Student Solutions Manual
to accompany

ORGANIC CHEMISTRY

seventh edition

Robert C. Atkins
James Madison University

Francis A. Carey
University of Virginia

Higher Education

Boston   Burr Ridge, IL   Dubuque, IA   New York   San Francisco   St. Louis
Bangkok   Bangota   Caracas   Kuala Lumpur   Lisbon   London   Madrid   Mexico City
Milan   Montreal   New Delhi   Santiago   Seoul   Singapore   Sydney   Taipei   Toronto

Higher Education

STUDENT SOLUTIONS MANUAL TO ACCOMPANY ORGANIC CHEMISTRY, SEVENTH EDITION

Published by McGraw-Hill, a business unit of The McGraw-Hill Companies, Inc., 1221 Avenue of the Americas, New York, NY 10020. Copyright © 2008 by The McGraw-Hill Companies, Inc. All rights reserved. Previously published under the title of *Study Guide to accompany Organic Chemistry*. Copyright © 1992, 1987 by The McGraw-Hill Companies, Inc. All rights reserved. No part of this publication may be reproduced or distributed in any form or by any means, or stored in a database or retrieval system, without the prior written consent of The McGraw-Hill Companies, Inc., including, but not limited to, in any network or other electronic storage or transmission, or broadcast for distance learning.

Some ancillaries, including electronic and print components, may not be available to customers outside the United States.

This book is printed on acid-free paper.

1 2 3 4 5 6 7 8 9 0 QPD/QPD 0 9 8 7

ISBN 978–0–07–304788–1
MHID 0–07–304788–0

Senior Sponsoring Editor: *Thomas D. Timp*
Developmental Editor: *Lorraine K. Buczek*
Marketing Manager: *Tamara Good-Hodge*
Lead Project Manager: *Peggy J. Selle*
Senior Production Supervisor: *Kara Kudronowicz*
Associate Design Coordinator*: Brenda Rolwes*
Cover designer: *Brenda Rolwes*
Compositor: *Techbooks*
Typeface: *10/12 Times Roman*
Printer: *Quebecor World Dubuque*

Library of Congress Control Number: 2006939224

www.mhhe.com

CONTENTS

PREFACE

It is our hope that in writing this *Solutions Manual* we will make the study of organic chemistry more meaningful and worthwhile. To be effective, a solutions manual should be more than just an answer book. What we present here was designed with that larger goal in mind.

The *Solutions Manual* contains detailed solutions to all the problems in the text. Learning how to solve a problem is, in our view, more important than merely knowing the correct answer. To that end we have included solutions sufficiently detailed to provide the student with the steps leading to the solution of each problem.

In addition, the Self-Test at the conclusion of each chapter is designed to test the student's mastery of the material. Both fill-in and multiple-choice questions have been included to truly test the student's understanding. Answers to the self-test questions may be found in the Appendix at the back of this book.

The completion of this manual was made possible through the time and talents of numerous people. Our thanks and appreciation also go to the many users of the previous editions who have provided us with helpful suggestions, comments, and corrections. We also wish to acknowledge the assistance and understanding of Thomas Timp, Jodi Rhomberg, and Peggy Selle of McGraw-Hill. Last, we thank our wives and families for their understanding of the long hours invested in this work.

Robert C. Atkins
Francis A. Carey

TO THE STUDENT

Before beginning the study of organic chemistry, a few words about "how to do it" are in order. You've probably heard that organic chemistry is difficult; there's no denying that. It need not be overwhelming, though, when approached with the right frame of mind and with sustained effort.

First of all you should realize that organic chemistry tends to "build" on itself. That is, once you have learned a reaction or concept, you will find it being used again and again later on. In this way it is quite different from general chemistry, which tends to be much more compartmentalized. In organic chemistry you will continually find previously learned material cropping up and being used to explain and to help you understand new topics. Often, for example, you will see the preparation of one class of compounds using reactions of other classes of compounds studied earlier in the year.

How to keep track of everything? It might be possible to memorize every bit of information presented to you, but you would still lack a fundamental understanding of the subject. It is far better to *generalize* as much as possible.

You will find that the early chapters of the text will emphasize concepts of *reaction theory*. These will be used, as the various classes of organic molecules are presented, to describe *mechanisms* of organic reactions. A relatively few fundamental mechanisms suffice to describe almost every reaction you will encounter. Once learned and understood, these mechanisms provide a valuable means of categorizing the reactions of organic molecules.

You will need to learn numerous facts in the course of the year, however. For example, chemical reagents necessary to carry out specific reactions must be learned. You might find a study aid known as *flash cards* helpful. These take many forms, but one idea is to use 3×5 index cards. As an example of how the cards might be used, consider the reduction of alkenes (compounds with carbon–carbon double bonds) to alkanes (compounds containing only carbon–carbon single bonds). The front of the card might look like this:

$$\text{Alkenes} \xrightarrow{\quad ? \quad} \text{alkanes}$$

The reverse of the card would show the reagents necessary for this reaction:

$$\boxed{H_2, \text{ Pt or Pd catalyst}}$$

The card can actually be studied in two ways. You may ask yourself: What reagents will convert alkenes into alkanes? Or, using the back of the card: What chemical reaction is carried out with hydrogen and a platinum or palladium catalyst? This is by no means the only way to use the cards—be creative! Just making up the cards will help you to study.

Although study aids such as flash cards will prove helpful, there is only one way to truly master the subject matter in organic chemistry—*do the problems*! The more you work, the more you will learn. Almost certainly the grade you receive will be a reflection of your ability to solve problems. Don't just think over the problems, either; write them out as if you were handing them in to be graded. Also, be careful of how you use the *Solutions Manual*. The solutions contained in this book are intended to provide explanations to help you understand the problem. Be sure to write out *your* solution to the problem first and only then look it up to see if you have done it correctly.

Students frequently feel that they understand the material but don't do as well as expected on tests. One way to overcome this is to "test" yourself. Each chapter in the *Solutions Manual* has a self-test at the end. Work the problems in these tests *without* looking up how to solve them in the text. You'll find it is much harder this way, but it is also a closer approximation to what will be expected of you when taking a test in class.

Success in organic chemistry depends on skills in analytical reasoning. Many of the problems you will be asked to solve require you to proceed through a series of logical steps to the correct answer. Most of the individual concepts of organic chemistry are fairly simple; stringing them together in a coherent fashion is where the challenge lies. By doing exercises conscientiously you should see a significant increase in your overall reasoning ability. Enhancement of their analytical powers is just one fringe benefit enjoyed by those students who are fully involved in the course rather than simply attending it.

Gaining a mastery of organic chemistry is hard work. We hope that the hints and suggestions outlined here will be helpful to you and that you will find your efforts rewarded with a knowledge and understanding of an important area of science.

Robert C. Atkins
Francis A. Carey

CHAPTER 1

Structure Determines Properties

SOLUTIONS TO TEXT PROBLEMS

1.1 The number of electrons in an element is equal to the number of protons in its nucleus and is given by its atomic number Z. For carbon, $Z = 6$. Therefore, carbon has 6 electrons. Of these, only the two $2s$ and two $2p$ electrons are valence electrons. The two $1s$ electrons are not valence electrons.

 Silicon, like carbon, is in group 4 of the periodic table. The number of valence electrons of main-group elements is the same as the group number. Therefore, silicon has 4 valence electrons.

1.2 Electron configurations of elements are derived by applying the following principles:

(*a*) The number of electrons in a neutral atom is equal to its atomic number Z.

(*b*) The maximum number of electrons in any orbital is 2.

(*c*) Electrons are added to orbitals in order of increasing energy, filling the $1s$ orbital before any electrons occupy the $2s$ level. The $2s$ orbital is filled before any of the $2p$ orbitals, and the $3s$ orbital is filled before any of the $3p$ orbitals.

(*d*) All the $2p$ orbitals ($2p_x, 2p_y, 2p_z$) are of equal energy, and each is singly occupied before any is doubly occupied. The same holds for the $3p$ orbitals.

 With this as background, the electron configuration of the third-row elements is derived as follows $[2p^6 = 2p_x^2 2p_y^2 2p_z^2]$:

Na	($Z = 11$)	$1s^2 2s^2 2p^6 3s^1$
Mg	($Z = 12$)	$1s^2 2s^2 2p^6 3s^2$
Al	($Z = 13$)	$1s^2 2s^2 2p^6 3s^2 3p_x^1$
Si	($Z = 14$)	$1s^2 2s^2 2p^6 3s^2 3p_x^1 3p_y^1$
P	($Z = 15$)	$1s^2 2s^2 2p^6 3s^2 3p_x^1 3p_y^1 3p_z^1$

S	$(Z = 16)$	$1s^2 2s^2 2p^6 3s^2 3p_x^2 3p_y^1 3p_z^1$
Cl	$(Z = 17)$	$1s^2 2s^2 2p^6 3s^2 3p_x^2 3p_y^2 3p_z^1$
Ar	$(Z = 18)$	$1s^2 2s^2 2p^6 3s^2 3p_x^2 3p_y^2 3p_z^2$

1.3   A sodium atom (Na) has 11 electrons. A sodium ion (Na^+) has one less, or 10 electrons. The +2 ion that contains 10 electrons must have 12 protons in its nucleus. The element with $Z = 12$ is magnesium. Mg^{2+} is isoelectronic with Na^+.

A –2 ion with 10 electrons must have 8 protons in its nucleus. The element with $Z = 8$ is oxygen. O^{2-} is isoelectronic with Na^+.

1.4   The atomic number of chlorine is 17. The ion Cl^- has one more electron than the 17 protons in its nucleus. Therefore, Cl^- has 18 electrons. In order to have a charge of −2, an ion that is isoelectronic with Cl^- must have 16 protons in its nucleus. The ion is S^{2-}.

1.5   The electron configurations of the designated ions are:

	Ion	Z	Number of Electrons in Ion	Electron Configuration of Ion
(b)	He^+	2	1	$1s^1$
(c)	H^-	1	2	$1s^2$
(d)	O^-	8	9	$1s^2 2s^2 2p^5$
(e)	F^-	9	10	$1s^2 2s^2 2p^6$
(f)	Ca^{2+}	20	18	$1s^2 2s^2 2p^6 3s^2 3p^6$

Those with a noble gas configuration are H^-, F^-, and Ca^{2+}. K^+ and Ca^{2+} are the only isoelectronic ions. Both have a total of 18 electrons.

1.6   A positively charged ion is formed when an electron is removed from a neutral atom. The following equation represents the ionization of carbon and the electron configurations of the neutral atom and the ion:

$$C \longrightarrow C^+ + e^-$$

$$1s^2 2s^2 2p_x^1 2p_y^1 \qquad 1s^2 2s^2 2p_x^1$$

A negatively charged carbon is formed when an electron is added to a carbon atom. The additional electron enters the $2p_z$ orbital.

$$C + e^- \longrightarrow C^-$$

$$1s^2 2s^2 2p_x^1 2p_y^1 \qquad 1s^2 2s^2 2p_x^1 p_y^1 2p_z^1$$

Neither C^+ nor C^- has a noble gas electron configuration.

1.7   (b) In order to have four bonds to carbon, CH_4O must have a carbon–oxygen bond.

The 14 valence electrons of CH_4O are distributed among five covalent bonds and two unshared pairs of oxygen.

(c) Three of the four bonds involving carbon are to hydrogen atoms; the fourth is to fluorine. Carbon contributes 4 electrons, fluorine 7, and the three hydrogens contribute one each. The total number of valence electrons in CH_3F is 14.

Combine carbon, fluorine and three hydrogens   to write a Lewis structure for methyl fluoride

The 14 valence electrons of CH_3F are distributed among four covalent bonds and three unshared pairs of fluorine.

(d) The only possible structure for C_2H_5F in which each carbon has four bonds must contain a carbon–carbon bond.

Combine two carbons, one fluorine, and five hydrogens   to write a Lewis structure for ethyl fluoride

The number of valence electrons in C_2H_5F is 20; eight contributed by two carbons, seven by fluorine, and a total of five by the hydrogens. These 20 valence electrons are distributed among 7 covalent bonds plus three unshared electron pairs of fluorine in C_2H_5F.

1.8 (b) Hydrogen cyanide (HCN) has 10 valence electrons; one from hydrogen, four from carbon, and five from nitrogen.

Combine   to give

In order to satisfy the octet rule, pair the remaining unpaired electrons of carbon with those of nitrogen so as to give a triple bond in:

1.9 The partial positive charge of hydrogen is greatest when the atom it is attached to is most electronegative. Among the atoms to which hydrogen is bonded in the compounds CH_4, NH_3, H_2O, SiH_4, and H_2S, oxygen is the most electronegative. Therefore, $\delta+$ is largest for hydrogen in H_2O.

Hydrogen will bear a partial negative charge when the atom to which it is bonded is less electronegative than hydrogen. The electronegativity of hydrogen is 2.1 versus 1.8 for silicon (text Table 1.3). Therefore, hydrogen bears a partial negative charge in SiH_4.

1.10 (b) First count the valence electrons in C_3H_7Cl.

carbon: 3 atoms $\times$ 4 valence electrons/atom = 12 valence electrons

hydrogen: 7 atoms $\times$ 1 valence electron/atom = 7 valence electrons

chlorine: 1 atom $\times$ 7 valence electrons/atom = 7 valence electrons

total = 26 valence electrons

Because chlorine can be bonded only to one other atom, only two connectivities are possible.

$$-\overset{|}{\underset{|}{C}}-\overset{|}{\underset{|}{C}}-\overset{|}{\underset{|}{C}}-Cl \quad \text{and} \quad -\overset{|}{\underset{|}{C}}-\overset{|}{\underset{Cl}{C}}-\overset{|}{\underset{|}{C}}-$$

Place one hydrogen on each of the bonds available to carbon.

$$H-\overset{H}{\underset{H}{\overset{|}{C}}}-\overset{H}{\underset{H}{\overset{|}{C}}}-\overset{H}{\underset{H}{\overset{|}{C}}}-Cl \quad \text{and} \quad H-\overset{H}{\underset{H}{\overset{|}{C}}}-\overset{H}{\underset{Cl}{\overset{|}{C}}}-\overset{H}{\underset{H}{\overset{|}{C}}}-H$$

Each structural formula has 10 bonds, which account for 20 electrons. Place the remaining 6 electrons as unshared pairs on chlorine to complete its octet.

$$H-\overset{H}{\underset{H}{\overset{|}{C}}}-\overset{H}{\underset{H}{\overset{|}{C}}}-\overset{H}{\underset{H}{\overset{|}{C}}}-\overset{\cdot\cdot}{\underset{\cdot\cdot}{Cl}}: \quad \text{and} \quad H-\overset{H}{\underset{H}{\overset{|}{C}}}-\overset{H}{\underset{:Cl:}{\overset{|}{C}}}-\overset{H}{\underset{H}{\overset{|}{C}}}-H$$

There are two constitutional isomers having the formula C_3H_7Cl.

(*c*) Proceed as in parts (*a*) and (*b*) of this problem.

The number of valence electrons contained in a molecular formula of C_3H_8O is 26 (12 contributed by three carbons, 8 contributed by eight hydrogens, and 6 contributed by one oxygen).

Three connectivities are possible in which carbon has four bonds. Two of these connectivities have a continuous chain of three carbons. The third has the atoms connected in the order CCOC.

$$-\overset{|}{\underset{|}{C}}-\overset{|}{\underset{|}{C}}-\overset{|}{\underset{|}{C}}-O-H \qquad -\overset{|}{\underset{|}{C}}-\overset{|}{\underset{\overset{|}{O}}{C}}-\overset{|}{\underset{|}{C}}- \qquad -\overset{|}{\underset{|}{C}}-\overset{|}{\underset{|}{C}}-O-\overset{|}{\underset{|}{C}}-$$
$$\underset{H}{}$$

Complete the number of available C—H bonds; there are seven in the first two structures, eight in the third.

$$H-\overset{H}{\underset{H}{\overset{|}{C}}}-\overset{H}{\underset{H}{\overset{|}{C}}}-\overset{H}{\underset{H}{\overset{|}{C}}}-O-H \qquad H-\overset{H}{\underset{H}{\overset{|}{C}}}-\overset{H}{\underset{\overset{|}{O}}{\overset{|}{C}}}-\overset{H}{\underset{H}{\overset{|}{C}}}-H \qquad H-\overset{H}{\underset{H}{\overset{|}{C}}}-\overset{H}{\underset{H}{\overset{|}{C}}}-O-\overset{H}{\underset{H}{\overset{|}{C}}}-H$$
$$\underset{H}{}$$

Each structural formula has 11 bonds, accounting for 22 electrons. The remaining four electrons are assigned to oxygen as two unshared pairs. The octet rule is satisfied.

$$H-\overset{H}{\underset{H}{\overset{|}{C}}}-\overset{H}{\underset{H}{\overset{|}{C}}}-\overset{H}{\underset{H}{\overset{|}{C}}}-\overset{\cdot\cdot}{\underset{\cdot\cdot}{O}}-H \qquad H-\overset{H}{\underset{H}{\overset{|}{C}}}-\overset{H}{\underset{:O:}{\overset{|}{C}}}-\overset{H}{\underset{H}{\overset{|}{C}}}-H \qquad H-\overset{H}{\underset{H}{\overset{|}{C}}}-\overset{H}{\underset{H}{\overset{|}{C}}}-\overset{\cdot\cdot}{\underset{\cdot\cdot}{O}}-\overset{H}{\underset{H}{\overset{|}{C}}}-H$$
$$\underset{H}{}$$

There are three constitutional isomers having the formula C_3H_8O.

1.11 (*b*) The compound $(CH_3)_3CH$ has a central carbon to which are attached three CH_3 groups and a hydrogen.

$$\begin{array}{c} H \\ | \\ H-C-H \\ \quad\; H \;|\; H \\ |\;\;\;|\;\;\;| \\ H-C-C-C-H \\ |\;\;\;|\;\;\;| \\ H \;\; H \;\; H \end{array}$$

Four carbons and ten hydrogens contribute 26 valence electrons. The structure shown has 13 covalent bonds, and so all the valence electrons are accounted for. The molecule has no unshared electron pairs.

(*c*) The number of valence electrons in $ClCH_2CH_2Cl$ is 26 (2Cl = 14; 4H = 4; 2C = 8). The constitution at the left shows seven covalent bonds accounting for 14 electrons. The remaining 12 electrons are divided equally between the two chlorines as unshared electron pairs. The octet rule is satisfied for both carbon and chlorine in the structure at the right.

$$\begin{array}{c} H \;\; H \\ |\;\;\;| \\ Cl-C-C-Cl \\ |\;\;\;| \\ H \;\; H \end{array} \qquad \begin{array}{c} H \;\; H \\ |\;\;\;| \\ :\ddot{C}l-C-C-\ddot{C}l: \\ |\;\;\;| \\ H \;\; H \end{array}$$

(*d*) This compound has the same molecular formula as the compound in part (*c*), but a different structure. It is a constitutional isomer of the compound in part (*c*). It, too, has 26 valence electrons, and again only chlorine has unshared pairs.

$$\begin{array}{c} H \;\; H \\ |\;\;\;| \\ H-C-C-\ddot{C}l: \\ |\;\;\;| \\ H \; :\ddot{C}l: \end{array}$$

(*e*) The connectivity of $CH_3NHCH_2CH_3$ is shown (on the left). There are 26 valence electrons, and 24 of them are accounted for by the covalent bonds in the structural formula. The remaining two electrons complete the octet of nitrogen as an unshared pair (on the right).

$$\begin{array}{c} H \;\;\;\;\; H \;\; H \\ |\;\;\;\;\;\;\;|\;\;\;| \\ H-C-N-C-C-H \\ |\;\;\;\;\;\;\;|\;\;\;| \\ H \;\; H \;\; H \;\; H \end{array} \qquad \begin{array}{c} H \;\;\;\;\; H \;\; H \\ |\;\;\;\;\;\;\;|\;\;\;| \\ H-C-\ddot{N}-C-C-H \\ |\;\;\;\;\;\;\;|\;\;\;| \\ H \;\; H \;\; H \;\; H \end{array}$$

(*f*) Oxygen has two unshared pairs in $(CH_3)_2CHCH{=}O$.

$$\begin{array}{c} H \\ | \\ H-C-H \\ \;H\;| \\ \;|\;\;\;| \\ H-C-C-C{=}\ddot{O}: \\ |\;\;\;|\;\;\;| \\ H \;\; H \;\; H \end{array}$$

1.12 (*b*) This compound has a four-carbon chain to which two other carbons are attached.

is equivalent to $\begin{array}{c} CH_3 \; H \\ |\;\;\;\;\; | \\ H_3C-C-C-CH_3 \\ |\;\;\;\;\; | \\ H \;\; CH_3 \end{array}$ which may be rewritten as $(CH_3)_2CHCH(CH_3)_2$

(c) The carbon skeleton is the same as that of the compound in part (b), but one of the terminal carbons bears an OH group in place of one of its hydrogens.

HO⟍ ... is equivalent to

$$H_3C-\underset{\underset{H}{|}}{\overset{\overset{H}{|}}{C}}-\underset{\underset{CH_3}{|}}{\overset{\overset{HO-\overset{\overset{H}{|}}{C}-H}{|}}{C}}-CH_3$$

which may be rewritten as

$$\underset{CH_3CHCH(CH_3)_2}{\overset{CH_2OH}{|}}$$

(d) The compound is a six-membered ring that bears a —$C(CH_3)_3$ substituent.

is equivalent to

which may be rewritten as

—$C(CH_3)_3$

1.13 (b) The formal charges in sulfuric acid are calculated as follows:

	Valence Electrons in Neutral Atom	Electron Count	Formal Charge
Hydrogen:	1	$\frac{1}{2}(2)=1$	0
Oxygen (of OH):	6	$\frac{1}{2}(4)+4=6$	0
Oxygen:	6	$\frac{1}{2}(2)+6=7$	-1
Sulfur:	6	$\frac{1}{2}(8)+0=4$	$+2$

$$H-\overset{..}{\underset{..}{O}}-\overset{|2+}{\underset{\underset{..}{\overset{..}{\underset{O}{}}}}{S}}-\overset{..}{\underset{..}{O}}-H$$

(c) The formal charges in nitrous acid are calculated as follows:

	Valence Electrons in Neutral Atom	Electron Count	Formal Charge
Hydrogen:	1	$\frac{1}{2}(2)=1$	0
Oxygen (of OH):	6	$\frac{1}{2}(4)+4=6$	0
Oxygen:	6	$\frac{1}{2}(4)+4=6$	0
Nitrogen:	5	$\frac{1}{2}(6)+2=5$	0

$$H-\overset{..}{\underset{..}{O}}-\overset{..}{N}=\overset{..}{\underset{..}{O}}:$$

1.14 The electron counts of nitrogen in ammonium ion and boron in borohydride ion are both four (one-half of eight electrons in covalent bonds).

Ammonium ion   Borohydride ion

Because a neutral nitrogen has five electrons in its valence shell, an electron count of 4 gives it a formal charge of +1. A neutral boron has three valence electrons, and so an electron count of 4 in borohydride ion corresponds to a formal charge of −1.

1.15 The two terminal nitrogens each have an electron count (6) that is one more than a neutral atom and thus each has a formal charge of −1. The central N has an electron count (4) that is one less than a neutral nitrogen; it has a formal charge of +1. The net charge on the species is (−1 + 1 − 1), or −1.

1.16 There are 18 valence electrons in NO_2^-. Nitrogen contributes 5, the two oxygens together contribute 12, and 1 electron is added for the negative charge. A structural formula can be written in which the octet rule is satisfied for both oxygens, but not for nitrogen.

Use an unshared electron pair from oxygen to form a double bond to nitrogen. There are two possibilities. The two are resonance forms of each other and contribute equally to the resonance hybrid.

1.17 Curved arrows guide us in writing the various resonance structures of carbonate ion.

Each oxygen is doubly bonded to carbon in one structure and singly bonded to carbon in the other two. All oxygens have, on average, two-thirds of a negative charge.

1.18 (*b*) Carbon has only six valence electrons and is positively charged in the original structural formula. Move the unshared electron pair from nitrogen to a shared position between nitrogen and carbon. The double bond becomes a triple bond.

The new structure has one more bond than the original structure, so is more stable and makes a greater contribution to the resonance hybrid. The octet rule is satisfied for both carbon and nitrogen in the new Lewis structure.

(*c*) Moving an unshared pair from oxygen to between carbon and oxygen converts C—O to C=O. Moving a pair of electrons to nitrogen converts C=N to C—N.

The number of bonds is the same in both the original and final structural formulas. The original structure, which has its negative charge on oxygen, is more stable than the second one, which has its negative charge on nitrogen. Oxygen is more electronegative than nitrogen and can bear a negative charge better.

(*d*) Reorganizing the electrons as indicated by the curved arrows changes the electron counts and formal charges of the two oxygens.

$$H-\ddot{\ddot{O}}-N=\ddot{O}: \longleftrightarrow H-\overset{+}{\ddot{O}}=N-\ddot{\ddot{O}}:^{-}$$

Positive and negatively charged atoms are separated from each other in the new structure. The original structure is more stable because it has no charge separation.

1.19   There are four B—H bonds in BH_4^-. The four electron pairs surround boron in a tetrahedral orientation. The H—B—H angles are 109.5°.

1.20   (*b*) Nitrogen in ammonium ion is surrounded by eight electrons in four covalent bonds. These four bonds are directed toward the corners of a tetrahedron.

$$H-\overset{H}{\underset{H}{\overset{|}{N^+}}}-H \qquad \text{Each HNH angle is 109.5°.}$$

(*c*) Double bonds are treated as a single unit when deducing the shape of a molecule using the VSEPR model. Thus azide ion is linear.

$$^-:\ddot{N}=\overset{+}{N}=\ddot{N}:^- \qquad \text{The NNN angle is 180°.}$$

(*d*) Because the double bond in carbonate ion is treated as if it were a single unit, the three sets of electrons are arranged in a trigonal planar arrangement around carbon.

$$^-:\ddot{O} \overset{\overset{\textstyle :\ddot{O}:}{\|}}{\underset{}{C}} \ddot{O}:^- \qquad \text{The OCO angle is 120°.}$$

1.21   (*b*) Water is a bent molecule, and so the individual O—H bond dipole moments do not cancel. Water has a dipole moment.

Individual OH bond   Direction of net
moments in water   dipole moment

(*c*) Methane, CH_4, is perfectly tetrahedral, and so the individual (small) C—H bond dipole moments cancel. Methane has no dipole moment.

(*d*) Methyl chloride has a dipole moment.

Directions of bond dipole   Direction of molecular
moments in CH_3Cl dipole moment

(e) Oxygen is more electronegative than carbon and attracts electrons from it. Formaldehyde has a dipole moment.

Direction of bond dipole moments in formaldehyde

Direction of molecular dipole moment

(f) Nitrogen is more electronegative than carbon. Hydrogen cyanide has a dipole moment.

Directions of bond dipole moments in HCN

Direction of molecular dipole moment

1.22 (b) The carbon–bromine bond breaks. The pair of electrons in that bond are retained by bromine, which becomes a bromide ion in the product. The organic product is a cation, having a formal charge of +1 on the central carbon.

Charge is conserved. A neutral molecule dissociates to a +1 ion and a –1 ion.

1.23 (b) In the reverse process an unshared pair of bromide ion becomes the pair of electrons in a C—Br bond.

1.24 The problem states that NaSH is converted to H_2S. Therefore, a bond must form between the sulfur of HS^- and a hydrogen of H_3O^+. Electrons flow from the negatively charged sulfur to one of the hydrogens. The pair of electrons in the O—H bond that is broken becomes an unshared pair of oxygen.

The net charge is zero on both sides of the equation.

1.25 The strength of an acid can be expressed by its pK_a, given by the expression:

$$pK_a = -\log K_a$$

The K_a of salicylic acid is 1.06×10^{-3}; its pK_a is 2.97.

1.26 Because the pK_a of HCN is given as 9.1, its $K_a = 10^{-9.1}$. In more conventional notation, $K_a = 8 \times 10^{-10}$.

1.27 (*b*) In an acid-base reaction, a proton is transferred from the Brønsted acid, in this case HCl, to the Brønsted base, in this case $(CH_3)_3N{:}$. Remember to use curved arrows to track the movement of electrons, not atoms.

| Trimethylamine
(base) | Hydrogen chloride
(acid) | Trimethylammonium
ion
(conjugate acid) | Chloride ion
(conjugate base) |

1.28 Hydride ion is a strong base and will remove a proton from water. The conjugate acid of hydride ion is hydrogen (H_2).

$$\overset{..}{H}{:}^{-} \quad + \quad H\!-\!\overset{..}{\underset{..}{O}}H \quad \rightleftharpoons \quad H_2 \quad + \quad {:}\overset{..}{\underset{..}{O}}H^{-}$$

| Hydride ion
(base) | Water
(acid) | Hydrogen
(conjugate acid) | Hydroxide ion
(conjugate base) |

1.29 (*b, c*) You can determine the relative strength of two bases by comparing the pK_as of their respective conjugate acids. Remember that the stronger base is derived from the weaker conjugate acid.

Base:	$HC\equiv C{:}^{-}$	$H_2\overset{..}{N}{:}^{-}$	$CH_3CH_2\overset{..}{\underset{..}{O}}{:}^{-}$
Conjugate acid:	$HC\equiv CH$	NH_3	CH_3CH_2OH
pK_a of conjugate acid:	26	36	16

Recall that the smaller pK_a is associated with the stronger acid. Ammonia is a weaker acid than acetylene ($HC\equiv CH$), therefore amide ion ($H_2\overset{..}{N}{:}^{-}$) is a stronger base than acetylide ion ($HC\equiv C{:}^{-}$). Similar reasoning leads to the conclusion that because acetylene is a weaker acid than ethanol (CH_3CH_2OH), acetylide ion is a stronger base than ethoxide ion ($CH_3CH_2\overset{..}{\underset{..}{O}}{:}^{-}$).

1.30 Bond strength weakens going down a group in the periodic table. Because sulfur lies below oxygen, the H—S bond is weaker than the H—O bond. We would expect H_2S to be a stronger acid than H_2O, and this prediction is borne out by their respective pK_as:

	H_2S	H_2O
pK_a	9	15.7

Because the stronger acid forms the weaker conjugate base, HS^- is a weaker base than HO^-.

1.31 (*b*) In part (*a*) of the problem you determined that $(CH_3)_2\overset{+}{\underset{..}{O}}H$ is a stronger acid than $(CH_3)_3\overset{+}{N}H$. Recalling that the weaker acid forms the stronger conjugate base, $(CH_3)_3N{:}$ is a stronger base than $(CH_3)_2\overset{..}{\underset{..}{O}}{:}$.

1.32 To compare the strength of bases, look up the pK_as of their conjugate acids. Ethanol has a pK_a of 16; *tert*-butyl alcohol has a pK_a of 18. Ethanol is a stronger acid than *tert*-butyl alcohol. Therefore *tert*-butoxide ion is a stronger base than ethoxide.

1.33 The Brønsted acid-base equilibria for the ionization of hypochlorous and hypobromous acid are:

| Hypochlorous acid | Water | | Hypochlorite ion | Hydronium ion |

| Hypobromous acid | Water | | Hypobromite ion | Hydronium ion |

The positive character (δ^+) of the proton in HOCl is greater than the proton in HOBr because Cl is more electronegative than Br. The greater electronegativity of Cl compared to Br also stabilizes ClO⁻ compared to BrO⁻ and makes the equilibrium for the ionization of HOCl more favorable. Therefore, hypochlorous acid is a stronger acid than hypobromous acid. (This prediction is borne out by the measured pK_a's: HOCl = 7.5; HOBr = 8.7. The trend continues with HOI which has a pK_a of 10.6.)

1.34 The three oxygens in nitrate ion share a total charge of −2. The average formal charge on each oxygen atom is −0.67.

1.35 Writing the two resonance forms of each conjugate base reveals that they are equivalent.

Oxygen is more electronegative than sulfur and will bear a greater share of the negative charge.

1.36 (b) Begin by writing the equation for the acid-base reaction between acetic acid and fluoride ion. Acetic acid will donate a proton to fluoride ion, converting acetic acid to its conjugate base (acetate ion) and fluoride ion to its conjugate acid, hydrogen fluoride.

Acetic acid	Fluoride ion		Acetate ion		Hydrogen fluoride
pK_a = 4.7					pK_a = 3.1
(weaker acid)					(stronger acid)

From the respective pK_as of the acids, you can see that the weaker acid (acetic acid) is on the left and the stronger acid (hydrogen fluoride) is on the right. Therefore, the equilibrium lies to the left. The equilibrium constant for the process is:

$$K_{eq} = \frac{10^{-pK_a} \text{ of acetic acid (reactant)}}{10^{-pK_a} \text{ of hydrogen fluoride (product)}} = \frac{10^{-4.7}}{10^{-3.1}} = 10^{-1.6}$$

1.37 Sulfuric acid is a strong acid; the pK_a of the first ionization of H_2SO_4 is −4.8. In an aqueous solution of a strong acid, the predominant acid species is H_3O^+. Hydrogen sulfate ion is a weak acid (p$K_a = 2.0$) and will only undergo ionization to a small extent. The relative amounts of the species present will be:

$$\left[H_3O^+\right] > \left[HSO_4^-\right] > \left[SO_4^{2-}\right] > \left[H_2SO_4\right]$$

1.38 Comparing the pK_as of phenol and water reveals that, indeed, the stronger acid (phenol) is on the left in the acid-base equation and the equilibrium constant is greater than 1.

$$C_6H_5OH \;+\; HO^- \;\rightleftharpoons\; C_6H_5O^- \;+\; H_2O$$

Phenol
pK_a = 10
(stronger acid)

Water
pK_a = 15.7
(weaker acid)

The same comparison for the reaction of phenol and hydrogen carbonate ion reveals that the stronger acid (carbonic acid) is on the right and the equilibrium constant is less than one.

$$C_6H_5OH \;+\; HO\overset{O}{\overset{\|}{C}}O^- \;\rightleftharpoons\; C_6H_5O^- \;+\; HO\overset{O}{\overset{\|}{C}}OH$$

Phenol
pK_a = 10
(weaker acid)

Carbonic acid
pK_a = 6.4
(stronger acid)

1.39 Boron trifluoride is a Lewis acid and will accept a pair of electrons from dimethyl sulfide to form a Lewis acid-Lewis base complex.

$$(CH_3)_2\ddot{S}\!: \;+\; BF_3 \;\rightleftharpoons\; (CH_3)_2\overset{+}{S}\!-\!\overset{-}{B}F_3$$

1.40 All these species are characterized by the formula $:X\equiv Y:$, and each atom has an electron count of 5.

$$:X\equiv Y:$$

Unshared electron pair contributes two electrons to electron count of X.

Unshared electron pair contributes two electrons to electron count of Y.

Triple bond contributes half of its six electrons, or three electrons each, to separate electron counts of X and Y.

Electron count X = electron count Y = 2 + 3 = 5.

(a) :N≡N: A neutral nitrogen atom has five valence electrons: therefore, each atom is electrically neutral in molecular nitrogen.

(b) :C≡N: Nitrogen, as before, is electrically neutral. A neutral carbon has four valence electrons, and so carbon in this species, with an electron count of 5, has a unit negative charge. The species is cyanide anion; its net charge is −1.

(c) :C≡C: There are two negatively charged carbon atoms in this species. It is a dianion; its net charge is −2.

(d) :N≡O: Here again is a species with a neutral nitrogen atom. Oxygen, with an electron count of 5, has one less electron in its valence shell than a neutral oxygen atom. Oxygen has a formal charge of +1; the net charge is +1.

(e) :C≡O: Carbon has a formal charge of −1; oxygen has a formal charge of +1. Carbon monoxide is a neutral molecule.

1.41 (a, b) The problem specifies that ionic bonding is present and that the anion is tetrahedral. The cations are the group 1 metals Na^+ and Li^+. Both boron and aluminum are group 3 elements, and thus have a formal charge of −1 in the tetrahedral anions BF_4^- and AlH_4^- respectively.

Sodium tetrafluoroborate   Lithium aluminum hydride

(c, d) Both of the tetrahedral anions have 32 valence electrons. Sulfur contributes six valence electrons and phosphorus five to the anions. Each oxygen contributes six electrons. The double negative charge in sulfate contributes two more, and the triple negative charge in phosphate contributes three more.

Potassium sulfate   Sodium phosphate

The formal charge on each oxygen in both ions is −1. The formal charge on sulfur in sulfate is +2; the charge on phosphorus is +1. The net charge of sulfate ion is −2; the net charge of phosphate ion is −3.

1.42 (a) The molecular formula OCS tells us that there are 16 valence electrons in carbon oxysulfide. Carbon contributes 4, oxygen 6, and sulfur 6. The connectivity O−C−S accounts for 4 electrons. Apportioning the remaining 12 equally between oxygen and sulfur completes the octet of both, but leaves carbon with only 4 electrons.

Use one of oxygen's unshared pairs to form a double bond to carbon. Do the same with one of sulfur's unshared pairs to give a Lewis structure in which all three atoms have complete octets.

(b) Double bonds are treated the same way as single bonds when applying VSEPR. The electrons in the O=C bond are farthest removed from the electrons in the C=S bond when the angle between them is 180°. Carbon oxysulfide is *linear*.

(*c*) According to Table 1.3 in the text, oxygen, with an electronegativity of 3.5, is the most electronegative atom of the three while the electronegativities of carbon and sulfur are both 2.5. Thus, electron density is drawn toward oxygen and away from the C=S group.

$$\overset{\longleftarrow +}{O=C=S}$$

1.43 Given that the compound $C_2H_6O_2$ must have an O—O bond, two connectivities are possible.

<pre>
    H H   H   H
    | | | |
H—C—C—O—O—H   and   H—C—O—O—C—H
    | | | |
    H H   H   H
</pre>

The nine bonds in each connectivity account for 18 of the 26 valence electrons in $C_2H_6O_2$. Add the remaining 8 electrons as unshared pairs on oxygen.

$$CH_3CH_2-\overset{..}{\underset{..}{O}}-\overset{..}{\underset{..}{O}}-H \quad \text{and} \quad CH_3-\overset{..}{\underset{..}{O}}-\overset{..}{\underset{..}{O}}-CH_3$$

1.44 (*a*) Each carbon has four valence electrons, each hydrogen one, and chlorine has seven. Hydrogen and chlorine each can form only one bond, and so the only stable structure must have a carbon–carbon bond. There are 18 valence electrons in C_2H_3Cl, and the framework of five single bonds accounts for only 10 electrons. Six of the remaining eight electrons are used to complete the octet of chlorine as three unshared pairs, and the last two are used to form a carbon–carbon double bond.

(*b*) All of the atoms except carbon (H, Br, Cl, and F) are monovalent; therefore, they can only be bonded to carbon. The problem states that all three fluorines are bonded to the same carbon, and so one of the carbons is present as a CF_3 group. The other carbon must be present as a CHBrCl group. Connect these groups together to give the structure of halothane.

<pre>
 ..   F H
:F: H | |
:F:C: C:Cl: or   F—C—C—Cl   (Unshared electron pairs omitted for clarity)
:F::Br: | |
 ..   F Br
</pre>

(*c*) As in part (*b*) all of the atoms except carbon are monovalent. Because each carbon bears one chlorine, two $ClCF_2$ groups must be bonded together.

<pre>
 .. ..   F F
:F: :F: | |
:Cl:C : C:Cl: or   Cl—C—C—Cl   (Unshared electron pairs omitted for clarity)
:F::F: | |
 .. ..   F F
</pre>

1.45 Place hydrogens on the given atoms so that carbon has four bonds, nitrogen three, and oxygen two. Place unshared electron pairs on nitrogen and oxygen so that nitrogen has an electron count of 5 and oxygen has an electron count of 6. These electron counts satisfy the octet rule when nitrogen has three bonds and oxygen two.

$$
(a) \quad H-\overset{\displaystyle H}{\underset{\displaystyle H}{\overset{|}{\underset{|}{C}}}}-\ddot{N}=\ddot{O}:
\qquad\qquad
(c) \quad H-\ddot{O}-\overset{}{C}=\underset{\displaystyle H}{\overset{|}{\ddot{N}}}-H
$$

$$
(b) \quad H-\overset{}{\underset{\displaystyle H}{\overset{|}{C}}}=\ddot{N}-\ddot{O}-H
\qquad\qquad
(d) \quad :\ddot{O}=\overset{}{\underset{\displaystyle H}{\overset{|}{C}}}-\underset{\displaystyle H}{\overset{|}{\ddot{N}}}-H
$$

1.46 (a) Species A, B, and C have the same molecular formula, the same atomic positions, and the same number of electrons. They differ only in the arrangement of their electrons. They are therefore resonance contributors of a single compound.

$$
\underset{H}{\overset{H}{\diagdown}}\overset{-}{\underset{}{:}}C-\overset{+}{N}\equiv N:
\qquad
\underset{H}{\overset{H}{\diagdown}}C=\overset{+}{N}=\ddot{\underset{}{N}}:^{-}
\qquad
\underset{H}{\overset{H}{\diagdown}}\overset{+}{C}-\ddot{N}=\ddot{\underset{}{N}}:^{-}
$$
$$
\qquad\quad A \qquad\qquad\qquad B \qquad\qquad\qquad C
$$

(b) Structure A has a formal charge of −1 on carbon.

(c) Structure C has a formal charge of +1 on carbon.

(d) Structures A and B have formal charges of +1 on the internal nitrogen.

(e) Structures B and C have a formal charge of −1 on the terminal nitrogen.

(f) All resonance contributors of a particular species must have the same net charge. In this case, the net charge on A, B, and C is 0.

(g) Both A and B have the same number of covalent bonds, but the negative charge is on a more electronegative atom in B (nitrogen) than it is in A (carbon). Structure B is more stable.

(h) Structure B is more stable than structure C. Structure B has one more covalent bond, all of its atoms have octets of electrons, and it has a lesser degree of charge separation than C. The carbon in structure C does not have an octet of electrons.

(i) The CNN unit is linear in A and B, but bent in C according to VSEPR. This is an example of how VSEPR can fail when comparing resonance structures.

1.47 The structures given and their calculated formal charges are:

$$
\overset{-1\ +1}{H-\ddot{C}=N=\ddot{O}:}
\qquad
\overset{\ \ +1\ \ \ \ -1}{H-C\equiv N-\ddot{O}:}
\qquad
H-C\equiv N=\ddot{O}:
\qquad
\overset{\ \ +1\ \ \ \ -1}{H-C=\ddot{N}-\ddot{O}:}
$$
$$
\quad A \qquad\qquad\qquad B \qquad\qquad\qquad C \qquad\qquad\qquad D
$$

(a) Structure D contains a positively charged carbon.

(b) Structures A and B contain a positively charged nitrogen.

(c) None of the structures contains a positively charged oxygen.

(d) Structure A contains a negatively charged carbon.

(e) None of the structures contains a negatively charged nitrogen.

(f) Structures B and D contain a negatively charged oxygen.

(g) All the structures are electrically neutral.

(h) Structure B is the most stable. All the atoms except hydrogen have octets of electrons, and the negative charge resides on the most electronegative element (oxygen).

(i) Structure C is the least stable. Nitrogen has five bonds (ten electrons), which violates the octet rule.

1.48 (a) These two structures are resonance contributors because they have the same atomic positions and the same number of electrons.

16 valence electrons   16 valence electrons
(net charge = –1)   (net charge = –1)

(b) The two structures have different numbers of electrons and, therefore, cannot be resonance contributors of the same species.

16 valence electrons   14 valence electrons
(net charge = –1)   (net charge = +1)

(c) These two structures have different numbers of electrons; they are not resonance contributors.

16 valence electrons   20 valence electrons
(net charge = –1)   (net charge = –5)

1.49 (a) Structure C has 5 covalent bonds to nitrogen (10 electrons). The octet rule limits nitrogen to a maximum of 8 electrons in its valence shell.

        A   B   C   D

(b) Of the three remaining structures, B and C have 4 bonds and each is more stable than A which has only three. B is more stable than D because its negative charge is on a more electronegative element (oxygen versus carbon).

(c) Use curved arrows to show movement of electrons among the resonance contributors.

        A   B   D

1.50 Notice that in all of the cases in which the more stable Lewis formula contains more bonds than the original one, the additional bond leads to a structure in which the octet rule is satisfied.

(a)

one more bond than
original Lewis formula

(b)

no separation
of opposite charges

(c)

one more bond and
no charge separation

(d)

one more bond and
no charge separation

(e)

one more bond and
no charge separation

(f)

negative charge on more
electronegative element

(g)

one more bond

(h)

one more bond

(i)

no separation
of opposite charges

1.51 (a) There are 26 valence electrons in $(CH_3)_2SO$. The connectivity given corresponds to nine bonds (18 electrons). The remaining 8 electrons can be placed so as to satisfy the octet rule as shown on the left:

(b) The structural formula on the right has one more bond and lacks the charge separation of the one on the left, but has 10 electrons in the valence shell of sulfur.

1.52 (*a*) To generate constitutionally isomeric structures having the molecular formula C_4H_{10}, you need to consider the various ways in which four carbon atoms can be bonded together. These are

$$C-C-C-C \quad \text{and} \quad \begin{array}{c} C-C-C \\ | \\ C \end{array}$$

Filling in the appropriate hydrogens gives the correct structures:

$$CH_3CH_2CH_2CH_3 \quad \text{and} \quad \begin{array}{c} CH_3CHCH_3 \\ | \\ CH_3 \end{array}$$

Continue with the remaining parts of the problem using the general approach outlined for part (*a*).

(*b*) C_5H_{12}

$$CH_3CH_2CH_2CH_2CH_3 \qquad \begin{array}{c} CH_3CHCH_2CH_3 \\ | \\ CH_3 \end{array} \qquad \begin{array}{c} CH_3 \\ | \\ H_3C-C-CH_3 \\ | \\ CH_3 \end{array}$$

(*c*) $C_2H_4Cl_2$

$$CH_3CHCl_2 \quad \text{and} \quad ClCH_2CH_2Cl$$

(*d*) C_4H_9Br

$$CH_3CH_2CH_2CH_2Br \qquad \begin{array}{c} CH_3CHCH_2CH_3 \\ | \\ Br \end{array} \qquad \begin{array}{c} CH_3CHCH_2Br \\ | \\ CH_3 \end{array} \qquad \begin{array}{c} CH_3 \\ | \\ H_3C-C-Br \\ | \\ CH_3 \end{array}$$

(*e*) C_3H_9N

$$CH_3CH_2CH_2NH_2 \qquad CH_3CH_2NHCH_3 \qquad \begin{array}{c} \quad\quad CH_3 \\ H_3C-N \\ \quad\quad CH_3 \end{array} \qquad \begin{array}{c} CH_3CHNH_2 \\ | \\ CH_3 \end{array}$$

Note that when the three carbons and the nitrogen are arranged in a ring, the molecular formula based on such a structure is C_3H_7N, not C_3H_9N as required.

$$\begin{array}{c} H_2C-CH_2 \\ | \quad\quad | \\ H_2C-NH \end{array}$$

(not an isomer)

1.53 (*a*) All three carbons must be bonded together, and each one has four bonds; therefore, the molecular formula C_3H_8 uniquely corresponds to:

$$\begin{array}{c} \quad H \quad H \quad H \\ \quad | \quad\; | \quad\; | \\ H-C-C-C-H \\ \quad | \quad\; | \quad\; | \\ \quad H \quad H \quad H \end{array} \qquad (CH_3CH_2CH_3)$$

(b) With two fewer hydrogen atoms than the preceding compound, C_3H_6 must either contain a carbon–carbon double bond or its carbons must be arranged in a ring; thus the following structures are constitutional isomers:

$$H_2C\!=\!CHCH_3 \quad \text{and} \quad \underset{\displaystyle CH_2}{H_2C\!-\!CH_2}$$

(c) The molecular formula C_3H_4 is satisfied by the structures

$$H_2C\!=\!C\!=\!CH_2 \quad HC\!\equiv\!CCH_3 \quad \underset{\displaystyle CH_2}{HC\!=\!CH}$$

1.54 (a) The connectivities of C_3H_6O that contain only single bonds must have a ring as part of their structure.

$$\underset{\displaystyle CH_2}{H_2C\!-\!CHOH} \qquad \underset{\displaystyle O}{H_2C\!-\!CHCH_3} \qquad \underset{\displaystyle O\!-\!CH_2}{H_2C\!-\!CH_2}$$

(b) Structures corresponding to C_3H_6O are possible in noncyclic compounds if they contain a carbon–carbon or carbon–oxygen double bond.

$$\overset{\displaystyle O}{\overset{\|}{CH_3CH_2CH}} \qquad \overset{\displaystyle O}{\overset{\|}{CH_3CCH_3}} \qquad CH_3CH\!=\!CHOH \qquad CH_3OCH\!=\!CH_2$$

$$\underset{\displaystyle OH}{CH_3C\!=\!CH_2} \qquad H_2C\!=\!CHCH_2OH$$

1.55 The direction of a bond dipole is governed by the electronegativity of the atoms it connects. In each of the parts to this problem, the more electronegative atom is partially negative and the less electronegative atom is partially positive. Electronegativities of the elements are given in Table 1.3 of the text.

(a) Chlorine is more electronegative than hydrogen.

$$\overset{\longmapsto}{H\!-\!Cl}$$

(b) Chlorine is more electronegative than iodine.

$$\overset{\longmapsto}{I\!-\!Cl}$$

(c) Iodine is more electronegative than hydrogen.

$$\overset{\longmapsto}{H\!-\!I}$$

(d) Oxygen is more electronegative than hydrogen.

$$\underset{\displaystyle H \qquad H}{\nearrow O \nwarrow}$$

(e) Oxygen is more electronegative than either hydrogen or chlorine.

$$\underset{\displaystyle H \qquad Cl}{\nearrow O \nwarrow}$$

1.56 The direction of a bond dipole is governed by the electronegativity of the atoms involved. Among the halogens the order of electronegativity is F > Cl > Br > I. Fluorine therefore attracts electrons away from chlorine in FCl, and chlorine attracts electrons away from iodine in ICl.

$$F\!\!-\!\!Cl \qquad\qquad I\!\!-\!\!Cl$$
$$\mu = 0.9\ D \qquad\qquad \mu = 0.7\ D$$

Chlorine is the positive end of the dipole in FCl and the negative end in ICl.

1.57 (a) Sodium chloride is ionic; it has a unit positive charge and a unit negative charge separated from each other. Hydrogen chloride has a polarized bond but is a covalent compound. Sodium chloride has a larger dipole moment. The measured values are as shown.

$$Na^+Cl^- \qquad \text{is more polar than} \qquad H\!\!-\!\!Cl$$
$$\mu = 9.4\ D \qquad\qquad\qquad\qquad \mu = 1.1\ D$$

(b) Fluorine is more electronegative than chlorine, and so its bond to hydrogen is more polar, as the measured dipole moments indicate.

$$H\!\!-\!\!F \qquad \text{is more polar than} \qquad H\!\!-\!\!Cl$$
$$\mu = 1.7\ D \qquad\qquad\qquad\qquad \mu = 1.1\ D$$

(c) Boron trifluoride is planar. Its individual B—F bond dipoles cancel. It has no dipole moment.

$$H\!\!-\!\!F \qquad\qquad \text{is more polar than}$$
$$\mu = 1.7\ D \qquad\qquad\qquad\qquad\qquad\qquad \mu = 0\ D$$

(d) A carbon-chlorine bond is strongly polar; carbon-hydrogen and carbon-carbon bonds are only weakly polar.

$$\text{is more polar than}$$
$$\mu = 2.1\ D \qquad\qquad\qquad\qquad\qquad\qquad \mu = 0.1\ D$$

(e) A carbon-fluorine bond in CCl$_3$F opposes the polarizing effect of the chlorines. The carbon-hydrogen bond in CHCl$_3$ reinforces it. CHCl$_3$ therefore has a larger dipole moment.

$$\text{is more polar than}$$
$$\mu = 1.0\ D \qquad\qquad\qquad\qquad\qquad\qquad \mu = 0.5\ D$$

(f) Oxygen is more electronegative than nitrogen; its bonds to carbon and hydrogen are more polar than the corresponding bonds formed by nitrogen.

$$\text{is more polar than}$$
$$\mu = 1.7\ D \qquad\qquad\qquad\qquad\qquad\qquad \mu = 1.3\ D$$

(g) The Lewis structure for CH_3NO_2 has a formal charge of +1 on nitrogen, making it more electron-attracting than the uncharged nitrogen of CH_3NH_2.

$$H_3C-\overset{+}{N}\overset{O}{\underset{O^-}{\diagup}} \qquad \text{is more polar than} \qquad H_3C-\overset{..}{N}\overset{H}{\underset{H}{\diagdown}}$$

$$\mu = 3.1\ D \qquad\qquad\qquad\qquad \mu = 1.3\ D$$

1.58 The structures, written in a form that indicates hydrogens and unshared electrons, are as shown. Remember: A neutral carbon has four bonds, a neutral nitrogen has three bonds plus one unshared electron pair, and a neutral oxygen has two bonds plus two unshared electron pairs. Halogen substituents have one bond and three unshared electron pairs.

(a) is equivalent to $(CH_3)_3CCH_2CH(CH_3)_2$

(b) is equivalent to $(CH_3)_2C=CHCH_2CH_2\overset{\overset{\displaystyle CH_2}{||}}{C}CH=CH_2$

(c) is equivalent to

(d) OH is equivalent to $CH_3\overset{..}{\underset{|}{O}}HCH_2CH_2CH_2CH_2CH_3$

(d) OH is equivalent to $CH_3CHCH_2CH_2CH_2CH_2CH_3$ with $:\ddot{O}H$ above.

(e) O is equivalent to $CH_3\overset{\overset{\displaystyle \ddot{O}:}{||}}{C}CH_2CH_2CH_2CH_2CH_3$

(f) is equivalent to

(g) is equivalent to

(*h*) is equivalent to

(*i*) is equivalent to

(*j*) is equivalent to

1.59 (*a*) C_8H_{18}　　　　　　(*e*) $C_7H_{14}O$　　　　　　(*i*) $C_{10}H_{14}N_2$

(*b*) $C_{10}H_{16}$　　　　　　(*f*) C_6H_6　　　　　　(*j*) $C_{16}H_8Br_2N_2O_2$

(*c*) $C_{10}H_{16}$　　　　　　(*g*) $C_{10}H_8$

(*d*) $C_7H_{16}O$　　　　　　(*h*) $C_9H_8O_4$

Isomers are different compounds that have the same molecular formula. Two of these compounds, (*b*) and (*c*), have the same molecular formula and are isomers of each other.

1.60 (*a*) The molecular weight of CHN is 27 or one-fifth the molecular weight of adenine (135). Multiply the empirical formula CHN by 5 to give the molecular formula $C_5H_5N_5$ for adenine.

(*b*) The molecular weight of C_3H_5 is 41. Multiply C_3H_5 by 10 to obtain a molecular formula of $C_{30}H_{50}$ for squalene which has a molecular weight of 410.

(*c*) The molecular weight of C_5H_7 is 67 or one-eighth the molecular weight of β-carotene (536). Multiply the empirical formula C_5H_7 by 8 to give the molecular formula $C_{40}H_{56}$.

(*d*) The empirical formula CH_2O has a molecular weight of 30. This value is one-sixth the molecular weight of D-fructose. Multiply CH_2O by six to give the molecular formula $C_6H_{12}O_6$.

1.61 Since 10^{-pK_a} equals K_a the ratio is $\dfrac{10^{-2}}{10^{-8}} = 10^6$. K_a for the acid with a pK_a of 2 is 1,000,000 times larger than for the acid with a pK_a of 8.

1.62 (*a*) This problem reviews the relationship between logarithms and exponential numbers. We need to determine K_a, given pK_a. The equation that relates the two is

$$pK_a = -\log_{10} K_a$$

Therefore

$$K_a = 10^{-pK_a}$$
$$= 10^{-3.48}$$
$$= 3.3 \times 10^{-4}$$

(b) As described in part (a), $K_a = 10^{-pK_a}$, therefore K_a for vitamin C is given by the expression:

$$K_a = 10^{-4.17}$$
$$= 6.7 \times 10^{-5}$$

(c) Similarly, $K_a = 1.8 \times 10^{-4}$ for formic acid (pK_a 3.75).

(d) $K_a = 6.5 \times 10^{-2}$ for oxalic acid (pK_a 1.19).

In ranking the acids in order of decreasing acidity, remember that the larger the equilibrium constant K_a, the stronger the acid; and the lower the pK_a value, the stronger the acid.

Acid	K_a	pK_a
Oxalic (strongest)	6.5×10^{-2}	1.19
Aspirin	3.3×10^{-4}	3.48
Formic acid	1.8×10^{-4}	3.75
Vitamin C (weakest)	6.7×10^{-5}	4.17

1.63 By comparison with examples found in Table 1.8 the relative acidities of these four compounds and species can be determined. The two cations are more acidic than the neutral compounds, and because oxygen is more electronegative than nitrogen, the oxygen cation is more acidic than the nitrogen cation. Likewise, the neutral compound containing oxygen is more acidic than the one containing nitrogen.

1.64   As in the previous problem, the species given in the problem can be compared with examples in Table 1.8. The best way to determine the basicity of an anion is to consider the acidity of its conjugate acid, remembering that the weaker the acid the stronger the conjugate base.

Most basic $CH_3CH_2CH_2C \equiv C\overset{\cdot\cdot}{:}^-$ is the conjugate base of $CH_3CH_2CH_2C \equiv CH$ $pK_a = 26$

$CH_3CH_2CH_2CH_2\overset{\cdot\cdot}{\underset{\cdot\cdot}{O}}:^-$ is the conjugate base of $CH_3CH_2CH_2CH_2OH$ $pK_a = 16$

$CH_3CH_2CH_2CH_2\overset{\cdot\cdot}{\underset{\cdot\cdot}{S}}:^-$ is the conjugate base of $CH_3CH_2CH_2CH_2SH$ $pK_a = 10.7$

Least basic $CH_3CH_2CH_2C\overset{\overset{\cdot\cdot}{O}:}{\underset{\underset{\cdot\cdot}{O}:^-}{\Big\backslash}}$ is the conjugate base of $CH_3CH_2CH_2C\overset{\overset{O}{\diagup\diagup}}{\underset{OH}{\diagdown}}$ $pK_a = 4.7$

1.65   When evaluating each statement, keep in mind that a strong acid is one that is stronger than hydronium ion (H_3O^+; $pK_a = -1.7$). A strong base is one that is stronger than hydroxide ion (HO^-). Hydroxide is the conjugate base of water ($pK_a = 15.7$), so for a base to be strong its conjugate acid must have a pK_a greater than 15.7.

(*a*)   HF is a weak acid ($pK_a = 3.1$). However its conjugate base, fluoride ion, is weak because it is less basic than hydroxide. The statement is false.

(*b*)   As noted in part (*a*), fluoride ion is a weak base. Its conjugate acid, hydrogen fluoride, is less acidic than hydronium ion and is a weak acid. The statement is false.

(*c*)   Ammonia (NH_3) is a weak acid ($pK_a = 36$). Because ammonia is a weaker acid than water, its conjugate base is strong. The statement is true.

(*d*)   Although ammonia is a weak base, its conjugate acid ammonium ion (NH_4^+, $pK_a = 9.3$) is not as strong as hydronium ion and is a weak acid. The statement is false.

1.66   In each case the more basic solution will be the one that contains the stronger base. Only the anions need be considered; sodium ion is a spectator ion and can be ignored. Remember that the stronger base is the one with the weaker conjugate acid.

(*a*)   HCN ($pK_a = 9.1$) is a weaker acid than HF ($pK_a = 3.1$), and a solution of CN^- (from NaCN) will be more basic than a solution of F^- (from NaF).

(*b*)   Carbonic acid ($pK_a = 6.4$) is a weaker acid than acetic acid ($pK_a = 4.7$). A solution of sodium carbonate will be more basic than a solution of sodium acetate.

(*c*)   Methanethiol, (CH_3SH, $pK_a = 10.7$) is the conjugate acid of methanethiolate ion. Methanethiol is a weaker acid than sulfuric acid ($pK_a = -4.8$), and a solution of sodium methanethiolate will be more basic than a solution of sodium sulfate.

1.67   (*a*)   Comparing two atoms as we go down a group in the periodic table, the bond to hydrogen becomes weaker and acid strength increases. Thus $(CH_3)_3\overset{+}{P}H$ is a stronger acid than $(CH_3)_3\overset{+}{N}H$.

(*b*)   The weaker acid forms the stronger conjugate base, so $(CH_3)_3\overset{\cdot\cdot}{N}:$ is a stronger base than $(CH_3)_3\overset{\cdot\cdot}{P}:$.

1.68 (*a*) Water will remove the proton from HCN in a Brønsted acid-base reaction.

$$H_3C-C\equiv N: \quad + \quad \text{...}$$

Water   +   H—C≡N: ⇌   Hydronium ion   +   :C≡N:

(*b*) Water uses one of its unshared electron pairs to remove a proton from nitrogen.

Water   +   ⇌   Hydronium ion   +

(*c*) The proton on the positively charged oxygen is the one that is transferred to water.

H_3C-C ... $:OCH_3$ +   Water   ⇌ H_3C-C ... $:OCH_3$ +   Hydronium ion

1.69 Water can act as both a base, as in the previous problem, or as an acid as in this problem. Water donates one of its protons and forms hydroxide ion in each case. Remember to use curved arrows to track electron movement, not the movement of atoms.

(*a*) $H_3C-C\equiv C:^-$ +   Water   ⇌ $H_3C-C\equiv C-H$ + $H-O:^-$ Hydroxide ion

(*b*)   ...   +   Water   ⇌   ...   +   Hydroxide ion

(*c*) H_3C-C ... N—H   +   Water   ⇌ H_3C-C ... $^+$N—H   + $H-O:^-$ Hydroxide ion

1.70 The problem states that the equilibrium in each part of this problem lies to the right. The stronger acid is on the left in each case.

(*a*) The negatively charged oxygen uses one of its unshared electron pairs to remove the proton from sulfur.

$(CH_3)_3C-O:^-$ + $H-S-CH_3$ ⇌ $(CH_3)_3C-O-H$ + $^-:S-CH_3$

Base   Acid   Conjugate acid   Conjugate base

$pK_a = 10.7$ $pK_a = 18$

The equilibrium constant is calculated from the pK_as.

$$K_{eq} = \frac{10^{-pK_a} \text{ of reactant acid}}{10^{-pK_a} \text{ of product acid}}$$

$$K_{eq} = \frac{10^{-10.7}}{10^{-18}} = 10^{7.3} = 2 \times 10^7$$

(*b*) One of the negatively charged oxygens removes the proton on oxygen. Notice that because the base (carbonate ion) has a −2 charge, its conjugate acid (hydrogen carbonate) bears a −1 charge.

Acid   +   Base   ⇌   Conjugate base   +   Conjugate acid

pK_a = 4.7   pK_a = 10.2

Calculate the equilibrium constant as in part (*a*).

$$K_{eq} = \frac{10^{-pK_a} \text{ of reactant acid}}{10^{-pK_a} \text{ of product acid}}$$

$$K_{eq} = \frac{10^{-4.7}}{10^{-10.2}} = 10^{5.5} = 3.2 \times 10^5$$

(*c*) One of the unshared electron pairs on nitrogen removes the most acidic proton of the acid, the one bound to oxygen.

Base   +   Acid   ⇌   Conjugate acid   +   Conjugate base

pK_a = 17   pK_a = 36

Calculate the equilibrium constant as in parts (*a*) and (*b*).

$$K_{eq} = \frac{10^{-pK_a} \text{ of reactant acid}}{10^{-pK_a} \text{ of product acid}}$$

$$K_{eq} = \frac{10^{-17}}{10^{-36}} = 10^{19}$$

1.71 (*a*) The acid on the left is acetylene. Amide ion is the base which abstracts a proton from acetylene. Their pK_as are shown in the equation.

$HC\equiv C-H$ + $:NH_2^-$ ⇌ $HC\equiv C:^-$ + $H-NH_2$

pK_a = 26   pK_a = 36

Acetylene is a stronger acid than ammonia. The equilibrium lies to the right. The equilibrium constant is:

$$K_{eq} = \frac{10^{-pK_a} \text{ of reactant acid}}{10^{-pK_a} \text{ of product acid}}$$

$$K_{eq} = \frac{10^{-26}}{10^{-36}} = 10^{10}$$

(b) Contrast this part of the problem with part (a). Methoxide ion is a much weaker base than the amide ion of part (a).

$$HC\equiv C\!-\!H \;+\; {}^{-}\!\ddot{\underset{\cdot\cdot}{O}}\!-\!CH_3 \;\rightleftharpoons\; HC\equiv C\!:^{-} \;+\; H\!-\!\ddot{\underset{\cdot\cdot}{O}}\!-\!CH_3$$

pK_a = 26   pK_a = 15.2

Here, the weaker acid is on the left. The equilibrium lies to the left. The equilibrium constant is more than 10^{20} times smaller than that of part (a):

$$K_{eq} = \frac{10^{-26}}{10^{-15.2}} = 10^{-10.8} = 1.6 \times 10^{-11}$$

(c) Table 1.8 tells us that the most acidic protons in the reactant (2,4-pentanedione) are those in the CH_2 group.

$$CH_3\overset{\overset{O}{\|}}{C}\overset{|}{\underset{\underset{H}{|}}{CH}}\overset{\overset{O}{\|}}{C}CH_3 \;+\; {}^{-}\!\ddot{\underset{\cdot\cdot}{O}}\!-\!CH_3 \;\rightleftharpoons\; CH_3\overset{\overset{O}{\|}}{C}\overset{}{\underset{\cdot\cdot}{CH}}\overset{\overset{O}{\|}}{C}CH_3 \;+\; H\!-\!\ddot{\underset{\cdot\cdot}{O}}\!-\!CH_3$$

pK_a = 9   pK_a = 15.2

2,4-Pentanedione is a stronger acid than methanol. The equilibrium lies to the right and the equilibrium constant is:

$$K_{eq} = \frac{10^{-9}}{10^{-15.2}} = 10^{5.2} = 1.6 \times 10^{5}$$

(d) The acidic proton in part (c) was part of a CH_2 flanked by two C=O groups. In methyl butanoate (given in Table 1.8), the CH_2 is flanked by only one C=O group and is far less acidic.

$$CH_3CH_2\overset{|}{\underset{\underset{H}{|}}{CH}}\overset{\overset{O}{\|}}{C}OCH_3 \;+\; {}^{-}\!\ddot{\underset{\cdot\cdot}{O}}\!-\!CH_3 \;\rightleftharpoons\; CH_3CH_2\overset{}{\underset{\cdot\cdot}{CH}}\overset{\overset{O}{\|}}{C}CH_3 \;+\; H\!-\!\ddot{\underset{\cdot\cdot}{O}}\!-\!CH_3$$

pK_a = 22   pK_a = 15.2

The stronger acid is on the right. Therefore, the equilibrium lies to the left and the equilibrium constant is:

$$K_{eq} = \frac{10^{-22}}{10^{-15.2}} = 10^{-6.8} = 1.6 \times 10^{-7}$$

(e) As in part (d), methyl butanoate is the reactant acid. The base in this case is far stronger than the base in part (d), however.

$pK_a = 22$ $pK_a = 36$

The stronger acid is on the left. The equilibrium lies far to the right.

$$K_{eq} = \frac{10^{-22}}{10^{-36}} = 10^{14}$$

1.72 The Conjugate base of squaric acid has the structure:

Move electron pairs are indicated, beginning at the negatively charged oxygen.

1.73 In structure A, the only contributing stricture that can be written which satisfies the octet rule is A:

A   A′

Contributing structure A′ is less stable than A because it has more charge separation than A.

In structure B, electron delocalization disperses the charge without increasing charge separation. All three resonance forms are equally stable.

B   B′   B″

Structure B is stabilized by electron delocalization more than A.

ANSWERS TO INTERPRETIVE PROBLEMS

1.74 D;   **1.75** B;   **1.76** A;   **1.77** C;   **1.78**   A

SELF-TEST

PART A

A-1. Write the electron configuration for each of the following:

(*a*) Phosphorus
(*b*) Sulfide ion in Na_2S

A-2. Determine the formal charge of each atom and the net charge for each of the following species:

(*a*) $:\ddot{N}=C=\ddot{S}:$
(*b*) $:O\equiv N-\ddot{\underset{..}{O}}:$
(*c*)

A-3. Write a second Lewis structure that satisfies the octet rule for each of the species in Problem A-2, and determine the formal charge of each atom. Which of the Lewis structures for each species in this and Problem A-2 is more stable?

A-4. Write a correct Lewis structure for each of the following. Be sure to show explicitly any unshared pairs of electrons.

(*a*) Methylamine, CH_3NH_2

(*b*) Acetaldehyde, C_2H_4O (The connectivity is CCO; all the hydrogens are connected to carbon.)

A-5. What is the molecular formula of each of the structures shown? Clearly draw any unshared electron pairs that are present.

(*a*)

(*b*)

(*c*)

(*d*)

A-6. Write the products of the acid-base reaction that follows, and identify the acid, base, conjugate acid, and conjugate base. What is the value of the equilibrium constant, K_{eq}? The approximate pK_a of NH_3 is 36; that of CH_3CH_2OH is 16.

$$CH_3CH_2O^- \quad + \quad NH_3 \quad \rightleftharpoons$$

Show all unshared electron pairs and formal charges, and use curved arrows to track electron movement.

A-7. Account for the fact that all three sulfur–oxygen bonds in SO_3 are the same by drawing the appropriate Lewis structure(s).

A-8. The cyanate ion contains 16 valence electrons, and its three atoms are connected in the order OCN. Write the most stable Lewis structure for this species, and assign a formal charge to each atom. What is the net charge of the ion?

A-9. Using the VSEPR method,

    (*a*) Describe the geometry at each carbon atom and the oxygen atom in the following molecule: $CH_3OCH{=}CHCH_3$.

    (*b*) Deduce the shape of NCl_3, and draw a three-dimensional representation of the molecule. Is NCl_3 polar?

A-10. Assign the shape of each of the following as either linear or bent.

    (*a*) CO_2 (*b*) $NO_2{}^{+}$ (*c*) $NO_2{}^{-}$

A-11. Consider structures A, B, C, and D:

        A   B   C   D

    (*a*) Which structure (or structures) contains a positively charged carbon?

    (*b*) Which structure (or structures) contains a positively charged nitrogen?

    (*c*) Which structure (or structures) contains a positively charged oxygen?

    (*d*) Which structure (or structures) contains a negatively charged carbon?

    (*e*) Which structure (or structures) contains a negatively charged nitrogen?

    (*f*) Which structure (or structures) contains a negatively charged oxygen?

    (*g*) Which structure is the most stable?

    (*h*) Which structure is the least stable?

A-12. Given the following information, write a Lewis structure for urea, CH_4N_2O. The oxygen atom and both nitrogen atoms are bonded to carbon, there is a carbon-oxygen double bond, and none of the atoms bears a formal charge. Be sure to include all unshared electron pairs.

A-13. Draw a second resonance contributor for each of the species shown. Be sure to show all unshared electron pairs, and indicate the formal charge (if any) on each atom. Is the first or the second structure more stable?

    (*a*) $\overset{-}{:\!\ddot{O}\!:}{-}C{\equiv}N:$ (*b*) $:\!\ddot{O}\!={N}\!-\overset{-}{\ddot{C}H_2}$

PART B

B-1. Which one of the following is most likely to have ionic bonds?

    (*a*) HCl   (*b*) Na_2O (*c*) N_2O (*d*) NCl_3

B-2. Which of the following is *not* an electron configuration for an atom in its ground state?

    (*a*) $1s^2 2s^2 2p_x^2 2p_y^1 2p_z^1$ (*c*) $1s^2 2s^2 2p_x^2 2p_y^2 2p_z^1$

    (*b*) $1s^2 2s^2 2p_x^2 2p_y^2 2p_z^0$ (*d*) $1s^2 2s^2 2p_x^2 2p_y^2 2p_z^2$

B-3. The formal charge on phosphorus in $(CH_3)_4P$ is

(a) 0   (b) −1   (c) +1   (d) +2

B-4. Which of the following is an isomer of compound 1?

$$H_2C-CHCH_3 \qquad CH_3CH_2\overset{\overset{\displaystyle O}{\|}}{C}H \qquad CH_3\overset{\overset{\displaystyle O}{\|}}{C}CH_3 \qquad CH_3CH=\underset{\underset{\displaystyle OH}{|}}{CH}$$
$$\underset{O}{\diagdown\diagup}$$

       1   2   3   4

(a) 2   (b) 4   (c) 2 and 3   (d) All are isomers

B-5. In which of the following is oxygen the positive end of the bond dipole?

(a) O—F   (b) O—N   (c) O—S   (d) O—H

B-6. What two structural formulas are resonance contributors of the same species?

(a) $H-C\equiv\overset{+}{N}-\overset{..}{\underset{..}{O}}:^-$ and $H-\overset{..}{\underset{..}{O}}-C\equiv N:$

(b) $H-\overset{+}{\underset{..}{O}}=C=\overset{..}{N}:^-$ and $H-\overset{..}{\underset{..}{O}}-C\equiv N:$

(c) $H-C\equiv\overset{+}{N}-\overset{..}{\underset{..}{O}}:^-$ and $H-\overset{\overset{\displaystyle :O:}{\|}}{C}-\overset{..}{N}:$

(d) $H-\overset{..}{\underset{..}{O}}-C\equiv N:$ and $H-\overset{..}{N}=C=\overset{..}{\underset{..}{O}}:$

B-7. The total number of *unshared pairs* of electrons in the molecule

is

(a) 0   (b) 1   (c) 2   (d) 3

B-8. Which of the following contains a triple bond?

(a) SO_2 (b) HCN   (c) C_2H_4 (d) NH_3

B-9. Which one of the compounds shown is *not* an isomer of two of the others?

     (a)   (b)   (c)   (d)

B-10. Which one of the following is the most stable Lewis structure? The answer must be correct in terms of bonds, unshared pairs of electrons, and formal charges.

(a) $\overset{-}{:\!\ddot{N}}\!-\!\overset{..}{\underset{..}{N}}\!-\!\overset{+}{\text{CH}_2}$

(b) $\overset{-}{:\!\ddot{N}}\!-\!\overset{..}{\underset{..}{N}}\!=\!\text{CH}_2$

(c) $:\!\ddot{N}\!=\!\overset{..}{\underset{..}{N}}\!-\!\overset{-}{\ddot{C}H_2}$

(d) $:\text{N}\!\equiv\!\overset{+}{N}\!-\!\text{CH}_2$

(e) $\overset{-}{:\!\ddot{N}}\!=\!\overset{+}{N}\!=\!\text{CH}_2$

B-11. Which of the following molecules would you expect to be *nonpolar*?

1. CH_2F_2 2. CO_2 3. CF_4 4. CH_3OCH_3

(a) 1 and 2   (b) 1 and 3   (c) 1 and 4   (d) 2 and 3   (e) 2, 3, and 4

The following two questions refer to the hypothetical compounds:

$$\text{A}-\text{B}-\text{A} \qquad \text{A}=\overset{..}{\text{B}}-\text{A} \qquad \text{A}=\text{B}=\text{A} \qquad \text{A}-\overset{..}{\underset{|}{\text{B}}}-\text{A}$$
$$\text{A}$$
$$\qquad\quad 1 \qquad\qquad\qquad 2 \qquad\qquad\qquad 3 \qquad\qquad\qquad 4$$

B-12. Which substance(s) is (are) linear?

(a) 1 only   (b) 1 and 3   (c) 1 and 2   (d) 3 only

B-13. Assuming A is more electronegative than B, which substance(s) is (are) polar?

(a) 1 and 3   (b) 2 only   (c) 4 only   (d) 2 and 4

B-14. Identify each species in the following equilibrium according to the code:
SA = stronger acid; SB = stronger base; WA = weaker acid; WB = weaker base.
The pK_a of $(CH_3)_2NH$ is 36; the pK_a of CH_3OH is 15.2.

$$CH_3OH \quad + \quad (CH_3)_2NK \quad \rightleftharpoons$$
$$\qquad\quad 1 \qquad\qquad\qquad\quad 2$$

	1	2		1	2
(a)	WA	WB	(c)	SA	SB
(b)	WB	WA	(d)	SB	SA
			(e)	WA	WA

B-15. What is the value of the equilibrium constant for the reaction in the previous problem?

(a) $10^{-20.8}$ (b) $10^{51.2}$ (c) $10^{20.8}$ (d) $10^{-51.2}$

CHAPTER 2

Alkanes and Cycloalkanes: Introduction to Hydrocarbons

SOLUTIONS TO TEXT PROBLEMS

2.1 An orbital diagram of sp^3-hybridized nitrogen is the same as for sp^3-hybridized carbon, except nitrogen has one more electron.

$$2p \ \uparrow \ \uparrow \ \uparrow \qquad 2sp^3 \ \uparrow\downarrow \ \uparrow \ \uparrow \ \uparrow$$

$$2s \ \uparrow\downarrow$$

Energy →

Ground electronic
state of nitrogen
(a)

sp^3 hybrid
state of nitrogen
(b)

The unshared electron pair in ammonia ($:NH_3$) occupies an sp^3-hybridized orbital of nitrogen. Each N—H bond corresponds to overlap of a half-filled sp^3 hybrid orbital of nitrogen and a half-filled $1s$ orbital of hydrogen.

2.2 A carbon atom is sp^3-hybridized when it is directly bonded to four atoms. Each of the three carbons of propane is bonded to four atoms, either carbon or hydrogen:

$$\begin{array}{ccc} H & H & H \\ | & | & | \\ H-C-C-C-H \\ | & | & | \\ H & H & H \end{array}$$

Each carbon–carbon σ bond arises from overlap of two half-filled sp^3 hybrid orbitals, one from each carbon atom. The eight carbon–hydrogen bonds are sp^3-$1s$ σ bonds.

2.3 An unbranched alkane (*n*-alkane) of 28 carbons has 26 methylene (CH_2) groups flanked by a methyl (CH_3) group at each end. The condensed formula is $CH_3(CH_2)_{26}CH_3$.

2.4   The alkane represented by the bond-line formula has 11 carbons. The general formula for an alkane is C_nH_{2n+2}, and thus there are 24 hydrogens. The molecular formula is $C_{11}H_{24}$; the condensed structural formula is $CH_3(CH_2)_9CH_3$.

2.5   In addition to $CH_3(CH_2)_4CH_3$ and $(CH_3)_2CHCH_2CH_2CH_3$, there are three more isomers. One has a five-carbon chain with a one-carbon (methyl) branch:

$$\underset{\qquad\qquad\;\; |}{\overset{\qquad\qquad CH_3}{CH_3CH_2CHCH_2CH_3}} \quad \text{or}$$

The remaining two isomers have two methyl branches on a four-carbon chain.

$$\underset{\underset{\qquad\;\; CH_3}{\quad|}}{\overset{\overset{\quad\; CH_3}{\quad|}}{CH_3CHCHCH_3}} \quad \text{or} \qquad\qquad \underset{\qquad\quad CH_3}{\overset{\qquad\quad CH_3}{CH_3CH_2CCH_3}} \quad \text{or}$$

2.6   (*b*) Octacosane is not listed in Table 2.2, but its structure can be deduced from its systematic name. The suffix -*cosane* pertains to alkanes that contain 20–29 carbons in their longest continuous chain. The prefix *octa*- means "eight." *Octacosane* is therefore the unbranched alkane having 28 carbon atoms. It is $CH_3(CH_2)_{26}CH_3$.

(*c*) The alkane in Problem 2.4 has an unbranched chain of 11 carbon atoms and is named *undecane*.

2.7   The ending -*hexadecane* reveals that the longest continuous carbon chain has 16 carbon atoms.

There are four methyl groups (represented by *tetramethyl*-), and they are located at carbons 2, 6, 10, and 14.

2,6,10,14-Tetramethylhexadecane
(phytane)

2.8   (*b*) The systematic name of the unbranched C_5H_{12} isomer is *pentane* (Table 2.2).

$$CH_3CH_2CH_2CH_2CH_3$$

IUPAC name: **pentane**

A second isomer, $(CH_3)_2CHCH_2CH_3$, has four carbons in the longest continuous chain and so is named as a derivative of butane. Because it has a methyl group at C-2, it is *2-methylbutane*.

$$\underset{\qquad\; CH_3}{\overset{}{CH_3CHCH_2CH_3}}$$

IUPAC name: **2-methylbutane**

The remaining isomer, $(CH_3)_4C$, has three carbons in its longest continuous chain and so is named as a derivative of propane. There are two methyl groups at C-2, and so it is a 2,2-dimethyl derivative of propane.

$$
\begin{array}{c}
CH_3 \\
| \\
CH_3CCH_3 \\
| \\
CH_3
\end{array}
$$

IUPAC name: **2,2-dimethylpropane**

(*c*) First write out the structure in more detail, and identify the longest continuous carbon chain.

$$
H_3C-\underset{\underset{CH_3}{|}}{\overset{\overset{CH_3}{|}}{C}}-CH_2-\underset{\underset{CH_3}{|}}{\overset{\overset{H}{|}}{C}}-CH_3
$$

There are five carbon atoms in the longest chain, and so the compound is named as a derivative of pentane. This five-carbon chain has three methyl substituents attached to it, making it a trimethyl derivative of pentane. Number the chain in the direction that gives the lowest numbers to the substituents at the first point of difference.

$$
H_3\overset{1}{C}-\underset{\underset{CH_3}{|}}{\overset{\overset{CH_3}{2|}}{C}}-\overset{3}{CH_2}-\underset{\underset{CH_3}{|}}{\overset{\overset{H}{4|}}{C}}-\overset{5}{CH_3} \quad not \quad H_3\overset{5}{C}-\underset{\underset{CH_3}{|}}{\overset{\overset{CH_3}{4|}}{C}}-\overset{3}{CH_2}-\underset{\underset{CH_3}{|}}{\overset{\overset{H}{2|}}{C}}-\overset{1}{CH_3}
$$

2,2,4-Trimethylpentane (correct)   2,4,4-Trimethylpentane (incorrect)

(*d*) The longest continuous chain in $(CH_3)_3CC(CH_3)_3$ contains four carbon atoms.

$$
H_3C-\underset{\underset{CH_3}{|}}{\overset{\overset{CH_3}{|}}{C}}-\underset{\underset{CH_3}{|}}{\overset{\overset{CH_3}{|}}{C}}-CH_3
$$

The compound is named as a tetramethyl derivative of butane; it is *2,2,3,3-tetramethylbutane*.

2.9   There are three C_5H_{11} alkyl groups with unbranched carbon chains. One is primary, and two are secondary. The IUPAC name of each group is given beneath the structure. Remember to number the alkyl groups from the potential point of attachment.

$$
CH_3CH_2CH_2CH_2CH_2- \qquad \overset{4}{C}H_3\overset{3}{C}H_2\overset{2}{C}H_2\overset{}{\underset{|}{C}}HCH_3 \qquad \overset{3}{C}H_3\overset{2}{C}H_2\overset{1}{\underset{|}{C}}HCH_2CH_3
$$

Pentyl group (primary)   1-Methylbutyl group (secondary)   1-Ethylpropyl group (secondary)

Four alkyl groups are derived from $(CH_3)_2CHCH_2CH_3$. Two are primary, one is secondary, and one is tertiary.

$$\overset{CH_3}{\underset{4|321}{CH_3CHCH_2CH_2-}}$$

3-Methylbutyl group (primary)

$$\overset{CH_3}{\underset{1|234}{-CH_2CHCH_2CH_3}}$$

2-Methylbutyl group (primary)

$$\overset{CH_3}{\underset{1|23}{CH_3CCH_2CH_3}}$$
$$|$$

1,1-Dimethylpropyl group (tertiary)

$$\overset{CH_3}{\underset{3|21}{CH_3CHCHCH_3}}$$
$$|$$

1,2-Dimethylpropyl group (secondary)

2.10 (*b*) Begin by writing the structure in more detail, showing each of the groups written in parentheses. The compound is named as a derivative of hexane, because it has six carbons in its longest continuous chain.

$$\underset{CH_3CH_2}{\overset{654321}{CH_3CH_2CHCH_2CHCH_3}}$$
$$CH_3CH_2CH_3$$

The chain is numbered so as to give the lowest number to the substituent that appears closest to the end of the chain. In this case it is numbered right to left so that the substituents are located at C-2 and C-4 rather than at C-3 and C-5. In alphabetical order the groups are ethyl and methyl; they are listed in alphabetical order in the name. The compound is *4-ethyl-2-methylhexane*.

(*c*) The longest continuous chain is shown in the structure; it contains ten carbon atoms. The structure also shows the numbering scheme that gives the lowest number to the substituent nearest the end of the chain.

$$\overset{CH_3CH_3}{\underset{10987|654|}{CH_3CH_2CHCH_2CHCH_2CHCHCH_3}}$$
$$CH_2CH_3CH_2CHCH_3$$
$$|32$$
$$|1$$
$$CH_3$$

In alphabetical order, the substituents are ethyl (at C-8), isopropyl (at C-4), and two methyl groups (at C-2 and C-6). The alkane is *8-ethyl-4-isopropyl-2,6-dimethyldecane*. The systematic name for the isopropyl group (1-methylethyl) may also be used, and the name becomes *8-ethyl-2,6-dimethyl-4-(1-methylethyl)decane*. Notice in both names that the "di" prefix is ignored when alphabetizing.

2.11 (*b*) There are ten carbon atoms in the ring in this cycloalkane, thus it is named as a derivative of cyclodecane.

Cyclodecane

The numbering pattern of the ring is chosen so as to give the lowest number to the substituent at the first point of difference between them. Thus, the carbon bearing two methyl groups is C-1, and the ring is numbered counterclockwise, placing the isopropyl group on C-4 (numbering clockwise would place the isopropyl on C-8). Listing the substituent groups in alphabetical order, the correct name is *4-isopropyl-1,1-dimethylcyclodecane*. Alternatively, the systematic name for isopropyl (1-methylethyl) could be used, and the name would become *1,1-dimethyl-4-(1-methylethyl)cyclodecane*.

(*c*) When two cycloalkyl groups are attached by a single bond, the compound is named as a cycloalkyl-substituted cycloalkane. This compound is *cyclohexylcyclohexane*.

2.12 The alkane that has the most carbons (nonane) has the highest boiling point (151°C). Among the others, all of which have eight carbons, the unbranched isomer (octane) has the highest boiling point (126°C) and the most branched one (2,2,3,3-tetramethylbutane) the lowest (106°C). The remaining alkane, 2-methylheptane, boils at 116°C.

2.13 All hydrocarbons burn in air to give carbon dioxide and water. To balance the equation for the combustion of cyclohexane (C_6H_{12}), first balance the carbons and the hydrogens on the right side. Then balance the oxygens on the left side.

2.14 (*b*) Icosane (Table 2.2) is $C_{20}H_{42}$. It has four more methylene (CH_2) groups than hexadecane, the last unbranched alkane in Table 2.3. Its calculated heat of combustion is therefore (4×653 kJ/mol) higher.

Heat of combustion of icosane = heat of combustion of hexadecane + (4×653 kJ/mol)

$$= 10,701 \text{ kJ/mol} + 2612 \text{ kJ/mol}$$

$$= 13,313 \text{ kJ/mol}$$

2.15 Two factors that influence the heats of combustion of alkanes are, in order of decreasing importance, (1) the number of carbon atoms and (2) the extent of chain branching. Pentane, 2-methylbutane, and 2,2-dimethylpropane are all C_5H_{12}; hexane is C_6H_{14}. Hexane has the largest heat of combustion. Branching leads to a lower heat of combustion; 2,2-dimethylpropane is the most branched and has the lowest heat of combustion.

Hexane	$CH_3(CH_2)_4CH_3$	Heat of combustion 4163 kJ/mol (995.0 kcal/mol)
Pentane	$CH_3CH_2CH_2CH_2CH_3$	Heat of combustion 3536 kJ/mol (845.3 kcal/mol)
2-Methylbutane	$(CH_3)_2CHCH_2CH_3$	Heat of combustion 3529 kJ/mol (843.4 kcal/mol)
2,2-Dimethylpropane	$(CH_3)_4C$	Heat of combustion 3514 kJ/mol (839.9 kcal/mol)

2.16 An oxidation number for carbon can be calculated from the molecular formula of ethanol (C_2H_6O) by first assigning oxidation numbers to hydrogen (+1) and to oxygen (–2). The sum of the oxidation numbers of the atoms in a neutral compound equals zero, so by this method both carbons have an oxidation number of –2. As described in Table 2.5, the oxidation numbers of the carbon atoms in ethanol are –3 for the CH_3 carbon and –1 for the carbon of the CH_2OH group. The difference between the two methods is that a calculation based on the molecular formula gives the *average* oxidation state for the two carbons.

2.17 In the first reaction a carbon–oxygen bond is replaced by a carbon–chlorine bond.

$$(CH_3)_3C{-}OH \ + \ HCl \ \longrightarrow \ (CH_3)_3C{-}Cl \ + \ H_2O$$

Both oxygen and chlorine are more electronegative than carbon, so the oxidation number of carbon does not change in this reaction. In the second reaction, a carbon–hydrogen bond is replaced by a carbon–bromine bond.

$$(CH_3)_3C{-}H \ + \ Br_2 \ \longrightarrow \ (CH_3)_3C{-}Br \ + \ HBr$$

Bromine is more electronegative than carbon; hydrogen is less electronegative. The oxidation number of carbon has increased, and carbon has been oxidized. Reduction of bromine has also occurred; the oxidation number of bromine in Br_2 is 0, in HBr it is –1.

2.18 The carbons of the double bond are both sp^2-hybridized. The carbon of the methyl group is sp^3-hybridized. The bond indicated with the arrow is formed by overlap of an sp^2 and an sp^3 hybrid orbital. It is a σ bond, as are all single bonds.

$$H_2C{=}CH{-}CH_3$$

sp^2 - sp^3 σ bond

2.19 The carbons of the double bond are both sp^2-hybridized. The carbons of the triple bond are sp-hybridized. The σ bond indicated with the arrow is formed by overlap of an sp^2 and an sp hybrid orbital.

$$H_2C{=}CH{-}C{\equiv}CH$$

sp^2 - sp σ bond

Vinylacetylene

There are a total of seven σ bonds in vinylacetylene: three sp^2-$1s$ C—H bonds, one sp-$1s$ C—H bond, one sp^2-sp^2 C—C bond (the σ component of the double bond), one sp-sp C—C bond (the σ component of the triple bond), and the sp^2-sp C—C bond described earlier. In addition there are three π bonds: one in the double bond and two in the triple bond.

2.20 Each ring decreases the number of hydrogens in the general molecular formula of the hydrocarbon by two.

(*a*) Cycloalkanes have the general formula C_nH_{2n}. Compare hexane (C_6H_{14}) with cyclohexane (C_6H_{12}), for example:

CH₃CH₂CH₂CH₂CH₂CH₃

Hexane
(C_6H_{14})

Cyclohexane
(C_6H_{12})

(b) The general formula of an alkene has two fewer hydrogens than that of an alkane and is C_nH_{2n}. A typical example is 1-hexene (C_6H_{12}):

$$H_2C=CHCH_2CH_2CH_2CH_3$$

1-Hexene
(C_6H_{12})

(c) A triple bond decreases the general molecular formula by four hydrogens; the general molecular formula of an alkyne is C_nH_{2n-2}. An example is 1-hexyne (C_6H_{10}).

$$HC\equiv CCH_2CH_2CH_2CH_3$$

1-Hexyne
(C_6H_{10})

(d) A compound that has both a ring and a double bond would have the general molecular formula C_nH_{2n-2}. Cyclohexene is an example:

Cyclohexene
(C_6H_{10})

2.21 A hydrocarbon with a molecular formula C_5H_8 fits the general formula C_nH_{2n-2}. Therefore the compound must have two rings, two double bonds, one double bond and one ring, or one triple bond.

(a) The hydrocarbon cannot be a cycloalkane because a cycloalkane with five carbon atoms would have the formula C_5H_{10}.

(b–d) Each of the remaining possibilities (one ring and one double bond, two double bonds, and one triple bond) are plausible for a formula C_5H_8.

2.22 (a) Isomers are different compounds that have the same molecular formula. Butane and 2-methylpropane both have the molecular formula C_4H_{10} and are isomers. The molecular formula of cyclobutane is C_4H_8 and that of 2-methylbutane is C_5H_{12}.

Butane	2-Methylpropane	Cyclobutane	2-Methylbutane
(C_4H_{10})	(C_4H_{10})	(C_4H_8)	(C_5H_{12})

(b) The general molecular formula for a compound with either a ring or a double bond is C_nH_{2n}. Cyclopentane and 1-pentene are isomers; the molecular formula of both compounds is C_5H_{10}.

Cyclopentane	1-Pentene	2,2-Dimethylpropane	Cyclopentene
(C_5H_{10})	(C_5H_{10})	(C_5H_{12})	(C_5H_8)

(*c*) Any combination of four rings or double bonds will give a compound with the general molecular formula C_nH_{2n-6}. A compound with two triple bonds will also have the same molecular formula. All four compounds shown are isomers because they all have the molecular formula C_7H_8.

$-CH_3$ $=CH_2$ $HC \equiv CCH_2CH_2C \equiv CCH_3$

One ring; three   One ring; three   One ring; three   Two triple bonds
double bonds   double bonds   double bonds

2.23 It is best to approach problems of this type systematically. Because the problem requires all the isomers of C_7H_{16} to be written, begin with the unbranched isomer heptane.

$CH_3CH_2CH_2CH_2CH_2CH_2CH_3$

Heptane

Two isomers have six carbons in their longest continuous chain. One bears a methyl substituent at C-2, the other a methyl substituent at C-3.

$(CH_3)_2CHCH_2CH_2CH_2CH_3$ $CH_3CH_2CHCH_2CH_2CH_3$
                                                    $|$
                                                    CH_3

2-Methylhexane   3-Methylhexane

Now consider all the isomers that have two methyl groups as substituents on a five-carbon continuous chain.

$(CH_3)_3CCH_2CH_2CH_3$ $(CH_3CH_2)_2C(CH_3)_2$

2,2-Dimethylpentane   3,3-Dimethylpentane

$(CH_3)_2CHCHCH_2CH_3$ $(CH_3)_2CHCH_2CH(CH_3)_2$
              $|$
              CH_3

2,3-Dimethylpentane   2,4-Dimethylpentane

There is one isomer characterized by an ethyl substituent on a five-carbon chain:

$(CH_3CH_2)_3CH$

3-Ethylpentane

The remaining isomer has three methyl substituents attached to a four-carbon chain.

$(CH_3)_3CCH(CH_3)_2$

2,2,3-Trimethylbutane

2.24 In the course of doing this problem, you will write and name the 17 alkanes that, in addition to octane, $CH_3(CH_2)_6CH_3$, comprise the 18 constitutional isomers of C_8H_{18}.

(*a*) The easiest way to attack this part of the exercise is to draw a bond-line depiction of heptane and add a methyl branch to the various positions.

        2-Methylheptane   3-Methylheptane   4-Methylheptane

Other structures bearing a continuous chain of seven carbons would be duplicates of these isomers rather than unique isomers. "5-Methylheptane," for example, is an incorrect name for 3-methylheptane, and "6-methylheptane" is an incorrect name for 2-methylheptane.

(*b*) Six of the isomers named as derivatives of hexane contain two methyl branches on a continuous chain of six carbons.

    2,2-Dimethylhexane   2,3-Dimethylhexane   2,4-Dimethylhexane   2,5-Dimethylhexane

          3,3-Dimethylhexane   3,4-Dimethylhexane

One isomer bears an ethyl substituent:

                3-Ethylhexane

(*c*) Four isomers are trimethyl-substituted derivatives of pentane:

   2,2,3-Trimethylpentane   2,3,3-Trimethylpentane   2,2,4-Trimethylpentane   2,3,4-Trimethylpentane

Two bear an ethyl group and a methyl group on a continuous chain of five carbons:

       3-Ethyl-2-methylpentane   3-Ethyl-3-methylpentane

(d) Only one isomer is named as a derivative of butane:

2,2,3,3-Tetramethylbutane

2.25 The IUPAC name for pristane reveals that the longest chain contains 15 carbon atoms (as indicated by *-pentadecane*). The chain is substituted with four methyl groups at the positions indicated in the name.

Pristane
(2,6,10,14-tetramethylpentadecane)

2.26 (a) The alkane contains 13 carbons. Because all alkanes have the molecular formula C_nH_{2n+2}, the molecular formula must be $C_{13}H_{28}$.

(b) The longest continuous chain is indicated and numbered as shown.

$$CH_3CHCH_2CH_2CHCHCH_3$$
with substituents CH_3 and CH_2CH_3 (top) and $CH_2CH_2CH_3$

In alphabetical order, the substituents are ethyl (at C-5), methyl (at C-2), methyl (at C-6). The IUPAC name is *5-ethyl-2,6-dimethylnonane*.

(c) Fill in the hydrogens in the alkane to identify the various kinds of groups present. There are five *methyl* (CH_3) groups, five *methylene* (CH_2) groups, and three *methine* (CH) groups in the molecule.

(d) A primary carbon is attached to one other carbon. There are five primary carbons (the carbons of the five CH_3 groups). A secondary carbon is attached to two other carbons, and there are five of these (the carbons of the five CH_2 groups). A tertiary carbon is attached to three other carbons, and there are three of these (the carbons of the three methine groups). A quaternary carbon is attached to four other carbons. None of the carbons is a quaternary carbon.

2.27 (a) This compound is an unbranched alkane with 27 carbons. Table 2.2 in the text indicates that alkanes with 20–29 carbons have names ending in *-cosane*. Thus, we add the prefix *hepta-* ("seven") to *-cosane* to name the alkane $CH_3(CH_2)_{25}CH_3$ as *heptacosane*.

(b) The alkane $(CH_3)_2CHCH_2(CH_2)_{14}CH_3$ has 18 carbons in its longest continuous chain. It is named as a derivative of *octadecane*. There is a single substituent, a methyl group at C-2. The compound is *2-methyloctadecane*.

(c) Write the structure out in more detail to reveal that it is *3,3,4-triethylhexane*.

$(CH_3CH_2)_3CCH(CH_2CH_3)_2$ is rewritten as $CH_3CH_2\overset{2}{C}H_2\overset{3}{C}\text{---}\overset{4}{C}HCH_2CH_3$ with CH_3CH_2 and CH_2CH_3 groups above and CH_3CH_2 below

(d) Each line of a bond-line formula represents a bond between two carbon atoms. Hydrogens are added so that the number of bonds to each carbon atom totals four.

is the same as $CH_3CH_2CHCH_2C(CH_3)_3$ with CH_2CH_3

The IUPAC name is *4-ethyl-2,2-dimethylhexane*.

(e) is the same as $CH_3CH_2CHCH_2CHCH_2CH_3$ with CH_3 and CH_3

The IUPAC name is *3,5-dimethylheptane*.

(f) Number the chain in the direction shown to give *3-ethyl-4,5,6-trimethyloctane*. When numbered in the opposite direction, the locants are also 3, 4, 5, and 6. In the case of ties, however, choose the direction that gives the lower number to the substituent that appears first in the name. "Ethyl" precedes "methyl" alphabetically.

2.28 (a) The longest continuous chain contains nine carbon atoms. Begin the problem by writing and numbering the carbon skeleton of nonane.

Now add two methyl groups (one to C-2 and the other to C-3) and an isopropyl group (to C-6) to give a structural formula for 6-isopropyl-2,3-dimethylnonane.

or $CH_3CHCHCH_2CH_2CHCH_2CH_2CH_3$ with CH_3, CH_3, $CH(CH_3)_2$

(b) To the carbon skeleton of heptane (seven carbons) add a *tert*-butyl group to C-4 and a methyl group to C-3 to give 4-*tert*-butyl-3-methylheptane.

or $CH_3CH_2CHCHCH_2CH_2CH_3$ with $C(CH_3)_3$ and CH_3

(c) An isobutyl group is $-CH_2CH(CH_3)_2$. The structure of 4-isobutyl-1,1-dimethylcyclohexane is as shown.

or $-CH_2CH(CH_3)_2$

(*d*) A *sec*-butyl group is $CH_3CHCH_2CH_3$. *sec*-Butylcycloheptane has a *sec*-butyl group on a seven-

membered ring.

$CH_3CHCH_2CH_3$

or

(*e*) A cyclobutyl group is a substituent on a five-membered ring in cyclobutylcyclopentane.

2.29 (*a*) The group $CH_3(CH_2)_{10}CH_2-$ is an unbranched alkyl group with 12 carbons. It is a *dodecyl group*. The carbon at the point of attachment is directly attached to only one other carbon. It is a primary alkyl group.

(*b*) The longest continuous chain from the point of attachment is six carbons; it is a hexyl group bearing an ethyl substituent at C-3. The group is a *3-ethylhexyl group*. It is a primary alkyl group.

$$-\overset{1}{C}H_2\overset{2}{C}H_2\overset{3}{C}H\overset{4}{C}H_2\overset{5}{C}H_2\overset{6}{C}H_3$$
$$\underset{CH_2CH_3}{|}$$

(*c*) By writing the structural formula of this alkyl group in more detail, we see that the longest continuous chain from the point of attachment contains three carbons. It is a *1,1-diethylpropyl* group. Because the carbon at the point of attachment is directly bonded to three other carbons, it is a tertiary alkyl group.

$-C(CH_2CH_3)_3$ is rewritten as $-\overset{1}{\underset{CH_2CH_3}{\overset{CH_2CH_3}{C}}}\overset{2}{C}H_2\overset{3}{C}H_3$

(*d*) This group contains four carbons in its longest continuous chain. It is named as a butyl group with a cyclopropyl substituent at C-1. It is a *1-cyclopropylbutyl* group and is a secondary alkyl group.

$$-\overset{1}{C}H\overset{2}{C}H_2\overset{3}{C}H_2\overset{4}{C}H_3$$

(*e*, *f*) A two-carbon group that bears a cyclohexyl substituent is a *cyclohexylethyl* group. Number from the potential point of attachment when assigning a locant to the cyclohexyl group.

$-\overset{2}{C}H_2\overset{1}{C}H_2-$ $-\overset{1}{C}H-$
                                     $\underset{CH_3}{\overset{2|}{|}}$

2-Cyclohexylethyl   1-Cyclohexylethyl
(primary)   (secondary)

2.30 The C_4H_9 alkyl groups are named according to the 2004 IUPAC recommendations as follows:

$$\overset{4}{C}H_3\overset{3}{C}H_2\overset{2}{C}H_2\overset{1}{C}H_2- \qquad \overset{1}{C}H_3\overset{2}{C}H\overset{3}{C}H_2\overset{4}{C}H_3 \qquad \overset{3}{C}H_3\overset{2}{C}H\overset{1}{C}H_2- \qquad$$

Butan-1-yl   Butan-2-yl   2-Methylpropan-1-yl   2-Methylpropan-2-yl

2.31 (a) Methylene groups are $-CH_2-$. $ClCH_2CH_2CH_2CH_2Cl$ is therefore the $C_4H_8Cl_2$ isomer in which all the carbons belong to methylene groups.

(b) The $C_4H_8Cl_2$ isomers that lack methylene groups are

$$(CH_3)_2CHCHCl_2 \quad \text{and} \quad CH_3CHCHCH_3$$
$$\overset{}{\underset{Cl}{|}} \overset{}{\underset{Cl}{|}}$$

2.32 Because it is an alkane, the sex attractant of the tiger moth has a molecular formula of C_nH_{2n+2}. The number of carbons and hydrogens may be calculated from its molecular weight.

$$\text{(atomic weight of carbon)}n + \text{(atomic weight of hydrogen)}(2n+2) = 254$$
$$12n + 1(2n+2) = 254$$
$$14n = 252$$
$$n = 18$$

The molecular formula of the alkane is $C_{18}H_{38}$. In the problem the sex attractant is stated as a 2-methyl-branched alkane. It is therefore 2-methylheptadecane, $(CH_3)_2CHCH_2(CH_2)_{13}CH_3$.

2.33 When any hydrocarbon is burned in air, the products of combustion are carbon dioxide and water.

(a) $CH_3(CH_2)_8CH_3 \quad + \quad \frac{31}{2}O_2 \quad \longrightarrow \quad 10CO_2 \quad + \quad 11H_2O$

Decane   Oxygen   Carbon   Water
$(C_{10}H_{22})$ dioxide

(b) $+ \quad 15O_2 \quad \longrightarrow \quad 10CO_2 \quad + \quad 10H_2O$

Cyclodecane   Oxygen   Carbon   Water
$(C_{10}H_{20})$ dioxide

(c) $-CH_3 \quad + \quad 15O_2 \quad \longrightarrow \quad 10CO_2 \quad + \quad 10H_2O$

Methylcyclononane   Oxygen   Carbon   Water
$(C_{10}H_{20})$ dioxide

(d) $+ \quad \frac{29}{2}O_2 \quad \longrightarrow \quad 10CO_2 \quad + \quad 9H_2O$

Cyclopentylcyclopentane   Oxygen   Carbon   Water
$(C_{10}H_{18})$ dioxide

2.34 To determine the quantity of heat evolved per unit mass of material, divide the heat of combustion by the molecular weight.

> Methane   Heat of combustion = 890 kJ/mol (212.8 kcal/mol)
> Molecular weight = 16.0 g/mol
> Heat evolved per gram = 55.6 kJ/g (13.3 kcal/g)
>
> Butane   Heat of combustion = 2876 kJ/mol (687.4 kcal/mol)
> Molecular weight = 58.0 g/mol
> Heat evolved per gram = 49.6 kJ/g (11.8 kcal/g)

When equal masses of methane and butane are compared, methane evolves more heat when it is burned. Equal volumes of gases contain an equal number of moles, so that when equal volumes of methane and butane are compared, the one with the greater heat of combustion in kilojoules (or kilocalories) per mole gives off more heat. Butane evolves more heat when it is burned than does an equal volume of methane.

2.35 When comparing heats of combustion of alkanes, two factors are important:

1. The heats of combustion of alkanes increase as the number of carbon atoms increases.

2. An unbranched alkane has a greater heat of combustion than a branched isomer.

(*a*) In the group hexane, heptane, and octane, three unbranched alkanes are being compared. Octane (C_8H_{18}) has the most carbons and has the greatest heat of combustion. Hexane (C_6H_{14}) has the fewest carbons and the lowest heat of combustion. The measured values in this group are as follows:

> Hexane   Heat of combustion 4163 kJ/mol (995.0 kcal/mol)
> Heptane   Heat of combustion 4817 kJ/mol (1151.3 kcal/mol)
> Octane   Heat of combustion 5471 kJ/mol (1307.5 kcal/mol)

(*b*) 2-Methylpropane has fewer carbons than either pentane or 2-methylbutane and so is the member of the group with the lowest heat of combustion. 2-Methylbutane is a 2-methyl-branched isomer of pentane and so has a lower heat of combustion. Pentane has the highest heat of combustion among these compounds.

2-Methylpropane	$(CH_3)_3CH$	Heat of combustion 2868 kJ/mol (685.4 kcal/mol)
2-Methylbutane	$(CH_3)_2CHCH_2CH_3$	Heat of combustion 3529 kJ/mol (843.4 kcal/mol)
Pentane	$CH_3CH_2CH_2CH_2CH_3$	Heat of combustion 3536 kJ/mol (845.3 kcal/mol)

(*c*) 2-Methylbutane and 2,2-dimethylpropane each have fewer carbons than 2-methylpentane, which therefore has the greatest heat of combustion. 2,2-Dimethylpropane is more highly branched than 2-methylbutane; 2,2-dimethylpropane has the lowest heat of combustion.

2,2-Dimethylpropane	$(CH_3)_4C$	Heat of combustion 3514 kJ/mol (839.9 kcal/mol)
2-Methylbutane	$(CH_3)_2CHCH_2CH_3$	Heat of combustion 3529 kJ/mol (843.4 kcal/mol)
2-Methylpentane	$(CH_3)_2CHCH_2CH_2CH_3$	Heat of combustion 4157 kJ/mol (993.6 kcal/mol)

(d) Chain branching has a small effect on heat of combustion; the number of carbons has a much larger effect. The alkane with the most carbons in this group is 3,3-dimethylpentane; it has the greatest heat of combustion. Pentane has the fewest carbons in this group and has the smallest heat of combustion.

Pentane	$CH_3CH_2CH_2CH_2CH_3$	Heat of combustion 3527 kJ/mol (845.3 kcal/mol)
3-Methylpentane	$(CH_3CH_2)_2CHCH_3$	Heat of combustion 4159 kJ/mol (994.1 kcal/mol)
3,3-Dimethylpentane	$(CH_3CH_2)_2C(CH_3)_2$	Heat of combustion 4804 kJ/mol (1148.3 kcal/mol)

(e) In this series the heat of combustion increases with increasing number of carbons. Ethylcyclopentane has the lowest heat of combustion; ethylcycloheptane has the greatest.

Ethylcyclopentane
4592 kJ/mol
(1097.5 kcal/mol)

Ethylcyclohexane
5222 kJ/mol
(1248.2 kcal/mol)

Ethylcycloheptane
(combustion data not available)

2.36 (a) The equation for the hydrogenation of ethylene is given by the sum of the following three reactions:

(1) $H_2(g) + \frac{1}{2}O_2(g) \longrightarrow H_2O(l)$ $\Delta H^\circ = -286$ kJ

(2) $H_2C{=}CH_2(g) + 3O_2(g) \longrightarrow 2CO_2(g) + 2H_2O(l)$ $\Delta H^\circ = -1410$ kJ

(3) $3H_2O(l) + 2CO_2(g) \longrightarrow CH_3CH_3(g) + \frac{7}{2}O_2(g)$ $\Delta H^\circ = +1560$ kJ

Sum: $H_2C{=}CH_2(g) + H_2(g) \longrightarrow CH_3CH_3(g)$ $\Delta H^\circ = -136$ kJ

Equations (1) and (2) are the combustion of hydrogen and ethylene, respectively, and ΔH° values for these reactions are given in the statement of the problem. Equation (3) is the reverse of the combustion of ethane, and its value of ΔH° is the negative of the heat of combustion of ethane.

(b) Again we need to collect equations of reactions for which the ΔH° values are known.

(1) $H_2(g) + \frac{1}{2}O_2(g) \longrightarrow H_2O(l)$ $\Delta H^\circ = -286$ kJ

(2) $HC{\equiv}CH(g) + \frac{5}{2}O_2(g) \longrightarrow 2CO_2(g) + 2H_2O(l)$ $\Delta H^\circ = -1300$ kJ

(3) $2CO_2(g) + 2H_2O(l) \longrightarrow H_2C{=}CH_2(g) + 3O_2(g)$ $\Delta H^\circ = +1410$ kJ

Sum: $HC{\equiv}CH(g) + H_2(g) \longrightarrow H_2C{=}CH_2(g)$ $\Delta H^\circ = -176$ kJ

Equations (1) and (2) are the combustion of hydrogen and acetylene, respectively. Equation (3) is the reverse of the combustion of ethylene, and its value of ΔH° is the negative of the heat of combustion of ethylene.

The value of ΔH° for the hydrogenation of acetylene to ethane is equal to the sum of the two reactions just calculated:

$HC{\equiv}CH(g) + H_2(g) \longrightarrow H_2C{=}CH_2(g)$ $\Delta H^\circ = -176$ kJ

$H_2C{=}CH_2(g) + H_2(g) \longrightarrow CH_3CH_3(g)$ $\Delta H^\circ = -136$ kJ

Sum: $HC{\equiv}CH(g) + 2H_2(g) \longrightarrow CH_3CH_3(g)$ $\Delta H^\circ = -312$ kJ

(c) We use the equations for the combustion of ethane, ethylene, and acetylene as shown.

(1) $2H_2C{=}CH_2(g) + 6O_2(g) \longrightarrow 4CO_2(g) + 4H_2O(l)$ $\Delta H° = -2820$ kJ

(2) $2CO_2(g) + H_2O(l) \longrightarrow HC{\equiv}CH(g) + \frac{5}{2}O_2(g)$ $\Delta H° = +1300$ kJ

(3) $3H_2O(l) + 2CO_2(g) \longrightarrow CH_3CH_3(g) + \frac{7}{2}O_2(g)$ $\Delta H° = +1560$ kJ

Sum: $2H_2C{=}CH_2(g) \longrightarrow CH_3CH_3(g) + HC{\equiv}CH(g)$ $\Delta H° = +40$ kJ

The value of $\Delta H°$ for reaction (1) is twice that for the combustion of ethylene because 2 moles of ethylene are involved.

2.37 Greater stability is *not* the reason branched isomers have lower boiling points. Branching affects boiling point because of the effect on a molecule's shape. A branched isomer is more spherical, and has a smaller surface than an unbranched isomer. The smaller surface area results in fewer induced-dipole/induced-dipole intermolecular attractions, and thus a lower boiling point.

Branched isomers are more stable because of stronger intramolecular van der Waals forces. Although the trends in boiling point and stability are parallel, there is not a cause and effect relationship between them. One trend is not the reason for the other.

2.38 Branched alkanes actually give off less heat, not more, on combustion. Branched isomers of an alkane are more stable than unbranched ones. A compound that is more stable has a lower heat of combustion and will give off less heat than a less stable isomer.

2.39 (a) In the reaction

$$2CH_3Cl + Si \longrightarrow (CH_3)_2SiCl_2$$

bonds between carbon and an atom more electronegative than itself (chlorine) are replaced by bonds between carbon and an atom less electronegative than itself (silicon). Carbon is reduced; silicon is oxidized.

(b) Silicon has a tetrahedral shape in $(CH_3)_2SiCl_2$. This shape is best described by the sp^3-hybridization bonding model. Silicon is in the third row of the periodic table so the principal quantum number, n, equals 3.

2.40 (a) The reaction between pentane and fluorine can be described by the equation:

$$CH_3CH_2CH_2CH_2CH_3 + 16F_2 \longrightarrow 5CF_4 + 12HF$$

     Pentane   Fluorine   Carbon   Hydrogen
                                             tetrafluoride   fluoride

(b) Each carbon of pentane becomes bonded to four fluorines in the reaction shown. Fluorine is more electronegative than carbon, and each carbon atom has become oxidized.

2.41 Two atoms appear in their elemental state: Na on the left and H_2 on the right. The oxidation state of an atom in its elemental state is 0. Assign an oxidation state of +1 to the hydrogen in the OH group of CH_3CH_2OH. H goes from +1 on the left to 0 on the right; it is reduced. Na goes from 0 on the left to +1 on the right; it is oxidized.

$$2CH_3CH_2\overset{+1}{O}H + 2\overset{0}{Na} \longrightarrow 2CH_3CH_2O\overset{+1}{Na} + \overset{0}{H_2}$$

2.42 (*a, b*) An oxidizing agent is required when the reaction to be carried out is an oxidation. Carbon is oxidized when the oxygen content of the molecule has increased, as in the top reaction. A reducing agent brings about a reduction. Reduction occurs when the hydrogen content of the molecule increases or the oxygen content (or bonds to oxygen) decreases, or both. The second and third reactions are reductions.

$$CH_3\overset{\overset{\displaystyle O}{\|}}{C}C(CH_3)_3 \quad \text{Compound A}$$

oxidation $\longrightarrow$

$$CH_3\overset{\overset{\displaystyle O}{\|}}{C}OC(CH_3)_3$$

reduction $\longrightarrow$

$$CH_3CH_2C(CH_3)_3$$

reduction $\longrightarrow$

$$CH_3\underset{\underset{\displaystyle OH}{|}}{CH}C(CH_3)_3$$

2.43 (*a*) The hydrogen content increases in going from $CH_3C{\equiv}CH$ to $CH_3CH{=}CH_2$. The organic compound $CH_3C{\equiv}CH$ is *reduced*.

(*b*) *Oxidation* occurs because a C—O bond has replaced a C—H bond in going from starting material to product.

(*c*) There are two carbon–oxygen bonds in the starting material and four carbon–oxygen bonds in the products. *Oxidation* occurs.

$$HO-CH_2CH_2-OH \longrightarrow 2H_2C{=}O$$

Two C—O bonds   Four C—O bonds

(*d*) Although the oxidation state of carbon is unchanged in the process

overall, *reduction* of the organic compound has occurred. Its hydrogen content has increased and its oxygen content has decreased. Nitrogen is reduced.

2.44 A bonding interaction exists when two orbitals overlap "in phase" with each other, that is, when the algebraic signs of their wave functions are the same in the region of overlap. The following orbital is a bonding orbital. It involves overlap of an *s* orbital with the lobe of a *p* orbital of the same sign.

On the other hand, the overlap of an *s* orbital with the lobe of a *p* orbital of opposite sign is antibonding.

Overlap in the manner shown next is nonbonding. Both the positive lobe and the negative lobe of the p orbital overlap with the spherically symmetrical s orbital. The bonding overlap between the s orbital and one lobe of the p orbital is exactly canceled by an antibonding interaction between the s orbital and the lobe of opposite sign.

2.45 The end-to-end overlap of two p orbitals corresponds to a σ bond as the overlap is symmetrical along the internuclear axis. The side-to-side overlap of two parallel p orbitals gives rise to a π bond.

ANSWERS TO INTERPRETIVE PROBLEMS

2.46 C; **2.47** B; **2.48** C; **2.49** C

SELF-TEST

PART A

A-1. Write the structure of each of the four-carbon alkyl groups. Give the common name and the systematic name for each.

A-2. How many σ bonds are present in each of the following?

  (*a*) Nonane   (*b*) Cyclononane

A-3. Classify each of the following reactions according to whether the organic substrate is oxidized, reduced, or neither.

  (*a*) CH_3CH_3 + Br_2 $\xrightarrow{\text{light}}$ CH_3CH_2Br + HBr

  (*b*) CH_3CH_2Br + HO^- $\longrightarrow$ CH_3CH_2OH + Br^-

  (*c*) CH_3CH_2OH $\xrightarrow[\text{heat}]{H_2SO_4}$ $H_2C{=}CH_2$ + H_2O

  (*d*) $H_2C{=}CH_2$ + H_2 $\xrightarrow{\text{Pt}}$ CH_3CH_3

A-4. (*a*) Write a structural formula for 3-isopropyl-2,4-dimethylpentane.

  (*b*) How many methyl groups are there in this compound? How many isopropyl groups?

A-5. Give the IUPAC name for each of the following substances:

A-6. The compounds in each part of the previous question contain _____ primary carbon(s), _____ secondary carbon(s), and _____ tertiary carbon(s).

A-7. Using the method outlined in text Section 2.13, give an IUPAC name for each of the following alkyl groups, and classify each one as primary, secondary, or tertiary.

(*a*) $(CH_3)_2CHCH_2CHCH_3$
|

(*b*) $(CH_3CH_2)_3C$—

(*c*) $(CH_3CH_2)_3CCH_2$—

A-8. Write a balanced chemical equation for the complete combustion of 2,3-dimethylpentane.

A-9. Write structural formulas, and give the names of all the constitutional isomers of C_5H_{10} that contain a ring.

A-10. Each of the following names is incorrect. Give the correct name for each compound.

(*a*) 2,3-Diethylhexane

(*c*) 2,3-Dimethyl-3-propylpentane

(*b*) (2-Ethylpropyl)cyclohexane

A-11. Which C_8H_{18} isomer

(*a*) Has the highest boiling point?

(*b*) Has the lowest boiling point?

(*c*) Has the greatest number of tertiary carbons?

(*d*) Has only primary and quaternary carbons?

A-12. Draw the constitutional isomers of C_7H_{16} that have five carbons in their longest chain, and give an IUPAC name for each of them.

A-13. Given the following heats of combustion (in kilojoules per mole) for the homologous series of unbranched alkanes: hexane (4163), heptane (4817), octane (5471), nonane (6125), estimate the heat of combustion (in kilojoules per mole) for **pentadecane**.

A-14. How many σ and π bonds are present in each of the following?

(*a*) CH_3CH=$CHCH_3$

(*c*)

(*b*) HC≡CCH_2CH_3

(*d*)

A-15. Give the hybridization of each carbon atom in the preceding problem.

PART B

B-1. Choose the response that best describes the following compounds:

1   2   3   4

(*a*) 1, 3, and 4 represent the same compound.

(*b*) 1 and 3 are isomers of 2 and 4.

(*c*) 1 and 4 are isomers of 2 and 3.

(*d*) All the structures represent the same compound.

B-2. Which of the following is a correct name according to the IUPAC rules?

(*a*) 2-Methylcyclohexane   (*c*) 2-Ethyl-2-methylpentane

(*b*) 3,4-Dimethylpentane   (*d*) 3-Ethyl-2-methylpentane

B-3. Following are the structures of four isomers of hexane. Which of the names given correctly identifies a fifth isomer?

$$CH_3CH_2CH_2CH_2CH_2CH_3$$
$$(CH_3)_2CHCH_2CH_2CH_3$$

$$(CH_3)_3CCH_2CH_3$$
$$(CH_3)_2CHCH(CH_3)_2$$

(*a*) 2-Methylpentane   (*c*) 2-Ethylbutane

(*b*) 2,3-Dimethylbutane   (*d*) 3-Methylpentane

B-4. Which of the following is cyclohexylcyclohexane?

$$CH_2CH_2CH_2CH_2CH_2CH_3$$

(*a*)   (*c*)

(*b*)   (*d*)

B-5. Which of the following is a 3-methylbutyl group?

(*a*) $CH_3CH_2CH_2CH_2CH_2$—   (*c*) $(CH_3CH_2)_2CH$—

(*b*) $(CH_3)_2CHCH_2CH_2$—   (*d*) $(CH_3)_3CCH_2$—

B-6. Rank the following substances in decreasing order of heats of combustion (most exothermic → least exothermic).

                         1   2   3

(*a*) 2 > 1 > 3   (*c*) 3 > 1 > 2

(*b*) 2 > 3 > 1   (*d*) 3 > 2 > 1

B-7. What is the total number of σ bonds present in the molecule shown?

(a) 18   (b) 26   (c) 27   (d) 30

B-8. Which of the following substances is *not* an isomer of 3-ethyl-2-methylpentane?

(a)

(c)

(b)

(d) None of these (all are isomers)

B-9. Which alkane has the highest boiling point?

(a) Hexane

(d) 2,3-Dimethylbutane

(b) 2,2-Dimethylbutane

(e) 3-Methylpentane

(c) 2-Methylpentane

B-10. Which of the following is a correct IUPAC name of the alkyl group shown?

$$\overset{\displaystyle CH_2CH_3}{\overset{|}{-CHCH_2CH(CH_3)_2}}$$

(a) 1-Ethyl-3-methylbutyl

(c) 4-Ethyl-2-methylbutyl

(b) 1-Ethyl-3,3-dimethylpropyl

(d) 5-Methylhexyl

B-11. Which of the following compounds is *not* a constitutional isomer of three others?

(a) Methylcyclohexane

(d) 1,1,2-Trimethylcyclobutane

(b) Cyclopropylcyclobutane

(e) Cycloheptane

(c) Ethylcyclopentane

B-12. The correct IUPAC name for the compound shown is

(a) 2-Ethyl-5-isobutyl-3-methylhexane

(d) 2-Ethyl-3,5,7-trimethyloctane

(b) 5-*sec*-Butyl-2-ethyl-3-methylhexane

(e) 2,4,6,7-Tetramethylnonane

(c) 2-Isobutyl-4,5-dimethylheptane

B-13. The heats of combustion of two isomers, A and B, are 4817 kJ/mol and 4812 kJ/mol, respectively. From this information it may be determined that

(*a*) Isomer A is 5 kJ/mol more stable

(*b*) Isomer B is 5 kJ/mol less stable

(*c*) Isomer B has 5 kJ/mol more potential energy

(*d*) Isomer A is 5 kJ/mol less stable

B-14. Which of the following reactions requires an *oxidizing agent*?

(*a*) $RCH_2OH \longrightarrow RCH_2Cl$

(*b*) $RCH=CH_2 \longrightarrow RCH_2CH_3$

(*c*) $RCH_2Cl \longrightarrow RCH_3$

(*d*) $RCH_2OH \longrightarrow RCH=O$

(*e*) None of these

B-15. The bond identified (with the arrow) in the structure shown is best described as:

$$\overset{\overset{\textstyle O}{\underset{\textstyle \|}{}}}{CH_3C}-CH=CH_2$$

(*a*) $sp\text{-}sp^2 \ \sigma$

(*b*) $p\text{-}p \ \pi$

(*c*) $sp^2\text{-}sp^3 \ \sigma$

(*d*) $sp^2\text{-}sp^2 \ \sigma$

(*e*) $p\text{-}p \ \sigma$

B-16. The compound shown has _____ $sp^2\text{-}sp^2 \ \sigma$ (sigma) bonds.

(*a*) 1   (*b*) 3   (*c*) 4   (*d*) 5   (*e*) 6

B-17. Select the response that correctly identifies the number of carbon atoms of each hybridization in the compound shown.

$$O=CH-CH_2-CH=C=CH_2$$

	sp^3	sp^2	sp
(*a*)	1	3	1
(*b*)	1	4	0
(*c*)	2	3	0
(*d*)	2	2	1
(*e*)	0	4	1

CHAPTER 3

Alkanes and Cycloalkanes: Conformations and cis-trans Stereoisomers

SOLUTIONS TO TEXT PROBLEMS

3.1 (*b*) The sawhorse formula contains four carbon atoms in an unbranched chain. The compound is butane, $CH_3CH_2CH_2CH_3$.

3.2 Red circles gauche: 60° and 300°. Red circles anti: 180°. Gauche and anti relationships occur only in staggered conformations; therefore, ignore the eclipsed conformations (0°, 120°, 240°, 360°).

3.3 All the staggered conformations of propane are equivalent to one another, and all its eclipsed conformations are equivalent to one another. The energy diagram resembles that of ethane in that it is a symmetrical one.

The activation energy for bond rotation in propane is expected to be somewhat higher than that in ethane because of van der Waals strain between the methyl group and a hydrogen in the eclipsed conformation. This strain is, however, less than the van der Waals strain between the methyl groups of butane, which makes the activation energy for bond rotation less for propane than for butane.

3.4   Ethylcyclopropane and methylcyclobutane are isomers (both are C_5H_{10}). The less stable isomer has the higher heat of combustion. Ethylcyclopropane has more angle strain and is less stable (has higher potential energy) than methylcyclobutane.

	Less stable	More stable
Heat of combustion:	3384 kJ/mol	3352 kJ/mol
	(808.8 kcal/mol)	(801.2 kcal/mol)

3.5   (*b*) To be gauche, substituents X and A must be related by a 60° torsion angle. If A is axial as specified in the problem, X must therefore be equatorial.

X and A are gauche.

(*c*) For substituent X at C-1 to be anti to C-3, it must be equatorial.

(*d*) When X is axial at C-1, it is gauche to C-3.

3.6   (*b*) According to the numbering scheme given in the problem, a methyl group is axial when it is "up" at C-1 but is equatorial when it is up at C-4. Because substituents are more stable when they occupy equatorial rather than axial sites, a methyl group that is up at C-1 is less stable than one that is up at C-4.

(*c*) An alkyl substituent is more stable in the equatorial position. An equatorial substituent at C-3 is "down."

3.7   A *tert*-butyl group is much larger than a methyl group and has a greater preference for the equatorial position. The most stable conformation of 1-*tert*-butyl-1-methylcyclohexane has an axial methyl group and an equatorial *tert*-butyl group.

1-*tert*-Butyl-1-methylcyclohexane

3.8   The four constitutional isomers of *cis*- and *trans*-1,2-dimethylcyclopropane that do not contain double bonds are

1,1-Dimethylcyclopropane

Ethylcyclopropane

Methylcyclobutane

Cyclopentane

3.9   When comparing two stereoisomeric cyclohexane derivatives, the more stable stereoisomer is the one with the greater number of its substituents in equatorial orientations. Rewrite the structures as chair conformations to see which substituents are axial and which are equatorial.

cis-1,3,5-Trimethylcyclohexane

All methyl groups are equatorial in *cis*-1,3,5-trimethylcyclohexane. It is more stable than *trans*-1,3,5-trimethylcyclohexane (shown in the following), which has one axial methyl group in its most stable conformation.

trans-1,3,5-Trimethylcyclohexane

3.10   In each of these problems, a *tert*-butyl group is the larger substituent and will be equatorial in the most stable conformation. Draw a chair conformation of cyclohexane, add an equatorial *tert*-butyl group, and then add the remaining substituent so as to give the required cis or trans relationship to the *tert*-butyl group.

(*b*) Begin by drawing a chair cyclohexane with an equatorial *tert*-butyl group. In *cis*-1-*tert*-butyl-3-methylcyclohexane the C-3 methyl group is equatorial.

(*c*) In *trans*-1-*tert*-butyl-4-methylcyclohexane both the *tert*-butyl and the C-4 methyl group are equatorial.

(*d*) Again the *tert*-butyl group is equatorial; however, in *cis*-1-*tert*-butyl-4-methylcyclohexane the methyl group on C-4 is axial.

3.11   The simplest approach is to break the four bonds that belong to two rings (the ring fusions, in this case). This leaves one large ring.

Hopane

4 disconnections

Cleaving any of the bonds in the large ring gives a noncyclic species. Therefore, a total of five bond disconnections is needed. Hopane is pentacyclic.

3.12   Isomers are different compounds that have the same molecular formula. Compare the molecular formulas of the compounds given to the molecular formula of spiro[2.2]pentane.

Spiro[2.2]pentane (C_5H_8) C_5H_8 C_6H_{10} C_5H_8 C_5H_{10}

Only the two compounds that have the molecular formula C_5H_8 are isomers of spiro[2.2]pentane.

3.13 Hydrocarbon A is a spiro compound, encompassing 10 carbons. It is, therefore, a *spirodecane*. Four carbons are unique to the cyclopentane ring, five to the cyclohexane ring, making it a substituted *spiro[4.5]decane*. Numbering proceeds through the smaller ring to the larger and substituents are listed in alphabetical order.

2-Isopropyl-6,10-dimethylspiro[4.5]decane

3.14 The two bond cleavages shown convert camphene to a noncyclic species; therefore, camphene is bicyclic. (Other pairs of bond cleavages are possible and lead to the same conclusion.)

3.15 (*b*) The structure of bicyclo[2.2.1]heptane was given in part (*a*) of this problem in the text. Numbering begins at a bridgehead position and continues in the direction of the largest bridge. The smallest bridge is numbered last. 1,7,7-Trimethylbicyclo[2.2.1]heptane has the structure shown at the right.

(*c*) The three bridges in bicyclo[3.1.1]heptane contain three carbons, one carbon, and one carbon. The structure can be written in a form that shows the actual shape of the molecule or one that simply emphasizes its constitution.

Three-carbon bridge

One-carbon bridge

One-carbon bridge

3.16 Because the two conformations are of approximately equal stability when R = H, it is reasonable to expect that the most stable conformation when R = CH$_3$ will have the CH$_3$ group equatorial.

R = H: both conformations similar in energy
R = CH$_3$: most stable conformation has CH$_3$ equatorial

3.17 Naming these alkanes is simplified if you first translate the Newman or sawhorse projection into a more conventional structural formula.

(a)

$H_3C-\underset{\underset{CH_3}{|}}{\overset{\overset{CH_3}{|}}{C}}-CH_3$ 2,2-Dimethylpropane

(b)

$H_3C-\underset{\underset{H}{|}}{\overset{\overset{CH_3}{|}}{C}}-CH_2\underset{\underset{CH_3}{|}}{CH}CH_3$ 2,4-Dimethylpentane

(c)

$H_3C-\underset{\underset{CH_3}{|}}{\overset{\overset{CH_3}{|}}{C}}-CH_2-\underset{\underset{CH_3}{|}}{\overset{\overset{CH_3}{|}}{CH}}-CH_2CH_3$ 2,2,4-Trimethylhexane

3.18 (a) First write out the structural formula of 2,2-dimethylbutane in order to identify the substituent groups attached to C-2 and C-3. As shown at left, C-2 bears three methyl groups, and C-3 bears two hydrogens and a methyl group. The most stable conformation is the staggered one shown at right. All other staggered conformations are equivalent to this one.

(b) The constitution of 2-methylbutane and its two most stable conformations are shown.

Both conformations are staggered. In one Newman projection (left), the methyl group at the back is gauche to the two methyl groups in the front. In the other (right), it is gauche to one and anti to the other.

(c) The hydrogens at C-2 and C-3 may be gauche to one another (left), or they may be anti (right).

3.19 The 2-methylbutane conformation with one gauche $CH_3\cdots CH_3$ and one anti $CH_3\cdots CH_3$ relationship is more stable than the one with two gauche $CH_3\cdots CH_3$ relationships. The more stable conformation has less van der Waals strain.

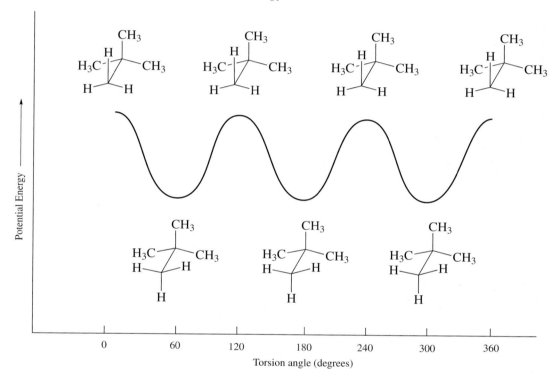

More stable   Less stable

3.20 All the staggered conformations about the C-2—C-3 bond of 2,2-dimethylpropane are equivalent to one another and of equal energy; they represent potential energy minima. All the eclipsed conformations are equivalent and represent potential energy maxima.

The shape of the potential energy profile for internal rotation in 2,2-dimethylpropane more closely resembles that of ethane than that of butane.

3.21 The potential energy diagram of 2-methylbutane more closely resembles that of butane than that of ethane in that the three staggered forms are not all of the same energy. Similarly, not all of the eclipsed forms are of equal energy.

3.22 Bromine is attached to C-1. Sight down the C-1—C-2 bond and convert the structural drawing to a Newman projection.

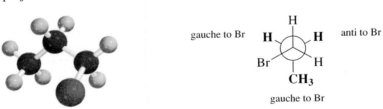

3.23 The structure shown in the text is not the most stable conformation because the bonds of the methyl group are eclipsed with those of the ring carbon to which it is attached. The most stable conformation has the bonds of the methyl group and its attached carbon in a staggered relationship.

Bonds of methyl group eclipsed   Bonds of methyl group staggered
with those of attached carbon   with those of attached carbon

3.24 Conformation B is more stable than A. The methyl groups are rather close together in A, resulting in van der Waals strain between them. In B, the methyl groups are farther apart.

Van der Waals strain between cis methyl groups.

Methyl groups remain cis, but are far apart.

A

B

3.25 (*a*) By rewriting the structures in a form that shows the order of their atomic connections, it is apparent that the two structures are constitutional isomers.

is equivalent to

$$CH_3CCH_3$$ with CH_3 above and CH_3 below   (2,2-Dimethylpropane)

is equivalent to $CH_3CHCH_2CH_3$ with CH_3 above   (2-Methylbutane)

(*b*) Both models represent alkanes of molecular formula C_6H_{14}. In each one the carbon chain is unbranched. The two models are different conformations of the same compound, $CH_3CH_2CH_2CH_2CH_2CH_3$ (hexane).

(*c*) The two compounds have the same constitution; both are $(CH_3)_2CHCH(CH_3)_2$. The Newman projections represent different staggered conformations of the same molecule: in one the hydrogens are anti to each other, whereas in the other they are gauche.

and

are different conformations of 2,3-dimethylbutane

Hydrogens at C-2 and C-3 are anti.

Hydrogens at C-2 and C-3 are gauche.

(*d*) The compounds differ in their connectivity. They are constitutional isomers. Although the compounds have different stereochemistry (one is cis, the other trans), they are not stereoisomers. Stereoisomers must have the same constitution.

cis-1,2-Dimethylcyclopentane   *trans*-1,3-Dimethylcyclopentane

(e) Both structures are *cis*-1-ethyl-4-methylcyclohexane (the methyl and ethyl groups are both "up"). In the structure on the left, the methyl is axial and the ethyl equatorial. The orientations are opposite to these in the structure on the right. The two structures are ring-flipped forms of each other—different conformations of the same compound.

(f) The methyl and the ethyl groups are cis in the first structure but trans in the second. The two compounds are stereoisomers; they have the same constitution but differ in the arrangement of their atoms in space.

cis-1-Ethyl-4-methylcyclohexane (both alkyl groups are up)	*trans*-1-Ethyl-4-methylcyclohexane (ethyl group is down; methyl group is up)

Do not be deceived because the six-membered rings look like ring-flipped forms. Remember, chair–chair interconversion converts all the equatorial bonds to axial and vice versa. Here the ethyl group is equatorial in both structures.

(g) The two structures have the same constitution but differ in the arrangement of their atoms in space; they are stereoisomers. They are not different conformations of the same compound, because they are not related by rotation about C—C bonds. In the first structure as shown here the methyl group is trans to the indicated bonds, whereas in the second it is cis to these bonds.

Methyl is trans to these bonds.	Methyl is cis to these bonds.

3.26. (a) Isomers are different compounds that have the same molecular formula. 2,2-Dimethylpentane has a molecular formula of C_7H_{16} and is not an isomer of 2,2-dimethylhexane or 2,2,3-trimethylpentane, both of which have the molecular formula C_8H_{18}.

$(CH_3)_3CCH_2CH_2CH_3$ $(CH_3)_3CCH_2CH_2CH_2CH_3$ $(CH_3)_2CHCHCH_2CH_3$
$\overset{|}{C}H_3$

2,2-Dimethylpentane (C_7H_{16})	2,2-Dimethylhexane (C_8H_{18})	2,2,3-Trimethylpentane (C_8H_{18})

The two C_8H_{18} alkanes differ in the order in which their atoms are connected. They are constitutional isomers.

(b) The alkane in the middle is the same as the one at the left, simply written in the opposite direction. Both are 2-methylpentane. 3-Methylpentane is a constitutional isomer of 2-methylpentane.

2-Methylpentane	2-Methylpentane	3-Methylpentane

(*c*) Pentane has a molecular formula of C_5H_{12}; it is not an isomer of cyclopentane or 1,1-dimethylcyclopropane which have molecular formulas of C_5H_{10} and are constitutional isomers of each other.

Pentane   Cyclopentane   1,1-Dimethylcyclopropane

(*d*) *cis-* and *trans*-1,2-Dimethylcyclobutane are stereoisomers. They have the same constitution but differ in the arrangement of atoms in space. Both are constitutional isomers of *trans*-1,3-dimethylcyclobutane.

cis-1,2-Dimethylcyclobutane   *trans*-1,3-Dimethylcyclobutane   *trans*-1,2-Dimethylcyclobutane

(*e*) The relationships here are the same as in part (*d*) except the compounds are the stereoisomeric *cis-* and *trans*-1,3-dimethylcyclohexanes and the constitutionally isomeric *cis*-1,2-dimethylcyclohexane.

cis-1,2-Dimethylcyclohexane   *trans*-1,3-Dimethylcyclohexane   *cis*-1,3-Dimethylcyclohexane

3.27 (*a*) Three isomers of C_5H_8 contain two rings and have no alkyl substituents:

Spiro[2.2]pentane   Bicyclo[2.1.0]pentane   Bicyclo[1.1.1]pentane

(*b*) Five isomers of C_6H_{10} contain two rings and have no alkyl substituents:

Spiro[2.3]hexane   Bicyclo[2.2.0]hexane   Bicyclo[3.1.0]hexane

Bicyclo[2.1.1]hexane   Cyclopropylcyclopropane

3.28 (*a*) The heat of combustion is highest for the hydrocarbon with the greatest number of carbons. Thus, cyclopropane, even though it is more strained than cyclobutane or cyclopentane, has the lowest heat of combustion.

		Cyclopentane	Heat of combustion 3291 kJ/mol (786.6 kcal/mol)
		Cyclobutane	Heat of combustion 2721 kJ/mol (650.3 kcal/mol)
		Cyclopropane	Heat of combustion 2091 kJ/mol (499.8 kcal/mol)

A comparison of heats of combustion can only be used to assess relative stability when the compounds are isomers.

(*b*) All these compounds have the molecular formula C_7H_{14}. They are isomers, and so the one with the most strain will have the highest heat of combustion.

1,1,2,2-Tetramethylcyclopropane (high in angle strain; bonds are eclipsed; van der Waals strain between cis methyl groups)   Heat of combustion 4635 kJ/mol (1107.9 kcal/mol)

cis-1,2-Dimethylcyclopentane (low angle strain; some torsional strain; van der Waals strain between cis methyl groups)   Heat of combustion 4590 kJ/mol (1097.1 kcal/mol)

Methylcyclohexane (minimal angle, torsional, and van der Waals strain)   Heat of combustion 4565 kJ/mol (1091.1 kcal/mol)

(*c*) These hydrocarbons all have different molecular formulas. Their heats of combustion decrease with decreasing number of carbons, and comparisons of relative stability cannot be made.

Cyclopropylcyclopropane (C_6H_{10})   Heat of combustion 3886 kJ/mol (928.8 kcal/mol)

Spiro[2.2]pentane (C_5H_8)   Heat of combustion 3296 kJ/mol (787.8 kcal/mol)

Bicyclo[1.1.0]butane (C_4H_6)   Heat of combustion 2648 kJ/mol (633.0 kcal/mol) (*d*) Bicyclo[3.3.0]octane and bicyclo[5.1.0]octane are isomers, and their heats of combustion can be compared on the basis of their relative stabilities. The three-membered ring in bicyclo[5.1.0]-octane imparts a significant amount of angle strain to this isomer, making it less stable than bicyclo[3.3.0]octane. The third hydrocarbon, bicyclo[4.3.0]nonane, has a greater number of carbons than either of the others and has the largest heat of combustion.

	Bicyclo[4.3.0]nonane (C_9H_{16})	Heat of combustion 5562 kJ/mol (1350.9 kcal/mol)
	Bicyclo[5.1.0]octane (C_8H_{14})	Heat of combustion 5089 kJ/mol (1216.3 kcal/mol)
	Bicyclo[3.3.0]octane (C_8H_{14})	Heat of combustion 5016 kJ/mol (1198.9 kcal/mol)

3.29 (*a*) The structural formula of 2,2,5,5-tetramethylhexane is $(CH_3)_3CCH_2CH_2C(CH_3)_3$. Both C-3 and C-4 have two hydrogens and a *tert*-butyl group attached. The most stable conformation has the large *tert*-butyl groups anti to each other.

Anti conformation of
2,2,5,5-tetramethylhexane

(*b*) The zigzag conformation of 2,2,5,5-tetramethylhexane is an alternative way of expressing the same conformation implied in the Newman projection of part (*a*). It is more complete, however, in that it also shows the spatial arrangement of the substituents attached to the main chain.

2,2,5,5-Tetramethylhexane

(*c*) An isopropyl group is bulkier than a methyl group and will have a greater preference for an equatorial orientation in the most stable conformation of *cis*-1-isopropyl-3-methylcyclohexane. Draw a chair conformation of cyclohexane, and place an isopropyl group in an equatorial position.

Notice that the equatorial isopropyl group is down on the carbon atom to which it is attached. Add a methyl group to C-3 so that it is also down.

Both substituents are equatorial in the most stable conformation of *cis*-1-isopropyl-3-methylcyclohexane.

(*d*) One substituent is up and the other is down in the most stable conformation of *trans*-1-isopropyl-3-methylcyclohexane. Begin as in part (*c*) by placing an isopropyl group in an equatorial orientation on a chair conformation of cyclohexane.

To be trans to the C-1 isopropyl group, the C-3 methyl group must be up.

The bulkier isopropyl group is equatorial and the methyl group axial in the most stable conformation.

(*e*) To be cis to each other, one substituent must be axial and the other equatorial when they are located at positions 1 and 4 on a cyclohexane ring.

Place the larger substituent (the *tert*-butyl group) at the equatorial site and the smaller substituent (the ethyl group) at the axial one.

(*f*) First write a chair conformation of cyclohexane, then add two methyl groups at C-1, and draw in the axial and equatorial bonds at C-3 and C-4. Next, add methyl groups to C-3 and C-4 so that they are cis to each other. This can be accomplished two different ways: either the C-3 and C-4 methyl groups are both up or they are both down.

More stable chair conformation: C-3 methyl group is equatorial; no van der Waals strain between axial C-1 methyl group and C-3 methyl group

Less stable chair conformation: C-3 methyl group is axial; strong van der Waals strain between axial C-1 and C-3 methyl groups

(*g*) Draw the projection formula as a chair conformation.

Check to see if this is the most stable conformation by writing its ring-flipped form.

Less stable conformation: two axial methyl groups

More stable conformation: one axial methyl group

The ring-flipped form, with two equatorial methyl groups and one axial methyl group, is more stable than the originally drawn conformation, with two axial methyl groups and one equatorial methyl group.

3.30 Begin by writing each of the compounds in its most stable conformation. Compare them by examining their conformations for sources of strain, particularly van der Waals strain arising from groups located too close together in space.

(*a*) Its axial methyl group makes the cis stereoisomer of 1-isopropyl-2-methylcyclohexane less stable than the trans.

cis-1-Isopropyl-2-methylcyclohexane
(less stable stereoisomer)

trans-1-Isopropyl-2-methylcyclohexane
(more stable stereoisomer)

The axial methyl group in the cis stereoisomer is involved in unfavorable repulsions with the C-4 and C-6 axial hydrogens indicated in the drawing.

(*b*) Both groups are equatorial in the cis stereoisomer of 1-isopropyl-3-methylcyclohexane; cis is more stable than trans in 1,3-disubstituted cyclohexanes.

cis-1-Isopropyl-3-methylcyclohexane
(more stable stereoisomer; both
groups are equatorial)

trans-1-Isopropyl-3-methylcyclohexane
(less stable stereoisomer; methyl group
is axial and involved in repulsions
with axial hydrogens at C-1 and C-5)

(*c*) The more stable stereoisomer of 1,4-disubstituted cyclohexanes is the trans; both alkyl groups are equatorial in *trans*-1-isopropyl-4-methylcyclohexane.

cis-1-Isopropyl-4-methylcyclohexane
(less stable stereoisomer; methyl
group is axial and involved in
repulsions with axial
hydrogens at C-2 and C-6)

trans-1-Isopropyl-4-methylcyclohexane
(more stable stereoisomer; both
groups are equatorial)

(*d*) The first stereoisomer of 1,2,4-trimethylcyclohexane is the more stable one. All its methyl groups are equatorial in its most stable conformation. The most stable conformation of the second stereoisomer has one axial and two equatorial methyl groups.

More stable stereoisomer

All methyl groups equatorial in
most stable conformation

Less stable stereoisomer

One axial methyl group in most
stable conformation

(*e*) The first stereoisomer of 1,2,4-trimethylcyclohexane is the more stable one here, as it was in part (*d*). All its methyl groups are equatorial, but one of the methyl groups is axial in the most stable conformation of the second stereoisomer.

More stable stereoisomer

All methyl groups equatorial in most stable conformation

Less stable stereoisomer

One axial methyl group in most stable conformation

(*f*) Each stereoisomer of 2,3-dimethylbicyclo[3.2.1]octane has one axial and one equatorial methyl group. The first one, however, has a close contact between its axial methyl group and both methylene groups of the two-carbon bridge. The second stereoisomer has repulsions with only one axial methylene group; it is more stable.

Less stable stereoisomer
(more van der Waals strain)

More stable stereoisomer
(less van der Waals strain)

3.31 First write structural formulas showing the relative stereochemistries and the preferred conformations of the two stereoisomers of 1,1,3,5-tetramethylcyclohexane.

written in its most stable conformation as

cis-1,1,3,5-Tetramethylcyclohexane

written in its most stable conformation as

trans-1,1,3,5-Tetramethylcyclohexane

The cis stereoisomer is more stable than the trans. It exists in a conformation with only one axial methyl group, while the trans stereoisomer has two axial methyl groups in close contact with each other. The trans stereoisomer is destabilized by van der Waals strain.

3.32 When comparing isomers, one that is less stable has a higher heat of combustion than one that is more stable. Both methyl groups can adopt an equatorial conformation in *trans*-1,2-dimethylcyclohexane, and thus trans is the more stable stereoisomer. Notice, however, that the two methyl groups are gauche to one another in the 1,2 isomer. Van der Waals strain between the gauche methyl groups is the reason for the higher heat of combustion compared with the more stable stereoisomers of the 1,3 and 1,4 isomers.

These methyl groups
are gauche to each other

trans-1,2-Dimethylcyclohexane
(more stable stereoisomer)

cis-1,3-Dimethylcyclohexane   *trans*-1,4-Dimethylcyclohexane
(more stable stereoisomer)   (more stable stereoisomer)

The same reasoning applies to the less stable stereoisomer, *cis*-1,2-dimethylcyclohexane. One methyl group is axial while the other is equatorial in the cis stereoisomer, but once again the methyl groups are gauche to each other and there is van der Waals strain between them.

These methyl groups
are gauche to each other

cis-1,2-Dimethylcyclohexane
(less stable stereoisomer)

3.33 Both structures have approximately the same degree of angle strain and of torsional strain. Structure B has more van der Waals strain than A because two pairs of hydrogens (shown here) approach each other at distances that are rather close.

A:
More stable stereoisomer

Van der Waals strain
destabilizes B

3.34 Five bond cleavages are required to convert cubane to a noncyclic skeleton; cubane is pentacyclic.

3.35 (*a*) The reactant is a spiro hydrocarbon. Its IUPAC name is spiro[2.5]octane.

Alcohol A   Alcohol B   Alcohol C

(b) Carbons 4, 5, and 6 are oxidized in the formation of alcohols A, B, and C, respectively.

(c) The alcohols are constitutional isomers.

3.36 Conformational representations of the two different forms of glucose are drawn in the usual way. An oxygen atom is present in the six-membered ring, and we are told in the problem that the ring exists in a chair conformation.

written in its most stable conformation as

One axial substituent

written in its most stable conformation as

All substituents equatorial

The two structures are not interconvertible by ring flipping; therefore they are not different conformations of the same molecule. Remember, ring flipping transforms all axial substituents to equatorial ones and vice versa. The two structures differ with respect to only one substituent; they are *stereoisomers* of each other.

3.37 This problem is primarily an exercise in correctly locating equatorial and axial positions in cyclohexane rings that are joined together into a steroid skeleton. Parts (a) through (e) are concerned with positions 1, 4, 7, 11, and 12 in that order. The following diagram shows the orientation of axial and equatorial bonds at each of those positions.

Both methyl groups are up.

(a) At C-1 the bond that is cis to the methyl groups is equatorial (up).

(b) At C-4 the bond that is cis to the methyl groups is axial (up).

(c) At C-7 the bond that is trans to the methyl groups is axial (down).

(d) At C-11 the bond that is trans to the methyl groups is equatorial (down).

(e) At C-12 the bond that is cis to the methyl groups is equatorial (up).

3.38 Analyze this problem in exactly the same way as the preceding one by locating the axial and equatorial bonds at each position. It will be seen that the only differences are those at C-1 and C-4.

Both methyl groups are up.

(*a*) At C-1 the bond that is cis to the methyl groups is axial (up).

(*b*) At C-4 the bond that is cis to the methyl groups is equatorial (up).

(*c*) At C-7 the bond that is trans to the methyl groups is axial (down).

(*d*) At C-11 the bond that is trans to the methyl groups is equatorial (down).

(*e*) At C-12 the bond that is cis to the methyl groups is equatorial (up).

3.39 (*a*) The torsion angle between chlorine substituents is 60° in the gauche conformation and 180° in the anti conformation of $ClCH_2CH_2Cl$.

Gauche
(can have a dipole moment)

Anti
(cannot have a dipole moment)

(*b*) All the individual bond dipole moments cancel in the anti conformation of $ClCH_2CH_2Cl$, and this conformation has no dipole moment. Because $ClCH_2CH_2Cl$ has a dipole moment of 1.12 D, it can exist entirely in the gauche conformation or it can be a mixture of anti and gauche conformations, but it cannot exist entirely in the anti conformation. Statement 1 is false.

ANSWERS TO INTERPRETIVE PROBLEMS

3.40 B;   **3.41** D;   **3.42** A;   **3.43** B

SELF-TEST

PART A

A-1. Draw Newman projections for both the gauche and the anti conformations of 1-chloropropane, $CH_3CH_2CH_2Cl$. Sight along the C-1, C-2 bond (the chlorine is attached to C-1).

A-2. Write Newman projection formulas for

(*a*) The least stable conformation of butane

(*b*) Two different staggered conformations of $Cl_2CHCHCl_2$

4.34 The functional groups to which the four oxygens belong are classified as:

4.35 One nitrogen belongs to a tertiary amine function; the other is an amide nitrogen.

4.36 *Uscharidin* has the structure shown.

(*a*) There are two alcohol groups, one aldehyde group, one ketone group, and one ester functionality.

(*b*) Uscharidin contains ten methylene groups (CH_2). They are indicated in the structure by small squares.

(*c*) The primary carbons in uscharidin are the carbons of the two methyl groups.

4.37 This problem illustrates the reactions of a primary alcohol with the reagents described in the chapter.

(*a*) $CH_3CH_2CH_2CH_2OH$ + $NaNH_2$ $\longrightarrow$ $CH_3CH_2CH_2CH_2O^- Na^+$ + NH_3

Sodium butoxide

(*b*) $CH_3CH_2CH_2CH_2OH$ $\xrightarrow[\text{heat}]{\text{HBr}}$ $CH_3CH_2CH_2CH_2Br$

1-Bromobutane

(*c*) $CH_3CH_2CH_2CH_2OH$ $\xrightarrow[\text{heat}]{\text{NaBr, H}_2\text{SO}_4}$ $CH_3CH_2CH_2CH_2Br$

1-Bromobutane

(d) $CH_3CH_2CH_2CH_2OH$ $\xrightarrow{PBr_3}$ $CH_3CH_2CH_2CH_2Br$

1-Bromobutane

(e) $CH_3CH_2CH_2CH_2OH$ $\xrightarrow{SOCl_2}$ $CH_3CH_2CH_2CH_2Cl$

1-Chlorobutane

4.38 (a) This reaction was used to convert the primary alcohol to the corresponding bromide in 60% yield.

(b) Thionyl chloride treatment of this secondary alcohol gave the chloro derivative in 59% yield.

(c) The starting material is a tertiary alcohol and reacted readily with hydrogen chloride to form the corresponding chloride in 67% yield.

(d) Both primary alcohol functional groups were converted to primary bromides; the yield was 88%.

(e) This molecule is called adamantane. It has six equivalent CH_2 groups and four equivalent CH groups. Bromination is selective for tertiary hydrogens, so a hydrogen of one of the CH groups is replaced. The product shown was isolated in 76% yield.

$C_{10}H_{15}Br$

4.39 The order of reactivity of alcohols with hydrogen halides is tertiary > secondary > primary.

$$ROH \ + \ HBr \ \longrightarrow \ RBr \ + \ H_2O$$

Reactivity of Alcohols with Hydrogen Bromide:

Part	More reactive	Less reactive
(a)	$CH_3CHCH_2CH_3$ $\mid$ OH 2-Butanol: secondary	$CH_3CH_2CH_2CH_2OH$ 1-Butanol: primary
(b)	$CH_3CHCH_2CH_3$ $\mid$ OH 2-Butanol: secondary	$CH_3CH_2CHCH_2OH$ $\mid$ CH_3 2-Methyl-1-butanol: primary
(c)	$(CH_3)_2CCH_2CH_3$ $\mid$ OH 2-Methyl-2-butanol: tertiary	$CH_3CHCH_2CH_3$ $\mid$ OH 2-Butanol: secondary
(d)	$CH_3CHCH_2CH_3$ $\mid$ OH 2-Butanol	$CH_3CHCH_2CH_3$ $\mid$ CH_3 2-Methylbutane: not an alcohol; does not react with HBr
(e)	 1-Methylcyclopentanol tertiary	 Cyclohexanol: secondary
(f)	 1-Methylcyclopentanol tertiary	 *trans*-2-Methylcyclopentanol: secondary
(g)	 1-Ethylcyclopentanol tertiary	 1-Cyclopentylethanol: secondary

4.40 The unimolecular step in the reaction of 2-methyl-2-butanol with hydrogen chloride to give 2-chloro-2-methylbutane is the dissociation of the alkyloxonium ion to a carbocation.

$$\overset{+\,\cdot\cdot}{\underset{\underset{CH_3}{|}}{\overset{\overset{\cdot OH_2}{|}}{CH_3CCH_2CH_3}}} \longrightarrow \overset{+}{\underset{\underset{CH_3}{|}}{CH_3CCH_2CH_3}} + \quad H\overset{\cdot\cdot}{\underset{H}{O}}$$

1,1-Dimethylpropyl-   1,1-Dimethylpropyl   Water
oxonium ion   cation

4.41 The nucleophile that attacks the oxonium ion in the reaction of 1-hexanol with hydrogen bromide is bromide ion.

$$:\!\overset{\cdot\cdot}{\underset{\cdot\cdot}{Br}}\!:^{-} + \quad CH_3(CH_2)_4CH_2\overset{+}{\underset{\diagdown H}{\overset{/}{O}}} \quad \xrightarrow{\text{slow}} \quad CH_3(CH_2)_4CH_2\!-\!\overset{\cdot\cdot}{\underset{\cdot\cdot}{Br}}\!: + \quad :\!\overset{H}{\underset{H}{\overset{/}{O}}}$$

Bromide ion Hexyloxonium ion   1-Bromohexane   Water

4.42 Methanol cannot react by the S_N1 mechanism because CH_3^+ is far too unstable to be an intermediate in a chemical reaction. The mechanism is S_N2. The first step is proton transfer from HBr to methanol. In the second step, bromide acts as a nucleophile to displace water from methyloxonium ion.

Step 1: $\underset{H}{\overset{H_3C}{\diagdown}}\overset{\cdot\cdot}{O}\overset{\diagup}{}\!: \quad + \quad H\!-\!\overset{\cdot\cdot}{\underset{\cdot\cdot}{Br}}\!: \quad \underset{\xrightarrow{\text{fast}}}{\rightleftharpoons} \quad \underset{H}{\overset{H_3C}{\diagdown}}\overset{+}{O}\overset{\diagup}{}\!-\!H \quad + \quad :\!\overset{\cdot\cdot}{\underset{\cdot\cdot}{Br}}\!:^{-}$

Methanol Hydrogen bromide Methyloxonium ion Bromide ion

Step 2: $:\!\overset{\cdot\cdot}{\underset{\cdot\cdot}{Br}}\!:^{-} \quad + \quad \underset{H}{\overset{H_3C}{\diagdown}}\overset{\cdot\cdot}{\underset{+}{O}}\!-\!H \quad \xrightarrow{\text{slow}} \quad :\!\overset{\cdot\cdot}{\underset{\cdot\cdot}{Br}}\!-\!CH_3 \quad + \quad \overset{\cdot\cdot}{\underset{H}{O}}\!-\!H$

Bromide ion Methyloxonium ion   Methyl bromide   Water

4.43 (*a*) Both the methyl group and the hydroxyl group are equatorial in the most stable conformation of *trans*-4-methylcyclohexanol.

trans-4-Methylcyclohexanol

(*b*) The positively charged carbon in the carbocation intermediate is sp^2-hybridized with a planar arrangement of bonds.

$$H_3C\diagdown\!\!\!\bigwedge\!\!\!\bigwedge\overset{H}{\underset{+}{}}$$

Carbocation intermediate

(c) Bromide ion attacks the carbocation from both above and below, giving rise to two stereoisomers, *cis-* and *trans*-1-bromo-4-methylcyclohexane.

cis-1-Bromo-4-methylcyclohexane   *trans*-1-Bromo-4-methylcyclohexane

4.44 (a) For the propagation step:

$$CH_4 \ + \ \cdot \ddot{Br}\!: \ \longrightarrow \ \cdot CH_3 \ + \ H\!-\!\ddot{Br}\!:$$

Energy required to break C—H bond of methane: 435 kJ/mol = 104 kcal/mol

Energy given off when H—Br bond forms: 366 kJ/mol = 87.5 kcal/mol

Net energy change for equation: $\Delta H° = (435–366)$ kJ = +69 kJ (+17.5 kcal)

(b)

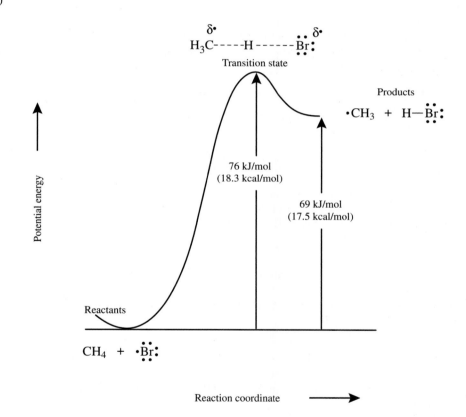

(c) This reaction step is endothermic; the transition state lies closer to the products than to the reactants. The structure of the transition state more closely resembles the products than the reactants.

4.45 Examine the equations to ascertain which bonds are made and which are broken. Then use the bond dissociation enthalpies in text Table 4.3 to calculate $\Delta H°$ for each reaction.

(a) $(CH_3)_2CH-OH$ + $H-F$ $\longrightarrow$ $(CH_3)_2CH-F$ + $H-OH$

385 kJ/mol (92 kcal/mol)   568 kJ/mol (136 kcal/mol)   439 kJ/mol (105 kcal/mol)   497 kJ/mol (119 kcal/mol)

Bond breaking: 953 kJ/mol (228 kcal/mol)   Bond making: 936 kJ/mol (224 kcal/mol)

$\Delta H°$ = energy cost of breaking bonds – energy given off in making bonds
= 953 kJ/mol – 936 kJ/mol (228 kcal/mol – 224 kcal/mol)
= +17 kJ/mol (+4 kcal/mol)

The reaction of isopropyl alcohol with hydrogen fluoride is endothermic.

(b) $(CH_3)_2CH-OH$ + $H-Cl$ $\longrightarrow$ $(CH_3)_2CH-Cl$ + $H-OH$

385 kJ/mol (92 kcal/mol)   431 kJ/mol (103 kcal/mol)   339 kJ/mol (81 kcal/mol)   497 kJ/mol (119 kcal/mol)

Bond breaking: 816 kJ/mol (195 kcal/mol)   Bond making: 836 kJ/mol (200 kcal/mol)

$\Delta H°$ = energy cost of breaking bonds – energy given off in making bonds
= 816 kJ/mol – 836 kJ/mol (195 kcal/mol – 200 kcal/mol)
= –20 kJ/mol (–5 kcal/mol)

The reaction of isopropyl alcohol with hydrogen chloride is exothermic.

(c) CH_3CHCH_3 + $H-Cl$ $\longrightarrow$ CH_3CHCH_3 + $H-H$
   | |
   H   Cl

397 kJ/mol (95 kcal/mol)   431 kJ/mol (103 kcal/mol)   339 kJ/mol (81 kcal/mol)   435 kJ/mol (104 kcal/mol)

Bond breaking: 828 kJ/mol (198 kcal/mol)   Bond making: 774 kJ/mol (185 kcal/mol)

$\Delta H°$ = energy cost of breaking bonds – energy given off in making bonds
= 828 kJ/mol – 774 kJ/mol (198 kcal/mol – 185 kcal/mol)
= +54 kJ/mol (+13 kcal/mol)

The reaction of propane with hydrogen chloride is endothermic.

4.46 In the statement of the problem you are told that the starting material is 2,2-dimethylpropane, that the reaction is fluorination, meaning that F_2 is a reactant, and that the product is $(CF_3)_4C$. You need to complete the equation by realizing that HF is also formed in the fluorination of alkanes. The balanced equation is therefore:

$$(CH_3)_4C + 12F_2 \longrightarrow (CF_3)_4C + 12HF$$

4.47 The reaction is free-radical chlorination, and substitution occurs at all possible positions that bear a replaceable hydrogen. Write the structure of the starting material, and identify the nonequivalent hydrogens.

$$\begin{array}{c} \quad\;\; F \;\;\; Cl \\ \quad\;\; | \;\;\;\; | \\ Cl-C-C-CH_3 \\ \quad\;\; | \;\;\;\; | \\ \quad\;\; F \;\;\; H \end{array}$$

1,2-Dichloro-1,1-difluoropropane

The problem states that one of the products is 1,2,3-trichloro-1,1-difluoropropane. This compound arises by substitution of one of the methyl hydrogens by chlorine. We are told that the other product is an isomer of 1,2,3-trichloro-1,1-difluoropropane; therefore, it must be formed by replacement of the hydrogen at C-2.

$$
\begin{array}{cc}
\overset{\displaystyle F}{\underset{\displaystyle F}{\text{Cl}-\text{C}}}-\overset{\displaystyle Cl}{\underset{\displaystyle H}{\text{C}}}-\text{CH}_2\text{Cl}
&
\overset{\displaystyle F}{\underset{\displaystyle F}{\text{Cl}-\text{C}}}-\overset{\displaystyle Cl}{\underset{\displaystyle Cl}{\text{C}}}-\text{CH}_3
\end{array}
$$

1,2,3-Trichloro-1,1-difluoropropane   1,2,2-Trichloro-1,1-difluoropropane

4.48 Free-radical chlorination leads to substitution at each carbon that bears a hydrogen. This problem essentially requires you to recognize structures that possess various numbers of nonequivalent hydrogens. The easiest way to determine the number of constitutional isomers that can be formed by chlorination of a particular compound is to replace one hydrogen with chlorine and assign an IUPAC name to the product. Continue by replacing one hydrogen on each carbon in the compound, and compare names to identify duplicates.

(*a*) 2,2-Dimethylpropane is the C_5H_{12} isomer that gives a single monochloride, since all the hydrogens are equivalent.

$$
\underset{\displaystyle \text{CH}_3}{\overset{\displaystyle \text{CH}_3}{\text{CH}_3\text{CCH}_3}}
\quad\xrightarrow[\text{light}]{\text{Cl}_2}\quad
\underset{\displaystyle \text{CH}_3}{\overset{\displaystyle \text{CH}_3}{\text{CH}_3\text{CCH}_2\text{Cl}}}
$$

2,2-Dimethylpropane   1-Chloro-2,2-dimethylpropane

(*b*) The C_5H_{12} isomer that has three nonequivalent sets of hydrogens is pentane. It yields three isomeric monochlorides on free-radical chlorination.

$$
\text{CH}_3\text{CH}_2\text{CH}_2\text{CH}_2\text{CH}_3 \xrightarrow{\text{Cl}_2}
$$

Pentane

ClCH$_2$CH$_2$CH$_2$CH$_2$CH$_3$

1-Chloropentane

$$\text{CH}_3\underset{\displaystyle \text{Cl}}{\text{CH}}\text{CH}_2\text{CH}_2\text{CH}_3$$

2-Chloropentane

$$\text{CH}_3\text{CH}_2\underset{\displaystyle \text{Cl}}{\text{CH}}\text{CH}_2\text{CH}_3$$

3-Chloropentane

(*c*) 2-Methylbutane forms four different monochlorides.

(*d*) For only two dichlorides to be formed, the starting alkane must have a structure that is rather symmetrical; that is, one in which most (or all) of the hydrogens are equivalent. 2,2-Dimethylpropane satisfies this requirement.

$$
\begin{array}{ccc}
\underset{\text{2,2-Dimethylpropane}}{\overset{\displaystyle CH_3 \atop |}{CH_3\underset{\displaystyle |}{\overset{\displaystyle |}{C}}CH_3} \atop \displaystyle CH_3} & \xrightarrow[\text{light}]{2Cl_2} & \underset{\substack{\text{1,1-Dichloro-2,2-}\\ \text{dimethylpropane}}}{\overset{\displaystyle CH_3 \atop |}{CH_3\underset{\displaystyle |}{\overset{\displaystyle |}{C}}CHCl_2} \atop \displaystyle CH_3} \;+\; \underset{\substack{\text{1,3-Dichloro-2,2-}\\ \text{dimethylpropane}}}{\overset{\displaystyle CH_3 \atop |}{ClCH_2\underset{\displaystyle |}{\overset{\displaystyle |}{C}}CH_2Cl} \atop \displaystyle CH_3}
\end{array}
$$

4.49 (*a*) Heptane has five methylene groups, which on chlorination together contribute 85% of the total monochlorinated product.

$$CH_3(CH_2)_5CH_3 \longrightarrow CH_3(CH_2)_5CH_2Cl \;+\; (\text{2-chloro} + \text{3-chloro} + \text{4-chloro})$$

Heptane   15%   85%

Because the problem specifies that attack at each methylene group is equally probable, the five methylene groups each give rise to 85/5, or 17%, of the monochloride product.

Because C-2 and C-6 of heptane are equivalent, we calculate that 2-chloroheptane will constitute 34% of the monochloride fraction. Similarly, C-3 and C-5 are equivalent, and so there should be 34% 3-chloroheptane. The remainder, 17%, is 4-chloroheptane.

These predictions are very close to the observed proportions.

	Calculated, %	Observed, %
2-Chloro	34	35
3-Chloro	34	34
4-Chloro	17	16

(b) There are a total of 20 methylene hydrogens in dodecane, $CH_3(CH_2)_{10}CH_3$. The 19% 2-chloro-dodecane that is formed arises by substitution of any of the four equivalent methylene hydrogens at C-2 and C-11. The total amount of substitution of methylene hydrogens must therefore be:

$$\frac{20}{4} \times 19\% = 95\%$$

The remaining 5% corresponds to substitution of methyl hydrogens at C-1 and C-12. The proportion of 1-chlorododecane in the monochloride fraction is 5%.

4.50 (a) Two of the monochlorides derived from chlorination of 2,2,4-trimethylpentane are primary chlorides:

$$\underset{\substack{|\\ CH_3}}{\overset{\substack{CH_3\\ |}}{ClCH_2CCH_2CHCH_3}}\underset{CH_3}{}$$
1-Chloro-2,2,4-trimethylpentane

$$\underset{\substack{|\quad\;\;|\\ CH_3\;\;CH_3}}{\overset{\substack{CH_3\\ |}}{CH_3CCH_2CHCH_2Cl}}$$
1-Chloro-2,4,4-trimethylpentane

The two remaining isomers are a secondary chloride and a tertiary chloride:

$$\underset{\substack{|\quad\;\;|\quad\;|\\ CH_3\;\;Cl\;\;CH_3}}{\overset{\substack{CH_3\\ |}}{CH_3C\!-\!CHCHCH_3}}$$
3-Chloro-2,2,4-trimethylpentane

$$\underset{\substack{|\quad\;\;|\\ CH_3\;\;CH_3}}{\overset{\substack{CH_3\;\;Cl\\ |\quad\;|}}{CH_3CCH_2CCH_3}}$$
2-Chloro-2,4,4-trimethylpentane

(b) Substitution of any one of the nine hydrogens designated as x in the structural diagram yields 1-chloro-2,2,4-trimethylpentane. Substitution of any one of the six hydrogens designated as y gives 1-chloro-2,4,4-trimethylpentane.

$$\underset{\substack{|\quad\;\;|\\ \underset{x}{CH_3}\;\;\underset{y}{CH_3}}}{\overset{\substack{\overset{x}{CH_3}\\ |}}{\overset{x}{CH_3}C CH_2 CH\overset{y}{CH_3}}}$$

Assuming equal reactivity of a single x hydrogen and a single y hydrogen, the ratio of the two isomers is then expected to be 9:6. Because together the two primary chlorides total 65% of the monochloride fraction, there will be 39% 1-chloro-2,2,4-trimethylpentane (substitution of x) and 26% 1-chloro-2,4,4-trimethylpentane (substitution of y).

4.51 The three monochlorides are shown in the equation

$$CH_3CH_2CH_2CH_2CH_3 \xrightarrow[\text{light}]{Cl_2} CH_3CH_2CH_2CH_2CH_2Cl + \underset{\substack{|\\ Cl}}{CH_3CHCH_2CH_2CH_3} + \underset{\substack{|\\ Cl}}{CH_3CH_2CHCH_2CH_3}$$

Pentane   1-Chloropentane   2-Chloropentane   3-Chloropentane

Pentane has six primary hydrogens (two CH_3 groups) and six secondary hydrogens (three CH_2 groups). Because a single secondary hydrogen is abstracted three times faster than a single primary hydrogen and there are equal numbers of secondary and primary hydrogens, the product mixture should contain three times as much of the secondary chloride isomers as the primary chloride. The primary chloride 1-chloropentane, therefore, is expected to constitute 25% of the product mixture. The secondary chlorides 2-chloropentane and 3-chloropentane are not formed in equal amounts. Rather, 2-chloropentane may be formed by replacement of a hydrogen at C-2 or at C-4, whereas 3-chloropentane is formed only when a C-3 hydrogen is replaced. The amount of 2-chloropentane is therefore 50%, and that of 3-chloropentane is 25%. We predict the major product to be 2-chloropentane, and the predicted proportion of 50% corresponds closely to the observed 46%.

4.52 The equation for the reaction is

Cyclopropane   Chlorine   Cyclopropyl chloride   Hydrogen chloride

The reaction begins with the initiation step in which a chlorine molecule dissociates to two chlorine atoms.

Chlorine   2 Chlorine atoms

A chlorine atom abstracts a hydrogen atom from cyclopropane in the first propagation step.

Cyclopropane   Chlorine atom   Cyclopropyl radical   Hydrogen chloride

Cyclopropyl radical reacts with Cl_2 in the next propagation step.

Cyclopropyl radical   Chlorine   Cyclopropyl chloride   Chlorine atom

4.53 (*a*) Acid-catalyzed hydrogen–deuterium exchange takes place by a pair of Brønsted acid–base reactions.

(*b*) Base-catalyzed hydrogen–deuterium exchange occurs by a different pair of Brønsted acid–base equilibria.

ANSWERS TO INTERPRETIVE PROBLEMS

4.54 C;   **4.55** D;   **4.56** A;   **4.57** A;   **4.58** C;   **4.59** A;   **4.60** A;   **4.61** B

SELF-TEST

PART A

A-1. The compound shown is an example of the broad class of organic compounds known as **steroids**. What functional groups does the molecule contain?

A-2. Give the correct substitutive IUPAC name for each of the following compounds:

(*a*)

(*b*) $CH_3CH_2CHCH_2CHCH_3$ with CH_2OH and CH_2CH_3 substituents

A-3. Draw the structures of the following substances:

(*a*) 2-Chloro-1-iodo-2-methylheptane

(*b*) *cis*-3-Isopropylcyclohexanol

A-4. Give both a functional class and a substitutive IUPAC name for each of the following compounds:

(*a*)

(*b*)

A-5. Supply the missing component for each of the following reactions:

(a) $CH_3CH_2CH_2OH \xrightarrow{SOCl_2}$?

(b) ? $\xrightarrow{HBr}$ $CH_3CH_2\overset{\overset{\displaystyle Br}{\displaystyle |}}{C}(CH_3)_2$

A-6. (a) How many different free radicals can be produced in the reaction between chlorine atoms and 2,4-dimethylpentane?

(b) Write their structures.

(c) Which is the most stable? Which is the least stable?

A-7. Write a balanced chemical equation for the reaction of chlorine with the pentane isomer that gives only one product on monochlorination.

A-8. Write the propagation steps for the light-initiated reaction of bromine with methylcyclohexane.

A-9. Using the data in text Table 4.3, calculate the heat of reaction ($\Delta H°$) for the light-initiated reaction of bromine (Br_2) with 2-methylpropane to give 2-bromo-2-methylpropane and hydrogen bromide.

A-10. (a) Write out each of the elementary steps in the reaction of *tert*-butyl alcohol with hydrogen bromide. Use curved arrows to show electron movement in each step.

(b) Draw the structure of the transition state representing the unimolecular dissociation of the alkyloxonium ion in the preceding reaction.

(c) How does the mechanism of the reaction between 1-butanol and hydrogen bromide differ from the reaction in part (a)?

A-11. (Choose the correct response for each part.) Which species or compound:

(a) Reacts faster with sodium bromide and sulfuric acid?

2-methyl-3-pentanol   or   3-methyl-3-pentanol

(b) Reacts more vigorously with cyclohexane?

Fluorine   or   iodine

(c) Has an odd number of electrons?

Ethoxide ion   or   ethyl radical

(d) Undergoes bond cleavage in the initiation step in the reaction by which methane is converted to chloromethane?

CH_4 or Cl_2

PART B

B-1. A certain alcohol has the functional class IUPAC name *1-ethyl-3-methylbutyl alcohol*. What is its substitutive name?

(a) 1-Ethyl-3-methyl-1-butanol   (d) 2-Methyl-4-hexanol

(b) 2-Methyl-1-hexanol   (e) 5-Methyl-3-hexanol

(c) 3-Methyl-1-hexanol

B-2. Rank the following substances in order of increasing boiling point (lowest → highest):

$$CH_3CH_2CH_2CH_2OH \quad (CH_3)_2CHOCH_3 \quad (CH_3)_3COH \quad (CH_3)_4C$$

$$\qquad\qquad 1 \qquad\qquad\qquad 2 \qquad\qquad\qquad 3 \qquad\qquad 4$$

(*a*) 1 < 3 < 2 < 4   (*c*) 4 < 2 < 3 < 1   (*e*) 4 < 3 < 2 < 1

(*b*) 2 < 4 < 3 < 1   (*d*) 2 < 3 < 1 < 4

B-3. Which one of the following reacts with HBr at the fastest rate?

(*a*)

(*b*)

(*c*)

(*d*)

(*e*)

B-4. What is the decreasing stability order (most stable → least stable) of the following carbocations?

$$1 \qquad\qquad 2 \qquad\qquad 3 \qquad\qquad 4 \qquad\qquad 5$$

(*a*) 3 > 2 > 1 > 4 > 5   (*c*) 3 > 2 ≈ 5 > 1 ≈ 4

(*b*) 1 ≈ 4 > 2 ≈ 5 > 3   (*d*) 3 > 1 ≈ 4 > 2 ≈ 5

B-5. Rank the bond dissociation enthalpies (BDEs) of the bonds indicated with the arrows from smallest to largest.

(*a*) 1 < 2 < 3   (*b*) 3 < 2 < 1   (*c*) 2 < 3 < 1   (*d*) 1 < 3 < 2   (*e*) 3 < 1 < 2

B-6. What are the chain-propagating steps in the free-radical chlorination of methane?

1. $Cl_2 \longrightarrow 2Cl\bullet$

2. $Cl\bullet + CH_4 \longrightarrow CH_3Cl + H\bullet$

3. $Cl\bullet + CH_4 \longrightarrow \bullet CH_3 + HCl$

4. $H\bullet + Cl_2 \longrightarrow HCl + Cl\bullet$

5. $\bullet CH_3 + Cl_2 \longrightarrow CH_3Cl + Cl\bullet$

6. $\bullet CH_3 + CH_4 \longrightarrow CH_4 + \bullet CH_3$

(*a*) 2, 4   (*c*) 3, 5

(*b*) 1, 2   (*d*) 1, 3, 5

B-7. Which of the following is *least* able to serve as a nucleophile in a chemical reaction?

   (*a*) Br^- (*b*) OH^- (*c*) NH_3 (*d*) CH_3^+

B-8. Thiols are alcohol analogs in which the oxygen has been replaced by sulfur (e.g., CH_3SH). Given the fact that the S—H bond is less polar than the O—H bond, which of the following statements comparing thiols and alcohols is correct?

   (*a*) Hydrogen bonding forces are weaker in thiols.

   (*b*) Hydrogen bonding forces are stronger in thiols.

   (*c*) Hydrogen bonding forces would be the same.

   (*d*) No comparison can be made without additional information.

B-9. Rank the *transition states* that occur during the following reaction steps in order of increasing stability (least → most stable):

   1. $H_3C-\overset{+}{O}H_2$ $\longrightarrow$ CH_3^+ + H_2O

   2. $(CH_3)_3C-\overset{+}{O}H_2$ $\longrightarrow$ $(CH_3)_3C^+$ + H_2O

   3. $(CH_3)_2CH-\overset{+}{O}H_2$ $\longrightarrow$ $(CH_3)_2CH^+$ + H_2O

   (*a*) $1 < 2 < 3$ (*b*) $2 < 3 < 1$ (*c*) $1 < 3 < 2$ (*d*) $2 < 1 < 3$

B-10. Using the data from text Table 4.3, calculate the heat of reaction $\Delta H°$ for the following:

   $$CH_3CH_2\cdot + HBr \longrightarrow CH_3CH_3 + Br\cdot$$

   (*a*) +69 kJ/mol (+16.5 kcal/mol)   (*c*) +44 kJ/mol (+10.5 kcal/mol)

   (*b*) −69 kJ/mol (−16.5 kcal/mol)   (*d*) −44 kJ/mol (−10.5 kcal/mol)

B-11. An alkane with a molecular formula C_6H_{14} reacts with chlorine in the presence of light and heat to give four constitutionally isomeric monochlorides of molecular formula $C_6H_{13}Cl$. What is the most reasonable structure for the starting alkane?

   (*a*) $CH_3CH_2CH_2CH_2CH_2CH_3$ (*d*) $(CH_3)_3CCH_2CH_3$

   (*b*) $(CH_3)_2CHCH_2CH_2CH_3$ (*e*) $(CH_3)_2CHCH(CH_3)_2$

   (*c*) $CH_3CH(CH_2CH_3)_2$

B-12. The species shown in the box represents ___ of the reaction between isopropyl alcohol and hydrogen bromide.

$$(CH_3)_2\overset{\delta+}{C}H - - -\overset{\delta+}{O}H_2$$

   (*a*) the alkyloxonium ion intermediate

   (*b*) the transition state of the bimolecular proton transfer step

   (*c*) the transition state of the capture of the carbocation by a nucleophile

   (*d*) the carbocation intermediate

   (*e*) the transition state of the unimolecular dissociation step

For the remaining four questions, consider the following free-radical reaction:

+ X_2 $\xrightarrow{\text{light}}$ monohalogenation product

B-13. Light is involved in which of the following reaction steps?

(*a*) Initiation only

(*b*) Propagation only

(*c*) Termination only

(*d*) Initiation and propagation

B-14. Which of the following statements about the reaction is *not* true?

(*a*) Halogen atoms are consumed in the first propagation step.

(*b*) Halogen atoms are regenerated in the second propagation step.

(*c*) Hydrogen atoms are produced in the first propagation step.

(*d*) Chain termination occurs when two radicals react with each other.

B-15. How many monohalogenation products are possible? (Do not consider stereoisomers.)

(*a*) 2   (*b*) 3   (*c*) 4   (*d*) 5

B-16. Which halogen (X_2) will give the best yield of a single monohalogenation product?

(*a*) F_2 (*b*) Cl_2 (*c*) Br_2 (*d*) I_2

CHAPTER 5

Structure and Preparation of Alkenes: Elimination Reactions

SOLUTIONS TO TEXT PROBLEMS

5.1   (*b*) Writing the structure in more detail, we see that the longest continuous chain contains four carbon atoms.

$$\overset{\displaystyle CH_3}{\underset{\displaystyle CH_3}{H_3\overset{4}{C}-\overset{3}{\underset{|}{\overset{|}{C}}}-\overset{2}{C}H=\overset{1}{C}H_2}}$$

The double bond is located at the end of the chain, and so we name the alkene as a derivative of 1-butene. Two methyl groups are substituents at C-3. The IUPAC name is 3,3-dimethyl-1-butene or 3,3-dimethylbut-1-ene.

  (*c*) Expanding the structural formula reveals the molecule to be a methyl-substituted derivative of hexene.

$$\underset{\displaystyle CH_3}{H_3\overset{1}{C}-\overset{2}{\underset{|}{C}}=\overset{3}{C}H\overset{4}{C}H_2\overset{5}{C}H_2\overset{6}{C}H_3}$$

2-Methyl-2-hexene
or
2-Methylhex-2-ene

(*d*) In compounds containing a double bond and a halogen, the double bond takes precedence in numbering the longest carbon chain.

$$\overset{1}{H_2}C{=}\overset{2}{C}H\overset{3}{C}H_2\overset{4}{C}H\overset{5}{C}H_3$$
$$|$$
$$Cl$$

4-Chloro-1-pentene
or
4-Chloropent-1-ene

(*e*) When a hydroxyl group is present in a compound containing a double bond, the hydroxyl takes precedence over the double bond in numbering the longest carbon chain.

$$\overset{5}{H_2}C{=}\overset{4}{C}H\overset{3}{C}H_2\overset{2}{C}H\overset{1}{C}H_3$$
$$|$$
$$OH$$

4-Penten-2-ol
or
Pent-4-en-2-ol

5.2 There are three sets of nonequivalent positions on a cyclopentene ring, identified as *a*, *b*, and *c* on the cyclopentene structure shown:

Thus, there are three different monochloro-substituted derivatives of cyclopentene. The carbons that bear the double bond are numbered C-1 and C-2 in each isomer, and the other positions are numbered in sequence in the direction that gives the chlorine-bearing carbon its lower number.

1-Chlorocyclopentene   3-Chlorocyclopentene   4-Chlorocyclopentene

5.3 (*b*) The alkene is a derivative of 3-hexene regardless of whether the chain is numbered from left to right or from right to left. Number it in the direction that gives the lower number to the substituent.

3-Ethyl-3-hexene

(*c*) There are only *two sp*^{2}-hybridized carbons, the two connected by the double bond. The other six carbons are *sp*^{3}-hybridized.

(*d*) There are three *sp*^{2}–*sp*^{3} σ bonds and three *sp*^{3}–*sp*^{3} σ bonds.

5.4 Consider first the C_5H_{10} alkenes that have an unbranched carbon chain:

1-Pentene   *cis*-2-Pentene   *trans*-2-Pentene

There are three additional isomers. These have a four-carbon chain with a methyl substituent.

2-Methyl-1-butene   2-Methyl-2-butene   3-Methyl-1-butene

5.5 No. They are constitutional isomers. The double bond connects C-2 and C-3 in *cis*-2-hexene and C-3 and C-4 in *trans*-3-hexene. Stereoisomers have the same constitution but differ in the arrangement of atoms in space.

cis-2-Hexene   trans-3-Hexene

5.6 First, identify the constitution of 9-tricosene. Referring back to Table 2.2 in Section 2.11 of the text, we see that tricosane is the unbranched alkane containing 23 carbon atoms. 9-Tricosene, therefore, contains an unbranched chain of 23 carbons with a double bond between C-9 and C-10. Because the problem specifies that the pheromone has the cis configuration, the first 8 carbons and the last 13 must be on the same side of the C-9–C-10 double bond.

$$CH_3(CH_2)_7 \quad (CH_2)_{12}CH_3$$
$$C=C$$
$$H \qquad H$$

cis-9-Tricosene

5.7 (*b*) One of the carbons of the double bond bears a methyl group and a hydrogen; methyl is of higher rank than hydrogen. The other doubly bonded carbon bears the groups —CH_2CH_2F and —$CH_2CH_2CH_2CH_3$. At the first point of difference between these two, fluorine is of higher atomic number than carbon, and so —CH_2CH_2F is of higher precedence.

Higher H_3C CH_2CH_2F Higher
              C=C
Lower H $CH_2CH_2CH_2CH_3$ Lower

Higher ranked substituents are on the same side of the double bond; the alkene has the *Z* configuration.

(*c*) One of the carbons of the double bond bears a methyl group and a hydrogen; as we have seen, methyl is of higher rank. The other doubly bonded carbon bears —CH_2CH_2OH and —$C(CH_3)_3$. Let's analyze these two groups to determine their order of precedence.

—CH_2CH_2OH —$C(CH_3)_3$

—C(C,H,H)   —C(C,C,C)

Lower priority   Higher priority

We examine the atoms one by one at the point of attachment before proceeding down the chain. Therefore, —$C(CH_3)_3$ outranks —CH_2CH_2OH.

Higher H_3C CH_2CH_2OH Lower
              C=C
Lower H $C(CH_3)_3$ Higher

Higher ranked groups are on opposite sides; the configuration of the alkene is *E*.

(*d*) The cyclopropyl ring is attached to the double bond by a carbon that bears the atoms (C,C,H) and is therefore of higher precedence than an ethyl group —C(C,H,H).

Higher ranked groups are on opposite sides; the configuration of the alkene is *E*.

5.8 The dipole moments of (*E*)-2,3-dichloro-2-butene cancel; the molecule is nonpolar. The dipole moments of *trans*-1-chloropropene are additive, as are the dipole moments of 1,1-dichloro-2-methylpropene. The latter compound has twice as many dipoles and is the most polar.

(*E*)-2,3-Dichloro-2-butene;
least polar

trans-1-Chloropropene

1,1-Dichloro-2-methylpropene;
most polar

5.9 A trisubstituted alkene has three carbons directly attached to the doubly bonded carbons. There are three trisubstituted C₆H₁₂ isomers, two of which are stereoisomers.

2-Methyl-2-pentene

(*E*)-3-Methyl-2-pentene

(*Z*)-3-Methyl-2-pentene

5.10 The most stable C₆H₁₂ alkene has a tetrasubstituted double bond:

2,3-Dimethyl-2-butene

5.11 Apply the two general rules for alkene stability to rank these compounds. First, more highly substituted double bonds are more stable than less substituted ones. Second, when two double bonds are similarly constituted, the trans stereoisomer is more stable than the cis. The predicted order of decreasing stability is therefore:

2-Methyl-2-butene
(trisubstituted):
most stable

(*E*)-2-Pentene
(disubstituted)

(*Z*)-2-Pentene
(disubstituted)

1-Pentene
(monosubstituted):
least stable

5.12 Begin by writing the structural formula corresponding to the IUPAC name given in the problem. A bond-line depiction is useful here.

3,4-Di-*tert*-butyl-2,2,5,5-tetramethyl-3-hexene

The alkene is extremely crowded and destabilized by van der Waals strain. Bulky *tert*-butyl groups are cis to one another on each side of the double bond. Highly strained compounds are often quite difficult to synthesize, and this alkene is a good example.

5.13 Use the zigzag arrangement of bonds in the parent skeleton figure to place *E* and *Z* bonds as appropriate for each part of the problem. From the sample solution to parts (*a*) and (*b*), the ring carbons have the

higher priorities. Thus, an *E* double bond will have ring carbons arranged ⌐⌐ and a *Z* double bond

.

(*c*) (Z)-3-Methylcyclodecene

(*e*) (Z)-5-Methylcyclodecene

(*d*) (E)-3-Methylcyclodecene

(*f*) (E)-5-Methylcyclodecene

5.14 Write out the structure of the alcohol, recognizing that the alkene is formed by loss of a hydrogen and a hydroxyl group from adjacent carbons.

(*b*, *c*) Both 1-propanol and 2-propanol give propene on acid-catalyzed dehydration.

$$\overset{\beta}{C}H_3\overset{\alpha}{C}H_2CH_2OH \quad \xrightarrow[\text{heat}]{H^+} \quad CH_3CH{=}CH_2 \quad \xleftarrow[\text{heat}]{H^+} \quad \overset{\beta}{C}H_3\overset{\alpha}{C}H\overset{\beta}{C}H_3$$
$$\qquad\qquad\qquad\qquad\qquad\qquad\qquad\qquad\qquad\qquad\qquad\qquad\qquad\qquad\qquad\quad |$$
$$\qquad\qquad\qquad\qquad\qquad\qquad\qquad\qquad\qquad\qquad\qquad\qquad\qquad\qquad\qquad\quad OH$$

1-Propanol   Propene   2-Propanol

(*d*) Carbon-3 has no hydrogens in 2,3,3-trimethyl-2-butanol. Elimination can involve only the hydroxyl group at C-2 and a hydrogen at C-1.

2,3,3-Trimethyl-2-butanol   2,3,3-Trimethyl-1-butene

5.15 (*b*) Elimination can involve loss of a hydrogen from the methyl group or from C-2 of the ring in 1-methylcyclohexanol.

| 1-Methylcyclohexanol | Methylenecyclohexane (a disubstituted alkene; minor product) | 1-Methylcyclohexene (a trisubstituted alkene; major product) |

According to the Zaitsev rule, the major alkene is the one corresponding to loss of a hydrogen from the alkyl group that has the smaller number of hydrogens. Thus hydrogen is removed from the methylene group in the ring rather than from the methyl group, and 1-methylcyclohexene is formed in greater amounts than methylenecyclohexane.

(*c*) The two alkenes formed are as shown in the equation.

Compound has a trisubstituted double bond

Compound has a tetrasubstituted double bond; more stable

The more highly substituted alkene is formed in greater amounts, as predicted by Zaitsev's rule.

5.16 2-Pentanol can undergo dehydration in two different directions, giving either 1-pentene or 2-pentene. 2-Pentene is formed as a mixture of the cis and trans stereoisomers.

| 2-Pentanol | 1-Pentene | *cis*-2-Pentene | *trans*-2-Pentene |

5.17 (*b*) The site of positive charge in the carbocation is the carbon atom that bears the hydroxyl group in the starting alcohol.

1-Methylcyclohexanol

Water may remove a proton from the methyl group, as shown in the following equation:

Methylenecyclohexane

Loss of a proton from the ring gives the major product 1-methylcyclohexene.

1-Methylcyclohexene

(c) Loss of the hydroxyl group under conditions of acid catalysis yields a tertiary carbocation.

Water may remove a proton from an adjacent methylene group to give a trisubstituted alkene.

Removal of the methine proton gives a tetrasubstituted alkene.

5.18 In writing mechanisms for acid-catalyzed dehydration of alcohols, begin with formation of the carbocation intermediate:

2,2-Dimethylcyclohexanol   2,2-Dimethylcyclohexyl cation

This secondary carbocation can rearrange to a more stable tertiary carbocation by a methyl group shift.

2,2-Dimethylcyclohexyl cation   1,2-Dimethylcyclohexyl cation
(secondary)   (tertiary)

Loss of a proton from the 1,2-dimethylcyclohexyl cation intermediate yields 1,2-dimethylcyclohexene.

1,2-Dimethylcyclohexyl
cation

1,2-Dimethylcyclohexene

5.19 (*b*) All the hydrogens of *tert*-butyl chloride are equivalent. Loss of any of these hydrogens along with the chlorine yields 2-methylpropene as the only alkene.

tert-Butyl chloride

2-Methylpropene

(*c*) All the β hydrogens of 3-bromo-3-ethylpentane are equivalent. Therefore, β elimination can give only 3-ethyl-2-pentene.

3-Bromo-3-ethylpentane

3-Ethyl-2-pentene

(*d*) There are two possible modes of β elimination from 2-bromo-3-methylbutane. Elimination in one direction provides 3-methyl-1-butene; elimination in the other gives 2-methyl-2-butene.

$$\overset{\beta}{C}H_3\overset{\beta}{C}HCH(CH_3)_2 \longrightarrow H_2C=CHCH(CH_3)_2 + CH_3CH=C(CH_3)_2$$

2-Bromo-3-methylbutane

3-Methyl-1-butene
(monosubstituted)

2-Methyl-2-butene
(trisubstituted)

The major product is the more highly substituted alkene, 2-methyl-2-butene. It is the more stable alkene and corresponds to removal of a hydrogen from the carbon that has the fewer hydrogens.

(*e*) Regioselectivity is not an issue here, because 3-methyl-1-butene is the only alkene that can be formed by β elimination from 1-bromo-3-methylbutane.

$$Br\overset{\beta}{C}H_2CH_2CH(CH_3)_2 \longrightarrow H_2C=CHCH(CH_3)_2$$

1-Bromo-3-methylbutane

3-Methyl-1-butene

(*f*) Two alkenes may be formed here. The more highly substituted one is 1-methylcyclohexene, and this is predicted to be the major product in accordance with Zaitsev's rule.

1-Iodo-1-methylcyclohexane   Methylenecyclohexane   1-Methylcyclohexene
                                              (disubstituted)   (trisubstituted;
                                                                                 major product)

5.20 Elimination in 2-bromobutane can take place between C-1 and C-2 or between C-2 and C-3. Three alkenes are capable of being formed: 1-butene and the stereoisomers *cis*-2-butene and *trans*-2-butene.

2-Bromobutane   1-Butene   *cis*-2-Butene   *trans*-2-Butene

As predicted by Zaitsev's rule, the most stable alkene predominates. The major product is *trans*-2-butene.

5.21 An unshared electron pair of the base methoxide (CH_3O^-) abstracts a proton from carbon. The pair of electrons in this $C-H$ bond becomes the π component of the double bond of the alkene. The pair of electrons in the $C-Cl$ bond becomes an unshared electron pair of chloride ion.

5.22 The most stable conformation of *cis*-4-*tert*-butylcyclohexyl bromide has the bromine substituent in an axial orientation. The hydrogen that is removed by the base is an axial proton at C-2. This hydrogen and the bromine are anti coplanar to each other in the most stable conformation.

5.23 These problems relate to isotope effects on the rate of elimination by the E2 mechanism.

(*b*) As in part (*a*) the β $C-H$ bond of $(CH_3)_2CHCD_2Br$ breaks faster than the β $C-D$ bond of $(CH_3)_2CDCH_2Br$.

(*c*) As in parts (*a*) and (*b*) the alkyl halide that has H at the β carbon reacts faster than the one with D at its β carbon.

$$
\underset{\underset{\text{H at } \beta \text{ carbon}}{}}{\overset{\overset{CD_3}{|}}{CD_3CD_2\underset{|}{C}CH_2Br}} \quad \text{faster rate of E2 than} \quad \underset{\underset{\text{D at } \beta \text{ carbon}}{}}{\overset{\overset{CH_3}{|}}{CH_3CH_2\underset{|}{C}CH_2Br}}
$$

5.24 The rate-determining step in the E1 elimination of 2-bromo-2-methylbutane is ionization to form a carbocation. A proton is lost from the β-carbon in the next step, which is fast.

$$
\underset{:\ddot{Br}:}{\overset{\overset{CH_3}{|}}{CH_3\underset{|}{C}CH_2CH_3}} \xrightarrow[\text{slow}]{-:\ddot{Br}:^-} \overset{\overset{CH_3}{|}}{\underset{H_3C\ \ \overset{+}{}\ \ CH_2CH_3}{C}} \xrightarrow[\text{fast}]{-H^+} \overset{\overset{CH_2}{||}}{\underset{H_3C\diagdown\ \diagup CH_2CH_3}{C}} + \overset{\overset{CH_3}{|}}{\underset{H_3C\diagdown\ \diagup CHCH_3}{C}}
$$

In the E1 mechanism shown, hydrogen is lost after the rate-determining step, so no isotope effect is expected.

5.25 (*a*) 1-Heptene is $H_2C{=}CH(CH_2)_4CH_3$

(*b*) 3-Ethyl-2-pentene is $CH_3CH{=}C(CH_2CH_3)_2$

(*c*) *cis*-3-Octene is
$$
\underset{H}{\overset{CH_3CH_2}{\diagdown}}C{=}C\underset{H}{\overset{CH_2CH_2CH_2CH_3}{\diagup}}
$$

(*d*) *trans*-1,4-Dichloro-2-butene is
$$
\underset{H}{\overset{ClCH_2}{\diagdown}}C{=}C\underset{CH_2Cl}{\overset{H}{\diagup}}
$$

(*e*) (*Z*)-3-Methyl-2-hexene is
$$
\underset{H}{\overset{H_3C}{\diagdown}}C{=}C\underset{CH_3}{\overset{CH_2CH_2CH_3}{\diagup}}
$$

(*f*) (*E*)-3-Chloro-2-hexene is
$$
\underset{H}{\overset{H_3C}{\diagdown}}C{=}C\underset{Cl}{\overset{CH_2CH_2CH_3}{\diagup}}
$$

(*g*) 1-Bromo-3-methylcyclohexene is

(*h*) 1-Bromo-6-methylcyclohexene is

(*i*) 4-Methyl-4-penten-2-ol is
$$
\underset{\overset{|}{OH}}{CH_3\overset{}{C}HCH_2}\overset{\overset{CH_3}{|}}{C}{=}CH_2
$$

(*j*) A vinyl group is —CH=CH$_2$. Vinylcycloheptane is

(*k*) An allyl group is —CH$_2$CH=CH$_2$. 1,1-Diallylcyclopropane is

(*l*) An isopropenyl substituent is $\underset{\underset{\displaystyle \text{CH}_3}{|}}{—\text{C}}$=CH$_2$. *trans*-1-Isopropenyl-3-methylcyclohexane is

or

5.26 Alkenes with tetrasubstituted double bonds have four alkyl groups attached to the doubly bonded carbons. There is only one alkene of molecular formula C$_7$H$_{14}$ that has a tetrasubstituted double bond. Acceptable IUPAC names are 2,3-dimethyl-2-pentene or 2,3-dimethylpent-2-ene.

$$\underset{\text{H}_3\text{C}}{\overset{\text{H}_3\text{C}}{>}}\text{C}=\text{C}\underset{\text{CH}_2\text{CH}_3}{\overset{\text{CH}_3}{<}}$$

2,3-Dimethyl-2-pentene
or
2,3-Dimethylpent-2-ene

5.27 (*a*) The longest chain that includes the double bond in (CH$_3$CH$_2$)$_2$C=CHCH$_3$ contains five carbon atoms, and so the parent alkene is a pentene. The numbering scheme that gives the double bond the lowest number is

The compound is named 3-ethyl-2-pentene or 3-ethylpent-2-ene.

(*b*) Write out the structure in detail, and identify the longest continuous chain that includes the double bond.

$$\underset{\text{CH}_3\text{CH}_2}{\overset{\overset{1}{\text{CH}_3}\overset{2}{\text{CH}_2}}{}}\underset{3}{\text{C}}=\underset{4}{\text{C}}\underset{\text{CH}_2\text{CH}_3}{\overset{\overset{5}{\text{CH}_2}\overset{6}{\text{CH}_3}}{}}$$

The longest chain contains six carbon atoms, and the double bond is between C-3 and C-4. The compound is named as a derivative of 3-hexene. There are ethyl substituents at C-3 and C-4. Acceptable IUPAC names are 3,4-diethyl-3-hexene or 3,4-diethylhex-3-ene.

(c) Write out the structure completely.

The longest carbon chain contains four carbons. Number the chain so as to give the lowest numbers to the doubly bonded carbons, and list the substituents in alphabetical order. This compound may be named 1,1-dichloro-3,3-dimethyl-1-butene or 1,1-dichloro-3,3-dimethylbut-1-ene.

(d) The longest chain has five carbon atoms, the double bond is at C-1, and there are two methyl substituents. Acceptable IUPAC names are 4,4-dimethyl-1-pentene or 4,4-dimethylpent-1-ene.

(e) We number this trimethylcyclobutene derivative so as to provide the lowest number for the substituent at the first point of difference, remembering also that the double bond is between C-1 and C-2. We therefore number

rather than

Acceptable IUPAC names are 1,4,4-trimethylcyclobutene or 1,4,4-trimethylcyclobut-1-ene.

(f) The cyclohexane ring has a 1,2-cis arrangement of vinyl substituents. The compound is cis-1,2-divinylcyclohexane.

(g) Name this compound as a derivative of cyclohexene. It may be named 1,2-divinylcyclohexene or 1,2-divinylcyclohex-1-ene.

5.28 (a) Go to the end of the name, because this tells you how many carbon atoms are present in the longest chain. In the hydrocarbon name 2,6,10,14-tetramethyl-2-pentadecene, the suffix "2-pentadecene" reveals that the longest continuous chain has 15 carbon atoms and that there is a double bond between C-2 and C-3. The rest of the name provides the information that there are four methyl groups and that they are located at C-2, C-6, C-10, and C-14.

2,6,10,14-Tetramethyl-2-pentadecene

(b) An allyl group is $H_2C=CHCH_2-$. Allyl isothiocyanate is therefore $H_2C=CHCH_2N=C=S$.

(*c*) Replacing the generalized abbreviation R by an isopropenyl group gives the structure of grandisol.

5.29 (*a*) The *E* configuration means that the higher priority groups are on opposite sides of the double bond.

(*E*)-6-Nonen-1-ol

(*b*) Geraniol has two double bonds, but only one of them, the one between C-2 and C-3, is capable of stereochemical variation. Of the groups at C-2, CH_2OH is of higher priority than H. At C-3, CH_2CH_2 outranks CH_3. Higher priority groups are on opposite sides of the double bond in the *E* isomer; hence geraniol has the structure shown.

Geraniol

(*c*) Because nerol is a stereoisomer of geraniol, it has the same constitution and differs from geraniol only in having the *Z* configuration of the double bond.

Nerol

(*d*) Beginning at the C-6, C-7 double bond, we see that the propyl group is of higher priority than the methyl group at C-7. Because the C-6, C-7 double bond is *E*, the propyl group must be on the opposite side of the higher priority group at C-6, where the CH_2 fragment has a higher priority than hydrogen. We therefore write the stereochemistry of the C-6, C-7 double bond as:

E

At C-2, CH_2OH is of higher priority than H; and at C-3, CH_2CH_2C— is of higher priority than CH_2CH_3. The double-bond configuration at C-2 is Z. Therefore

Combining the two partial structures, we obtain for the full structure of the codling moth's sex pheromone

The compound is (2Z,6E)-3-ethyl-7-methyl-2,6-decadien-1-ol.

(e) The sex pheromone of the honeybee is (E)-9-oxo-2-decenoic acid, with the structure

(f) Looking first at the C-2, C-3 double bond of the cecropia moth's growth hormone

we find that its configuration is E, because the higher priority groups are on opposite sides of the double bond.

The configuration of the C-6, C-7 double bond is also E.

5.30 The alkenes are listed as follows in order of decreasing heat of combustion:

(e)   2,4,4-Trimethyl-2-pentene; 5293 kJ/mol (1264.9 kcal/mol). Highest heat of combustion because it is C_8H_{16}; all others are C_7H_{14}.

(a)   1-Heptene; 4658 kJ/mol (1113.4 kcal/mol). Monosubstituted double bond; therefore least stable C_7H_{14} isomer.

(d)   (Z)-4,4-Dimethyl-2-pentene; 4650 kJ/mol (1111.4 kcal/mol). Disubstituted double bond, but destabilized by van der Waals strain.

(b)   2,4-Dimethyl-1-pentene; 4638 kJ/mol (1108.6 kcal/mol). Disubstituted double bond.

(c)   2,4-Dimethyl-2-pentene; 4632 kJ/mol (1107.1 kcal/mol). Trisubstituted double bond.

5.31 (*a*) 1-Methylcyclohexene is more stable; it contains a *trisubstituted* double bond, whereas 3-methylcyclohexene has only a disubstituted double bond.

more stable than

      1-Methylcyclohexene   3-Methylcyclohexene

(*b*) Both isopropenyl and allyl are three-carbon alkenyl groups: isopropenyl is $H_2C=CCH_3$, allyl is

$H_2C=CHCH_2-$.

         Isopropenylcyclopentane   Allylcyclopentane

Isopropenylcyclopentane has a disubstituted double bond and so is predicted to be more stable than allylcyclopentane, in which the double bond is monosubstituted.

(*c*) A double bond in a six-membered ring is less strained than a double bond in a four-membered ring; therefore bicyclo[4.2.0]oct-3-ene is more stable.

more stable than

       Bicyclo[4.2.0]oct-3-ene   Bicyclo[4.2.0]oct-7-ene

(*d*) Cis double bonds are more stable than trans double bonds when the ring is smaller than 11-membered. (*Z*)-Cyclononene has a cis double bond in a nine-membered ring, and is therefore more stable than (*E*)-cyclononene.

more stable than

         (*Z*)-Cyclononene   (*E*)-Cyclononene

(*e*) Trans double bonds are more stable than cis when the ring is large. Here the rings are 18-membered. Therefore, (*E*)-cyclooctadecene is more stable than (*Z*)-cyclooctadecene.

more stable than

      (*E*)-Cyclooctadecene   (*Z*)-Cyclooctadecene

5.32 (*a*) Carbon atoms that are involved in double bonds are sp^2-hybridized, with ideal bond angles of 120°. Incorporating an sp^2-hybridized carbon into a three-membered ring leads to more angle strain than incorporation of an sp^3-hybridized carbon with an ideal bond angle of 109°. 1-Methylcyclopropene has two sp^2-hybridized carbons in a three-membered ring and so has substantially more angle strain than methylenecyclopropane.

$$\triangleright\!\!-\!\!CH_3 \qquad\qquad \triangleright\!\!=\!\!CH_2$$

1-Methylcyclopropene Methylenecyclopropane

The higher degree of substitution at the double bond in 1-methylcyclopropene is not sufficient to offset the increased angle strain, and so 1-methylcyclopropene is less stable than methylenecyclopropane.

(*b*) 3-Methylcyclopropene has a disubstituted double bond and two sp^2-hybridized carbons in its three-membered ring. It is the least stable of the isomers.

$$\triangleright\!\!-\!\!CH_3$$

3-Methylcyclopropene

5.33 In all parts of this exercise, write the structure of the alkyl halide in sufficient detail to identify the carbon that bears the halogen and the β-carbon atoms that bear at least one hydrogen. These are the carbons that become doubly bonded in the alkene product.

(*a*) 1-Bromohexane can give only 1-hexene under conditions of E2 elimination.

$$\overset{\beta}{Br}CH_2\overset{\beta}{CH_2}CH_2CH_2CH_2CH_3 \quad\xrightarrow[\text{E2}]{\text{base}}\quad H_2C=CHCH_2CH_2CH_2CH_3$$

1-Bromohexane   1-Hexene (only alkene)

(*b*) 2-Bromohexane can give both 1-hexene and 2-hexene on dehydrobromination. The 2-hexene fraction is a mixture of cis and trans stereoisomers.

2-Bromohexane   1-Hexene   cis-2-Hexene   trans-2-Hexene

(*c*) Both a cis–trans pair of 2-hexenes and a cis–trans pair of 3-hexenes are capable of being formed from 3-bromohexane.

cis-2-Hexene   trans-2-Hexene

cis-3-Hexene   trans-3-Hexene

3-Bromohexane

(*d*) Dehydrobromination of 2-bromo-2-methylpentane can involve one of the hydrogens of either a methyl group (C-1) or a methylene group (C-3).

2-Bromo-2-methylpentane   2-Methyl-1-pentene   2-Methyl-2-pentene

Neither alkene is capable of existing in stereoisomeric forms, and so these two are the only products of E2 elimination.

(*e*) 2-Bromo-3-methylpentane can undergo dehydrohalogenation by loss of a proton from either C-1 or C-3. Loss of a proton from C-1 gives 3-methyl-1-pentene.

2-Bromo-3-methylpentane   3-Methyl-1-pentene

Loss of a proton from C-3 gives a mixture of (*E*)- and (*Z*)-3-methyl-2-pentene.

2-Bromo-3-methylpentane   (*E*)-3-Methyl-2-pentene   (*Z*)-3-Methyl-2-pentene

(*f*) Three alkenes are possible from 3-bromo-2-methylpentane. Loss of the C-2 proton gives 2-methyl-2-pentene.

3-Bromo-2-methylpentane   2-Methyl-2-pentene

Abstraction of a proton from C-4 can yield either (*E*)- or (*Z*)-4-methyl-2-pentene.

3-Bromo-2-methylpentane   (*E*)-4-Methyl-2-pentene   (*Z*)-4-Methyl-2-pentene

(*g*) Proton abstraction from the C-3 methyl group of 3-bromo-3-methylpentane yields 2-ethyl-1-butene.

3-Bromo-3-methylpentane   2-Ethyl-1-butene

Stereoisomeric 3-methyl-2-pentenes are formed by proton abstraction from C-2.

3-Bromo-3-methylpentane   (E)-3-Methyl-2-pentene   (Z)-3-Methyl-2-pentene

(h) Only 3,3-dimethyl-1-butene may be formed under conditions of E2 elimination from 3-bromo-2,2-dimethylbutane.

3-Bromo-2,2-dimethylbutane   3,3-Dimethyl-1-butene

5.34 (a) The reaction that takes place with 1-bromo-3,3-dimethylbutane is an E2 elimination involving loss of the bromine at C-1 and abstraction of the proton at C-2 by the strong base potassium *tert*-butoxide, yielding a single alkene.

1-Bromo-3,3-dimethylbutane   3,3-Dimethyl-1-butene

(b) Two alkenes are capable of being formed in this β elimination reaction.

1-Methylcyclopentyl chloride   Methylenecyclopentane   1-Methylcyclopentene

The more highly substituted alkene is 1-methylcyclopentene; it is the major product of this reaction. According to Zaitsev's rule, the major alkene is formed by proton removal from the β carbon that has the fewest hydrogens.

(c) Acid-catalyzed dehydration of 3-methyl-3-pentanol can lead either to 2-ethyl-1-butene or to a mixture of (E)- and (Z)-3-methyl-2-pentene.

3-Methyl-3-pentanol   2-Ethyl-1-butene   (E)-3-Methyl-2-pentene   (Z)-3-Methyl-2-pentene

The major product is a mixture of the trisubstituted alkenes, (E)- and (Z)-3-methyl-2-pentene. Of these two stereoisomers the E isomer is slightly more stable and is expected to predominate.

(d) Acid-catalyzed dehydration of 2,3-dimethyl-2-butanol can proceed in either of two directions.

$$H_3C-\underset{\underset{OH}{|}}{\overset{\overset{H_3C \quad CH_3}{|\quad\quad|}}{\underset{\beta}{C}}}-\underset{\beta}{CHCH_3} \quad \xrightarrow[\substack{120°C \\ -H_2O}]{H_3PO_4} \quad H_2C=C\overset{CH_3}{\underset{CH(CH_3)_2}{\diagdown}} \quad + \quad (CH_3)_2C=C(CH_3)_2$$

| 2,3-Dimethyl-2-butanol | 2,3-Dimethyl-1-butene (disubstituted) | 2,3-Dimethyl-2-butene (tetrasubstituted) |

The major alkene is the one with the more highly substituted double bond, 2,3-dimethyl-2-butene. Its formation corresponds to Zaitsev's rule in that a proton is lost from the β carbon that has the fewest hydrogens.

(e) Only a single alkene is capable of being formed on E2 elimination from this alkyl iodide. Stereoisomeric alkenes are not possible, and because both β hydrogens are equivalent, regioisomers cannot be formed either.

$$\underset{\underset{CH_3 \quad CH_3}{|\quad\quad|}}{CH_3\overset{\overset{\quad\quad I}{\quad\quad|}}{\underset{\beta}{C}}HCH\underset{\beta}{CHCH_3}} \quad \xrightarrow[\substack{ethanol \\ 70°C}]{NaOCH_2CH_3} \quad (CH_3)_2C=CHCH(CH_3)_2$$

| 3-Iodo-2,4-dimethylpentane | 2,4-Dimethyl-2-pentene |

(f) Despite the structural similarity of this alcohol to the alkyl halide in the preceding part of this problem, its dehydration is more complicated. The initially formed carbocation is secondary and can rearrange to a more stable tertiary carbocation by a hydride shift.

$$\underset{\underset{CH_3 \quad CH_3}{|\quad\quad|}}{CH_3\overset{\overset{OH}{|}}{C}HCHCHCH_3} \quad \xrightarrow[-H_2O]{H^+} \quad \underset{\underset{CH_3 \quad CH_3}{|\quad\quad|}}{CH_3\overset{\overset{H}{|}}{C}-CHCHCH_3} \quad \xrightarrow[shift]{hydride} \quad \underset{\underset{CH_3 \quad CH_3}{|\quad\quad|}}{CH_3\overset{+}{C}CH_2CHCH_3}$$

| 2,4-Dimethyl-3-pentanol | Secondary carbocation (less stable) | Tertiary carbocation (more stable) |

The tertiary carbocation, once formed, can give either 2,4-dimethyl-1-pentene or 2,4-dimethyl-2-pentene by loss of a proton.

$$\underset{\underset{CH_3 \quad CH_3}{|\quad\quad|}}{CH_3\overset{+}{C}CH_2CHCH_3} \quad \longrightarrow \quad \underset{\underset{CH_3}{|}}{H_2C=CCH_2CH(CH_3)_2} \quad + \quad (CH_3)_2C=CHCH(CH_3)_2$$

| 2,4-Dimethyl-1-pentene (disubstituted) | 2,4-Dimethyl-2-pentene (trisubstituted) |

The proton is lost from the methylene group in preference to the methyl group. The major alkene is the more highly substituted one, 2,4-dimethyl-2-pentene.

5.35 In all parts of this problem you need to reason backward from an alkene to an alkyl bromide of molecular formula $C_7H_{13}Br$ that gives *only* the desired alkene under E2 elimination conditions.

Recall that the carbon–carbon double bond is formed by loss of a proton from one of the carbons that becomes doubly bonded and a bromine from the other.

(a) Cycloheptene is the only alkene formed by an E2 elimination reaction of cycloheptyl bromide.

Cycloheptyl bromide   Cycloheptene

(b) (Bromomethyl)cyclohexane is the correct answer. It gives methylenecyclohexane as the *only* alkene under E2 conditions.

(Bromomethyl)cyclohexane   Methylenecyclohexane

1-Bromo-1-methylcyclohexane is not correct. It gives a mixture of 1-methylcyclohexene and methylenecyclohexane on elimination.

1-Bromo-1-methylcyclohexane   Methylenecyclohexane   +   1-Methylcyclohexene

(c) In order for 4-methylcyclohexene to be the only alkene, the starting alkyl bromide must be 1-bromo-4-methylcyclohexane. Either the cis or the trans isomer may be used, although the cis will react more readily because its more stable conformation (equatorial methyl) has an axial bromine.

cis- or trans-1-Bromo-4-methylcyclohexane   4-Methylcyclohexene

1-Bromo-3-methylcyclohexane is incorrect; its dehydrobromination yields a mixture of 3-methylcyclohexene and 4-methylcyclohexene.

1-Bromo-3-methylcyclohexene   3-Methylcyclohexene   +   4-Methylcyclohexene

(d) The alkyl bromide must be primary in order for the desired alkene to be the only product of E2 elimination.

2-Cyclopentylethyl bromide   Vinylcyclopentane

If 1-cyclopentylethyl bromide were used, a mixture of regioisomeric alkenes would be formed, with the desired vinylcyclopentane being the minor component of the mixture.

| 1-Cyclopentylethyl bromide | Ethylidenecyclopentane (major product) | Vinylcyclopentane (minor product) |

(*e*) Either *cis*- or *trans*-1-bromo-3-isopropylcyclobutane would be appropriate here.

cis- or *trans*-1-Bromo-3-isopropylcyclobutane   3-Isopropylcyclobutene

(*f*) The desired alkene is the exclusive product formed on E2 elimination from 1-bromo-1-*tert*-butylcyclopropane.

1-Bromo-1-*tert*-butylcyclopropane   1-*tert*-Butylcyclopropene

5.36 (*a*) Both 1-bromopropane and 2-bromopropane yield propene as the exclusive product of E2 elimination.

$$CH_3CH_2CH_2Br \quad \text{or} \quad CH_3CHCH_3 \quad \xrightarrow[\text{E2}]{\text{base}} \quad CH_3CH{=}CH_2$$
$$\text{|}$$
$$Br$$

1-Bromopropane   2-Bromopropane   Propene

(*b*) 2-Methylpropene is formed on dehydrobromination of either *tert*-butyl bromide or isobutyl bromide.

$$(CH_3)_3CBr \quad \text{or} \quad (CH_3)_2CHCH_2Br \quad \xrightarrow[\text{E2}]{\text{base}} \quad (CH_3)_2C{=}CH_2$$

tert-Butyl bromide   Isobutyl bromide   2-Methylpropene

(*c*) A tetrabromoalkane is required as the starting material to form a tribromoalkene under E2 elimination conditions. Either 1,1,2,2-tetrabromoethane or 1,1,1,2-tetrabromoethane is satisfactory.

$$Br_2CHCHBr_2 \quad \text{or} \quad BrCH_2CBr_3 \quad \xrightarrow[\text{E2}]{\text{base}} \quad BrCH{=}CBr_2$$

1,1,2,2-Tetrabromoethane   1,1,1,2-Tetrabromoethane   1,1,2-Tribromoethene

(*d*) The bromine substituent may be at either C-2 or C-3.

2-Bromo-1,1-dimethylcyclobutane   3-Bromo-1,1-dimethylcyclobutane   3,3-Dimethylcyclobutene

5.37 (*a*) The isomeric alkyl bromides having the molecular formula $C_5H_{11}Br$ are:

$CH_3CH_2CH_2CH_2CH_2Br$ $\underset{\underset{Br}{|}}{CH_3CH_2CH_2CHCH_3}$ $\underset{\underset{Br}{|}}{CH_3CH_2CHCH_2CH_3}$

     1-Bromopentane   2-Bromopentane   3-Bromopentane

$\underset{\underset{CH_3}{|}}{CH_3CH_2CHCH_2Br}$ $\underset{\underset{CH_3}{|}}{CH_3CHCH_2CH_2Br}$ $\overset{\overset{Br}{|}}{\underset{\underset{CH_3}{|}}{CH_3CHCHCH_3}}$

1-Bromo-2-methylbutane   1-Bromo-3-methylbutane   2-Bromo-3-methylbutane

$\overset{\overset{Br}{|}}{\underset{\underset{CH_3}{|}}{CH_3CCH_2CH_3}}$ $\overset{\overset{CH_3}{|}}{\underset{\underset{CH_3}{|}}{CH_3CCH_2Br}}$

     2-Bromo-2-methylbutane   1-Bromo-2,2-dimethylpropane

(*b*) The order of reactivity toward E1 elimination parallels carbocation stability and is tertiary > secondary > primary. The tertiary bromide 2-bromo-2-methylbutane will undergo E1 elimination at the fastest rate.

(*c*) 1-Bromo-2,2-dimethylpropane has no hydrogens on the β carbon and so cannot form an alkene by an E2 process.

$\overset{\overset{CH_3}{|}}{\underset{\underset{CH_3}{|}}{CH_3C\text{—}CH_2Br}}$ β carbon has no hydrogens

(*d*) Only the primary bromides will give a single alkene on E2 elimination.

$CH_3CH_2CH_2CH_2CH_2Br$ $\xrightarrow[\text{E2}]{\text{base}}$ $CH_3CH_2CH_2CH{=}CH_2$

     1-Bromopentane   1-Pentene

$\underset{\underset{CH_3}{|}}{CH_3CH_2CHCH_2Br}$ $\xrightarrow[\text{E2}]{\text{base}}$ $\underset{\underset{CH_3}{|}}{CH_3CH_2C{=}CH_2}$

     1-Bromo-2-methylbutane   2-Methyl-1-butene

$\underset{\underset{CH_3}{|}}{CH_3CHCH_2CH_2Br}$ $\xrightarrow[\text{E2}]{\text{base}}$ $\underset{\underset{CH_3}{|}}{CH_3CHCH{=}CH_2}$

     1-Bromo-3-methylbutane   3-Methyl-1-butene

(*e*) Elimination in 3-bromopentane will give the stereoisomers (*E*)- and (*Z*)-2-pentene.

(*f*) Three alkenes can be formed from 2-bromopentane.

5.38 (*a*) The isomeric $C_5H_{12}O$ alcohols are:

$$CH_3CH_2CH_2CH_2CH_2OH$$

1-Pentanol

$$CH_3CH_2CH_2\underset{\underset{OH}{|}}{C}HCH_3$$

2-Pentanol

$$CH_3CH_2\underset{\underset{OH}{|}}{C}HCH_2CH_3$$

3-Pentanol

$$CH_3CH_2\underset{\underset{CH_3}{|}}{C}HCH_2OH$$

2-Methyl-1-butanol

$$CH_3\underset{\underset{CH_3}{|}}{C}HCH_2CH_2OH$$

3-Methyl-1-butanol

$$CH_3\overset{\overset{OH}{|}}{C}H\underset{\underset{CH_3}{|}}{C}HCH_3$$

3-Methyl-2-butanol

$$CH_3\overset{\overset{OH}{|}}{\underset{\underset{CH_3}{|}}{C}}CH_2CH_3$$

2-Methyl-2-butanol

$$CH_3\overset{\overset{CH_3}{|}}{\underset{\underset{CH_3}{|}}{C}}CH_2OH$$

2,2-Dimethyl-1-propanol

(*b*) The order of reactivity in alcohol dehydration parallels carbocation stability and is tertiary > secondary > primary. The only tertiary alcohol in the group is 2-methyl-2-butanol. It will dehydrate fastest.

(*c*) The most stable C_5H_{11} carbocation is the tertiary carbocation.

$$CH_3\overset{\overset{+}{}}{\underset{\underset{CH_3}{|}}{C}}CH_2CH_3$$

1,1-Dimethylpropyl cation

(*d*) A proton may be lost from C-1 or C-3:

$$\underset{\underset{+}{}}{CH_3\overset{\overset{CH_3}{|}}{C}CH_2CH_3} \longrightarrow H_2C=\overset{\overset{CH_3}{|}}{C}CH_2CH_3 \quad + \quad CH_3\overset{\overset{CH_3}{|}}{C}=CHCH_3$$

1,1-Dimethylpropyl cation

2-Methyl-1-butene
(minor alkene)

2-Methyl-2-butene
(major alkene)

(e) For the 1,1-dimethylpropyl cation to be formed by a process involving a hydride shift, the starting alcohol must have the same carbon skeleton as the 1,1-dimethylpropyl cation.

$$CH_3\overset{+}{\underset{\underset{CH_3}{|}}{C}}CH_2CH_3 \quad \text{has the same carbon skeleton as} \quad HOCH_2\underset{\underset{CH_3}{|}}{\overset{\overset{H}{|}}{C}}CH_2CH_3 \quad \text{and} \quad CH_3\underset{\underset{CH_3}{|}}{\overset{\overset{OH}{|}}{CH}}CHCH_3$$

2-Methyl-1-butanol

3-Methyl-2-butanol

Although the same carbon skeleton is necessary, it alone is not sufficient; the alcohol must also have its hydroxyl group on the carbon atom adjacent to the carbon that bears the migrating hydrogen. Thus, 3-methyl-1-butanol cannot form a tertiary carbocation by a single hydride shift. It requires two sequential hydride shifts.

3-Methyl-1-butanol

(f) 2,2-Dimethyl-1-propanol can yield a tertiary carbocation by a process involving a methyl shift.

2,2-Dimethyl-1-propanol

5.39 (a) Heating an alcohol in the presence of an acid catalyst ($KHSO_4$) leads to dehydration with formation of an alkene. In this alcohol, elimination can occur in only one direction to give a mixture of cis and trans alkenes.

Cis–trans mixture

(b) Alkyl halides undergo E2 elimination on being heated with potassium *tert*-butoxide.

$$ICH_2CH(OCH_2CH_3)_2 \xrightarrow[\substack{(CH_3)_3COH \\ heat}]{KOC(CH_3)_3} H_2C{=}C(OCH_2CH_3)_2$$

(*c*) This reaction is highly regioselective and gives only 1,2-dimethylcyclohexene.

trans-1,2-Dimethylcyclohexyl bromide   1,2-Dimethylcyclohexene
(100%)

(*d*) β-Elimination can occur only in one direction, to give the alkene shown.

(*e*) The reaction is a conventional one of alcohol dehydration and proceeds as written in 76–78% yield.

(*f*) Dehydration of citric acid occurs, giving aconitic acid.

Citric acid   Aconitic acid

(*g*) Sequential double dehydrohalogenation gives the diene.

Bornylene (83%)

(*h*) This example has been reported in the chemical literature, and in spite of the complexity of the starting material, elimination proceeds in the usual way.

(84%)

(*i*) Again, we have a fairly complicated substrate, but notice that it is well disposed toward E2 elimination of the axial bromine.

(*j*) In the most stable conformation of this compound, chlorine occupies an axial site, and so it is ideally situated to undergo an E2 elimination reaction by way of an anti arrangement in the transition state.

4-*tert*-Butyl-
1-methylcyclohexene (95%)

4-*tert*-Butyl-
(methylene)cyclohexane (5%)

The major product is the more highly substituted isomer, in which the double bond is trisubstituted.

5.40 First identify the base as the amide ion (H_2N^-) portion of potassium amide (KNH_2). Amide ion is a strong base and uses an unshared electron pair to abstract a proton from the β carbon of the alkyl halide. The pair of electrons in the C—H bond becomes the π component of the double bond as the C—Br bond breaks. The electrons in the C—Br bond become an unshared electron pair of bromide ion.

5.41 The problem states that the reaction is first order in $(CH_3)_3CCl$ (*tert*-butyl chloride) and first order in $NaSCH_2CH_3$ (sodium ethanethiolate). It therefore exhibits the kinetic behavior (overall second order) of a reaction that proceeds by the E2 mechanism. The base that abstracts the proton from carbon is the anion $CH_3CH_2S^-$.

5.42 The two starting materials are stereoisomers of each other, and so it is reasonable to begin by examining each one in more stereochemical detail. First, write the most stable conformation of each isomer, keeping in mind that isopropyl is the bulkiest of the three substituents and has the greatest preference for an equatorial orientation.

Menthyl chloride

Most stable conformation of menthyl chloride: none of the three β protons is anti to chlorine

Neomenthyl chloride

Most stable conformation of neomenthyl chloride: each β carbon has a proton that is anti to chlorine

The anti coplanar relationship of halide and proton can be achieved only when the chlorine is axial; this corresponds to the most stable conformation of neomenthyl chloride. Menthyl chloride, however, must undergo appreciable distortion of its ring to achieve an anti coplanar Cl—C—C—H geometry. Strain increases substantially in going to the transition state for E2 elimination in menthyl chloride but not in neomenthyl chloride. Neomenthyl chloride undergoes E2 elimination at the faster rate.

5.43 The proton that is removed by the base must be anti to bromine. Thus, the alkyl groups must be gauche to one another in the conformation that leads to *cis*-4-nonene and anti to one another in the one that leads to *trans*-4-nonene.

Gauche conformation of 5-bromononane

E2 transition state

cis-4-nonene

Anti conformation of 5-bromononane

E2 transition state

trans-4-nonene

The alkyl groups move closer together (van der Waals strain increases) as the transition state for formation of *cis*-4-nonene is approached. No comparable increase in strain is involved in going to the transition state for formation of the trans stereoisomer.

5.44 Choose B and treat it with a strong base such as $NaOCH_2CH_3$ to ensure that the reaction follows an E2 mechanism. The only possible E2 product is C.

Dehydration of alcohol A would proceed through a carbocation, which could rearrange prior to loss of a proton and yield a mixture of products.

5.45 Begin by writing chemical equations for the processes specified in the problem. First consider rearrangement by way of a hydride shift:

Rearrangement by way of a methyl group shift is as follows:

A hydride shift gives a tertiary carbocation; a methyl migration gives a secondary carbocation. It is reasonable to expect that rearrangement will occur so as to produce the more stable of these two carbocations because the transition state has carbocation character at the carbon that bears the migrating group. We predict that rearrangement proceeds by a hydride shift rather than a methyl shift.

5.46 Rearrangement proceeds by migration of a hydrogen or an alkyl group from the carbon atom adjacent to the positively charged carbon.

(a) A propyl cation is primary and rearranges to an isopropyl cation, which is secondary, by migration of a hydrogen with its pair of electrons.

$$\underset{\substack{\text{Propyl cation} \\ \text{(primary, less stable)}}}{CH_3\overset{H}{\overset{|}{CH}}-\overset{+}{CH_2}} \longrightarrow \underset{\substack{\text{Isopropyl cation} \\ \text{(secondary, more stable)}}}{CH_3\overset{+}{CH}CH_3}$$

(b) A hydride shift transforms the secondary carbocation to a tertiary one.

$$\underset{\substack{\text{1,2-Dimethylpropyl cation} \\ \text{(secondary, less stable)}}}{\underset{\underset{CH_3}{|}}{CH_3\overset{H}{\overset{|}{C}}-\overset{+}{C}HCH_3}} \longrightarrow \underset{\substack{\text{1,1-Dimethylpropyl cation} \\ \text{(tertiary, more stable)}}}{\underset{\underset{CH_3}{|}}{CH_3\overset{+}{C}CH_2CH_3}}$$

This hydride shift occurs in preference to methyl migration, which would produce the same carbocation as the original one. (You can verify this by writing the appropriate structural formulas.)

(c) Migration of a methyl group converts this secondary carbocation to a tertiary one.

$$\underset{\substack{\text{1,2,2-Trimethylpropyl cation} \\ \text{(secondary, less stable)}}}{\underset{\underset{CH_3}{|}}{CH_3\overset{CH_3}{\overset{|}{C}}-\overset{+}{C}HCH_3}} \longrightarrow \underset{\substack{\text{1,1,2-Trimethylpropyl cation} \\ \text{(tertiary, more stable)}}}{\underset{\underset{CH_3}{|}}{CH_3\overset{+}{C}-\overset{CH_3}{\overset{|}{C}}HCH_3}}$$

(d) The group that shifts in this case is the entire ethyl group.

$$\underset{\substack{\text{2,2-Diethylbutyl cation} \\ \text{(primary, less stable)}}}{\underset{\underset{CH_3CH_2}{|}}{CH_3CH_2\overset{CH_3CH_2}{\overset{|}{C}}-\overset{+}{C}H_2}} \longrightarrow \underset{\substack{\text{1,1-Diethylbutyl cation} \\ \text{(tertiary, more stable)}}}{\underset{\underset{CH_3CH_2}{|}}{CH_3CH_2\overset{+}{C}-CH_2-CH_2CH_3}}$$

(e) Migration of a hydride from the ring carbon that bears the methyl group produces a tertiary carbocation.

2-Methylcyclopentyl cation   1-Methylcyclopentyl cation
(secondary, less stable)   (tertiary, more stable)

5.47 (*a*) Note that the starting material is an alcohol and that it is treated with an acid. The product is an alkene but its carbon skeleton is different from that of the starting alcohol. The reaction is one of alcohol dehydration accompanied by rearrangement at the carbocation stage. Begin by writing the step in which the alcohol is converted to a carbocation.

The carbocation is tertiary and relatively stable. Migration of a methyl group from the *tert*-butyl substituent, however, converts it to an isomeric carbocation, which is also tertiary.

Loss of a proton from this carbocation gives the observed product.

(*b*) Here also we have an alcohol dehydration reaction accompanied by rearrangement. The initially formed carbocation is secondary.

This cation can rearrange to a tertiary carbocation by an alkyl group shift.

Loss of a proton from the tertiary carbocation gives the observed alkene.

(c) The reaction begins as a normal alcohol dehydration in which the hydroxyl group is protonated by the acid catalyst and then loses water from the oxonium ion to give a carbocation.

4-Methylcamphenilol   Secondary carbocation

We see that the final product, 1-methylsantene, has a rearranged carbon skeleton corresponding to a methyl shift, and so we consider the rearrangement of the initially formed secondary carbocation to a tertiary ion.

Secondary carbocation   Tertiary carbocation   1-Methylsantene

Deprotonation of the tertiary carbocation yields 1-methylsantene.

5.48   The secondary carbocation can, as we have seen, rearrange by a methyl shift (Problem 5.18). It can also rearrange by migration of one of the ring bonds.

Secondary carbocation   Tertiary carbocation

The tertiary carbocation formed by this rearrangement can lose a proton to give the observed byproduct.

Isopropylidenecyclopentane

5.49   Let's do both part (a) and part (b) together by reasoning mechanistically. The first step in any acid-catalyzed alcohol dehydration is proton transfer to the OH group.

But notice that because this alcohol does not have any hydrogens on its β carbon, it cannot dehydrate directly. Any alkenes that are formed must arise by rearrangement processes. Consider, for example, migration of either of the two equivalent methyl groups at C-2.

The resulting carbocation can lose a proton in three different directions.

$$\overset{+}{CH_3CH_2CH_2CH_2\overset{\underset{|}{CH_3}}{C}HCH_2CH_3} \xrightarrow{-H^+}$$

$$\underset{\underset{CH_3}{|}}{CH_3CH_2CH_2CH}=CCH_2CH_3 \quad + \quad \underset{\underset{CH_2}{\|}}{CH_3CH_2CH_2CH_2CCH_2CH_3} \quad + \quad \underset{\underset{CH_3}{|}}{CH_3CH_2CH_2CH_2C}=CHCH_3$$

| 3-Methyl-3-heptene | 2-Ethyl-1-hexene | 3-Methyl-2-heptene |
| (mixture of *E* and *Z*) | | (mixture of *E* and *Z*) |

The alkene mixture shown in the preceding equation constitutes part of the answer to part (*b*). None of the alkenes arising from methyl migration is 2-methyl-2-heptene, the answer to part (*a*), however.

What other group can migrate? The other group attached to the β carbon is a butyl group. Consider its migration.

$$CH_3CH_2CH_2CH_2-\overset{\overset{CH_3}{|}}{\underset{\underset{CH_3}{|}}{C}}\overset{+}{-}CH_2-\overset{\cdot\cdot}{\underset{\underset{H}{\backslash}}{O}}\overset{/}{\underset{}{H}} \longrightarrow \underset{H_3C}{\overset{H_3C}{\diagdown}}\overset{}{\underset{}{}}^+CCH_2CH_2CH_2CH_2CH_3 \quad + \quad :\overset{\cdot\cdot}{O}\overset{\diagup H}{\underset{\diagdown H}{}}$$

Loss of a proton from the carbocation gives the alkene in part (*a*).

$$\underset{H_3C}{\overset{H_3C}{\diagdown}}{}^+CCH_2CH_2CH_2CH_2CH_3 \xrightarrow{-H^+} \underset{H_3C}{\overset{H_3C}{\diagdown}}C=CHCH_2CH_2CH_2CH_3$$

2-Methyl-2-heptene

A proton can also be lost from one of the methyl groups to give 2-methyl-1-heptene. This is the last alkene constituting the answer to part (*b*).

$$\underset{H_3C}{\overset{H_3C}{\diagdown}}{}^+CCH_2CH_2CH_2CH_2CH_3 \xrightarrow{-H^+} \underset{H_3C}{\overset{H_2C}{\diagdown\diagdown}}CCH_2CH_2CH_2CH_2CH_3$$

2-Methyl-1-heptene

5.50 Only two alkanes have the molecular formula C_4H_{10}: butane and isobutane (2-methylpropane)—both of which give two monochlorides on free-radical chlorination. However, dehydrochlorination of one of the monochlorides derived from butane yields a mixture of alkenes.

$$\underset{\underset{Cl}{|}}{CH_3CHCH_2CH_3} \xrightarrow[\substack{\text{dimethyl} \\ \text{sulfoxide}}]{KOC(CH_3)_3} H_2C=CHCH_2CH_3 \quad + \quad CH_3CH=CHCH_3$$

| 2-Chlorobutane | 1-Butene | 2-Butene (*cis* + *trans*) |

Both monochlorides derived from 2-methylpropane yield only 2-methylpropene under conditions of E2 elimination.

$$(CH_3)_3CCl \quad \text{or} \quad (CH_3)_2CHCH_2Cl \xrightarrow[\substack{\text{dimethyl} \\ \text{sulfoxide}}]{KOC(CH_3)_3} (CH_3)_2C{=}CH_2$$

tert-Butyl chloride   Isobutyl chloride   2-Methylpropene

Compound A is therefore 2-methylpropane, the two alkyl chlorides are *tert*-butyl chloride and isobutyl chloride, and alkene B is 2-methylpropene.

5.51   The key to this problem is the fact that one of the alkyl chlorides of molecular formula $C_6H_{13}Cl$ does not undergo E2 elimination. It must therefore have a structure in which the carbon atom that is β to the chlorine bears no hydrogens. This $C_6H_{13}Cl$ isomer is 1-chloro-2,2-dimethylbutane.

$$\begin{array}{c} CH_3 \\ | \\ CH_3CH_2CCH_2Cl \\ | \\ CH_3 \end{array}$$

1-Chloro-2,2-dimethylbutane
(cannot form an alkene)

Identifying this monochloride derivative gives us the carbon skeleton. The starting alkane (compound A) must be 2,2-dimethylbutane. Its free-radical halogenation gives three different monochlorides:

$$\begin{array}{c} CH_3 \\ | \\ CH_3CH_2CCH_3 \\ | \\ CH_3 \end{array} \xrightarrow[\text{light}]{Cl_2} \begin{array}{c} CH_3 \\ | \\ CH_3CH_2CCH_2Cl \\ | \\ CH_3 \end{array} + \begin{array}{c} CH_3 \\ | \\ CH_3CHCCH_3 \\ | \quad | \\ Cl \;\; CH_3 \end{array} + \begin{array}{c} CH_3 \\ | \\ ClCH_2CH_2CCH_3 \\ | \\ CH_3 \end{array}$$

2,2-Dimethylbutane   1-Chloro-2,2-dimethylbutane   3-Chloro-2,2-dimethylbutane   1-Chloro-3,3-dimethylbutane
(compound A)

Both 3-chloro-2,2-dimethylbutane and 1-chloro-3,3-dimethylbutane give only 3,3-dimethyl-1-butene on E2 elimination.

$$\begin{array}{c} CH_3 \\ | \\ CH_3CHCCH_3 \\ | \quad | \\ Cl \;\; CH_3 \end{array} \quad \text{or} \quad \begin{array}{c} CH_3 \\ | \\ ClCH_2CH_2CCH_3 \\ | \\ CH_3 \end{array} \xrightarrow[(CH_3)_3COH]{KOC(CH_3)_3} \begin{array}{c} CH_3 \\ | \\ H_2C{=}CHCCH_3 \\ | \\ CH_3 \end{array}$$

3-Chloro-2,2-   1-Chloro-3,3-   3,3-Dimethyl-1-butene
dimethylbutane   dimethylbutane   (alkene B)

ANSWERS TO INTERPRETIVE PROBLEMS

5.52 A;   **5.53** A;   **5.54** D;   **5.55** B;   **5.56** D;   **5.57** D;   **5.58** B;   **5.59** D

SELF-TEST

PART A

A-1. Give two acceptable IUPAC names for each of the following:

(a) $(CH_3)_3CCH{=}C(CH_3)_2$

(b)

(c)

(d)

A-2. Each of the following is an incorrect name for an alkene. Write the structure and give the correct name for each.

(a) 2-Ethyl-3-methyl-2-butene

(b) 2-Chloro-5-methyl-5-hexene

(c) 2,3-Dimethylcyclohexene

(d) 2-Methyl-1-penten-4-ol

A-3. (a) Write the structures of all the alkenes of molecular formula C_5H_{10}.

(b) Which isomer is the most stable?

(c) Which isomers are the least stable?

(d) Which isomers are a pair of stereoisomers?

A-4. How many carbon atoms are sp^2-hybridized in 2-methyl-2-pentene? How many are sp^3-hybridized? How many σ bonds are of the sp^2–sp^3 type?

A-5. Write the structure, clearly indicating the stereochemistry, of each of the following:

(a) (Z)-4-Ethyl-3-methyl-3-heptene

(b) (E)-1,2-Dichloro-3-methyl-2-hexene

(c) (E)-3-Methyl-3-penten-1-ol

A-6. Write structural formulas for two alkenes of molecular formula C_7H_{14} that are stereoisomers of each other and have a trisubstituted double bond. Give systematic names for each.

A-7. Write structural formulas for the reactant or product(s) omitted from each of the following. If more than one product is formed, indicate the major one.

(a) $(CH_3)_2CCH_2CH_2CH_3 \ \underset{\text{heat}}{\overset{H_2SO_4}{\longrightarrow}} \ ?$
   $\qquad\quad |$
   $\qquad\quad OH$

(b) $\underset{\text{CH}_3\text{OH}}{\overset{\text{NaOCH}_3}{\longrightarrow}} \ ?$

(c) ? $\xrightarrow[\text{CH}_3\text{CH}_2\text{OH}]{\text{KOCH}_2\text{CH}_3}$ (only alkene formed)

(d) $(\text{CH}_3)_2\underset{\underset{\text{OH}}{|}}{\text{C}}\text{C}(\text{CH}_3)_3$ $\xrightarrow{\text{H}_3\text{PO}_4}$?

A-8.   Write the structure of the $\text{C}_6\text{H}_{13}\text{Br}$ isomer that is *not* capable of undergoing E2 elimination.

A-9.   Write a stepwise mechanism for the formation of 2-methyl-2-butene from the dehydration of 2-methyl-2-butanol in sulfuric acid.

A-10.   Draw the structures of all the alkenes, including stereoisomers, that can be formed from the E2 elimination of 3-bromo-2,3-dimethylpentane with sodium ethoxide ($\text{NaOCH}_2\text{CH}_3$) in ethanol. Which of these would you expect to be the major product?

A-11.   Using curved arrows and perspective drawings (of chair cyclohexanes), explain the formation of the indicated product from the following reaction:

A-12.   Compare the relative rate of reaction of *cis-* and *trans*-1-chloro-3-isopropylcyclohexane with sodium methoxide in methanol by the E2 mechanism.

A-13.   Outline a mechanism for the following reaction:

                 $\xrightarrow[\text{heat}]{\text{H}_3\text{PO}_4}$ +   other alkenes

A-14.   Compound A, on reaction with bromine in the presence of light, gave as the major product compound B ($\text{C}_9\text{H}_{19}\text{Br}$). Reaction of B with sodium ethoxide in ethanol gave 3-ethyl-4,4-dimethyl-2-pentene as the only alkene. Identify compounds A and B.

PART B

B-1.   Which one of the alkenes shown below has the *Z* configuration of its double bond?

        (*a*)   (*b*)   (*c*)   (*d*)

B-2.   Carbon–carbon double bonds do not undergo rotation as do carbon–carbon single bonds. The reason is that

   (*a*) The double bond is much stronger and thus more difficult to rotate

   (*b*) Overlap of the sp^2 orbitals of the carbon–carbon σ bond would be lost

   (*c*) Overlap of the *p* orbitals of the carbon–carbon π bond would be lost

   (*d*) The shorter bond length of the double bond makes it more difficult for the attached groups to pass one another

B-3. Rank the following substituent groups in order of decreasing priority according to the Cahn–Ingold–Prelog system:

         —CH(CH₃)₂   —CH₂Br   —CH₂CH₂Br

             1   2   3

(*a*) 2 > 3 > 1   (*b*) 1 > 3 > 2   (*c*) 3 > 1 > 2   (*d*) 2 > 1 > 3

B-4. The heats of combustion for the four C_6H_{12} isomers shown are (not necessarily in order): 955.3, 953.6, 950.6, and 949.7 (all in kilocalories per mole). Which of these values is most likely the heat of combustion of isomer 1?

      1   2   3   4

(*a*) 955.3 kcal/mol   (*c*) 950.6 kcal/mol

(*b*) 953.6 kcal/mol   (*d*) 949.7 kcal/mol

B-5. Referring to the structures in the previous question, what can be said about isomers 3 and 4?

(*a*) 3 is more stable by 1.7 kcal/mol.   (*c*) 3 is more stable by 3.0 kcal/mol.

(*b*) 4 is more stable by 1.7 kcal/mol.   (*d*) 3 is more stable by 0.9 kcal/mol.

B-6. The structure of (*E*)-1-chloro-3-methyl-3-hexene is

(*a*)   (*c*)

(*b*)   (*d*) None of these

B-7. In the dehydrohalogenation of 2-bromobutane, which conformation leads to the formation of *cis*-2-butene?

      (*a*)   (*b*)   (*c*)   (*d*)

(*e*) None of these is the correct conformation.

B-8. Which of the following alcohols would be *most* likely to undergo dehydration with rearrangement by a process involving a methyl migration (methyl shift)?

(*a*)   (*c*)   (*e*)

(*b*)   (*d*)

B-9. Rank the following alcohols in order of decreasing reactivity (fastest → slowest) toward dehydration with 85% H_3PO_4:

$(CH_3)_2CHCH_2CH_2OH$

1

OH
|
$(CH_3)_2CCH_2CH_3$

2

OH
|
$(CH_3)_2CHCHCH_3$

3

(a) $2 > 3 > 1$ (b) $1 > 3 > 2$ (c) $2 > 1 > 3$ (d) $1 > 2 > 3$

B-10. Consider the following reaction:

$$\text{(structure)} \xrightarrow[\text{heat}]{H_3PO_4} \text{(structure)} + \text{(structure)}$$

Which response contains all the correct statements about this process and no incorrect ones?

1. Dehydration   4. Most stable carbocation forms
2. E2 mechanism   5. Single-step reaction
3. Carbon skeleton migration

(a) 1, 3   (b) 1, 2, 3   (c) 1, 2, 5   (d) 1, 3, 4

B-11. Select the formula or statement representing the major product(s) of the following reaction:

$$\text{(structure)} \xrightarrow[\text{CH}_3\text{CH}_2\text{OH}]{\text{KOCH}_2\text{CH}_3} \ ?$$

(a) (structure)   (c) $H_2C{=}CHCHCH_2CH_3$ with CH_3 substituent

(b) (structure)   (d) Both (a) and (b) form in approximately equal amounts.

B-12. Which one of the following statements concerning E2 reactions of alkyl halides is true?

(a) The rate of an E2 reaction depends only on the concentration of the alkyl halide.
(b) The rate of an E2 reaction depends only on the concentration of the base.
(c) The C—H bond and the C—X (X = halogen) bond are broken in the same step.
(d) Alkyl chlorides generally react faster than alkyl bromides.

B-13. Which alkyl halide undergoes E2 elimination at the fastest rate?

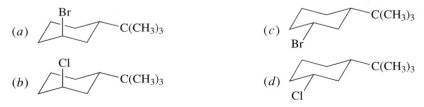

(a) (structure) —$C(CH_3)_3$ with Br   (c) (structure) —$C(CH_3)_3$ with Br

(b) (structure) —$C(CH_3)_3$ with Cl   (d) (structure) —$C(CH_3)_3$ with Cl

B-14. What is the relationship between the pair of compounds shown?

(*a*) Identical: superimposable without bond rotations

(*b*) Conformations

(*c*) Stereoisomers

(*d*) Constitutional isomers

B-15. Which one of the following will give 2-methyl-1-butene as the *only* alkene on treatment with $KOC(CH_3)_3$ in dimethyl sulfoxide?

(*a*) 1-Bromo-2-methylbutane   (*c*) 2-Bromo-2-methylbutane

(*b*) 2-Methyl-1-butanol   (*d*) 2-Methyl-2-butanol

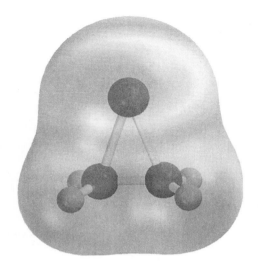

CHAPTER 6

Addition Reactions of Alkenes

SOLUTIONS TO TEXT PROBLEMS

6.1 Catalytic hydrogenation converts an alkene to an alkane having the same carbon skeleton. Because 2-methylbutane is the product of hydrogenation, all three alkenes must have a four-carbon chain with a one-carbon branch. The three alkenes are therefore:

6.2 The most highly substituted double bond is the most stable and has the smallest heat of hydrogenation.

2-Methyl-2-butene: most stable (trisubstituted)	2-Methyl-1-butene (disubstituted)	3-Methyl-1-butene (monosubstituted)

Heat of hydrogenation:	112 kJ/mol (26.7 kcal/mol)	118 kJ/mol (28.2 kcal/mol)	126 kJ/mol (30.2 kcal/mol)

6.3   Both *cis*-2-butene and *trans*-2-butene react with HBr to give the same product (2-bromobutane).

$$CH_3CH{=}CHCH_3 \quad + \quad HBr \quad \longrightarrow \quad \underset{\underset{Br}{|}}{CH_3CH_2CHCH_3}$$

cis-2-butene: $\Delta H = -77$ kJ (-18.4 kcal)

trans-2-butene: $\Delta H = -72$ kJ (-17.3 kcal)

Therefore, the difference in the heats of reaction is due entirely to the difference in their respective energies. *cis*-2-Butene gives off 5 kJ/mol (1.1 kcal/mol) more heat when it reacts with HBr than *trans*-2-butene does, which makes *cis*-2-butene 5 kJ/mol (1.1 kcal/mol) higher in energy (less stable) than *trans*-2-butene. This energy difference is approximately the same as that obtained from their heats of hydrogenation [4 kJ/mol (1 kcal/mol)] and heats of combustion [3 kJ/mol (0.7 kcal/mol)].

6.4   (*b*) Begin by writing out the structure of the starting alkene. Identify the doubly bonded carbon that has the greater number of hydrogens; this is the one to which the proton of hydrogen chloride adds. Chlorine adds to the carbon atom of the double bond that has the fewer attached hydrogens.

2-Methyl-1-butene   2-Chloro-2-methylbutane

By applying Markovnikov's rule, we see that the major product is 2-chloro-2-methylbutane.

(*c*) Regioselectivity of addition is not an issue here, because the two carbons of the double bond are equivalent in *cis*-2-butene. Hydrogen chloride adds to *cis*-2-butene to give 2-chlorobutane.

cis-2-Butene   2-Chlorobutane

(*d*) One end of the double bond has no hydrogens, but the other end has one. In accordance with Markovnikov's rule, the proton of hydrogen chloride adds to the carbon that already has one hydrogen. The product is 1-chloro-1-ethylcyclohexane.

Ethylidenecyclohexane   1-Chloro-1-ethylcyclohexane

6.5   (*b*) A proton is transferred to C-1 of 2-methyl-1-butene to produce a tertiary carbocation.

2-Methyl-1-butene   Hydrogen chloride   Tertiary carbocation   Chloride ion

This is the carbocation that leads to the observed product, 2-chloro-2-methylbutane.

(*c*) A secondary carbocation is an intermediate in the reaction of *cis*-2-butene with hydrogen chloride.

| | | | |
| *cis*-2-Butene | Hydrogen chloride | Secondary carbocation | Chloride ion |

Capture of this carbocation by chloride gives 2-chlorobutane.

(*d*) A tertiary carbocation is formed by protonation of the double bond.

| | | | |
| Ethylidenecyclohexane | Hydrogen chloride | Tertiary carbocation | Chloride ion |

This carbocation is captured by chloride to give the observed product, 1-chloro-1-ethylcyclohexane.

6.6   The carbocation formed by protonation of the double bond of 3,3-dimethyl-1-butene is secondary. Methyl migration can occur to give a more stable tertiary carbocation.

The two chlorides are 3-chloro-2,2-dimethylbutane and 2-chloro-2,3-dimethylbutane.

6.7   The structure of allyl bromide (3-bromo-1-propene) is $H_2C=CHCH_2Br$. Its reaction with hydrogen bromide in accordance with Markovnikov's rule proceeds by addition of a proton to the doubly bonded carbon that has the greater number of hydrogens.

Addition according to Markovnikov's rule:

$$H_2C=CHCH_2Br \ + \ HBr \longrightarrow \underset{\underset{Br}{|}}{CH_3CHCH_2Br}$$

| | | |
| Allyl bromide | Hydrogen bromide | 1,2-Dibromopropane |

Addition of hydrogen bromide opposite to Markovnikov's rule leads to 1,3-dibromopropane.

Addition opposite to Markovnikov's rule:

$$H_2C=CHCH_2Br \ + \ HBr \ \longrightarrow \ BrCH_2CH_2CH_2Br$$

    Allyl bromide   Hydrogen   1,3-Dibromopropane
                      bromide

6.8 (*b*) Hydrogen bromide adds to 2-methyl-1-butene in accordance with Markovnikov's rule when peroxides are absent. The product is 2-bromo-2-methylbutane.

    2-Methyl-1-butene   Hydrogen   2-Bromo-2-methylbutane
                            bromide

The opposite regioselectivity is observed when peroxides are present. The product is 1-bromo-2-methylbutane.

    2-Methyl-1-butene   Hydrogen   1-Bromo-2-methylbutane
                            bromide

(*c*) Both ends of the double bond in *cis*-2-butene are equivalently substituted. Therefore, the same product (2-bromobutane) is formed by hydrogen bromide addition regardless of whether the reaction is carried out in the presence of peroxides or in their absence.

    cis-2-butene   Hydrogen   2-Bromobutane
                    bromide

(*d*) A tertiary bromide is formed on addition of hydrogen bromide to ethylidenecyclohexane in the absence of peroxides.

    Ethylidenecyclohexane   Hydrogen   1-Bromo-1-ethylcyclohexane
                            bromide

The regioselectivity of addition is reversed in the presence of peroxides, and the product is (1-bromoethyl)cyclohexane.

    Ethylidenecyclohexane   Hydrogen   (1-Bromoethyl)cyclohexane
                            bromide

6.9 The three alkenes give the same alkyl hydrogen sulfate because they give the same carbocation on protonation.

sec-Butyl cation   *sec*-Butyl hydrogen sulfate

6.10 The presence of hydroxide ion in the second step is incompatible with the acidic medium in which the reaction is carried out. The reaction as shown in step 1

$$(CH_3)_2C{=}CH_2 \;+\; H_3O^+ \longrightarrow (CH_3)_3C^+ \;+\; H_2O$$

is performed in acidic solution. There are, for all practical purposes, no hydroxide ions in aqueous acid, the strongest base present being water itself. It is quite important to pay attention to the species that are actually present in the reaction medium whenever you formulate a reaction mechanism.

6.11 The more stable the carbocation, the faster it is formed. The more reactive alkene gives a tertiary carbocation in the rate-determining step.

Tertiary carbocation

Protonation of $\triangleright{-}CH{=}CHCH_3$ gives a secondary carbocation.

6.12 The mechanism of electrophilic addition of hydrogen chloride to 2-methylpropene as outlined in text Section 6.6 proceeds through a carbocation intermediate. This mechanism is the reverse of the E1 elimination. The E2 mechanism is concerted, meaning it occurs in a single step.

6.13 Water is one of the products in the reaction of *tert*-butyl alcohol with HCl.

$$(CH_3)_3COH \;+\; HCl \rightleftharpoons (CH_3)_3CCl \;+\; H_2O$$

tert-Butyl alcohol   Hydrogen chloride   *tert*-Butyl chloride   Water

According to Le Châtelier's principle, in order to maximize the yield of *tert*-butyl chloride we need to minimize the amount of water. A concentrated solution is a better choice than a dilute solution of HCl in water. Excess water will cause the position of the equilibrium to shift to the left.

6.14 In a reversible reaction, the equilibrium constant of the reverse reaction is the reciprocal of the forward reaction. In other words, $K_{reverse} = \dfrac{1}{K_{forward}}$. If $K_{forward} = 9$ for the reaction

$$(CH_3)_2C{=}CH_2(g) \;+\; H_2O(l) \rightleftharpoons (CH_3)_3COH(l)$$

then $K_{reverse} = 0.11$.

6.15 A catalyst affects the rate of a reaction, but not its equilibrium constant, by speeding up both the forward and reverse reactions. The usual hydrogenation catalysts (Pd, Pt, Ni) allow the ethane $\rightleftharpoons$ ethylene equilibrium to be established more quickly but do not affect the relative amounts of reactant and product.

6.16 (*b*) The carbon–carbon double bond is symmetrically substituted in *cis*-2-butene. Therefore, the regioselectivity of hydroboration–oxidation is not an issue. Hydration of the double bond gives 2-butanol.

cis-2-Butene   2-Butanol

(*c*) Hydroboration–oxidation of alkenes leads to hydration of the double bond with a regioselectivity opposite to Markovnikov's rule.

Methylenecyclobutane   Cyclobutylmethanol

(*d*) Hydroboration–oxidation of cyclopentene gives cyclopentanol.

Cyclopentene   Cyclopentanol

(*e*) When alkenes are converted to alcohols by hydroboration–oxidation, the hydroxyl group is introduced at the less substituted carbon of the double bond.

$$CH_3CH=C(CH_2CH_3)_2 \xrightarrow[\text{2. oxidation}]{\text{1. hydroboration}} \underset{\underset{OH}{|}}{CH_3CHCH(CH_2CH_3)_2}$$

3-Ethyl-2-pentene   3-Ethyl-2-pentanol

(*f*) The less substituted carbon of the double bond in 3-ethyl-1-pentene is at the end of the chain. It is this carbon that bears the hydroxyl group in the product of hydroboration–oxidation.

$$H_2C=CHCH(CH_2CH_3)_2 \xrightarrow[\text{2. oxidation}]{\text{1. hydroboration}} HOCH_2CH_2CH(CH_2CH_3)_2$$

3-Ethyl-1-pentene   3-Ethyl-1-pentanol

6.17 The bottom face of the double bond of α-pinene is less hindered than the top face.

Syn addition of H and OH takes place and with a regioselectivity opposite to that of Markovnikov's rule.

6.18 Bromine adds anti to the double bond of 1-bromocyclohexene to give 1,1,2-tribromocyclohexane. The radioactive bromines (^{82}Br) are vicinal and trans to each other.

1-Bromocyclohexene   Bromine   1,1,2-Tribromocyclohexane

6.19 Alkyl substituents on the double bond increase the reactivity of the alkene toward addition of bromine.

2-Methyl-2-butene   2-Methyl-1-butene   3-Methyl-1-butene
(trisubstituted double bond; most reactive)   (disubstituted double bond)   (monosubstituted double bond; least reactive)

6.20 (*b*) Bromine becomes bonded to the less highly substituted carbon of the double bond, the hydroxyl group to the more highly substituted one.

$$(CH_3)_2C{=}CHCH_3 \xrightarrow[\text{H}_2\text{O}]{\text{Br}_2} (CH_3)_2\overset{\text{Br}}{\underset{\text{HO}}{C}}CHCH_3$$

2-Methyl-2-butene   3-Bromo-2-methyl-2-butanol

(*c*) $(CH_3)_2CHCH{=}CH_2 \xrightarrow[\text{H}_2\text{O}]{\text{Br}_2} (CH_3)_2CH\underset{\text{OH}}{C}HCH_2Br$

3-Methyl-1-butene   1-Bromo-3-methyl-2-butanol

(*d*) Anti addition occurs.

1-Methylcyclopentene   *trans*-2-Bromo-1-methylcyclopentanol

6.21 The structure of disparlure is as shown.

Its longest continuous chain contains 18 carbon atoms, and so it is named as an epoxy derivative of octadecane. Number the chain in the direction that gives the lowest numbers to the carbons that bear oxygen. Thus, disparlure is *cis*-7,8-epoxy-2-methyloctadecane.

6.22 Disparlure can be prepared by epoxidation of the corresponding alkene. Cis alkenes yield cis epoxides upon epoxidation. *cis*-2-Methyl-7-octadecene is therefore the alkene chosen to prepare disparlure by epoxidation.

cis-2-Methyl-7-octadecene

Disparlure

6.23 The products of ozonolysis are formaldehyde and 4,4-dimethyl-2-pentanone.

Formaldehyde   4,4-Dimethyl-2-pentanone

The two carbons that were doubly bonded to each other in the alkene become the carbons that are doubly bonded to oxygen in the products of ozonolysis. Therefore, mentally remove the oxygens and connect these two carbons by a double bond to reveal the structure of the starting alkene.

2,4,4-Trimethyl-1-pentene

6.24 From the structural formula of the desired product, we see that it is a **vicinal bromohydrin**. Vicinal bromohydrins are made from alkenes by reaction with bromine in water.

$$BrCH_2C(CH_3)_2 \quad \text{is made from} \quad H_2C{=}C(CH_3)_2$$
$$\underset{OH}{|}$$

1-Bromo-2-methyl-2-propanol   2-Methylpropene

Because the starting material given is *tert*-butyl bromide, a practical synthesis is:

$$(CH_3)_3CBr \xrightarrow[\substack{CH_3CH_2OH \\ \text{heat}}]{NaOCH_2CH_3} (CH_3)_2C{=}CH_2 \xrightarrow[H_2O]{Br_2} (CH_3)_2\underset{OH}{\overset{}{C}}CH_2Br$$

tert-Butyl bromide   2-Methylpropene   1-Bromo-2-methyl-2-propanol

6.25 Catalytic hydrogenation of the double bond converts 2,4,4-trimethyl-1-pentene and 2,4,4-trimethyl-2-pentene to 2,2,4-trimethylpentane.

2,4,4-Trimethyl-1-pentene   2,4,4-Trimethyl-2-pentene   2,2,4-Trimethylpentane

6.26 This problem illustrates the reactions of alkenes with various reagents and requires application of Markovnikov's rule to the addition of unsymmetrical electrophiles.

(*a*) Addition of hydrogen chloride to 1-pentene will give 2-chloropentane.

$$H_2C{=}CHCH_2CH_2CH_3 \ + \ HCl \ \longrightarrow \ CH_3\underset{\underset{Cl}{|}}{C}HCH_2CH_2CH_3$$

1-Pentene   2-Chloropentane

(*b*) Electrophilic addition of hydrogen bromide will give 2-bromopentane.

$$H_2C{=}CHCH_2CH_2CH_3 \ + \ HBr \ \longrightarrow \ CH_3\underset{\underset{Br}{|}}{C}HCH_2CH_2CH_3$$

2-Bromopentane

(*c*) The presence of peroxides will cause free-radical addition of hydrogen bromide, and regioselective addition opposite to Markovnikov's rule will be observed.

$$H_2C{=}CHCH_2CH_2CH_3 \ + \ HBr \ \xrightarrow{\text{peroxides}} \ BrCH_2CH_2CH_2CH_2CH_3$$

1-Bromopentane

(*d*) Hydrogen iodide will add according to Markovnikov's rule.

$$H_2C{=}CHCH_2CH_2CH_3 \ + \ HI \ \longrightarrow \ CH_3\underset{\underset{I}{|}}{C}HCH_2CH_2CH_3$$

2-Iodopentane

(*e*) Dilute sulfuric acid will cause hydration of the double bond with regioselectivity in accord with Markovnikov's rule.

$$H_2C{=}CHCH_2CH_2CH_3 \ + \ H_2O \ \xrightarrow{H_2SO_4} \ CH_3\underset{\underset{OH}{|}}{C}HCH_2CH_2CH_3$$

2-Pentanol

(*f*) Hydroboration–oxidation of an alkene brings about hydration of the double bond opposite to Markovnikov's rule; 1-pentanol will be the product.

$$H_2C{=}CHCH_2CH_2CH_3 \ \xrightarrow[\text{2. } H_2O_2,\ HO^-]{\text{1. } B_2H_6} \ HOCH_2CH_2CH_2CH_2CH_3$$

1-Pentanol

(*g*) Bromine adds across the double bond to give a vicinal dibromide.

$$H_2C{=}CHCH_2CH_2CH_3 \ + \ Br_2 \ \xrightarrow{CCl_4} \ BrCH_2\underset{\underset{Br}{|}}{C}HCH_2CH_2CH_3$$

1,2-Dibromopentane

(*h*) Vicinal bromohydrins are formed when bromine in water adds to alkenes. Br adds to the less substituted carbon, OH to the more substituted one.

$$H_2C{=}CHCH_2CH_2CH_3 \ + \ Br_2 \ \xrightarrow{\ H_2O\ } \ \underset{\underset{\textstyle OH}{|}}{Br}CH_2CHCH_2CH_2CH_3$$

1-Bromo-2-pentanol

(*i*) Epoxidation of the alkene occurs on treatment with peroxy acids.

$$H_2C{=}CHCH_2CH_2CH_3 \ + \ CH_3\overset{\overset{\textstyle O}{||}}{C}OOH \ \longrightarrow \ H_2C\overset{O}{\underset{}{\diagdown\!\!\diagup}}CHCH_2CH_2CH_3 \ + \ CH_3CO_2H$$

1,2-Epoxypentane   Acetic acid

(*j*) Ozone reacts with alkenes to give ozonides.

$$H_2C{=}CHCH_2CH_2CH_3 \ + \ O_3 \ \longrightarrow \ H_2C\overset{O}{\underset{O-O}{\diagup\quad\diagdown}}CHCH_2CH_2CH_3$$

Ozonide

(*k*) When the ozonide in part (*j*) is hydrolyzed in the presence of zinc, formaldehyde and butanal are formed.

$$H_2C\overset{O}{\underset{O-O}{\diagup\quad\diagdown}}CHCH_2CH_2CH_3 \ \xrightarrow[\ Zn\ ]{\ H_2O\ } \ \overset{\overset{\textstyle O}{||}}{H}CH \ + \ \overset{\overset{\textstyle O}{||}}{H}CCH_2CH_2CH_3$$

Formaldehyde   Butanal

(*l*) As an alternative to treating an ozonide with H_2O and Zn, dimethyl sulfide $(CH_3)_2S$ is often used. The products are the same as in part (*k*) of this problem. $(CH_3)_2SO$ (dimethyl sulfoxide) is also formed.

6.27 When we compare the reactions of 2-methyl-2-butene with the analogous reactions of 1-pentene, we find that the reactions proceed in a similar manner.

(*a*) $(CH_3)_2C{=}CHCH_3 \ + \ HCl \ \longrightarrow \ (CH_3)_2\underset{\underset{\textstyle Cl}{|}}{C}CH_2CH_3$

2-Methyl-2-butene   2-Chloro-2-methylbutane

(*b*) $(CH_3)_2C{=}CHCH_3 \ + \ HBr \ \longrightarrow \ (CH_3)_2\underset{\underset{\textstyle Br}{|}}{C}CH_2CH_3$

2-Bromo-2-methylbutane

(*c*) $(CH_3)_2C{=}CHCH_3 \ + \ HBr \ \xrightarrow{\ peroxides\ } \ (CH_3)_2CH\underset{\underset{\textstyle Br}{|}}{C}HCH_3$

2-Bromo-3-methylbutane

(*d*) $(CH_3)_2C{=}CHCH_3 \ + \ HI \ \longrightarrow \ (CH_3)_2\underset{\underset{\textstyle I}{|}}{C}CH_2CH_3$

2-Iodo-2-methylbutane

(e) $(CH_3)_2C=CHCH_3$ + H_2O $\xrightarrow{H_2SO_4}$ $(CH_3)_2CCH_2CH_3$
$\underset{OH}{|}$

2-Methyl-2-butanol

(f) $(CH_3)_2C=CHCH_3$ $\xrightarrow[\text{2. } H_2O_2,\ HO^-]{\text{1. } B_2H_6}$ $(CH_3)_2CHCHCH_3$
$\underset{OH}{|}$

3-Methyl-2-butanol

(g) $(CH_3)_2C=CHCH_3$ + Br_2 $\xrightarrow{CCl_4}$ $(CH_3)_2\overset{Br}{\underset{Br}{C}}CHCH_3$

2,3-Dibromo-2-methylbutane

(h) $(CH_3)_2C=CHCH_3$ + Br_2 $\xrightarrow{H_2O}$ $(CH_3)_2\overset{Br}{\underset{OH}{C}}CHCH_3$

3-Bromo-2-methyl-2-butanol

(i) $(CH_3)_2C=CHCH_3$ + $CH_3\overset{O}{\overset{||}{C}}OOH$ $\longrightarrow$ $(CH_3)_2\underset{O}{C}{-}CHCH_3$ + CH_3CO_2H

2,3-Epoxy-2-methylbutane

(j) $(CH_3)_2C=CHCH_3$ + O_3 $\longrightarrow$

Ozonide

(k)

$\xrightarrow[Zn]{H_2O}$ $CH_3\overset{O}{\overset{||}{C}}CH_3$ + $H\overset{O}{\overset{||}{C}}CH_3$

Acetone   Acetaldehyde

(l) As an alternative to treating an ozonide with H_2O and Zn, dimethyl sulfide $(CH_3)_2S$ is often used. The products are the same as in part (k) of this problem. $(CH_3)_2SO$ (dimethyl sulfoxide) is also formed.

6.28 Cycloalkenes undergo the same kinds of reactions as do noncyclic alkenes.

(a)

1-Methylcyclohexene   +   HCl $\longrightarrow$ 1-Chloro-1-methylcyclohexane

(b) + HBr ⟶

1-Bromo-1-methylcyclohexane

(c) + HBr $\xrightarrow{\text{peroxides}}$

1-Bromo-2-methylcyclohexane
(mixture of cis and trans)

(d) + HI ⟶

1-Iodo-1-methylcyclohexane

(e) + H_2O $\xrightarrow{H_2SO_4}$

1-Methylcyclohexanol

(f) $\xrightarrow[\text{2. } H_2O_2, \text{ HO}^-]{\text{1. } B_2H_6}$

trans-2-Methylcyclohexanol

(g) + Br_2 $\xrightarrow{CCl_4}$

trans-1,2-Dibromo-1-
methylcyclohexane

(h) + Br_2 $\xrightarrow{H_2O}$

trans-2-Bromo-1-
methylcyclohexanol

(i) + CH$_3$C(=O)OOH ⟶ + CH_3CO_2H

1,2-Epoxy-1-methylcyclohexane

(*j*) [cyclohexene with CH$_3$] + O$_3$ ⟶ [ozonide with CH$_3$] Ozonide

(*k*) [bicyclic ozonide with CH$_3$] $\xrightarrow[\text{Zn}]{\text{H}_2\text{O}}$ [open-chain keto-aldehyde with CH$_3$] ≡ $\text{CH}_3\overset{\displaystyle O}{\overset{\|}{C}}\text{CH}_2\text{CH}_2\text{CH}_2\text{CH}_2\overset{\displaystyle O}{\overset{\|}{C}}\text{H}$ 6-Oxoheptanal

(*l*) As an alternative to treating an ozonide with H$_2$O and Zn, dimethyl sulfide (CH$_3$)$_2$S is often used. The products are the same as in part (*k*) of this problem. (CH$_3$)$_2$SO (dimethyl sulfoxide) is also formed.

6.29 We need first to write out the structures in more detail to evaluate the substitution patterns at the double bonds.

(*a*)   1-Pentene   [structure]   Monosubstituted

(*b*)   (*E*)-4,4-Dimethyl-2-pentene   [structure]   trans-Disubstituted

(*c*)   (*Z*)-4-Methyl-2-pentene   [structure]   cis-Disubstituted

(*d*)   (*Z*)-2,2,5,5-Tetramethyl-3-hexene   [structure]   Two *tert*-butyl groups cis

(*e*)   2,4-Dimethyl-2-pentene   [structure]   Trisubstituted

Compound (*d*), having two cis *tert*-butyl groups, should have the least stable (highest energy) double bond. The remaining alkenes are arranged in order of increasing stability (decreasing heats of hydrogenation) according to the degree of substitution of the double bond: monosubstituted, cis-disubstituted, trans-disubstituted, trisubstituted. The heats of hydrogenation are therefore:

(*d*)   [structure]   151 kJ/mol (36.2 kcal/mol)

(*a*)   [structure]   122 kJ/mol (29.3 kcal/mol)

(*c*)   [structure]   114 kJ/mol (27.3 kcal/mol)

(b)   111 kJ/mol (26.5 kcal/mol)

(e)   105 kJ/mol (25.1 kcal/mol)

6.30 In all parts of this exercise we deduce the carbon skeleton on the basis of the alkane formed on hydrogenation of an alkene and then determine what carbon atoms may be connected by a double bond in that skeleton. Problems of this type are best done by using carbon skeleton formulas.

(a) Product is 2,2,3,4,4-pentamethylpentane.   The only possible alkene precursor is

(b) Product is 2,3-dimethylbutane.   May be formed by hydrogenation of

or

(c) Product is methylcyclobutane.   May be formed by hydrogenation of

or   or

6.31 Hydrogenation of the alkenes shown will give a mixture of *cis*- and *trans*-1,4,-dimethylcyclohexane.

H_2C⚌———CH_3

or

H_3C———⬡———CH_3

$\xrightarrow[\text{catalyst}]{H_2}$

H_3C◼⬡◼CH_3 + H_3C⋯⬡◼CH_3

cis-1,4-Dimethylcyclohexane   *trans*-1,4-Dimethylcyclohexane

Only when the methyl groups are cis in the starting alkene will the cis stereoisomer be the sole product following hydrogenation. Hydrogenation of *cis*-3,6-dimethylcyclohexane will yield exclusively *cis*-1,4-dimethylcyclohexane.

H_3C◼⬡◼CH_3 $\xrightarrow[\text{catalyst}]{H_2}$ H_3C◼⬡◼CH_3

cis-3,6-Dimethylcyclohexane   *cis*-1,4-Dimethylcyclohexane

6.32 (*a*) The desired transformation is the conversion of an alkene to a vicinal dibromide.

$$CH_3CH{=}C(CH_2CH_3)_2 \xrightarrow[CCl_4]{Br_2} CH_3\underset{|}{\overset{|}{C}}H\underset{|}{\overset{|}{C}}(CH_2CH_3)_2$$
$$\qquad\qquad\qquad\qquad\qquad\qquad\qquad\qquad Br\ Br$$

3-Ethyl-2-pentene   2,3-Dibromo-3-ethylpentane

(*b*) Markovnikov addition of hydrogen chloride is indicated.

$$CH_3CH{=}C(CH_2CH_3)_2 \xrightarrow{HCl} CH_3CH_2\underset{|}{\overset{}{C}}(CH_2CH_3)_2$$
$$\qquad\qquad\qquad\qquad\qquad\qquad\qquad\qquad Cl$$

3-Chloro-3-ethylpentane

(*c*) Free-radical addition of hydrogen bromide opposite to Markovnikov's rule will give the required regiochemistry.

$$CH_3CH{=}C(CH_2CH_3)_2 \xrightarrow[peroxides]{HBr} CH_3\underset{|}{\overset{}{C}}HCH(CH_2CH_3)_2$$
$$\qquad\qquad\qquad\qquad\qquad\qquad\qquad\qquad Br$$

2-Bromo-3-ethylpentane

(*d*) Acid-catalyzed hydration will occur in accordance with Markovnikov's rule to yield the desired tertiary alcohol.

$$CH_3CH{=}C(CH_2CH_3)_2 \xrightarrow[H_2SO_4]{H_2O} CH_3CH_2\underset{|}{\overset{}{C}}(CH_2CH_3)_2$$
$$\qquad\qquad\qquad\qquad\qquad\qquad\qquad\qquad OH$$

3-Ethyl-3-pentanol

(*e*) Hydroboration–oxidation results in hydration of alkenes with a regioselectivity opposite to that of Markovnikov's rule.

$$CH_3CH{=}C(CH_2CH_3)_2 \xrightarrow[2.\ H_2O_2,\ HO^-]{1.\ B_2H_6} CH_3\underset{|}{\overset{}{C}}HCH(CH_2CH_3)_2$$
$$\qquad\qquad\qquad\qquad\qquad\qquad\qquad\qquad OH$$

3-Ethyl-2-pentanol

(*f*) A peroxy acid will convert an alkene to an epoxide.

$$CH_3CH{=}C(CH_2CH_3)_2 \xrightarrow{CH_3\overset{O}{\overset{\|}{C}}OOH}$$

3-Ethyl-2-pentanol → (epoxide structure)

2,3-Epoxy-3-ethylpentane

(*g*) Hydrogenation of alkenes converts them to alkanes.

$$CH_3CH{=}C(CH_2CH_3)_2 \xrightarrow[Pt]{H_2} CH_3CH_2CH(CH_2CH_3)_2$$

3-Ethylpentane

6.33 (*a*) Four primary alcohols have the molecular formula $C_5H_{12}O$:

$$CH_3CH_2CH_2CH_2CH_2OH \qquad \underset{\underset{CH_3}{|}}{CH_3CH_2CHCH_2OH} \qquad (CH_3)_2CHCH_2CH_2OH \qquad (CH_3)_3CCH_2OH$$

      1-Pentanol   2-Methyl-1-butanol   3-Methyl-1-butanol   2,2-Dimethyl-1-propanol

2,2-Dimethyl-1-propanol cannot be prepared by hydroboration–oxidation of an alkene, because no alkene can have this carbon skeleton.

(*b*) Hydroboration–oxidation of alkenes is the method of choice for converting terminal alkenes to primary alcohols.

$$CH_3CH_2CH_2CH{=}CH_2 \xrightarrow[\text{2. } H_2O_2,\ HO^-]{\text{1. } B_2H_6} CH_3CH_2CH_2CH_2CH_2OH$$

    1-Pentene   1-Pentanol

$$\underset{\underset{CH_3}{|}}{CH_3CH_2C{=}CH_2} \xrightarrow[\text{2. } H_2O_2,\ HO^-]{\text{1. } B_2H_6} \underset{\underset{CH_3}{|}}{CH_3CH_2CHCH_2OH}$$

  2-Methyl-1-butene   2-Methyl-1-butanol

$$(CH_3)_2CHCH{=}CH_2 \xrightarrow[\text{2. } H_2O_2,\ HO^-]{\text{1. } B_2H_6} (CH_3)_2CHCH_2CH_2OH$$

  3-Methyl-1-butene   3-Methyl-1-butanol

(*c*) The only tertiary alcohol is 2-methyl-2-butanol. It can be made by acid-catalyzed hydration of 2-methyl-1-butene or 2-methyl-2-butene.

$$\underset{\underset{CH_3}{|}}{H_2C{=}CCH_2CH_3} \xrightarrow{H_2O,\ H_2SO_4} \underset{\underset{OH}{|}}{(CH_3)_2CCH_2CH_3}$$

  2-Methyl-1-butene   2-Methyl-2-butanol

$$(CH_3)_2C{=}CHCH_3 \xrightarrow{H_2O,\ H_2SO_4} \underset{\underset{OH}{|}}{(CH_3)_2CCH_2CH_3}$$

  2-Methyl-2-butene   2-Methyl-2-butanol

6.34 (*a*) Because the double bond is symmetrically substituted, the same addition product is formed under either ionic or free-radical conditions. Peroxides are absent, and so addition takes place by an ionic mechanism to give 3-bromohexane. (It does not matter whether the starting material is *cis*- or *trans*-3-hexene; both give the same product.)

$$CH_3CH_2CH{=}CHCH_2CH_3 \ + \ HBr \xrightarrow{\text{no peroxides}} \underset{\underset{Br}{|}}{CH_3CH_2CH_2CHCH_2CH_3}$$

      3-Hexene   Hydrogen   3-Bromohexane
                            bromide   (observed yield 76%)

(*b*) In the presence of peroxides, hydrogen bromide adds with a regioselectivity opposite to that predicted by Markovnikov's rule. The product is the corresponding primary bromide.

$$(CH_3)_2CHCH_2CH_2CH_2CH{=}CH_2 \xrightarrow[\text{peroxides}]{\text{HBr}} (CH_3)_2CHCH_2CH_2CH_2CH_2CH_2Br$$

6-Methyl-1-heptene

1-Bromo-6-methylheptane
(observed yield 92%)

(*c*) Hydroboration–oxidation of alkenes leads to hydration of the double bond with a regioselectivity opposite to Markovnikov's rule and without rearrangement of the carbon skeleton.

2-*tert*-Butyl-3,3-dimethyl-1-butene

2-*tert*-Butyl-3,3-dimethyl-1-butanol
(observed yield 65%)

(*d*) Hydroboration–oxidation of an alkene leads to syn hydration of the double bond.

1,2-Dimethylcyclohexene

cis-1,2-Dimethylcyclohexanol
(observed yield 82%)

(*e*) Bromine adds across the double bond of alkenes to give vicinal dibromides.

2-Methyl-1-pentene

1,2-Dibromo-2-methylpentane
(observed yield 60%)

(*f*) In aqueous solution bromine reacts with alkenes to give bromohydrins. Bromine is the electrophile in this reaction and adds to the carbon that has the greater number of attached hydrogens.

2-Methyl-2-butene   Bromine

3-Bromo-2-methyl-2-butanol
(observed yield 77%)

(*g*) An aqueous solution of chlorine will react with 1-methylcyclopentene by anti addition. Chlorine is the electrophile and adds to the less substituted end of the double bond.

1-Methylcyclopentene

trans-2-Chloro-1-methylcyclopentanol

(*h*) Compounds of the type $RCOOH$ are peroxy acids and react with alkenes to give epoxides.

$$(CH_3)_2C{=}C(CH_3)_2 \quad + \quad CH_3COOH \longrightarrow (CH_3)_2C{-}C(CH_3)_2 \quad + \quad CH_3COH$$

| 2,3-Dimethyl-2-butene | Peroxyacetic acid | 2,3-Epoxy-2,3-dimethylbutane (observed yield 70–80%) | Acetic acid |

(*i*) The double bond is cleaved by ozonolysis. Each of the doubly bonded carbons becomes doubly bonded to oxygen in the product.

Cyclodecane-1,6-dione
(observed yield 45%)

6.35 The product is epoxide B.

A   B

Major product;
formed faster

Epoxidation is an electrophilic addition; oxygen is transferred to the more electron-rich, more highly substituted double bond. A tetrasubstituted double bond reacts faster than a disubstituted one.

6.36 (*a*) No direct, one-step transformation moves a hydroxyl group from one carbon to another, and so it is not possible to convert 2-propanol to 1-propanol in a single reaction. Analyze the problem by reasoning backward. 1-Propanol is a primary alcohol. What reactions are available for the preparation of primary alcohols? One way is by the hydroboration–oxidation of terminal alkenes.

$$CH_3CH{=}CH_2 \xrightarrow{\text{hydroboration–oxidation}} CH_3CH_2CH_2OH$$

Propene   1-Propanol

The problem now becomes the preparation of propene from 2-propanol. The simplest way is by acid-catalyzed dehydration.

$$CH_3CHCH_3 \xrightarrow{\text{H}^+,\ \text{heat}} CH_3CH{=}CH_2$$
$$\underset{OH}{\mid}$$

2-Propanol   Propene

After analyzing the problem in terms of overall strategy, present the synthesis in detail showing the reagents required in each step. Thus, the answer is:

$$CH_3CHCH_3 \xrightarrow[\text{heat}]{H_2SO_4} CH_3CH=CH_2 \xrightarrow[\text{2. } H_2O_2, \text{ } HO^-]{\text{1. } B_2H_6} CH_3CH_2CH_2OH$$
$$\quad\quad | $$
$$\quad\quad OH$$

　　　　2-Propanol　　　　　　　　　　　　　Propene　　　　　　　　　　　1-Propanol

(b) We analyze this synthetic exercise in a manner similar to the preceding one. There is no direct way to move a bromine from C-2 in 2-bromopropane to C-1 in 1-bromopropane. We can, however, prepare 1-bromopropane from propene by free-radical addition of hydrogen bromide in the presence of peroxides.

$$CH_3CH=CH_2 \quad + \quad HBr \xrightarrow{\text{peroxides}} CH_3CH_2CH_2Br$$

　　　　Propene　　　　　Hydrogen　　　　　　　　1-Bromopropane
　　　　　　　　　　　　　bromide

We prepare propene from 2-bromopropane by dehydrohalogenation.

$$CH_3CHCH_3 \xrightarrow{E2} CH_3CH=CH_2$$
$$\quad | $$
$$\quad Br$$

　　　　2-Bromopropane　　　　　　　　　　　Propene

Sodium ethoxide in ethanol is a suitable base-solvent system for this conversion. Sodium methoxide in methanol or potassium *tert*-butoxide in *tert*-butyl alcohol could also be used, as could potassium hydroxide in ethanol.

Combining these two transformations gives the complete synthesis.

$$CH_3CHCH_3 \xrightarrow[\text{CH}_3\text{CH}_2\text{OH, heat}]{NaOCH_2CH_3} CH_3CH=CH_2 \xrightarrow[\text{peroxides}]{HBr} CH_3CH_2CH_2Br$$
$$\quad | $$
$$\quad Br$$

　　　2-Bromopropane　　　　　　　　　　Propene　　　　　　　　　1-Bromopropane

(c) Planning your strategy in a forward direction can lead to problems when the conversion of 2-bromopropane to 1,2-dibromopropane is considered. There is a temptation to try to simply add the second bromine by free-radical halogenation.

$$CH_3CHCH_3 \xrightarrow{Br_2, \text{ light and heat}} CH_3CHCH_2Br$$
$$\quad | \quad\quad\quad\quad\quad\quad\quad\quad\quad\quad | $$
$$\quad Br \quad\quad\quad\quad\quad\quad\quad\quad\quad\quad Br$$

　　　2-Bromopropane　　　　　　　　1,2-Dibromopropane

This is *incorrect*! There is no reason to believe that the second bromine will be introduced exclusively at C-1. In fact, the selectivity rules for bromination tell us that 2,2-dibromopropane is the expected major product.

　　　The best approach is to reason backward. 1,2-Dibromopropane is a vicinal dibromide, and we prepare vicinal dibromides by adding elemental bromine to alkenes.

$$CH_3CH=CH_2 \quad + \quad Br_2 \longrightarrow CH_3CHCH_2Br$$
$$\quad\quad\quad\quad\quad\quad\quad\quad\quad\quad\quad\quad\quad\quad\quad | $$
$$\quad\quad\quad\quad\quad\quad\quad\quad\quad\quad\quad\quad\quad\quad\quad Br$$

　　　　Propene　　　　　Bromine　　　　　1,2-Dibromopropane

As described in part (*b*), we prepare propene from 2-bromopropane by E2 elimination. The correct synthesis is therefore

$$CH_3\underset{\underset{Br}{|}}{C}HCH_3 \xrightarrow[CH_3CH_2OH]{NaOCH_2CH_3} CH_3CH{=}CH_2 \xrightarrow{Br_2} CH_3\underset{\underset{Br}{|}}{C}HCH_2Br$$

2-Bromopropane   Propene   1,2-Dibromopropane

(*d*) Do not attempt to reason forward and convert 2-propanol to 1-bromo-2-propanol by free radical bromination. Reason backward! The desired compound is a vicinal bromohydrin, and vicinal bromohydrins are prepared by adding bromine to alkenes in aqueous solution. The correct solution is

$$CH_3\underset{\underset{OH}{|}}{C}HCH_3 \xrightarrow[heat]{H_2SO_4} CH_3CH{=}CH_2 \xrightarrow[H_2O]{Br_2} CH_3\underset{\underset{OH}{|}}{C}HCH_2Br$$

2-Propanol   Propene   1,2-Bromo-2-propanol

(*e*) Here we have another problem where reasoning forward can lead to trouble. If we try to conserve the oxygen of 2-propanol so that it becomes the oxygen of 1,2-epoxypropane, we need a reaction in which this oxygen becomes bonded to C-1.

$$CH_3\underset{\underset{OH}{|}}{C}HCH_3 \longrightarrow CH_3CH{-}CH_2 \atop \diagdown O \diagup$$

2-Propanol   1,2-Epoxypropane

This will not work because no synthetic method for such a single-step transformation exists!

By reasoning backward, recalling that epoxides are made from alkenes by reaction with peroxy acids, we develop a proper synthesis.

$$CH_3\underset{\underset{OH}{|}}{C}HCH_3 \xrightarrow[heat]{H_2SO_4} CH_3CH{=}CH_2 \xrightarrow{\overset{\overset{O}{\|}}{CH_3COOH}} CH_3CH{-}CH_2 \atop \diagdown O \diagup$$

2-Propanol   Propene   1,2-Epoxypropane

(*f*) *tert*-Butyl alcohol and isobutyl alcohol have the same carbon skeleton; all that is required is to move the hydroxyl group from C-1 to C-2. As pointed out in part (*a*) of this problem, we cannot do that directly but we can do it in two efficient steps through a synthesis that involves hydration of an alkene.

$$(CH_3)_2CHCH_2OH \xrightarrow[heat]{H_2SO_4} (CH_3)_2C{=}CH_2 \xrightarrow{H_2O,\ H_2SO_4} (CH_3)_3COH$$

Isobutyl alcohol   2-Methylpropene   *tert*-Butyl alcohol

Acid-catalyzed hydration of the alkene gives the desired regioselectivity.

(*g*) The strategy of this exercise is similar to that of the preceding one. Convert the starting material to an alkene by elimination, followed by electrophilic addition to the double bond.

$$(CH_3)_2CHCH_2I \xrightarrow[(CH_3)_3COH,\ heat]{KOC(CH_3)_3} (CH_3)_2C{=}CH_2 \xrightarrow{HI} (CH_3)_3CI$$

Isobutyl iodide   2-Methylpropene   *tert*-Butyl iodide (*h*) This problem is similar to the one in part (*d*) in that it requires the preparation of a vicinal halohydrin from an alkyl halide. The strategy is the same. Convert the alkyl halide to an alkene, and then form the halohydrin by treatment with the appropriate halogen in aqueous solution.

Cyclohexyl chloride　　　　　　　　　　Cyclohexene　　　　　　　*trans*-2-Chlorocyclohexanol

(*i*) Halogenation of an alkane is required here. Iodination of alkanes, however, is not a feasible reaction. We can make alkyl iodides from alcohols or from alkenes by treatment with HI. A reasonable synthesis using reactions that have been presented to this point proceeds as shown:

Cyclopentane　　　　　Cyclopentyl chloride　　　　　Cyclopentene　　　　　Cyclopentyl iodide

(*j*) Dichlorination of cyclopentane under free-radical conditions is not a realistic approach to the introduction of two chlorines in a trans-1,2 relationship without contamination by isomeric dichlorides. Vicinal dichlorides are prepared by electrophilic addition of chlorine to alkenes. The stereochemistry of addition is anti.

Cyclopentane　　　　　Cyclopentyl chloride　　　　　Cyclopentene　　　　*trans*-1,2-Dichlorocyclopentane

(*k*) The desired compound contains all five carbon atoms of cyclopentane but is not cyclic. Two aldehyde functions are present. We know that cleavage of carbon–carbon double bonds by ozonolysis leads to two carbonyl groups, which suggests the synthesis shown in the following equation:

Cyclopentanol　　　　　　　　Cyclopentene　　　　　　　　　　　　　　　　Pentanedial

6.37 (*a*) When peroxides are present, hydrogen bromide adds to alkenes by a free-radical chain mechanism.

Initiation:

Peroxide dissociation is the first part of the initiation stage. The alkoxy radicals so generated then abstract a hydrogen atom from HBr to produce bromine atoms.

RO—OR → 2RO·

A peroxide   Alkoxy radical

RO· + H—Br: → RO—H + ·Br:

Alkoxy radical   Hydrogen bromide   Alcohol   Bromine atom

Propagation:

The propagation phase begins with a bromine atom adding to the double bond in the direction that produces the more stable alkyl radical. This radical then abstracts a hydrogen atom from HBr to give the alkyl bromide and a bromine atom

:Br· +   1-Octene   → :Br   1-Bromomethylheptyl radical

Bromine atom

:Br (1-Bromomethylheptyl radical) + H—Br: Hydrogen bromide → :Br   H   1-Bromooctane + ·Br: Bromine atom

The bromine atom produced in the second propagation reaction adds to a second 1-octene molecule and the two propagation steps repeat over and over.

(*b*) Phosphoric acid protonates cyclohexene giving cyclohexyl cation. This carbocation reacts with iodide ion to produce iodocyclohexane. The overall reaction corresponds to electrophilic addition of HI.

Cyclohexene + H—O—P(=O)(OH)—OH (Phosphoric acid) → (Cyclohexyl cation)⁺ + :O—P(=O)(OH)—OH (Dihydrogen phosphate ion)

Cyclohexyl cation⁺ + :I:⁻ (Iodide ion) → Iodocyclohexane (with I)

(c) This reaction follows the conventional mechanism for the acid-catalyzed hydration of alkenes. Water can attack the carbocation intermediate from either side giving a mixture of syn and anti addition.

| 1,2-Dimethylcyclohexene | Hydronium ion | 1,2-Dimethylcyclohexyl cation | Water |

| 1,2-Dimethylcyclohexyl ion | Water | 1,2-Dimethylcyclohexyloxonium ion |

| 1,2-Dimethylcyclohexyloxonium ion | Water | 1,2-Dimethylcyclohexanol | Hydronium ion |

(d) This mechanism is the same as that for acid-catalyzed hydration of an alkene except that the nucleophile which attacks the carbocation is methanol rather than water.

| Methyloxonium ion | 2-Methylpropene | Methanol | tert-Butyl cation |

| tert-Butyl cation | Methanol | Conjugate acid of tert-butyl methyl ether |

| Conjugate acid of tert-butyl methyl ether | Methanol | tert-Butyl methyl ether | Methyloxonium ion |

(*e*) The product, a bromohydrin, is formed when water attacks a bromonium ion.

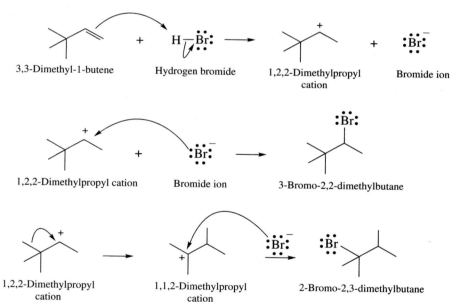

(*f*) The reaction is electrophilic addition of HBr to the double bond, accompanied by carbocation rearrangement.

6.38 The two products formed by addition of hydrogen bromide to 1,2-dimethylcyclohexene cannot be regioisomers. Stereoisomers are possible, however.

1,2-Dimethylcyclohexene	*cis*-1,2-Dimethylcyclohexyl bromide	*trans*-1,2-Dimethylcyclohexyl bromide

The same two products are formed from 1,6-dimethylcyclohexene because addition of hydrogen bromide follows Markovnikov's rule in the absence of peroxides.

1,6-Dimethylcyclohexene	*cis*-1,2-Dimethylcyclohexyl bromide	*trans*-1,2-Dimethylcyclohexyl bromide

6.39 The problem presents the following experimental observation:

4-*tert*-Butyl(methylene)-cyclohexane	*cis*-1-*tert*-Butyl-4-methylcyclohexane (88%)	*trans*-1-*tert*-Butyl-4-methylcyclohexane (12%)

This observation tells us that the predominant mode of hydrogen addition to the double bond is from the equatorial direction. Equatorial addition is the less hindered approach and thus occurs faster.

(*a*) Epoxidation should therefore give the following products:

Major product	Minor product

The major product is the stereoisomer that corresponds to transfer of oxygen from the equatorial direction.

(*b*) Hydroboration–oxidation occurs from the equatorial direction.

Major product   Minor product

6.40 The methyl group in compound B shields one face of the double bond from the catalyst surface, therefore hydrogen can be transferred only to the bottom face of the double bond. The methyl group in compound A does not interfere with hydrogen transfer to the double bond.

Top face of double bond

Compound A   Compound B

Thus, hydrogenation of A is faster than that of B because B contains a more sterically hindered double bond.

6.41 Hydrogen can add to the double bond of 1,4-dimethylcyclopentene either from the same side as the C-4 methyl group or from the opposite side. The two possible products are *cis*- and *trans*-1,3-dimethylcyclopentane.

1,4-Dimethylcyclopentene   *cis*-1,3-Dimethylcyclopentane   *trans*-1,3-Dimethylcyclopentane

Hydrogen transfer occurs to the less hindered face of the double bond, that is, trans to the C-4 methyl group. Thus, the major product is *cis*-1,3-dimethylcyclopentane.

6.42 Hydrogen can add to either the top face or the bottom face of the double bond. Syn addition to the double bond requires that the methyl groups in the product be cis.

6.43 3-Carene can in theory undergo hydrogenation to give either *cis*-carane or *trans*-carane.

cis-Carane (98%)   *trans*-Carane

The exclusive product is *cis*-carane, because it corresponds to transfer of hydrogen from the less hindered side.

cis-Carane

6.44

Reaction is	Sign of $\Delta H°$	$\Delta S°$
(*a*) exergonic at all temperatures	–	+
(*b*) exergonic at low temperature; endergonic at high temperature	–	–
(*c*) endergonic at all temperatures	+	–
(*d*) endergonic at low temperature; exergonic at high temperature	+	+

(*a*) A reaction will be exergonic at all temperatures when $\Delta G°$ is negative at all temperatures. Therefore, $\Delta H°$ should have a negative sign. $\Delta S°$ should have a positive sign to ensure that the $-T\Delta S°$ term is also negative. A reaction that is exergonic at all temperatures is one which is exothermic and proceeds with an increase in entropy.

(*b*) This case is one in which an increase in temperature reverses an exergonic reaction to an endergonic one. It must be a reaction in which $\Delta S°$ has a negative sign, so that $-T\Delta S°$ becomes positive as T increases and overcomes a negative value for $\Delta H°$.

(*c*) Endergonic at all temperatures is the "worst" case—an endothermic reaction in which entropy decreases.

(*d*) Entropy must increase in order for an endothermic reaction (endergonic at low temperature) to become exergonic at high temperature.

6.45 (*a*) $\Delta G°$ for the iodination of ethylene at 25°

$$H_2C=CH_2(g) \; + \; I_2(g) \; \rightleftharpoons \; ICH_2CH_2I(g)$$

can be calculated from the values given for $\Delta H°$ and $\Delta S°$.

$\Delta G° = \Delta H° - T\Delta S°$

$\Delta G° = -48 \text{ kJ} - (298 \text{ K})(-0.13 \text{ kJ/K})$

$\Delta G° = -48 \text{ kJ} + 38.7 \text{ kJ}$

$\Delta G° = -9 \text{ kJ}$

The equilibrium constant K is calculated from $\Delta G°$.

$$\Delta G° = -RT \ln K$$

$$-(9 \text{ kJ})(1000 \text{ J/kJ}) = -(8.314 \text{ J/K})(298 \text{ K}) \ln K$$

$$-9000 \text{ J} = -(8.314 \text{ J/K})(298 \text{ K}) \ln K$$

$$\ln K = \frac{-9000 \text{ J}}{-(8.314 \text{ J/K})(298 \text{ K})} = 3.6$$

$$K = e^{3.6} = 38$$

(b) The sign of $\Delta G°$ is negative; the reaction is exergonic.

(c) The sign of $\Delta S°$ is negative. Therefore $(-T\Delta S°)$ becomes increasingly positive as the temperature is raised, $\Delta G°$ becomes less negative, and K decreases.

6.46 Ethylene and propene react with concentrated sulfuric acid to form alkyl hydrogen sulfates. Addition of water hydrolyzes the alkyl hydrogen sulfates to the corresponding alcohols.

$$H_2C{=}CH_2 \xrightarrow{H_2SO_4} CH_3CH_2OSO_2OH \xrightarrow[\text{heat}]{H_2O} CH_3CH_2OH$$

| Ethylene | Ethyl hydrogen sulfate | Ethanol |

$$H_2C{=}CHCH_3 \xrightarrow{H_2SO_4} \underset{\underset{OSO_2OH}{|}}{CH_3CHCH_3} \xrightarrow[\text{heat}]{H_2O} \underset{\underset{OH}{|}}{CH_3CHCH_3}$$

| Propene | Isopropyl hydrogen sulfate | Isopropyl alcohol |

Recall that alkyl substituents on the double bond increase the reactivity of alkenes toward electrophilic addition. Propene therefore reacts faster than ethylene with sulfuric acid, and the mixture of alkyl hydrogen sulfates is mainly isopropyl hydrogen sulfate, and the alcohol obtained on hydrolysis is isopropyl alcohol.

6.47 The first step in the mechanism of acid-catalyzed hydration of alkenes is protonation of the double bond to give a carbocation intermediate.

| Hydronium ion | 3-Methyl-1-butene | Water | 1,2-Dimethylpropyl cation (secondary) |

The carbocation formed in this step is secondary and capable of rearranging to a more stable tertiary carbocation by a hydride shift.

| 1,2-Dimethylpropyl cation (secondary) | | 1,1-Dimethylpropyl cation (tertiary) |

The alcohol that is formed when water reacts with the tertiary carbocation is 2-methyl-2-butanol, not 3-methyl-2-butanol.

Water   1,1-Dimethylpropyl cation   2-Methyl-2-butanol

6.48 In the presence of sulfuric acid, the carbon–carbon double bond of 2-methyl-1-butene is protonated and a carbocation is formed.

Hydronium ion   2-Methyl-1-butene   Water   1,1-Dimethylpropyl cation

This carbocation can then lose a proton from its CH_2 group to form 2-methyl-2-butene.

Water   1,1-Dimethylpropyl cation   Hydronium ion   2-Methyl-2-butene

6.49 The first step in the reaction of an alkene with bromine is the formation of a bromonium ion.

This bromonium ion can react with Br_2 to form 1,2-dibromohexane, or it can be attacked by methanol.

1-Bromo-2-methoxyhexane

Attack on the bromonium ion by methanol is analogous to the attack by water in the mechanism of bromohydrin formation.

6.50 The problem states that a bridged sulfonium ion is an intermediate. Therefore, use the π electrons of the double bond to attack one of the sulfur atoms of thiocyanogen and cleave the S—S bond in a manner analogous to cleavage of a Br—Br bond in the reaction of bromine with an alkene.

The sulfonium ion is then attacked by thiocyanate (NCS^-) to give the observed product, which has the trans stereochemistry.

6.51 Alkenes of molecular formula $C_{12}H_{24}$ are *trimers* of 2-methylpropene (C_4H_8). The first molecule of 2-methylpropene is protonated to form *tert*-butyl cation, which reacts with a second molecule of 2-methylpropene to give a tertiary carbocation having eight carbons.

tert-Butyl cation   2-Methylpropene   1,1,3,3-Tetramethylbutyl cation

This carbocation reacts with a third molecule of 2-methylpropene to give a 12-carbon tertiary carbocation.

1,1,3,3-Tetramethylbutyl cation   2-Methylpropene   1,1,3,3,5,5-Hexamethylhexyl cation

The 12-carbon carbocation can lose a proton in either of two directions to give the alkenes shown.

1,1,3,3,5,5-Hexamethylhexyl cation   2,4,4,6,6-Pentamethyl-1-heptene   2,4,4,6,6-Pentamethyl-2-heptene

6.52 The carbon skeleton is revealed by the hydrogenation experiment. Compounds B and C must have the same carbon skeleton as 3-ethylpentane.

Three alkyl bromides have this carbon skeleton, namely, 1-bromo-3-ethylpentane, 2-bromo-3-ethylpentane, and 3-bromo-3-ethylpentane. Of these three only 2-bromo-3-ethylpentane will give two alkenes on dehydrobromination.

1-Bromo-3-ethylpentane

3-Bromo-3-ethylpentane

2-Bromo-3-ethylpentane   3-Ethyl-1-pentene   3-Ethyl-2-pentene

Compound A must therefore be 2-bromo-3-ethylpentane. Dehydrobromination of A will follow Zaitsev's rule, so that the major alkene (compound B) is 3-ethyl-2-pentene and the minor alkene (compound C) is 3-ethyl-1-pentene.

6.53 The information that compound B gives 2,4-dimethylpentane on catalytic hydrogenation establishes its carbon skeleton.

2,4-Dimethylpentane

Compound B is an alkene derived from compound A—an alkyl bromide of molecular formula $C_7H_{15}Br$. We are told that compound A is not a primary alkyl bromide. Compound A can therefore be only:

Because compound A gives a single alkene on being treated with sodium ethoxide in ethanol, it can only be 3-bromo-2,4-dimethylpentane, and compound B must be 2,4-dimethyl-2-pentene.

3-Bromo-2,4-dimethylpentane   2,4-Dimethyl-2-pentene
(compound A)   (compound B)

6.54 Alkene C must have the same carbon skeleton as its hydrogenation product, 2,3,3,4-tetramethylpentane.

2,3,3,4-Tetramethylpentane

The only alkene with this carbon skeleton is 2,3,3,4-tetramethyl-1-pentene, alkene C. The two alkyl bromides, compounds A and B, that give this alkene on dehydrobromination have their bromine substituents at C-1 and C-2, respectively.

1-Bromo-2,3,3,4-tetramethylpentane

KOC(CH₃)₃, dimethyl sulfoxide

2,3,3,4-tetramethyl-1-pentene

2-Bromo-2,3,3,4-tetramethylpentane

KOC(CH₃)₃, dimethyl sulfoxide

6.55 The only alcohol (compound A) that can undergo acid-catalyzed dehydration to alkene B without rearrangement is the one shown in the equation.

Alcohol A

$\xrightarrow[\substack{\text{heat} \\ -H_2O}]{\text{KHSO}_4}$

Alkene B

Dehydration of alcohol A also yields an isomeric alkene under these conditions.

Alcohol A

$\xrightarrow[\substack{\text{heat} \\ -H_2O}]{\text{KHSO}_4}$

Alkene C

6.56 First decide what three alkenes can be formed by dehydration of 1,2-dimethylcyclohexanol.

A   +   B   +   C

Notice that the three alkenes differ in the regioselectivity of dehydration and that each will give 1,2-dimethylcyclohexane on catalytic hydrogenation. Assigning structures to the alkenes on the basis of Zaitsev's rule gives A,B, and C as indicated in the equation. A has the least substituted double bond and is formed in the smallest amount. C has the most substituted double bond and is formed in greatest amount. A is the least stable alkene and is present in the smallest amount at equilibrium. C has the most stable double bond and predominates at equilibrium.

6.57 Electrophilic addition of hydrogen iodide should occur in accordance with Markovnikov's rule.

$$H_2C{=}CHC(CH_3)_3 \quad \underset{\text{KOH, } n\text{-PrOH}}{\overset{\text{HI}}{\rightleftarrows}} \quad CH_3\underset{\underset{I}{|}}{C}HC(CH_3)_3$$

3,3-Dimethyl-1-butene   3-Iodo-2,2-dimethylbutane

Treatment of 3-iodo-2,2-dimethylbutane with alcoholic potassium hydroxide should bring about E2 elimination to regenerate the starting alkene. Hence, compound A is 3-iodo-2,2-dimethylbutane.

The carbocation intermediate formed in the addition of hydrogen iodide to the alkene is one that can rearrange by a methyl group migration.

$$H_2C{=}CHC(CH_3)_3 \quad + \quad H^+ \quad \longrightarrow \quad H_3C{-}\overset{+}{CH}{-}\underset{\underset{CH_3}{|}}{\overset{\overset{CH_3}{|}}{C}}CH_3 \quad \xrightarrow[\text{migration}]{\text{methyl}} \quad H_3C{-}CH{-}\underset{\underset{CH_3}{|}}{\overset{\overset{CH_3}{|}}{\overset{+}{C}}}CH_3$$

$$\Big\downarrow I^- \qquad\qquad\qquad\qquad\qquad \Big\downarrow I^-$$

$$CH_3\underset{\underset{I}{|}}{CH}{-}C(CH_3)_3 \qquad\qquad\qquad (CH_3)_2CH{-}\underset{\underset{I}{|}}{C}(CH_3)_2$$

Compound A   Compound B
                                   (2-iodo-2,3-dimethylbutane)

A likely candidate for compound B is therefore the one with a rearranged carbon skeleton, 2-iodo-2,3-dimethylbutane. This is confirmed by the fact that compound B undergoes elimination to give 2,3-dimethyl-2-butene.

$$(CH_3)_2CH{-}\underset{\underset{I}{|}}{C}(CH_3)_2 \quad \xrightarrow{\text{E2}} \quad (CH_3)_2C{=}C(CH_3)_2$$

Compound B   2,3-Dimethyl-2-butene

6.58 The ozonolysis data are useful in quickly identifying alkenes A and B.

$$\text{Compound A} \quad \longrightarrow \quad \overset{O}{\overset{||}{H}}CH \quad + \quad (CH_3)_3C\overset{\overset{O}{||}}{C}C(CH_3)_3$$

Compound A is therefore 2-*tert*-butyl-3,3-dimethyl-1-butene.

$$(CH_3)_3C\overset{\overset{CH_2}{||}}{C}C(CH_3)_3$$

Compound A

$$\text{Compound B} \quad \longrightarrow \quad \overset{O}{\overset{||}{H}}CH \quad + \quad CH_3\overset{\overset{O}{||}}{C}{-}\underset{\underset{CH_3}{|}}{\overset{\overset{CH_3}{|}}{C}}{-}C(CH_3)_3$$

Compound B is therefore 2,3,3,4,4-pentamethyl-1-pentene.

$$CH_3\overset{\overset{H_2C}{||}}{C}{-}\underset{\underset{CH_3}{|}}{\overset{\overset{CH_3}{|}}{C}}C(CH_3)_3$$

Compound B

Compound B has a carbon skeleton different from the alcohol that produced it by dehydration. We are therefore led to consider a carbocation rearrangement.

Compound A

methyl migration

Compound B

6.59 The important clue to deducing the structures of A and B is the ozonolysis product C. Remembering that the two carbonyl carbons of C must have been joined by a double bond in the precursor B, we write

These two carbons must have been connected by a double bond.

Compound C

Compound B

The tertiary bromide that gives compound B on dehydrobromination is 1-methylcyclohexyl bromide.

Compound A

Compound B

When tertiary halides are treated with base, they undergo E2 elimination. The regioselectivity of elimination of tertiary halides follows the Zaitsev rule.

6.60 Because santene and 1,3-diacetylcyclopentane (compound A) contain the same number of carbon atoms, the two carbonyl carbons of the diketone must have been connected by a double bond in santene. The structure of santene must therefore be

more appropriately represented as

6.61   (*a*) Compound A contains nine of the ten carbons and 14 of the 16 hydrogens of sabinene. Ozonolysis has led to the separation of one carbon and two hydrogens from the rest of the molecule. The carbon and the two hydrogens must have been lost as formaldehyde, $H_2C=O$. This H_2C unit was originally doubly bonded to the carbonyl carbon of compound A. Sabinene must therefore have the structure shown in the equation representing its ozonolysis:

           Sabinene   Compound A   Formaldehyde

    (*b*) Compound B contains all ten of the carbons and all 16 of the hydrogens of Δ^3-carene. The two carbonyl carbons of compound B must have been linked by a double bond in Δ^3-carene.

           Δ^3-Carene   Compound B

6.62   The sex attractant of the female housefly consumes one mole of hydrogen on catalytic hydrogenation (the molecular formula changes from $C_{23}H_{46}$ to $C_{23}H_{48}$). Thus, the molecule has one double bond. The position of the double bond is revealed by the ozonolysis data.

An unbranched 9-carbon unit and an unbranched 14-carbon unit make up the carbon skeleton, and these two units must be connected by a double bond. The housefly sex attractant therefore has the constitution:

$$CH_3(CH_2)_7CH=CH(CH_2)_{12}CH_3$$

9-Tricosene

The data cited in the problem do not permit the stereochemistry of this natural product to be determined.

6.63   The hydrogenation data tell us that $C_{19}H_{38}$ contains one double bond and has the same carbon skeleton as 2,6,10,14-tetramethylpentadecane. We locate the double bond at C-2 on the basis of the fact that acetone, $(CH_3)_2C=O$, is obtained on ozonolysis. The structures of the natural product and the aldehyde produced on its ozonolysis are as follows:

Ozonolysis cleaves molecule here.   Aldehyde obtained on ozonolysis

6.64 Because $\overset{\overset{O}{\|}}{H}C\overset{\overset{O}{\|}}{CH_2C}H$ is one of the products of its ozonolysis, the sex attractant of the arctiid moth must contain the unit $=CHCH_2CH=$. This unit must be bonded to an unbranched 12-carbon unit at one end and an unbranched 6-carbon unit at the other in order to give $CH_3(CH_2)_{10}CH=O$ and $CH_3(CH_2)_4CH=O$ on ozonolysis.

$$CH_3(CH_2)_{10}CH \overset{\xi}{=} CHCH_2CH \overset{\xi}{=} CH(CH_2)_4CH_3$$

Sex attractant of arctiid moth
(wavy lines show positions of cleavage on ozonolysis)

1. O_3
2. H_2O, Zn

$$CH_3(CH_2)_{10}\overset{\overset{O}{\|}}{C}H \ + \ \overset{\overset{O}{\|}}{H}C\overset{\overset{O}{\|}}{CH_2C}H \ + \ \overset{\overset{O}{\|}}{H}C(CH_2)_4CH_3$$

The stereochemistry of the double bonds cannot be determined on the basis of the available information.

ANSWERS TO INTERPRETIVE PROBLEMS

6.65 A; **6.66** D; **6.67** D; **6.68** B; **6.69** A; **6.70** C

SELF-TEST

PART A

A-1. How many different alkenes will yield 2,3-dimethylpentane on catalytic hydrogenation? Draw their structures, and name them.

A-2. Write structural formulas for the reactant, reagents, or product omitted from each of the following:

(*a*) $(CH_3)_2C=CHCH_3 \xrightarrow{H_2SO_4 \text{ (dilute)}}$?

(*b*) $=CH_2 \xrightarrow{?}$ $-CH_2Br$

(*c*) ? $(C_{10}H_{16}) \xrightarrow[\text{2. } H_2O, \text{ Zn}]{\text{1. } O_3}$

(*d*) $-CH_3$ + $Br_2 \xrightarrow{H_2O}$?

A-3. Provide a sequence of reactions to carry out the following conversions. More than one synthetic step is necessary for each. Write the structure of the product of each synthetic step.

(*a*)

(b) $CH_3CH_2\underset{\underset{\text{Cl}}{|}}{CH}CH(CH_3)_2 \longrightarrow CH_3CH_2CH\overset{\overset{\text{O}}{\diagup\!\!\diagdown}}{-}C(CH_3)_2$

(c) $(CH_3)_3CCHCH_3 \longrightarrow (CH_3)_3CCH_2CH_2Br$
  $\underset{\underset{\text{Br}}{|}}{}$

A-4. Provide a detailed mechanism describing the elementary steps in the reaction of 1-butene with HBr in the presence of peroxides.

A-5. Chlorine reacts with an alkene to give the 2,3-dichlorobutane isomer whose structure is shown. What are the structure and name of the alkene? Outline a mechanism for the reaction.

A-6. Write a structural formula, including stereochemistry, for the compound formed from *cis*-3-hexene on treatment with peroxyacetic acid.

A-7. Give a mechanism describing the elementary steps in the reaction of 2-methyl-1-butene with hydrogen chloride. Use curved arrows to show the flow of electrons.

A-8. What two alkenes give 2-chloro-2-methylbutane on reaction with hydrogen chloride?

A-9. Give the major organic product formed from the following sequence of reactions.

$$\underset{\text{light}}{\overset{\text{Br}_2}{\longrightarrow}} \quad \underset{CH_3CH_2OH}{\overset{NaOCH_2CH_3}{\longrightarrow}} \quad \underset{\text{2. H}_2O_2, \text{HO}^-}{\overset{\text{1. B}_2H_6}{\longrightarrow}} \quad ?$$

A-10. The reaction of 3-methyl-1-butene with hydrogen chloride gives two alkyl halide products; one is a secondary alkyl chloride and the other is tertiary. Write the structures of the products, and provide a mechanism explaining their formation.

A-11. A hydrocarbon A (C_6H_{12}) undergoes reaction with HBr to yield compound B ($C_6H_{13}Br$). Treatment of B with sodium ethoxide in ethanol yields C, an isomer of A. Reaction of C with ozone followed by treatment with water and zinc gives acetone, $(CH_3)_2C=O$, as the only organic product. Provide structures for A, B, and C, and outline the reaction pathway.

PART B

B-1. Rank the following alkenes in order of decreasing heats of hydrogenation (largest first)

           1   2   3   4

(a) 2 > 3 > 4 > 1   (d) 2 > 4 > 3 > 1

(b) 1 > 3 > 4 > 2   (e) 1 > 2 > 3 > 4

(c) 1 > 4 > 3 > 2

B-2. The product from the reaction of 1-pentene with Cl_2 in H_2O is named:

(a) 1-Chloro-2-pentanol   (c) 1-Chloro-1-pentanol

(b) 2-Chloro-2-pentanol   (d) 2-Chloro-1-pentanol

B-3. In the reaction of hydrogen bromide with an alkene (in the absence of peroxides), the first step of the reaction is the _____ to the alkene.

(a) Fast addition of an electrophile

(b) Fast addition of a nucleophile

(c) Slow addition of an electrophile

(d) Slow addition of a nucleophile

B-4. The major product of the following reaction sequence is

B-5. Which, if any, of the following alcohols *cannot* be prepared from an alkene?

(a) [structure: pentanol with OH]

(b) [structure with OH]

(c) [structure with OH]

(d) [structure with OH]

(e) None of these—all of the alcohols shown can be prepared from an alkene.

B-6. Which of the species shown is the most stable form of the intermediate in the electrophilic addition of Cl_2 in water to cyclohexene to form a halohydrin? Electron pairs have been omitted for convenience, and their absence should not be considered as part of the problem.

B-7. Treatment of 2-methyl-2-butene with HBr in the presence of a peroxide yields

(a) A primary alkyl bromide

(b) A secondary alkyl bromide

(c) A tertiary alkyl bromide

(d) A vicinal dibromide

B-8. The reaction

$$(CH_3)_2C=CH_2 \ + \ Br\bullet \ \longrightarrow \ (CH_3)_2\overset{\bullet}{C}-CH_2Br$$

is an example of a(n) _____ step in a radical chain reaction.

(*a*) Initiation   (*c*) Termination

(*b*) Propagation   (*d*) Heterolytic cleavage

B-9. To which point on the potential energy diagram for the reaction of 2-methylpropene with hydrogen chloride does the figure shown at the right correspond?

B-10. Which of the following most accurately describes the first step in the reaction of hydrogen chloride with 1-butene?

(*a*) $Cl-H$ ... $\longrightarrow$ ... $+ \ Cl\bullet$

(*b*) $Cl-H$ ... $\longrightarrow$ ... $+ \ Cl^-$

(*c*) $Cl-H$ ... $\longrightarrow$ ... $+ \ Cl^-$

(*d*) $H-Cl$ ... $\longrightarrow$ ... $+ \ H^-$

B-11. Which of the following best describes the flow of electrons in the acid-catalyzed dimerization of $(CH_3)_2C=CH_2$?

(*a*) ...

(*b*) ...

(*c*) ...

(*d*) ...

B-12. Which one of the following compounds gives acetone $(CH_3)_2C=O$ as one of the products of its ozonolysis?

(a) (b) (c) (d) (e)

B-13. Addition of HCl to 3,3-dimethyl-1-butene yields two products, one of which has a rearranged carbon skeleton. Which of the following cations are intermediates in that reaction?

$(CH_3)_3\overset{+}{C}CHCH_2Cl$ $(CH_3)_3\overset{+}{C}CHCH_3$ $(CH_3)_2\overset{+}{C}\underset{\underset{Cl}{|}}{C}(CH_3)_2$ $(CH_3)_2\overset{+}{C}CH(CH_3)_2$

1   2   3   4

(a) 1, 2   (b) 1, 3   (c) 1, 4   (d) 2, 3   (e) 2, 4

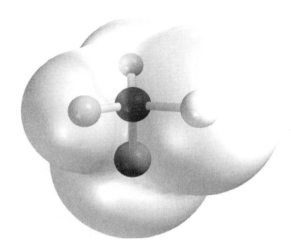

CHAPTER 7

Stereochemistry

SOLUTIONS TO TEXT PROBLEMS

7.1   (*c*) Carbon-2 is a chirality center in 1-bromo-2-methylbutane. The four groups attached to it: H, CH_3, CH_3CH_2, and $BrCH_2$, are different from each other.

$$BrCH_2-\overset{\overset{\displaystyle H}{|}}{\underset{\underset{\displaystyle CH_3}{|}}{C}}-CH_2CH_3$$

        (*d*) There are no chirality centers in 2-bromo-2-methylbutane.

$$H_3C-\overset{\overset{\displaystyle Br}{|}}{\underset{\underset{\displaystyle CH_3}{|}}{C}}-CH_2CH_3$$

7.2   (*b*) Carbon-2 is a chirality center in 1,1,2-trimethylcyclobutane.

A chirality center; the four groups to which it is directly bonded [H, CH_3, CH_2, and $C(CH_3)_2$] are all different from one another.

1,1,3-Trimethylcyclobutane however, has no chirality centers.

Not a chirality center; two of its groups are the same.

7.3 (*b*) (Z)-1,2-Dichloroethene has *two* planes of symmetry, of which one is the plane of the molecule and the second bisects the carbon–carbon bond. The molecule is achiral.

(*c*) A plane of symmetry in *cis*-1,2-dichlorocyclopropane bisects the C-1—C-2 bond and passes through C-3. The molecule is achiral.

(*d*) *trans*-1,2-Dichlorocyclopropane lacks a plane of symmetry. Its two mirror images cannot be superimposed on each other. The molecule is chiral.

Cl■——▲——■H   and   H■——▲——■Cl   are nonsuperimposable
  H   Cl   Cl   H   mirror images

7.4 The equation relating specific rotation $[\alpha]$ to observed rotation α is

$$[\alpha] = \frac{100\alpha}{cl}$$

The concentration c is expressed in grams per 100 mL and the length l of the polarimeter tube in decimeters. Because the problem specifies the concentration as 0.3 g/15 mL and the path length as 10 cm, the specific rotation $[\alpha]$ is:

$$[\alpha] = \frac{100(-0.78°)}{100\left(\dfrac{0.3\ \text{g}}{15\ \text{mL}}\right)\left(\dfrac{10\ \text{cm}}{10\ \text{cm/dm}}\right)}$$

$$= -39°$$

7.5 From the previous problem, the specific rotation of natural cholesterol is $[\alpha] = -39°$. The mixture of natural (−)-cholesterol and synthetic (+)-cholesterol specified in this problem has a specific rotation $[\alpha]$ of −13°. The optical purity is 33.3% $[(-13°\,/-39°)\,(100)]$.

Optical purity = %(−)-cholesterol − %(+)-cholesterol

33.3% = %(−)-cholesterol − [100 − %(−)-cholesterol]

133.3% = 2[%(−)-cholesterol]

66.7% = %(−)-cholesterol

The mixture is two-thirds natural (−)-cholesterol and one-third synthetic (+)-cholesterol.

7.6   Draw the molecular model so that it is in the same format as the drawings of (+) and (−)-2-butanol in the text.

Reorient the molecule so that it can be compared with the drawings of (+) and (−)-2-butanol.

Reorient $\begin{array}{c} \text{HO} \quad \text{H} \\ \diagdown \text{C} \diagup \\ \text{H}_3\text{C} \quad \text{CH}_2\text{CH}_3 \end{array}$ which becomes $\begin{array}{c} \text{CH}_3\text{CH}_2 \quad \text{H} \\ \diagdown \text{C} \diagdown \text{OH} \\ \text{H}_3\text{C} \end{array}$

The molecular model when redrawn matches the text's drawing of (+)-2-butanol.

7.7   (*b*) The solution to this problem is analogous to the sample solution given in the text to part (*a*).

$$\begin{array}{c} \text{H}_3\text{C} \quad \text{H} \\ \diagdown \text{C} \text{—CH}_2\text{F} \\ \diagup \\ \text{CH}_3\text{CH}_2 \end{array}$$

(+)-1-Fluoro-2-methylbutane

Order of precedence: $CH_2F > CH_3CH_2 > CH_3 > H$

The lowest ranked atom (H) at the chirality center points away from us in the drawing. The three higher ranked substituents trace a clockwise path from CH_2F to CH_2CH_3 to CH_3.

$$\begin{array}{c} \text{H}_3\text{C} \diagdown \quad \diagup \text{CH}_2\text{F} \\ \text{C} \\ | \\ \text{CH}_2\text{CH}_3 \end{array}$$

The absolute configuration is *R*; the compound is (*R*)-(+)-1-fluoro-2-methylbutane.

(*c*) The highest ranked substituent at the chirality center of 1-bromo-2-methylbutane is CH_2Br, and the lowest ranked atom is H. Of the remaining two, ethyl outranks methyl.

Order of precedence: $CH_2Br > CH_2CH_3 > CH_3 > H$

The lowest ranked atom (H) is directed toward you in the drawing, and therefore the molecule needs to be reoriented so that H points in the opposite direction.

$$\begin{array}{c} \text{CH}_3 \\ \text{H} \diagdown \text{C} \text{—CH}_2\text{Br} \\ \diagup \\ \text{CH}_3\text{CH}_2 \end{array} \qquad \bigcirc \qquad \begin{array}{c} \text{H} \quad \text{CH}_3 \\ \text{BrCH}_2 \text{—C} \\ \diagdown \\ \text{CH}_2\text{CH}_3 \end{array}$$

turn 180°

(+)-1-Bromo-2-methylbutane

The three highest ranking substituents trace a counterclockwise path when the lowest ranked atom (H) is held away from you.

$$BrCH_2 \quad CH_3$$
$$C$$
$$CH_2CH_3$$

The absolute configuration is S, and thus the compound is (S)-$(+)$-1-bromo-2-methylbutane.

(*d*) The highest ranked substituent at the chirality center of 3-buten-2-ol is the hydroxyl group, and the lowest ranked atom is H. Of the remaining two, vinyl outranks methyl.

Order of precedence: $\quad HO > H_2C{=}CH > CH_3 > H$

The lowest ranked atom (H) is directed away from you in the drawing. We see that the order of decreasing precedence appears in a counterclockwise manner.

$$H_3C \quad H$$
$$C{-}CH{=}CH_2 \qquad H_3C \quad CH{=}CH_2$$
$$HO \qquad\qquad\qquad OH$$

(+)-3-Buten-2-ol

The absolute configuration is S, and the compound is (S)-$(+)$-3-buten-2-ol.

7.8 (*b*) The chirality center is the carbon that bears the methyl group. Its substituents are:

$$-CF_2CH_2 > -CH_2CF_2 > CH_3 > H$$

Highest   Lowest
rank   rank

When the lowest ranked substituent points away from you, the remaining three must appear in descending order of precedence in a counterclockwise fashion in the S enantiomer. (S)-1,1-difluoro-2-methylcyclopropane is therefore

$$H \quad CH_3$$
$$F$$
$$F$$

7.9 Begin by converting the Fischer projection to a perspective (three-dimensional) representation, remembering that the vertical bonds at the chirality center are directed away from you and the horizontal bonds are directed toward you.

$$CH_2OH \qquad\qquad CH_2OH$$
$$H{-}{-}OH \quad \text{is the} \quad H{-}{-}OH$$
$$CH_2CH_3 \quad \text{same as} \quad CH_2CH_3$$

Now rotate the three-dimensional representation so that the lowest ranked atom (H) points away from you. You can then see that the order of decreasing precedence $OH > CH_2OH > CH_2CH_3 > H$ traces a clockwise path. The absolute configuration is R.

7.10 (b) As in part (a), exchanging any two groups in a Fischer projection converts it to the mirror image. Thus exchanging the positions of CH_3 and CH_2CH_3 in the Fischer projection of (R)-2-butanol converts the configuration of the chirality center to S.

(R)-2-Butanol   (S)-2-Butanol

(c) Switching the positions of three groups gives a Fischer projection that looks different, but leaves the configuration of the chirality center unchanged.

(R)-2-Butanol   (R)-2-Butanol

(d, e) Switching H with OH and CH_3 and CH_2CH_3 has the same effect as rotating the Fischer projection 180°. The resulting projection has the same configuration at the chirality center as the original.

(R)-2-Butanol   (R)-2-Butanol

7.11 Begin by using the molecular model of thalidomide in the text to draw a perspective view, paying particular attention to the configuration of the chirality center.

Thalidomide

The four groups attached to the chirality center, in order of decreasing precedence, are:

$$N- \;>\; C=O \;>\; CH_2 \;>\; H-$$

These groups trace a clockwise path with the lowest ranked group (H) pointing away from us. The molecular model of thalidomide is the R enantiomer.

7.12 The top and bottom faces of the double bond in (E)-3-methyl-2-hexene are enantiotopic. Addition of hydrogen to either the top or the bottom face occurs at the same rate, so a racemic (1:1) mixture of the two enantiomers of 3-methylhexane is formed.

(E)-3-Methyl-2-hexene
(achiral)

3-Methylhexane
(chiral, but racemic)

7.13 The two carboxyl groups at the ends of the five-carbon chain in citric acid are enantiotopic.

These carboxyl groups
are enantiotopic

Replacing one of them will give an enantiomer of the structure obtained by replacing the other.

is the enantiomer of

7.14 In order of decreasing precedence, the substituents attached to the chirality center in lactic acid are —OH, —CO$_2$H, —CH$_3$, and —H. The Fischer projection given for (+)-lactic acid (a) corresponds to the three-dimensional representation (b), which can be reoriented as in (c). When (c) is viewed from the side opposite the lowest ranked atom (H), the order of decreasing precedence is counterclockwise, as shown in (d). (+)-Lactic acid has the S configuration.

(a)   (b)   (c)   (d)

7.15 The erythro stereoisomers are characterized by Fischer projections in which analogous substituents, in this case OH and NH_2, are on the same side when the carbon chain is vertical. There are two erythro stereoisomers that are enantiomers of each other:

$$
\begin{array}{cc}
CH_3 & CH_3 \\
H-\!\!-OH & HO-\!\!-H \\
H-\!\!-NH_2 & H_2N-\!\!-H \\
CH_3 & CH_3 \\
\text{Erythro} & \text{Erythro}
\end{array}
$$

Analogous substituents are on opposite sides in the threo isomer:

$$
\begin{array}{cc}
CH_3 & CH_3 \\
H-\!\!-OH & HO-\!\!-H \\
H_2N-\!\!-H & H-\!\!-NH_2 \\
CH_3 & CH_3 \\
\text{Threo} & \text{Threo}
\end{array}
$$

7.16 3-Amino-2-butanol has four stereoisomeric forms

(2R,3R) and its enantiomer (2S,3S)
(2R,3S) and its enantiomer (2S,3R)

In the text we are told that the (2R,3R) stereoisomer is a liquid. Its enantiomer (2S,3S) has the same physical properties and so must also be a liquid. The text notes that the (2R,3S) stereoisomer is a solid (mp 49°C). Its enantiomer (2S,3R) must therefore be the other stereoisomer that is a crystalline solid.

7.17 Examine the structural formula of each compound for equivalently substituted chirality centers. The only one capable of existing in a meso form is 2,4-dibromopentane.

Equivalently substituted
chirality centers

$$
CH_3\overset{\downarrow}{C}HCH_2\overset{\downarrow}{C}HCH_3
$$
$$
|\qquad\quad |
$$
$$
Br\qquad Br
$$

$$
\begin{array}{c}
CH_3 \\
H-\!\!-Br \\
H-\!\!-H \quad \text{- - -} \leftarrow \text{Plane of symmetry}\\
H-\!\!-Br \\
CH_3
\end{array}
$$

2,4-Dibromopentane

Fischer projection of
meso-2,4-dibromopentane

None of the other compounds has equivalently substituted chirality centers. No meso forms are possible for:

$$
\underset{\underset{Br\ Br}{|\ \ |}}{CH_3CHCHCH_2CH_3}
\qquad
\overset{\overset{OH}{|}}{\underset{\underset{Br}{|}}{CH_3CHCHCH_2CH_3}}
\qquad
\underset{\underset{OH\quad Br}{|\qquad |}}{CH_3CHCH_2CHCH_3}
$$

2,3-Dibromopentane   3-Bromo-2-pentanol   4-Bromo-2-pentanol

7.18 The cis stereoisomer of 1,3-dimethylcyclohexane has a plane of symmetry, and so it is an achiral substance—it is a meso form.

Plane of symmetry passes through
C-2 and C-5 and bisects the ring.

The trans stereoisomer is chiral. It is not a meso form.

7.19 A molecule with three chirality centers has eight (2^3) stereoisomers. The eight combinations of R and S chirality centers are:

	Chirality center		Chirality center
	1 2 3		**1 2 3**
Isomer 1	$R\,R\,R$	Isomer 5	$S\,S\,S$
Isomer 2	$R\,R\,S$	Isomer 6	$S\,S\,R$
Isomer 3	$R\,S\,R$	Isomer 7	$S\,R\,S$
Isomer 4	$S\,R\,R$	Isomer 8	$R\,S\,S$

7.20 2-Ketohexoses have three chirality centers. They are marked with asterisks in the structural formula.

$$\underset{\underset{\text{OH}}{|}\ \underset{\text{OH}}{|}}{\overset{\overset{\text{O}}{||}\ \overset{\text{OH}}{|}}{\text{HOCH}_2\text{CCHCHCHCH}_2\text{OH}}}$$

No meso forms are possible, and so there are a total of eight (2^3) stereoisomeric 2-ketohexoses.

7.21 Epoxidation of (Z)-2-butene gives the meso (achiral) epoxide. Oxygen transfer from the peroxy acid can occur at either face of the double bond, but the product formed is the same because the two mirror-image forms of the epoxide are superimposable.

meso-2,3-Epoxybutane   *meso*-2,3-Epoxybutane

Epoxidation of (E)-2-butene gives a racemic mixture of two enantiomeric epoxides.

(2R,3R)-2,3-Epoxybutane   (2S,3S)-2,3-Epoxybutane

7.22 The observed product mixture (68% *cis*-1,2-dimethylcyclohexane: 32% *trans*-1,2-dimethylcyclohexane) contains more of the less stable cis stereoisomer than the trans. The relative stabilities of the products therefore play no role in determining the stereoselectivity of this reaction.

7.23 The tartaric acids incorporate two equivalently substituted chirality centers. (+)-Tartaric acid, as noted in the text, is the 2R,3R stereoisomer. There will be two additional stereoisomers, the enantiomeric (−)-tartaric acid (2S,3S) and an optically inactive meso form.

(2S,3S)-Tartaric acid   *meso*-Tartaric acid
(optically active)   (optically inactive)
(mp 170°C, [α]$_D$ − 12°)   (mp 140°C)

7.24 No. Pasteur separated an optically inactive racemic mixture into two optically active enantiomers. A meso form is achiral, is identical to its mirror image, and is incapable of being separated into optically active forms.

7.25 The more soluble salt must have the opposite configuration at the chirality center of 1-phenylethyl-amine, that is, the S configuration. The malic acid used in the resolution is a single enantiomer, S. In this particular case the more soluble salt is therefore (S)-1-phenylethylammonium (S)-malate.

7.26 In an earlier exercise (Problem 4.25) the structures of all the isomeric $C_5H_{12}O$ alcohols were presented. Those that lack a chirality center and thus are *achiral* are

$CH_3CH_2CH_2CH_2CH_2OH$ $CH_3CHCH_2CH_2OH$ $(CH_3)_3CCH_2OH$
                                                                 |
                                                               CH_3

1-Pentanol   3-Methyl-1-butanol   2,2-Dimethyl-1-propanol

                                                                          CH_3
                                                                            |
            $CH_3CH_2CHCH_2CH_3$ CH_3CH_2COH
                            | |
                           OH CH_3

                    3-Pentanol   2-Methyl-2-butanol

The chiral isomers are characterized by carbons that bear four different groups. These are:

CH₃CHCH₂CH₂CH₃ $\qquad$ CH₃CHCH(CH₃)₂ $\qquad$ CH₃CH₂CHCH₂OH
$\quad$|$\qquad\qquad\qquad\qquad$|$\qquad\qquad\qquad\qquad\quad$|
$\quad$OH$\qquad\qquad\qquad\qquad\;OH\qquad\qquad\qquad\qquad\;$CH₃

$\qquad\;$2-Pentanol$\qquad\qquad$3-Methyl-2-butanol$\qquad\qquad$2-Methyl-1-butanol

7.27 The isomers of trichlorocyclopropane are

$\quad$Enantiomeric forms of 1,1,2-trichlorocyclopropane $\qquad$ *cis*-1,2,3-Trichlorocyclopropane $\qquad$ *trans*-1,2,3-Trichlorocyclopropane
$\qquad\qquad\qquad$(both chiral) $\qquad\qquad\qquad\qquad$ (achiral—contains a plane of symmetry) (achiral—contains a plane of symmetry)

7.28 (*a*) Carbon-2 is a chirality center in 3-chloro-1,2-propanediol. Carbon-2 has two equivalent substituents in 2-chloro-1,3-propanediol, and is not a chirality center.

$\qquad\qquad\qquad$ClCH₂CHCH₂OH $\qquad\qquad$ HOCH₂CHCH₂OH
$\qquad\qquad\qquad\qquad\quad$|$\qquad\qquad\qquad\qquad\qquad$|
$\qquad\qquad\qquad\qquad\;OH\qquad\qquad\qquad\qquad\qquad$Cl

$\qquad\qquad\qquad\;$3-Chloro-1,2-propanediol $\qquad\;$ 2-Chloro-1,3-propanediol
$\qquad\qquad\qquad\qquad\qquad$Chiral $\qquad\qquad\qquad\qquad\qquad\;$Achiral

(*b*) The primary bromide is achiral; the secondary bromide contains a chirality center and is chiral.

$\qquad\qquad\qquad$CH₃CH=CHCH₂Br $\qquad\qquad$ CH₃CHCH=CH₂
$\qquad\qquad\qquad\qquad\qquad\qquad\qquad\qquad\qquad\qquad\quad$|
$\qquad\qquad\qquad\qquad\qquad\qquad\qquad\qquad\qquad\qquad\;$Br

$\qquad\qquad\qquad\qquad\;$Achiral $\qquad\qquad\qquad\qquad\qquad$Chiral

(*c*) Both stereoisomers have two equivalently substituted chirality centers, and so we must be alert for the possibility of a meso stereoisomer. The structure at the left is chiral. The one at the right has a plane of symmetry and is the achiral meso stereoisomer.

$\qquad\qquad\qquad\qquad$CH₃ $\qquad\qquad\qquad\qquad$ CH₃
$\qquad\qquad\qquad\qquad\;$| $\qquad\qquad\qquad\qquad\quad$| $\qquad$ Plane of symmetry
$\qquad\qquad$H₂N——H $\qquad\qquad$ H——NH₂
$\qquad\qquad\qquad\qquad\;$| $\qquad\qquad\qquad\qquad\quad$|
$\qquad\qquad$H——NH₂ $\qquad\qquad$ H——NH₂
$\qquad\qquad\qquad\qquad\;$| $\qquad\qquad\qquad\qquad\quad$|
$\qquad\qquad\qquad\qquad$CH₃ $\qquad\qquad\qquad\qquad$ CH₃

$\qquad\qquad\qquad\qquad$Chiral $\qquad\qquad\qquad$ Meso: achiral

(*d*) The first structure is achiral; it has a plane of symmetry.

Plane of symmetry passes through C-1, C-4, and C-7.

The second structure cannot be superimposed on its mirror image; it is chiral.

Reference structure   Mirror image   Reoriented mirror image

7.29 There are four stereoisomers of 2,3-pentanediol, represented by the Fischer projections shown. All are chiral.

Enantiomeric erythro isomers   Enantiomeric threo isomers

2,4-Pentanediol has three stereoisomers. The meso form is achiral; both threo forms are chiral.

meso-2,4-Pentanediol   Enantiomeric threo isomers

7.30 Among the atoms attached to the chiralty center, the order of decreasing precedence is Br > Cl > F > H. When the molecule is viewed with the hydrogen pointing away from us, the order Br → Cl → F appears clockwise in the R enantiomer, counterclockwise in the S enantiomer.

R-(−)   *S*-(+)   *R*-(−)   *S*-(+)

7.31 (*a*) The order of substituent precedence at the chirality center is

$$HO > CH_2CH_2 > CH_3 > H$$

The molecule is oriented so that the lowest ranking substituent is directed away from you and the order of decreasing precedence is clockwise. (−)-2-Octanol has the R configuration.

(b) In order of decreasing sequence rule precedence, the four substituents at the chirality center of monosodium L-glutamate are

$$\overset{+}{N}H_3 > CO_2^- > CH_2 > H$$

is the same as

When the molecule is oriented so that the lowest ranking atom (hydrogen) is directed away from you, the other three substituents are arranged as shown.

The order of decreasing rank is counterclockwise; the absolute configuration is *S*.

7.32 (a) Among the isotopes of hydrogen, T has the highest mass number (3), D next (2), and H lowest (1). Thus, the order of rank at the chirality center in the reactant is $CH_3 > T > D > H$. The order of rank in the product is $HO > CH_3 > T > D$.

Orient with lowest ranked substituent away from you.

The order of decreasing precedence in the reactant is counterclockwise; the configuration is *S*. The order of decreasing precedence in the product is clockwise; the configuration is *R*.

(b) Retention of configuration means that the three-dimensional arrangement of bonds at the chirality center is the same in the reactant and the product. The *R* and *S* descriptors change because the order of precedence changes in going from reactant to product; for example, CH_3 is the highest ranked substituent in the reactant, but becomes the second-highest ranked in the product.

7.33 Two compounds can be stereoisomers only if they have the *same* constitution. Thus, you should compare first the constitution of the two structures and then their stereochemistry. One way to compare constitutions is to assign a systematic (IUPAC) name to each molecule. Also remember that enantiomers are nonsuperimposable mirror images, and diastereomers are stereoisomers that are not enantiomers.

(a) The two compounds are constitutional isomers. Their IUPAC names clearly reflect this difference.

(b) The two structures have the same constitution. Test them for superimposability. To do this we need to place them in comparable orientations.

and

The two are nonsuperimposable mirror images of each other. They are enantiomers.

To check this conclusion, work out the absolute configuration of each using the Cahn–Ingold–Prelog system.

(*S*)-2-Bromobutane   (*R*)-2-Bromobutane

(c) Again, place the structures in comparable orientations, and examine them for superimposability.

and

The two structures represent the same compound, because they are superimposable. (As a check, notice that both have the *S* configuration.)

(d) If we reorient the first structure,

which is the enantiomer of

As a check, the first structure is seen to have the S configuration, and the second has the R configuration.

(*e*) As drawn, the two structures are mirror images of each other; however, they represent an achiral molecule. The two structures are superimposable mirror images and are not stereoisomers but identical.

$$\begin{array}{ccc} & \text{CH}_2\text{OH} & & \text{CH}_2\text{OH} \\ \text{H}\!\!-\!\!\!&\!\!\!-\!\!\text{OH} & \text{and} & \text{HO}\!\!-\!\!\!-\!\!\!&\!\!\!-\!\!\text{H} \\ & \text{CH}_2\text{OH} & & \text{CH}_2\text{OH} \end{array} \quad \text{are identical}$$

(*f*) The two structures—one cis, the other trans—are stereoisomers that are not mirror images; they are diastereomers.

trans-1-Chloro-2-methylcyclopropane   *cis*-1-Chloro-2-methylcyclopropane

(*g*) The two structures are enantiomers because they are nonsuperimposable mirror images. Checking their absolute configurations reveals one to be R, the other S. Both have the E configuration at the double bond.

HO H   H OH

(2*R*,3*E*)-3-Penten-2-ol   (2*S*,3*E*)-3-Penten-2-ol

(*h*) These two structures are identical; both have the E configuration at the double bond and the R configuration at the chirality center.

Alternatively, we can show their superimposability by rotating the second structure 180° about an axis passing through the doubly bonded carbons.

HO H   H OH   HO H

Reference structure   Rotate 180° around this axis   Identical to reference structure

(*i*) One structure has a cis double bond, the other a trans double bond; therefore, the two are diastereomers. Even though one chirality center is R and the other is S, the two structures are *not* enantiomers. The mirror image of a cis (or Z) double bond is cis, and that of a trans (or E) double bond is trans.

(2*R*,3*E*)-3-Penten-2-ol   (2*S*,3*Z*)-3-Penten-2-ol

(*j*) Here it will be helpful to reorient the second structure so that it may be more readily compared with the first.

| Reference structure | | | Enantiomer of reference structure |

The two compounds are enantiomers, and examining their absolute configurations will confirm this.

(*R*)-3-Hydroxymethyl-2-cyclopenten-1-ol   (*S*)-3-Hydroxymethyl-2-cyclopenten-1-ol

(*k*) These two compounds differ in the order in which their atoms are joined together; they are constitutional isomers.

3-Hydroxymethyl-2-cyclopenten-1-ol   3-Hydroxymethyl-3-cyclopenten-1-ol

(*l*) To better compare these two structures, place them both in the same format.

which is equivalent to

The two are enantiomers.

(*m*) Because *cis*-1,3-dimethylcyclopentane has a plane of symmetry, it is achiral and cannot have an enantiomer. The two structures given in the problem are identical.

Plane of symmetry

(*n*) These structures are diastereomers, that is, stereoisomers that are not mirror images. They have the same configuration at C-3 but opposite configurations at C-2.

2R,3R   2S,3R

(*o*) To compare these compounds, reorient the first structure so that it may be drawn as a Fischer projection. The first step in the reorientation consists of a 180° rotation about an axis passing through the midpoint of the C-2—C-3 bond.

becomes

Thus

is

Reference structure

Now rotate the "back" carbon of the reoriented structure to give the necessary alignment for a Fischer projection.

becomes   which is the same as

This reveals that the original two structures in the problem are equivalent.

(*p*) These two structures are nonsuperimposable mirror images of a molecule with two nonequivalent chirality centers; they are enantiomers.

2R,3R   2S,3S

(*q*) The two structures are stereoisomers that are not mirror images; they are diastereomers.

cis-3-Methylcyclohexanol

trans-3-Methylcyclohexanol

(*r*) These two structures, *cis-* and *trans*-4-*tert*-butylcyclohexyl iodide, are diastereomers.

Trans

Cis

(*s*) The two structures are nonsuperimposable mirror images; they are enantiomers.

Reference structure

is equivalent to

Enantiomer of reference structure

(*t*) The two structures are identical.

Reference structure

is equivalent to

Identical to reference structure

(*u*) As represented, the two structures are mirror images of each other, but because the molecule is achiral (it has a plane of symmetry), the two must be superimposable. They represent the same compound.

Reference structure

is equivalent to

Identical to reference structure

The plane of symmetry passes through C-7 and bisects the C-2—C-3 bond and the C-5—C-6 bond.

(v) The structures are stereoisomers but not enantiomers; they are diastereomers. (Both are achiral and so cannot have enantiomers.)

and   are stereoisomers but not mirror images

Achiral   Achiral

7.34 Examine the structural formula of phytol and count the number of structural units capable of stereochemical variation.

$$HOCH_2CH{=}CCH_2CH_2CH_2CHCH_2CH_2CH_2CHCH_2CH_2CH_2CHCH_3$$

with CH₃ groups:

Phytol

Phytol has two chirality centers (C-7 and C-11) and one double bond. The chirality centers may be either R or S, and the double bond may be either E or Z. Eight stereoisomers are possible.

	Isomer							
	1	**2**	**3**	**4**	**5**	**6**	**7**	**8**
Double bond	E	E	E	E	Z	Z	Z	Z
Carbon-7	R	S	R	S	R	S	R	S
Carbon-11	R	S	S	R	R	S	S	R

7.35 (*a*) Muscarine has three chirality centers, and so eight (2^3) stereoisomers have this constitution.

(*b*) The three substituents on the ring (at C-2, C-3, and C-5) can be thought of as being either up (U) or down (D) in a perspective drawing. Thus the eight possibilities are:

UUU, UUD, UDU, DUU, UDD, DUD, DDU, DDD

Of these, six have one substituent trans to the other two.

(*c*) Muscarine is

7.36 To write a stereochemically accurate representation of ectocarpene, it is best to begin with the configuration of the chirality center, which we are told is S.

Clearly, hydrogen is the lowest ranked substituent; among the other three substituents, two are part of the ring and the third is the four-carbon side chain. The sequence rule precedence of these groups is determined by systematically working along the chain.

The substituents

$$—CH{=}CHCH_2CH{=}CHCH_2 \qquad —CH{=}CHCH_2CH_3 \qquad —CH_2CH{=}CHCH_2CH{=}CH$$

(Ring)   (Side Chain)   (Ring)

Are considered as if they were

(Ring)   (Side Chain)   (Ring)

Orienting the molecule with the hydrogen away from you

we place the double bonds in the ring so that the order of decreasing sequence rule precedence is counterclockwise:

Finally, because all the double bonds are cis, the complete structure becomes:

7.37 (a) Multifidene has two chirality centers and three double bonds. Neither the ring double bond nor the double bond of the vinyl substituent can give rise to stereoisomers, but the butenyl side chain can be either E or Z. Eight (2^3) stereoisomers are therefore possible. We can rationalize them as

Stereoisomer	C-3	C-4	Butenyl double bond	
1	R	R	E	enantiomers
2	S	S	E	
3	R	R	Z	enantiomers
4	S	S	Z	
5	R	S	E	enantiomers
6	S	R	E	
7	R	S	Z	enantiomers
8	S	R	Z	

(b) Given the information that the alkenyl substituents are cis to each other, the number of stereoisomers is reduced by half. Four stereoisomers are therefore possible.

(c) Knowing that the butenyl group has a Z double bond reduces the number of possibilities by half. Two stereoisomers are possible.

(d) The two stereoisomers are

and

(e) These two stereoisomers are enantiomers. They are nonsuperimposable mirror images.

Mirror plane

7.38 In a substance with more than one chirality center, each center is independently specified as R or S. Streptimidone has two chirality centers and two double bonds. Only the internal double bond is capable of stereoisomerism.

The three stereochemical variables give rise to eight (2^3) stereoisomers, of which one is streptimidone and a second is the enantiomer of streptimidone. The remaining six stereoisomers are diastereomers of streptimidone.

7.39 (*a*) The first step is to write the constitution of menthol, which we are told is 2-isopropyl-5-methylcyclohexanol.

2-Isopropyl-5-methylcyclohexanol

Because the configuration at C-1 is *R* in (−)-menthol, the hydroxyl group must be "up" in our drawing.

Because menthol is the most stable stereoisomer of this constitution, all three of its substituents must be equatorial. We therefore draw the chair form of the preceding structure, which has the hydroxyl group equatorial and up, placing isopropyl and methyl groups so as to preserve the *R* configuration at C-1.

(−)-Menthol

(*b*) To transform the structure of (−)-menthol to that of (+)-isomenthol, the configuration at C-5 must remain the same, whereas those at C-1 and C-2 are inverted.

(−)-Menthol   (+)-Isomenthol

(+)-Isomenthol is represented here in its correct configuration, but the conformation with two axial substituents is not the most stable one. The ring-flipped form is the preferred conformation of (+)-isomenthol:

Most stable conformation of
(+)-isomenthol

7.40 Because the only information available about the compound is its optical activity, examine the two structures for chirality, recalling that only chiral substances can be optically active.

The structure with the six-membered ring has a plane of symmetry passing through C-1 and C-4. It is achiral and cannot be optically active.

Achiral; $[\alpha]_D$ 0°   Chiral; can be optically active

The open-chain structure has neither a plane of symmetry nor a center of symmetry; it is not superimposable on its mirror image and so is chiral. It can be optically active and is more likely to be the correct choice.

7.41 Compound B has a center of symmetry, is achiral, and thus cannot be optically active.

Compound B: not optically active
(center of symmetry is midpoint of C-16—C-17 bond)

The diol in the problem is optically active, and so it must be chiral. Compound A is the naturally occurring diol.

7.42 (*a*) The equation that relates specific rotation $[\alpha]_D$ to observed rotation α is

$$[\alpha]_D = \frac{100\alpha}{cl}$$

where c is concentration in grams per 100 mL and l is path length in decimeters.

$$[\alpha]_D = \frac{100(-5.20°)}{(2.0 \text{ g} / 100 \text{ mL})(2 \text{ dm})}$$
$$= -130°$$

(*b*) The optical purity of the resulting solution is 10/15, or 66.7%, because 10 g of optically pure fructose has been mixed with 5 g of racemic fructose. The specific rotation will therefore be two thirds (10/15) of the specific rotation of optically pure fructose:

$$[\alpha]_D = \frac{2}{3}(-130°) = -87°$$

7.43 (*a*) The four methylene hydrogens of 1,2-dibromoethane are enantiotopic. Replacement of any one of them by some atom or group produces two enantiomers. Light-initated free-radical chlorination will replace one of the hydrogens with a chlorine. The enantiomers are chiral and are formed in equal amounts as a racemic mixture.

$$BrCH_2CH_2Br \ + \ Cl_2 \ \xrightarrow{\text{light}} \ BrCH_2\overset{\overset{\displaystyle Cl}{|}}{C}HBr \ + \ HCl$$

The product is a racemic mixture of

(*b*) Tertiary alcohols react with hydrogen chloride to produce alkyl chlorides by way of a carbocation intermediate. In this case, chloride ion can approach the carbocation from either an axial direction or an equatorial one. The two approaches are not equivalent, nor are the alkyl chloride products. Two diastereomers, both of which are achiral, are formed in unequal amounts. (On the basis of the information given, we cannot predict which one is the major product.)

(*c*) The starting material is a chiral tertiary alcohol, shown as a single enantiomer. As in part (*b*), the reaction proceeds through a tertiary carbocation and gives a mixture of two alkyl chlorides.

The two products are diastereomers, and both are chiral. Because the axial and equatorial approach routes of chloride to the positively charged carbon are not equivalent, it is likely that they are formed in unequal amounts. [As in part (*b*) we cannot predict which one predominates, however.]

7.44 (*a*) The reaction of 1-butene with hydrogen iodide is one of electrophilic addition. It follows Markovnikov's rule and yields a racemic mixture of (*R*)- and (*S*)-2-iodobutane.

(b) Bromine adds anti to carbon–carbon double bonds to give vicinal dibromides.

H₃C, H / H, CH₂CH₃ →(Br₂ / CCl₄)→

Br / H—CH₂CH₃ / H₃C—H / Br   +   CH₂CH₃ / H—Br / Br—H / CH₃

(E)-2-Pentene   (2R,3S)-2,3-Dibromopentane   (2S,3R)-2,3-Dibromopentane

The two stereoisomers are enantiomers and are formed in equal amounts.

(c) Two enantiomers are formed in equal amounts in this reaction, involving electrophilic addition of bromine to (Z)-2-pentene. These two are diastereomers of those formed in part (b).

H₃C, CH₂CH₃ / H, H →(Br₂ / CCl₄)→

Br / CH₃CH₂—H / H₃C—H / Br   +   CH₃CH₂—H / Br/Br / H₃C—H

(Z)-2-Pentene   (2R,3R)-2,3-Dibromopentane   (2S,3S)-2,3-Dibromopentane

(d) Epoxidation of 1-butene yields a racemic epoxide mixture.

$CH_3CH_2CH=CH_2$ →(CH₃COOH (O) / CH₂Cl₂)→

CH₃CH₂ (epoxide with H, O)   +   H (epoxide with O) / CH₃CH₂

1-Butene   (S)-1,2-Epoxybutane   (R)-1,2-Epoxybutane

(e) Two enantiomeric epoxides are formed in equal amounts on epoxidation of (Z)-2-pentene.

H, H / H₃C, CH₂CH₃ →(CH₃COOH (O) / CH₂Cl₂)→

H, H / H₃C—epoxide—CH₂CH₃   +   O epoxide / H···H / H₃C, CH₂CH₃

(Z)-2-Pentene   (2S,3R)-2,3-Epoxypentane   (2R,3S)-2,3-Epoxypentane

The reaction is a stereospecific syn addition. The cis alkyl groups in the starting alkene remain cis in the product epoxide.

(f) The starting material is achiral, so even though a chiral product is formed, it is a racemic mixture of enantiomers and is optically inactive.

(cyclopentene with CH₃, CH₃, CH₃) →(H₂ / Pt)→

(cyclopentane with CH₃, CH₃, H₃C, H)   +   (cyclopentane with CH₃, CH₃, H, CH₃)

1,5,5-Trimethylcyclopentene   (R)-1,1,2-Trimethylcyclopentane   (S)-1,1,2-Trimethylcyclopentane

(g) Recall that hydroboration–oxidation leads to hydration of the double bond opposite to Markovnikov's rule.

1,5,5-Trimethylcyclopentene   (1S,2S)-2,3,3-Trimethylcyclopentanol   (1R,2R)-2,3,3-Trimethylcyclopentanol

The product has two chirality centers. It is formed as a racemic mixture of enantiomers.

7.45 Hydration of the double bond of aconitic acid (shown in the center) can occur in two regiochemically distinct ways:

Chiral
(isocitric acid)

Aconitic acid

Achiral
(citric acid)

One of the hydration products lacks a chirality center. It must be citric acid, the achiral, optically inactive isomer. The other one has two different chirality centers and must be isocitric acid, the optically active isomer.

7.46 (a) Structures A and B are chiral. Structure C has a plane of symmetry and is an achiral meso form.

(b) Ozonolysis of the starting material proceeds with the stereochemistry shown. Compound B is the product of the reaction.

Compound B

(c) If the methyl groups were cis to each other in the cycloalkene, they would be on the same side of the Fischer projection in the product. Compound C would be formed.

7.47 (a) The E2 transition state requires that the bromine and the hydrogen that is lost be anti and coplanar to each other. Examination of the compound given in the problem reveals that loss of bromine and the deuterium will yield *trans*-2-butene, whereas loss of the bromine and the hydrogen on C-3 will yield *cis*-2-butene.

The *trans*-2-butene that forms does not contain deuterium, but *cis*-2-butene does. 1-Butene also contains deuterium.

$$\text{Br} \atop CH_3\overset{|}{C}HCHCH_3 \underset{D}{\overset{|}{}} \xrightarrow{\text{E2}} H_2C=CHCHCH_3 \atop \underset{D}{\overset{|}{}}$$

1-Butene

(b) The starting material in part (a) is the erythro isomer. The relative positions of the H and D at C-3 are reversed in the threo isomer. The erythro and threo isomers can be drawn using Fischer projections:

Erythro:
```
        CH3
   Br ---+--- H
    D ---+--- H
        CH3
```
Erythro

Threo:
```
        CH3
   Br ---+--- H
    H ---+--- D
        CH3
```
Threo

Because the positions of the H and D on C-3 in the threo isomer are opposite that in the erythro, the deuterium content of *cis*- and *trans*-2-butene would be reversed. *trans*-2-Butene obtained from the threo isomer would contain deuterium, and *cis*-2-butene would not. 1-Butene obtained from the threo isomer would also contain deuterium.

7.48 Bromine adds to the unknown compound, suggesting the presence of a double bond in addition to the five-membered ring. The following are possible structures for the unknown:

Methylenecyclopentane   1-Methylcyclopentene   3-Methylcyclopentene   4-Methylcyclopentene

Which of these form diastereomeric dibromides on anti addition of bromine?

Methylenecyclopentane

1-Methylcyclopentene

3-Methylcyclopentene

4-Methylcyclopentene

We are told in the problem that two diastereomeric bromides were formed, thus the compound must be 3-methylcyclopentene.

7.49 Dehydration of this tertiary alcohol can yield 2,3-dimethyl-1-pentene or 2,3-dimethyl-2-pentene. Only the terminal alkene in this case is chiral.

CH₃CH₂CHC(CH₃)₂
| |
H₃C   OH

2,3-Dimethyl-2-pentanol
(chiral, optically pure)

2,3-Dimethyl-1-pentene
(chiral, optically pure)

2,3-Dimethyl-2-pentene
(achiral, optically inactive)

H₂, Pt

H₂, Pt

2,3-Dimethylpentane
(chiral, optically pure)

2,3-Dimethylpentane
(chiral, racemic)

The 2,3-dimethyl-1-pentene formed in the dehydration reaction must be optically pure because it arises from optically pure alcohol by a reaction that does not involve any of the bonds to the chirality center. When optically pure 2,3-dimethyl-1-pentene is hydrogenated, it must yield optically pure 2,3-dimethylpentane—again, no bonds to the chirality center are involved in this step.

The 2,3-dimethyl-2-pentene formed in the dehydration reaction is achiral and must yield racemic 2,3-dimethylpentane on hydrogenation.

Because the alkane is 50% optically pure, the alkene fraction must have contained equal amounts of optically pure 2,3-dimethyl-1-pentene and its achiral isomer 2,3-dimethyl-2-pentene.

7.50 (*a*) Oxygen may be transferred to either the front face or the back face of the double bond when (*R*)-3-buten-2-ol reacts with a peroxy acid. The structure of the minor stereoisomer was given in the problem. The major stereoisomer results from addition to the opposite face of the double bond.

(*R*)-3-Buten-2-ol   Minor stereoisomer   Major stereoisomer

(*b*) The two epoxides have the same configuration (*R*) at the secondary alcohol carbon, but opposite configurations at the chirality center of the epoxide ring. They are diastereomers.

(*c*) In addition to the two diastereomeric epoxides whose structures are shown in the solution to part (*a*), the enantiomers of each will be formed when racemic 3-buten-2-ol is epoxidized. The relative amounts of the four products will be:

Enantiomeric forms of minor stereoisomer, totaling 40%

20%   20%

Enantiomeric forms of major stereoisomer, totaling 60%

30%   30%

ANSWERS TO INTERPRETIVE PROBLEMS

7.51 C; **7.52** B; **7.53** B; **7.54** B; **7.55** B; **7.56** B; **7.57** D

SELF-TEST

PART A

A-1. For each of the following pairs of drawings, identify the molecules as chiral or achiral and tell whether each pair represents molecules that are enantiomers, diastereomers, or identical.

(a)   and

(1)   (2)

(b)   and

(3)   (4)

(c)   and

(5)   (6)

(d)   and

(7)   (8)

(e)   and

(9)   (10)

A-2. Specify the configuration of each chirality center in the preceding problem, using the Cahn–Ingold–Prelog *R–S* system.

A-3. Predict the number of stereoisomers possible for each of the following constitutions. For which of these will meso forms be possible?

(a)

(c) H_2C=CHCHCH=CHCH$_3$
       |
       OH

(b) CH$_3$CHCHCHCH$_3$ with Br Br Cl above

A-4. Using the skeletons provided as a guide,

(a) Draw a perspective view of (2R,3R)-3-chloro-2-butanol.

(b) Draw a sawhorse diagram of (R)-2-bromobutane.

(c) Draw Fischer projections of both these compounds.

A-5. Draw Fischer projections of each stereoisomer of 2,3-dichlorobutane. Identify each chirality center as R or S. Which stereoisomers are chiral? Which are not? Why?

A-6. (a) The specific rotation of pure (−)-cholesterol is −39°. What is the specific rotation of a sample of cholesterol containing 10% (+)-cholesterol and 90% (−)-cholesterol.

(b) If the rotation of optically pure (R)-2-octanol is −10°, what is the percentage of the S enantiomer in a sample of 2-octanol that has a rotation of −4°?

A-7. Write the organic product(s) expected from each of the following reactions. Show each stereoisomer if more than one forms.

(a) 1,5,5-Trimethylcyclopentene and hydrogen bromide
(b) (E)-2-Butene and chlorine (Cl$_2$)
(c) (Z)-2-Pentene and peroxyacetic acid

A-8. Give the IUPAC name, including stereochemistry, for the following:

(a)

(b)

A-9. How many stereoisomeric products are obtained from the reaction of (S)-3-chloro-1-butene with hydrogen bromide? What is their relationship (enantiomers, diastereomers)?

A-10. Write the final product of the following reaction sequence, clearly showing its stereochemistry. Is the product achiral, a meso compound, optically active, or a racemic mixture?

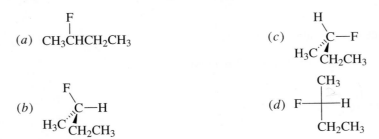

PART B

B-1. The structure of (S)-2-fluorobutane is best represented by

(a)
$$\begin{array}{c} F \\ | \\ CH_3CHCH_2CH_3 \end{array}$$

(c)
$$\begin{array}{c} H \\ | \\ C\!-\!F \\ H_3C^{\prime\prime\prime} \;\; CH_2CH_3 \end{array}$$

(b)
$$\begin{array}{c} F \\ \backslash \\ C\!-\!H \\ H_3C^{\prime\prime\prime} \;\; CH_2CH_3 \end{array}$$

(d)
$$\begin{array}{c} CH_3 \\ | \\ F\!-\!\!\!\!-\!\!\!\!-H \\ | \\ CH_2CH_3 \end{array}$$

B-2. Which one of the following is chiral?

(a) 1,1-Dibromo-1-chloropropane

(b) 1,1-Dibromo-3-chloropropane

(c) 1,3-Dibromo-1-chloropropane

(d) 1,3-Dibromo-2-chloropropane

B-3. Which of the following compounds are meso forms?

1   2   3

(a) 1 only   (b) 3 only   (c) 1 and 2   (d) 2 and 3

B-4. The 2,3-dichloropentane whose structure is shown is

$$\begin{array}{c} CH_3 \\ H\!-\!\!\!\!-\!\!\!\!-Cl \\ Cl\!-\!\!\!\!-\!\!\!\!-H \\ CH_2CH_3 \end{array}$$

(a) 2R,3R   (b) 2R,3S   (c) 2S,3R   (d) 2S,3S

B-5. The separation of a racemic mixture into the pure enantiomers is termed

(a) Racemization

(b) Resolution

(c) Isomerization

(d) Equilibration

B-6. Rank the following groups in order of R,S precedence (4 is highest):

—CH(CH$_3$)$_2$ —CH$_2$CH$_2$Br   —CH$_2$Br   —C(CH$_3$)$_3$

          A   B   C   D

	4	3	2	1
(a)	C	B	D	A
(b)	A	D	B	C
(c)	C	D	A	B
(d)	C	D	B	A

B-7. A meso compound

(a) Is an achiral molecule that contains chirality centers.

(b) Contains a plane of symmetry or a center of symmetry.

(c) Is optically inactive.

(d) Is characterized by all of these.

B-8. The S enantiomer of ibuprofen is responsible for its pain-relieving properties. Which one of the structures shown is (S)-ibuprofen?

(a)

(c)

(b)

(d)

B-9. Which one of the following is a diastereomer of (R)-4-bromo-*cis*-2-hexene?

(a) (S)-4-bromo-*cis*-2-hexene

(b) (R)-4-bromo-*trans*-2-hexene

(c) (R)-5-bromo-*cis*-2-hexene

(d) (S)-5-bromo-*trans*-2-hexene

(e) (R)-5-bromo-*trans*-2-hexene

B-10. The reaction sequence

will yield:

(a) A pair of products that are enantiomers

(b) A single product that is optically active

(c) A pair of products that are diastereomers

(d) A pair of products one of which is meso

B-11. Which of the following depict the same stereoisomer?

$$
\begin{array}{ccc}
CH_2CH_3 & CH_3 & CH_2CH_3 \\
Br\!-\!\overset{\vdots}{C}\!\!-\!\!Cl & CH_3CH_2\!-\!\overset{\vdots}{C}\!\!-\!\!Cl & H_3C\!-\!\overset{\vdots}{C}\!\!-\!\!Br \\
CH_3 & Br & Cl \\
1 & 2 & 3
\end{array}
$$

(*a*) 1 and 2   (*b*) 1 and 3   (*c*) 2 and 3   (*d*) 1, 2, and 3

B-12. A naturally occurring substance has the constitution shown. How many stereoisomers may have this constitution?

(*a*) 2   (*b*) 8   (*c*) 16   (*d*) 64   (*e*) 128

B-13. Acid-catalyzed hydration of an unknown compound X, C_6H_{12}, yielded as the major product a racemic mixture Y, $C_6H_{14}O$. Which (if any) of the following is (are) likely candidate(s) for X?

$$
\begin{array}{ccc}
1 & 2 & 3
\end{array}
$$

(*a*) 3 only   (*b*) 2 only   (*c*) 1 and 3   (*d*) 2 and 3   (*e*) None of these

B-14. The major product(s) from the reaction of Br_2 with (Z)-3-hexene is (are)

(*a*) Optically pure
(*b*) A racemic mixture of enantiomers
(*c*) The meso form
(*d*) Both the racemic mixture and the meso form

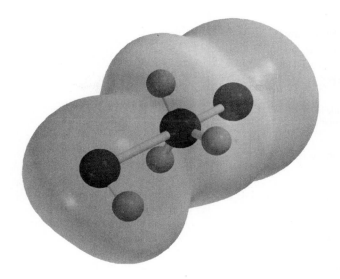

CHAPTER 8

Nucleophilic Substitution

SOLUTIONS TO TEXT PROBLEMS

8.1 Identify the nucleophilic anion in each reactant. The nucleophilic anion replaces bromine as a substituent on carbon.

(*b*) Potassium ethoxide is a source of the nucleophilic anion $CH_3CH_2O^-$.

$$CH_3CH_2\overset{..}{\underset{..}{O}}:^- \quad + \quad CH_3\overset{..}{\underset{..}{Br}}: \quad \longrightarrow \quad CH_3CH_2\overset{..}{\underset{..}{O}}CH_3 \quad + \quad :\overset{..}{\underset{..}{Br}}:^-$$

Ethoxide ion   Methyl bromide   Ethyl methyl ether   Bromide ion
(nucleophile)   (product)

(*c*)

Benzoate ion   Methyl bromide   Methyl benzoate   Bromide ion

(*d*) Lithium azide is a source of azide ion.

$$:\overset{-}{\underset{..}{N}}=\overset{+}{N}=\overset{-}{\underset{..}{N}}:$$

It reacts with methyl bromide to give methyl azide.

$$:\overset{-}{\underset{..}{N}}=\overset{+}{N}=\overset{-}{\underset{..}{N}}: \quad + \quad CH_3\overset{..}{\underset{..}{Br}}: \quad \longrightarrow \quad CH_3\overset{}{N}=\overset{+}{N}=\overset{-}{\underset{..}{N}}: \quad + \quad :\overset{..}{\underset{..}{Br}}:^-$$

Azide ion   Methyl bromide   Methyl azide   Bromide ion
(nucleophile)   (product)

(*e*) The nucleophilic anion in KCN is cyanide ($:\overset{-}{\underset{}{C}}{\equiv}N:$). The carbon atom is negatively charged and is normally the site of nucleophilic reactivity.

$$:N{\equiv}C:^- \quad + \quad CH_3\ddot{B}r: \quad \longrightarrow \quad CH_3C{\equiv}N: \quad + \quad :\ddot{B}r:^-$$

Cyanide ion   Methyl bromide   Methyl cyanide   Bromide ion
(nucleophile)   (product)

(*f*) The anion in sodium hydrogen sulfide (NaSH) is $^-:\ddot{S}H$.

$$H\ddot{S}:^- \quad + \quad CH_3\ddot{B}r: \quad \longrightarrow \quad CH_3\ddot{S}H \quad + \quad :\ddot{B}r:^-$$

Hydrogen   Methyl bromide   Methanethiol   Bromide ion
sulfide ion

(*g*) Sodium iodide is a source of the nucleophilic anion iodide ion, $:\ddot{\underset{..}{I}}:^-$. The reaction of sodium iodide with alkyl bromides is usually carried out in acetone to precipitate the sodium bromide formed.

$$:\ddot{\underset{..}{I}}:^- \quad + \quad CH_3\ddot{B}r: \quad \xrightarrow{\text{acetone}} \quad CH_3\ddot{\underset{..}{I}}: \quad + \quad :\ddot{B}r:^-$$

Iodide ion   Methyl bromide   Methyl iodide   Bromide ion

8.2   Write out the structure of the starting material. Notice that it contains a primary bromide and a primary chloride. Bromide is a better leaving group than chloride and is the one that is displaced faster by the nucleophilic cyanide ion.

$$ClCH_2CH_2CH_2Br \quad \xrightarrow[\text{ethanol–water}]{NaCN} \quad ClCH_2CH_2CH_2C{\equiv}N$$

1-Bromo-3-chloropropane   4-Chlorobutanenitrite

8.3   No, the two-step sequence is not consistent with the observed behavior for the hydrolysis of methyl bromide. The rate-determining step in the two-step sequence shown is the first step, ionization of methyl bromide to give methyl cation.

1. $CH_3Br \quad \xrightarrow{\text{slow}} \quad CH_3^+ \quad + \quad Br^-$

2. $CH_3^+ \ + \ HO^- \quad \xrightarrow{\text{fast}} \quad CH_3OH$

In such a sequence the nucleophile would not participate in the reaction until after the rate-determining step is past, and the reaction rate would depend only on the concentration of methyl bromide and be independent of the concentration of hydroxide ion.

$$\text{Rate} = k[CH_3Br]$$

The predicted kinetic behavior is first order. Second-order kinetic behavior is actually observed for methyl bromide hydrolysis, so the proposed mechanism cannot be correct.

8.4   Inversion of configuration occurs at the chirality center. When shown in a Fischer projection, this corresponds to replacing the leaving group on one side by the nucleophile on the opposite side.

(*S*)-(+)-2-Bromooctane   (*R*)-(–)-2-Octanol

8.5   The example given in the text illustrates inversion of configuration in the S_N2 hydrolysis of (S)-(+)-2-bromooctane, which yields (R)-(−)-2-octanol. The hydrolysis of (R)-(−)-2-bromooctane exactly mirrors that of its enantiomer and yields (S)-(+)-2-octanol.

Hydrolysis of racemic 2-bromooctane gives racemic 2-octanol. Remember, optically inactive reactants must yield optically inactive products.

8.6   C-2 of isopropyl bromide is partially bonded to both Br and I in the S_N2 transition state. Iodide attacks from the side opposite the C—Br bond.

| Iodide ion | Isopropyl bromide | S_N2 Transition state | Isopropyl iodide | Bromide ion |

8.7   Sodium iodide in acetone is a reagent that converts alkyl chlorides and bromides into alkyl iodides by an S_N2 mechanism. Pick the alkyl halide in each pair that is more reactive toward S_N2 displacement.

(b)   The less crowded alkyl halide reacts faster in an S_N2 reaction. 1-Bromopentane is a primary alkyl halide and so is more reactive than 3-bromopentane, which is secondary.

$$BrCH_2CH_2CH_2CH_2CH_3 \qquad\qquad CH_3CH_2CHCH_2CH_3$$
$$| $$
$$Br$$

1-Bromopentane
(primary; more reactive in S_N2)

3-Bromopentane
(secondary; less reactive in S_N2)

(c)   Both halides are secondary, but fluoride is a poor leaving group in nucleophilic substitutions. Alkyl chlorides are more reactive than alkyl fluorides.

$$CH_3CHCH_2CH_2CH_3 \qquad\qquad CH_3CHCH_2CH_2CH_3$$
$$| \qquad\qquad\qquad\qquad\qquad | $$
$$Cl \qquad\qquad\qquad\qquad\qquad F$$

2-Chloropentane
(more reactive)

2-Fluoropentane
(less reactive)

(d)   A secondary alkyl bromide reacts faster under S_N2 conditions than a tertiary one.

$$\qquad\qquad\qquad\qquad\qquad\qquad\qquad CH_3$$
$$\qquad\qquad\qquad\qquad\qquad\qquad\qquad |$$
$$CH_3CHCH_2CH_2CHCH_3 \qquad CH_3CCH_2CH_2CH_2CH_3$$
$$| \qquad\qquad | \qquad\qquad\qquad\qquad |$$
$$Br \qquad\quad CH_3 \qquad\qquad\qquad\quad Br$$

2-Bromo-5-methylhexane
(secondary; more reactive in S_N2)

2-Bromo-2-methylhexane
(tertiary; less reactive in S_N2)

(e)   The number of carbons does not matter as much as the degree of substitution at the reaction site. The primary alkyl bromide is more reactive than the secondary.

$$BrCH_2(CH_2)_8CH_3 \qquad\qquad CH_3CHCH_3$$
$$|$$
$$Br$$

1-Bromodecane
(primary; more reactive in S_N2)

2-Bromopropane
(secondary; less reactive in S_N2)

8.8   Begin by writing the balanced equation for the reaction of ethyl bromide with methanol.

$$CH_3CH_2Br + 2CH_3OH \longrightarrow CH_3CH_2OCH_3 + CH_3\overset{+}{O}H_2 + Br^-$$

Ethyl bromide   Methanol   Ethyl methyl ether   Methyloxonium ion   Bromide ion

The first step of the reaction is nucleophilic substitution. It is the rate-determining step of the process.

Methanol   Ethyl bromide   Ethylmethyloxonium ion   Bromide ion

The second step is a fast Brønsted acid-base reaction.

Methanol   Ethylmethyloxonium ion   Methyloxonium ion   Ethyl methyl ether

8.9   Solvolysis of alkyl halides in alcohols yields ethers as the products of reaction.

$$(CH_3)_3CBr + 2CH_3OH \longrightarrow (CH_3)_3COCH_3 + CH_3\overset{+}{O}H_2 + Br^-$$

tert-Butyl   Methanol   *tert*-Butyl methyl   Methyloxonium ion   Bromide ion
bromide   ether

The reaction proceeds by an S_N1 mechanism.

tert-Butyl   *tert*-Butyl   Bromide
bromide   cation   ion

tert-Butyl   Methanol   *tert*-Butylmethyloxonium
cation   ion

tert-Butylmethyloxonium ion   Methanol   *tert*-Butyl   Methyloxonium
                                                                  methyl ether   ion

8.10   The reactivity of an alkyl halide in an S_N1 reaction is dictated by the ease with which it ionizes to form a carbocation. A more stable carbocation is formed faster than a less stable one. Tertiary alkyl halides are the most reactive, methyl halides the least reactive.

(*b*) Cyclopentyl iodide ionizes to form a secondary carbocation, while the carbocation from 1-methylcyclopentyl iodide is tertiary. The tertiary halide is more reactive.

1-Methylcyclopentyl iodide
(tertiary; more reactive in S$_N$1)

Cyclopentyl iodide
(secondary; less reactive in S$_N$1)

(*c*) Cyclopentyl bromide ionizes to a secondary carbocation. 1-Bromo-2,2-dimethylpropane is a primary alkyl halide and is therefore less reactive.

$(CH_3)_3CCH_2Br$

Cyclopentyl bromide
(secondary; more reactive in S$_N$1)

1-Bromo-2,2-dimethylpropane
(primary; less reactive in S$_N$1)

(*d*) Iodide is a better leaving group than chloride in both S$_N$1 and S$_N$2 reactions.

$(CH_3)_3CI$

tert-Butyl iodide
(more reactive)

$(CH_3)_3CCl$

tert-Butyl chloride
(less reactive)

8.11 The alkyl halide is tertiary and so undergoes hydrolysis by an S$_N$1 mechanism. The carbocation can be captured by water at either face. A mixture of the axial and the equatorial alcohols is formed.

cis-1,4-Dimethylcyclohexyl bromide

trans-1,4-Dimethylcyclohexanol   Carbocation intermediate   *cis*-1,4-Dimethylcyclohexanol

The same two substitution products are formed from *trans*-1,4-dimethylcyclohexyl bromide because it undergoes hydrolysis via the same carbocation intermediate.

8.12 Write chemical equations illustrating each rearrangement process.

Hydride shift:

$$H_3C-\overset{+}{\underset{\underset{H}{|}}{C}}-\underset{|}{\overset{H\quad CH_3}{\overset{|}{C}}}CH_3 \longrightarrow H_3C-\underset{\underset{H}{|}}{\overset{H}{\overset{|}{C}}}-\overset{+}{C}\overset{CH_3}{\underset{CH_3}{}}$$

Tertiary carbocation

Methyl shift:

$$H_3C-\overset{+}{\underset{}{C}}-\underset{\underset{CH_3}{|}}{\overset{H\quad H}{\overset{|}{C}}}-CH_3 \longrightarrow H_3C-\underset{\underset{CH_3}{|}}{\overset{H}{\overset{|}{C}}}-\overset{+}{C}\overset{H}{\underset{CH_3}{}}$$

Secondary carbocation

Rearrangement by a hydride shift is observed because it converts a secondary carbocation to a more stable tertiary one. A methyl shift gives a secondary carbocation—in this case the same carbocation as the one that existed prior to rearrangement.

8.13 (b) Ethyl bromide is a primary alkyl halide and reacts with the potassium salt of cyclohexanol by substitution.

$$CH_3CH_2Br \ + \ \langle\rangle-OK \longrightarrow \langle\rangle-OCH_2CH_3$$

Ethyl bromide   Potassium   Cyclohexyl ethyl ether
                     cyclohexanolate

(c) No strong base is present in this reaction; the nucleophile is methanol itself, not methoxide. It reacts with sec-butyl bromide by substitution, not elimination.

$$CH_3\underset{\underset{Br}{|}}{CH}CH_2CH_3 \xrightarrow{CH_3OH} CH_3\underset{\underset{OCH_3}{|}}{CH}CH_2CH_3$$

sec-Butyl bromide   sec-Butyl methyl ether

(d) Secondary alkyl halides react with alkoxide bases by E2 elimination.

$$CH_3\underset{\underset{Br}{|}}{CH}CH_2CH_3 \xrightarrow[CH_3OH]{NaOCH_3} CH_3CH=CHCH_3 \ + \ H_2C=CHCH_2CH_3$$

sec-Butyl bromide   2-Butene   1-Butene
                         (major product; mixture
                         of cis and trans)

8.14 There are two possible combinations of alkoxide and alkyl halide (shown as the bromide), both of which involve secondary alkyl halides.

$$CH_3CH_2\underset{\underset{CH_3}{|}}{CH}-\overset{..}{\underset{..}{O}}{:}^- \ + \ Br-CH(CH_3)_2$$

$$CH_3CH_2\underset{\underset{CH_3}{|}}{CH}-Br \ + \ {:}\overset{..}{\underset{..}{O}}^--CH(CH_3)_2$$

Secondary alkyl halides undergo *elimination* (E2) rather than substitution (S_N2) with alkoxide bases. Alkenes are the major products of the combinations shown.

$$CH_3CH_2\underset{\underset{CH_3}{|}}{CH}-\ddot{\underset{..}{O}}{:}^- \quad + \quad Br-CH(CH_3)_2 \longrightarrow H_2C{=}CHCH_3 \quad + \quad CH_3CH_2\underset{\underset{CH_3}{|}}{CH}-OH$$

Propene

$$CH_3CH_2\underset{\underset{CH_3}{|}}{CH}-Br \quad + \quad {:}\ddot{\underset{..}{O}}-CH(CH_3)_2 \longrightarrow CH_3CH{=}CHCH_3 \quad + \quad HO-CH(CH_3)_2$$

2-Butene
(mixture of cis and trans)

8.15 Alkyl *p*-toluenesulfonates are prepared from alcohols and *p*-toluenesulfonyl chloride.

$$CH_3(CH_2)_{16}CH_2OH \quad + \quad H_3C\text{—}\!\!\!\bigcirc\!\!\!\text{—}\overset{\overset{O}{\|}}{\underset{\underset{O}{\|}}{S}}Cl \quad \xrightarrow{\text{pyridine}} \quad CH_3(CH_2)_{16}CH_2O\overset{\overset{O}{\|}}{\underset{\underset{O}{\|}}{S}}\text{—}\!\!\!\bigcirc\!\!\!\text{—}CH_3 \quad + \quad HCl$$

1-Octadecanol   *p*-Toluenesulfonyl   Octadecyl *p*-toluenesulfonate   Hydrogen
                  cholride   chloride

8.16 As in part (*a*), identify the nucleophilic anion in each part. The nucleophile replaces the *p*-toluenesulfonate (tosylate) leaving group by an S_N2 process. The tosylate group is abbreviated as OTs.

(*b*) $:\ddot{\underset{..}{I}}{:}^- \quad + \quad CH_3(CH_2)_{16}CH_2\ddot{\underset{..}{O}}Ts \longrightarrow CH_3(CH_2)_{16}CH_2\ddot{\underset{..}{I}}{:} \quad + \quad Ts\ddot{\underset{..}{O}}{:}^-$

Iodide   Octadecyl   Octadecyl iodide   *p*-Toluenesulfonate
ion   *p*-toluenesulfonate   anion

(*c*) $:\bar{C}{\equiv}N{:} \quad + \quad CH_3(CH_2)_{16}CH_2\ddot{\underset{..}{O}}Ts \longrightarrow CH_3(CH_2)_{16}CH_2C{\equiv}N{:} \quad + \quad Ts\ddot{\underset{..}{O}}{:}^-$

Cyanide   Octadecyl *p*-toluenesulfonate   Octadecyl cyanide   *p*-Toluenesulfonate
ion   anion

(*d*) $H\ddot{\underset{..}{S}}{:}^- \quad + \quad CH_3(CH_2)_{16}CH_2\ddot{\underset{..}{O}}Ts \longrightarrow CH_3(CH_2)_{16}CH_2\ddot{\underset{..}{S}}H \quad + \quad Ts\ddot{\underset{..}{O}}{:}^-$

Hydrogen   Octadecyl *p*-toluenesulfonate   1-Octadecanethiol   *p*-Toluenesulfonate
sulfide ion   anion

(*e*)

$$CH_3CH_2CH_2CH_2\ddot{\underset{..}{S}}{:}^- \quad + \quad CH_3(CH_2)_{16}CH_2\ddot{\underset{..}{O}}Ts \longrightarrow CH_3(CH_2)_{16}CH_2\ddot{\underset{..}{S}}CH_2CH_2CH_2CH_3 \quad + \quad Ts\ddot{\underset{..}{O}}{:}^-$$

Butanethiolate ion   Octadecyl *p*-toluenesulfonate   Butyl octadecyl sulfide   *p*-Toluenesulfonate
                                                                                        anion

8.17 The hydrolysis of (*S*)-(+)-1-methylheptyl *p*-toluenesulfonate proceeds with inversion of configuration, giving the *R* enantiomer of 2-octanol.

$$\underset{\underset{H_3C}{\diagdown}}{CH_3(CH_2)_5}\overset{H}{\underset{}{\diagup}}C-OTs \quad \xrightarrow{H_2O} \quad HO-C\overset{\overset{H}{\diagup}(CH_2)_5CH_3}{\underset{CH_3}{\diagdown}}$$

(*S*)-(+)-1-Methylheptyl   (*R*)-(−)-2-Octanol
p-toluenesulfonate

In Section 8.12 of the text we are told that optically pure (S)-(+)-1-methylheptyl p-toluenesulfonate is prepared from optically pure (S)-(+)-2-octanol having a specific rotation $[\alpha]_D^{25} + 9.9°$. The conversion of an alcohol to a p-toluenesulfonate proceeds with complete *retention* of configuration. Hydrolysis of this p-toluenesulfonate with *inversion* of configuration therefore yields optically pure (R)-(−)-2-octanol of $[\alpha]_D^{25} - 9.9°$.

8.18 Protonation of 3-methyl-2-butanol and dissociation of the alkyloxonium ion gives a secondary carbocation. A hydride shift yields a tertiary, and thus more stable, carbocation. Capture of this carbocation by chloride ion gives the major product, 2-chloro-2-methylbutane.

$$CH_3CHCH(CH_3)_2 \xrightarrow{HCl} CH_3\overset{+}{C}HCH(CH_3)_2 \xrightarrow[\text{shift}]{\text{hydride}} CH_3CH_2\overset{+}{C}(CH_3)_2$$
$$\underset{OH}{|}$$

3-Methyl-2-butanol   Secondary carbocation;   Tertiary carbocation;
                                             less stable   more stable

$$\Big\downarrow Cl^-\qquad\qquad\qquad\qquad\qquad \Big\downarrow Cl^-$$

$$CH_3CHCH(CH_3)_2 \qquad\qquad\qquad CH_3CH_2C(CH_3)_2$$
$$\underset{Cl}{|} \qquad\qquad\qquad\qquad\qquad \underset{Cl}{|}$$

2-Chloro-3-methylbutane   2-Chloro-2-methylbutane
       (trace)   (major product; ≈97%)

8.19 1-Bromopropane is a primary alkyl halide, and so it will undergo predominantly S_N2 displacement regardless of the basicity of the nucleophile.

(a) $CH_3CH_2CH_2Br \xrightarrow[\text{acetone}]{NaI} CH_3CH_2CH_2I$

1-Bromopropane   1-Iodopropane

(b) $CH_3CH_2CH_2Br \xrightarrow[\text{acetic acid}]{\overset{\overset{O}{\|}}{CH_3CONa}} CH_3CH_2CH_2O\overset{\overset{O}{\|}}{C}CH_3$

Propyl acetate

(c) $CH_3CH_2CH_2Br \xrightarrow[\text{ethanol}]{NaOCH_2CH_3} CH_3CH_2CH_2OCH_2CH_3$

Ethyl propyl ether

(d) $CH_3CH_2CH_2Br \xrightarrow[\text{DMSO}]{NaCN} CH_3CH_2CH_2CN$

Butanenitrile

(e) $CH_3CH_2CH_2Br \xrightarrow[\text{ethanol–water}]{NaN_3} CH_3CH_2CH_2N_3$

1-Azidopropane

(f) $CH_3CH_2CH_2Br$ $\xrightarrow[\text{ethanol}]{\text{NaSH}}$ $CH_3CH_2CH_2SH$

1-Propanethiol

(g) $CH_3CH_2CH_2Br$ $\xrightarrow[\text{ethanol}]{\text{NaSCH}_3}$ $CH_3CH_2CH_2SCH_3$

Methyl propyl sulfide

8.20 Elimination is the major product when secondary halides react with anions as basic as or more basic than hydroxide ion. Alkoxide ions have a basicity comparable with hydroxide ion and react with secondary halides to give predominantly elimination products. Thus ethoxide ion [part (c)] will react with 2-bromopropane to give mainly propene.

$\underset{\underset{Br}{|}}{CH_3CHCH_3}$ $\xrightarrow{\text{NaOCH}_2\text{CH}_3}$ $CH_3CH{=}CH_2$

Propene

8.21 (a) The substrate is a primary alkyl bromide and reacts with sodium iodide in acetone to give the corresponding iodide.

$\underset{\text{Ethyl bromoacetate}}{BrCH_2\overset{\overset{\text{O}}{\|}}{C}OCH_3CH_3}$ $\xrightarrow[\text{acetone}]{\text{NaI}}$ $\underset{\text{Ethyl iodoacetate (89\%)}}{ICH_2\overset{\overset{\text{O}}{\|}}{C}OCH_2CH_3}$

(b) Primary alkyl chlorides react with sodium acetate to yield the corresponding acetate esters.

$O_2N-\langle\text{benzene}\rangle-CH_2Cl$ $\xrightarrow[\text{acetic acid}]{CH_3\overset{\overset{\text{O}}{\|}}{C}ONa}$ $O_2N-\langle\text{benzene}\rangle-CH_2O\overset{\overset{\text{O}}{\|}}{C}CH_3$

p-Nitrobenzyl chloride p-Nitrobenzyl acetate (78–82%)

(c) The only leaving group in the substrate is bromide. Neither of the carbon–oxygen bonds is susceptible to cleavage by nucleophilic attack.

$CH_3CH_2OCH_2CH_2Br$ $\xrightarrow[\text{ethanol–water}]{\text{NaCN}}$ $CH_3CH_2OCH_2CH_2CN$

2-Bromoethyl ethyl ether   2-Cyanoethyl ethyl ether
(52–58%)

(d) Hydrolysis of the primary chloride yields the corresponding alcohol.

$NC-\langle\text{benzene}\rangle-CH_2Cl$ $\xrightarrow{H_2O, HO^-}$ $NC-\langle\text{benzene}\rangle-CH_2OH$

p-Cyanobenzyl chloride p-Cyanobenzyl alcohol (85%)

(e) The substrate is a primary chloride.

$\underset{\text{\textit{tert}-Butyl chloroacetate}}{ClCH_2\overset{\overset{\text{O}}{\|}}{C}OC(CH_3)_3}$ $\xrightarrow[\text{acetone–water}]{\text{NaN}_3}$ $\underset{\text{\textit{tert}-Butyl azidoacetate (92\%)}}{N_3CH_2\overset{\overset{\text{O}}{\|}}{C}OC(CH_3)_3}$

(*f*) Primary alkyl tosylates yield iodides on treatment with sodium iodide in acetone.

(2,2-Dimethyl-1,3-dioxolan-4-yl)-
methyl *p*-toluenesulfonate

2,2-Dimethyl-5-(iodomethyl)-
1,3-dioxolane (60%)

(*g*) Sulfur displaces bromide from ethyl bromide.

Sodium (2-furyl)-
methanethiolate

Ethyl bromide

Ethyl (2-furyl)methyl
sulfide (80%)

(*h*) The first reaction is one in which a substituted alcohol is converted to a *p*-toluenesulfonate. This is followed by an S_N2 displacement with lithium iodide.

4-(2,3,4-Trimethoxyphenyl)-1-butanol

LiI, acetone (88%)

4-(2,3,4-Trimethoxyphenyl)butyl iodide

8.22 The two products are diastereomers. They are formed by bimolecular nucleophilic substitution (S_N2). In each case, a good nucleophile ($C_6H_5S^-$) displaces chloride from a secondary carbon with inversion of configuration.

(*a*) The trans chloride yields a cis substitution product.

trans-4-*tert*-Butylcyclohexyl
chloride

Sodium
benzenethiolate

cis-4-*tert*-Butylcyclohexyl phenyl sulfide

(*b*) The cis chloride yields a trans substitution product.

cis-4-*tert*-Butylcyclohexyl chloride	Sodium benzenethiolate	*trans*-4-*tert*-Butylcyclohexyl phenyl sulfide

8.23 The isomers of C_4H_9Cl are:

1-Chlorobutane (*n*-butyl chloride)	1-Chloro-2-methylpropane (isobutyl chloride)
2-Chlorobutane (*sec*-butyl chloride)	2-Chloro-2-methylpropane (*tert*-butyl chloride)

The reaction conditions (sodium iodide in acetone) are typical for an S_N2 process.

The order of S_N2 reactivity is primary > secondary > tertiary, and branching of the chain close to the site of substitution hinders reaction. The unbranched primary halide *n*-butyl chloride will be the most reactive and the tertiary halide *tert*-butyl chloride the least. The order of reactivity is therefore: 1-chlorobutane > 1-chloro-2-methylpropane > 2-chlorobutane > 2-chloro-2-methylpropane.

8.24 1-Chlorohexane is a primary alkyl halide; 2-chlorohexane and 3-chlorohexane are secondary.

$CH_3CH_2CH_2CH_2CH_2CH_2Cl$ $CH_3CHCH_2CH_2CH_2CH_3$ $CH_3CH_2CHCH_2CH_2CH_3$

1-Chlorohexane (primary)	2-Chlorohexane (secondary)	3-Chlorohexane (secondary)

Primary and secondary alkyl halides react with potassium iodide in acetone by an S_N2 mechanism, and the rate depends on steric hindrance to attack on the alkyl halide by the nucleophile.

(*a*) Primary alkyl halides are more reactive than secondary alkyl halides in S_N2 reactions. 1-Chlorohexane is the most reactive isomer.

(*b*) Substituents at the carbon adjacent to the one that bears the leaving group slow down the rate of nucleophilic displacement. In 2-chlorohexane the group adjacent to the point of attack is CH_3. In 3-chlorohexane the group adjacent to the point of attack is CH_2CH_3. 2-Chlorohexane has been observed to be more reactive than 3-chlorohexane by a factor of 2.

8.25 (*a*) Iodide is a better leaving group than bromide, and so 1-iodobutane should undergo S_N2 attack by cyanide faster than 1-bromobutane.

(*b*) The reaction conditions are typical for an S_N2 process. The methyl branch in 1-chloro-2-methylbutane sterically hinders attack at C-1. The unbranched isomer, 1-chloropentane, reacts faster.

$$\underset{\substack{\text{CH}_3 \\ |}}{\text{CH}_3\text{CH}_2\text{CHCH}_2\text{Cl}}$$

$$\text{CH}_3\text{CH}_2\text{CH}_2\text{CH}_2\text{CH}_2\text{Cl}$$

1-Chloro-2-methylbutane
is more sterically
hindered, therefore
less reactive.

1-Chloropentane is less
sterically hindered,
therefore more
reactive.

(*c*) Hexyl chloride is a primary alkyl halide, and cyclohexyl chloride is secondary. Azide ion is a good nucleophile, and so the S_N2 reactivity rules apply; primary is more reactive than secondary.

$$\text{CH}_3\text{CH}_2\text{CH}_2\text{CH}_2\text{CH}_2\text{CH}_2\text{Cl}$$

Hexyl chloride is primary,
therefore more reactive in S_N2

Cyclohexyl chloride is secondary,
therefore less reactive in S_N2

(*d*) 1-Bromo-2,2-dimethylpropane is too hindered to react with the weakly nucleophilic ethanol by an S_N2 reaction, and since it is a primary alkyl halide, it is less reactive in S_N1 reactions. *tert*-Butyl bromide will react with ethanol by an S_N1 mechanism at a reasonable rate owing to formation of a tertiary carbocation.

$$(\text{CH}_3)_3\text{CBr}$$

$$(\text{CH}_3)_3\text{CCH}_2\text{Br}$$

tert-Butyl bromide;
more reactive in
S_N1 solvolysis

1-Bromo-2,2-dimethylpropane;
relatively unreactive in nucleophilic
substitution reactions

(*e*) Solvolysis of alkyl halides in aqueous formic acid is faster for those that form carbocations readily. The S_N1 reactivity order applies here: secondary > primary.

$$\underset{\substack{| \\ \text{Br}}}{\text{CH}_3\text{CHCH}_2\text{CH}_3}$$

$$(\text{CH}_3)_2\text{CHCH}_2\text{Br}$$

sec-Butyl bromide; secondary,
therefore more reactive in S_N1

Isobutyl bromide; primary,
therefore less reactive in S_N1

(*f*) 1-Chlorobutane is a primary alkyl halide and so should react by an S_N2 mechanism. Sodium methoxide is more basic than sodium acetate and is a better nucleophile. Reaction will occur faster with sodium methoxide than with sodium acetate.

(*g*) Azide ion is a very good nucleophile, whereas *p*-toluenesulfonate is a very good leaving group but a very poor nucleophile. In an S_N2 reaction with 1-chlorobutane, sodium azide will react faster than sodium *p*-toluenesulfonate.

8.26 There are only two possible products from free-radical chlorination of the starting alkane:

$$(\text{CH}_3)_3\text{CCH}_2\text{C}(\text{CH}_3)_3 \xrightarrow[h\nu]{\text{Cl}_2} \underset{\substack{| \\ \text{CH}_2\text{Cl}}}{(\text{CH}_3)_3\text{CCH}_2\text{C}(\text{CH}_3)_2} + \underset{\substack{| \\ \text{Cl}}}{(\text{CH}_3)_3\text{CCHC}(\text{CH}_3)_3}$$

2,2,4,4-Tetramethylpentane

1-Chloro-2,2,4,4-
tetramethylpentane
(primary)

3-Chloro-2,2,4,4-
tetramethylpentane
(secondary)

As revealed by their structural formulas, one isomer is a primary alkyl chloride, the other is secondary. The problem states that the major product (compound A) undergoes S_N1 hydrolysis much more slowly than the minor product (compound B). Because secondary halides are much more reactive than primary halides under S_N1 conditions, the major (unreactive) product is the primary alkyl halide 1-chloro-2,2,4,4-tetramethylpentane (compound A) and the minor (reactive) product is the secondary alkyl halide 3-chloro-2,2,4,4-tetramethylpentane (compound B).

8.27 (*a*) The two most stable Lewis structures (resonance contributors) of thiocyanate are:

$$\ddot{\ddot{S}}-C\equiv N: \quad \longleftrightarrow \quad :\ddot{S}=C=\ddot{N}:^-$$

(*b*) The two Lewis structures indicate that the negative charge is shared by two atoms: S and N. Thus thiocyanate ion has two potentially nucleophilic sites, and the two possible products are

$$CH_3CH_2CH_2CH_2Br \xrightarrow[\text{DMF}]{\text{KSCN}} CH_3CH_2CH_2CH_2-\ddot{S}-C\equiv N: \quad + \quad CH_3CH_2CH_2CH_2-\ddot{N}=C=\ddot{S}:$$

1-Bromobutane   Butyl thiocyanate   Butyl isothiocyanate

8.28 Nitrite ion has two potentially nucleophilic sites, oxygen and nitrogen.

Nitrite ion   Alkyl iodide   Alkyl nitrite   Iodide ion

Nitrite ion   Alkyl iodide   Nitroalkane   Iodide ion

Thus, an alkyl iodide can yield either an alkyl nitrite or a nitroalkane depending on whether the oxygen or the nitrogen of nitrite ion attacks carbon. Both do, and the product from 2-iodooctane is a mixture of

$$\underset{\underset{\displaystyle ONO}{|}}{CH_3CH}(CH_2)_5CH_3 \quad \text{and} \quad \underset{\underset{\displaystyle NO_2}{|}}{CH_3CH}(CH_2)_5CH_3$$

8.29 Using the unshared electron pair on its nitrogen, triethylamine acts as a nucleophile in an S_N2 reaction toward ethyl iodide.

Triethylamine   Ethyl iodide   Tetraethylammonium iodide

The product of the reaction is a salt and has the structure shown. The properties given in the problem (soluble in polar solvents, high melting point) are typical of those of an ionic compound.

8.30 This reaction has been reported in the chemical literature and proceeds as shown (91% yield):

$$CH_3CH_2\overset{\overset{H}{\vdots}}{\underset{H_3C}{C}}-CH_2Br \xrightarrow[\text{acetone}]{\text{NaI}} CH_3CH_2\overset{\overset{H}{\vdots}}{\underset{H_3C}{C}}-CH_2I$$

(*S*)-1-Bromo-2-methylbutane   (*S*)-1-Iodo-2-methylbutane

Notice that the configuration of the product is the *same* as the configuration of the reactant. This is because the chirality center is not involved in the reaction. When we say that S_N2 reactions proceed with inversion of configuration we refer only to the carbon at which substitution takes place, not a chirality center somewhere else in the molecule.

8.31 (*a*) The starting material incorporates both a primary chloride and a secondary chloride. The nucleophile (iodide) attacks the less hindered primary position.

$$\underset{\underset{Cl}{|}}{ClCH_2CH_2CHCH_2CH_3} \xrightarrow[\text{acetone}]{\text{NaI}} \underset{\underset{Cl}{|}}{ICH_2CH_2CHCH_2CH_3}$$

1,3-Dichloropentane   3-Chloro-1-iodopentane
($C_5H_{10}ClI$)

(*b*) Nucleophilic substitution of the first bromide by sulfur occurs in the usual way.

$$\overset{..}{\underset{..}{:}}SCH_2CH_2\overset{..}{\underset{..}{S}}\overset{\frown}{:} \quad + \quad \underset{\underset{CH_2-Br}{|}}{CH_2\overset{\frown}{-}Br} \longrightarrow \overset{..}{\underset{..}{:}}SCH_2CH_2\overset{..}{\underset{..}{S}}CH_2CH_2Br$$

The product of this step cyclizes by way of an intramolecular nucleophilic substitution.

$$\underset{\underset{\underset{\frown Br}{|}}{CH_2-CH_2}}{\overset{CH_2-CH_2}{\overset{..}{\underset{..}{S}}\diagdown\diagup\overset{..}{\underset{..}{S}}:}} \longrightarrow \overset{..}{\underset{..}{:}}S\underset{}{\bigcirc}\overset{..}{\underset{..}{S}}:$$

1,4-Dithiane ($C_4H_8S_2$)

(*c*) The nucleophile is a dianion (S^{2-}). Two nucleophilic substitutions take place; the second of the two leads to intramolecular cyclization.

$$\overset{..}{\underset{..}{:}}S^{2-} \quad + \quad \underset{\underset{CH_2CH_2CH_2Cl}{|}}{CH_2\overset{\frown}{-}Cl} \longrightarrow \underset{\underset{CH_2-CH_2}{}}{\overset{\overset{\overset{..}{\underset{..}{S}}:^-}{|}}{CH_2}\diagdown CH_2\overset{\frown}{-}Cl} \longrightarrow \overset{..}{\underset{..}{:}}S\underset{}{\bigcirc}$$

Thiolane (C_4H_8S)

8.32 (*a*) Methyl halides are unhindered and react rapidly by the S_N2 mechanism.

(*b*) Sodium ethoxide is a good nucleophile and will react with unhindered primary alkyl halides by the S_N2 mechanism.

(*c*) Cyclohexyl bromide is a secondary halide and will react with a strong base (sodium ethoxide) predominantly by the E2 mechanism.

(*d*) The tertiary halide *tert*-butyl bromide will undergo solvolysis by the S_N1 mechanism.

(*e*) The presence of the strong base sodium ethoxide will cause the E2 mechanism to predominate.

(f) Concerted reactions are those that occur in a single step. The bimolecular mechanisms S_N2 and E2 represent concerted processes.

(g) In a stereospecific reaction, stereoisomeric reactants yield products that are stereoisomers of each other. Reactions that occur by the S_N2 and E2 mechanisms are stereospecific.

(h) The unimolecular mechanisms S_N1 and E1 involve the formation of carbocation intermediates.

(i) Rearrangements are possible when carbocations are intermediates in a reaction. Thus reactions occurring by the S_N1 and E1 mechanisms are most likely to have a rearranged carbon skeleton.

(j) Iodide is a better leaving group than bromide, and alkyl iodides react faster than alkyl bromides by any of the four mechanisms S_N1, S_N2, E1, and E2.

8.33 (a) Cyclopentyl cyanide can be prepared from a cyclopentyl halide by a nucleophilic substitution reaction. The first task, therefore, is to convert cyclopentane to a cyclopentyl halide.

| Cyclopentane | Chlorine | | Cyclopentyl chloride | Hydrogen chloride |

| Cyclopentyl chloride | Sodium cyanide | | Cyclopentyl cyanide | Sodium chloride |

An analogous sequence involving cyclopentyl bromide could be used.

(b) Cyclopentene can serve as a precursor to a cyclopentyl halide.

| Cyclopentene | Hydrogen bromide | | Cyclopentyl bromide |

Once cyclopentyl bromide has been prepared, it is converted to cyclopentyl cyanide by nucleophilic substitution, as shown in part (a).

(c) Reaction of cyclopentanol with hydrogen bromide gives cyclopentyl bromide. Then cyclopentyl bromide can be converted to cyclopentyl cyanide, as shown in part (a).

(d) Two cyano groups are required here, both of which must be introduced in nucleophilic substitution reactions. The substrate in the key reaction is $BrCH_2CH_2Br$.

$$BrCH_2CH_2Br \ + \ 2NaCN \ \xrightarrow{\text{ethanol–water or}} \ N{\equiv}CCH_2CH_2C{\equiv}N$$

| 1,2-Dibromoethane | Sodium cyanide | | 1,2-Dicyanoethane |

1,2-Dibromoethane is prepared from ethylene. The overall synthesis from ethyl alcohol is therefore formulated as shown:

$$CH_3CH_2OH \xrightarrow[\text{heat}]{H_2SO_4} H_2C{=}CH_2 \xrightarrow{Br_2} BrCH_2CH_2Br \xrightarrow{NaCN} NCCH_2CH_2CN$$

Ethyl alcohol   Ethylene   1,2-Dibromoethane   1,2-Dicyanoethane

(*e*) In this synthesis a primary alkyl chloride must be converted to a primary alkyl iodide. This is precisely the kind of transformation for which sodium iodide in acetone is used.

$$(CH_3)CHCH_2Cl \quad + \quad NaI \xrightarrow{\text{acetone}} (CH_3)_2CHCH_2I \quad + \quad NaCl$$

Isobutyl chloride   Sodium   Isobutyl iodide   Sodium
                   iodide   chloride

(*f*) First convert *tert*-butyl chloride into an isobutyl halide.

$$(CH_3)_3CCl \xrightarrow{NaOCH_3} (CH_3)_2C{=}CH_2 \xrightarrow[\text{peroxides}]{HBr} (CH_3)_2CHCH_2Br$$

tert-Butyl   2-Methylpropene   Isobutyl bromide
chloride

Treating isobutyl bromide with sodium iodide in acetone converts it to isobutyl iodide.

$$(CH_3)_2CHCH_2Br \xrightarrow[\text{acetone}]{NaI} (CH_3)_2CHCH_2I$$

Isobutyl bromide   Isobutyl iodide

A second approach is by way of isobutyl alcohol.

$$(CH_3)_3CCl \xrightarrow{NaOCH_3} (CH_3)_2C{=}CH_2 \xrightarrow[\text{2. } H_2O_2,\ HO^-]{\text{1. } B_2H_6} (CH_3)_2CHCH_2OH$$

tert-Butyl   2-Methylpropene   Isobutyl alcohol
chloride

Isobutyl alcohol is then converted to its *p*-toluenesulfonate, which reacts with sodium iodide in acetone in a manner analogous to that of isobutyl bromide.

(*g*) First introduce a leaving group into the molecule by converting isopropyl alcohol to an isopropyl halide. Then convert the resulting isopropyl halide to isopropyl azide by nucleophilic substitution.

$$\underset{\underset{OH}{|}}{CH_3CHCH_3} \xrightarrow{HBr} \underset{\underset{Br}{|}}{CH_3CHCH_3} \xrightarrow{NaN_3} \underset{\underset{N_3}{|}}{CH_3CHCH_3}$$

Isopropyl alcohol   Isopropyl bromide   Isopropyl azide

(*h*) In this synthesis 1-propanol must be first converted to an isopropyl halide.

$$CH_3CH_2CH_2OH \xrightarrow[\text{heat}]{H_2SO_4} CH_3CH{=}CH_2 \xrightarrow{HBr} \underset{\underset{Br}{|}}{CH_3CHCH_3}$$

1-Propanol   Propene   Isopropyl bromide

After an isopropyl halide has been obtained, it can be treated with sodium azide as in part (*g*).

(*i*) First write out the structure of the starting material and of the product so as to determine their relationship in three dimensions.

(*R*)-*sec*-Butyl alcohol   (*S*)-*sec*-Butyl azide

The hydroxyl group must be replaced by azide with inversion of configuration. First, however, a leaving group must be introduced, and it must be introduced in such a way that the configuration at the chirality center is not altered. The best way to do this is to convert (*R*)-*sec*-butyl alcohol to its corresponding *p*-toluenesulfonate.

(*R*)-*sec*-Butyl alcohol   (*R*)-*sec*-Butyl *p*-toluenesulfonate

Next, convert the *p*-toluenesulfonate to the desired azide by an S_N2 reaction.

(*R*)-*sec*-Butyl *p*-toluenesulfonate   (*S*)-*sec*-Butyl azide

(*j*) This problem is carried out in exactly the same way as the preceding one, except that the nucleophile in the second step is HS⁻.

(*R*)-*sec*-Butyl alcohol   (*R*)-*sec*-Butyl *p*-toluenesulfonate   (*S*)-2-Butanethiol

8.34 (*a*) The two possible combinations of alkyl bromide and alkoxide ion that might yield *tert*-butyl methyl ether are

1. CH_3Br + $(CH_3)_3CO^-$ $\xrightarrow{\text{fast}}$ $(CH_3)_3COCH_3$

Methyl bromide   *tert*-Butoxide   *tert*-Butyl methyl ether
                          ion

2. CH_3O^- + $(CH_3)_3CBr$ $\xrightarrow{\text{slow}}$ $(CH_3)_3COCH_3$

Methoxide ion   *tert*-Butyl bromide   *tert*-Butyl methyl ether

We choose the first approach because it is an S_N2 reaction on the unhindered substrate, methyl bromide. The second approach requires an S_N2 reaction on a hindered tertiary alkyl halide, a very poor choice. Indeed, we expect that the reaction of methoxide ion with *tert*-butyl bromide would not give any ether at all but would proceed entirely by E2 elimination:

$$CH_3O^- + (CH_3)_3CBr \xrightarrow{fast} CH_3OH + H_2C=C(CH_3)_2$$

Methanol   2-Methylpropene

(*b*) Again, the better alternative is to choose the less hindered alkyl halide to permit substitution to predominate over elimination.

⬠—OK + CH₃Br ⟶ ⬠—OCH₃

Potassium   Methyl   Cyclopentyl
cyclopentoxide   bromide   methyl ether

An attempt to prepare this compound by the reaction

⬠—Cl + CH₃ONa ⟶ ⬠—OCH₃

Chlorocyclopentane   Sodium   Cyclopentyl
          methoxide   methyl ether

gave cyclopentyl methyl ether in only 24% yield. Cyclopentene was isolated in 31% yield.

(*c*) A 2,2-dimethylpropyl halide is too sterically hindered to be a good candidate for this synthesis. The only practical method is

$$(CH_3)_3CCH_2OK + CH_3CH_2Br \longrightarrow (CH_3)_3CCH_2OCH_2CH_3$$

Potassium   Bromoethane   Ethyl 2,2-dimethylpropyl
2,2-dimethylpropoxide   ether

8.35 (*a*) The problem states that the reaction type is nucleophilic substitution. Sodium acetylide is therefore the nucleophile and must be treated with an alkyl halide to give the desired product.

$$Na^+ :\bar{C}{\equiv}CH + CH_3CH_2Br \longrightarrow CH_3CH_2C{\equiv}CH + NaBr$$

Sodium acetylide   Ethyl bromide   1-Butyne   Sodium bromide

(*b*) The acidity data given in the problem for acetylene tell us that $HC{\equiv}CH$ is a very weak acid ($pK_a = 26$), so that sodium acetylide must be a very strong base—stronger than hydroxide ion. *Elimination* by the E2 mechanism rather than S_N2 substitution is therefore expected to be the principal (probably the exclusive) reaction observed with secondary and tertiary alkyl halides. The substitution reaction will work well with primary alkyl halides but will likely fail for secondary and tertiary ones. Alkynes such as $(CH_3)_2CHC{\equiv}CH$ and $(CH_3)_3CC{\equiv}CH$ could not be prepared by this method.

8.36 The compound that reacts with *trans*-4-*tert*-butylcyclohexanol is a sulfonyl chloride and converts the alcohol to the corresponding sulfonate.

(CH₃)₃C—⬡—OH + O₂N—⬡—SO₂Cl $\xrightarrow{pyridine}$ (CH₃)₃C—⬡—OSO₂—⬡—NO₂

Compound A

Reaction of compound A with lithium bromide in acetone effects displacement of the sulfonate leaving group by bromide with inversion of configuration.

Compound A

cis-1-Bromo-4-*tert*-butylcyclohexane
Compound B

8.37 (*a*) To convert *trans*-2-methylcyclopentanol to *cis*-2-methylcyclopentyl acetate the hydroxyl group must be replaced by acetate with inversion of configuration. Hydroxide is a poor leaving group and so must first be converted to a good leaving group. The best choice is *p*-toluenesulfonate, because this can be prepared by a reaction that alters none of the bonds to the chirality center.

trans-2-Methylcyclopentanol

p-toluenesulfonyl chloride, pyridine

trans-2-Methylcyclopentyl *p*-toluenesulfonate

Treatment of the *p*-toluenesulfonate with potassium acetate in acetic acid will proceed with inversion of configuration to give the desired product.

trans-2-Methylcyclopentyl *p*-toluenesulfonate

cis-2-Methylcyclopentyl acetate

(*b*) To decide on the best sequence of reactions, we must begin by writing structural formulas to determine what kinds of transformations are required.

1-Methylcyclopentanol

cis-2-Methylcyclopentyl acetate

We already know from part (*a*) how to convert *trans*-2-methylcyclopentanol to *cis*-2-methylcyclopentyl acetate. So it is necessary is to design a synthesis of *trans*-2-methylcyclopentanol. Therefore,

1-Methylcyclopentanol

H_2SO_4 heat

1-Methylcyclopentene

1. B_2H_6
2. H_2O_2, HO^-

trans-2-Methylcyclopentanol

Hydroboration–oxidation converts 1-methylcyclopentene to the desired alcohol by anti-Markovnikov syn hydration of the double bond. The resulting alcohol is then converted to its *p*-toluenesulfonate and treated with acetate ion as in part (*a*) to give *cis*-2-methylcyclopentyl acetate.

8.38 (*a*) The reaction of an alcohol with a sulfonyl chloride gives a sulfonate. The oxygen of the alcohol remains in place and is the atom to which the sulfonyl group becomes attached.

$$\text{H}-\overset{\displaystyle \text{CH}_3}{\underset{\displaystyle \text{CH}_2\text{CH}_3}{|}}-\text{OH} \quad \xrightarrow[\text{pyridine}]{\text{CH}_3\text{SO}_2\text{Cl}} \quad \text{H}-\overset{\displaystyle \text{CH}_3}{\underset{\displaystyle \text{CH}_2\text{CH}_3}{|}}-\text{OSO}_2\text{CH}_3$$

(*S*)-(+)-2-Butanol   (*S*)-*sec*-Butyl methanesulfonate

(*b*) Sulfonate is similar to iodide in its leaving-group behavior. The product in part (*a*) is attacked by NaSCH$_2$CH$_3$ in an S$_N$2 reaction. Inversion of configuration occurs at the chirality center.

$$\text{H}-\overset{\displaystyle \text{CH}_3}{\underset{\displaystyle \text{CH}_2\text{CH}_3}{|}}-\text{OSO}_2\text{CH}_3 \quad \xrightarrow{\text{NaSCH}_2\text{CH}_3} \quad \text{CH}_3\text{CH}_2\text{S}-\overset{\displaystyle \text{CH}_3}{\underset{\displaystyle \text{CH}_2\text{CH}_3}{|}}-\text{H}$$

(*S*)-*sec*-Butyl methanesulfonate   (*R*)-(−)-*sec*-Butyl ethyl sulfide

(*c*) In this part of the problem we deduce the stereochemical outcome of the reaction of 2-butanol with PBr$_3$. We know the absolute configuration of (+)-2-butanol (*S*) from the statement of the problem and the configuration of (−)-*sec*-butyl ethyl sulfide (*R*) from part (*b*). We are told that the sulfide formed from (+)-2-butanol via the bromide has a positive rotation. It must therefore have the opposite configuration of the product in part (*b*).

$$\text{H}-\overset{\displaystyle \text{CH}_3}{\underset{\displaystyle \text{CH}_2\text{CH}_3}{|}}-\text{OH} \quad \xrightarrow{\text{PBr}_3} \quad \text{CH}_3\underset{\displaystyle \underset{\displaystyle \text{Br}}{|}}{\text{CHCH}_2\text{CH}_3} \quad \xrightarrow{\text{NaSCH}_2\text{CH}_3} \quad \text{H}-\overset{\displaystyle \text{CH}_3}{\underset{\displaystyle \text{CH}_2\text{CH}_3}{|}}-\text{SCH}_2\text{CH}_3$$

(*S*)-(+)-2-Butanol   2-Bromobutane   (*S*)-(+)-*sec*-Butyl ethyl sulfide

Because the reaction of the bromide with NaSCH$_2$CH$_3$ proceeds with inversion of configuration at the chirality center and because the final product has the same configuration as the starting alcohol, the conversion of the alcohol to the bromide must proceed with inversion of configuration.

$$\text{H}-\overset{\displaystyle \text{CH}_3}{\underset{\displaystyle \text{CH}_2\text{CH}_3}{|}}-\text{OH} \quad \xrightarrow{\text{PBr}_3} \quad \text{Br}-\overset{\displaystyle \text{CH}_3}{\underset{\displaystyle \text{CH}_2\text{CH}_3}{|}}-\text{H}$$

(*S*)-(+)-2-Butanol   (*R*)-(−)-2-Bromobutane

(*d*) The conversion of 2-butanol to *sec*-butyl methanesulfonate does not involve any of the bonds to the chirality center, and so it must proceed with 100% retention of configuration. Assuming that the reaction of the methanesulfonate with NaSCH$_2$CH$_3$ proceeds with 100% inversion of configuration, we conclude that the maximum rotation of *sec*-butyl ethyl sulfide is the value given in the statement of part (*b*), that is, ±25°. Because the sulfide produced in part (*c*) has a rotation of +23°, it is 92% optically pure. It is reasonable to assume that the loss of optical purity occurred in the conversion of the alcohol to the bromide, rather than in the reaction of the bromide with NaSCH$_2$CH$_3$. If the bromide is 92% optically pure and has a rotation of −38°, optically pure

2-bromobutane therefore has a rotation of $\dfrac{38}{0.92}$, or ±41°.

8.39 (*a*) Tertiary alkyl halides undergo nucleophilic substitution only by way of carbocations: S_N1 is the most likely mechanism for solvolysis of the 2-halo-2-methylbutanes.

$$\begin{array}{c} X \\ | \\ CH_3\overset{}{C}CH_2CH_3 \\ | \\ CH_3 \end{array}$$

2-Halo-2-methylbutanes are
tertiary alkyl halides.

(*b*) Tertiary alkyl halides can undergo either E1 or E2 elimination. Because no alkoxide base is present, solvolytic elimination most likely occurs by an E1 mechanism.

(*c, d*) Iodides react faster than bromides in substitution and elimination reactions irrespective of whether the mechanism is E1, E2, S_N1, or S_N2.

(*e*) Solvolysis in aqueous ethanol can give rise to an alcohol or an ether as product, depending on whether the carbocation is captured by water or ethanol.

$$\underset{X}{(CH_3)_2CCH_2CH_3} \xrightarrow[\text{H}_2\text{O}]{\text{CH}_3\text{CH}_2\text{OH}} \underset{OH}{(CH_3)_2CCH_2CH_3} + \underset{OCH_2CH_3}{(CH_3)_2CCH_2CH_3}$$

2-Methyl-2-butanol   Ethyl 1,1-
dimethylpropyl ether

(*f*) Elimination can yield either of two isomeric alkenes.

$$\underset{X}{(CH_3)_2CCH_2CH_3} \longrightarrow \underset{CH_3}{H_2C=CCH_2CH_3} + (CH_3)_2C=CHCH_3$$

2-Methyl-1-butene   2-Methyl-2-butene

Zaitsev's rule predicts that 2-methyl-2-butene should be the major alkene.

(*g*) The product distribution is determined by what happens to the carbocation intermediate. If the carbocation is free of its leaving group, its fate will be the same no matter whether the leaving group is bromide or iodide.

8.40 Both aspects of this reaction—its slow rate and the formation of a rearranged product—have their origin in the positive character developed at a primary carbon. The alcohol is protonated and the carbon–oxygen bond of the resulting alkyloxonium ion begins to break:

2,2-Dimethyl-1-propanol

As positive character develops at the primary carbon, a methyl group migrates. Rearrangement gives a tertiary carbocation, which is captured by bromide to give the product.

2-Bromo-2-methylbutane

8.41 The substrate is a tertiary alkyl bromide and can undergo S_N1 substitution and E1 elimination under these reaction conditions. Elimination in either of two directions to give regioisomeric alkenes can also occur.

2-Bromo-2-methylbutane	1,1-Dimethylpropyl acetate	2-Methyl-1-butene	2-Methyl-2-butene

8.42 Solvolysis of 1,2-dimethylpropyl p-toluenesulfonate in acetic acid is expected to give one substitution product and two alkenes.

$$(CH_3)_2CHCHCH_3 \xrightarrow{CH_3CO_2H} (CH_3)_2CHCHCH_3 + (CH_3)_2C=CHCH_3 + (CH_3)_2CHCH=CH_2$$

1,2-Dimethylpropyl p-toluenesulfonate	1,2-Dimethylpropyl acetate	2-Methyl-2-butene	3-Methyl-1-butene

Because five products are formed, we are led to consider the possibility of carbocation rearrangements in S_N1 and E1 solvolysis.

$$(CH_3)_2CH\overset{+}{C}HCH_3 \longrightarrow (CH_3)_2\overset{+}{C}CH_2CH_3 \longrightarrow (CH_3)_2CCH_2CH_3 + (CH_3)_2C=CHCH_3 + H_2C=CCH_2CH_3$$

1,2-Dimethylpropyl cation (secondary)	1,1-Dimethylpropyl cation (tertiary)	1,1-Dimethylpropyl acetate	2-Methyl-2-butene	2-Methyl-1-butene

Because 2-methyl-2-butene is a product common to both carbocation intermediates, a total of five different products are accounted for. There are two substitution products:

$$(CH_3)_2CHCHCH_3 \quad (CH_3)_2CCH_2CH_3$$

1,2-Dimethylpropyl acetate	1,1-Dimethylpropyl acetate

and three elimination products:

$$(CH_3)_2C=CHCH_3 \quad (CH_3)_2CHCH=CH_2 \quad H_2C=CCH_2CH_3$$

2-Methyl-2-butene	3-Methyl-1-butene	2-Methyl-1-butene

8.43 Solution A contains both acetate ion and methanol as nucleophiles. Acetate is more nucleophilic than methanol, and so the major observed reaction is:

$$CH_3I \quad + \quad CH_3\overset{\overset{\displaystyle O}{\|}}{C}O^- \quad \xrightarrow{\text{methanol}} \quad CH_3O\overset{\overset{\displaystyle O}{\|}}{C}CH_3$$

Methyl iodide	Acetate	Methyl acetate

Solution B prepared by adding potassium methoxide to acetic acid rapidly undergoes an acid–base reaction:

$$CH_3O^- \quad + \quad CH_3\overset{\overset{\displaystyle O}{\|}}{C}OH \quad \longrightarrow \quad CH_3OH \quad + \quad CH_3\overset{\overset{\displaystyle O}{\|}}{C}O^-$$

Methoxide (stronger base)	Acetic acid (stronger acid)	Methanol (weaker acid)	Acetate (weaker base)

Thus the major base present is not methoxide but acetate. Methyl iodide therefore reacts with acetate anion in solution B to give methyl acetate.

8.44 Alkyl chlorides arise by the reaction sequence:

Primary alcohol	p-Toluenesulfonyl chloride	Pyridine	Primary alkyl p-toluenesulfonate	Pyridinium chloride

Primary alkyl p-toluenesulfonate		Primary alkyl chloride

The reaction proceeds to form the alkyl p-toluenesulfonate as expected, but the chloride anion formed in this step subsequently acts as a nucleophile and displaces p-toluenesulfonate from RCH_2OTs.

8.45 (a) If each act of exchange (substitution) occurred with retention of configuration, there would be no observable racemization; $k_{rac} = 0$.

(R)-(−)-2-Iodooctane (*indicates radioactive label)		[(R)-(−)-2-Iodooctane]*

Therefore $\dfrac{k_{rac}}{k_{exch}} = 0.$

(*b*) If each act of exchange proceeds with inversion of configuration, (*R*)-(−)-2-iodooctane will be transformed to radioactively labeled (*S*)-(+)-2-iodooctane.

$$H_3C \overset{H}{\underset{CH_3(CH_2)_5}{\diagdown}} C{-}I \; + \; [I^*]^- \longrightarrow \; I^*{-}\overset{H}{\underset{(CH_2)_5CH_3}{C}}CH_3 \; + \; I^-$$

(*R*)-(−)-2-Iodooctane   [(*S*)-(+)-2-Iodooctane]*

Starting with 100 molecules of (*R*)-(−)-2-iodooctane, the compound will be completely racemized when 50 molecules have become radioactive. Therefore,

$$\frac{k_{rac}}{k_{exch}} = 2$$

(*c*) If radioactivity is incorporated in a stereorandom fashion, then 2-iodooctane will be 50% racemized when 50% of it has reacted. Therefore,

$$\frac{k_{rac}}{k_{exch}} = 1$$

In fact, Hughes found that the rate of racemization was twice the rate of incorporation of radioactive iodide. This experiment provided strong evidence for the belief that bimolecular nucleophilic substitution proceeds stereospecifically with inversion of configuration.

8.46 Iodide ion reacts with (*R*)-2-chlorobutane with inversion of configuration by an S_N2 process to give (*S*)-2-iodobutane.

$$I^- \; + \; H_3C\overset{H}{\underset{CH_3CH_2}{\diagdown}}C{-}Cl \; \xrightarrow[\text{acetone}]{\text{slow}} \; I{-}\overset{H}{\underset{CH_2CH_3}{C}}CH_3 \; + \; Cl^-$$

Iodide ion   (*R*)-2-Chlorobutane   (*S*)-2-Iodobutane   Chloride ion

When some of the (*R*)-2-chlorobutane has reacted, the solution contains both (*R*)-2-chlorobutane and (*S*)-2-iodobutane. Iodide is a better leaving group than chloride, so the (*S*)-2-iodobutane that is formed reacts with iodide faster than the starting (*R*)-2-chlorobutane does.

$$I{-}\overset{H}{\underset{CH_2CH_3}{C}}CH_3 \; + \; I^- \; \underset{\text{acetone}}{\overset{\text{fast}}{\rightleftharpoons}} \; I^- \; + \; H_3C\overset{H}{\underset{CH_3CH_2}{\diagdown}}C{-}I$$

(*S*)-2-Iodobutane   Iodide ion   Iodide ion   (*R*)-2-Iodobutane

Thus, the (*S*)-2-iodobutane undergoes racemization faster than the (*R*)-2-chlorobutane reacts, and the 2-iodobutane that is isolated after all the (*R*)-2-chlorobutane has reacted is racemic.

8.47 Iodide ion is both a better nucleophile than cyanide and a better leaving group than bromide. The two reactions shown are therefore faster than the reaction of cyclopentyl bromide with sodium cyanide alone.

Cyclopentyl bromide   Cyclopentyl iodide   Cyclopentyl cyanide

ANSWERS TO INTERPRETIVE PROBLEMS

8.48 D;   **8.49** A;   **8.50** C;   **8.51** D;   **8.52** B;   **8.53** B;   **8.54** C

SELF-TEST

PART A

A-1. Write the correct structure of the reactant or product omitted from each of the following. Clearly indicate stereochemistry where it is important.

(*a*) $CH_3CH_2CH_2CH_2Br$ $\xrightarrow[CH_3CH_2OH]{CH_3CH_2ONa}$?

(*b*) ? $\xrightarrow{NaCN}$

(*c*) 1-Chloro-3-methylbutane + sodium iodide $\xrightarrow{acetone}$?

(*d*) $\xrightarrow[CH_3OH]{CH_3ONa}$? (major)

(*e*)   + NaN_3 $\longrightarrow$?

(*f*)   + $NaSCH_3$ $\longrightarrow$?

(*g*) $\xrightarrow{NaSH}$?

A-2. Choose the best pair of reactants to form the following ether by an S_N2 reaction:

$$(CH_3)_2CHOCH_2CH_2CH_3$$

A-3. Outline the chemical steps necessary to convert:

(a)
$$H-\overset{\overset{\displaystyle CH_3}{|}}{\underset{\underset{\displaystyle CH_2CH_3}{|}}{C}}-OTs \quad \text{to} \quad H-\overset{\overset{\displaystyle CH_3}{|}}{\underset{\underset{\displaystyle CH_2CH_3}{|}}{C}}-CN$$

(b) (S)-2-Pentanol to (R)-$CH_3\overset{\overset{\displaystyle SCH_3}{|}}{C}HCH_2CH_2CH_3$

A-4. Hydrolysis of 3-chloro-2,2-dimethylbutane yields 2,3-dimethyl-2-butanol as the major product. Explain this observation, using structural formulas to outline the mechanism of the reaction.

A-5. Identify the class of reaction (e.g., E2), and write the kinetic and chemical equations for:

(a) The solvolysis of *tert*-butyl bromide in methanol

(b) The reaction of chlorocyclohexane with sodium azide (NaN_3).

A-6. (a) Provide a brief explanation why the halogen exchange reaction shown is an acceptable synthetic method:

$$CH_3\overset{\overset{\displaystyle Br}{|}}{C}HCH_3 + NaI \xrightarrow{\text{acetone}} CH_3\overset{\overset{\displaystyle I}{|}}{C}HCH_3 + NaBr$$

(b) Briefly explain why the reaction of 1-bromobutane with sodium azide occurs faster in dimethyl sulfoxide [$(CH_3)_2S{=}O$] than in water.

A-7. Write chemical structures for compounds A through D in the following sequence of reactions. Compounds A and C are alcohols.

$$A \xrightarrow{\text{NaNH}_2} B$$

$$C \xrightarrow{\text{HBr, heat}} D$$

$$B + D \longrightarrow CH_3CH_2O{-}\bigcirc$$

A-8. Write a mechanism describing the solvolysis (S_N1) of 1-bromo-1-methylcyclohexane in ethanol.

A-9. Solvolysis of the compound shown occurs with carbocation rearrangement and yields an alcohol as the major product. Write the structure of this product, and give a mechanism to explain its formation.

$$CH_3CH_2CH_2\overset{\overset{\displaystyle }{|}}{C}HCH(CH_3)_2 \xrightarrow{\text{H}_2\text{O}} ?$$
$$\underset{\displaystyle Br}{|}$$

PART B

B-1. The bimolecular substitution reaction

$$CH_3Br + OH^- \longrightarrow CH_3OH + Br^-$$

is represented by the kinetic equation:

(a) Rate = $k[CH_3Br]^2$

(c) Rate = $k[CH_3Br] + k[OH^-]$

(b) Rate = $k[CH_3Br][OH^-]$

(d) Rate = $\dfrac{k}{[CH_3Br][OH^-]}$

B-2. Which compound undergoes nucleophilic substitution with NaCN at the fastest rate?

(a) Br

(d) Br

(b) Br

(e) Br

(c) Br

B-3. For the reaction

the major product is formed by

(a) An S_N1 reaction

(c) An E1 reaction

(b) An S_N2 reaction

(d) An E2 reaction

B-4. Which of the following statements pertaining to an S_N2 reaction are true?

1. The rate of reaction is independent of the concentration of the nucleophile.
2. The nucleophile attacks carbon on the side of the molecule opposite the group being displaced.
3. The reaction proceeds with simultaneous bond formation and bond rupture.
4. Partial racemization of an optically active substrate results.

(a) 1, 4   (b) 1, 3, 4   (c) 2, 3   (d) All

B-5. Which one of the following alkyl halides would be expected to give the *highest* substitution-to-elimination ratio (most substitution, least elimination) on treatment with sodium ethoxide in ethanol?

(a) $(CH_3)_3C$ — Br, CH$_2$CH$_3$

(c) $(CH_3)_3C$ — CHCH$_3$, Br

(b) $(CH_3)_3C$ — CH$_2$CH$_3$, Br

(d) $(CH_3)_3C$ — CH$_2$CH$_2$Br

B-6. Which of the following phrases is (are) *not* correctly associated with an S_N1 reaction?

1. Rearrangement is possible.
2. Rate is affected by solvent polarity.
3. The strength of the nucleophile is important in determining rate.

4. The reactivity series is tertiary > secondary > primary.

5. Proceeds with complete inversion of configuration.

(a) 3, 5   (b) 5 only   (c) 2, 3, 5   (d) 3 only

B-7. Rank the following in order of decreasing rate of solvolysis with aqueous ethanol (fastest → slowest):

$$H_2C=\overset{\overset{CH_3}{|}}{C}-Br \qquad \qquad CH_3\overset{\overset{Br}{|}}{C}HCH_2CH(CH_3)_2$$

1   2   3

(a) 2 > 1 > 3   (b) 1 > 2 > 3   (c) 2 > 3 > 1   (d) 1 > 3 > 2

B-8. Rank the following species in order of decreasing nucleophilicity in a polar protic solvent (most → least nucleophilic):

$$CH_3CH_2CH_2O^- \qquad CH_3CH_2CH_2S^- \qquad CH_3CH_2\overset{\overset{O}{||}}{C}O^-$$

1   2   3

(a) 3 > 1 > 2   (b) 2 > 3 > 1   (c) 1 > 3 > 2   (d) 2 > 1 > 3

B-9. From each of the following pairs select the compound that will react faster with sodium iodide in acetone.

2-Chloropropane   or   2-bromopropane
1   2

1-Bromobutane   or   2-bromobutane
3   4

(a) 1, 3   (b) 1, 4   (c) 2, 3   (d) 2, 4

B-10. Select the reagent that will yield the greater amount of substitution on reaction with 1-bromobutane.

(a) CH_3CH_2OK in dimethyl sulfoxide (DMSO)

(b) $(CH_3)_3COK$ in dimethyl sulfoxide (DMSO)

(c) Both (a) and (b) will give comparable amounts of substitution.

(d) Neither (a) nor (b) will give any appreciable amount of substitution.

B-11. The reaction of (R)-1-chloro-3-methylpentane with sodium iodide in acetone will yield 1-iodo-3-methylpentane that is

(a) R   (b) S   (c) A mixture of R and S   (d) Meso   (e) None of these

B-12. What is the principal product of the following reaction?

$$
\begin{array}{c}
CH_3 \\
H{-}{\mid}{-}Br \\
H{-}{\mid}{-}H \quad + \quad NaN_3 \longrightarrow \\
H{-}{\mid}{-}Cl \\
CH_3
\end{array}
$$

(a)	(b)	(c)	(d)

B-13. Which of the following statements is true?

(a) $CH_3CH_2S^-$ is both a stronger base and more nucleophilic than $CH_3CH_2O^-$.

(b) $CH_3CH_2S^-$ is a stronger base but is less nucleophilic than $CH_3CH_2O^-$.

(c) $CH_3CH_2S^-$ is a weaker base but is more nucleophilic than $CH_3CH_2O^-$.

(d) $CH_3CH_2S^-$ is both a weaker base and less nucleophilic than $CH_3CH_2O^-$.

B-14. Which of the following alkyl halides would be most likely to give a rearranged product under S_N1 conditions?

(a)　　　(b)　　　(c)　　　(d)

(e) None of these. Rearrangements only occur under S_N2 conditions.

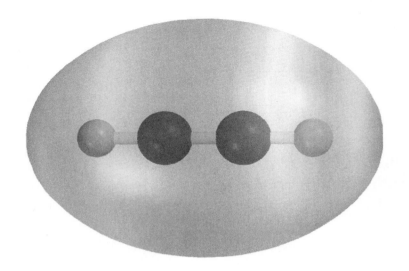

CHAPTER 9

Alkynes

SOLUTIONS TO TEXT PROBLEMS

9.1 The reaction is an acid–base process; water is the proton donor. Two separate proton-transfer steps are involved.

$$:\bar{C}\equiv\bar{C}: \quad + \quad H-\ddot{O}-H \quad \longrightarrow \quad :\bar{C}\equiv C-H \quad + \quad {}^-\!:\ddot{O}-H$$

Carbide ion   Water   Acetylide ion   Hydroxide ion

$$H-\ddot{O}-H \quad + \quad :\bar{C}\equiv C-H \quad \longrightarrow \quad H-\ddot{O}:^- \quad + \quad H-C\equiv C-H$$

Water   Acetylide ion   Hydroxide ion   Acetylene

9.2 A triple bond may connect C-1 and C-2 or C-2 and C-3 in an unbranched chain of five carbons.

$$CH_3CH_2CH_2C\equiv CH \qquad CH_3CH_2C\equiv CCH_3$$

1-Pentyne   2-Pentyne
or   or
Pent-1-yne   Pent-2-yne

One of the C_5H_8 isomers has a branched carbon chain.

$$CH_3CHC\equiv CH$$
$$|$$
$$CH_3$$

3-Methyl-1-butyne
or
3-Methylbut-1-yne

9.3 The bonds become shorter and stronger in the series as the electronegativity increases.

	NH_3	H_2O	HF
Electronegativity:	N (3.0)	O (3.5)	F (4.0)
Bond distance (pm):	N—H (101)	O—H (95)	F—H (92)
Bond dissociation enthalpy (kJ/mol):	N—H (435)	O—H (497)	F—H (568)
Bond dissociation enthalpy (kcal/mol):	N—H (104)	O—H (119)	F—H (136)

9.4 (*b*) A proton is transferred from acetylene to ethyl anion.

$$HC\equiv C-H \ + \ :\bar{C}H_2CH_3 \ \rightleftharpoons \ HC\equiv C: \ + \ CH_3CH_3$$

Acetylene	Ethyl anion	Acetylide ion	Ethane
(stronger acid)	(stronger base)	(weaker base)	(weaker acid)
(pK_a 26)			($pK_a \approx 62$)

The position of equilibrium lies to the right. Ethyl anion is a very powerful base and deprotonates acetylene quantitatively.

(*c*) Amide ion is not a strong enough base to remove a proton from ethylene. The equilibrium lies to the left.

$$H_2C=CH-H \ + \ :\ddot{N}H_2 \ \rightleftharpoons \ H_2C=\ddot{C}H \ + \ :NH_3$$

Ethylene	Amide anion	Vinyl anion	Ammonia
(weaker acid)	(weaker base)	(stronger base)	(stronger acid)
($pK_a \approx 45$)			(pK_a 36)

(*d*) Alcohols are stronger acids than ammonia; the position of equilibrium lies to the right.

$$CH_3C\equiv CCH_2\ddot{O}-H \ + \ :\ddot{N}H_2 \ \rightleftharpoons \ CH_3C\equiv CCH_2\ddot{O}: \ + \ :NH_3$$

2-Butyn-1-ol	Amide ion	2-Butyn-1-olate anion	Ammonia
(stronger acid)	(stronger base)	(weaker base)	(weaker acid)
($pK_a \approx 16$–20)			(pK_a 36)

9.5 (*b*) The desired alkyne has a methyl group and a butyl group attached to a —C≡C— unit. Two alkylations of acetylene are therefore required: one with a methyl halide, the other with a butyl halide.

$$HC\equiv CH \xrightarrow[\text{2. } CH_3Br]{\text{1. } NaNH_2,\ NH_3} CH_3C\equiv CH \xrightarrow[\text{2. } CH_3CH_2CH_2CH_2Br]{\text{1. } NaNH_2,\ NH_3} CH_3C\equiv CCH_2CH_2CH_2CH_3$$

Acetylene	Propyne	2-Heptyne

It does not matter whether the methyl group or the butyl group is introduced first; the order of steps shown in this synthetic scheme may be inverted.

(*c*) An ethyl group and a propyl group need to be introduced as substituents on a —C≡C— unit. As in part (*b*), it does not matter which of the two is introduced first.

$$HC\equiv CH \xrightarrow[\text{2. } CH_3CH_2CH_2Br]{\text{1. } NaNH_2,\ NH_3} CH_3CH_2CH_2C\equiv CH \xrightarrow[\text{2. } CH_3CH_2Br]{\text{1. } NaNH_2,\ NH_3} CH_3CH_2CH_2C\equiv CCH_2CH_3$$

Acetylene	1-Pentyne	3-Heptyne

9.6 Both 1-pentyne and 2-pentyne can be prepared by alkylating acetylene. All the alkylation steps involve nucleophilic substitution of a methyl or primary alkyl halide.

$$HC\equiv CH \xrightarrow[\text{2. } CH_3CH_2CH_2Br]{\text{1. } NaNH_2, NH_3} CH_3CH_2CH_2C\equiv CH$$

Acetylene   1-Pentyne

$$HC\equiv CH \xrightarrow[\text{2. } CH_3CH_2Br]{\text{1. } NaNH_2, NH_3} CH_3CH_2C\equiv CH \xrightarrow[\text{2. } CH_3Br]{\text{1. } NaNH_2, NH_3} CH_3CH_2C\equiv CCH_3$$

Acetylene   1-Butyne   2-Pentyne

A third isomer, 3-methyl-1-butyne, cannot be prepared by alkylation of acetylene, because it requires a secondary alkyl halide as the alkylating agent. The reaction that takes place is elimination, not substitution.

$$HC\equiv \overset{\cdot\cdot}{C}{:}^{-} + CH_3\underset{\underset{Br}{|}}{C}HCH_3 \xrightarrow{E2} HC\equiv CH + H_2C=CHCH_3$$

Acetylide   Isopropyl   Acetylene   Propene
ion   bromide

9.7 Each of the dibromides shown yields 3,3-dimethyl-1-butyne when subjected to double dehydrohalogenation with strong base.

$$(CH_3)_3\underset{\underset{Br}{|}}{\overset{\overset{Br}{|}}{C}}CCH_3 \quad\text{or}\quad (CH_3)_3CCH_2CHBr_2 \quad\text{or}\quad (CH_3)_3C\underset{\underset{Br}{|}}{C}HCH_2Br \xrightarrow[\text{2. } H_2O]{\text{1. } 3NaNH_2} (CH_3)_3CC\equiv CH$$

2,2-Dibromo-3,3-   1,1-Dibromo-3,3-   1,2-Dibromo-3,3-   3,3-Dimethyl-1-butyne
dimethylbutane   dimethylbutane   dimethylbutane

9.8 (*b*) The first task is to convert 1-propanol to propene:

$$CH_3CH_2CH_2OH \xrightarrow[\text{heat}]{H_2SO_4} CH_3CH=CH_2$$

1-Propanol   Propene

After propene is available, it is converted to 1,2-dibromopropane and then to propyne as described in the sample solution for part (*a*).

(*c*) Treat isopropyl bromide with a base to effect dehydrohalogenation.

$$(CH_3)_2CHBr \xrightarrow{NaOCH_2CH_3} CH_3CH=CH_2$$

Isopropyl bromide   Propene

Next, convert propene to propyne as in parts (*a*) and (*b*).

(*d*) The starting material contains only two carbon atoms, and so an alkylation step is needed at some point. Propyne arises by alkylation of acetylene, and so the last step in the synthesis is

$$HC\equiv CH \xrightarrow[\text{2. } CH_3Br]{\text{1. } NaNH_2, NH_3} CH_3C\equiv CH$$

Acetylene   Propyne

The designated starting material, 1,1-dichloroethane, is a geminal dihalide and can be used to prepare acetylene by a double dehydrohalogenation.

$$CH_3CHCl_2 \xrightarrow[\text{2. H}_2\text{O}]{\text{1. NaNH}_2, \text{NH}_3} HC\equiv CH$$

1,1-Dichloroethane   Acetylene

(*e*) The first task is to convert ethyl alcohol to acetylene. Once acetylene is prepared it can be alkylated with a methyl halide.

$$CH_3CH_2OH \xrightarrow[\text{heat}]{\text{H}_2\text{SO}_4} H_2C=CH_2 \xrightarrow{\text{Br}_2} BrCH_2CH_2Br \xrightarrow[\text{2. H}_2\text{O}]{\text{1. NaNH}_2, \text{NH}_3} HC\equiv CH \xrightarrow[\text{2. CH}_3\text{Br}]{\text{1. NaNH}_2, \text{NH}_3} CH_3C\equiv CH$$

Ethyl alcohol   Ethylene   1,2-Dibromoethane   Acetylene   Propyne

9.9 The first task is to assemble a carbon chain containing eight carbons. Acetylene has two carbon atoms and can be alkylated via its sodium salt to 1-octyne. Hydrogenation over platinum converts 1-octyne to octane.

$$HC\equiv CH \xrightarrow[\text{NH}_3]{\text{NaNH}_2} HC\equiv CNa \xrightarrow{BrCH_2(CH_2)_4CH_3} HC\equiv CCH_2(CH_2)_4CH_3 \xrightarrow[\text{Pt}]{\text{H}_2 \text{ (2 moles)}} CH_3CH_2CH_2(CH_2)_4CH_3$$

Acetylene   Sodium acetylide   1-Octyne   Octane

Alternatively, two successive alkylations of acetylene with $CH_3CH_2CH_2Br$ could be carried out to give 4-octyne ($CH_3CH_2CH_2C\equiv CCH_2CH_2CH_3$), which could then be hydrogenated to octane.

9.10 As the equation preceding the problem in the text shows, *cis*-5-decene can be prepared by reduction of 5-decyne using the Lindlar catalyst.

$$CH_3(CH_2)_3C\equiv C(CH_2)_3CH_3 \xrightarrow[\text{Lindlar Pd}]{\text{H}_2}$$

5-Decyne   *cis*-5-Decene

The task, then, is to prepare 5-decyne from acetylene. This can be accomplished by two successive alkylations using 1-bromobutane.

$$HC\equiv CH \xrightarrow[\text{2. CH}_3(\text{CH}_2)_3\text{Br}]{\text{1. NaNH}_2, \text{NH}_3} HC\equiv C(CH_2)_3CH_3 \xrightarrow[\text{2. CH}_3(\text{CH}_2)_3\text{Br}]{\text{1. NaNH}_2, \text{NH}_3} CH_3(CH_2)_3C\equiv C(CH_2)_3CH_3$$

Acetylene   1-Hexyne   5-Decyne

9.11 The desired trans stereochemistry of the product may be achieved by metal–ammonia reduction of 2-heptyne.

$$CH_3C\equiv C(CH_2)_3CH_3 \xrightarrow[\text{NH}_3]{\text{Na}}$$

2-Heptyne   *trans*-2-Heptene

2-Heptyne is available from propyne by alkylation with 1-bromobutane.

$$CH_3C\equiv CH \xrightarrow[\text{2. CH}_3(\text{CH}_2)_3\text{Br}]{\text{1. NaNH}_2, \text{NH}_3} CH_3C\equiv C(CH_2)_3CH_3$$

Propyne   2-Heptyne

9.12 (*b*) Addition of hydrogen chloride to vinyl chloride gives the geminal dichloride 1,1-dichloroethane.

$$H_2C=CHCl \xrightarrow{\text{HCl}} CH_3CHCl_2$$

Vinyl chloride   1,1-Dichloroethane

(*c*) Because 1,1-dichloroethane can be prepared by adding 2 moles of hydrogen chloride to acetylene as shown in the sample solution to part (*a*), first convert 1,1-dibromoethane to acetylene by dehydrohalogenation.

$$CH_3CHBr_2 \xrightarrow[\text{2. } H_2O]{\text{1. } NaNH_2, NH_3} HC\equiv CH \xrightarrow{\text{2HCl}} CH_3CHCl_2$$

1,1-Dibromoethane   Acetylene   1,1-Dichloroethane

9.13 The enol arises by addition of water to the triple bond.

$$CH_3C\equiv CCH_3 + H_2O \longrightarrow \underset{\underset{OH}{|}}{CH_3C=CHCH_3} \longrightarrow CH_3\overset{\overset{O}{\|}}{C}CH_2CH_3$$

2-Butyne   2-Buten-2-ol (enol form)   2-Butanone

The mechanism described in the textbook (Mechanism 9.2) is adapted to the case of 2-butyne hydration as shown:

Hydronium Ion   2-Buten-2-ol   Water   Carbocation

Carbocation   Water   2-Butanone   Hydronium ion

9.14 Hydration of 1-octyne gives 2-octanone according to the equation that immediately precedes this problem in the text. Prepare 1-octyne as described in the solution to Problem 9.9, and then carry out its hydration in the presence of mercury(II) sulfate and sulfuric acid.

Hydration of 4-octyne gives 4-octanone. Prepare 4-octyne as described in the solution to Problem 9.9.

9.15 Each of the carbons that are part of —CO_2H groups was once part of a —$C\equiv C$— unit. The two fragments $CH_3(CH_2)_4CO_2H$ and $HO_2CCH_2CH_2CO_2H$ account for only 10 of the original 16 carbons. The full complement of carbons can be accommodated by assuming that two molecules of $CH_3(CH_2)_4CO_2H$ are formed, along with one molecule of $HO_2CCH_2CH_2CO_2H$. The starting alkyne is therefore deduced from the ozonolysis data to be as shown:

$$CH_3(CH_2)_4C\equiv CCH_2CH_2C\equiv C(CH_2)_4CH_3$$

$$CH_3(CH_2)_4CO_2H \quad HO_2CCH_2CH_2CO_2H \quad HO_2C(CH_2)_4CH_3$$

9.16 Three isomers have unbranched carbon chains:

$$CH_3CH_2CH_2CH_2C \equiv CH \qquad CH_3CH_2CH_2C \equiv CCH_3 \qquad CH_3CH_2C \equiv CCH_2CH_3$$

1-Hexyne   2-Hexyne   3-Hexyne

Next consider all the alkynes with a single methyl branch:

$$\underset{CH_3}{CH_3CHCH_2C} \equiv CH \qquad \underset{CH_3}{CH_3CH_2CHC} \equiv CH \qquad \underset{CH_3}{CH_3CHC} \equiv CCH_3$$

4-Methyl-1-pentyne   3-Methyl-1-pentyne   4-Methyl-2-pentyne

One isomer has two methyl branches. None is possible with an ethyl branch.

$$\overset{CH_3}{\underset{CH_3}{CH_3CC}} \equiv CH$$

3,3-Dimethyl-1-butyne

9.17 (a) $\overset{5}{C}H_3\overset{4}{C}H_2\overset{3}{C}H_2\overset{2}{C} \equiv \overset{1}{C}H$ is 1 pentyne

(b) $\overset{5}{C}H_3\overset{4}{C}H_2\overset{3}{C} \equiv \overset{2}{C}\overset{1}{C}H_3$ is 2-pentyne

(c) $\overset{1}{C}H_3\overset{2}{C} \equiv \overset{3}{C}\overset{4}{C}H\overset{5}{C}H\overset{6}{C}H_3$ is 4,5-dimethyl-2-hexyne
$\qquad\qquad \underset{H_3C \quad CH_3}{|\quad\quad|}$

(d) $\triangleright\!\!-\overset{5}{C}H_2\overset{4}{C}H_2\overset{3}{C}H_2\overset{2}{C} \equiv \overset{1}{C}H$ is 5-cyclopropyl-1-pentyne

(e) $\overset{13}{C}H_2\overset{1}{C} \equiv \overset{2}{C}\overset{3}{C}H_2$ is cyclotridecyne

(f) $CH_3CH_2CH_2CH_2\overset{4}{C}H\overset{5}{C}H_2\overset{6}{C}H_2\overset{7}{C}H_2\overset{8}{C}H_2\overset{9}{C}H_3$ is 4-butyl-2-nonyne
$\qquad\qquad\qquad\qquad \underset{\overset{3}{C} \equiv \overset{2}{C}\overset{1}{C}H_3}{|}$

(Parent chain must contain the triple bond.)

(g) $\overset{1}{C}H_3\overset{2}{\underset{CH_3}{\overset{CH_3}{C}}}\overset{3}{C} \equiv \overset{4}{C}\overset{5}{\underset{CH_3}{\overset{CH_3}{C}}}\overset{6}{C}H_3$ is 2,2,5,5-tetramethyl-3-hexyne

9.18 (a) 1-Octyne is $HC \equiv CCH_2CH_2CH_2CH_2CH_2CH_3$

(b) 2-Octyne is $CH_3C \equiv CCH_2CH_2CH_2CH_2CH_3$

(c) 3-Octyne is $CH_3CH_2C \equiv CCH_2CH_2CH_2CH_3$

(d) 4-Octyne is $CH_3CH_2CH_2C \equiv CCH_2CH_2CH_3$

(e) 2,5-Dimethyl-3-hexyne is $\underset{CH_3}{CH_3CHC} \equiv \underset{CH_3}{CCHCH_3}$

(*f*)   4-Ethyl-1-hexyne is $CH_3CH_2CHCH_2C{\equiv}CH$
                           $|$
                         CH_2CH_3

(*g*)   Ethynylcyclohexane is   ⬡—$C{\equiv}CH$

(*h*)   3-Ethyl-3-methyl-1-pentyne is $\overset{\displaystyle CH_3}{\underset{\displaystyle CH_2CH_3}{CH_3CH_2\overset{|}{\underset{|}{C}}C{\equiv}CH}}$

9.19   Ethynylcyclohexane has the molecular formula C_8H_{12}. All the other compounds are C_8H_{14}.

9.20   Only alkynes with the carbon skeletons shown can give 3-ethylhexane on catalytic hydrogenation.

| 3-Ethyl-1-hexyne | | 4-Ethyl-1-hexyne | | 4-Ethyl-2-hexyne | $\xrightarrow[\text{Pt}]{\text{H}_2\ (2\ \text{moles})}$ | 3-Ethylhexane |

9.21   The carbon skeleton of the unknown acetylenic amino acid must be the same as that of homoleucine. The structure of homoleucine is such that there is only one possible location for a carbon–carbon triple bond in an acetylenic precursor.

$$HC{\equiv}CCHCH_2\overset{\displaystyle O}{\overset{\|}{C}}CO^- \xrightarrow[\text{Pt}]{\text{H}_2\ (2\ \text{moles})} CH_3CH_2CHCH_2\overset{\displaystyle O}{\overset{\|}{C}}CO^-$$

            CH_3 $^+NH_3$ CH_3 $^+NH_3$

            $C_7H_{11}NO_2$ Homoleucine

9.22   (*a*) $CH_3CH_2CH_2CH_2CH_2CHCl_2 \xrightarrow[\text{2. H}_2\text{O}]{\text{1. NaNH}_2,\ \text{NH}_3} CH_3CH_2CH_2CH_2C{\equiv}CH$

               1,1-Dichlorohexane   1-Hexyne

(*b*)

$$CH_3CH_2CH_2CH_2CH{=}CH_2 \xrightarrow[\text{CCl}_4]{\text{Br}_2} CH_3CH_2CH_2CH_2\underset{\displaystyle Br}{\overset{|}{C}}HCH_2Br \xrightarrow[\text{2. H}_2\text{O}]{\text{1. NaNH}_2,\ \text{NH}_3} CH_3CH_2CH_2CH_2C{\equiv}CH$$

          1-Hexene   1,2-Dibromohexane   1-Hexyne

(*c*) $HC{\equiv}CH \xrightarrow[\text{NH}_3]{\text{NaNH}_2} HC{\equiv}C{:}^-\ Na^+ \xrightarrow{CH_3CH_2CH_2CH_2Br} CH_3CH_2CH_2CH_2C{\equiv}CH$

        Acetylene   1-Hexyne

(*d*) $CH_3CH_2CH_2CH_2CH_2CH_2I \xrightarrow[\text{DMSO}]{\text{KOC(CH}_3)_3} CH_3CH_2CH_2CH_2CH{=}CH_2$

               1-Iodohexane   1-Hexene

1-Hexene is then converted to 1-hexyne as in part (*b*).

9.23 (*a*) Working backward from the final product, it can be seen that preparation of 1-butyne will allow the desired carbon skeleton to be constructed.

$$CH_3CH_2C\equiv CCH_2CH_3 \quad \text{prepared from} \quad CH_3CH_2C\equiv C:^- \; + \; BrCH_2CH_3$$

3-Hexyne

The desired intermediate, 1-butyne, is available by halogenation followed by double dehydrohalogenation of 1-butene.

$$CH_3CH_2CH{=}CH_2 \; + \; Br_2 \longrightarrow \overset{\overset{\displaystyle Br}{|}}{CH_3CH_2CHCH_2Br} \; \xrightarrow[\text{2. H}_2\text{O}]{\text{1. NaNH}_2,\ \text{NH}_3} \; CH_3CH_2C\equiv CH$$

1-Butene   1-Butyne

Reaction of the anion of 1-butyne with ethyl bromide completes the synthesis.

$$CH_3CH_2C\equiv CH \xrightarrow[\text{NH}_3]{\text{NaNH}_2} CH_3CH_2C\equiv C:^- \; Na^+ \xrightarrow{CH_3CH_2Br} CH_3CH_2C\equiv CCH_2CH_3$$

1-Butyne   3-Hexyne

(*b*) Double dehydrohalogenation of 1,1-dichlorobutane yields 1-butyne. The synthesis is completed as in part (*a*).

$$CH_3CH_2CH_2CHCl_2 \xrightarrow[\text{2. H}_2\text{O}]{\text{1. NaNH}_2,\ \text{NH}_3} CH_3CH_2C\equiv CH$$

1,1-Dichlorobutane   1-Butyne

(*c*) 1-Butyne can be prepared from acetylene as follows.

$$HC\equiv CH \xrightarrow[\text{NH}_3]{\text{NaNH}_2} HC\equiv C:^- \; Na^+ \xrightarrow{CH_3CH_2Br} HC\equiv CCH_2CH_3$$

Acetylene   1-Butyne

1-Butyne is then converted to 3-hexyne as in part (*a*).

9.24 A single dehydrobromination step occurs in the conversion of 1,2-dibromodecane to $C_{10}H_{19}Br$. Bromine may be lost from C-1 to give 2-bromo-1-decene.

$$\underset{\overset{\displaystyle |}{Br}}{BrCH_2CH}(CH_2)_7CH_3 \xrightarrow[\text{ethanol–water}]{\text{KOH}} \underset{\overset{\displaystyle |}{Br}}{H_2C{=}C}(CH_2)_7CH_3$$

1,2-Dibromodecane   2-Bromo-1-decene

Loss of bromine from C-2 gives (*E*)- and (*Z*)-1-bromo-1-decene.

$$\underset{\overset{\displaystyle |}{Br}}{BrCH_2CH}(CH_2)_7CH_3 \xrightarrow[\text{ethanol–water}]{\text{KOH}} \quad \overset{H}{\underset{Br}{}}C{=}C\overset{(CH_2)_7CH_3}{\underset{H}{}} \; + \; \overset{Br}{\underset{H}{}}C{=}C\overset{(CH_2)_7CH_3}{\underset{H}{}}$$

1,2-Dibromodecane   (*E*)-1-Bromo-1-decene   (*Z*)-1-Bromo-1-decene

9.25 (*a*) $CH_3CH_2CH_2CH_2C\equiv CH$ + $2H_2$ $\xrightarrow{\text{Pt}}$ $CH_3CH_2CH_2CH_2CH_2CH_3$

1-Hexyne   Hexane

(*b*) $CH_3CH_2CH_2CH_2C\equiv CH$ + H_2 $\xrightarrow{\text{Lindlar Pd}}$ $CH_3CH_2CH_2CH_2CH=CH_2$

1-Hexyne   1-Hexene

(*c*) $CH_3CH_2CH_2CH_2C\equiv CH$ $\xrightarrow[\text{NH}_3]{\text{Li}}$ $CH_3CH_2CH_2CH_2CH=CH_2$

1-Hexyne   1-Hexene

(*d*) $CH_3CH_2CH_2CH_2C\equiv CH$ $\xrightarrow[\text{NH}_3]{\text{NaNH}_2}$ $CH_3CH_2CH_2CH_2C\equiv C\colon^-$ Na^+

1-Hexyne   Sodium 1-hexynide

(*e*)
$CH_3CH_2CH_2CH_2C\equiv C\colon^-$ Na^+ + $CH_3CH_2CH_2CH_2Br$ $\longrightarrow$ $CH_3CH_2CH_2CH_2C\equiv CCH_2CH_2CH_2CH_3$

Sodium 1-hexynide   1-Bromobutane   5-Decyne

(*f*)
$CH_3CH_2CH_2CH_2C\equiv C\colon^-$ Na^+ + $(CH_3)_3CBr$ $\longrightarrow$ $CH_3CH_2CH_2CH_2C\equiv CH$ + $(CH_3)_2C=CH_2$

Sodium 1-hexynide   *tert*-Butyl   1-Hexyne   2-Methylpropene
                                                    bromide

(*g*) $CH_3CH_2CH_2CH_2C\equiv CH$ $\xrightarrow[\text{(1 mol)}]{\text{HCl}}$ $CH_3CH_2CH_2CH_2\underset{\underset{Cl}{|}}{C}=CH_2$

1-Hexyne   2-Chloro-1-hexene

(*h*) $CH_3CH_2CH_2CH_2C\equiv CH$ $\xrightarrow[\text{(2 mol)}]{\text{HCl}}$ $CH_3CH_2CH_2CH_2\overset{\overset{Cl}{|}}{\underset{\underset{Cl}{|}}{C}}CH_3$

1-Hexyne   2,2-Dichlorohexane

(*i*) $CH_3CH_2CH_2CH_2C\equiv CH$ $\xrightarrow[\text{(1 mol)}]{\text{Cl}_2}$ $\underset{Cl}{\overset{CH_3CH_2CH_2CH_2}{C}}=\underset{H}{\overset{Cl}{C}}$

1-Hexyne   (*E*)-1,2-Dichloro-1-hexene

(*j*) $CH_3CH_2CH_2CH_2C\equiv CH$ $\xrightarrow[\text{(2 mol)}]{\text{Cl}_2}$ $CH_3CH_2CH_2CH_2\overset{\overset{Cl}{|}}{\underset{\underset{Cl}{|}}{C}}CHCl_2$

1-Hexyne   1,1,2,2-Tetrachlorohexane

(k) $CH_3CH_2CH_2CH_2C\equiv CH$ $\xrightarrow[\text{HgSO}_4]{\text{H}_2\text{O, H}_2\text{SO}_4}$ $CH_3CH_2CH_2CH_2\overset{\displaystyle O}{\overset{\|}{C}}CH_3$

1-Hexyne

2-Hexanone

(l) $CH_3CH_2CH_2CH_2C\equiv CH$ $\xrightarrow[\text{2. H}_2\text{O}]{\text{1. O}_3}$ $CH_3CH_2CH_2CH_2\overset{\displaystyle O}{\overset{\|}{C}}OH$ + $HO\overset{\displaystyle O}{\overset{\|}{C}}OH$

1-Hexyne

Pentanoic acid

Carbonic acid

9.26 (a) $CH_3CH_2C\equiv CCH_2CH_3$ + $2H_2$ $\xrightarrow{\text{Pt}}$ $CH_3CH_2CH_2CH_2CH_2CH_3$

3-Hexyne

Hexane

(b) $CH_3CH_2C\equiv CCH_2CH_3$ + H_2 $\xrightarrow{\text{Lindlar Pd}}$

$$\underset{H}{\overset{CH_3CH_2}{>}}C=C\underset{H}{\overset{CH_2CH_3}{<}}$$

3-Hexyne

(Z)-3-Hexene

(c) $CH_3CH_2C\equiv CCH_2CH_3$ $\xrightarrow[\text{NH}_3]{\text{Li}}$

$$\underset{H}{\overset{CH_3CH_2}{>}}C=C\underset{CH_2CH_3}{\overset{H}{<}}$$

3-Hexyne

(E)-3-Hexene

(d) $CH_3CH_2C\equiv CCH_2CH_3$ $\xrightarrow[\text{(1 mol)}]{\text{HCl}}$

$$\underset{Cl}{\overset{CH_3CH_2}{>}}C=C\underset{CH_2CH_3}{\overset{H}{<}}$$

3-Hexyne

(Z)-3-Chloro-3-hexene

(e) $CH_3CH_2C\equiv CCH_2CH_3$ $\xrightarrow[\text{(2 mol)}]{\text{HCl}}$ $CH_3CH_2\overset{\displaystyle Cl}{\underset{\displaystyle Cl}{\overset{|}{\underset{|}{C}}}}CH_2CH_2CH_3$

3-Hexyne

3,3-Dichlorohexane

(f) $CH_3CH_2C\equiv CCH_2CH_3$ $\xrightarrow[\text{(1 mol)}]{\text{Cl}_2}$

$$\underset{Cl}{\overset{CH_3CH_2}{>}}C=C\underset{CH_2CH_3}{\overset{Cl}{<}}$$

3-Hexyne

(E)-3,4-Dichloro-3-hexene

(g) $CH_3CH_2C\equiv CCH_2CH_3$ $\xrightarrow[\text{(2 mol)}]{\text{Cl}_2}$ $CH_3CH_2\overset{\displaystyle Cl\ \ Cl}{\underset{\displaystyle Cl\ \ Cl}{\overset{|\ \ \ |}{\underset{|\ \ \ |}{C-C}}}}CH_2CH_3$

3-Hexyne

3,3,4,4-Tetrachlorohexane

(h) $CH_3CH_2C{\equiv}CCH_2CH_3$ $\xrightarrow[\text{HgSO}_4]{\text{H}_2\text{O, H}_2\text{SO}_4}$ $CH_3CH_2\overset{\overset{\text{O}}{\|}}{C}CH_2CH_2CH_3$

3-Hexyne   3-Hexanone

(i) $CH_3CH_2C{\equiv}CCH_2CH_3$ $\xrightarrow[\text{2. H}_2\text{O}]{\text{1. O}_3}$ $2CH_3CH_2\overset{\overset{\text{O}}{\|}}{C}OH$

3-Hexyne   Propanoic acid

9.27 The two carbons of the triple bond are similarly but not identically substituted in 2-heptyne, $CH_3C{\equiv}CCH_2CH_2CH_2CH_3$. Two regioisomeric enols are formed, each of which gives a different ketone.

$CH_3C{\equiv}CCH_2CH_2CH_2CH_3$ $\xrightarrow[\text{HgSO}_4]{\text{H}_2\text{O, H}_2\text{SO}_4}$ $CH_3\overset{\overset{\text{OH}}{|}}{C}{=}CHCH_2CH_2CH_2CH_3$ + $CH_3CH{=}\overset{\overset{\text{OH}}{|}}{C}CH_2CH_2CH_2CH_3$

2-Heptyne   2-Hepten-2-ol   2-Hepten-3-ol

$CH_3\overset{\overset{\text{O}}{\|}}{C}CH_2CH_2CH_2CH_2CH_3$ $CH_3CH_2\overset{\overset{\text{O}}{\|}}{C}CH_2CH_2CH_2CH_3$

2-Heptanone   3-Heptanone

9.28 The alkane formed by hydrogenation of (S)-3-methyl-1-pentyne is achiral; it cannot be optically active.

$\underset{\text{H}_3\text{C}}{\overset{\text{CH}_3\text{CH}_2}{}}\overset{\text{H}}{\underset{}{C}}{-}C{\equiv}CH$ $\xrightarrow[\text{Pt}]{\text{H}_2}$ $CH_3CH_2\underset{\underset{\text{CH}_3}{|}}{CH}CH_2CH_3$

(S)-3-Methyl-1-pentyne   3-Methylpentane
                                                               (does not have a chirality
                                                               center; optically inactive)

The product of hydrogenation of (S)-4-methyl-1-hexyne is optically active because a chirality center is present in the starting material and is carried through to the product.

$\underset{\text{H}_3\text{C}}{\overset{\text{CH}_3\text{CH}_2}{}}\overset{\text{H}}{\underset{}{C}}{-}CH_2C{\equiv}CH$ $\xrightarrow[\text{Pt}]{\text{H}_2}$ $\underset{\text{H}_3\text{C}}{\overset{\text{CH}_3\text{CH}_2}{}}\overset{\text{H}}{\underset{}{C}}CH_2CH_2CH_3$

(S)-4-Methyl-1-hexyne   (S)-3-Methylhexane

Both (S)-3-methyl-1-pentyne and (S)-4-methyl-1-hexyne yield optically active products when their triple bonds are reduced to double bonds.

9.29 (a) The dihaloalkane contains both a primary alkyl chloride and a primary alkyl iodide functional group. Iodide is a better leaving group than chloride and is the one replaced by acetylide.

$NaC{\equiv}CH$ + $ClCH_2CH_2CH_2CH_2CH_2CH_2I$ $\longrightarrow$ $ClCH_2CH_2CH_2CH_2CH_2CH_2C{\equiv}CH$

Sodium   1-Chloro-6-iodohexane   8-Chloro-1-octyne
acetylide

(*b*) Both vicinal dibromide functions are converted to alkyne units on treatment with excess sodium amide.

$$BrCH_2CHCH_2CH_2CHCH_2Br \xrightarrow[\text{2. H}_2\text{O}]{\text{1. excess NaNH}_2,\ \text{NH}_3} HC{\equiv}CCH_2CH_2C{\equiv}CH$$
$$\underset{Br}{|} \qquad\qquad \underset{Br}{|}$$

1,2,5,6-Tetrabromohexane   1,5-Hexadiyne

(*c*) The starting material is a geminal dichloride. Potassium *tert*-butoxide in dimethyl sulfoxide is a sufficiently strong base to convert it to an alkyne.

1,1-Dichloro-1-   Ethynylcyclopropane
cyclopropylethane

(*d*) Alkyl *p*-toluenesulfonates react similarly to alkyl halides in nucleophilic substitutions. The alkynide nucleophile displaces the *p*-toluenesulfonate leaving group from ethyl *p*-toluenesulfonate.

Phenylacetylide ion   Ethyl *p*-toluenesulfonate   1-Phenyl-1-butyne

(*e*) Both carbons of a —C≡C— unit are converted to carboxyl groups (—CO_2H) on ozonolysis.

Cyclodecyne   Decanedioic acid

(*f*) Ozonolysis cleaves the carbon–carbon triple bond.

1-Ethynylcyclohexanol   1-Hydroxycyclohexane-   Carbonic acid
                                carboxylic acid

(g) Hydration of a terminal carbon–carbon triple bond converts it to a $-\overset{\overset{\text{O}}{\|}}{\text{C}}\text{CH}_3$ group.

$$\underset{\underset{\text{CH}_3}{|}}{\text{CH}_3\text{CHCH}_2}\underset{\underset{\text{CH}_3}{|}}{\overset{\overset{\text{OH}}{|}}{\text{C}}}\text{C}\equiv\text{CH} \quad \xrightarrow[\text{HgO}]{\text{H}_2\text{O, H}_2\text{SO}_4} \quad \underset{\underset{\text{CH}_3}{|}}{\text{CH}_3\text{CHCH}_2}\underset{\underset{\text{CH}_3}{|}}{\overset{\overset{\text{HO}}{|}}{\text{C}}}-\overset{\overset{\text{O}}{\|}}{\text{C}}\text{CH}_3$$

3,5-Dimethyl-1-hexyn-3-ol   3-Hydroxy-3,5-dimethyl-2-hexanone

(h) Sodium-in-ammonia reduction of an alkyne yields a trans alkene. The stereochemistry of a double bond that is already present in the molecule is not altered during the process.

$$\underset{\text{H}}{\overset{\text{CH}_3\text{CH}_2\text{CH}_2\text{CH}_2}{\diagdown}}\text{C}=\text{C}\underset{\text{H}}{\overset{\diagup \text{CH}_2(\text{CH}_2)_7\text{C}\equiv\text{CCH}_2\text{CH}_2\text{OH}}{}}$$

(Z)-13-Octadecen-3-yn-1-ol

1. Na, NH$_3$
2. H$_2$O

$$\underset{\text{H}}{\overset{\text{CH}_3\text{CH}_2\text{CH}_2\text{CH}_2}{\diagdown}}\text{C}=\text{C}\underset{\text{H}}{\overset{\diagup \text{CH}_2(\text{CH}_2)_7}{}}\quad \underset{\text{H}}{\overset{}{}}\text{C}=\text{C}\underset{\text{CH}_2\text{CH}_2\text{OH}}{\overset{\diagup \text{H}}{}}$$

(3E,13Z)-3,13-Octadecadien-1-ol

(i) The primary chloride leaving group is displaced by the alkynide nucleophile.

$+ \text{ NaC}\equiv\text{CCH}_2\text{CH}_2\text{CH}_2\text{CH}_3 \longrightarrow$

8-Chlorooctyl   Sodium 1-hexynide   9-Tetradecyn-1-yl tetrahydropyranyl ether
tetrahydropyranyl ether

(j) Hydrogenation of the triple bond over the Lindlar catalyst converts the compound to a cis alkene.

O(CH$_2$)$_8$C≡CCH$_2$CH$_2$CH$_2$CH$_3$ $\xrightarrow[\text{Lindlar Pd}]{\text{H}_2}$

9-Tetradecyn-1-yl tetrahydropyranyl ether   (Z)-9-Tetradecen-1-yl tetrahydropyranyl ether

9.30 (a) Hydrogenation over Lindlar palladium converts an alkyne to a cis alkene. Oleic acid therefore has the structure indicated in the following equation:

$$\text{CH}_3(\text{CH}_2)_7\text{C}\equiv\text{C}(\text{CH}_2)_7\text{CO}_2\text{H} \quad \xrightarrow[\text{Lindlar Pd}]{\text{H}_2} \quad \underset{\text{H}}{\overset{\text{CH}_3(\text{CH}_2)_7}{\diagdown}}\text{C}=\text{C}\underset{\text{H}}{\overset{\diagup (\text{CH}_2)_7\text{CO}_2\text{H}}{}}$$

Stearolic acid   Oleic acid

Hydrogenation of alkynes over platinum leads to alkanes.

$$CH_3(CH_2)_7C\equiv C(CH_2)_7CO_2H \xrightarrow[Pt]{2H_2} CH_3(CH_2)_{16}CO_2H$$

Stearolic acid Stearic acid

(b) Alkynes are converted to trans alkenes on reduction with sodium in liquid ammonia.

$$CH_3(CH_2)_7C\equiv C(CH_2)_7CO_2H \xrightarrow[2.\ H_3O^+]{1.\ Na,\ NH_3}$$

$$\underset{H}{\overset{CH_3(CH_2)_7}{}}C=C\underset{(CH_2)_7CO_2H}{\overset{H}{}}$$

Stearolic acid Elaidic acid

9.31 Ketones such as 2-heptanone may be readily prepared by hydration of terminal alkynes. Thus, if we had 1-heptyne, it could be converted to 2-heptanone.

$$HC\equiv C(CH_2)_4CH_3 \xrightarrow[H_2SO_4,\ HgSO_4]{H_2O} CH_3\overset{O}{\overset{\|}{C}}(CH_2)_4CH_3$$

1-Heptyne 2-Heptanone

Acetylene, as we have seen in earlier problems, can be converted to 1-heptyne by alkylation.

$$HC\equiv CH \xrightarrow[NH_3]{NaNH_2} HC\equiv C:^- \ Na^+$$

Acetylene Sodium acetylide

$$HC\equiv C:^- \ Na^+ + CH_3CH_2CH_2CH_2CH_2Br \longrightarrow HC\equiv C(CH_2)_4CH_3$$

Sodium acetylide 1-Bromopentane 1-Heptyne

9.32 Apply the technique of reasoning backward to gain a clue to how to attack this synthesis problem. A reasonable final step is the formation of the Z double bond by hydrogenation of an alkyne over Lindlar palladium.

$$CH_3(CH_2)_7C\equiv C(CH_2)_{12}CH_3 \xrightarrow[Lindlar\ Pd]{H_2}$$

$$\underset{H}{\overset{CH_3(CH_2)_7}{}}C=C\underset{H}{\overset{(CH_2)_{12}CH_3}{}}$$

9-Tricosyne (Z)-9-Tricosene

The necessary alkyne 9-tricosyne can be prepared by a double alkylation of acetylene.

$$HC\equiv CH \xrightarrow[2.\ CH_3(CH_2)_7Br]{1.\ NaNH_2,\ NH_3} CH_3(CH_2)_7C\equiv CH \xrightarrow[2.\ CH_3(CH_2)_{12}Br]{1.\ NaNH_2,\ NH_3} CH_3(CH_2)_7C\equiv C(CH_2)_{12}CH_3$$

Acetylene 1-Decyne 9-Tricosyne

It does not matter which alkyl group is introduced first.
The alkyl halides are prepared from the corresponding alcohols.

$$CH_3(CH_2)_7OH \xrightarrow[or\ PBr_3]{HBr} CH_3(CH_2)_7Br$$

1-Octanol 1-Bromooctane

$$CH_3(CH_2)_{12}OH \xrightarrow[or\ PBr_3]{HBr} CH_3(CH_2)_{12}Br$$

1-Tridecanol 1-Bromotridecane

9.33 (*a*) 2,2-Dibromopropane is prepared by addition of hydrogen bromide to propyne.

$$CH_3C{\equiv}CH \ + \ 2HBr \longrightarrow CH_3\overset{\overset{\displaystyle Br}{|}}{\underset{\underset{\displaystyle Br}{|}}{C}}CH_3$$

Propyne   Hydrogen   2,2-Dibromopropane
bromide

The designated starting material, 1,1-dibromopropane, is converted to propyne by a double dehydrohalogenation.

$$CH_3CH_2CHBr_2 \ \xrightarrow[\text{2. H}_2\text{O}]{\text{1. NaNH}_2,\ \text{NH}_3} \ CH_3C{\equiv}CH$$

1,1-Dibromopropane   Propyne

(*b*) As in part (*a*), first convert the designated starting material to propyne, and then add 2 moles of hydrogen bromide.

$$CH_3\underset{\underset{\displaystyle Br}{|}}{C}HCH_2Br \ \xrightarrow[\text{2. H}_2\text{O}]{\text{1. NaNH}_2,\ \text{NH}_3} \ CH_3C{\equiv}CH \ \xrightarrow{\text{2HBr}} \ CH_3\overset{\overset{\displaystyle Br}{|}}{\underset{\underset{\displaystyle Br}{|}}{C}}CH_3$$

1,2-Dibromopropane   Propyne   2,2-Dibromopropane

(*c*) Instead of trying to introduce two additional chlorines into 1,2-dichloropropane by free-radical substitution (a mixture of products would result), convert the vicinal dichloride to propyne, and then add 2 moles of Cl$_2$.

$$CH_3\underset{\underset{\displaystyle Cl}{|}}{C}HCH_2Cl \ \xrightarrow[\text{2. H}_2\text{O}]{\text{1. NaNH}_2,\ \text{NH}_3} \ CH_3C{\equiv}CH \ \xrightarrow{\text{2Cl}_2} \ CH_3\overset{\overset{\displaystyle Cl}{|}}{\underset{\underset{\displaystyle Cl}{|}}{C}}CHCl_2$$

1,2-Dichloropropane   Propyne   1,1,2,2-Tetrachloropropane

(*d*) The required carbon skeleton can be constructed by alkylating acetylene with ethyl bromide.

$$HC{\equiv}CH \ \xrightarrow[\text{NH}_3]{\text{NaNH}_2} \ HC{\equiv}C{:}^- \ Na^+ \ \xrightarrow{\text{CH}_3\text{CH}_2\text{Br}} \ HC{\equiv}CCH_2CH_3$$

Acetylene   Sodium acetylide   1-Butyne

Addition of 2 moles of hydrogen iodide to 1-butyne gives 2,2-diiodobutane.

$$HC{\equiv}CCH_2CH_3 \ + \ 2HI \longrightarrow CH_3\overset{\overset{\displaystyle I}{|}}{\underset{\underset{\displaystyle I}{|}}{C}}CH_2CH_3$$

1-Butyne   Hydrogen   2,2-Diiodobutane
iodide

(*e*) The six-carbon chain is available by alkylation of acetylene with 1-bromobutane.

$$HC{\equiv}CH \ \xrightarrow[\text{2. CH}_3\text{CH}_2\text{CH}_2\text{CH}_2\text{Br}]{\text{1. NaNH}_2,\ \text{NH}_3} \ HC{\equiv}CCH_2CH_2CH_2CH_3$$

Acetylene   1-Hexyne

The alkylating agent, 1-bromobutane, is prepared from 1-butene by free-radical (anti-Markovnikov) addition of hydrogen bromide.

$$CH_3CH_2CH{=}CH_2 \quad + \quad HBr \quad \xrightarrow{\text{peroxides}} \quad CH_3CH_2CH_2CH_2Br$$

1-Butene   Hydrogen bromide   1-Bromobutane

Once 1-hexyne is prepared, it can be converted to 1-hexene by hydrogenation over Lindlar palladium or by sodium–ammonia reduction.

$$CH_3CH_2CH_2CH_2C{\equiv}CH \quad \xrightarrow[\text{or Na, NH}_3]{H_2,\ \text{Lindlar Pd}} \quad CH_3CH_2CH_2CH_2CH{=}CH_2$$

1-Hexyne   1-Hexene

(*f*) Dialkylation of acetylene with 1-bromobutane, prepared in part (*e*), gives the necessary ten-carbon chain.

$$HC{\equiv}CH \xrightarrow[\text{2. CH}_3\text{CH}_2\text{CH}_2\text{CH}_2\text{Br}]{\text{1. NaNH}_2,\ \text{NH}_3} CH_3CH_2CH_2CH_2C{\equiv}CH \xrightarrow[\text{2. CH}_3\text{CH}_2\text{CH}_2\text{CH}_2\text{Br}]{\text{1. NaNH}_2,\ \text{NH}_3} CH_3CH_2CH_2CH_2C{\equiv}CCH_2CH_2CH_2CH_3$$

Acetylene   1-Hexyne   5-Decyne

Hydrogenation of 5-decyne yields decane.

$$CH_3(CH_2)_3C{\equiv}C(CH_2)_3CH_3 \quad \xrightarrow[\text{Pt}]{2H_2} \quad CH_3(CH_2)_3CH_2CH_2(CH_2)_3CH_3$$

5-Decyne   Decane

(*g*) A standard method for converting alkenes to alkynes is to add Br_2 and then carry out a double dehydrohalogenation.

Cyclopentadecene   1,2-Dibromocyclopentadecane   Cyclopentadecyne

(*h*) Alkylation of the triple bond gives the required carbon skeleton.

1-Ethynylcyclohexene   1-(1-Propynyl)cyclohexene

Hydrogenation over the Lindlar catalyst converts the carbon–carbon triple bond to a cis double bond.

1-(1-Propynyl)cyclohexene   (Z)-1-(1-Propenyl)cyclohexene (*i*) The stereochemistry of *meso*-2,3-dibromobutane is most easily seen with a Fischer projection:

Recalling that the addition of Br_2 to alkenes occurs with anti stereochemistry, rotate the sawhorse diagram so that the bromines are anti to each other:

Thus, the starting alkene must be *trans*-2-butene. *trans*-2-Butene is available from 2-butyne by metal–ammonia reduction:

    2-Butyne   *trans*-2-Butene   *meso*-2,3-Dibromobutane

9.34   Attack this problem by first planning a synthesis of 4-methyl-2-pentyne from any starting material in a single step. Two different alkyne alkylations suggest themselves:

$$CH_3C\equiv CCH(CH_3)_2 \quad \begin{cases} (a) & \text{from } CH_3C\equiv C\overset{\bullet\bullet}{C}{}^- \text{ and } BrCH(CH_3)_2 \\ (b) & \text{from } CH_3I \text{ and } {}^-\!\!:\!C\equiv CCH(CH_3)_2 \end{cases}$$

4-Methyl-2-pentyne

Isopropyl bromide is a secondary alkyl halide and cannot be used to alkylate $CH_3C\equiv C\overset{\bullet\bullet}{C}{}^-$ according to reaction (*a*); elimination would occur. A reasonable last step is therefore the alkylation of $(CH_3)_2CHC\equiv CH$ via reaction of its anion with methyl iodide.

    The next question that arises from this analysis is the origin of $(CH_3)_2CHC\equiv CH$. One of the available starting materials is 1,1-dichloro-3-methylbutane. It can be converted to $(CH_3)_2CHC\equiv CH$ by a double dehydrohalogenation. The complete synthesis is therefore:

$$(CH_3)_2CHCH_2CHCl_2 \xrightarrow[\text{2. } H_2O]{\text{1. } NaNH_2,\ NH_3} (CH_3)_2CHC\equiv CH \xrightarrow[\text{2. } CH_3I]{\text{1. } NaNH_2} (CH_3)_2CHC\equiv CCH_3$$

1,1-Dichloro-3-methylbutane   3-Methyl-1-butyne   4-Methyl-2-pentyne

9.35 The reaction that produces compound A is reasonably straightforward. Compound A is 14-bromo-1-tetradecyne.

$$NaC\equiv CH \quad + \quad Br(CH_2)_{12}Br \quad \longrightarrow \quad Br(CH_2)_{12}C\equiv CH$$

Sodium acetylide   1,12-Dibromododecane   Compound A ($C_{14}H_{25}Br$)

Treatment of compound A with sodium amide converts it to compound B. Compound B on ozonolysis gives a diacid that retains all the carbon atoms of B. Compound B must therefore be a cyclic alkyne, formed by an intramolecular alkylation.

Compound A   Compound B

Compound B is cyclotetradecyne.

Hydrogenation of compound B over Lindlar palladium yields *cis*-cyclotetradecene (compound C).

Compound C ($C_{14}H_{26}$)

Hydrogenation over platinum gives cyclotetradecane (compound D).

Compound D ($C_{14}H_{28}$)

Sodium–ammonia reduction of compound B yields *trans*-cyclotetradecene.

Compound E ($C_{14}H_{26}$)

The cis and trans isomers of cyclotetradecene are both converted to $O=CH(CH_2)_{12}CH=O$ on ozonolysis, whereas cyclotetradecane does not react with ozone.

ANSWERS TO INTERPRETIVE PROBLEMS

9.36 D;   **9.37** B;   **9.38** A;   **9.39** C;   **9.40** B

SELF-TEST

PART A

A-1. Provide the IUPAC names for the following:

(a) $CH_3C{\equiv}CCHCH(CH_3)_2$
          |
          CH_3

(c)

(b) CH_2CH_3
             |
  $CH_3CH_2CH_2CHCHC{\equiv}CH$
                 |
             $CH_2CH_2CH_3$

A-2. Give the structure of the reactant, reagent, or product omitted from each of the following reactions.

(a) $CH_3CH_2CH_2C{\equiv}CH$ $\xrightarrow{\text{HCl (1 mol)}}$?

(b) $CH_3CH_2CH_2C{\equiv}CH$ $\xrightarrow{\text{HCl (2 mol)}}$?

(c) $CH_3CH_2CH_2C{\equiv}CH$ $\xrightarrow{\text{?}}$ $CH_3CH_2CH_2\overset{\displaystyle O}{\overset{\displaystyle \|}{C}}CH_3$

(d) $CH_3C{\equiv}CCH_3$ $\xrightarrow[\text{Lindlar Pd}]{\text{H}_2}$?

(e) ? $\xrightarrow[\text{2. CH}_3\text{CH}_2\text{Br}]{\text{1. NaNH}_2}$ $(CH_3)_2CHC{\equiv}CCH_2CH_3$

(f) $CH_3C{\equiv}CCH_2CH_3$ $\xrightarrow{\text{?}}$ (E)-2-pentene

(g) $CH_3C{\equiv}CCH_2CH_3$ $\xrightarrow{\text{Cl}_2\text{ (1 mol)}}$?

(h) $CH_3CH_2CH_2CHC{\equiv}CCH_2CH_3$ $\xrightarrow[\text{2. H}_2\text{O}]{\text{1. O}_3}$?
              |
            CH_3

A-3. Which one of the following two reactions is effective in the synthesis of 4-methyl-2-hexyne? Why is the other not effective?

1.   Br
            |
    $CH_3CH_2CHCH_3$ + $CH_3C{\equiv}CNa$ $\longrightarrow$

2. CH_3
            |
    $CH_3CH_2CHC{\equiv}CNa$ + CH_3I $\longrightarrow$

A-4. Outline a series of steps, using any necessary organic and inorganic reagents, for the preparation of:

(a) 1-Butyne from ethyl bromide as the source of all carbon atoms

(b) 3-Hexyne from 1-butyne

(c) 3-Hexyne from 1-butene

(d) $CH_3\overset{O}{\overset{\|}{C}}CH_2CH(CH_3)_2$ from acetylene

A-5. Treatment of propyne in successive steps with sodium amide, 1-bromobutane, and sodium in liquid ammonia yields as the final product _____.

A-6. Give the structures of compounds A through D in the following series of equations.

$$A \xrightarrow{\text{NaNH}_2,\text{NH}_3} B$$

$$C \xrightarrow{\text{HBr, heat}} D$$

$$B + D \longrightarrow CH_3CH_2CH_2C{\equiv}CC(CH_3)_3$$

A-7. What are the structures of compounds E and F in the following sequence of reactions?

$$\text{Compound E} \xrightarrow[\text{2. CH}_3\text{CH}_2\text{Br}]{\text{1. NaNH}_2,\text{ NH}_3} \text{Compound F} \xrightarrow[\text{HgSO}_4]{\text{H}_2\text{O, H}_2\text{SO}_4} CH_3CH_2\overset{O}{\overset{\|}{C}}CH_2CH_2CH_3$$

A-8. Give the reagents that would be suitable for carrying out the following transformation. Two or more reaction steps are necessary.

PART B

B-1. An acceptable IUPAC name for the compound shown is

$$\underset{CH_3CHCH_2C{\equiv}CCH(CH_3)_2}{\overset{CH_2CH_3}{\overset{|}{}}}$$

(a) 2,6-Dimethyl-3-octyne   (c) 2-Ethylpropyl isopropyl acetylene

(b) 6-Ethyl-2-methyl-3-heptyne   (d) 2-Ethyl-6-methyl-4-heptyne

B-2. Which of the following statements best explains the greater acidity of terminal alkynes ($RC{\equiv}CH$) compared with monosubstituted alkenes ($RCH{=}CH_2$)?

(a) The sp-hybridized carbons of the alkyne are less electronegative than the sp^2 carbons of the alkene.

(b) The two π bonds of the alkyne are better able to stabilize the negative charge of the anion by resonance.

(c) The sp-hybridized carbons of the alkyne are more electronegative than the sp^2 carbons of the alkene.

(d) The question is incorrect—alkenes are more acidic than alkynes.

B-3. Referring to the following equilibrium (R = alkyl group)

$$RCH_2CH_3 + RC{\equiv}C\!:^{-} \;\rightleftharpoons\; RCH_2\overset{\cdot\,\cdot}{C}H_2{}^{-} + RC{\equiv}C{-}H$$

(a) $K < 1$; the equilibrium would lie to the left.

(b) $K > 1$; the equilibrium would lie to the right.

(c) $K = 1$; equal amounts of all species would be present.

(d) Not enough information is given; the structure of R must be known.

B-4. Which of the following is an effective way to prepare 1-pentyne?

(a) 1-Pentene $\xrightarrow[\text{2. NaNH}_2,\text{ heat}]{\text{1. Cl}_2}$

(b) Acetylene $\xrightarrow[\text{2. CH}_3\text{CH}_2\text{CH}_2\text{Br}]{\text{1. NaNH}_2}$

(c) 1,1-Dichloropentane $\xrightarrow[\text{2. H}_2\text{O}]{\text{1. NaNH}_2,\text{ NH}_3}$

(d) All these are effective.

B-5. Which alkyne yields butanoic acid ($CH_3CH_2CH_2CO_2H$) as the only organic product on treatment with ozone followed by hydrolysis?

(a) 1-Butyne   (b) 4-Octyne   (c) 1-Pentyne   (d) 2-Hexyne

B-6. Which of the following produces a significant amount of acetylide ion on reaction with acetylene?

(a) Conjugate base of CH_3OH ($pK_a = 16$)   (c) Conjugate base of H_2O ($pK_a = 16$)

(b) Conjugate base of H_2 ($pK_a = 35$)   (d) Both (a) and (c).

B-7. Which of the following is the product of the reaction of 1-hexyne with 1 mole of Br_2?

(a)   (d)

(b)   (e)

(c)

B-8. Choose the sequence of steps that describes the best synthesis of 1-butene from ethanol.

(a) (1) NaC≡CH;
    (2) H_2, Lindlar Pd

(b) (1) NaC≡CH;
    (2) Na, NH_3

(c) (1) HBr, heat; (2) NaC≡CH;
    (3) H_2, Lindlar Pd

(d) (1) HBr, heat; (2) KOC(CH_3)$_3$, DMSO;
    (3) NaC≡CH; (4) H_2, Lindlar Pd

B-9. What is (are) the major product(s) of the following reaction?

$$(CH_3)_3CBr + HC{\equiv}C{:}^- \ Na^+ \longrightarrow \ ?$$

(a) $(CH_3)_3CC{\equiv}CH$

(c) H_3C—△—CH_3 with H_3C

(b) $H_2C{=}\overset{\underset{|}{CH_3}}{C}CH_3 + HC{\equiv}CH$

(d) $HC{\equiv}CCH_2CH(CH_3)_2$

B-10. Which would be the best sequence of reactions to use to prepare *cis*-3-nonene from 1-butyne?

(a) (1) $NaNH_2$ in NH_3; (2) 1-bromopentane; (3) H_2, Lindlar Pd

(b) (1) $NaNH_2$ in NH_3; (2) 1-bromopentane; (3) Na, NH_3

(c) (1) H_2, Lindlar Pd; (2) $NaNH_2$ in NH_3; (3) 1-bromopentane

(d) (1) Na, NH_3; (2) $NaNH_2$ in NH_3; (3) 1-bromopentane

B-11. Which one of the following is the intermediate in the preparation of a ketone by hydration of an alkyne in the presence of sulfuric acid and mercury(II) sulfate?

(a)   (b)   (c)   (d)   (e)

B-12. Which combination is best for preparing the compound shown in the box?

$$\boxed{\begin{array}{c} H\overset{\overset{\textstyle CH_3}{|}}{\underset{\underset{\textstyle CH_3CH_2}{\diagup}}{C}}{-}CH_2CH_2CH_2C{\equiv}CH \end{array}}$$

(a) $H_3C{\cdots}\overset{\overset{\textstyle H}{\vdots}}{\underset{\underset{\textstyle CH_3CH_2}{\diagup}}{C}}{-}CH_2CH_2CH_2Br \xrightarrow{\text{NaC}{\equiv}\text{CH}}$

(b) $H{\cdots}\overset{\overset{\textstyle CH_3}{\vdots}}{\underset{\underset{\textstyle CH_3CH_2}{\diagup}}{C}}{-}CH_2CH_2CH_2Br \xrightarrow{\text{NaC}{\equiv}\text{CH}}$

(c) $H_3C{\cdots}\overset{\overset{\textstyle H}{\vdots}}{\underset{\underset{\textstyle CH_3CH_2}{\diagup}}{C}}{-}Br \xrightarrow[\text{2. } BrCH_2CH_2CH_2C{\equiv}CH]{\text{1. } NaNH_2, NH_3}$

(d) $H{\cdots}\overset{\overset{\textstyle CH_3}{\vdots}}{\underset{\underset{\textstyle CH_3CH_2}{\diagup}}{C}}{-}Br \xrightarrow[\text{2. } BrCH_2CH_2CH_2C{\equiv}CH]{\text{1. } NaNH_2, NH_3}$

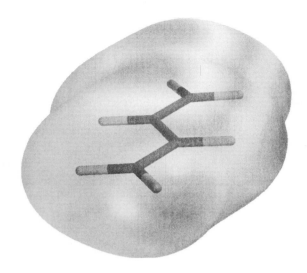

CHAPTER 10

Conjugation in Alkadienes and Allylic Systems

SOLUTIONS TO TEXT PROBLEMS

10.1 As noted in the sample solution to part (*a*), a pair of electrons is moved from the double bond toward the positively charged carbon.

(*b*) $H_2C\!\!=\!\!\overset{+}{C}\!\!-\!\!\overset{+}{C}H_2$ $\longleftrightarrow$ $H_2\overset{+}{C}\!\!-\!\!C\!\!=\!\!CH_2$
            $|$ $|$
            CH_3 CH_3

The two resonance structures are equivalent, and the positive charge is shared equally by the two allylic carbons.

(*c*) 〈 〉$=\!C(CH_3)_2$ $\longleftrightarrow$ 〈 〉$-\!\overset{+}{C}(CH_3)_2$

The two resonance structures are not equivalent, therefore the positive charge is not shared equally. In the first structure the positive charge is on a secondary carbon; in the second structure the carbon bearing the positive charge is tertiary. The second structure contributes more to the resonance hybrid than the first.

10.2 For two isomeric halides to yield the same carbocation on ionization, they must have the same carbon skeleton. They may have their leaving group at a different location, but the carbocations must become equivalent by allylic resonance.

3-Bromo-1-
methylcyclohexene

3-Chloro-3-
methylcyclohexene

Not an allylic carbocation

4-Bromo-1-
methylcyclohexene

Not an allylic carbocation

5-Chloro-1-
methylcyclohexene

Not an allylic carbocation

1-Bromo-3-
methylcyclohexene

10.3 3-Chloro-1-pentene is a secondary allylic halide with the potential to react by an $S_N 1$ mechanism with allylic rearrangement.

$$CH_3CH_2CH-CH=CH_2 \longrightarrow CH_3CH_2\overset{+}{CH}-CH=CH_2 \longleftrightarrow CH_3CH_2CH=CH-\overset{+}{CH_2}$$
$$:\overset{..}{\underset{..}{Cl}}:$$

3-Chloro-1-pentene   Resonance contributors to allylic carbocation

The positive charge is shared between C-1 and C-3. Phenoxide can bond to either to give a mixture of isomers.

1-Ethyl-2-propenyl phenyl ether   2-Pentenyl phenyl ether

10.4 The allylic hydrogens are the ones shown in the structural formulas.

(b)   1-Methylcyclohexene

(c)   2,3,3-Trimethyl-1-butene

(d)   1-Octene

10.5 The statement of the problem specifies that in allylic brominations using *N*-bromosuccinimide the active reagent is Br_2. Thus, the equation for the overall reaction is

Cyclohexene   + Br_2 ⟶   3-Bromocyclohexene   +   HBr

Bromine   Hydrogen bromide

The propagation steps are analogous to those of other free-radical brominations. An allylic hydrogen is removed by a bromine atom in the first step.

Cyclohexene   +   Bromine atom   ⟶   2-Cyclohexenyl radical   +   Hydrogen bromide

The allylic radical formed in the first step abstracts a bromine atom from Br_2 in the second propagation step.

2-Cyclohexenyl radical   +   Bromine   ⟶   3-Bromocyclohexene   +   Bromine atom

10.6 Write both resonance forms of the allylic radicals produced by hydrogen atom abstraction from the alkene.

2,3,3-Trimethyl-1-butene

Both resonance forms are equivalent, and so 2,3,3-trimethyl-1-butene gives a single bromide on treatment with *N*-bromosuccinimide (NBS).

2,3,3-Trimethyl-1-butene $\xrightarrow{\text{NBS}}$ 2-(Bromomethyl)-3,3-dimethyl-1-butene

10.7 The allylic hydrogens in this hydrocarbon are the most acidic. Therefore, remove a proton from the methyl group to form the conjugate base.

$$\text{Base:} + \text{H}-\text{CH}_2-\text{CH}=\text{CH}-\text{CH}=\text{CH}_2 \longrightarrow \text{Base}-\text{H} + \text{H}_2\ddot{\text{C}}-\text{CH}=\text{CH}-\text{CH}=\text{CH}_2$$

Two additional contributors can be generated by allylic resonance of the original structure shown for the conjugate base.

$$\text{H}_2\ddot{\text{C}}-\text{CH}=\text{CH}-\text{CH}=\text{CH}_2 \longleftrightarrow \text{H}_2\text{C}=\text{CH}-\ddot{\text{C}}\text{H}-\text{CH}=\text{CH}_2 \longleftrightarrow \text{H}_2\text{C}=\text{CH}-\text{CH}=\text{CH}-\ddot{\text{C}}\text{H}_2$$

10.8 Drawing the structure of 1,4-cyclooctadiene reveals that the double bonds are separated by an sp^3 carbon atom. 1,4-Cyclooctadiene is an isolated diene.

sp^3 carbon atom

1,4-Cyclooctadiene

10.9 (*b*) The C-1 and C-3 double bonds of cembrene are conjugated with each other.

Cembrene

The double bonds at C-6 and C-10 are isolated from each other and from the conjugated diene system.

(*c*) The sex attractant of the dried-bean beetle has a cumulated diene system involving C-4, C-5, and C-6. This allenic system is conjugated with the C-2 double bond.

$$\text{CH}_3(\text{CH}_2)_6\text{CH}_2\overset{6}{\text{C}}\text{H}=\overset{5}{\text{C}}=\overset{4}{\text{C}}\text{H}\overset{3}{\text{C}}\text{H}=\overset{2}{\text{C}}\text{H}\overset{1}{\text{C}}\text{O}_2\text{CH}_3$$

10.10 The more stable the isomer, the lower its heat of combustion. The conjugated diene is the most stable and has the lowest heat of combustion. The cumulated diene is the least stable and has the highest heat of combustion.

(*E*)-1,3-Pentadiene

Most stable
3186 kJ/mol

$$\text{H}_2\text{C}=\text{CHCH}_2\text{CH}=\text{CH}_2$$

1,4-Pentadiene

3217 kJ/mol

$$\text{H}_2\text{C}=\text{C}=\text{CHCH}_2\text{CH}_3$$

1,2-Pentadiene

Least stable
3251 kJ/mol

10.11 Compare the mirror-image forms of each compound for superimposability. For 2-methyl-2,3-pentadiene,

2-Methyl-2,3-pentadiene

Reference structure   Mirror image

Rotation of the mirror image 180° around an axis passing through the three carbons of the C=C=C unit demonstrates that the reference structure and its mirror image are superimposable.

Mirror image   Reoriented mirror image

2-Methyl-2,3-pentadiene is an achiral allene.

Comparison of the mirror-image forms of 2-chloro-2,3-pentadiene reveals that they are not superimposable. 2-Chloro-2,3-pentadiene is a chiral allene.

2-Chloro-2,3-pentadiene

Reference structure   Mirror image   Reoriented mirror image

10.12 Both starting materials undergo β elimination to give a conjugated diene system. Two minor products result, both of which have isolated double bonds.

CH_3

$H_2C{=}CHCH_2\overset{|}{\underset{|}{C}}CH_2CH_3$

X

X = OH   3-Methyl-5-hexen-3-ol
X = Br   4-Bromo-4-methyl-1-hexene

Faster   Slower

4-Methyl-1,3-hexadiene   4-Methyl-1,4-hexadiene   2-Ethyl-1,4-pentadiene
(mixture of *E* and *Z* isomers;   (mixture of *E* and *Z* isomers;   (minor product)
major product)   minor product)

10.13 Using HCl as a representative acid, we can write the protonation of C-2 of 1,3-cyclopentadiene as:

1.3-Cyclopentadiene   3-Cyclopentenyl cation

The carbocation that is formed is not allylic; therefore, cannot be stabilized by allylic resonance.

10.14 The numbers 1,2 and 1,4 refer to the carbons of the conjugated diene unit, not to the entire carbon chain. Therefore:

As can be seen by the structural formulas, the product of 1,2-addition is 4-chloro-2-hexene while the product of 1,4-addition is 2-chloro-3-hexene.

10.15 The two double bonds of 2-methyl-1,3-butadiene are not equivalent, and so two different products of 1,2-addition are possible, along with one 1,4-addition product.

2-Methyl-1,3-butadiene   3,4-Dibromo-3-methyl-1-butene (1,2-addition)   3,4-Dibromo-2-methyl-1-butene (1,2-addition)   1,4-Dibromo-2-methyl-2-butene (1,4-addition)

10.16 The molecular formula of the product, $C_{10}H_9ClO_2$, is that of a 1:1 Diels–Alder adduct between 2-chloro-1,3-butadiene and benzoquinone.

2-Chloro-1,3-butadiene   Benzoquinone $C_{10}H_9ClO_2$

10.17 "Unravel" the Diels–Alder adduct as described in the sample solution to part (*a*).

(*b*)

Diels–Alder adduct   is prepared from   Diene   Dienophile
(cyano groups are cis)

(*c*)

is prepared from   Diene   Dienophile

10.18 Two stereoisomeric Diels–Alder adducts are possible from the reaction of 1,3-cyclopentadiene and methyl acrylate. In one stereoisomer the CO_2CH_3 group is syn to the $HC\!=\!CH$ bridge, and is called the *endo* isomer. In the other stereoisomer the CO_2CH_3 group is anti to the $HC\!=\!CH$ bridge and is called the *exo* isomer.

1,3-Cyclopentadiene   Methyl acrylate   Endo isomer (75%)   Exo isomer (25%)

(Stereoisomeric forms of methyl
bicyclo[2.2.1]hept-5-ene-2-carboxylate)

10.19 An electrophile is by definition an *electron-seeker*. When an electrophile attacks ethylene, it interacts with the π orbital because this is the orbital that contains electrons. The π^* orbital of ethylene is unoccupied.

10.20 Analyze the reaction of two butadiene molecules by examining the symmetry properties of the highest occupied molecular orbital (HOMO) of one diene and the lowest unoccupied molecular orbital (LUMO) of the other.

HOMO   Bonding

Antibonding   LUMO

This reaction is symmetry forbidden. Both interactions involving the ends of the dienes need to be bonding for concerted cycloaddition to take place. Here, one is bonding and the other is antibonding.

10.21 Dienes and trienes are given IUPAC names by replacing the *-ane* ending of the alkane with *-adiene* or *-atriene* and locating the positions of the double bonds by number. The stereoisomers are identified as *E* or *Z* according to the rules established in Chapter 5.

(*a*) 3,4-Octadiene: $CH_3CH_2CH=C=CHCH_2CH_2CH_3$

(*b*) (*E,E*)-3,5-Octadiene:

(*c*) (*Z,Z*)-1,3-Cyclooctadiene:

(*d*) (*Z,Z*)-1,4-Cyclooctadiene:

(*e*) (*E,E*)-1,5-Cyclooctadiene:

(*f*) (2*E*,4*Z*,6*E*)-2,4,6-Octatriene:

(*g*) 5-Allyl-1,3-cyclopentadiene:

(*h*) *trans*-1,2-Divinylcyclopropane:

(*i*) 2,4-Dimethyl-1,3-pentadiene: $H_2C=CCH=CCH_3$
with CH_3 substituents

10.22 (*a*) $H_2C=CH(CH_2)_5CH=CH_2$ 1,8-Nonadiene

(*b*) 2,3,4,5-Tetramethyl-2,4-hexadiene

(*c*) $H_2C=CH-CH-CH=CH_2$ 3-Vinyl-1,4-pentadiene
with $CH=CH_2$ substituent

(*d*) 3-Isopropenyl-1,4-cyclohexadiene

(*e*)   (1Z,3E,5Z)-1,6-Dichloro-1,3,5-hexatriene

(*f*) $H_2C=C=CHCH=CHCH_3$ 1,2,4-Hexatriene

(*g*)   (1E,5E,9E)-1,5,9-Cyclododecatriene

(*h*)   (*E*)-3-Ethyl-4-methyl-1,3-hexadiene

10.23 (*a*) Because the product is 2,3-dimethylbutane, we know that the carbon skeleton of the starting material must be

$$
\begin{array}{c}
C-C-C-C \\
\;\;\;|\;\;\;| \\
\;\;\;C\;\;\;C
\end{array}
$$

Because 2,3-dimethylbutane is C_6H_{14} and the starting material is C_6H_{10}, *two* molecules of H_2 must have been taken up and the starting material must have two double bonds. The starting material can only be 2,3-dimethyl-1,3-butadiene.

$$
\underset{\displaystyle \underset{CH_3\;\;CH_3}{|\;\;\;\;\;|}}{H_2C=C-C=CH_2} + 2H_2 \quad \xrightarrow{\;Pt\;} \quad (CH_3)_2CHCH(CH_3)_2
$$

(*b*) Write the carbon skeleton corresponding to 2,2,6,6-tetramethylheptane.

Compounds of molecular formula $C_{11}H_{20}$ have two double bonds or one triple bond. The only compounds with the proper carbon skeleton are the alkyne and the allene shown.

$(CH_3)_3CC≡CCH_2C(CH_3)_3$ $(CH_3)_3CCH=C=CHC(CH_3)_3$

2,2,6,6-Tetramethyl-3-heptyne   2,2,6,6-Tetramethyl-3,4-heptadiene

10.24 The dienes that give 2,4-dimethylpentane on catalytic hydrogenation must have the same carbon skeleton as that alkane.

(*a*)	(*b*)	(*c*)	
2,4-Dimethyl-1,3-pentadiene (conjugated diene)	2,4-Dimethyl-1,4-pentadiene (isolated diene)	2,4-Dimethyl-2,3-pentadiene (cumulated diene)	2,4-Dimethylpentane

10.25 The important piece of information that allows us to complete the structure properly is that the ant repellent is an *allenic* substance. The allenic unit cannot be incorporated into the ring, because the three carbons must be collinear. The only possible constitution is therefore

10.26 (*a*) Allylic halogenation of propene with *N*-bromosuccinimide (NBS) gives allyl bromide.

$$H_2C{=}CHCH_3 \xrightarrow[\text{CCl}_4, \text{ heat}]{N\text{-bromosuccinimide (NBS)}} H_2C{=}CHCH_2Br$$

Propene   Allyl bromide

(*b*) Electrophilic addition of bromine to the double bond of propene gives 1,2-dibromopropane.

$$H_2C{=}CHCH_3 \xrightarrow{Br_2} \underset{\underset{Br}{|}}{BrCH_2CHCH_3}$$

Propene   1,2-Dibromopropane

(*c*) 1,3-Dibromopropane is made from allyl bromide from part (*a*) by free-radical addition of hydrogen bromide.

$$H_2C{=}CHCH_2Br \xrightarrow[\text{peroxides}]{HBr} BrCH_2CH_2CH_2Br$$

Allyl bromide   1,3-Dibromopropane

(*d*) Addition of hydrogen chloride to allyl bromide proceeds in accordance with Markovnikov's rule.

$$H_2C{=}CHCH_2Br \xrightarrow{HCl} \underset{\underset{Cl}{|}}{CH_3CHCH_2Br}$$

Allyl bromide   1-Bromo-2-chloropropane

(*e*) Addition of bromine to allyl bromide gives 1,2,3-tribromopropane.

$$H_2C{=}CHCH_2Br \xrightarrow{Br_2} \underset{\underset{Br}{|}}{BrCH_2CHCH_2Br}$$

Allyl bromide   1,2,3-Tribromopropane

(*f*) Nucleophilic substitution by hydroxide on allyl bromide gives allyl alcohol.

$$H_2C{=}CHCH_2Br \xrightarrow{NaOH} H_2C{=}CHCH_2OH$$

Allyl bromide   Allyl alcohol

(*g*) Alkylation of sodium acetylide using allyl bromide gives the desired pent-1-en-4-yne.

$$H_2C{=}CHCH_2Br \xrightarrow{NaC{\equiv}CH} H_2C{=}CHCH_2C{\equiv}CH$$

Allyl bromide   Pent-1-en-4-yne

(*h*) Sodium–ammonia reduction of pent-1-en-4-yne reduces the triple bond but leaves the double bond intact. Hydrogenation over Lindlar palladium could also be used.

$$H_2C=CHCH_2C\equiv CH \xrightarrow[\text{or } H_2, \text{ Lindlar Pd}]{\text{Na, NH}_3} H_2C=CHCH_2CH=CH_2$$

Pent-1-en-4-yne   1,4-Pentadiene

10.27 (*a*) The desired allylic alcohol can be prepared by hydrolysis of an allylic halide. Cyclopentene can be converted to an allylic bromide by free-radical bromination with *N*-bromosuccinimide (NBS).

Cyclopentene   3-Bromocyclopentene   2-Cyclopentenol

(*b*) Reaction of the allylic bromide from part (*a*) with sodium iodide in acetone converts it to the corresponding iodide.

3-Bromocyclopentene   3-Iodocyclopentene

(*c*) Nucleophilic substitution by cyanide converts the allylic bromide to 3-cyanocyclopentene.

3-Bromocyclopentene   3-Cyanocyclopentene

(*d*) Reaction of the allylic bromide with a strong base will yield cyclopentadiene by an E2 elimination.

3-Bromocyclopentene   1,3-Cyclopentadiene

(*e*) Cyclopentadiene formed in part (*d*) is needed in order to form the required Diels–Alder adduct.

1,3-Cyclopentadiene   Dimethyl bicyclo[2.2.1]heptadiene-
                                         2,3-dicarboxylate

10.28 The starting material in all cases is 2,3-dimethyl-1,3-butadiene.

2,3-Dimethyl-1,3-butadiene

(*a*) Hydrogenation of both double bonds yields 2,3-dimethylbutane.

$$\text{H}_2\text{C}=\overset{\underset{\displaystyle \text{CH}_3}{|}}{\text{C}}-\overset{\overset{\displaystyle \text{CH}_3}{|}}{\text{C}}=\text{CH}_2 \quad \xrightarrow[\text{Pt}]{2\text{H}_2} \quad (\text{CH}_3)_2\text{CHCH}(\text{CH}_3)_2$$

(*b*) 1,2-Addition of 1 mole of hydrogen chloride will give the product of Markovnikov addition to one of the double bonds, 3-chloro-2,3-dimethyl-1-butene.

$$\text{H}_2\text{C}=\overset{\underset{\displaystyle \text{CH}_3}{|}}{\text{C}}-\overset{\overset{\displaystyle \text{CH}_3}{|}}{\text{C}}=\text{CH}_2 \quad \xrightarrow{\text{HCl}} \quad (\text{CH}_3)_2\overset{\overset{\displaystyle \text{CH}_3}{|}}{\underset{\underset{\displaystyle \text{Cl}}{|}}{\text{C}}}\text{C}=\text{CH}_2$$

(*c*) 1,4-Addition will lead to double bond migration and produce 1-chloro-2,3-dimethyl-2-butene.

$$\text{H}_2\text{C}=\overset{\underset{\displaystyle \text{CH}_3}{|}}{\text{C}}-\overset{\overset{\displaystyle \text{CH}_3}{|}}{\text{C}}=\text{CH}_2 \quad \xrightarrow{\text{HCl}} \quad (\text{CH}_3)_2\text{C}=\overset{\overset{\displaystyle \text{CH}_3}{|}}{\text{C}}\text{CH}_2\text{Cl}$$

(*d*) The 1,2-addition product is 3,4-dibromo-2,3-dimethyl-1-butene.

$$\text{H}_2\text{C}=\overset{\underset{\displaystyle \text{CH}_3}{|}}{\text{C}}-\overset{\overset{\displaystyle \text{CH}_3}{|}}{\text{C}}=\text{CH}_2 \quad \xrightarrow{\text{Br}_2} \quad \text{BrCH}_2\overset{\overset{\displaystyle \text{CH}_3}{|}}{\underset{\underset{\displaystyle \text{Br}}{|}}{\text{C}}}-\overset{\overset{\displaystyle \text{CH}_3}{|}}{\text{C}}=\text{CH}_2$$

(*e*) The 1,4-addition product will be 1,4-dibromo-2,3-dimethyl-2-butene.

$$\text{H}_2\text{C}=\overset{\underset{\displaystyle \text{CH}_3}{|}}{\text{C}}-\overset{\overset{\displaystyle \text{CH}_3}{|}}{\text{C}}=\text{CH}_2 \quad \xrightarrow{\text{Br}_2} \quad \text{BrCH}_2\overset{\overset{\displaystyle \text{CH}_3}{|}}{\text{C}}=\overset{\underset{\displaystyle \text{CH}_3}{|}}{\text{C}}\text{CH}_2\text{Br}$$

(*f*) Bromination of both double bonds will lead to 1,2,3,4-tetrabromo-2,3-dimethylbutane irrespective of whether the first addition step occurs by 1,2- or 1,4-addition.

$$\text{H}_2\text{C}=\overset{\underset{\displaystyle \text{CH}_3}{|}}{\text{C}}-\overset{\overset{\displaystyle \text{CH}_3}{|}}{\text{C}}=\text{CH}_2 \quad \xrightarrow{2\text{Br}_2} \quad \text{BrCH}_2\overset{\overset{\displaystyle \text{CH}_3}{|}}{\underset{\underset{\displaystyle \text{Br}}{|}}{\text{C}}}-\overset{\overset{\displaystyle \text{CH}_3}{|}}{\underset{\underset{\displaystyle \text{Br}}{|}}{\text{C}}}\text{CH}_2\text{Br}$$

(*g*) The reaction of a diene with maleic anhydride is a Diels–Alder reaction.

10.29 The starting material in all cases is 1,3-cyclohexadiene.

(*a*) Cyclohexane will be the product of hydrogenation of 1,3-cyclohexadiene:

(*b*) 1,2-Addition will occur according to Markovnikov's rule to give 3-chlorocyclohexene.

(*c*) The product of 1,4-addition is 3-chlorocyclohexene also. 1,2-Addition and 1,4-addition of hydrogen chloride to 1,3-cyclohexadiene give the same product.

(*d*) Bromine can add to one of the double bonds to give 3,4-dibromocyclohexene:

(*e*) 1,4-Addition of bromine will give 3,6-dibromocyclohexene:

(*f*) Addition of 2 moles of bromine will yield 1,2,3,4-tetrabromocyclohexane.

(*g*) The constitution of the Diels–Alder adduct of 1,3-cyclohexadiene and maleic anhydride will have a bicyclo[2.2.2]octyl carbon skeleton.

10.30 Bond formation takes place at the end of the diene system to give a bridged bicyclic ring system.

1,3-Cyclohexadiene   Dimethyl acetylenedicarboxylate   Dimethyl bicyclo[2.2.2]octa-2,5-diene-2,3-dicarboxylate

10.31 The two Diels–Alder adducts formed in the reaction of 1,3-pentadiene with acrolein arise by the two alignments shown:

3-Methylcyclohexene-4-carboxaldehyde   3-Methylcyclohexene-5-carboxaldehyde

10.32 Compound B arises by way of a Diels–Alder reaction between compound A and dimethyl acetylenedicarboxylate. Compound A must therefore have a conjugated diene system.

Compound A $CH_3O_2CC\equiv CCO_2CH_3$ Compound B

10.33 The reaction is a nucleophilic substitution in which the nucleophile ($C_6H_5S^-$) becomes attached to the carbon that bore the chloride leaving group. Allylic rearrangement is not observed; therefore, it is reasonable to conclude that an allylic carbocation is *not* involved. The mechanism is S_N2.

$CH_3CH=CHCH_2Cl$ +   ⬡—SNa $\xrightarrow{\text{ethanol}}$ $CH_3CH=CHCH_2S$—⬡

1-Chloro-2-butene   Sodium benzenethiolate   2-Butenyl phenyl sulfide

10.34 (*a*) Solvolysis of $(CH_3)_2C=CHCH_2Cl$ in ethanol proceeds by an S_N1 mechanism and involves a carbocation intermediate.

$$(CH_3)_2C=CHCH_2Cl \xrightarrow{-Cl^-} \boxed{(CH_3)_2C \overset{\frown}{=} CH \overset{+}{-} CH_2 \longleftrightarrow (CH_3)_2\overset{+}{C}-CH=CH_2}$$

1-Chloro-3-methyl-2-butene

This carbocation has some of the character of a tertiary carbocation. It is more stable and is therefore formed faster than allyl cation, $H_2C=CH\overset{+}{C}H_2$.

(*b*) An allylic carbocation is formed from the alcohol in the presence of an acid catalyst.

$$H_2C=CHCHCH_3 \xrightarrow[H_2O]{H_2SO_4} H_2C=CHCHCH_3 \longrightarrow H_2C=CH\overset{+}{C}HCH_3 + H_2O$$

3-Buten-2-ol

This carbocation is a delocalized one and can be captured at either end of the allylic system by water acting as a nucleophile.

$$\boxed{\begin{array}{c} H_2C=CH-\overset{+}{C}HCH_3 \\ \updownarrow \\ H_2\overset{+}{C}-CH=CHCH_3 \end{array}}$$

$\xrightarrow{H_2O}$

$H_2C=CHCHCH_3 \xrightarrow{-H^+} H_2C=CHCHCH_3$ (with $\overset{+}{O}H_2$ and OH respectively)

3-Buten-2-ol

$H_2\overset{+}{O}-CH_2CH=CHCH_3 \xrightarrow{-H^+} HOCH_2CH=CHCH_3$

2-Buten-1-ol

(*c*) Hydrogen bromide converts the alcohol to an allylic carbocation. Bromide ion captures this carbocation at either end of the delocalized allylic system.

$$CH_3CH=CHCH_2OH \xrightarrow{HBr} CH_3CH=CHCH_2-\overset{+}{O}(H)(H) \xrightarrow{-H_2O} CH_3CH=CH\overset{+}{C}H_2$$

2-Buten-1-ol

$$\boxed{\begin{array}{c} CH_3CH=CH-\overset{+}{C}H_2 \\ \updownarrow \\ CH_3\overset{+}{C}H-CH=CH_2 \end{array}}$$

$\xrightarrow{Br^-}$

$CH_3CH=CHCH_2Br$

1-Bromo-2-butene

$CH_3CHCH=CH_2$ (with Br)

3-Bromo-1-butene

(*d*) The same delocalized carbocation is formed from 3-buten-2-ol as from 2-buten-1-ol.

$$CH_3CHCH=CH_2 \xrightarrow{HBr} CH_3\overset{+}{C}H-CH=CH_2 \longleftrightarrow CH_3CH=CH-\overset{+}{C}H_2$$
(with OH)

3-Buten-2-ol

Because this carbocation is the same as the one formed in part (*c*), it gives the same mixture of products when it reacts with bromide.

(*e*) We are told that the major product is 1-bromo-2-butene, not 3-bromo-1-butene.

$$CH_3CH=CHCH_2Br \qquad CH_3\underset{\overset{|}{Br}}{C}HCH=CH_2$$

1-Bromo-2-butene
(major)

3-Bromo-1-butene
(minor)

The major product is the more stable one. It is a primary rather than a secondary halide and contains a more substituted double bond. The reaction is therefore governed by thermodynamic control.

10.35 1,2-Addition can occur at either of the two double bonds of the diene system. Protonation of C-1 gives an allylic cation for which the most stable resonance form is a tertiary carbocation. Protonation of C-4 would give a less stable allylic carbocation for which the most stable resonance form is a secondary carbocation. Thus the major 1,2-addition product arises by addition to the C-1, C-2 double bond of the diene.

$$H_2C=\underset{\overset{|}{CH_3}}{C}CH=CH_2 \xrightarrow{HCl} \qquad \longrightarrow \qquad (CH_3)_2\underset{\overset{|}{Cl}}{C}CH=CH_2$$

2-Methyl-1,3-butadiene

3-Chloro-3-methyl-1-butene (major product)

10.36 Because both products of reaction of hydrogen chloride with vinylacetylene are chloro-substituted dienes, the first step in addition must involve the triple bond. The carbocation produced is an allylic vinyl cation for which two Lewis structures may be written. Capture of this cation gives the products of 1,2 and 1,4 addition. The 1,2 addition product is more stable because of its conjugated system. The observations of the experiment tell us that the 1,4 addition product is formed faster, although we could not have predicted that.

$$HC{\equiv}C-CH=CH_2 \xrightarrow{HCl} \qquad H_2C=\overset{+}{C}-CH=CH_2 \longleftrightarrow H_2C=C=CH-\overset{+}{C}H_2$$

Vinylacetylene

$$H_2C=\underset{\overset{|}{Cl}}{C}-CH=CH_2 \quad + \quad H_2C=C=CH-CH_2Cl$$

2-Chloro-1,3-butadiene
(1,2 addition)

4-Chloro-1,2-butadiene
(1,4 addition)

10.37 Compare the mirror-image forms of each compound for superimposability.

(*a*)

2-Methyl-2,3-hexadiene

Reference structure   Mirror image

Rotation of the mirror image 180° around an axis passing through the three carbons of the C=C=C unit demonstrates that the reference structure and its mirror image are superimposable.

Mirror image   Rotate 180°   Reoriented mirror image

2-Methyl-2,3-hexadiene is an achiral allene.

(*b*) The two mirror-image forms of 4-methyl-2,3-hexadiene are as shown:

Reference structure   and   Mirror image   Rotate 180°   Reoriented mirror image

The two structures cannot be superimposed. 4-Methyl-2,3-hexadiene is chiral. Rotation of either representation 180° around an axis that passes through the three carbons of the C=C=C unit leads to superposition of the groups at the "bottom" carbon but not at the "top."

(*c*) 2,4-Dimethyl-2,3-pentadiene is achiral. Its two mirror-image forms are superimposable.

Reference structure   Mirror image

The molecule has two planes of symmetry defined by the three carbons of each CH_3CCH_3 unit.

10.38 (*a*) Carbons 2 and 3 of 1,2,3-butatriene are *sp*-hybridized, and the bonding is an extended version of that seen in allene. Allene is nonplanar; its two CH$_2$ units must be in perpendicular planes in order to maximize overlap with the two mutually perpendicular *p* orbitals at C-2. With one more *sp*-hybridized carbon, 1,2,3-butatriene has an "extra turn" in its carbon chain, making the molecule planar.

Nonplanar geometry of allene

All atoms of 1,2,3-butatriene
lie in the same plane.

(*b*) The planar geometry of the cumulated triene system leads to the situation where cis and trans stereoisomers are possible for 2,3,4-hexatriene (CH$_3$CH=C=C=CHCH$_3$). Cis–trans stereoisomers are diastereomers of each other.

cis-2,3,4-Hexatriene   *trans*-2,3,4-Hexatriene

10.39 Reaction (*a*) is an electrophilic addition of bromine to an alkene; the appropriate reagent is bromine in carbon tetrachloride.

(74%)

Reaction (*b*) is an epoxidation of an alkene, for which almost any peroxy acid could be used. Peroxybenzoic acid was actually used.

(69%)

Reaction (*c*) is an elimination reaction of a vicinal dibromide to give a conjugated diene and requires E2 conditions. Sodium methoxide in methanol was used.

(80%)

Reaction (*d*) is a Diels–Alder reaction in which the dienophile is maleic anhydride. The dienophile adds from the side opposite that of the epoxide ring.

10.40 To predict the constitution of the Diels–Alder adducts, we can ignore the substituents and simply remember that the fundamental process is

(a)

(b)

(c)

10.41 Think mechanistically. The reaction of alkenes with *N*-bromosuccinimide is a free-radical process proceeding by way of an allylic radical.

This allylic radical is delocalized with the odd electron shared by two carbons.

Resonance forms of free-radical intermediate

The two isomeric bromides are formed from the same allylic radical.

and

3-Bromo-1,5-cyclooctadiene   6-Bromo-1,4-cyclooctadiene

10.42 (*a*) Because allyl cation is positively charged, examine the process in which electrons flow from the HOMO of ethylene to the LUMO of allyl cation.

This reaction is forbidden. The symmetries of the orbitals are such that one interaction is bonding and the other is antibonding. The same answer is obtained if the HOMO of allyl cation and the LUMO of ethylene are examined.

(*b*) In this part of the exercise we consider the LUMO of allyl cation and the HOMO of 1,3-butadiene.

This reaction is allowed by the Woodward–Hoffmann rules. Both interactions are bonding. The same prediction would be arrived at if the HOMO of allyl cation and LUMO of 1,3-butadiene were the orbitals considered.

10.43 Because oxygen has two unpaired electrons, it can abstract a hydrogen atom from the allylic position of cyclohexene to give a free-radical intermediate.

The cyclohexenyl radical is resonance-stabilized. It reacts further via the following two propagation steps:

ANSWERS TO INTERPRETIVE PROBLEMS

10.44 C;   **10.45** A;   **10.46** B;   **10.47** C;   **10.48** A;   **10.49** D

SELF-TEST

PART A

A-1. Give the structures of all the constitutionally isomeric alkadienes of molecular formula C_5H_8. Indicate which are conjugated and which are allenes.

A-2. Provide an acceptable IUPAC name for each of the conjugated dienes of the previous problem, *including stereoisomers.*

A-3. Hydrolysis of 3-bromo-3-methylcyclohexene yields two isomeric alcohols. Draw their structures and the structure of the intermediate that leads to their formation.

A-4. Give the chemical structure of the reactant, reagent, or product omitted from each of the following:

(*a*) $CH_3CH{=}CHCH{=}CHCH_3$ $\xrightarrow{\text{Br}_2}$?   (two products)

(*b*) $H_2C{=}CHCH{=}CH_2$ $\xrightarrow{\text{HCl (1 mol)}}$?   (two products)

(*c*) ? $\xrightarrow{\text{Diels–Alder}}$

(*d*)

(*e*)

A-5. One of the isomeric conjugated dienes having the formula C_6H_8 is not able to react with a dienophile in a Diels–Alder reaction. Draw the structure of this compound.

A-6. Draw the structure of the carbocation formed on ionization of the compound shown. A constitutional isomer of this compound gives the same carbocation; draw its structure.

A-7. Give the structures of compounds A and B in the following reaction scheme.

A-8. Give the reagents necessary to carry out the following conversion. Note that more than one reaction step is necessary.

PART B

B-1. 2,3-Pentadiene, $CH_3CH=C=CHCH_3$, is

(a) A planar substance

(c) A conjugated diene

(b) An allene

(d) A substance capable of cis–trans isomerism

B-2. Rank the following carbocations in order of increasing stability (least → most):

$$CH_3\overset{+}{C}HCH_3 \qquad CH_3\overset{+}{C}HCH=CHCH_3 \qquad (CH_3)_3C\overset{+}{C}H_2$$

1   2   3

(a) 1 < 2 < 3   (b) 2 < 3 < 1   (c) 3 < 1 < 2   (d) 2 < 1 < 3

B-3. Hydrogenation of cyclohexene releases 120 kJ/mol (28.6 kcal/mol) of heat. Which of the following most likely represents the observed heat of hydrogenation of 1,3-cyclohexadiene?

(a) 232 kJ/mol (55.4 kcal/mol)

(c) 247 kJ/mol (59.0 kcal/mol)

(b) 240 kJ/mol (57.2 kcal/mol)

(d) 120 kJ/mol (28.6 kcal/mol)

B-4. Which of the following compounds give the *same* carbocation on ionization?

1   2   3   4

(a) 1 and 3   (b) 2 and 4   (c) 1 and 2   (d) 1 and 4

B-5. For the following reactions the major products are shown:

$$H_2C=CH-CH=CH_2 \quad \xrightarrow[0°C]{HBr} \quad H_2C=CHCHCH_3 \quad \xrightarrow{25°C} \quad BrCH_2CH=CHCH_3$$
$$\underset{Br}{|}$$

These provide an example of __1__ control at low temperature and __2__ control at higher temperature.

1	2
(a) kinetic	thermodynamic
(b) thermodynamic	kinetic
(c) kinetic	kinetic
(d) thermodynamic	thermodynamic

B-6. Which of the following C—H bonds would have the smallest bond dissociation enthalpy?

(a) $CH_3\overset{\overset{\displaystyle H}{|}}{C}HCH_3$

(c) $CH_3CH_2\overset{\overset{\displaystyle H}{|}}{C}H_2$

(b) $CH_3CH=\overset{\overset{\displaystyle H}{|}}{C}H$

(d) $H_2C=CH-\overset{\overset{\displaystyle H}{|}}{C}HCH_3$

B-7. Which of the following compounds would undergo solvolysis (S_N1) most rapidly in aqueous ethanol?

(a)   (b)   (c)   (d)

B-8. What is the product of 1,4-addition in the reaction shown?

(a)   (d)

(b)   (e)

(c)

B-9. Which of the following compounds will undergo hydrolysis (S_N1) to give a mixture of two alcohols that are constitutional isomers?

(a)   (b)   (c)   (d)

B-10. What hydrocarbon reacts with the compound shown (on heating) to give the indicated product?

(a) 2-Methyl-1-butene

(b) 2-Methyl-2-butene

(c) 3-Methyl-1-butyne

(d) 2-Methyl-1,3-butadiene

(e) 1,3-Pentadiene

CHAPTER 11

Arenes and Aromaticity

SOLUTIONS TO TEXT PROBLEMS

11.1 Toluene is $C_6H_5CH_3$; it has a methyl group attached to a benzene ring.

Kekulé forms of toluene

Robinson symbol
for toluene

Benzoic acid has a —CO_2H substituent on the benzene ring.

Kekulé forms of benzoic acid

Robinson symbol
for benzoic acid

11.2 Given

Cycloheptene + H₂ $\xrightarrow{\Delta H° = -110\ \text{kJ}\ (-26.3\ \text{kcal})}$ Cycloheptane and assuming that there is no resonance stabilization in 1,3,5-cycloheptatriene, we predict that its heat of hydrogenation would be three times that of cycloheptene, or 330 kJ/mol (78.9 kcal/mol). The measured heat of hydrogenation is

1,3,5-Cycloheptatriene   Cycloheptane

$+$ $3H_2$ $\xrightarrow{\Delta H^\circ = -305 \text{ kJ } (-73.0 \text{ kcal})}$

Therefore

Resonance energy = 330 kJ/mol (predicted for no delocalization) – 305 kJ/mol (observed)
            = 25 kJ/mol (5.9 kcal/mol)

The value given in the text for the resonance energy of benzene (152 kJ/mol) is six times larger than this. 1,3,5-Cycloheptatriene is *not* aromatic.

11.3 (*b*) The parent compound is styrene, $C_6H_5CH=CH_2$. The desired compound has a chlorine in the meta position.

m-Chlorostyrene

(*c*) The parent compound is aniline, $C_6H_5NH_2$. *p*-Nitroaniline is therefore

p-Nitroaniline

11.4 There are only three monochloro derivatives of anthracene. Any other monochloro derivative is equivalent to one of these. Carbons at the ring junction positions already have four bonds, so cannot bear a substituent.

1-Chloroanthracene   2-Chloroanthracene   9-Chloroanthracene

11.5 The most stable resonance form is the one that has the greatest number of rings that are Kekulé formulations of benzene. A resonance form for chrysene can be drawn in which all four rings correspond to Kekulé structures.

Chrysene

11.6 Birch reductions of monosubstituted arenes yield 1,4-cyclohexadiene derivatives in which the alkyl group is a substituent on the double bond. With *p*-xylene, both methyl groups are double-bond substituents in the product.

p-Xylene

1,4-Dimethyl-1,4-cyclohexadiene

11.7 By drawing the product obtained if chlorine became attached to a ring carbon it becomes clear that the product is *not aromatic*. The aromatic product obtained by substitution at the benzylic carbon is much more stable than any nonaromatic product that would be formed from addition to the ring.

Observed product
(aromatic)

Not observed
(not aromatic)

11.8 (*b*) Only the benzylic hydrogen is replaced by bromine in the reaction of 4-methyl-3-nitroanisole with *N*-bromosuccinimide (NBS).

Only these hydrogens
are benzylic.

11.9 The molecular formula of the product is $C_{12}H_{14}O_4$. Because it contains four oxygens, the product must have two —CO_2H groups. None of the hydrogens of a *tert*-butyl substituent on a benzene ring is benzylic, and so this group is inert to oxidation. Only the benzylic methyl groups of 4-*tert*-butyl-1,2-dimethylbenzene are susceptible to oxidation; therefore, the product is 4-*tert*-butylbenzene-1,2-dicarboxylic acid.

11.10 Beginning with the structure having the positive charge at the benzylic carbon (left), move electron pairs as indicated by the curved arrow to generate a resonance form with a positively charged ortho carbon (middle) and another with a positively charged para carbon.

11.11 Each of these reactions involves nucleophilic substitution of the S_N2 type at the benzylic position of benzyl bromide.

(e)

Iodide ion   Benzyl bromide   Benzyl iodide

11.12 The dihydronaphthalene in which the double bond is conjugated with the aromatic ring is more stable; thus 1,2-dihydronaphthalene has a lower heat of hydrogenation than 1,4-dihydronaphthalene.

1,2-Dihydronaphthalene
Heat of hydrogenation
101 kJ/mol (24.1 kcal/mol)

1,4-Dihydronaphthalene
Heat of hydrogenation
113 kJ/mol (27.1 kcal/mol)

11.13 (b) The regioselectivity of alcohol formation by hydroboration–oxidation is opposite to that predicted by Markovnikov's rule.

2-Phenylpropene

2-Phenyl-1-propanol (92%)

(c) Bromine adds to alkenes in aqueous solution to give bromohydrins. A water molecule acts as a nucleophile, attacking the bromonium ion at the carbon that can bear most of the positive charge, which in this case is the benzylic carbon.

Styrene

2-Bromo-1-phenylethanol (82%)

(d) Peroxy acids convert alkenes to epoxides.

Styrene   Peroxybenzoic acid   Epoxystyrene   Benzoic acid
(69–75%)

11.14 Styrene contains a benzene ring and will be appreciably stabilized by resonance, which makes it lower in energy than cyclooctatetraene.

Structure contains an
aromatic ring

Styrene: heat of
combustion 4393 kJ/mol
(1050 kcal/mol)

Cyclooctatetraene (not aromatic):
heat of combustion 4543 kJ/mol
(1086 kcal/mol)

11.15 The dimerization of cyclobutadiene is a Diels–Alder reaction in which one molecule of cyclobutadiene acts as a diene and the other as a dienophile.

Diene   Dienophile   Diels–Alder adduct

11.16 (*b*) Compound B is not aromatic for two reasons. Although the compound has six π electrons, they are not all part of a ring. Secondly, the *sp*³-hybridized carbon present in the ring acts as an insulator and prevents cyclic delocalization of the π electrons.

These π electrons
are not part of
the ring.

*sp*³ carbon atom

(*c*) Compound C is aromatic because it contains two benzene rings. Hückel's rule should only be applied to *monocyclic* compounds.

Compound C
(biphenyl)

11.17 (*b*) [12]Annulene is a 12-membered ring with six conjugated double bonds. As in part (*a*), inscribe a polygon with 12 sides in a circle with its vertex downward. Place an orbital at each point where the polygon contacts the circle. Ten of the π electrons fill the bonding orbitals in pairs. The remaining two electrons are in the nonbonding orbitals. Following Hund's rule, these electrons are in separate orbitals with parallel spin. [12]Annulene is a highly unstable diradical species, and is not aromatic.

[12]Annulene   Frost's circle   π Electron configuration

Antibonding

Bonding

11.18 One way to evaluate the relationship between heats of combustion and structure for compounds that are not isomers is to divide the heat of combustion by the number of carbons so that heats of combustion are compared on a "per carbon" basis.

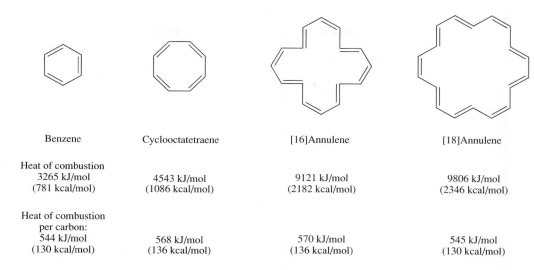

Benzene	Cyclooctatetraene	[16]Annulene	[18]Annulene

Heat of combustion			
3265 kJ/mol (781 kcal/mol)	4543 kJ/mol (1086 kcal/mol)	9121 kJ/mol (2182 kcal/mol)	9806 kJ/mol (2346 kcal/mol)

Heat of combustion per carbon:			
544 kJ/mol (130 kcal/mol)	568 kJ/mol (136 kcal/mol)	570 kJ/mol (136 kcal/mol)	545 kJ/mol (130 kcal/mol)

As the data indicate (within experimental error), the heats of combustion *per carbon* of the two aromatic hydrocarbons, benzene and [18]annulene, are equal. Similarly, the heats of combustion per carbon of the two nonaromatic hydrocarbons, cyclooctatetraene and [16]annulene, are equal. The two aromatic hydrocarbons have heats of combustion per carbon that are less than those of the nonaromatic hydrocarbons. On a per carbon basis, the aromatic hydrocarbons have lower potential energy (are more stable) than the nonaromatic hydrocarbons.

11.19 As in Problem 11.17, an energy-level diagram can be generated for cycloheptatrienyl cation by inscribing a heptagon in a circle with a point down. The cation has six π electrons that fill the bonding orbitals, yielding an energy-level diagram similar to Figure 11.13 in the text.

Cycloheptatrienyl cation	Frost's circle	π Electron configuration

11.20 Although cycloheptatrienyl radical contains a cyclic, completely conjugated π system, it is not aromatic. There are seven π electrons in the radical, a number that does not satisfy Hückel's rule. The seventh electron occupies an antibonding orbital.

Cycloheptatrienyl radical

To be antiaromatic a species must contain $4n$ π electrons, where n is a whole number. Cycloheptatrienyl radical has seven π electrons, and is neither aromatic nor antiaromatic.

11.21 The seven resonance forms for tropylium cation (cycloheptatrienyl cation) may be generated by moving π electrons in pairs toward the positive charge. The resonance forms are simply a succession of allylic carbocations.

11.22 As in previous problems, inscribe a polygon (a pentagon in this case) in a circle with a point down. The six π electrons of cyclopentadienyl anion fill the bonding orbitals, as shown in text Figure 11.14.

Cyclopentadienyl anion   Frost's circle   π Electron configuration

11.23 Resonance structures are generated for cyclopentadienide anion by moving the unshared electron pair from the carbon to which it is attached to a position where it becomes a shared electron pair in a π bond.

11.24 (*b*) Cyclononatetraenide anion has ten π electrons; it is aromatic. The ten π electrons are most easily seen by writing a Lewis structure for the anion: there are two π electrons for each of four double bonds, and the negatively charged carbon contributes two.

11.25 Indole is more stable than isoindole. Although the bonding patterns in both five-membered rings are the same, the six-membered ring in indole has a pattern of bonds identical to benzene and so is highly stabilized. The six-membered ring in isoindole is not of the benzene type.

Six-membered ring corresponds to benzene.

Six-membered ring does not have same pattern of bonds as benzene.

Indole more stable

Isoindole less stable

11.26 The prefix *benz-* in benzimidazole (structure given in text) signifies that a benzene ring is fused to an imidazole ring. By analogy, benzoxazole has a benzene ring fused to oxazole. Similarly, benzothiazole has a benzene ring fused to thiazole.

Benzimidazole   Benzoxazole   Benzothiazole

11.27 The original structure of pyrrole (left) has no formal charges. Reorganizing electrons as shown generates resonance forms in which nitrogen has a formal charge of +1. The two structures indicated have their negative charge at carbon-3 (middle) and carbon-5 (right). By continuing the electron reorganization, two more structures may be written. One has a negative charge at carbon-2, the other at carbon-4.

11.28 The problem states that pyrrole is 10^7–10^9 weaker in basicity than pyridine. Therefore, its conjugate acid is 10^7–10^9 times stronger. Subtract 7–9 from the pK_a of the conjugate acid of pyridine.

pK_a of [Pyrrole—H^+] = pK_a of [Pyridine—H^+] – (7–9)

pK_a of [Pyrrole—H^+] = 5.2 – (7–9)

pK_a of [Pyrrole—H^+] = –1.8 to –3.8

The conjugate acid of pyrrole is a strong acid. Its pK_a of –1.8 to –3.8 is smaller than that of hydronium ion (–1.7). The conjugate acid of pyridine is a weak acid. Its pK_a is greater than that of hydronium ion.

11.29 Move electrons in pairs from electron rich sites toward positive charge to convert the original Lewis structure to a resonance form in which the positive charge is on the other nitrogen.

Resonance in imidazolium ion

11.30 Because the problem requires that the benzene ring be monosubstituted, all that needs to be examined are the various isomeric forms of the C_4H_9 substituent.

Butylbenzene
(1-phenylbutane)

sec-Butylbenzene
(2-phenylbutane)

Isobutylbenzene
(2-methyl-1-phenylpropane)

tert-Butylbenzene
(2-methyl-2-phenylpropane)

These are the four constitutional isomers. *sec*-Butylbenzene is chiral and so exists in enantiomeric *R* and *S* forms.

11.31 (*a*) An allyl substituent is —$CH_2CH=CH_2$.

Allylbenzene

(*b*) The constitution of 1-phenyl-1-butene is $C_6H_5CH=CHCH_2CH_3$. The *E* stereoisomer is

(*E*)-1-Phenyl-1-butene

The two higher ranked substituents, phenyl and ethyl, are on opposite sides of the double bond.

(*c*) The constitution of 2-phenyl-2-butene is $CH_3C=CHCH_3$. The *Z* stereoisomer is
$\quad\quad C_6H_5$

(*Z*)-2-Phenyl-2-butene

The two higher ranked substituents, phenyl and methyl, are on the same side of the double bond.

(*d*) 1-Phenylethanol is chiral and has the constitution $CH_3CHC_6H_5$. Among the substituents attached
$\quad\quad\quad OH$
to the chirality center, the order of decreasing precedence is

$$HO > C_6H_5 > CH_3 > H$$

In the *R* enantiomer the order of decreasing precedence of the three highest ranked substituents must appear clockwise when the lowest ranked substituent is directed away from you.

(*R*)-1-Phenylethanol

(*e*) A benzyl group is $C_6H_5CH_2-$. Benzyl alcohol is therefore $C_6H_5CH_2OH$ and *o*-chlorobenzyl alcohol is

(*f*) In *p*-chlorophenol the benzene ring bears chlorine and hydroxyl substituents in a 1,4-substitution pattern.

p-Chlorophenol

(*g*) Benzenecarboxylic acid is an alternative IUPAC name for benzoic acid.

2-Nitrobenzenecarboxylic acid

(*h*) Two isopropyl groups are in a 1,4 relationship in *p*-diisopropylbenzene.

p-Diisopropylbenzene

(*i*) Aniline is $C_6H_5NH_2$. Therefore

2,4,6-Tribromoaniline

(*j*) Acetophenone (from text Table 11.1) is $C_6H_5\overset{\overset{\displaystyle O}{\|}}{C}CH_3$. Therefore

m-Nitroacetophenone

(*k*) Styrene is $C_6H_5CH=CH_2$, and numbering of the ring begins at the carbon that bears the alkenyl side chain.

4-Bromo-3-ethylstyrene

11.32 (*a*) Anisole is the name for $C_6H_5OCH_3$, and allyl is an acceptable name for the group $H_2C=CHCH_2-$. Number the ring beginning with the carbon that bears the methoxy group.

(*b*) Phenol is the name for C_6H_5OH. The ring is numbered beginning at the carbon that bears the hydroxyl group, and the substituents are listed in alphabetical order.

(*c*) Aniline is the name given to $C_6H_5NH_2$. This compound is named as a dimethyl derivative of aniline. Number the ring sequentially beginning with the carbon that bears the amino group.

(*a*)   Estragole
4-Allylanisole

(*b*)   Diosphenol
2,6-Diiodo-4-nitrophenol

(*c*)   *m*-Xylidine
2,6-Dimethylaniline

11.33 (*a*) There are three isomeric nitrotoluenes because the nitro group can be ortho, meta, or para to the methyl group.

o-Nitrotoluene
(2-nitrotoluene)

m-Nitrotoluene
(3-nitrotoluene)

p-Nitrotoluene
(4-nitrotoluene)

(*b*) Benzoic acid is $C_6H_5CO_2H$. In the isomeric dichlorobenzoic acids, two of the ring hydrogens of benzoic acid have been replaced by chlorines. The isomeric dichlorobenzoic acids are

2,3-Dichlorobenzoic
acid

2,4-Dichlorobenzoic
acid

2,5-Dichlorobenzoic
acid

2,6-Dichlorobenzoic
acid

3,4-Dichlorobenzoic
acid

3,5-Dichlorobenzoic
acid

The prefixes *o*-, *m*-, and *p*- may not be used in trisubstituted arenes; numerical prefixes are used. Note also that *benzenecarboxylic* may be used in place of *benzoic*.

(*c*) In the various tribromophenols, we are dealing with tetrasubstitution on a benzene ring. Again, *o*-, *m*-, and *p*- are not valid prefixes. The hydroxyl group is assigned position 1 because the base name is phenol.

2,3,4-Tribromophenol   2,3,5-Tribromophenol   2,3,6-Tribromophenol

2,4,5-Tribromophenol   2,4,6-Tribromophenol   3,4,5-Tribromophenol

(*d*) There are only three tetrafluorobenzenes. The two hydrogens may be ortho, meta, or para to each other.

1,2,3,4-Tetrafluorobenzene   1,2,3,5-Tetrafluorobenzene   1,2,4,5-Tetrafluorobenzene

(*e*) There are only two naphthalenecarboxylic acids.

Naphthalene-1-   Naphthalene-2-
carboxylic acid   carboxylic acid

11.34 (*a*) In the structure shown for naphthalene, one ring but not the other corresponds to a Kekulé form of benzene. We say that one ring is *benzenoid*, and the other is not.

This six-membered ring is not benzenoid (does not correspond to Kekulé form of benzene).

This six-membered ring is benzenoid (corresponds to a Kekulé form of benzene).

By rewriting the benzenoid ring in its alternative Kekulé form, *both* rings become benzenoid.

Both rings
are benzenoid.

(*b*) Here a cyclobutadiene ring is fused to benzene. By writing the alternative resonance form of cyclobutadiene, the six-membered ring becomes benzenoid.

(c) The structure portrayed for phenanthrene contains two terminal benzenoid rings and a nonbenzenoid central ring. All three rings may be represented in benzenoid forms by converting one of the terminal six-membered rings to its alternative Kekulé form as shown:

Central ring
not benzenoid

All three rings
benzenoid

(d) Neither of the six-membered rings is benzenoid in the structure shown. By writing the cyclooctatetraene portion of the molecule in its alternative representation, the two six-membered rings become benzenoid.

Six-membered rings
are not benzenoid.

Six-membered rings
are benzenoid.

(e) Move π electron pairs as shown so that both six-membered rings have an arrangement of bonds that corresponds to benzene.

Less stable

More stable

11.35 (a) Hydrogenation of isopropylbenzene converts the benzene ring to a cyclohexane unit.

Isopropylbenzene

Isopropylcyclohexane

(b) Sodium and ethanol in liquid ammonia is the combination of reagents that brings about Birch reduction of benzene rings. The 1,4-cyclohexadiene that is formed has its isopropyl group as a substituent on one of the double bonds.

Isopropylbenzene

1-Isopropyl-1,4-cyclohexadiene

(c) Oxidation of the isopropyl side chain occurs. The benzene ring remains intact.

Isopropylbenzene

Benzoic acid

(*d*) *N*-Bromosuccinimide (NBS) is a reagent effective for the substitution of a benzylic hydrogen.

Isopropylbenzene   2-Bromo-2-phenylpropane

(*e*) The tertiary bromide undergoes E2 elimination to give a carbon–carbon double bond.

2-Bromo-2-phenylpropane   2-Phenylpropene

11.36 All the specific reactions in this problem have been reported in the chemical literature with results as indicated.

(*a*) Hydroboration–oxidation of alkenes leads to syn anti-Markovnikov hydration of the double bond.

1-Phenylcyclobutene   *trans*-2-Phenyl-
cyclobutanol (82%)

(*b*) The compound contains a substituted benzene ring and an alkene-like double bond. When hydrogenation of this compound was carried out, the alkene-like double bond was hydrogenated cleanly.

1-Ethylindene   1-Ethylindan (80%)

(*c*) Free-radical chlorination will lead to substitution of benzylic hydrogens. The starting material contains four benzylic hydrogens, all of which may eventually be replaced.

(65%)

(*d*) Epoxidation of alkenes is stereospecific. Groups that are trans in the alkene remain trans in the epoxide.

(*E*)-1,2-Diphenylethene

trans-1,2-Epoxy-1, 2-diphenylethane
(78–83%)

(*e*) The reaction is one of acid-catalyzed alcohol dehydration.

cis-4-Methyl-1-phenylcyclohexanol

4-Methyl-1-
phenylcyclohexene (81%)

(*f*) This reaction illustrates identical reactivity at two equivalent sites in a molecule. Both alcohol functions are tertiary and benzylic and undergo acid-catalyzed dehydration readily.

1,4-Di-(1-hydroxy-1-methylethyl)-
benzene

1,4-Diisopropenylbenzene (68%)

(*g*) The compound shown is DDT (standing for the nonsystematic name *dichlorodiphenyltrichloroethane*). It undergoes β elimination to form an alkene.

(100%)

(*h*) Alkyl side chains on naphthalene undergo reactions analogous to those of alkyl groups on benzene.

1-Methylnaphthalene

1-(Bromomethyl)-
naphthalene (46%)

(*i*) Potassium carbonate is a weak base. Hydrolysis of the primary benzylic halide converts it to an alcohol.

$$NC-\!\!\!\!\bigcirc\!\!\!\!-CH_2Cl \xrightarrow[H_2O]{K_2CO_3} NC-\!\!\!\!\bigcirc\!\!\!\!-CH_2OH$$

p-Cyanobenzyl chloride   *p*-Cyanobenzyl alcohol (85%)

11.37 Only benzylic (or allylic) hydrogens are replaced by *N*-bromosuccinimide (NBS). Among the four bromines in 3,4,5-tribromobenzyl bromide, three are substituents on the ring and are incapable of being introduced by benzylic bromination. The starting material must therefore have these three bromines already in place.

3,4,5-Tribromotoluene   3,4,5-Tribromobenzyl bromide
Compound A

11.38 2,3,5-Trimethoxybenzoic acid has the structure shown. The three methoxy groups occupy the same positions in this oxidation product that they did in the unknown compound. The carboxylic acid function must have arisen by oxidation of the $-CH_2CH=C(CH_3)_2$ side chain. Therefore

$(C_{14}H_{20}O_3)$ 2,3,5-Trimethoxybenzoic acid

11.39 Hydroboration–oxidation leads to stereospecific syn addition of H and OH to the double bond. The regiochemistry of addition is opposite to that predicted by Markovnikov's rule. Hydroboration–oxidation of the *E* alkene gives alcohol A.

(*E*)-2-(*p*-Anisyl)-2-butene   2*S*,3*R*   2*R*,3*S*

$$An = CH_3O-\!\!\!\!\bigcirc\!\!\!\!-$$

Alcohol A is a racemic mixture of the 2*S*,3*R* and 2*R*,3*S* enantiomers of 3-(*p*-anisyl)-2-butanol. Hydroboration–oxidation of the *Z* alkene gives alcohol B.

(*Z*)-2-(*p*-Anisyl)-2-butene   (2*R*,3*R*)   (2*S*,3*S*)

Alcohol B is a racemic mixture of the 2R,3R and 2S,3S enantiomers of 3-(p-anisyl)-2-butanol. Alcohols A and B are stereoisomers that are not enantiomers; they are diastereomers.

11.40 Dehydrohalogenation of alkyl halides is stereospecific, requiring an anti arrangement between the hydrogen being lost and the leaving group in the transition state. (Z)-1,2-Diphenylpropene must therefore be formed from the diastereomer shown.

(1S,2S)-1-Chloro-1,2-diphenylpropane

(Z)-1,2-Diphenylpropene (90%)

The mirror-image chloride, 1R,2R, will also give the Z alkene. In fact, the reaction was carried out on a racemic mixture of the 1R,2R and 1S,2S stereoisomers.

The E isomer is formed from either the 1R,2S or the 1S,2R chloride (or from a racemic mixture of the two).

(1R,2S)-1-Chloro-1,2-diphenylpropane

(E)-1,2-Diphenylpropene (87%)

11.41 (*a*) The conversion of ethylbenzene to 1-phenylethyl bromide is a benzylic bromination. It can be achieved by using either bromine or N-bromosuccinimide (NBS).

$$C_6H_5CH_2CH_3 \xrightarrow[\substack{or \\ NBS, heat}]{Br_2, light} C_6H_5\underset{\overset{|}{Br}}{CH}CH_3$$

Ethylbenzene   1-Phenylethyl bromide

(*b*) The conversion of 1-phenylethyl bromide to 1,2-dibromo-1-phenylethane

$$C_6H_5\underset{\overset{|}{Br}}{CH}CH_3 \longrightarrow C_6H_5\underset{\overset{|}{Br}}{CH}CH_2Br$$

cannot be achieved cleanly in a single step. We must reason backward from the target molecule, that is, determine how to make 1,2-dibromo-1-phenylethane in one step from any starting material. Vicinal dibromides are customarily prepared by addition of bromine to alkenes. This suggests that 1,2-dibromo-1-phenylethane can be prepared by the reaction

$$C_6H_5CH{=}CH_2 \ + \ Br_2 \longrightarrow C_6H_5\underset{\overset{|}{Br}}{CH}CH_2Br$$

Styrene   1,2-Dibromo-1-phenylethane

The necessary alkene, styrene, is available by dehydrohalogenation of the given starting material, 1-phenylethyl bromide.

$$C_6H_5CHCH_3 \xrightarrow[CH_3CH_2OH]{NaOCH_2CH_3} C_6H_5CH{=}CH_2$$
$$\underset{|}{\;}$$
$$Br$$

　　　1-Phenylethyl bromide　　　　　　　　　　　Styrene

Thus, by reasoning backward from the target molecule, the synthetic scheme becomes apparent.

$$C_6H_5CHCH_3 \xrightarrow[CH_3CH_2OH]{NaOCH_2CH_3} C_6H_5CH{=}CH_2 \xrightarrow{Br_2} C_6H_5CHCH_2Br$$
$$\underset{|}{\;} \qquad\qquad\qquad\qquad\qquad\qquad \underset{|}{\;}$$
$$Br \qquad\qquad\qquad\qquad\qquad\qquad\qquad\qquad Br$$

　　1-Phenylethyl　　　　　　　　　Styrene　　　　　　　　1,2-Dibromo-1-
　　　bromide　　　　　　　　　　　　　　　　　　　　　　phenylethane

(*c*) The conversion of styrene to phenylacetylene cannot be carried out in a single step. As was pointed out in Chapter 9, however, a standard sequence for converting terminal alkenes to alkynes consists of bromine addition followed by a double dehydrohalogenation in strong base.

$$C_6H_5CH{=}CH_2 \xrightarrow{Br_2} C_6H_5CHCH_2Br \xrightarrow[NH_3]{NaNH_2} C_6H_5C{\equiv}CH$$
$$\underset{|}{\;}$$
$$Br$$

　　　Styrene　　　　　　　　　　1,2-Dibromo-1-　　　　　Phenylacetylene
　　　　　　　　　　　　　　　　　　phenylethane

(*d*) The conversion of phenylacetylene to butylbenzene requires both a carbon–carbon bond formation step and a hydrogenation step. The acetylene function is essential for carbon–carbon bond formation by alkylation. The correct sequence is therefore:

$$C_6H_5C{\equiv}CH \xrightarrow[NH_3]{NaNH_2} C_6H_5C{\equiv}C\text{:}^- \; Na^+$$

　　　Phenylacetylene

$$C_6H_5C{\equiv}C\text{:}^- \; Na^+ + CH_3CH_2Br \longrightarrow C_6H_5C{\equiv}CCH_2CH_3$$

$$C_6H_5C{\equiv}CCH_2CH_3 \xrightarrow[Pt]{H_2} C_6H_5CH_2CH_2CH_2CH_3$$

　　　　　　　　　　　　　　　　　　　　　　　　Butylbenzene

(*e*) The transformation corresponds to alkylation of acetylene, and so the alcohol must first be converted to a species with a good leaving group such as its halide derivative.

$$C_6H_5CH_2CH_2OH \xrightarrow{PBr_3} C_6H_5CH_2CH_2Br$$

　　　　　2-Phenylethanol　　　　　　　　　　　　2-Phenylethyl bromide

$$C_6H_5CH_2CH_2Br + NaC{\equiv}CH \longrightarrow C_6H_5CH_2CH_2C{\equiv}CH$$

　　　2-Phenylethyl　　　　Sodium　　　　　　　　4-Phenyl-1-butyne
　　　　bromide　　　　　acetylide

(f) The target compound is a bromohydrin. Bromohydrins are formed by addition of bromine and water to alkenes.

$$C_6H_5CH_2CH_2Br \xrightarrow[\text{(CH}_3)_3\text{COH}]{\text{KOC(CH}_3)_3} C_6H_5CH=CH_2 \xrightarrow[\text{H}_2\text{O}]{\text{Br}_2} C_6H_5\underset{\underset{\text{OH}}{|}}{C}HCH_2Br$$

2-Phenylethyl bromide   Styrene   2-Bromo-1-phenylethanol

11.42 The stability of free radicals is reflected in their ease of formation. Toluene, which forms a benzyl radical, reacts with bromine 64,000 times faster than does ethane, which forms a primary alkyl radical. Ethylbenzene, which forms a secondary benzylic radical, reacts 1 million times faster than ethane.

Ethylbenzene
(most reactive)

Secondary benzylic
radical

Toluene

Primary benzylic
radical

$$CH_3CH_3 + Br\bullet \longrightarrow CH_3\overset{\bullet}{C}H_2 + HBr$$

Ethane
(least reactive)

Primary
radical

11.43 A good way to develop alternative resonance structures for carbocations is to move electron pairs toward sites of positive charge.

o-Methylbenzyl cation

Tertiary
carbocation

m-Methylbenzyl cation

One of the resonance structures shown for o-methylbenzyl cation is a tertiary carbocation. o-Methylbenzyl cation has tertiary carbocation character; m-methylbenzyl cation does not.

11.44 The rate-determining step in an S_N1 process is formation of the carbocation. The reactivity order can be determined by examining the stability of the carbocations formed remembering the general principle that the more stable a carbocation is, the faster it is formed. All three carbocations are secondary, however the one formed from 3-iodocyclopentene is stabilized by conjugation with the double bond and is formed fastest. The carbocation formed from 5-iodo-1,3-cyclopentadiene is also conjugated, but it is a cyclic, fully conjugated π system having four π electrons and thus is *antiaromatic*. It is the least stable, and is formed slowest.

3-Iodocyclopentene → Conjugated (allylic) carbocation; most stable and formed fastest

Iodocyclopentane → Secondary carbocation

5-Iodo-1,3-cyclopentadiene → Antiaromatic carbocation; least stable and formed slowest

11.45 The process is an acid–base reaction in which cyclopentadiene transfers a proton to amide ion (the base) to give the aromatic cyclopentadienide anion. The sodium ion (Na^+) has been omitted from the equation.

1,3-Cyclopentadiene   +   Amide ion   ⟶   Cyclopentadienide anion   +   Ammonia

11.46 The resonance structures for the cyclopentadienide anions formed by loss of a proton from 1-methyl-1,3-cyclopentadiene and 5-methyl-1,3-cyclopentadiene are equivalent.

1-Methyl-1,3-cyclopentadiene

5-Methyl-1,3-cyclopentadiene

11.47 Cyclooctatetraene is not aromatic. 1,2,3,4-Tetramethylcyclooctatetraene and 1,2,3,8-tetramethylcyclooctatetraene are constitutional isomers.

1,2,3,4-Tetramethyl-
cyclooctatetraene

1,2,3,8-Tetramethyl-
cyclooctatetraene

Leo A. Paquette at Ohio State University synthesized each of these compounds independently of the other and showed them to be stable enough to be stored separately without interconversion.

11.48 Cyclooctatetraene has eight π electrons and thus does not satisfy the $(4n + 2)$ π electron requirement of the Hückel rule.

Cyclooctatetraene.
Each double bond contributes
two π electrons to give a total of eight.

All of the exercises in this problem involve counting the number of π electrons in the various species derived from cyclooctatetraene and determining whether they satisfy the $(4n + 2)$ π electron rule.

(a) Adding one π electron gives a species ($C_8H_8^-$) with nine π electrons. $4n + 2$, where n is a whole number, can never equal nine. This species is therefore *not aromatic*.

(b) Adding two π electrons gives a species ($C_8H_8^{2-}$) with ten π electrons. $4n + 2 = 10$ when $n = 2$. The species ($C_8H_8^{2-}$) *is aromatic*.

(c) Removing one π electron gives a species ($C_8H_8^+$) with seven π electrons. $4n + 2$ cannot equal 7. The species ($C_8H_8^+$) *is not aromatic*.

(d) Removing two π electrons gives a species ($C_8H_8^{2+}$) with six π electrons. $4n + 2 = 6$ when $n = 1$. The species ($C_8H_8^{2+}$) *is aromatic*. (It has the same number of π electrons as benzene.)

11.49 (a, b) Cyclononatetraene does not have a continuous conjugated system of π electrons. Conjugation is incomplete because it is interrupted by a CH_2 group. Thus (a) adding one more π electron or (b) two more π electrons will *not* give an aromatic system.

sp^3 carbon in ring

(*c*) Removing a proton from the CH_2 group permits complete conjugation. The species produced has ten π electrons and is aromatic, because $4n + 2 = 10$ when $n = 2$.

Two π electrons for each double bond
+
two π electrons for unshared pair
= ten π electrons

(*d*) Removing a proton from one of the sp^2-hybridized carbons of the ring does not produce complete conjugation; the CH_2 group remains present to interrupt cyclic conjugation. The anion formed is *not* aromatic.

11.50 (*a*) Cycloundecapentaene is *not aromatic*. Its π system is not fully conjugated; it is interrupted by an sp^3-hybridized carbon.

sp^3-hybridized carbon;
not a completely conjugated
monocyclic π system

(*b*) Cycloundecapentaenyl radical is *not aromatic*. Its π system is completely conjugated and monocyclic but contains 11 π electrons—a number not equal to $(4n + 2)$ where *n* is an integer.

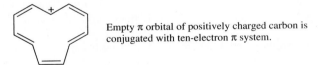

There are 11 electrons in the conjugated π system.
The five double bonds contribute ten π electrons;
the odd electron of the radical is the eleventh.

(*c*) Cycloundecapentaenyl cation is *aromatic*. It includes a completely conjugated π system which contains ten π electrons ($10 = 4n + 2$, where $n = 2$).

Empty π orbital of positively charged carbon is
conjugated with ten-electron π system.

(*d*) Cycloundecapentaenyl anion is *not aromatic*. It contains 12 π electrons and thus does not satisfy the $(4n + 2)$ rule. The presence of $4n$ π electrons ($n = 3$) suggests that the anion would be *antiaromatic*.

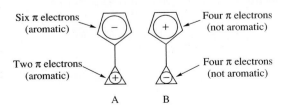

There are 12 π electrons. The five
double bonds contribute ten;
the anionic carbon contributes two.

11.51 (*a*) The more stable dipolar resonance structure is A because it has an aromatic cyclopentadienide anion bonded to an aromatic cyclopropenyl cation. In structure B neither ring is aromatic.

Six π electrons
(aromatic)

Four π electrons
(not aromatic)

Two π electrons
(aromatic)

Four π electrons
(not aromatic)

A   B

(b) Structure D can be stabilized by resonance involving a dipolar form.

Six π electrons
(aromatic)

Six π electrons
(aromatic)

D

Comparable stabilization is not possible in structure C because neither a cyclopropenyl system nor a cycloheptatrienyl system is aromatic in its anionic form. Both are aromatic as cations.

Eight π electrons
(not aromatic)

Six π electrons
(aromatic)

Two π electrons
(aromatic)

Four π electrons
(not aromatic)

C

11.52 (a) This molecule, called *oxepin*, is *not aromatic*. It is antiaromatic. The three double bonds each contribute two π electrons, and an oxygen atom contributes two π electrons to the conjugated system, giving a total of eight π electrons. Only one of the two unshared pairs on oxygen can contribute to the π system; the other unshared pair is in an sp^2-hybridized orbital and cannot interact with it.

p orbital aligned for overlap
with π system of ring

sp^2-hybridized orbital

(b) This compound, called *azonine*, has ten electrons in a completely conjugated planar monocyclic π system and therefore satisfies Hückel's rule for $(4n + 2)$ π electrons where $n = 2$. There are eight π electrons from the conjugated tetraene and two electrons contributed by the nitrogen unshared pair.

Two π electrons

Two π electrons

Unshared pair on nitrogen
is delocalized into π system
of ring.

Two π electrons

Two π electrons

(c) Borazole, sometimes called *inorganic benzene*, is *aromatic*. Six π electrons are contributed by the unshared pairs of the three nitrogen atoms. Each boron contributes a *p* orbital to maintain the conjugated system but no electrons.

(*d*) This compound has eight π electrons and is *not aromatic*. It is antiaromatic.

Two π electrons

Two π electrons

Two π electrons

Electrons in *sp*² orbital do not interact with the π system.

Electrons in *sp*² orbital do not interact with the π system.

Two π electrons

11.53 The structure and numbering system for pyridine are given in Section 11.23 of the text, where we were told that pyridine is aromatic. Oxidation of 3-methylpyridine is analogous to oxidation of toluene. The methyl side chain is oxidized to a carboxylic acid.

3-Methylpyridine   Niacin

11.54 The structure and numbering system for quinoline are given in Section 11.23 of the text. *Nitroxoline* has the structural formula:

5-Nitro-8-hydroxyquinoline

11.55 The ring system of *acridine* (C$_{13}$H$_9$N) is analogous to that of anthracene (i.e., tricyclic and linearly fused). Furthermore, the two most stable resonance forms are equivalent to each other. The nitrogen atom must therefore be in the central ring, and the structure of acridine is

The two resonance forms would not be equivalent if the nitrogen were present in one of the terminal rings. Can you see why?

ANSWERS TO INTERPRETIVE PROBLEMS

11.56 C;   **11.57** A;   **11.58** A;   **11.59** B;   **11.60** D;   **11.61** A;   **11.62** B

SELF-TEST

PART A

A-1. Give an acceptable IUPAC name for each of the following:

(a)

(b) $C_6H_5CHCHCH_3$

(c)

(d)

A-2. Draw the structure of each of the following:

(a) 3,5-Dichlorobenzoic acid

(b) *p*-Nitroanisole

(c) 2,4-Dimethylaniline

(d) *m*-Bromobenzyl chloride

A-3. Write a positive (+) or negative (−) charge at the appropriate position so that each of the following structures contains the proper number of π electrons to permit it to be considered an aromatic ion. For this problem ignore strain effects that might destabilize the molecule.

(a)

(b)

A-4. For each of the following, determine how many π electrons are counted toward satisfying Hückel's rule. Assuming the molecule can adopt a planar conformation, is it aromatic?

(a)

(b)

(c)

A-5. Azulene, shown in the following structure, is highly polar. Draw a dipolar resonance structure to explain this fact.

Azulene

A-6. Give the reactant, reagent, or product omitted from each of the following:

A-7. Provide two methods for the synthesis of 1-bromo-1-phenylpropane from an aromatic hydrocarbon.

A-8. Write the structures of the resonance forms that contribute to the stabilization of the intermediate in the reaction of styrene ($C_6H_5CH=CH_2$) with hydrogen bromide in the absence of peroxides.

A-9. Write one or more resonance structures that represent the delocalization of the following carbocation.

A-10. An unknown compound, $C_{12}H_{18}$, reacts with sodium dichromate ($Na_2Cr_2O_7$) in warm aqueous sulfuric acid to give *p-tert*-butylbenzoic acid. What is the structure of the unknown?

PART B

B-1. The number of possible dichloronitrobenzene isomers is

    (*a*) 3   (*b*) 4   (*c*) 6   (*d*) 8

B-2. Which of the following statements is correct concerning the class of reactions to be expected for benzene and cyclooctatetraene?

    (*a*) Both substances undergo addition reactions.

    (*b*) Both substances undergo substitution reactions.

    (*c*) Benzene undergoes substitution; cyclooctatetraene undergoes addition.

    (*d*) Benzene undergoes addition; cyclooctatetraene undergoes substitution.

B-3. Which, if any, of the following structures represents an aromatic species?

    (*a*)   (*b*)   (*c*)   (*d*) None of these is aromatic.

B-4. Which of the following compounds has a double bond that is conjugated with the π system of the benzene ring?

(a) p-Benzyltoluene   (c) 3-Phenylcyclohexene

(b) 2-Phenyl-1-decene   (d) 3-Phenyl-1,4-pentadiene

B-5. Rank the following compounds in order of increasing rate of solvolysis ($S_N 1$) in aqueous acetone (slowest → fastest):

$$(CH_3)_2CHCH_2CH_2Br \quad\quad (CH_3)_2CHCHCH_3 \quad\quad C_6H_5CHCH(CH_3)_2$$
$$\overset{|}{Br} \quad\quad\quad\quad\quad \overset{|}{Br}$$

1   2   3

(a) 1 < 2 < 3   (b) 2 < 1 < 3   (c) 3 < 2 < 1   (d) 1 < 3 < 2

B-6. When comparing the hydrogenation of benzene with that of a hypothetical 1,3,5-cyclohexatriene, benzene _____ than the cyclohexatriene.

(a) Absorbs 152 kJ/mol more heat   (c) Gives off 152 kJ/mol more heat

(b) Absorbs 152 kJ/mol less heat   (d) Gives off 152 kJ/mol less heat

B-7. The reaction

$$\xrightarrow{CH_3CH_2O^- \ Na^+} \quad ?$$

gives as the major elimination product

(a)   (c)   *Equal* amounts of (a) and (b)

(b)   (d)   Neither (a) nor (b)

B-8. Which one of the following is best classified as a *heterocyclic aromatic* compound?

(a) :N—H   (c)   N: (e)   —ṄH₂

(b)   N: (d)   N:

B-9. Which of the following has the smallest heat of combustion?

 (a)   (c)

 (b)   (d)

B-10. Which one of the following alcohols undergoes dehydration at the *fastest* rate on being heated with sulfuric acid? (The potential for rearrangement does not affect the rate.)

(a) ⬡—CH$_2$CH$_2$CH$_2$CH$_2$OH   (c) ⬡—CH$_2$CHCH$_2$CH$_3$
                                                      |
                                                      OH

(b) ⬡—CH$_2$CH$_2$CHCH$_3$ (d) ⬡—CHCH$_2$CH$_2$CH$_3$
              | |
              OH   OH

B-11. Ethylbenzene is treated with the reagents listed, in the order shown.

1. NBS, peroxides, heat; 2. CH$_3$CH$_2$O$^-$; 3. B$_2$H$_6$; 4. H$_2$O$_2$, HO$^-$.

The structure of the final product is:

(a) C$_6$H$_5$CH$_2$CH$_2$OH

         Br
         |
(b) C$_6$H$_5$CHCH$_2$OH

         OH
         |
(c) C$_6$H$_5$CHCH$_3$

         OH
         |
(d) C$_6$H$_5$CHCH$_2$OH

         OH
         |
(e) C$_6$H$_5$CHCH$_2$Br

B-12. Which of the following hydrogens is most easily abstracted (removed) on reaction with bromine atoms, Br• ?

CH$_3$CH$_2$—⬡—CH$_3$

(a) (b)   (c)   (d)

B-13. All the hydrocarbons shown are very weak acids. One, however, is far more acidic than the others. Which one is the strongest acid?

(a)   (b)   (c)   (d)   (e)

B-14. The compound shown is planar, and all the carbon–carbon bond lengths are the same. What can you deduce about the bonding of boron from these observations?

(*a*) The boron is sp^2-hybridized, and the *p* orbital contains an unshared pair of electrons.

(*b*) The boron is sp^3-hybridized, and a hybrid orbital contains an unshared pair of electrons.

(*c*) The boron is sp^3-hybridized, and a hybrid orbital is vacant.

(*d*) The boron is sp^2-hybridized, and the *p* orbital is vacant.

B-15. How many benzylic hydrogens are present in the hydrocarbon shown?

(*a*) 3   (*b*) 4   (*c*) 5   (*d*) 6   (*e*) 8

CHAPTER 12

Reactions of Arenes:
Electrophilic Aromatic Substitution

SOLUTIONS TO TEXT PROBLEMS

12.1 According to the mechanism of electrophilic aromatic substitution, formation of the cyclohexadienyl cation is rate-determining and endothermic. The transition state for an endothermic elementary step in a mechanism more closely resembles the products of that step, in this case the carbocation intermediate, than it resembles the reactants. Therefore, the transition state for electrophilic aromatic substitution more closely resembles the carbocation intermediate.

12.2 Electrophilic aromatic substitution leads to replacement of one of the hydrogens directly attached to the ring. All four of the ring hydrogens of *p*-xylene are equivalent; so it does not matter which one is replaced by the nitro group.

CH₃ —⟨benzene ring⟩— CH₃ $\xrightarrow[\text{H}_2\text{SO}_4]{\text{HNO}_3}$ CH₃ —⟨benzene ring with NO₂⟩— CH₃

p-Xylene   1,4-Dimethyl-2-
nitrobenzene

12.3 Following the pathway outlined in Mechanism 12.1, the aromatic ring reacts with nitronium ion to give the cyclohexadienyl cation intermediate.

p-Xylene   →(slow)→   Cyclohexadienyl
cation intermediate

Loss of a proton from the intermediate gives the product, 1,4-dimethyl-2-nitrobenzene.

Cyclohexadienyl
cation intermediate   1,4-Dimethyl-2-
nitrobenzene   Hydronium
ion

12.4   The aromatic ring of 1,2,4,5-tetramethylbenzene has two equivalent hydrogens. Sulfonation of the ring leads to replacement of one of them by —SO_3H.

1,2,4,5-Tetramethylbenzene   2,3,5,6-Tetramethylbenzene-
sulfonic acid

12.5   The major product is isopropylbenzene.

Benzene   1-Chloropropane   Propylbenzene
(20% yield)   Isopropylbenzene
(40% yield)

Aluminum chloride coordinates with 1-chloropropane to give a Lewis acid-Lewis base complex, which can be attacked by benzene to yield propylbenzene or can rearrange to produce isopropyl cation. Isopropylbenzene arises by reaction of isopropyl cation with benzene.

Isopropyl cation

12.6   The species that attacks the benzene ring is cyclohexyl cation, formed by protonation of cyclohexene.

Cyclohexene   Sulfuric acid   Cyclohexyl cation   Hydrogen sulfate ion

The mechanism for the reaction of cyclohexyl cation with benzene is analogous to the general mechanism for electrophilic aromatic substitution.

Benzene   Cyclohexyl   Cyclohexadienyl cation   Cyclohexylbenzene
                    cation   intermediate

12.7   The preparation of cyclohexylbenzene from cyclohexene and benzene was described in text Section 12.6. Cyclohexylbenzene is converted to 1-phenylcyclohexene by benzylic bromination, followed by dehydrohalogenation.

Benzene   Cyclohexene   Cyclohexylbenzene   1-Bromo-1-
                                                                                    phenylcyclohexane

1-Phenylcyclohexene

12.8   Treatment of 1,3,5-trimethoxybenzene with an acyl chloride and aluminum chloride brings about Friedel–Crafts acylation at one of the three equivalent positions available on the ring.

1,3,5-Trimethoxybenzene   3-Methylbutanoyl chloride   Isobutyl 2,4,6-trimethoxyphenyl ketone

12.9   Because the anhydride is cyclic, its structural units are not incorporated into a ketone and a carboxylic acid as two separate product molecules. Rather, they become part of a four-carbon unit attached to benzene by a ketone carbonyl. The acyl substituent terminates in a carboxylic acid functional group.

Benzene   Succinic   4-Oxo-4-phenylbutanoic acid
                 anhydride

12.10 (*b*) A Friedel–Crafts alkylation of benzene using 1-chloro-2,2-dimethylpropane would not be a satisfactory method to prepare (2,2-dimethylpropyl)benzene because of the likelihood of a carbocation rearrangement. The best way to prepare this compound is by Friedel–Crafts acylation followed by Clemmensen reduction.

| 2,2-Dimethylpropanoyl chloride | Benzene | | 2,2-Dimethyl-1-phenyl-1-propanone | | (2,2-Dimethylpropyl)benzene |

12.11 (*b*) Partial rate factors for nitration of toluene and *tert*-butylbenzene, relative to a single position of benzene, are as shown:

The sum of these partial rate factors is 147 for toluene, 90 for *tert*-butylbenzene. Toluene is 147/90, or 1.7, times more reactive than *tert*-butylbenzene.

(*c*) The product distribution for nitration of *tert*-butylbenzene is determined from the partial rate factors.

$$\text{Ortho:} \quad \frac{2(4.5)}{90} = 10\%$$

$$\text{Meta:} \quad \frac{2(3)}{90} = 6.7\%$$

$$\text{Para:} \quad \frac{75}{90} = 83.3\%$$

12.12 The compounds shown all undergo electrophilic aromatic substitution more slowly than benzene. Therefore, $-CH_2Cl$, $-CHCl_2$, and $-CCl_3$ are *deactivating* substituents.

| Benzyl chloride | (Dichloromethyl)benzene | (Trichloromethyl)benzene |

The electron-withdrawing power of these substituents, and their tendency to direct incoming electrophiles meta to themselves, will increase with the number of chlorines each contains. Thus, the substituent that gives 4% meta nitration (96% ortho + para) contains the fewest chlorine atoms ($-CH_2Cl$), and the one that gives 64% meta nitration contains the most ($-CCl_3$).

$-CH_2Cl$	$-CHCl_2$	$-CCl_3$
Deactivating, ortho, para-directing	Deactivating, ortho, para-directing	Deactivating, meta-directing

12.13 (*b*) Attack by bromine at the position meta to the amino group gives a cyclohexadienyl cation intermediate in which delocalization of the nitrogen lone pair cannot participate in dispersal of the positive charge.

(*c*) Attack at the position para to the amino group yields a cyclohexadienyl cation intermediate that is stabilized by delocalization of the electron pair of the amino group.

Most stable
resonance contributor

12.14 Electrophilic aromatic substitution in biphenyl is best understood by considering one ring as the functional group and the other as a substituent. An aryl substituent is ortho, para-directing. Nitration of biphenyl gives a mixture of *o*-nitrobiphenyl and *p*-nitrobiphenyl.

Biphenyl   *o*-Nitrobiphenyl   *p*-Nitrobiphenyl
                        (37%)   (63%)

12.15 (*b*) The carbonyl group attached directly to the ring is a signal that the substituent is a meta-directing group. Nitration of methyl benzoate yields methyl *m*-nitrobenzoate.

Methyl benzoate   Methyl *m*-nitrobenzoate
                        (isolated in 81–85% yield)

(*c*) The acyl group in 1-phenyl-1-propanone is meta-directing; the carbonyl is attached directly to the ring. The product is 1-(*m*-nitrophenyl)-1-propanone.

1-Phenyl-1-propanone   1-(*m*-Nitrophenyl)-1-propanone
                        (isolated in 60% yield)

12.16 Writing the structures out in more detail reveals that the substituent $\overset{+}{—N}(CH_3)_3$ lacks the unshared electron pair of $—\overset{..}{N}(CH_3)_2$.

This unshared pair is responsible for the powerful activating effect of an $—\overset{..}{N}(CH_3)_2$ group. On the other hand, the nitrogen in $—\overset{+}{N}(CH_3)_3$ is positively charged and in that respect resembles the nitrogen of a nitro group. We expect the substituent $—\overset{+}{N}(CH_3)_3$ to be deactivating and meta-directing.

12.17 The reaction is a Friedel–Crafts alkylation in which 4-chlorobenzyl chloride serves as the carbocation source and chlorobenzene is the aromatic substrate. Alkylation occurs at the positions ortho and para to the chlorine substituent of chlorobenzene.

Chlorobenzene   4-Chlorobenzyl chloride   1-Chloro-2-(4'-chlorobenzyl)-   1-Chloro-4-(4'-chlorobenzyl)-
                                                                              benzene   benzene

12.18 Bromine reacts with the aromatic ring to form the cyclohexadienyl cation intermediate.

4-Chloro-N-   Cyclohexadienyl   Bromide
methylaniline   cation intermediate   ion

Loss of a proton gives the product, 2-bromo-4-chloro-N-methylaniline.

Cyclohexadienyl   2-Bromo-4-chloro-   Hydrogen
cation intermediate   N-methylaniline   bromide

12.19 (*b*) Halogen substituents are ortho, para-directing, and the disposition in *m*-dichlorobenzene is such that their effects reinforce each other. The major product is 2,4-dichloro-1-nitrobenzene. Substitution at the position between the two chlorines is slow because it is a sterically hindered position.

Most reactive positions (arrows)
in electrophilic aromatic
substitution of *m*-dichlorobenzene

2,4-Dichloro-1-nitrobenzene
(major product of nitration)

(*c*) Nitro groups are meta-directing. Both nitro groups of *m*-dinitrobenzene direct an incoming electrophile to the same position in an electrophilic aromatic substitution reaction. Nitration of *m*-nitrobenzene yields 1,3,5-trinitrobenzene.

Both nitro groups of
m-dinitrobenzene direct
electrophile to same position (arrow).

1,3,5-Trinitrobenzene
(principal product of nitration
of *m*-dinitrobenzene)

(*d*) A methoxy group is ortho, para-directing, and a carbonyl group is meta-directing. The open positions of the ring that are activated by the methoxy group in *p*-methoxyacetophenone are also those that are meta to the carbonyl, so the directing effects of the two substituents reinforce each other. Nitration of *p*-methoxyacetophenone yields 4-methoxy-3-nitroacetophenone.

Positions ortho to the methoxy
group (arrows) are meta to
the carbonyl.

4-Methoxy-3-nitroacetophenone

(*e*) The methoxy group of *p*-methylanisole activates the positions that are ortho to it; the methyl activates those ortho to itself. Methoxy is a more powerful activating substituent than methyl, so nitration occurs ortho to the methoxy group.

Methyl activates C-3 and C-5
(light arrows); methoxy activates
C-2 and C-6 (heavy arrows).

4-Methyl-2-nitroanisole
(principal product of nitration)

(*f*) All the substituents in 2,6-dibromoanisole are ortho, para-directing, and their effects are felt at different positions. The methoxy group, however, is a far more powerful activating substituent than bromine, so it controls the regioselectivity of nitration.

Methoxy directs toward C-4
(heavy arrow); bromines direct
toward C-3 and C-5 (light arrows).

2,6-Dibromo-4-nitroanisole
(principal product of nitration)

12.20 The product obtained when benzene is subjected to bromination and nitration depends on the order in which the reactions are carried out. A nitro group is meta-directing, and so if it is introduced prior to the bromination step, *m*-bromonitrobenzene is obtained.

Benzene   Nitrobenzene   *m*-Bromonitrobenzene

Bromine is an ortho, para-directing group. If it is introduced first, nitration of the resulting bromobenzene yields a mixture of *o*-bromonitrobenzene and *p*-bromonitrobenzene.

Benzene   Bromobenzene   *o*-Bromonitrobenzene   *p*-Bromonitrobenzene

12.21 A straightforward approach to the synthesis of *m*-nitrobenzoic acid involves preparation of benzoic acid by oxidation of toluene, followed by nitration. The carboxyl group of benzoic acid is meta-directing. Nitration of toluene prior to oxidation would lead to a mixture of ortho and para products.

Toluene   Benzoic acid   *m*-Nitrobenzoic acid

12.22 The text points out that C-1 of naphthalene is more reactive than C-2 toward electrophilic aromatic substitution. Thus, of the two possible products of sulfonation, naphthalene-1-sulfonic acid should be formed faster and should be the major product under conditions of kinetic control. Because the problem states that the product under conditions of thermodynamic control is the other isomer, naphthalene-2-sulfonic acid is the major product at elevated temperature.

Naphthalene	Naphthalene-1-sulfonic acid major product at 0°C; formed faster	Naphthalene-2-sulfonic acid major product at 160°C; more stable

Naphthalene-2-sulfonic acid is the more stable isomer for steric reasons. The hydrogen at C-8 (the one shown in the equation) crowds the —SO_3H group in naphthalene-1-sulfonic acid.

12.23 The text states that electrophilic aromatic substitution in furan, thiophene, and pyrrole occurs at C-2. Sulfonation of thiophene gives thiophene-2-sulfonic acid.

Thiophene	Thiophene-2- sulfonic acid

12.24 (*a*) Nitration of benzene is the archetypical electrophilic aromatic substitution reaction.

Benzene	Nitrobenzene

(*b*) Nitrobenzene is much less reactive than benzene toward electrophilic aromatic substitution. The nitro group on the ring is a meta director.

Nitrobenzene	*m*-Dinitrobenzene

(*c*) Toluene is more reactive than benzene in electrophilic aromatic substitution. A methyl substituent is an ortho, para director.

Toluene	*o*-Bromotoluene	*p*-Bromotoluene

(*d*) Trifluoromethyl is deactivating and meta-directing.

(Trifluoromethyl)-benzene   *m*-Bromo(trifluoromethyl)-benzene

(*e*) Anisole is ortho, para-directing, strongly activated toward electrophilic aromatic substitution, and readily sulfonated in sulfuric acid.

Anisole   *o*-Methoxybenzene-sulfonic acid   *p*-Methoxybenzene-sulfonic acid

Sulfur trioxide could be added to the sulfuric acid to facilitate reaction. The para isomer is the predominant product.

(*f*) Acetanilide is similar to anisole in its behavior toward electrophilic aromatic substitution.

Acetanilide   *o*-Acetamidobenzene-sulfonic acid   *p*-Acetamidobenzene-sulfonic acid

(*g*) Bromobenzene is less reactive than benzene. A bromine substituent is ortho, para-directing.

Bromobenzene   *o*-Bromochloro-benzene   *p*-Bromochloro-benzene (*h*) Anisole is a reactive substrate toward Friedel–Crafts alkylation and yields a mixture of *o*- and *p*-benzylated products when treated with benzyl chloride and aluminum chloride.

Anisole   Benzyl chloride   *o*-Benzylanisole   *p*-Benzylanisole

(*i*) Benzene will undergo acylation with benzoyl chloride and aluminum chloride.

Benzene   Benzoyl   Benzophenone
                 chloride

(*j*) A benzoyl substituent is meta-directing and deactivating.

Benzophenone   *m*-Nitrobenzophenone

(*k*) Clemmensen reduction conditions involve treating a ketone with zinc amalgam and concentrated hydrochloric acid.

Benzophenone   Diphenylmethane

(*l*) Wolff–Kishner reduction utilizes hydrazine, a base, and a high-boiling alcohol solvent to reduce ketone functions to methylene groups.

Benzophenone   Diphenylmethane

12.25 (*a*) There are three principal resonance forms of the cyclohexadienyl cation intermediate formed by attack of bromine on *p*-xylene.

Tertiary carbocation

Any one of these resonance forms is a satisfactory answer to the question. Because of its tertiary carbocation character, this carbocation is more stable than the corresponding intermediate formed from benzene.

(*b*) Chlorination of *m*-xylene will give predominantly 4-chloro-1,3-dimethylbenzene.

m-Xylene   4-Chloro-1,3-dimethylbenzene   More stable cyclohexadienyl cation

The intermediate shown (or any of its resonance forms) is more stable for steric reasons than

Less stable cyclohexadienyl cation

The cyclohexadienyl cation intermediate leading to 4-chloro-1,3-dimethylbenzene is more stable and is formed faster than the intermediate leading to chlorobenzene because of its tertiary carbocation character.

more stable than

(c) The most stable carbocation intermediate formed during nitration of acetophenone is the one corresponding to meta attack.

An acyl group is electron-withdrawing and destabilizes a carbocation to which it is attached. The most stable carbocation intermediate in the nitration of acetophenone is less stable, and is formed more slowly, than is the corresponding carbocation formed during nitration of benzene.

(d) The methoxy group in anisole is strongly activating and ortho, para-directing. For steric reasons and because of inductive electron withdrawal by oxygen, the intermediate leading to para substitution is the most stable.

Of the various resonance forms for the most stable intermediate, the most stable one has eight electrons around each oxygen and carbon.

Most stable
resonance form

This intermediate is much more stable than the corresponding intermediate from acylation of benzene.

(*e*) An isopropyl group is an activating substituent and is ortho, para-directing. Attack at the ortho position is sterically hindered. The most stable intermediate is

$$CH(CH_3)_2$$

or any of its resonance forms. Because of its tertiary carbocation character, this cation is more stable than the corresponding cyclohexadienyl cation intermediate from benzene.

(*f*) A nitro substituent is deactivating and meta-directing. The most stable cyclohexadienyl cation formed in the bromination of nitrobenzene is

This ion is less stable than the cyclohexadienyl cation formed during bromination of benzene.

(*g*) Sulfonation of furan takes place at C-2. The cationic intermediate is more stable than the cyclohexadienyl cation formed from benzene because it is stabilized by electron release from oxygen.

Furan   Furan-2-sulfonic acid   via

(*h*) Pyridine reacts with electrophiles at C-3. It is less reactive than benzene, and the carbocation intermediate is less stable than the corresponding intermediate formed from benzene.

12.26 (*a*) Toluene is more reactive than chlorobenzene in electrophilic aromatic substitution reactions because a methyl substituent is activating but a halogen substituent is deactivating. Both are ortho, para-directing, however. Nitration of toluene is faster than nitration of chlorobenzene.

Faster:

Toluene $\xrightarrow[\text{H}_2\text{SO}_4]{\text{HNO}_3}$ *o*-Nitrotoluene   +   *p*-Nitrotoluene

Slower:

Chlorobenzene   o-Chloronitrobenzene   p-Chloronitrobenzene

(b) A fluorine substituent is not nearly as strongly deactivating as a trifluoromethyl group. The reaction that takes place is Friedel–Crafts alkylation of fluorobenzene.

o-Benzylfluorobenzene   p-Benzylfluorobenzene
(15%)   (85%)

Strongly deactivated aromatic compounds do not undergo Friedel–Crafts reactions.

(c) A carbonyl group directly bonded to a benzene ring strongly *deactivates* it toward electrophilic aromatic substitution. Methyl benzoate is much less reactive than benzene.

An oxygen attached directly to the ring strongly *activates* it toward electrophilic aromatic substitution. Phenyl acetate is much more reactive than benzene or methyl benzoate.

Phenyl acetate   o-Bromophenyl   p-Bromophenyl
                            acetate   acetate

Bromination of methyl benzoate requires more vigorous conditions; catalysis by iron(III) bromide is required for bromination of deactivated aromatic rings.

(*d*) Acetanilide is strongly activated toward electrophilic aromatic substitution and reacts faster than nitrobenzene, which is strongly deactivated.

Acetanilide
(Lone pair on nitrogen can stabilize cyclohexadienyl cation intermediate.)

Nitrobenzene
(Nitrogen is positively charged and is electron-withdrawing.)

Acetanilide

o-Acetamidobenzene-sulfonic acid

p-Acetamidobenzene-sulfonic acid

(*e*) Both substrates are of the type

R = alkyl

and are activated toward Friedel–Crafts acylation. Because electronic effects are comparable, we look to differences in steric factors and conclude that reaction will be faster for R = CH$_3$ than for R = (CH$_3$)$_3$C— .

p-Xylene

Acetyl chloride

2,5-Dimethylacetophenone

(*f*) A phenyl substituent is activating and ortho, para-directing. Biphenyl will undergo chlorination readily.

Biphenyl

o-Chlorobiphenyl

p-Chlorobiphenyl

Each benzene ring of benzophenone is deactivated by the carbonyl group.

Benzophenone

Benzophenone is much less reactive than biphenyl in electrophilic aromatic substitution reactions.

12.27 Reactivity toward electrophilic aromatic substitution increases with increasing number of electron-releasing substituents. Benzene, with no methyl substituents, is the least reactive, followed by toluene, with one methyl group. 1,3,5-Trimethylbenzene, with three methyl substituents, is the most reactive.

	Benzene	Toluene	1,3,5-Trimethylbenzene
Relative reactivity:	1	60	2×10^7

o-Xylene and *m*-xylene are intermediate in reactivity between toluene and 1,3,5-trimethylbenzene. Of the two, *m*-xylene is more reactive than *o*-xylene because the activating effects of the two methyl groups reinforce each other.

	o-Xylene (all positions somewhat activated)	*m*-Xylene (activating effects reinforce each other)
Relative reactivity:	5×10^2	5×10^4

12.28 (*a*) Chlorine is ortho, para-directing, carboxyl is meta-directing. The positions that are ortho to the chlorine are meta to the carboxyl, so that both substituents direct an incoming electrophile to the same position. Introduction of the second nitro group at the remaining position that is ortho to the chlorine puts it meta to the carboxyl and meta to the first nitro group.

p-Chlorobenzoic acid

$\xrightarrow[\text{H}_2\text{SO}_4, \text{heat}]{\text{HNO}_3}$

$\xrightarrow[\text{H}_2\text{SO}_4, \text{heat}]{\text{HNO}_3}$

4-Chloro-3,5-dinitrobenzoic acid (90%)

(b) An amino group is one of the strongest activating substituents. The para and both ortho positions are readily substituted in aniline. When aniline is treated with excess bromine, 2,4,6-tribromoaniline is formed in quantitative yield.

Aniline

2,4,6-Tribromoaniline
(100%)

(c) The positions ortho and para to the amino group in *o*-aminoacetophenone are the ones most activated toward electrophilic aromatic substitution.

o-Aminoacetophenone

2-Amino-3,5-
dibromoacetophenone (65%)

(d) The carboxyl group in benzoic acid is meta-directing, and so nitration gives *m*-nitrobenzoic acid. The second nitration step introduces a nitro group meta to both the carboxyl group and the first nitro group.

Benzoic acid

m-Nitrobenzoic acid

3,5-Dinitrobenzoic
acid (54–58%)

(e) Both bromine substituents are introduced ortho to the strongly activating hydroxyl group in *p*-nitrophenol.

p-Nitrophenol

2,6-Dibromo-4-
nitrophenol (96–98%)

(*f*) Friedel–Crafts alkylation occurs when biphenyl is treated with *tert*-butyl chloride and iron(III) chloride (a Lewis acid catalyst); the product of monosubstitution is *p-tert*-butylbiphenyl. All the positions of the ring that bears the *tert*-butyl group are sterically hindered, so the second alkylation step introduces a *tert*-butyl group at the para position of the second ring.

Biphenyl

C(CH$_3$)$_3$

$\dfrac{(CH_3)_3CCl}{FeCl_3}$

C(CH$_3$)$_3$

$\dfrac{(CH_3)_3CCl}{FeCl_3}$

C(CH$_3$)$_3$

C(CH$_3$)$_3$

4,4′-Di-*tert*-butylbiphenyl (70%)

(*g*) Disulfonation of phenol occurs at positions ortho and para to the hydroxyl group. The ortho, para product predominates over the ortho, ortho one.

OH

Phenol

$\xrightarrow{H_2SO_4}$

OH

SO$_3$H

SO$_3$H

2-Hydroxy-1,5-benzenedisulfonic acid

12.29 When carrying out each of the following syntheses, evaluate how the structure of the product differs from that of benzene or toluene; that is, determine which groups have been substituted on the benzene ring or altered in some way. The sequence of reaction steps when multiple substitution is desired is important; recall that some groups direct ortho, para and others meta.

(*a*) Isopropylbenzene may be prepared by a Friedel–Crafts alkylation of benzene with isopropyl chloride (or bromide, or iodide).

Benzene   +   (CH$_3$)$_2$CHCl $\xrightarrow{AlCl_3}$ —CH(CH$_3$)$_2$

Benzene   Isopropyl chloride   Isopropylbenzene

It would not be appropriate to use propyl chloride and trust that a rearrangement would lead to isopropylbenzene, because a mixture of propylbenzene and isopropylbenzene would be obtained.
     Isopropylbenzene may also be prepared by alkylation of benzene with propene in the presence of sulfuric acid.

Benzene   +   CH$_3$CH=CH$_2$ $\xrightarrow{H_2SO_4}$ —CH(CH$_3$)$_2$

Benzene   Propene   Isopropylbenzene

(b) Because the isopropyl and sulfonic acid groups are para to each other, the first group introduced on the ring must be the ortho, para director, that is, the isopropyl group. We may therefore use the product of part (a), isopropylbenzene, in this synthesis. An isopropyl group is a fairly bulky ortho, para director, and so sulfonation of isopropylbenzene gives mainly p-isopropylbenzenesulfonic acid.

$(CH_3)_2CH$—⬡ $\xrightarrow[\text{H}_2\text{SO}_4]{\text{SO}_3}$ $(CH_3)_2CH$—⬡—SO_3H

Isopropylbenzene   p-Isopropylbenzenesulfonic acid

A sulfonic acid group is meta-directing, so that the order of steps must be alkylation followed by sulfonation rather than the reverse.

(c) Free-radical halogenation of isopropylbenzene, from part (a), occurs with high regioselectivity at the benzylic position. N-Bromosuccinimide (NBS) is a good reagent to use for benzylic bromination.

⬡—$CH(CH_3)_2$ $\xrightarrow[\text{peroxides}]{\text{NBS}}$ ⬡—$\overset{\overset{\displaystyle Br}{|}}{\underset{\underset{\displaystyle CH_3}{|}}{C}}CH_3$

Isopropylbenzene   2-Bromo-2-phenylpropane

(d) Toluene is an obvious starting material for the preparation of 4-tert-butyl-2-nitrotoluene. Two possibilities, both involving nitration and alkylation of toluene, present themselves; the problem to be addressed is in what order to carry out the two steps. Friedel–Crafts alkylation must precede nitration.

CH_3-⬡ $\xrightarrow[\text{AlCl}_3]{(CH_3)_3CCl}$ CH_3-⬡-$C(CH_3)_3$ $\xrightarrow[\text{H}_2\text{SO}_4]{\text{HNO}_3}$ CH_3-⬡(NO_2)-$C(CH_3)_3$

Toluene   p-tert-Butyltoluene   4-tert-Butyl-2-nitrotoluene

Introduction of the nitro group as the first step is unsatisfactory because Friedel–Crafts reactions cannot be carried out on nitro-substituted aromatic compounds.

(e) Two electrophilic aromatic substitution reactions need to be performed: chlorination and Friedel–Crafts acylation. The order in which the reactions are carried out is important; chlorine is an ortho, para director, and the acetyl group is a meta director. Because the groups are meta in the desired compound, introduce the acetyl group first.

⬡ $\xrightarrow[\text{AlCl}_3]{\overset{\overset{\displaystyle O}{||}}{CH_3CCl}}$ ⬡—$\overset{\overset{\displaystyle O}{||}}{C}CH_3$ $\xrightarrow[\text{AlCl}_3]{\text{Cl}_2}$ ⬡(Cl)—$\overset{\overset{\displaystyle O}{||}}{C}CH_3$

Benzene   Acetophenone   m-Chloroacetophenone (*f*) Reverse the order of steps in part (*e*) to prepare *p*-chloroacetophenone.

Benzene   Chlorobenzene   *p*-Chloroacetophenone

Friedel–Crafts reactions can be carried out on halobenzenes but not on arenes that are more strongly deactivated.

(*g*) Here again the problem involves two successive electrophilic aromatic substitution reactions, in this case using toluene as the initial substrate. The proper sequence is Friedel–Crafts acylation first, followed by bromination of the ring.

Toluene   *p*-Methylacetophenone   3-Bromo-4-
methylacetophenone

If the sequence of steps had been reversed, with halogenation preceding acylation, the first intermediate would be *o*-bromotoluene, Friedel–Crafts acylation of which would give a complex mixture of products because both groups are ortho, para-directing. On the other hand, the orienting effects of the two groups in *p*-methylacetophenone reinforce each other, so that its bromination is highly regioselective and in the desired direction.

(*h*) Recalling that alkyl groups attached to the benzene ring by CH$_2$ may be prepared by reduction of the appropriate ketone, we may reduce 3-bromo-4-methylacetophenone, as prepared in part (*g*), by the Clemmensen or Wolff–Kishner procedure to give 2-bromo-4-ethyltoluene.

3-Bromo-4-
methylacetophenone   2-Bromo-4-
ethyltoluene

(*i*) This is relatively straightforward. Bromine is an ortho, para-directing substituent; nitro is meta-directing. Nitrate first, and then brominate to give 1-bromo-3-nitrobenzene.

Benzene   Nitrobenzene   1-Bromo-3-nitrobenzene

(*j*) Take advantage of the ortho, para-directing properties of bromine to prepare 1-bromo-2,4-dinitrobenzene. Brominate first, and then nitrate under conditions that lead to disubstitution. The nitro groups are introduced at positions ortho and para to the bromine and meta to each other.

Benzene Br_2, $FeBr_3$ Bromobenzene HNO_3 / H_2SO_4, heat   1-Bromo-2,4-dinitrobenzene

(*k*) Although bromo and nitro substituents are readily introduced by electrophilic aromatic substitution, the only method we have available so far to prepare carboxylic acids is by oxidation of alkyl side chains. Thus, use toluene as a starting material, planning to convert the methyl group to a carboxyl group by oxidation. Nitrate next; nitro and carboxyl are both meta-directing groups, so that the bromination in the last step occurs with the proper regioselectivity.

Toluene $Na_2Cr_2O_7$ / H_2O, H_2SO_4, heat   Benzoic acid HNO_3 / H_2SO_4 3-Nitrobenzoic acid Br_2 / $FeBr_3$ 3-Bromo-5-nitrobenzoic acid

If bromination is performed prior to nitration, the bromine substituent will direct an incoming electrophile to positions ortho and para to itself, giving the wrong orientation of substituents in the product.

(*l*) Again toluene is a suitable starting material, with its methyl group serving as the source of the carboxyl substituent. The orientation of the substituents in the final product requires that the methyl group be retained until the final step.

Toluene HNO_3 / H_2SO_4 *p*-Nitrotoluene Br_2 / $FeBr_3$ 2-Bromo-4-nitrotoluene $Na_2Cr_2O_7$ / H_2O, H_2SO_4, heat   2-Bromo-4-nitrobenzoic acid

Nitration must precede bromination, as in the previous part, in order to prevent formation of an undesired mixture of isomers.

(*m*) Friedel–Crafts alkylation of benzene with benzyl chloride (or benzyl bromide) is a satisfactory route to diphenylmethane.

Benzene   +   Benzyl chloride —CH$_2$Cl $AlCl_3$ Diphenylmethane —CH$_2$—

Benzyl chloride is prepared by free-radical chlorination of toluene.

$$\text{C}_6\text{H}_5\text{—CH}_3 \xrightarrow[\text{light or heat}]{\text{Cl}_2} \text{C}_6\text{H}_5\text{—CH}_2\text{Cl}$$

Toluene   Benzyl chloride

Alternatively, benzene could have been subjected to Friedel–Crafts acylation with benzoyl chloride to give benzophenone (see Problem 12.24 *i*). Clemmensen or Wolff–Kishner reduction of benzophenone would then furnish diphenylmethane (see Problems 12.24 *k* and *l*).

(*n*) 1-Phenyloctane cannot be prepared efficiently by direct alkylation of benzene, because of the probability that rearrangement will occur. Indeed, a mixture of 1-phenyloctane and 2-phenyloctane is formed under the usual Friedel–Crafts conditions, along with 3-phenyloctane.

$$\text{C}_6\text{H}_6 + \text{CH}_3(\text{CH}_2)_6\text{CH}_2\text{Br} \xrightarrow{\text{AlBr}_3} \text{C}_6\text{H}_5\text{CH}_2(\text{CH}_2)_6\text{CH}_3 + \underset{\overset{|}{\text{CH}_3}}{\text{C}_6\text{H}_5\text{CH}}(\text{CH}_2)_5\text{CH}_3 + \underset{\overset{|}{\text{CH}_2\text{CH}_3}}{\text{C}_6\text{H}_5\text{CH}}(\text{CH}_2)_4\text{CH}_3$$

Benzene   1-Bromooctane   1-Phenyloctane (40%)   2-Phenyloctane (30%)   3-Phenyloctane (30%)

A method that permits the synthesis of 1-phenyloctane free of isomeric compounds is acylation followed by reduction.

$$\text{C}_6\text{H}_6 + \text{CH}_3(\text{CH}_2)_6\overset{\overset{\text{O}}{\|}}{\text{C}}\text{Cl} \xrightarrow{\text{AlCl}_3} \text{C}_6\text{H}_5\overset{\overset{\text{O}}{\|}}{\text{C}}(\text{CH}_2)_6\text{CH}_3 \xrightarrow[\text{HCl}]{\text{Zn(Hg)}} + \text{C}_6\text{H}_5\text{CH}_2(\text{CH}_2)_6\text{CH}_3$$

Benzene   Octanoyl chloride   1-Phenyl-1-octanone   1-Phenyloctane

Alternatively, Wolff–Kishner conditions (hydrazine, potassium hydroxide, diethylene glycol, heat) could be used in the reduction step.

(*o*) Direct alkenylation of benzene under Friedel–Crafts reaction conditions does *not* take place, and so 1-phenyl-1-octene *cannot* be prepared by the reaction

$$\text{C}_6\text{H}_6 + \text{ClCH}=\text{CH}(\text{CH}_2)_5\text{CH}_3 \xrightarrow{\text{AlCl}_3} \text{C}_6\text{H}_5\text{CH}=\text{CH}(\text{CH}_2)_5\text{CH}_3$$

Benzene   1-Chloro-1-octene   1-Phenyl-1-octene

No! Reaction effective only with
alkyl halides, not 1-haloalkenes.

Having already prepared 1-phenyloctane in part (*n*), however, we can functionalize the benzylic position by bromination and then carry out a dehydrohalogenation to obtain the target compound.

$$\text{C}_6\text{H}_5\text{CH}_2(\text{CH}_2)_6\text{CH}_3 \xrightarrow[\text{or NBS, peroxides}]{\text{Br}_2, \text{ light}} \underset{\overset{|}{\text{Br}}}{\text{C}_6\text{H}_5\text{CH}}(\text{CH}_2)_6\text{CH}_3 \xrightarrow[\text{CH}_3\text{OH}]{\text{KOCH}_3} \text{C}_6\text{H}_5\text{CH}=\text{CH}(\text{CH}_2)_5\text{CH}_3$$

1-Phenyloctane   1-Bromo-1-phenyloctane   1-Phenyl-1-octene

(*p*) 1-Phenyl-1-octyne cannot be prepared in one step from benzene; 1-haloalkynes are unsuitable reactants for a Friedel–Crafts process. In Chapter 9, however, we learned that alkynes may be prepared from the corresponding alkene:

$$\text{RC}\equiv\text{CR} \text{ obtained from } \underset{\overset{|}{\text{Br}}}{\text{RCH}}\text{—}\underset{\overset{|}{\text{Br}}}{\text{CHR}} \text{ obtained from } \text{RCH}=\text{CHR}$$

Using the alkene prepared in part (o),

$$C_6H_5CH{=}CH(CH_2)_5CH_3 \xrightarrow{\text{Br}_2} C_6H_5\underset{\underset{\text{Br}}{|}}{C}H\underset{\underset{\text{Br}}{|}}{C}H(CH_2)_5CH_3 \xrightarrow[\text{NH}_3]{\text{NaNH}_2} C_6H_5C{\equiv}C(CH_2)_5CH_3$$

1-Phenyl-1-octene   1,2-Dibromo-1-phenyloctane   1-Phenyl-1-octyne

(q) Nonconjugated cyclohexadienes are prepared by Birch reduction of arenes. Thus the last step in the synthesis of 1,4-di-*tert*-butyl-1,4-cyclohexadiene is the Birch reduction of 1,4-di-*tert*-butylbenzene.

Benzene   *tert*-Butylbenzene   p-Di-*tert*-butylbenzene   1,4-Di-*tert*-butyl-1,4-cyclohexadiene

12.30 (a) Methoxy is an ortho, para-directing substituent. All that is required to prepare p-methoxy-benzenesulfonic acid is to sulfonate anisole.

Anisole   p-Methoxybenzene-sulfonic acid

(b) In reactions involving disubstitution of anisole, the better strategy is to introduce the para substituent first. The methoxy group is ortho, para-directing, but para substitution predominates.

Anisole   p-Nitroanisole   2-Bromo-4-nitroanisole

(c) Reversing the order of the steps used in part (b) yields 4-bromo-2-nitroanisole.

Anisole   p-Bromoanisole   4-Bromo-2-nitroanisole

(*d*) Direct introduction of a vinyl group onto an aromatic ring is not feasible. *p*-Methoxystyrene must be prepared indirectly by adding an ethyl side chain and then taking advantage of the reactivity of the benzylic position by bromination (e.g., with *N*-bromosuccinimide) and dehydrohalogenation.

    Anisole   *p*-Ethylanisole   *p*-(1-Bromoethyl)anisole   *p*-Methoxystyrene

12.31 (*a*) Methyl is an ortho, para-directing substituent, and toluene yields mainly *o*-nitrotoluene and *p*-nitrotoluene on mononitration. Some *m*-nitrotoluene is also formed.

    Toluene   *o*-Nitrotoluene   *m*-Nitrotoluene   *p*-Nitrotoluene

(*b*) There are six isomeric dinitrotoluenes:

    2,3-Dinitrotoluene   2,4-Dinitrotoluene   2,5-Dinitrotoluene

    2,6-Dinitrotoluene   3,5-Dinitrotoluene   3,4-Dinitrotoluene

The least likely product is 3,5-dinitrotoluene because neither of its nitro groups is ortho or para to the methyl group.

(*c*) There are six trinitrotoluene isomers:

2,4,6-Trinitrotoluene   2,3,4-Trinitrotoluene   2,3,5-Trinitrotoluene

2,3,6-Trinitrotoluene   3,4,5-Trinitrotoluene   2,4,5-Trinitrotoluene

The most likely major product is 2,4,6-trinitrotoluene because all the positions activated by the methyl group are substituted. This is, in fact, the compound commonly known as TNT.

12.32 From *o*-xylene:

o-Xylene   Acetyl
                  chloride   3,4-Dimethyl-
                                    acetophenone (94%)

From *m*-xylene:

m-Xylene   Acetyl
                  chloride   2,4-Dimethyl-
                                    acetophenone (86%)

From *p*-xylene:

p-Xylene	Acetyl chloride	2,5-Dimethyl-acetophenone (99%)

12.33 The ring that bears the nitrogen in benzanilide is activated toward electrophilic aromatic substitution. The ring that bears the C=O is strongly deactivated.

Benzanilide	*N*-(*o*-Chlorophenyl)benzamide	*N*-(*p*-Chlorophenyl)benzamide

12.34 (*a*) Nitration of the ring takes place para to the ortho, para-directing chlorine substituent; this position is also meta to the meta-directing carboxyl groups.

2-Chloro-1,3-benzenedicarboxylic acid	2-Chloro-5-nitro-1,3-benzenedicarboxylic acid (86%)

(*b*) Bromination of the ring occurs at the only available position activated by the amino group, a powerful activating substituent and an ortho, para director. This position is meta to the meta-directing trifluoromethyl group and to the meta-directing nitro group.

4-Nitro-2-(trifluoromethyl)-aniline	2-Bromo-4-nitro-6-(trifluoromethyl)aniline (81%)

(*c*) This may be approached as a problem in which there are two aromatic rings. One of them bears two activating substituents and so is more reactive than the other, which bears only one activating substituent. Of the two activating substituents (—OH and C₆H₅—), the hydroxyl substituent is the more powerful and controls the regioselectivity of substitution.

p-Phenylphenol	2-Bromo-4-phenylphenol

(*d*) Both substituents are activating and nitration occurs readily even in the absence of sulfuric acid; both are ortho, para-directing and comparable in activating power. The position at which substitution takes place is therefore

(*e*) Protonation of 1-octene yields a secondary carbocation, which attacks benzene.

Benzene   1-Octene   2-Phenyloctane (84%)

(*f*) The reaction that occurs with arenes and acid anhydrides in the presence of aluminum chloride is Friedel–Crafts acylation. The methoxy group is the more powerful activating substituent, so acylation occurs para to it.

o-Fluoroanisole   Acetic anhydride   3-Fluoro-4-methoxyacetophenone   Acetic
                                                                  (70–80%)   acid

(*g*) The isopropyl group is ortho, para-directing, and the nitro group is meta-directing. In this case their orientation effects reinforce each other. Electrophilic aromatic substitution takes place ortho to isopropyl and meta to nitro.

p-Nitroisopropylbenzene   2,4-Dinitroisopropylbenzene (96%)

(*h*) In the presence of an acid catalyst (H_2SO_4), 2-methylpropene is converted to *tert*-butyl cation, which then attacks the aromatic ring ortho to the strongly activating methoxy group.

$$(CH_3)_2C\text{=}CH_2 + H^+ \longrightarrow (CH_3)_3\overset{+}{C}$$

In this particular example, 2-*tert*-butyl-4-methylanisole was isolated in 98% yield.

(*i*) There are two things to consider in this problem: (1) In which ring does bromination occur, and (2) what is the orientation of substitution in that ring? Substitution will take place in the ring that bears the most powerful activating substituent, the hydroxyl group. Both positions ortho to the hydroxyl group are already substituted, so bromination takes place para to it. The product shown was isolated from the bromination reaction in 100% yield.

3-Benzyl-2,6-dimethylphenol

3-Benzyl-4-bromo-2,6-dimethylphenol (100%)

(*j*) Wolff–Kishner reduction converts benzophenone to diphenylmethane.

Benzophenone

Diphenylmethane (83%)

(*k*) Fluorine is an ortho, para-directing substituent. It undergoes Friedel–Crafts alkylation on being treated with benzyl chloride and aluminum chloride to give a mixture of *o*-fluorodiphenylmethane and *p*-fluorodiphenylmethane.

Fluorobenzene   Benzyl chloride

o-Fluorodiphenylmethane (15%)

p-Fluorodiphenylmethane (85%)

(l) The $-\overset{..}{N}H\overset{\overset{\displaystyle O}{\|}}{C}CH_3$ substituent is a more powerful activator than the ethyl group. It directs Friedel–Crafts acylation primarily to the position para to itself.

o-Ethylacetanilide

4-Acetamido-3-ethylacetophenone (57%)

(m) Clemmensen reduction converts the carbonyl group to a CH$_2$ unit.

2,4,6-Trimethylacetophenone

2-Ethyl-1,3,5-trimethylbenzene (74%)

(n) Bromination occurs at C-5 on thiophene-3-carboxylic acid. Reaction does not occur at C-2 because substitution at this position would place a carbocation adjacent to the electron-withdrawing carboxyl group.

Thiophene-3-carboxylic acid

2-Bromothiophene-4-carboxylic acid (69%)

12.35 In a Friedel–Crafts acylation reaction an acyl chloride or acid anhydride reacts with an arene to yield an aryl ketone.

$$ArH + R\overset{\overset{\displaystyle O}{\|}}{C}Cl \xrightarrow{AlCl_3} Ar\overset{\overset{\displaystyle O}{\|}}{C}R$$

or

$$ArH + R\overset{\overset{\displaystyle O}{\|}}{C}O\overset{\overset{\displaystyle O}{\|}}{C}R \xrightarrow{AlCl_3} Ar\overset{\overset{\displaystyle O}{\|}}{C}R + R\overset{\overset{\displaystyle O}{\|}}{C}OH$$

The ketone carbonyl is bonded directly to the ring. In each of these problems, therefore, you should identify the bond between the aromatic ring and the carbonyl group and realize that it arises as shown in this general reaction.

(*a*) The compound is derived from benzene and $C_6H_5CH_2\overset{\overset{\displaystyle O}{\|}}{C}Cl$. The observed yield in this reaction is 82%.

arises from   and

(*b*) The presence of the $Ar\overset{\overset{\displaystyle O}{\|}}{C}CH_2CH_2CO_2H$ unit suggests an acylation reaction using succinic anhydride.

arises from   and

Succinic anhydride

In practice, this reaction has been carried out in 55% yield.

(*c*) Two methods seem possible here but only one actually works. The only effective combination is

+ $\xrightarrow{\text{AlCl}_3}$

p-Nitrobenzoyl   Benzene   *p*-Nitrobenzophenone (87%)
chloride

The alternative combination

+ $\xrightarrow{\text{AlCl}_3}$ no reaction

fails because it requires a Friedel–Crafts reaction on a strongly deactivated aromatic ring (nitrobenzene).

(*d*) Here also two methods seem possible, but only one is successful in practice. The valid synthesis is

+ $\xrightarrow{\text{AlCl}_3}$

3,5-Dimethylbenzoyl   Benzene   3,5-Dimethylbenzophenone
chloride   (89%)

The alternative combination will not give 3,5-dimethylbenzophenone, because of the ortho, para-directing properties of the methyl substituents in *m*-xylene. The product will be 2,4-dimethyl-benzophenone.

| *m*-Xylene | Benzoyl chloride | 2,4-Dimethylbenzophenone |

(*e*) The combination that follows is not effective because it involves a Friedel–Crafts reaction on a deactivated aromatic ring.

| *p*-Methylbenzoyl chloride | Benzoic acid |

The following combination, utilizing toluene, therefore seems appropriate:

The actual sequence used a cyclic anhydride, phthalic anhydride, in a reaction analogous to that seen in part (*b*).

| Toluene | Phthalic anhydride | *o*-(4-Methylbenzoyl)benzoic acid (96%) |

12.36 (*a*) The problem to be confronted here is that two meta-directing groups are para to each other in the product. However, by recognizing that the carboxylic acid function can be prepared by oxidation of the isopropyl group

we have a reasonable last step in the synthesis. The key intermediate has its sulfonic acid group para to the ortho, para-directing isopropyl group, which suggests the following approach:

Isopropylbenzene   *p*-Isopropylbenzene-sulfonic acid   *p*-Carboxybenzene-sulfonic acid

(*b*) In this problem two methyl groups must be oxidized to carboxylic acid functions, and a *tert*-butyl group must be introduced, most likely by a Friedel–Crafts reaction. Because Friedel–Crafts alkylations cannot be performed on deactivated aromatic rings, oxidation must *follow*, not precede, alkylation. The following reaction sequence therefore seems appropriate:

o-Xylene   4-*tert*-Butyl-1,2-dimethylbenzene   4-*tert*-Butylbenzene-1,2-dicarboxylic acid

In practice, zinc chloride was used as the Lewis acid to catalyze the Friedel–Crafts reaction (64% yield). Oxidation of the methyl groups occurs preferentially because the *tert*-butyl group has no benzylic hydrogens.

(*c*) The carbonyl group is directly attached to the naphthalene unit in the starting material. Reduce it in the first step so that a Friedel–Crafts acylation can be accomplished on the naphthalene ring. An aromatic ring that bears a strongly electron-withdrawing group such as C=O does not undergo Friedel–Crafts reactions.

(*d*) *m*-Dimethoxybenzene is a strongly activated aromatic compound and so will undergo electrophilic aromatic substitution readily. The ring position between the two methoxy groups is sterically hindered and less reactive than the other activated positions.

Arrows indicate equivalent ring positions strongly activated by methoxy groups.

Because Friedel–Crafts reactions may not be performed on deactivated aromatic rings, the *tert*-butyl group must be introduced before the nitro group. The correct sequence is therefore

This is essentially the procedure actually followed. Alkylation was effected, however, not with *tert*-butyl chloride and aluminum chloride but with 2-methylpropene and phosphoric acid.

m-Dimethoxybenzene   2-Methylpropene   1-*tert*-Butyl-2,4-dimethoxybenzene (75%)

Nitration was carried out in the usual way. The orientation of nitration is controlled by the more powerfully activating methoxy groups rather than by the weakly activating *tert*-butyl.

1-*tert*-Butyl-2,4-dimethoxy-5-nitrobenzene

12.37 The first step is a Friedel–Crafts acylation reaction. The use of a cyclic anhydride introduces both the acyl and carboxyl groups into the molecule.

Benzene   Succinic anhydride   4-Oxo-4-phenylbutanoic acid

The second step is a reduction of the ketone carbonyl to a methylene group. A Clemmensen reduction is normally used for this step.

4-Oxo-4-phenylbutanoic acid   4-Phenylbutanoic acid

The cyclization phase of the process is an intramolecular Friedel–Crafts acylation reaction. It requires conversion of the carboxylic acid to the acyl chloride (thionyl chloride is a suitable reagent) followed by treatment with aluminum chloride.

4-Phenylbutanoic acid   4-Phenylbutanoyl chloride

12.38 Intramolecular Friedel–Crafts acylation reactions that produce five-membered or six-membered rings occur readily. Cyclization must take place at the position ortho to the reacting side chain.

(*a*) A five-membered cyclic ketone is formed here.

(46%)

(*b*) This intramolecular Friedel–Crafts acylation takes place to form a six-membered cyclic ketone in excellent yield.

(93%)

(*c*) In this case two aromatic rings are available for attack in the acylation reaction. The more reactive ring is the one that bears the two activating methoxy groups, and cyclization occurs on it.

Only product (78% yield)   Not observed

12.39 The partial rate factors given in the problem are greater than 1 and highest for the positions ortho and para to the substituent X. Thus X must be an activating, ortho, para-directing group. The most likely candidate for X is the only activating group given in the problem, *tert*-butyl [C(CH$_3$)$_3$].

12.40 (*a*) To determine the total rate of chlorination of biphenyl relative to that of benzene, we add up the partial rate factors for all the positions in each substrate and compare them.

Biphenyl
(sum = 2580)

Benzene
(sum = 6)

Relative rate of chlorination: $\dfrac{\text{Biphenyl}}{\text{Benzene}} = \dfrac{2580}{6} = \dfrac{430}{1}$

(*b*) The relative rate of attack at the para position compared with the ortho positions is given by the ratio of their partial rate factors.

$$\dfrac{\text{Para}}{\text{Ortho}} = \dfrac{1580}{1000} = \dfrac{1.58}{1}$$

Therefore, 15.8 g of *p*-chlorobiphenyl is formed for every 10 g of *o*-chlorobiphenyl.

12.41 The problem stipulates that the reactivity of various positions in *o*-bromotoluene can be estimated by multiplying the partial rate factors for the corresponding positions in toluene and bromobenzene. Therefore, given the partial rate factors:

and

the two are multiplied together to give the combined effects of the two substituents at the various ring positions.

or

The most reactive position is the one that is para to bromine. The predicted product is therefore 4-bromo-3-methylacetophenone. Indeed, this is what is observed experimentally.

o-Bromotoluene	Acetyl chloride	4-Bromo-3-methylacetophenone

This was first considered to be "anomalous" behavior on the part of *o*-bromotoluene, but, as can be seen, it is consistent with the individual directing properties of the two substituents.

12.42 The isomerization is triggered by protonation of the aromatic ring, an electrophilic attack by HCl catalyzed by $AlCl_3$.

2-Isopropyl-1,3,5-
trimethylbenzene

The carbocation then rearranges by a methyl shift, and the rearranged cyclohexadienyl cation loses a proton to form the isomeric product.

1-Isopropyl-2,4,5-
trimethylbenzene

The driving force for rearrangement is relief of steric strain between the isopropyl group and one of its adjacent methyl groups. Isomerization is acid-catalyzed. Protonation of the ring generates the necessary carbocation intermediate and rearomatization occurs by loss of a proton.

12.43 The relation of compound A to the starting material is

$$C_6H_5(CH_2)_5\overset{\overset{\displaystyle O}{\|}}{C}Cl \longrightarrow A + HCl$$

$$(C_{12}H_{15}ClO) \qquad\qquad (C_{12}H_{14}O)$$

The starting acyl chloride has lost the elements of HCl in the formation of A. Because A forms benzene-1,2-dicarboxylic acid on oxidation, it must have two carbon substituents ortho to each other.

These facts suggest the following process:

The reaction leading to compound A is an intramolecular Friedel–Crafts acylation. Because cyclization to form an eight-membered ring is difficult, it must be carried out in dilute solution to minimize competition with intermolecular acylation.

12.44 Although hexamethylbenzene has no positions available at which ordinary electrophilic aromatic *substitution* might occur, electrophilic *attack* on the ring can still take place to form a cyclohexadienyl cation.

Compound A is the tetrachloroaluminate ($AlCl_4^-$) salt of the carbocation shown. It undergoes deprotonation on being treated with aqueous sodium bicarbonate.

B

12.45 By examining the structure of the target molecule, compound C, we see that the bond indicated in the following structure joins two fragments that are related to the given starting materials A and B:

Compound C   A   B

The bond connecting the two fragments can be made by a Friedel–Crafts acylation–reduction sequence using the acyl chloride B.

The orientation is right; attack is para to one of the methoxy groups and ortho to the methyl. The substrate for the Friedel–Crafts acylation reaction, 3,4-dimethoxytoluene, is prepared from compound A by a Clemmensen or Wolff–Kishner reduction. Compound A cannot be acylated directly because it bears a strongly deactivating —CH substituent.

$$\overset{\|}{\underset{O}{}}$$

12.46 In the presence of aqueous sulfuric acid, the side-chain double bond of styrene undergoes protonation to form a benzylic carbocation.

$$C_6H_5CH{=}CH_2 + H^+ \longrightarrow C_6H_5\overset{+}{C}HCH_3$$

Styrene   1-Phenylethyl cation

This carbocation then reacts with a molecule of styrene in the manner we have seen earlier (Chapter 6) for alkene dimerization.

$$C_6H_5\overset{+}{C}HCH_3 + C_6H_5CH{=}CH_2 \longrightarrow C_6H_5\overset{+}{C}HCH_2\underset{\underset{CH_3}{|}}{C}HC_6H_5$$

The carbocation produced in this step can lose a proton to form 1,3-diphenyl-1-butene

$$C_6H_5\overset{+}{C}HCH_2\underset{\underset{CH_3}{|}}{C}HC_6H_5 \longrightarrow C_6H_5CH{=}CHCHC_6H_5 + H^+$$

$$\underset{CH_3}{|}$$

1,3-Diphenyl-1-butene

or it can undergo a cyclization reaction in what amounts to an intramolecular Friedel–Crafts alkylation.

1-Methyl-3-phenylindan

12.47 The alcohol is tertiary and benzylic. In the presence of sulfuric acid a carbocation is formed.

An intramolecular Friedel–Crafts alkylation reaction follows, in which the carbocation attacks the adjacent aromatic ring.

$C_{16}H_{16}$

ANSWERS TO INTERPRETIVE PROBLEMS

12.48 A;   **12.49** D;   **12.50** D;   **12.51** C;   **12.52** C;   **12.53** C;   **12.54** B;   **12.55** A

SELF-TEST

PART A

A-1. Write the three most stable resonance contributors to the cyclohexadienyl cation formed in the ortho bromination of toluene.

A-2. Give the major product(s) for each of the following reactions. Indicate whether the reaction proceeds faster or slower than the corresponding reaction of benzene.

(a) [benzene ring with NO_2] $\xrightarrow[\text{H}_2\text{SO}_4]{\text{HNO}_3}$?

(c) [benzene ring with Cl] $\xrightarrow[\text{H}_2\text{SO}_4]{\text{SO}_3}$?

(b) [benzene ring with CH_2CH_3] $\xrightarrow[\text{FeBr}_3]{\text{Br}_2}$?

A-3. Write the formula of the electrophilic reagent species present in each reaction of the preceding problem.

A-4. Provide the reactant, reagent, or product omitted from each of the following:

(a) ? $\xrightarrow[\text{HCl}]{\text{Zn(Hg)}}$ [benzene ring with $CH_2CH(CH_3)_2$ at top and $C(CH_3)_3$ at bottom]

(b) [benzene ring] $\xrightarrow{\text{AlCl}_3}$?

[reagent shown: $CH_2CH_2CH_2\overset{\text{Cl}}{C}(CH_3)_2$]

(c) [benzene ring with OCH_3] $\xrightarrow{?}$ [benzene ring with OCH_3 and $\overset{\text{O}}{\overset{\|}{C}}C_6H_5$] + [benzene ring with OCH_3 and $O=C-C_6H_5$]

(d) [benzene ring with $\overset{\text{O}}{\overset{\|}{C}}CH_2CH_3$ group] $\xrightarrow[\text{H}_2\text{SO}_4]{\text{HNO}_3}$?

(e) [thiophene ring with S] $\xrightarrow[\text{FeCl}_3]{\text{Cl}_2}$?

A-5. Draw the structure(s) of the major product(s) formed by reaction of each of the following compounds with Cl_2 and $FeCl_3$. If two products are formed in significant amounts, draw them both.

(a)

(b)

(c)

(d) CH_3CH_2—⟨benzene ring⟩—OH

A-6. Provide the necessary reagents for each of the following transformations. More than one step may be necessary.

(a) ⟨benzene ring⟩—$CH(CH_3)_2$ $\xrightarrow{\ ?\ }$ HO_3S—⟨benzene ring⟩—CO_2H

(b) ⟨benzene⟩ $\xrightarrow{\ ?\ }$ ⟨benzene ring with Cl⟩—$CH_2CH_2C_6H_5$

(c) $\xrightarrow{\ ?\ }$

(d) ⟨benzene⟩ $\xrightarrow{\ ?\ }$ ⟨benzene ring with Br and SO_3H⟩

(e) ⟨benzene⟩ $\xrightarrow{\ ?\ }$ $(CH_3)_2CH$—⟨benzene ring⟩—NO_2

A-7. Outline a reasonable synthesis of each of the following from either benzene or toluene and any necessary organic or inorganic reagents.

(a)

(b)

(c)

A-8. Outline a reasonable synthesis of the compound shown using anisole ($C_6H_5OCH_3$) and any necessary inorganic reagents.

$$CH_3O-\underset{\overset{|}{Br}}{\bigcirc}-NO_2$$

PART B

B-1. Consider the following statements concerning the effect of a trifluoromethyl group, $-CF_3$, on an electrophilic aromatic substitution.

1. The CF_3 group will activate the ring.

2. The CF_3 group will deactivate the ring.

3. The CF_3 group will be a meta director.

4. The CF_3 group will be an ortho, para director.

Which of these statements are correct?

(*a*) 1, 3   (*b*) 1, 4   (*c*) 2, 3   (*d*) 2, 4

B-2. Which of the following resonance structures is *not* a contributor to the cyclohexadienyl cation intermediate in the nitration of benzene?

(*a*)

(*c*)

(*b*)

(*d*) None of these (all are contributors)

B-3. All the following groups are activating ortho, para directors when attached to a benzene ring *except*

(*a*) $-OCH_3$ (*b*) $\overset{\overset{\displaystyle O}{\|}}{-NHCCH_3}$ (*c*) $-Cl$ (*d*) $-N(CH_3)_2$

B-4. Rank the following in terms of increasing reactivity toward nitration with HNO_3, H_2SO_4 (least → most):

1   2   3

(*a*) 1 < 2 < 3   (*b*) 2 < 1 < 3   (*c*) 3 < 1 < 2   (*d*) 3 < 2 < 1

B-5. For the reaction

$$? \longrightarrow$$

the best reactants are:

(a) $C_6H_5Br + HNO_3, H_2SO_4$

(c) $C_6H_5Br + H_2SO_4$, heat

(b) $C_6H_5NO_2 + Br_2, FeBr_3$

(d) $C_6H_5NO_2 + HBr$

B-6. For the reaction

$$? \longrightarrow$$

the best reactants are

(a) C_6H_5Cl + $C_6H_5\overset{O}{\overset{\|}{C}}Cl$, AlCl$_3$

(c) $C_6H_5CH_2C_6H_5 + Cl_2$, FeCl$_3$, followed by oxidation with chromic acid

(b) $C_6H_5\overset{O}{\overset{\|}{C}}C_6H_5$ + Cl_2, FeCl$_3$

(d) None of these yields the desired product.

B-7. The reaction

$$\xrightarrow{HNO_3/H_2SO_4}$$

gives as the major product:

(a)

(c)

(b)

(d)

B-8. Which one of the following compounds undergoes bromination of its aromatic ring (electrophilic aromatic substitution) at the *fastest* rate?

(a)   (b)   (c)   (d)

B-9. Which one of the following is the major resonance contributor?

(a)   (b)   (c)   (d)

B-10. The major product of the reaction

$$\underset{S}{\bigcirc} \quad \xrightarrow[\text{FeBr}_3]{\text{Br}_2} \quad ?$$

is

(a) (thiophene with Br at 2-position)

(c) An *equal* mixture of compound (a) and (b) would form.

(b) (thiophene with Br at 3-position)

(d) None of these; substitution would not occur.

B-11. What is the product of the following reaction?

$$\bigcirc\text{—OCH}_3 \quad + \quad \bigcirc\text{—}\overset{\overset{\text{CH}_3}{|}}{\text{C}}\text{=CH}_2 \quad \xrightarrow{\text{H}_2\text{SO}_4}$$

(a) (phenyl)—CHCH₂—(C₆H₄)—OCH₃ with CH₃ on the CH

(b) (phenyl)—CHCH₂—(C₆H₄) with OCH₃ and CH₃ on CH

(c) (CH₃)₂CH—(biphenyl)—OCH₃

(d) (CH₃)₂CH—(biphenyl with OCH₃)

(e) (phenyl)—C(CH₃)₂—(C₆H₄)—OCH₃

B-12. Partial rate factors are shown for nitration of a particular aromatic compound. Based on these data, the most reasonable choice for substituent X is:

$$X$$

0.003   0.003

0.001   0.001

0.1

(a)   —N(CH_3)_2   (c)   —Br   (e)   —CH=O

(b)   —SO_3H   (d)   —CH(CH_3)_2

B-13. Which reactants combine to give the species shown as a reactive intermediate?

CH(CH_3)_2

H   H

+

H   H

H   Br

(a) Benzene, isopropyl bromide, and HBr

(b) Bromobenzene, isopropyl chloride, and AlCl_3

(c) Isopropylbenzene, Br_2, and FeBr_3

(d) Isopropylbenzene, Br_2, light, and heat

(e) Isopropylbenzene, N-bromosuccinimide, benzoyl peroxide, and heat

B-14. Which sequence of steps describes the best synthesis of the compound shown?

—CH_2—

Br

(a) $\xrightarrow[\text{AlCl}_3]{\text{C}_6\text{H}_5\text{CH}_2\text{Cl}}$ $\xrightarrow[\text{FeBr}_3]{\text{Br}_2}$

(b) $\xrightarrow[\text{FeBr}_3]{\text{Br}_2}$ $\xrightarrow[\text{AlCl}_3]{\text{C}_6\text{H}_5\text{CH}_2\text{Cl}}$

(c) $\xrightarrow[\text{AlCl}_3]{\overset{\overset{\text{O}}{\|}}{\text{C}_6\text{H}_5\text{CCl}}}$ $\xrightarrow[\text{FeBr}_3]{\text{Br}_2}$ $\xrightarrow[\text{HCl}]{\text{Zn(Hg)}}$

(d) $\xrightarrow[\text{FeBr}_3]{\text{Br}_2}$ $\xrightarrow[\text{AlCl}_3]{\overset{\overset{\text{O}}{\|}}{\text{C}_6\text{H}_5\text{CCl}}}$ $\xrightarrow[\text{HCl}]{\text{Zn(Hg)}}$

B-15. Which one of the following is the best synthesis of 2-chloro-4-nitrobenzoic acid?

2-Chloro-4-
nitrobenzoic acid

(*a*) 1. Heat benzoic acid with HNO_3,
      H_2SO_4
    2. Cl_2, $FeCl_3$, heat

(*b*) 1. Treat toluene with HNO_3,
      H_2SO_4
    2. $K_2Cr_2O_7$, H_2O, H_2SO_4, heat
    3. Cl_2, $FeCl_3$, heat

(*c*) 1. Treat toluene with HNO_3, H_2SO_4
    2. Cl_2, $FeCl_3$, heat
    3. $K_2Cr_2O_7$, H_2O, H_2SO_4, heat

(*d*) 1. Treat nitrobenzene with Cl_2,
      $FeCl_3$, heat
    2. CH_3Cl, $AlCl_3$
    3. $K_2Cr_2O_7$, H_2O, H_2SO_4, heat

(*e*) 1. Treat chlorobenzene with
      HNO_3, H_2SO_4
    2. CH_3Cl, $AlCl_3$
    3. $K_2Cr_2O_7$, H_2O, H_2SO_4, heat

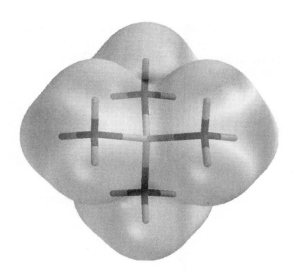

CHAPTER 13

Spectroscopy

SOLUTIONS TO TEXT PROBLEMS

13.1 The field strength of an NMR spectrometer magnet and the frequency of electromagnetic radiation used to observe an NMR spectrum are directly proportional. Thus, the ratio 4.7 T/200 MHz is the same as 1.41 T/60 MHz. The magnetic field strength of a 60-MHz NMR spectrometer is 1.41 T. The field strength of a 920 MHz instrument is 21.6 T.

13.2 The ratio of ^1H and ^{13}C resonance frequencies remains constant. When the ^1H frequency is 200 MHz, ^{13}C NMR spectra are recorded at 50.4 MHz. Thus, when the ^1H frequency is 100 MHz, ^{13}C NMR spectra will be observed at 25.2 MHz.

13.3 (a) Chemical shifts reported in parts per million (ppm) are independent of the field strength of the NMR spectrometer. Thus, to compare the ^1H NMR signal of bromoform ($CHBr_3$) recorded at 300 MHz with that of chloroform ($CHCl_3$) recorded at 200 MHz as given in the text, the chemical shift of bromoform must be converted from hertz to parts per million. The chemical shift for the proton in bromoform is

$$\delta = \frac{2065 \text{ Hz}}{300 \text{ MHz}} = 6.88 \text{ ppm}$$

(b) The chemical shift of the proton in bromoform (δ 6.88) is less than that of chloroform (δ 7.28). The proton signal of bromoform is farther upfield and thus is *more shielded* than the proton in chloroform.

13.4 (*b*) Electronegative atoms such as chlorine exert an electron-withdrawing effect on the protons that are separated from them by the fewest bonds. The effect is cumulative, and two chlorines will exert a greater effect than one. The proton on C-1 will be the least shielded and those on C-3 will be the most shielded.

Least shielded
(δ 5.9)
Most shielded
(δ 1.7)

Cl
|
Cl₂CHCHCH₃

(δ 4.3)

(*c*) The protons on the carbons adjacent to the oxygen of tetrahydrofuran will be less shielded than those separated from the oxygen by more bonds.

More shielded
(δ 1.8)

Less shielded
(δ 3.7)

O

13.5 The least shielded (most deshielded) protons are those directly attached to the double bond. The protons on the ring that are closer to the double bond are less shielded than those farther away.

=CH₂

δ 1.6   δ 4.8

δ 2.2

13.6 Protons attached to carbons adjacent to a carbonyl group are less shielded than those separated from the carbonyl by more bonds. Methyl protons are generally more shielded than methylene protons, thus the chemical shifts of each set of protons in 2-pentanone may be assigned.

O
||
CH₃CCH₂CH₂CH₃

δ 2.0   δ 2.3   δ 1.7   δ 1.1

13.7 The least shielded proton is the O—H proton of the carboxyl group. Recall (from Chapter 12) that nitro groups are strongly electron-withdrawing and that this effect is felt most strongly ortho and para to the group. Thus of the two sets of protons on the benzene ring, the less shielded of them is the set ortho to the nitro substituent. The most shielded protons in the spectrum are those of the methyl group.

H   H δ 4.0
O
||
O₂N—   —CHCOH ← δ 12.0
|
CH₃

H   H
δ 1.6

δ 8.2   δ 7.5

13.8 1,4-Dimethylbenzene has two types of protons: those attached directly to the benzene ring and those of the methyl groups. Aryl protons are significantly less shielded than alkyl protons. As shown in text Table 13.1 they are expected to give signals in the chemical shift range δ 6.5–8.5. Thus, the signal at δ 7.0 is due to the protons of the benzene ring. The signal at δ 2.2 is due to the methyl protons.

13.9 (*b*) Four nonequivalent sets of protons are bonded to carbon in 1-butanol as well as a fifth distinct type of proton, the one bonded to oxygen. There should be five signals in the ^{1}H NMR spectrum of 1-butanol.

$$CH_3CH_2CH_2CH_2OH$$

Five different proton environments
in 1-butanol; five signals

(*c*) Apply the "proton replacement" test to butane.

$$CH_3CH_2CH_2CH_3 \quad ClCH_2CH_2CH_2CH_3 \quad CH_3CHCH_2CH_3$$
$$|$$
$$Cl$$

Butane   1-Chlorobutane   2-Chlorobutane

$$CH_3CH_2CHCH_3 \quad CH_3CH_2CH_2CH_2Cl$$
$$|$$
$$Cl$$

2-Chlorobutane   1-Chlorobutane

Butane has *two* different types of protons; it will exhibit *two* signals in its ^{1}H NMR spectrum.

(*d*) Like butane, 1,4-dibromobutane has two different types of protons. This can be illustrated by using a chlorine atom as a test group.

$$\begin{array}{ccccc} & Cl & Cl & Cl & Cl \\ & | & | & | & | \\ BrCH_2CH_2CH_2CH_2Br & BrCHCH_2CH_2CH_2Br & BrCH_2CHCH_2CH_2Br & BrCH_2CH_2CHCH_2Br & BrCH_2CH_2CH_2CHBr \end{array}$$

1,4-Dibromobutane   1,4-Dibromo-1-chlorobutane 1,4-Dibromo-2-chlorobutane 1,4-Dibromo-2-chlorobutane 1,4-Dibromo-1-chlorobutane

The ^{1}H NMR spectrum of 1,4-dibromobutane is expected to consist of two signals.

(*e*) All the carbons in 2,2-dibromobutane are different from one another, and so protons attached to one carbon are not equivalent to the protons attached to any of the other carbons. This compound should have *three* signals in its ^{1}H NMR spectrum.

$$\begin{array}{c} Br \\ | \\ CH_3CCH_2CH_3 \\ | \\ Br \end{array}$$

2,2-Dibromobutane has three
nonequivalent sets of protons.

(f) All the protons in 2,2,3,3-tetrabromobutane are equivalent. Its ^{1}H NMR spectrum will consist of one signal.

$$\begin{array}{cc} Br & Br \\ | & | \\ CH_3C & -CCH_3 \\ | & | \\ Br & Br \end{array}$$

2,2,3,3-Tetrabromobutane

(g) There are *four* nonequivalent sets of protons in 1,1,4-tribromobutane. It will exhibit four signals in its ^{1}H NMR spectrum.

$$\begin{array}{c} Br \\ | \\ BrCCH_2CH_2CH_2Br \\ | \\ H \end{array}$$

1,1,4-Tribromobutane

(h) The seven protons of 1,1,1-tribromobutane belong to three nonequivalent sets, and hence the ^{1}H NMR spectrum will consist of three signals.

$$Br_3CCH_2CH_2CH_3$$

1,1,1-Tribromobutane

13.10 (b) Apply the replacement test to each of the protons of 1,1-dibromoethene.

| 1,1-Dibromoethene | 1,1-Dibromo-2-chloroethene | 1,1-Dibromo-2-chloroethene |

Replacement of one proton by a test group (Cl) gives exactly the same compound as replacement of the other. The two protons of 1,1-dibromoethene are equivalent, and there is only one signal in the ^{1}H NMR spectrum of this compound.

(c) The replacement test reveals that both protons of *cis*-1,2-dibromoethene are equivalent.

| *cis*-1,2-Dibromoethene | (Z)-1,2-Dibromo-1-chloroethene | (Z)-1,2-Dibromo-1-chloroethene |

Because both protons are equivalent, the ^{1}H NMR spectrum of *cis*-1,2-dibromoethene consists of one signal.

(d) Both protons of *trans*-1,2-dibromoethene are equivalent; each is cis to a bromine substituent.

trans-1,2-Dibromoethene
(one signal in the ^{1}H NMR spectrum)

(*e*) *Four* nonequivalent sets of protons occur in allyl bromide.

$$\begin{array}{ccc} H & & H \\ \diagdown & & \diagup \\ & C{=}C & \\ \diagup & & \diagdown \\ H & & CH_2Br \end{array}$$

Allyl bromide (four signals in
the ^{1}H NMR spectrum)

(*f*) The protons of a single methyl group are equivalent to one another, but all three methyl groups of 2-methyl-2-butene are nonequivalent. The vinyl proton is unique.

$$\begin{array}{ccc} H_3C & & CH_3 \\ \diagdown & & \diagup \\ & C{=}C & \\ \diagup & & \diagdown \\ H_3C & & H \end{array}$$

2-Methyl-2-butene (four signals in
the ^{1}H NMR spectrum)

13.11 (*b*) The three methyl protons of 1,1,1-trichloroethane (Cl_3CCH_3) are equivalent. They have the same chemical shift and do not split each other's signals. The ^{1}H NMR spectrum of Cl_3CCH_3 consists of a single sharp peak.

(*c*) Separate signals will be seen for the methylene (CH_2) protons and for the methine (CH) proton of 1,1,2-trichloroethane.

$$Cl_2C{-}CH_2Cl$$
$$\overset{\nearrow}{|}_{\,H} \qquad \overset{\nwarrow}{\big\backslash}$$

Triplet   Doublet

1,1,2-Trichloroethane

The methine proton splits the signal for the methylene protons into a doublet. The two methylene protons split the methine proton's signal into a triplet.

(*d*) Examine the structure of 1,2,2-trichloropropane.

$$\begin{array}{c} Cl \\ | \\ ClCH_2CCH_3 \\ | \\ Cl \end{array}$$

1,2,2-Trichloropropane

The ^{1}H NMR spectrum exhibits a signal for the two equivalent methylene protons and one for the three equivalent methyl protons. Both these signals are sharp singlets. The protons of the methyl group and the methylene group are separated by more than three bonds and do not split each other's signals.

(*e*) The methine proton of 1,1,1,2-tetrachloropropane splits the signal of the methyl protons into a doublet; its signal is split into a quartet by the three methyl protons.

$$\begin{array}{c} Cl \\ | \\ Cl_3CC{-}CH_3 \leftarrow \text{Doublet} \\ | \\ H \leftarrow \text{Quartet} \end{array}$$

1,1,1,2-Tetrachloropropane

13.12 (*b*) The ethyl group appears as a triplet–quartet pattern and the methyl group as a singlet.

$$CH_3CH_2OCH_3$$

Triplet   Quartet   Singlet; not vicinal to any
                          other protons in molecule

(*c*) The two ethyl groups of diethyl ether are equivalent to each other. The two methyl groups appear as one triplet and the two methylene groups as one quartet.

$$CH_3CH_2OCH_2CH_3$$

Triplet   Triplet
        Quartet   Quartet

(*d*) The two ethyl groups of *p*-diethylbenzene are equivalent to each other and give rise to a single triplet–quartet pattern.

Three signals:
CH₃ triplet;
CH₂ quartet;
aromatic H singlet

All four protons of the aromatic ring are equivalent, have the same chemical shift, and do not split either each other's signals or any of the signals of the ethyl group.

(*e*) Four nonequivalent sets of protons occur in this compound:

$$ClCH_2CH_2OCH_2CH_3$$

Triplet   Triplet
        Triplet   Quartet

Vicinal protons in the $ClCH_2CH_2O$ group split one another's signals, as do those in the CH_3CH_2O group.

13.13 (*b*) The signal of the proton at C-2 is split into a quartet by the methyl protons, and each line of this quartet is split into a doublet by the aldehyde proton. It appears as a doublet of quartets. (*Note:* It does not matter whether the splitting pattern is described as a doublet of quartets or a quartet of doublets. There is no substantive difference in the two descriptions.)

These three
protons split
the signal for
proton at C-2
into a quartet.

This proton splits the signal
for the proton at C-2
into a doublet.

13.14 The alcohol proton will be split into a triplet in a primary alcohol, a doublet in a secondary alcohol, and will not be split in a tertiary alcohol.

RCH₂—O—H   R₂CH—O—H   R₃C—O—H

Primary alcohol: Secondary alcohol: Tertiary alcohol: no
CH₂ group splits signal   CH group splits signal   protons vicinal to O—H proton;
for O—H proton into a triplet for O—H proton into a doublet   therefore, no splitting

13.15 (*b*) The two methyl carbons of the isopropyl group are equivalent.

Four different types of carbons occur in the aromatic ring and two different types are present in the isopropyl group. The ^{13}C NMR spectrum of isopropylbenzene contains *six* signals.

(*c*) The methyl substituent at C-2 is different from those at C-1 and C-3:

The four nonequivalent ring carbons and the two different types of methyl carbons give rise to a ^{13}C NMR spectrum that contains *six* signals.

(*d*) The three methyl carbons of 1,2,4-trimethylbenzene are different from one another:

Also, all the ring carbons are different from each other. The nine different carbons give rise to *nine* separate signals.

(*e*) All three methyl carbons of 1,3,5-trimethylbenzene are equivalent.

Because of its high symmetry 1,3,5-trimethylbenzene has only *three* signals in its ^{13}C NMR spectrum.

13.16 The two factors that affect chemical shift in a ^{13}C NMR spectrum are the electronegativity of the groups attached to carbon and the hybridization of carbon. The more electronegative chlorine will exert a greater deshielding effect than bromine in 1-bromo-3-chloropropane. The peaks can be assigned as follows:

$$BrCH_2CH_2CH_2Cl$$

$$(\delta\,35)\quad(\delta\,30)\quad(\delta\,43)$$

13.17 sp^3-Hybridized carbons are more shielded than sp^2-hybridized ones. Carbon x is the most shielded, and has a chemical shift of δ 20. The oxygen of the OCH_3 group decreased the shielding of carbon z; its chemical shift is δ 55. The least shielded is carbon y with a chemical shift of δ 157.

13.18 Carbon is more electronegative than hydrogen. Because the carbonyl of a ketone has two carbon atoms attached and an aldehyde only one, the carbonyl carbon of an aldehyde is expected to be more shielded. This effect can be illustrated by comparing the chemical shifts of the carbonyl carbons of the aldehyde butanal (δ 203) and the ketone 2-butanone (δ 208).

Butanal   2-Butanone

13.19 The ^{13}C NMR spectrum in Figure 13.25 shows nine signals and is the spectrum of 1,2,4-trimethyl-benzene from part (d) of Problem 13.15. Six of the signals, in the range δ 127–138, are due to the six nonequivalent carbons of the benzene ring. The three signals near δ 20 are due to the three nonequivalent methyl groups.

1,2,4-Trimethylbenzene

13.20 Absorption frequencies given in wavenumbers (cm^{-1}) are directly proportional to the energy difference between two adjacent vibrational states. Thus the energy difference is smaller for O—D stretching (2630 cm^{-1}) than for O—H stretching (3600 cm^{-1}).

13.21 In order for an absorption to be observable in an infrared spectrum a change in molecular dipole moment must occur. The double bond in ethylene, $H_2C{=}CH_2$, is symmetrically substituted. Thus the double bond stretching vibration does not produce a change in dipole moment and no peak is observed.

13.22 The ir spectrum shows absorption above 3000 cm^{-1} in the sp^2 C—H stretching region indicating a substituted benzene ring. The prominent peak in the 1700–1750 cm^{-1} region suggests the presence of a carbonyl group. The strong absorption at 1200 cm^{-1} due to C—O—C stretching allows the ester to be selected as the compound giving rise to the spectrum in text Figure 13.33.

Benzyl acetate; compound whose IR spectrum is shown in Figure 13.33

13.23 The energy of electromagnetic radiation is inversely proportional to its wavelength. Because excitation of an electron for the $\pi \rightarrow \pi^*$ transition of ethylene occurs at a shorter wavelength ($\lambda_{max} = 170$ nm) than that of *cis,trans*-1,3-cyclooctadiene ($\lambda_{max} = 230$ nm), the HOMO–LUMO energy difference in ethylene is *greater*.

13.24 Conjugation shifts λ_{max} to longer wavelengths in alkenes. The conjugated diene 2-methyl-1,3-butadiene has the longest wavelength absorption, $\lambda_{max} = 222$ nm. The isolated diene 1,4-pentadiene and the simple alkene cyclopentene both absorb below 200 nm.

2-Methyl-1,3-butadiene
($\lambda_{max} = 222$ nm)

13.25 (*b*) The distribution of molecular-ion peaks in *o*-dichlorobenzene is identical to that in the para isomer. As the sample solution to part (*a*) in the text describes, peaks at *m/z* 146, 148, and 150 are present for the molecular ion.

(*c*) The two isotopes of bromine are ^{79}Br and ^{81}Br. When both bromines of *p*-dibromobenzene are ^{79}Br, the molecular ion appears at *m/z* 234. When one is ^{79}Br and the other is ^{81}Br, *m/z* for the molecular ion is 236. When both bromines are ^{81}Br, *m/z* for the molecular ion is 238.

(*d*) The combinations of ^{35}Cl, ^{37}Cl, ^{79}Br, and ^{81}Br in *p*-bromochlorobenzene and the values of *m/z* for the corresponding molecular ion are as shown.

$$(^{35}\text{Cl}, \ ^{79}\text{Br}) \quad m/z = 190$$
$$(^{37}\text{Cl}, \ ^{79}\text{Br}) \ \text{or} \ (^{35}\text{Cl}, \ ^{81}\text{Br}) \quad m/z = 192$$
$$(^{37}\text{Cl}, \ ^{81}\text{Br}) \quad m/z = 194$$

13.26 The base peak in the mass spectrum of alkylbenzenes corresponds to carbon–carbon bond cleavage at the benzylic carbon.

Base peak: $C_9H_{11}^+$	Base peak: $C_8H_9^+$	Base peak: $C_9H_{11}^+$
m/z 119	*m/z* 105	*m/z* 119

13.27 (*b*) The index of hydrogen deficiency is given by the following formula:

$$\text{Index of hydrogen deficiency} = \tfrac{1}{2}(C_n H_{2n+2} - C_n H_x)$$

The compound given contains eight carbons (C_8H_8); therefore,

$$\text{Index of hydrogen deficiency} = \tfrac{1}{2}(C_8 H_{18} - C_8 H_8)$$
$$= 5$$

The problem specifies that the compound consumes 2 moles of hydrogen, and so it contains two double bonds (or one triple bond). Because the index of hydrogen deficiency is equal to 5, there must be three rings.

(c) Chlorine substituents are equivalent to hydrogens when calculating the index of hydrogen deficiency. Therefore, consider $C_8H_8Cl_2$ as equivalent to C_8H_{10}. Thus, the index of hydrogen deficiency of this compound is 4.

$$\text{Index of hydrogen deficiency} = \tfrac{1}{2}(C_8H_{18} - C_8H_{10})$$
$$= 4$$

Because the compound consumes 2 moles of hydrogen on catalytic hydrogenation, it must therefore contain two rings.

(d) Oxygen atoms are ignored when calculating the index of hydrogen deficiency. Thus, C_8H_8O is treated as if it were C_8H_8.

$$\text{Index of hydrogen deficiency} = \tfrac{1}{2}(C_8H_{18} - C_8H_8)$$
$$= 5$$

Because the problem specifies that 2 moles of hydrogen is consumed on catalytic hydrogenation, this compound contains three rings.

(e) Ignoring the oxygen atoms in $C_8H_{10}O_2$, we treat this compound as if it were C_8H_{10}.

$$\text{Index of hydrogen deficiency} = \tfrac{1}{2}(C_8H_{18} - C_8H_{10})$$
$$= 4$$

Because 2 moles of hydrogen is consumed on catalytic hydrogenation, there must be two rings.

(f) Ignore the oxygen, and treat the chlorine as if it were hydrogen. Thus, C_8H_9ClO is treated as if it were C_8H_{10}. Its index of hydrogen deficiency is 4, and it contains two rings.

13.28 Because each compound exhibits only a single peak in its ^1H NMR spectrum, all the hydrogens are equivalent in each one. Structures are assigned on the basis of their molecular formulas and chemical shifts.

(a) This compound has the molecular formula C_8H_{18} and so must be an alkane. The 18 hydrogens (δ 0.9) are contributed by six equivalent methyl groups.

$$(CH_3)_3CC(CH_3)_3$$

2,2,3,3-Tetramethylbutane

(b) A hydrocarbon with the molecular formula C_5H_{10} has an index of hydrogen deficiency of 1 and so is either a cycloalkane or an alkene. Because all ten hydrogens (δ 1.5) are equivalent, this compound must be cyclopentane.

Cyclopentane

(*c*) The chemical shift of the eight equivalent hydrogens in C_8H_8 is δ 5.8, which is consistent with protons attached to a carbon–carbon double bond.

1,3,5,7-Cyclooctatetraene

(*d*) The compound C_4H_9Br has no rings or double bonds. The nine hydrogens (δ 1.8) belong to three equivalent methyl groups.

$$(CH_3)_3CBr$$

tert-Butyl bromide

(*e*) The dichloride has no rings or double bonds (index of hydrogen deficiency = 0). The four equivalent hydrogens (δ 3.7) are present as two —CH_2Cl groups. Remember that equivalent protons on adjacent carbons do not couple.

$$ClCH_2CH_2Cl$$

1,2-Dichloroethane

(*f*) All three hydrogens (δ 2.7) in $C_2H_3Cl_3$ must be part of the same methyl group in order to be equivalent.

$$CH_3CCl_3$$

1,1,1-Trichloroethane

(*g*) This compound has no rings or double bonds. To have eight equivalent hydrogens (δ 3.7) it must have four equivalent methylene groups.

$$\begin{array}{c} CH_2Cl \\ | \\ ClCH_2CCH_2Cl \\ | \\ CH_2Cl \end{array}$$

1,3-Dichloro-2,2-di(chloromethyl)propane

(*h*) A compound with a molecular formula of $C_{12}H_{18}$ has an index of hydrogen deficiency of 4. A likely candidate for a compound with 18 equivalent hydrogens is one with six equivalent CH_3 groups. Thus, 6 of the 12 carbons belong to CH_3 groups, and the other 6 have no hydrogens. The compound is hexamethylbenzene.

A chemical shift of δ 2.2 is consistent with the fact that all of the protons are benzylic hydrogens.

(*i*)   The molecular formula of $C_3H_6Br_2$ tells us that the compound has no double bonds and no rings. All six hydrogens (δ 2.6) are equivalent, indicating two equivalent methyl groups. The compound is 2,2-dibromopropane, $(CH_3)_2CBr_2$.

13.29 In each of the parts to this problem, nonequivalent protons must *not* be bonded to adjacent carbons, because we are told that the two signals in each case are singlets.

(*a*) Each signal corresponds to four protons, and so each must result from two equivalent CH_2 groups. The four CH_2 groups account for four of the carbons of C_6H_8, leaving two carbons that bear no hydrogens. A molecular formula of C_6H_8 corresponds to an index of hydrogen deficiency of 3. A compound consistent with these requirements is

$$H_2C = \diamondsuit = CH_2$$

The signal at δ 5.6 is consistent with that expected for the four vinylic protons. The signal at δ 2.7 corresponds to that for the allylic protons of the ring.

(*b*) The compound has a molecular formula of $C_5H_{11}Br$ and therefore has no double bonds or rings. A nine-proton singlet at δ 1.1 indicates three equivalent methyl groups, and a two-proton singlet at δ 3.3 indicates a CH_2Br group. The correct structure is $(CH_3)_3CCH_2Br$.

(*c*) This compound ($C_6H_{12}O$) has three equivalent CH_3 groups, along with a fourth CH_3 group that is somewhat less shielded. Its molecular formula indicates that it can have either one double bond or one ring. This compound is $(CH_3)_3C\overset{\overset{\displaystyle O}{\|}}{C}CH_3$.

(*d*) A molecular formula of $C_6H_{10}O_2$ corresponds to an index of hydrogen deficiency of 2. The signal at δ 2.2 (6H) is likely due to two equivalent CH_3 groups, and the one at δ 2.7 (4H) to two equivalent CH_2 groups. The compound is $CH_3\overset{\overset{\displaystyle O}{\|}}{C}CH_2CH_2\overset{\overset{\displaystyle O}{\|}}{C}CH_3$. Remember that equivalent protons on adjacent carbons do not couple.

13.30 (*a*) A five-proton signal at δ 7.1 indicates a monosubstituted aromatic ring. With an index of hydrogen deficiency of 4, C_8H_{10} contains this monosubstituted aromatic ring and no other rings or multiple bonds. The triplet–quartet pattern at high field suggests an ethyl group.

Quartet   Triplet
δ 2.6 δ 1.2

Ethylbenzene

(*b*) The index of hydrogen deficiency of 4 and the five-proton multiplet at δ 7.0 to 7.5 are accommodated by a monosubstituted aromatic ring. The remaining four carbons and nine hydrogens are most reasonably a *tert*-butyl group, since all nine hydrogens are equivalent.

Singlet; δ 1.3

tert-Butylbenzene

(*c*) Its molecular formula requires that C_6H_{14} be an alkane. The doublet–septet pattern is consistent with an isopropyl group, and the total number of protons requires that two of these groups be present.

$(CH_3)_2CHCH(CH_3)_2$

Doublet δ 0.8   Septet δ 1.4

2,3-Dimethylbutane

Note that the methine (CH) protons do not split each other, because they are equivalent and have the same chemical shift.

(*d*) The molecular formula C_6H_{12} requires the presence of one double bond or ring. A peak at δ 5.1 is consistent with —C=CH, and so the compound is a noncyclic alkene. The vinyl proton gives a triplet signal, and so the group C=CHCH$_2$ is present. The ^{1}H NMR spectrum shows the presence of the following structural units:

δ 5.1 (Triplet)
δ 2.0 (Allylic)
CH_2CH_3 δ 0.9 (Triplet)
CH_3 δ 1.6 (Singlet; allylic)
CH_3 δ 1.7 (Singlet; allylic)

Putting all these fragments together yields a unique structure.

Singlet   Triplet   Singlet   Pentet   Triplet

2-Methyl-2-pentene

(*e*) The compound $C_4H_6Cl_4$ contains no double bonds or rings. There are no high-field peaks (δ 0.5 to 1.5), and so there are no methyl groups. At least one chlorine substituent must therefore be at each end of the chain. The most likely structure has the four chlorines divided into two groups of two.

$Cl_2CHCH_2CH_2CHCl_2$

δ 4.6 (Triplet) δ 3.9 (Doublet)

1,1,4,4-Tetrachlorobutane

(*f*) The molecular formula $C_4H_6Cl_2$ indicates the presence of one double bond or ring. A signal at δ 5.7 is consistent with a proton attached to a doubly bonded carbon. The following structural units are present:

$$C=C \begin{array}{c} H \\ CH_2 \end{array} \quad \longleftarrow \quad 5.7 \text{ (Triplet)}$$

$$C=CCH_3 \quad \longleftarrow \quad 2.2 \text{ (Allylic)}$$

For the methyl group to appear as a singlet and the methylene group to appear as a doublet, the chlorine substituents must be distributed as shown:

Singlet

$$\begin{array}{c} H_3C \\ \diagdown \\ Cl \diagup \end{array} C=CHCH_2Cl$$

Triplet   Doublet (δ 4.1)

1,3-Dichloro-2-butene

The stereochemistry of the double bond (E or Z) is not revealed by the ^1H NMR spectrum.

(*g*) A molecular formula of C_3H_7ClO is consistent with the absence of rings and multiple bonds (index of hydrogen deficiency = 0). None of the signals is equivalent to three protons, and so no methyl groups are present. Three methylene groups occur, all of which are different from one another. The compound is therefore:

$$ClCH_2CH_2CH_2OH \quad \longleftarrow \quad \delta \text{ 2.8 (Singlet)}$$

δ 3.7 or 3.8 δ 3.7 or 3.8
(Triplet) δ 2.0   (Triplet)
      (Pentet)

(*h*) The compound has a molecular formula of $C_{14}H_{14}$ and an index of hydrogen deficiency of 8. With a ten-proton signal at δ 7.1, a logical conclusion is that there are two monosubstituted benzene rings. The other four protons belong to two equivalent methylene groups.

$$\bigcirc\text{—}CH_2CH_2\text{—}\bigcirc$$

δ 2.9 (Singlet; benzylic)

1,2-Diphenylethane

13.31 The compounds of molecular formula C_4H_9Cl are the isomeric chlorides: *n*-butyl, isobutyl, *sec*-butyl, and *tert*-butyl chloride.

(*a*) All nine methyl protons of *tert*-butyl chloride $(CH_3)_3CCl$ are equivalent; its ^1H NMR spectrum has only one peak.

(*b*) A doublet at δ 3.4 indicates a —CH₂Cl group attached to a carbon that bears a single proton.

$$(CH_3)_2CHCH_2Cl$$

δ 3.4 (Doublet)

Isobutyl chloride

(*c*) A triplet at δ 3.5 means that a methylene group is attached to the carbon that bears the chlorine.

$$CH_3CH_2CH_2CH_2Cl$$

δ 3.5 (Triplet)

n-Butyl chloride

(*d*) This compound has two nonequivalent methyl groups.

δ 1.5 (Doublet)   δ 1.0 (Triplet)

$$CH_3CHCH_2CH_3$$
$$|$$
$$Cl$$

sec-Butyl chloride

13.32 Compounds with the molecular formula C_3H_5Br have either one ring or one double bond.

(*a*) The two peaks at δ 5.4 and 5.6 have chemical shifts consistent with the assumption that each peak is due to a vinyl proton (C=CH). The remaining three protons belong to an allylic methyl group (δ 2.3). The compound cannot be $CH_3CH=CHBr$, because the methyl signal would be split into a doublet. Isomer A can only be

$$H_2C=C\begin{array}{c} CH_3 \\ Br \end{array}$$

2-Bromopropene

(*b*) Two of the carbons of isomer B have chemical shifts characteristic of sp^2-hybridized carbon. One of these bears two protons (δ 118.8); the other bears one proton (δ 134.2). The remaining carbon is sp^3-hybridized and bears two hydrogens. Isomer B is allyl bromide.

$$H_2C=CHCH_2Br$$

δ 118.8   δ 134.2   δ 32.6

Allyl bromide

(*c*) All the carbons are sp^3-hybridized in this isomer. Two of the carbons belong to equivalent CH_2 groups, and the other bears only one hydrogen. Isomer C is cyclopropyl bromide.

δ 12.0   H
                    Br
δ 16.8

Cyclopropyl bromide

13.33 All these compounds have the molecular formula $C_4H_{10}O$. They have neither multiple bonds nor rings.

(*a*) Two equivalent CH_3 groups occur at δ 18.9. One carbon bears a single hydrogen. The least shielded carbon, presumably the one bonded to oxygen, has two hydrogen substituents. Putting all the information together reveals this compound to be isobutyl alcohol.

$(CH_3)_2CHCH_2OH$

δ 18.9   δ 30.8   δ 69.4

Isobutyl alcohol

(*b*) This compound has four distinct peaks, and so none of the four carbons is equivalent to any of the others. The signal for the least shielded carbon represents CH, and so the oxygen is attached to a secondary carbon. Only one carbon appears at low field; the compound is an alcohol, not an ether. Therefore;

δ 69.2

$CH_3CHCH_2CH_3$

δ 22.7 | δ 10.0
              OH
                   δ 32.0

sec-Butyl alcohol

(*c*) Signals for three equivalent CH_3 carbons indicate that this isomer is *tert*-butyl alcohol. This assignment is reinforced by the observation that the least shielded carbon has no hydrogens attached to it.

$(CH_3)_3COH$

δ 31.2   δ 68.9

tert-Butyl alcohol

13.34 The molecular formula of C_6H_{14} for each of these isomers requires that all of them be alkanes.

(*a*) This compound contains only CH_3 and CH carbons.

$(CH_3)_2CHCH(CH_3)_2$

δ 19.1   δ 33.9

2,3-Dimethylbutane

(*b*) This isomer has no CH carbons, and two different kinds of CH_2 groups.

$CH_3CH_2CH_2CH_2CH_2CH_3$

δ 13.7   δ 22.8   δ 31.9

Hexane

(*c*) CH$_3$, CH$_2$, and CH carbons are all present in this isomer. There are two different kinds of CH$_3$ groups.

$$\delta\, 29.1 \qquad\qquad \delta\, 36.4$$

$$\text{CH}_3\text{CH}_2\text{CHCH}_2\text{CH}_3$$
$$|$$
$$\text{CH}_3$$

$$\delta\, 11.1 \qquad\qquad \delta\, 18.4$$

3-Methylpentane

(*d*) This isomer contains a quaternary carbon in addition to a CH$_2$ group and two different kinds of CH$_3$ groups.

$$\delta\, 28.7 \qquad \text{CH}_3$$
$$|$$
$$\text{H}_3\text{C}-\text{C}-\text{CH}_2\text{CH}_3$$
$$|$$
$$\text{CH}_3 \qquad \delta\, 8.5$$

$$\delta\, 30.2 \qquad \delta\, 36.5$$

2,2-Dimethylbutane

(*e*) This isomer contains two different kinds of CH$_3$ groups, two different kinds of CH$_2$ groups, and a CH group.

$$\delta\, 27.6 \qquad\qquad \delta\, 20.5$$

$$\text{CH}_3\text{CHCH}_2\text{CH}_2\text{CH}_3$$
$$|$$
$$\text{CH}_3 \qquad \delta\, 14.0$$

$$\delta\, 22.4$$
$$\delta\, 41.6$$

2-Methylpentane

13.35 The index of hydrogen deficiency of the compound C$_4$H$_6$ is 2. It can have two double bonds, two rings, one ring and one double bond, or one triple bond. The chemical shift data indicate that two carbons are sp^3-hybridized and two are sp^2. The most reasonable structure that is consistent with ^{13}C NMR data is cyclobutene.

$\delta\, 136$ $\qquad$ $\delta\, 30.2$

Cyclobutene

The compound cannot be 1- or 2-methylcyclopropene. Neither of the carbon signals represents a methyl group.

13.36 Each of the carbons in the compound gives its ^{13}C NMR signal at relatively low field; it is likely that each one bears an electron-withdrawing substituent. The compound is

$$\delta\, 72.0$$

$$\text{ClCH}_2\text{CHCH}_2\text{OH}$$
$$|$$
$$\delta\, 46.8 \qquad \text{OH} \qquad \delta\, 63.5$$

3-Chloro-1,2-propanediol

The isomeric compound 2-chloro-1,3-propanediol

$$HOCH_2\underset{\underset{Cl}{|}}{C}HCH_2OH$$

cannot be correct. The C-1 and C-3 positions are equivalent; the ^{13}C NMR spectrum of this compound exhibits only two peaks, not three.

13.37 (*a*) All the hydrogens are equivalent in *p*-dichlorobenzene; therefore it has the simplest ^1H NMR spectrum of the three compounds chlorobenzene, *o*-dichlorobenzene, and *p*-dichlorobenzene.

Chlorobenzene
(three different kinds of protons)

o-Dichlorobenzene
(two different kinds of protons)

p-Dichlorobenzene
(all protons are equivalent)

(*b–d*) In addition to giving the simplest ^1H NMR spectrum, *p*-dichlorobenzene gives the simplest ^{13}C NMR spectrum. It has two peaks in its ^{13}C NMR spectrum, chlorobenzene has four, and *o*-dichlorobenzene has three.

Chlorobenzene
(four different kinds of carbon)

o-Dichlorobenzene
(three different kinds of carbon)

p-Dichlorobenzene
(two different kinds of carbon)

13.38 The hydrogens bonded to the sp^3 carbon in the five-membered ring of fluorene are relatively acidic. The reason for this can be seen by realizing that the central ring of fluorene resembles cyclopentadiene, and recalling from text Chapter 11 (Section 11.22) that cyclopentadiene can lose a proton to form a stable aromatic ion.

Fluorene

Cyclopentadiene

Cyclopentadienyl
anion (aromatic)

Fluorene is even more acidic than cyclopentadiene because its conjugate base is not only an aromatic anion, but it is stabilized by benzylic conjugation with the two benzene rings. A strong base such as $NaOCH_3$ can thus readily remove a proton from fluorene. Reprotonation in the presence of CH_3OD results in exchange of the protons (^1H) with deuterium ($^2H = D$).

13.39 Compounds A and B ($C_{10}H_{14}$) have an index of hydrogen deficiency of 4. Both have peaks in the δ 130–140 range of their ^{13}C NMR spectra, so that the index of hydrogen deficiency can be accommodated by a benzene ring.

The ^{13}C NMR spectrum of compound A shows only a single peak in the upfield region, at δ 20. Thus, the four remaining carbons, after accounting for the benzene ring, are four equivalent methyl groups. The benzene ring is symmetrically substituted as there are only two signals in the aromatic region at δ 132 and 135. Compound A is 1,2,4,5-tetramethylbenzene.

1,2,4,5-Tetramethylbenzene
(compound A)

In compound B the four methyl groups are divided into two pairs. Three different carbons occur in the benzene ring, as noted by the appearance of three signals in the aromatic region (δ 128–135). Compound B is 1,2,3,4-tetramethylbenzene.

1,2,3,4-Tetramethylbenzene
(compound B)

13.40 A hydrocarbon having a molecular ion, M^+, at m/z 102 has a molecular formula of C_8H_6. This formula corresponds to an index of hydrogen deficiency of 6, suggesting the presence of a phenyl group (C_6H_5) accounting for three double bonds and one ring. Infrared absorptions due to sp^2 C—H stretching can be seen in text Figure 13.44 at 3000 cm^{-1}.

The remaining atoms of the formula (C_2H) suggest an alkyne side-chain. The infrared spectrum supports this observation with absorptions at 2100 cm^{-1} (C≡C stretching) and 3300 cm^{-1} (sp C—H stretching). The compound is phenylacetylene.

Phenylacetylene

13.41 Because the compound has a five-proton signal at δ 7.4 and an index of hydrogen deficiency of 4, we conclude that six of its eight carbons belong to a monosubstituted benzene ring. The IR spectrum exhibits absorption at 3300 cm^{-1}, indicating the presence of a hydroxyl group. The compound is an alcohol. A three-proton doublet at δ 1.6, along with a one-proton quartet at δ 4.9, suggests the presence of a CH_3CH unit. The compound is 1-phenylethanol.

1-Phenylethanol

13.42 The peak at highest m/z in the mass spectrum of the compound $m/z = 134$; this is likely to correspond to the molecular ion. Among the possible molecular formulas, $C_{10}H_{14}$ correlates best with the information from the ^1H NMR spectrum. What is evident is that there is a signal due to aromatic protons, as well as a triplet–quartet pattern of an ethyl group. A molecular formula of $C_{10}H_{14}$ suggests a benzene ring that bears two ethyl groups. Because the signal for the aryl protons is so sharp, they are probably equivalent. The compound is p-diethylbenzene.

p-Diethylbenzene

That the benzene ring is para substituted is supported by the band at 820 cm^{-1} in the infrared spectrum.

13.43 There is a prominent peak in the IR spectrum of the compound near 1700 cm^{-1}, characteristic of C=O stretching vibrations. The ^{1}H NMR spectrum shows only two sets of signals, a triplet at δ 1.1 and a quartet at δ 2.4. The compound contains a CH_3CH_2 group as its only protons. Its ^{13}C NMR spectrum has three peaks, one of which is at very low field. The signal at δ 211 is in the region characteristic of carbons of C=O groups. If one assumes that the compound contains only carbon, hydrogen, and one oxygen atom and that the peak at highest m/z in its mass spectrum $m/z = 86$ corresponds to the molecular ion, then the compound has the molecular formula $C_5H_{10}O$. All the information points to the conclusion that the compound is 3-pentanone.

$$\underset{\text{3-Pentanone}}{CH_3CH_2\overset{\overset{\displaystyle O}{\|}}{C}CH_2CH_3}$$

13.44 [24]Annulene is expected to be antiaromatic because the number of π electrons equals $4n$, not the $4n + 2$ required for aromaticity according to Hückel's rule. As shown in text Figure 13.10b, the outer protons are expected to be more shielded in a compound that is antiaromatic.

[24]Annulene

13.45 (*a*) The nuclear spin of ^{19}F is $\pm\frac{1}{2}$, that is, the same as that of a proton. The splitting rules for ^{19}F–^{1}H couplings are the same as those for ^{1}H–^{1}H. Thus, the single fluorine atom of CH_3F splits the signal for the protons of the methyl group into a *doublet*.

(*b*) The set of three equivalent protons of CH_3F splits the signal for fluorine into a *quartet*.

(*c*) The proton signal in CH_3F is a doublet centered at δ 4.3. The separation between the two halves of this doublet is 45 Hz, which is equivalent to 0.225 ppm at 200 MHz (200 Hz = 1 ppm). Thus, one line of the doublet appears at δ (4.3 + 0.225) and the other at δ (4.3 − 0.225).

13.46 Because ^{31}P has a spin of $\pm\frac{1}{2}$, it is capable of splitting the ^1H NMR signal of protons in the same molecule. The problem stipulates that the methyl protons are coupled through three bonds to phosphorus in trimethyl phosphite.

$$CH_3O-P\underset{OCH_3}{\overset{OCH_3}{<}}$$

(*a*) The reciprocity of splitting requires that the protons split the ^{31}P signal of phosphorus. There are nine equivalent protons, and so the ^{31}P signal is split into ten peaks.

(*b*) Each peak in the ^{31}P multiplet is separated from the next by a value equal to the ^1H–^{31}P coupling constant of 12 Hz. There are nine such intervals in a ten-line multiplet, and so the separation is 108 Hz between the highest and lowest field peaks in the multiplet.

13.47 The trans and cis isomers of 1-bromo-4-*tert*-butylcyclohexane can be taken as models to estimate the chemical shift of the proton of the CHBr group when it is axial and equatorial, respectively, in the two chair conformations of bromocyclohexane. An axial proton is more shielded (δ 3.81 for *trans*-1-bromo-4-*tert*-butylcyclohexane) than an equatorial one (δ 4.62 for *cis*-1-bromo-4-*tert*-butylcyclohexane).

cis-1-Bromo-4-*tert*-butylcyclohexane;   *trans*-1-Bromo-4-*tert*-butylcyclohexane;   Bromocyclohexane
less shielded   more shielded

The difference in chemical shift between these stereoisomers is 0.81 ppm. The corresponding proton in bromocyclohexane is 0.67 ppm more shielded than the equatorial proton in *cis*-1-bromo-4-*tert*-butylcyclohexane. The proportion of bromocyclohexane that has an axial hydrogen is therefore $\frac{0.67}{0.81}$, or 83%. For bromocyclohexane, 83% of the molecules have an equatorial bromine, and 17% have an axial bromine.

13.48 The two staggered conformations of 1,2-dichloroethane are anti and gauche:

Anti   Gauche

The species present at low temperature (crystalline 1,2-dichloroethane) has a center of symmetry and is therefore the anti conformation. Liquid 1,2-dichloroethane is a mixture of the anti and the gauche conformations.

13.49 (*a*) Energy is proportional to frequency and inversely proportional to wavelength. The longer the wavelength, the lower the energy. Microwave photons have a wavelength in the range of 10^{-2} m, which is longer than that of infrared photons (on the order of 10^{-5} m). Thus, microwave radiation is lower in energy than IR radiation, and the separation between rotational energy levels (measured by microwave) is less than the separation between vibrational energy levels (measured by IR).

(*b*) Absorption of a photon occurs only when its energy matches the energy difference between two adjacent energy levels in a molecule. Microwave photons have energies that match the differences between the rotational energy levels of water. They are not sufficiently high in energy to excite a water molecule to a higher vibrational or electronic energy state.

13.50 A shift in the UV-VIS spectrum of acetone from 279 nm in hexane to 262 nm in water is a shift to shorter wavelength on going from a less polar solvent to a more polar one. This means that the energy difference between the starting electronic state (the ground state, n) and the excited electronic state (π^*) is *greater* in water than in hexane. Hexane as a solvent does not interact appreciably with either the ground or the excited state of acetone. Water is polar and solvates the ground state of acetone, lowering its energy. Because the energy gap between the ground state and the excited state increases, it must mean that the ground state is more solvated than the excited state and therefore more polar than the excited state.

Acetone in hexane   Acetone in water

ANSWERS TO INTERPRETIVE PROBLEMS

13.51 C;   **13.52** A;   **13.53** C;   **13.54** C;   **13.55** B;   **13.56** A;   **13.57** B;   **13.58** B

SELF-TEST

PART A

A-1. Complete the following table relating to ^{1}H NMR spectra by supplying the missing data.

	Spectrometer frequency	Chemical shift	
		ppm	**Hz**
(a)	60 MHz	——1——	336
(b)	300 MHz	4.35	——2——
(c)	——3——	3.50	700
(d)	100 MHz	——4——	of TMS

A-2. Indicate the number of signals to be expected and the multiplicity of each in the ^{1}H NMR spectrum of each of the following substances:

(a) $BrCH_2CH_2CH_2Br$

(b) $CH_3CH_2\overset{\displaystyle Cl}{\underset{\displaystyle Cl}{C}}CH_2CH_3$

(c) $CH_3OCH_2\overset{\displaystyle O}{\overset{\|}{C}}CH_3$

A-3. Two isomeric compounds having the molecular formula $C_6H_{12}O_2$ both gave ^{1}H NMR spectra consisting of only two singlets. Given the chemical shifts and integrations shown, identify both compounds.

Compound A: δ 1.45 (9H)   Compound B: δ 1.20 (9H)
            δ 1.95 (3H)   δ 3.70 (3H)

A-4. Identify each of the following compounds on the basis of the IR and ^{1}H NMR information provided.

(a) $C_{10}H_{12}O$:

IR: 1710 cm^{-1}

NMR: δ 1.0 (triplet, 3H)
δ 2.4 (quartet, 2H)
δ 3.7 (singlet, 2H)
δ 7.2 (singlet, 5H)

(b) $C_6H_{14}O_2$:

IR: 3400 cm^{-1}

NMR: δ 1.2 (singlet, 12H)
δ 2.0 (broad singlet, 2H)

(c) $C_{10}H_{16}O_6$:

IR: 1740 cm^{-1}

NMR: δ 1.3 (triplet, 9H)
δ 4.2 (quartet, 6H)
δ 4.4 (singlet, 1H)

(d) C_4H_7NO:

IR: 2240 cm^{-1}
3400 cm^{-1} (broad)

NMR: δ 1.65 (singlet, 6H)
δ 3.7 (singlet, 1H)

A-5. Predict the number of signals and their approximate chemical shifts in the ^{13}C NMR spectrum of the compound shown.

A-6. How many signals will appear in the ^{13}C NMR spectrum of each of the three C_5H_{12} isomers?

A-7. The ^{13}C NMR spectrum of an alkane of molecular formula C_6H_{14} exhibits two signals at δ 23 (4C) and δ 37 (2C). What is the structure of this alkane?

PART B

The following three problems refer to the ^{1}H NMR spectrum of $CH_3CH_2OCH_2OCH_2CH_3$.

B-1. How many signals are expected?

(a) 12　　　　(b) 5　　　　(c) 4　　　　(d) 3

B-2. The signal farthest downfield (relative to TMS) will be a

(a) Singlet　　　(b) Triplet　　　(c) Doublet　　　(d) Quartet

B-3. The signal farthest upfield (closest to TMS) will be a

(a) Singlet　　　(b) Triplet　　　(c) Doublet　　　(d) Quartet

B-4. The relationship between magnetic field strength and the energy difference between nuclear spin states is

(a) They are independent of each other.　　(c) They are inversely proportional.

(b) They are directly proportional.　　　　(d) The relationship varies from molecule to molecule.

B-5. An IR spectrum exhibits a broad band in the 3000–3500-cm^{-1} region and a strong peak at 1710 cm^{-1}. Which of the following substances best fits the data?

(*a*) $C_6H_5CH_2CH_2OH$

$$(c) \quad C_6H_5CH_2\overset{\overset{\textstyle O}{\|}}{C}CH_3$$

$$(b) \quad C_6H_5CH_2\overset{\overset{\textstyle O}{\|}}{C}OH$$

$$(d) \quad C_6H_5CH_2\overset{\overset{\textstyle O}{\|}}{C}OCH_3$$

B-6. Considering the ^1H NMR spectrum of the following substance, which set of protons appears farthest downfield relative to TMS?

B-7. Which of the following substances does *not* give a ^1H NMR spectrum consisting of only two peaks?

(*a*)

$$H_3C-\overset{\overset{\textstyle CH_3}{|}}{\underset{\underset{\textstyle CH_3}{|}}{C}}-OCH_3$$

(*c*) $H_3C-\!\!\left\langle\ \right\rangle\!\!-CH_3$

(*b*)

$$H_3C-\overset{\overset{\textstyle Br}{|}}{\underset{\underset{\textstyle CH_3}{|}}{C}}-\overset{\overset{\textstyle Br}{|}}{\underset{\underset{\textstyle CH_3}{|}}{C}}-CH_3$$

(*d*) None of these (all satisfy the spectrum)

B-8. The multiplicity of the *a* protons in the ^1H NMR spectrum of the following substance is

$$\underset{a}{(CH_3)_2}\overset{\overset{\textstyle OH}{|}}{C}\underset{b}{CH_2Cl}$$

(*a*) Singlet   (*b*) Doublet   (*c*) Triplet   (*d*) Quartet

B-9. An unknown compound C_4H_8O gave a strong IR absorption at 1710 cm^{-1}. The ^{13}C NMR spectrum exhibited four peaks at δ 9, 29, 37, and 209. The ^1H NMR spectrum had three signals at δ 1.1 (triplet), 2.1 (singlet), and 2.3 (quartet). Which, if any, of the following compounds is the unknown?

(*a*)   (*b*)   (*c*)   (*d*)   None of these (*e*)

B-10. How many signals are expected in the ^{13}C NMR spectrum of the following substance?

(*a*) 5   (*b*) 6   (*c*) 8   (*d*) 10

B-11. Which one of the following has the *greatest* number of signals in its ^{13}C NMR spectrum? (The spectrum is run under conditions in which splitting due to $^{13}C–^1H$ coupling is not observed.)

(*a*) Hexane   (*c*) 1-Hexene   (*e*) 1,5-Hexadiene

(*b*) 2-Methylpentane   (*d*) *cis*-3-Hexene

B-12. Which of the following C_9H_{12} isomers has the *fewest* signals in its ^{13}C NMR spectrum?

(*a*)   (*b*)   (*c*)   (*d*)

B-13. Which of the following compounds would best fit a ^{13}C NMR spectrum having peaks at δ 16, 21, 32, 36, 115, and 140?

(*a*) ⌇⌇⌇⌇   (*c*) ⌇⌇⌇⌇

(*b*)   (*d*) ⌇⌇⌇⌇

B-14. Which of the following compounds would have the *fewest* peaks in its ^{13}C NMR spectrum?

(*a*)   (*b*)   (*c*)   (*d*)   (*e*)

B-15. Which of the compounds in the previous problem would have the *most* peaks in its ^{13}C NMR spectrum?

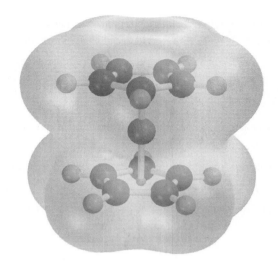

CHAPTER 14

Organometallic Compounds

SOLUTIONS TO TEXT PROBLEMS

14.1 (*b*) Magnesium bears a cyclohexyl substituent and a chlorine. Chlorine is named as an anion. The compound is cyclohexylmagnesium chloride.

14.2 (*b*) The alkyl bromide precursor to *sec*-butyllithium must be *sec*-butyl bromide.

$$CH_3CHCH_2CH_3 \ + \ 2Li \longrightarrow CH_3CHCH_2CH_3 \ + \ LiBr$$
$$\quad\quad | \quad\quad\quad\quad\quad\quad\quad\quad\quad\quad\quad\quad | $$
$$\quad\quad Br \quad\quad\quad\quad\quad\quad\quad\quad\quad\quad\quad\quad Li$$

2-Bromobutane   1-Methylpropyllithium
(*sec*-butyl bromide)   (*sec*-butyllithium)

14.3 (*b*) Allyl chloride is converted to allylmagnesium chloride on reaction with magnesium.

$$H_2C{=}CHCH_2Cl \xrightarrow[\text{diethyl ether}]{Mg} H_2C{=}CHCH_2MgCl$$

Allyl chloride   Allylmagnesium chloride

(*c*) The carbon–iodine bond of iodocyclobutane is replaced by a carbon–magnesium bond in the Grignard reagent.

Iodocyclobutane   Cyclobutylmagnesium
iodide

(*d*) Bromine is attached to an *sp*^{2}-hybridized carbon in 1-bromocyclohexene. The product of its reaction with magnesium has a carbon–magnesium bond in place of the carbon–bromine bond.

1-Bromocyclohexene   1-Cyclohexenylmagnesium bromide

14.4   Proton transfer from water to an organometallic compound is a rapid process.

Butyllithium   Water   Butane   Lithium hydroxide

Methyl alcohol will also transfer a proton quantitatively to an organometallic compound.

Phenylmagnesium bromide   Methanol   Benzene   Methoxymagnesium bromide

14.5   The Bronsted acid-base reaction involves transfer of the N—H proton of diisopropylamine to butyllithium.

$$[(CH_3)_2CH]_2\ddot{N}{-}H \; + \; {:}CH_2CH_2CH_2CH_3 \longrightarrow [(CH_3)_2CH]_2\ddot{N}{:}^- \; + \; CH_3CH_2CH_2CH_3$$

Diisopropylamine   Butyllithium   LDA   Butane
(acid)   (base)   (conjugate base)   (conjugate acid)

14.6   (*b*) Propylmagnesium bromide reacts with benzaldehyde by addition to the carbonyl group.

1-Phenyl-1-butanol

(*c*) Tertiary alcohols result from the reaction of Grignard reagents and ketones.

1-Propylcyclohexanol

(*d*) The starting material is a ketone and so reacts with a Grignard reagent to give a tertiary alcohol.

$CH_3CH_2CH_2$—MgBr

Propylmagnesium bromide
+ 2-butanone

3-Methyl-3-hexanol

14.7 Ethyl anion reacts as a Brønsted base to remove a proton from the alkyne. The proton at C-1 is removed because it is the most acidic, having a pK_a of approximately 26.

$CH_3\overset{..}{\underset{..}{C}}H_2^-$ + H—$C\equiv CCH_2CH_2CH_2CH_3$ ⟶ CH_3CH_3 + $:\bar{C}\equiv CCH_2CH_2CH_2CH_3$

Ethyl anion   1-Hexyne   Ethane   Conjugate base of 1-hexyne

14.8 (*b*) The target alcohol is tertiary and so is prepared by addition of a Grignard reagent to a ketone. The retrosynthetic transformations are:

Because two of the alkyl groups on the hydroxyl-bearing carbon are the same (methyl), only two, not three, different ketones are possible starting materials:

CH_3MgI +

$\dfrac{\text{1. diethyl ether}}{\text{2. H}_3\text{O}^+}$

Methylmagnesium iodide   Acetophenone   2-Phenyl-2-propanol

—MgBr + $CH_3\overset{O}{\overset{||}{C}}CH_3$ $\dfrac{\text{1. diethyl ether}}{\text{2. H}_3\text{O}^+}$

Phenylmagnesium bromide   Acetone   2-Phenyl-2-propanol

14.9 (*b*) Recall that the two identical groups bonded to the hydroxyl-bearing carbon of the alcohol arose from the Grignard reagent. That leads to the following retrosynthetic analysis:

$(C_6H_5)_2\underset{OH}{C}$ ◁ ⟹ ▷—$\overset{O}{\underset{||}{C}}OR$ + $2C_6H_5MgX$

Thus, the two phenyl substituents arise by addition of a phenyl Grignard reagent to an ester of cyclopropanecarboxylic acid.

Phenylmagnesium bromide

Methyl cyclopropanecarboxylate

Cyclopropyldiphenylmethanol

Methanol

14.10 (*b*) Of the three methyl groups of 1,3,3-trimethylcyclopentene, only the one connected to the double bond can be attached by way of an organocuprate reagent. Attachment of either of the other methyls would involve a tertiary carbon, a process that does not occur very efficiently.

Lithium dimethylcuprate

1-Bromo-3,3-dimethylcyclopentene

1,3,3-Trimethylcyclopentene

14.11 (*b*) Methylenecyclobutane is the appropriate precursor to the spirohexane shown.

Methylenecyclobutane

Spiro[3.2]hexane (22%)

14.12 Syn addition of dibromocarbene to *cis*-2-butene yields a cyclopropane derivative in which the methyl groups are cis.

cis-2-Butene

cis-1,1-Dibromo-2,3-dimethylcyclopropane

Conversely, the methyl groups in the cyclopropane derivative of *trans*-2-butene are trans to each other.

trans-2-Butene

trans-1,1-Dibromo-2,3-dimethylcyclopropane

14.13 Iron has an atomic number of 26 and an electron configuration of $[Ar]4s^2 3d^6$. Thus, it has 8 valence electrons and requires 10 more to satisfy the 18-electron rule. Five CO ligands, each providing two electrons, are therefore needed. The compound is $Fe(CO)_5$.

14.14 The chirality center in naproxen is introduced via enantioselective hydrogenation of a double bond. Therefore, retrosynthetically:

14.15 Cross-metathesis exchanges doubly bonded carbons. The products can be deduced graphically. They are 2-butene and 3-hexene. Both are mixtures of their cis and trans isomers.

$$CH_3CH{=}CHCH_2CH_3 \longrightarrow \begin{matrix} CH_3CH\text{-}\text{-}\text{-}CHCH_2CH_3 \\ | \qquad | \\ CH_3CH\text{-}\text{-}\text{-}CHCH_2CH_3 \end{matrix} \longrightarrow \begin{matrix} CH_3CH \\ \| \\ CH_3CH \end{matrix} + \begin{matrix} CHCH_2CH_3 \\ \| \\ CHCH_2CH_3 \end{matrix}$$

2-Pentene 2-Butene 3-Hexene

14.16 The molecular formula of the product ($C_{15}H_{19}NO_2$) differs from the starting material by C_2H_4, which is the same as the molecular formula of ethylene. This suggests that ring-closing metathesis has occurred.

$C_{17}H_{23}NO_2$ $C_{15}H_{19}NO_2$ Ethylene

14.17 (*a*) Cyclopentyllithium is

It has a carbon–lithium bond. It satisfies the requirement for classification as an organometallic compound.

(*b*) Ethoxymagnesium chloride does not have a carbon–metal bond. It is not an organometallic compound.

$$CH_3CH_2OMgCl \quad \text{or} \quad CH_3CH_2O^- \ Mg^{2+} \ Cl^-$$

(*c*) 2-Phenylethylmagnesium iodide is an example of a Grignard reagent. It is an organometallic compound.

(*d*) Lithium divinylcuprate has two vinyl groups bonded to copper. It is an organometallic compound.

$$Li^+ \ (H_2C{=}CH{-}\bar{C}u{-}CH{=}CH_2)$$

(*e*) Sodium carbonate, Na_2CO_3, can be represented by the Lewis structure.

$$Na^+ \quad :\overset{..}{\underset{..}{O}}-\overset{\overset{\overset{..}{O}:}{\|}}{C}-\overset{..}{\underset{..}{O}}:^- \quad Na^+$$

There is no carbon–metal bond, and sodium carbonate is not an organometallic compound.

(*f*) Benzylpotassium is represented as

It has a carbon–potassium bond and thus is an organometallic compound.

(*g*) The metal in lithium diisopropylamide (LDA) is bonded to nitrogen. Thus LDA is not an organometallic compound.

$$[(CH_3)_2CH]_2\overset{..}{N}:^- \; Li^+$$

LDA

14.18 The two alkyl groups attached to aluminum in $[(CH_3)_2CHCH_2]_2AlH$ are isobutyl groups. The hydrogen bonded to aluminum is named in a separate word as hydride. Thus, "dibal" is a shortened form of the systematic name *diisobutylaluminum hydride*.

14.19 (*a*) Grignard reagents such as pentylmagnesium iodide are prepared by reaction of magnesium with the corresponding alkyl halide.

$$CH_3CH_2CH_2CH_2CH_2I \; + \; Mg \; \xrightarrow{\text{diethyl ether}} \; CH_3CH_2CH_2CH_2CH_2MgI$$

1-Iodopentane   Pentylmagnesium iodide

(*b*) Acetylenic Grignard reagents are normally prepared by reaction of a terminal alkyne with a readily available Grignard reagent such as an ethylmagnesium halide. The reaction that takes place is an acid–base reaction in which the terminal alkyne acts as a proton donor.

$$CH_3CH_2C{\equiv}CH \; + \; CH_3CH_2MgI \; \xrightarrow{\text{diethyl ether}} \; CH_3CH_2C{\equiv}CMgI \; + \; CH_3CH_3$$

1-Butyne   Ethylmagnesium   1-Butynylmagnesium   Ethane
                    iodide   iodide

(*c*) Alkyllithiums are formed by reaction of lithium with an alkyl halide.

$$CH_3CH_2CH_2CH_2CH_2X \; + \; 2Li \; \longrightarrow \; CH_3CH_2CH_2CH_2CH_2Li \; + \; LiX$$

1-Halopentane   Pentyllithium
(X = Cl, Br, or I)

(*d*) Lithium dialkylcuprates arise by the reaction of an alkyllithium with a Cu(I) salt.

$$2CH_3CH_2CH_2CH_2CH_2Li \; + \; CuX \; \longrightarrow \; LiCu(CH_2CH_2CH_2CH_2CH_3)_2 \; + \; LiX$$

Pentyllithium, from part (*c*)   (X = Cl, Br, or I)   Lithium dipentylcuprate

14.20 The polarity of a covalent bond increases with an increase in the electronegativity difference between the connected atoms. Carbon has an electronegativity of 2.5 (text Figure 14.1). Metals are less electronegative than carbon. When comparing two metals, the less electronegative (or more electropositive) one therefore has the more polar bond to carbon.

(*a*) Text Figure 14.1 gives the electronegativity of lithium as 1.0, whereas that for aluminum is 1.5. The carbon–lithium bond in CH_3CH_2Li is more polar than the carbon–aluminum bond in $(CH_3CH_2)_3Al$.

(*b*) The electronegativity of magnesium (1.2) is less than that of zinc (1.6). $(CH_3)_2Mg$ therefore has a more polar carbon–metal bond than $(CH_3)_2Zn$.

(*c*) In this part of the problem two Grignard reagents are compared. Magnesium is the metal in both cases. The difference is the hybridization state of carbon. The *sp*-hybridized carbon in $HC\equiv CMgBr$ is more electronegative than the sp^3-hybridized carbon in CH_3CH_2MgBr, and $HC\equiv CMgBr$ has a more polar carbon–magnesium bond.

14.21 (*a*) $CH_3CH_2CH_2Br \ + \ 2Li \ \xrightarrow{\text{diethyl ether}} \ CH_3CH_2CH_2Li \ + \ LiBr$

      1-Bromopropane   Propyllithium

(*b*) $CH_3CH_2CH_2Br \ + \ Mg \ \xrightarrow{\text{diethyl ether}} \ CH_3CH_2CH_2MgBr$

      1-Bromopropane   Propylmagnesium bromide

(*c*) $\underset{\underset{I}{|}}{CH_3CHCH_3} \ + \ 2Li \ \xrightarrow{\text{diethyl ether}} \ \underset{\underset{Li}{|}}{CH_3CHCH_3} \ + \ LiI$

      2-Iodopropane   Isopropyllithium

(*d*) $\underset{\underset{I}{|}}{CH_3CHCH_3} \ + \ Mg \ \xrightarrow{\text{diethyl ether}} \ \underset{\underset{MgI}{|}}{CH_3CHCH_3}$

      2-Iodopropane   Isopropylmagnesium
                                               iodide

(*e*) $2CH_3CH_2CH_2Li \ + \ CuI \ \xrightarrow{\text{diethyl ether}} \ (CH_3CH_2CH_2)_2CuLi$

      Propyllithium   Lithium dipropylcuprate

(*f*) $(CH_3CH_2CH_2)_2CuLi \ + \ CH_3CH_2CH_2CH_2Br \ \longrightarrow \ CH_3CH_2CH_2CH_2CH_2CH_2CH_3$

      Lithium dipropylcuprate   1-Bromobutane   Heptane

(*g*) $(CH_3CH_2CH_2)_2CuLi \ + \ $ $\longrightarrow$

      Lithium dipropylcuprate   Iodobenzene   Propylbenzene

(*h*) $CH_3CH_2CH_2MgBr \ \xrightarrow[\text{DCl}]{\text{D}_2\text{O}} \ CH_3CH_2CH_2D$

      Propylmagnesium   1-Deuteriopropane
         bromide

(*i*)   CH$_3$CHCH$_3$ $\xrightarrow[\text{DCl}]{\text{D}_2\text{O}}$ CH$_3$CHCH$_3$
   | |
   Li   D

  Isopropyllithium   2-Deuteriopropane

(*j*)   CH$_3$CH$_2$CH$_2$Li + HCH ($\overset{O}{\overset{\|}{}}$) $\xrightarrow[\text{2. H}_3\text{O}^+]{\text{1. diethyl ether}}$ CH$_3$CH$_2$CH$_2$CH$_2$OH

  Propyllithium   1-Butanol

(*k*)   CH$_3$CH$_2$CH$_2$MgBr + [benzaldehyde] $\xrightarrow[\text{2. H}_3\text{O}^+]{\text{1. diethyl ether}}$ [phenyl]—CHCH$_2$CH$_2$CH$_3$
                          |
                          OH

 Propylmagnesium   Benzaldehyde   1-Phenyl-1-butanol
   bromide

(*l*)   CH$_3$CHCH$_3$ + [cycloheptanone] $\xrightarrow[\text{2. H}_3\text{O}^+]{\text{1. diethyl ether}}$ [cycloheptane ring]
   |
   Li   (CH$_3$)$_2$CH   OH

 Isopropyllithium Cycloheptanone   1-Isopropylcycloheptanol

(*m*)   CH$_3$CHCH$_3$ + CH$_3$CCH$_2$CH$_3$ ($\overset{O}{\overset{\|}{}}$) $\xrightarrow[\text{2. H}_3\text{O}^+]{\text{1. diethyl ether}}$ CH$_3$CH—CCH$_2$CH$_3$ ($\overset{OH}{\overset{|}{}}$)
   | | |
   MgI   CH$_3$ CH$_3$

 Isopropylmagnesium   2-Butanone   2,3-Dimethyl-3-pentanol
   iodide

(*n*)   2CH$_3$CH$_2$CH$_2$MgBr + C$_6$H$_5$COCH$_3$ ($\overset{O}{\overset{\|}{}}$) $\xrightarrow[\text{2. H}_3\text{O}^+]{\text{1. diethyl ether}}$ C$_6$H$_5$C(CH$_2$CH$_2$CH$_3$)$_2$ ($\overset{OH}{\overset{|}{}}$) + CH$_3$OH

 Propylmagnesium   Methyl benzoate   4-Phenyl-4-heptanol   Methanol
   bromide

(*o*)   H$_2$C=CH(CH$_2$)$_5$CH$_3$ $\xrightarrow[\text{Zu(Cu), diethyl ether}]{\text{CH}_2\text{I}_2}$ H$_2$C—CH(CH$_2$)$_5$CH$_3$
                          _ /
                          CH$_2$

  1-Octene   1-Cyclopropylhexane

(*p*)   [H$_3$C, H / C=C / H, (CH$_2$)$_6$CH$_3$] $\xrightarrow[\text{Zu(Cu), diethyl ether}]{\text{CH}_2\text{I}_2}$ [H$_3$C, H / cyclopropane / H, (CH$_2$)$_6$CH$_3$]

  (*E*)-2-Decene   *trans*-1-Heptyl-2-
                   methylcyclopropane

(q)

$$CH_3CH_2\overset{\displaystyle (CH_2)_5CH_3}{\underset{H}{\diagdown}}C=C\overset{\displaystyle (CH_2)_5CH_3}{\underset{H}{\diagup}} \quad \xrightarrow[\text{Zu(Cu), diethyl ether}]{CH_2I_2} \quad CH_3CH_2\!\!-\!\!\triangle\!\!-\!\!(CH_2)_5CH_3$$

(Z)-3-Decene

cis-1-Ethyl-2-
hexylcyclopropane

(r) $H_2C{=}CHCH_2CH_2CH_3 \quad \xrightarrow[\text{KOC(CH}_3)_3]{CHBr_3} \quad$

1-Pentene

1,1-Dibromo-2-propylcyclopropane

14.22 In the solutions to this problem, the Grignard reagent butylmagnesium bromide is used. In each case the use of butyllithium would be equally satisfactory.

(a) 1-Pentanol is a primary alcohol having one more carbon atom than 1-bromobutane. Retrosynthetic analysis suggests the reaction of a Grignard reagent with formaldehyde.

$$CH_3CH_2CH_2CH_2\text{---}CH_2OH \quad \Longrightarrow \quad CH_3CH_2CH_2CH_2MgX \quad + \quad H_2C{=}O$$

1-Pentanol

Butylmagnesium halide

Formaldehyde

An appropriate synthetic scheme is

$$CH_3CH_2CH_2CH_2Br \quad \xrightarrow[\text{diethyl ether}]{Mg} \quad CH_3CH_2CH_2CH_2MgBr \quad \xrightarrow[\text{2. } H_3O^+]{\overset{\overset{\displaystyle O}{\|}}{\text{1. HCH}}} \quad CH_3CH_2CH_2CH_2CH_2OH$$

1-Bromobutane

Butylmagnesium
bromide

1-Pentanol

(b) 2-Hexanol is a secondary alcohol having two more carbon atoms than 1-bromobutane. As revealed by retrosynthetic analysis, it may be prepared by reaction of ethanal (acetaldehyde) with butylmagnesium bromide.

$$CH_3CH_2CH_2CH_2\text{---}\underset{\underset{\displaystyle OH}{|}}{CHCH_3} \quad \Longrightarrow \quad CH_3CH_2CH_2CH_2MgX \quad + \quad \overset{\overset{\displaystyle O}{\|}}{CH_3CH}$$

2-Hexanol

Butylmagnesium halide

Ethanal
(acetaldehyde)

The correct reaction sequence is

$$CH_3CH_2CH_2CH_2Br \quad \xrightarrow[\text{diethyl ether}]{Mg} \quad CH_3CH_2CH_2CH_2MgBr \quad \xrightarrow[\text{2. } H_3O^+]{\overset{\overset{\displaystyle O}{\|}}{\text{1. } CH_3CH}} \quad CH_3CH_2CH_2CH_2\underset{\underset{\displaystyle OH}{|}}{CHCH_3}$$

1-Bromobutane

Butylmagnesium bromide

2-Hexanol

(*c*) 1-Phenyl-1-pentanol is a secondary alcohol. Disconnection suggests that it can be prepared from butylmagnesium bromide and an aldehyde; benzaldehyde is the appropriate aldehyde.

1-Phenyl-1-pentanol   Butylmagnesium halide   Benzaldehyde

Butylmagnesium bromide   Benzaldehyde   1-Phenyl-1-pentanol

(*d*) The target molecule 3-methyl-3-heptanol has the structure

$$CH_3CH_2CH_2CH_2-\underset{\underset{OH}{|}}{\overset{\overset{CH_3}{|}}{C}}-CH_2CH_3$$

By retrosynthetically disconnecting the butyl group from the carbon that bears the hydroxyl substituent, we see that the appropriate starting ketone is 2-butanone.

Butylmagnesium halide   2-Butanone

Therefore

Butylmagnesium bromide   2-Butanone   3-Methyl-3-heptanol

(*e*) 1-Butylcyclobutanol is a tertiary alcohol. The appropriate ketone is cyclobutanone.

Butylmagnesium bromide   Cyclobutanone   1-Butylcyclobutanol

14.23 In each part of this problem in which there is a change in the carbon skeleton, disconnect the phenyl group of the product to reveal the aldehyde or ketone precursor that reacts with the Grignard reagent derived from bromobenzene. Recall that reaction of a Grignard reagent with formaldehyde ($H_2C{=}O$) yields a primary alcohol, reaction with an aldehyde (other than formaldehyde) yields a secondary alcohol, and reaction with a ketone yields a tertiary alcohol.

(*a*) Conversion of bromobenzene to benzyl alcohol requires formation of the corresponding Grignard reagent and its reaction with formaldehyde. Retrosynthetically, this can be seen as

Therefore,

Bromobenzene   Phenylmagnesium bromide   Benzyl alcohol

(*b*) The product is a secondary alcohol and is formed by reaction of phenylmagnesium bromide with hexanal.

1-Phenyl-1-hexanol   Phenylmagnesium halide   Hexanal

Phenylmagnesium bromide   Hexanal   1-Phenyl-1-hexanol

(*c*) The desired product is a secondary alkyl *bromide*. A reasonable synthesis would be to first prepare the analogous secondary alcohol by reaction of phenylmagnesium bromide with benzaldehyde, followed by a conversion of the alcohol to the bromide. Retrosynthetically this can be seen as

Phenylmagnesium bromide   Benzaldehyde   Diphenylmethanol   Bromodiphenylmethane (*d*) The target molecule is a tertiary alcohol, which requires that phenylmagnesium bromide react with a ketone. By mentally disconnecting the phenyl group from the carbon that bears the hydroxyl group, we see that the appropriate ketone is 4-heptanone.

4-Phenyl-4-heptanol   4-Heptanone

The synthesis is therefore

Phenylmagnesium   4-Heptanone   4-Phenyl-4-heptanol
bromide

(*e*) Reaction of phenylmagnesium bromide with cyclooctanone will give the desired tertiary alcohol.

Phenylmagnesium   Cyclooctanone   1-Phenylcyclooctanol
bromide

(*f*) The 1-phenylcyclooctanol prepared in part (*e*) of this problem can be subjected to acid-catalyzed dehydration to give 1-phenylcyclooctene. Hydroboration–oxidation of 1-phenylcyclooctene gives *trans*-2-phenylcyclooctanol.

1-Phenylcyclooctanol   1-Phenylcyclooctene   *trans*-2-Phenylcyclooctanol

14.24 In these problems the principles of retrosynthetic analysis are applied. The alkyl groups attached to the carbon that bears the hydroxyl group are mentally disconnected to reveal the Grignard reagent and carbonyl compound.

(1)   (2)

(*a*)   CH$_3$CH$_2$┊CH┊CH$_2$(CH$_3$)$_2$
                                         |
                                        OH

5-Methyl-3-hexanol

(1)   (2)

$$\left[\; CH_3CH_2MgX \;+\; \underset{\underset{O}{||}}{HCCH_2CH(CH_3)_2} \;\right] \qquad \left[\; \underset{\underset{O}{||}}{CH_3CH_2CH} \;+\; XMgCH_2CH(CH_3)_2 \;\right]$$

Ethylmagnesium   3-Methylbutanal   Propanal   Isobutylmagnesium
halide   halide

(1)   (2)

(*b*)   ▷┊CH┊〈 〉—OCH$_3$
                        |
                       OH

1-Cyclopropyl-1-(*p*-anisyl)methanol

(1)   (2)

$$\left[\; \triangleright\!\!-MgX \;+\; \underset{\underset{O}{||}}{HC}\!\!-\!\!\langle\;\rangle\!\!-\!OCH_3 \;\right] \qquad \left[\; \triangleright\!\!-\underset{\underset{O}{||}}{CH} \;+\; XMg\!\!-\!\!\langle\;\rangle\!\!-\!OCH_3 \;\right]$$

Cyclopropyl-   *p*-Anisaldehyde   Cyclopropane-   *p*-Anisylmagnesium
magnesium halide   carbaldehyde   halide

O
||
(*c*) (CH$_3$)$_3$C┊CH$_2$OH ⟹ (CH$_3$)$_3$CMgX + HCH

2,2-Dimethyl-1-propanol   *tert*-Butylmagnesium   Formaldehyde
                                              halide

(1)   (2)

(*d*)   (CH$_3$)$_2$C=CHCH$_2$CH$_2$┊CH┊CH$_3$
                                            |
                                           OH

(1)   6-Methyl-5-hepten-2-ol   (2)

$$\left[\; (CH_3)_2C{=}CHCH_2CH_2MgX \;+\; \underset{\underset{O}{||}}{HCCH_3} \;\right] \qquad \left[\; (CH_3)_2C{=}CHCH_2CH_2\underset{\underset{O}{||}}{CH} \;+\; XMgCH_3 \;\right]$$

4-Methyl-3-hexen-1-ylmagnesium   Ethanal   5-Methyl-4-hexenal   Methylmagnesium
halide   halide

(*e*)

4-Ethyl-4-octanol

Propylmagnesium halide   3-Heptanone

4-Octanone   +   Ethylmagnesium halide

3-Hexanone   +   Butylmagnesium halide

14.25 (*a*) Meparfynol is a tertiary alcohol and so can be prepared by addition of a carbanionic species to a ketone. Use the same reasoning that applies to the synthesis of alcohols from Grignard reagents. On mentally disconnecting one of the bonds to the carbon bearing the hydroxyl group

we see that the addition of acetylide ion to 2-butanone will provide the target molecule.

$$HC{\equiv}CNa \ + \ CH_3CH_2\overset{O}{\overset{\|}{C}}CH_3 \quad \xrightarrow[\text{2. } H_3O^+]{\text{1. } NH_3} \quad CH_3CH_2\overset{OH}{\underset{CH_3}{C}}C{\equiv}CH$$

Sodium   2-Butanone   Meparfynol (94%)
acetylide

The alternative, reaction of a Grignard reagent with an alkynyl ketone, is not acceptable in this case. The acidic terminal alkyne C—H would transfer a proton to the Grignard reagent.

(*b*) Diphepanol is a tertiary alcohol and so may be prepared by reaction of a Grignard or organolithium reagent with a ketone. Retrosynthetically, two possibilities seem reasonable:

and

In principle, either strategy is acceptable; in practice, the one involving phenylmagnesium bromide is used.

Phenylmagnesium bromide

Diphepanol

(*c*) A reasonable last step in the synthesis of mestranol is the addition of sodium acetylide to the ketone shown.

Mestranol

Acetylide anion adds to the carbonyl from the less sterically hindered side. The methyl group shields the top face of the carbonyl, and so acetylide adds from the bottom.

14.26 (*a*) Sodium acetylide adds to ketones to give tertiary alcohols.

Benzophenone

1,1-Diphenyl-2-propyn-1-ol
(50%)

(*b*) The substrate is a ketone, which reacts with ethyllithium to yield a tertiary alcohol.

2-Adamantanone

2-Ethyl-2-adamantanol (83%)

(*c*) The first step is conversion of bromocyclopentene to the corresponding Grignard reagent, which then reacts with formaldehyde to give a primary alcohol.

1-Bromocyclopentene

1-Cyclopentenylmagnesium
bromide

1-Cyclopentenylmethanol
(53%)

(*d*) The reaction is one in which an alkene is converted to a cyclopropane through use of the Simmons–Smith reagent, iodomethylzinc iodide.

$$\text{Allylbenzene} \xrightarrow[\substack{\text{Zn(Cu)}\\\text{diethyl ether}}]{\text{CH}_2\text{I}_2} \text{Benzylcyclopropane (64\%)}$$

Allylbenzene   Benzylcyclopropane (64%)

(*e*) Methylene transfer using the Simmons–Smith reagent is stereospecific. The trans arrangement of substituents in the alkene is carried over to the cyclopropane product.

$$\text{(E)-1-Phenyl-2-butene} \xrightarrow[\text{Zn(Cu), ether}]{\text{CH}_2\text{I}_2} \text{trans-1-Benzyl-2-methylcyclopropane}$$

(*E*)-1-Phenyl-2-butene   *trans*-1-Benzyl-2-methylcyclopropane (50%)

(*f*) Lithium dimethylcuprate transfers a methyl group, which substitutes for iodine on the iodoalkene. Even halogens on sp^2-hybridized carbon are reactive in substitutions with lithium dialkylcuprates.

2-Iodo-8-methoxybenzonorbornadiene   + $\text{LiCu(CH}_3)_2$ ⟶   8-Methoxy-2-methylbenzonorbornadiene (73%)

(*g*) The starting material is a *p*-toluenesulfonate. *p*-Toluenesulfonates are similar to alkyl halides in their reactivity. Substitution occurs; a butyl group from lithium dibutylcuprate replaces *p*-toluenesulfonate.

(3-Furyl)methyl *p*-toluenesulfonate   + $\text{LiCu(CH}_2\text{CH}_2\text{CH}_2\text{CH}_3)_2$ ⟶   3-Pentylfuran

Lithium dibutylcuprate

14.27 Phenylmagnesium bromide reacts with 4-*tert*-butylcyclohexanone as shown.

$$\text{4-tert-Butylcyclohexanone} \xrightarrow[\substack{\text{1. C}_6\text{H}_5\text{MgBr, diethyl ether}\\\text{2. H}_3\text{O}^+}]{} \text{4-tert-Butyl-1-phenylcyclohexanol}$$

4-*tert*-Butylcyclohexanone   4-*tert*-Butyl-1-phenylcyclohexanol

The phenyl substituent can be introduced either cis or trans to the *tert*-butyl group. The two alcohols are therefore stereoisomers (diastereomers). Dehydration of either alcohol yields 4-*tert*-butyl-1-phenylcyclohexene.

4-*tert*-Butyl-1-phenylcyclohexene

14.28 (*a*) By working through the sequence of reactions that occur when ethyl formate reacts with a Grignard reagent, we can see that this combination leads to *secondary alcohols*.

| Grignard reagent | Ethyl formate | | Aldehyde | | | Secondary alcohol |

This is simply because the substituent on the carbonyl carbon of the ester, in this case a hydrogen, is carried through and becomes a substituent on the hydroxyl-bearing carbon of the alcohol.

(*b*) Diethyl carbonate has the potential to react with 3 moles of a Grignard reagent.

The tertiary alcohols that are formed by the reaction of diethyl carbonate with Grignard reagents have three identical R groups attached to the carbon that bears the hydroxyl substituent.

14.29 If we use the 2-bromobutane given, along with the information that the reaction occurs with net inversion of configuration, the stereochemical course of the reaction may be written as

The phenyl group becomes bonded to carbon from the opposite side of the leaving group. Applying the Cahn–Ingold–Prelog notational system described in text Section 7.6 to the product, the order of decreasing precedence is

$$C_6H_5 > CH_3CH_2 > CH_3 > H$$

Orienting the molecule so that the lowest ranked group (H) is away from us, we see that the order of decreasing precedence is clockwise.

The absolute configuration is *R*.

14.30 The substrates are secondary alkyl *p*-toluenesulfonates, and so we expect elimination to compete with substitution. Compound B is formed in both reactions and has the molecular formula of 4-*tert*-butylcyclohexene. Because the two *p*-toluenesulfonates are diastereomers, it is likely that compounds A and C are also diastereomers, especially since they have the same molecular formula. Assuming that the substitution reactions proceed with inversion of configuration, we conclude that the products are as shown.

trans-4-*tert*-Butylcyclohexyl
p-toluenesulfonate

cis-1-*tert*-Butyl-4-methylcyclohexane
(compound A, $C_{11}H_{22}$)

4-*tert*-Butylcyclohexene
(compound B, $C_{10}H_{18}$)

cis-4-*tert*-Butylcyclohexyl
p-toluenesulfonate

trans-1-*tert*-Butyl-4-methylcyclohexane
(compound C, $C_{11}H_{22}$)

Compound B

Inversion of configuration is borne out by the fact given in the problem that compound C is more stable than compound A. Both substituents are equatorial in C; the methyl group is axial in A.

14.31 Sharing the nitrogen unshared electron pair with carbon allows a resonance form to be written in which carbon's octet is satisfied.

14.32 We are told in the statement of the problem that the first step is conversion of the alcohol to the corresponding *p*-toluenesulfonate. This step is carried out as follows:

3,8-Epoxy-1-undecanol

p-Toluenesulfonyl
chloride (TsCl)

3,8-Epoxyundecyl
p-toluenesulfonate

Alkyl *p*-toluenesulfonates react with lithium dialkylcuprates in the same way that alkyl halides do. Treatment of the preceding *p*-toluenesulfonate with lithium dibutylcuprate gives the desired compound.

3,8-Epoxyundecyl
p-toluenesulfonate   4,9-Epoxypentadecane

As actually performed, a 91% yield of the desired product was obtained in the reaction of the *p*-toluenesulfonate with lithium dibutylcuprate.

14.33 Hydrogenation of a double bond must generate the chirality center. Therefore, the double bond cannot be part of the four-carbon side chain that becomes the isobutyl group. It must be part of the other substituent.

14.34 Each reactant contains a carbon–carbon double bond and the reaction is carried out in the presence of an olefin-metathesis catalyst. The molecular formula of the product is the sum of the molecular formulas of the reactants, less C_2H_4. These facts suggest cross-metathesis.

14.35 (*a*) Exaltolide is the hydrogenation product of Compound A. Compound A is formed by ring-closing metathesis of the reactant. To derive the structure of compound A, rewrite the reactant in a form that approximates the shape of the resulting ring.

$$H_2C=CHCH_2(CH_2)_7\overset{\overset{\displaystyle O}{\|}}{C}OCH_2(CH_2)_3CH=CH_2 \quad \text{is the same as}$$

ring-closing
metathesis
$-H_2C=CH_2$

H_2, Pd

Compound A
$(C_{15}H_{26}O_2)$

(*b*) Analyze this ring-closing metathesis in the same way as part (*a*).

$$H_2C=CHCH_2(CH_2)_2\overset{\overset{\displaystyle O}{\|}}{C}OCH_2(CH_2)_8CH=CH_2 \quad \text{is the same as}$$

ring-closing
metathesis
$-H_2C=CH_2$

H_2, Pd

14.36 Olefin metathesis between the reactants leads to double-bond formation between carbons a and c and between b and d.

$CHCH_2CH_2CH_2CH_3$

$CHCH_2CH_2CH_2CH_3$

14.37 (*a*) The desired 1-deuteriobutane can be obtained by reaction of D_2O with butyllithium or butylmagnesium bromide.

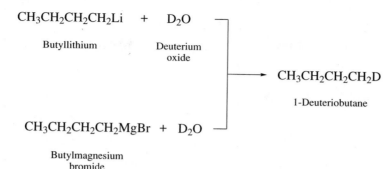

CH₃CH₂CH₂CH₂Li   +   D₂O

Butyllithium   Deuterium
                    oxide

⟶ CH₃CH₂CH₂CH₂D

1-Deuteriobutane

CH₃CH₂CH₂CH₂MgBr + D₂O

Butylmagnesium
bromide

Preparation of the organometallic compounds requires an alkyl bromide, which is synthesized from the corresponding alcohol.

$$\text{CH}_3\text{CH}_2\text{CH}_2\text{CH}_2\text{OH} \xrightarrow[\text{or HBr}]{\text{PBr}_3} \text{CH}_3\text{CH}_2\text{CH}_2\text{CH}_2\text{Br}$$

1-Butanol   1-Bromobutane

$$\text{CH}_3\text{CH}_2\text{CH}_2\text{CH}_2\text{Li} \xleftarrow[\text{diethyl ether}]{\text{Li}} \text{CH}_3\text{CH}_2\text{CH}_2\text{CH}_2\text{Br} \xrightarrow[\text{diethyl ether}]{\text{Mg}} \text{CH}_3\text{CH}_2\text{CH}_2\text{CH}_2\text{MgBr}$$

Butyllithium   1-Bromobutane   Butylmagnesium bromide

(*b*) In a sequence identical to that of part (*a*) in design but using 2-butanol as the starting material, 2-deuteriobutane may be prepared.

$$\underset{\overset{|}{\text{OH}}}{\text{CH}_3\text{CHCH}_2\text{CH}_3} \xrightarrow{\text{PBr}_3} \underset{\overset{|}{\text{Br}}}{\text{CH}_3\text{CHCH}_2\text{CH}_3} \xrightarrow[\text{diethyl ether}]{\text{Mg}} \underset{\overset{|}{\text{MgBr}}}{\text{CH}_3\text{CHCH}_2\text{CH}_3} \xrightarrow{\text{D}_2\text{O}} \underset{\overset{|}{\text{D}}}{\text{CH}_3\text{CHCH}_2\text{CH}_3}$$

2-Butanol   2-Bromobutane   *sec*-Butylmagnesium   2-Deuteriobutane
                                          bromide

An analogous procedure involving *sec*-butyllithium in place of the Grignard reagent can be used.

ANSWERS TO INTERPRETIVE PROBLEMS

14.38 C; **14.39** B; **14.40** C; **14.41** D; **14.42** A; **14.43** B

SELF-TEST

PART A

A-1. Give a method for the preparation of each of the following organometallic compounds, using appropriate starting materials:

(*a*) Cyclohexyllithium   (*b*) *tert*-Butylmagnesium bromide   (*c*) Lithium dibenzylcuprate

A-2. Give the structure of the product obtained by each of the following reaction schemes:

(a) $CH_3\overset{\displaystyle O}{\overset{\|}{C}}OCH_2CH_3$ $\xrightarrow[\text{2. } H_3O^+]{\text{1. } 2C_6H_5MgBr}$?

(d) $\xrightarrow[\text{2. } H_3O^+]{\text{1. } CH_3CH_2Li}$?

(b) $(CH_3)_2CHCH_2Li$ $\xrightarrow{D_2O}$?

(e) $+$ $CHBr_3$ $\xrightarrow[(CH_3)_3COH]{KOC(CH_3)_3}$?

(c) $H_3C-$$-Br$ $\xrightarrow[\text{3. } H_3O^+]{\substack{\text{1. Mg} \\ \text{2. } H_2C=O}}$?

A-3. Give two combinations of an organometallic reagent and a carbonyl compound that may be used for the preparation of each of the following:

(a) $C_6H_5\overset{\displaystyle OH}{\overset{|}{C}}HC(CH_3)_3$

(b) $CH_3CH_2CH_2\overset{\displaystyle OH}{\overset{|}{C}}HCH_2CH_2CH_2CH_3$

A-4. Give the structure of the organometallic reagent necessary to carry out each of the following:

(a) $\xrightarrow{?}$

(b) $C_6H_5CH_2\overset{\displaystyle O}{\overset{\|}{C}}OCH_3$ $\xrightarrow[\text{2. } H_3O^+]{\text{1. ?}}$ $(CH_3)_2CHC\overset{\displaystyle OH}{\overset{|}{\underset{\underset{\displaystyle CH_2C_6H_5}{|}}{C}}}H(CH_3)_2$ $+$ CH_3OH

(c) $\xrightarrow{?}$

A-5. Compounds A through F are some common organic solvents. Which ones would be suitable for use in the preparation of a Grignard reagent? For those that are not suitable, give a brief reason why.

$CH_3CH_2CH_2CH_2OCH_2CH_2CH_2CH_3$ $CH_3OCH_2CH_2OCH_3$ $HOCH_2CH_2OH$

         A   B   C

$CH_3\overset{\displaystyle O}{\overset{\|}{C}}OCH_2CH_3$ $CH_3\overset{\displaystyle O}{\overset{\|}{C}}OH$

        D   E   F

A-6. Show by a series of chemical equations how to prepare octane from 1-butanol as the source of all its carbon atoms.

A-7. Synthesis of the following alcohol is possible by three schemes using Grignard reagents. Give the reagents necessary to carry out each of them.

$$(CH_3)_2CH\overset{\displaystyle OH}{\overset{|}{C}}(CH_3)_2$$

A-8. Using ethylbenzene and any other necessary organic or inorganic reagents, outline a synthesis of 3-phenyl-2-butanol.

A-9. Give the structure of the final product of each of the following sequences of reactions.

(a)

$$\text{benzene} \xrightarrow[\text{FeBr}_3]{\text{Br}_2} \xrightarrow{\text{Mg}} \xrightarrow{\text{CH}_3\text{CCH}_2\text{CH}_3} \xrightarrow{\text{H}_3\text{O}^+} ?$$

(b) 1-Butene $\xrightarrow{\text{HCl}} \xrightarrow{\text{Mg}} \xrightarrow{\text{CH}_3\text{CH}} \xrightarrow{\text{H}_3\text{O}^+} ?$

(c) $\text{CH}_3\text{C}{\equiv}\text{CH} \xrightarrow{\text{NaNH}_2} \xrightarrow{\text{(cyclopentanone)}} \xrightarrow{\text{H}_3\text{O}^+} ?$

PART B

B-1. Which (if any) of the following would *not* be classified as an organometallic substance?

(a) Triethylaluminum

(c) Potassium *tert*-butoxide

(b) Ethylmagnesium iodide

(d) None of these (all are organometallic compounds)

B-2. Rank the following species in order of increasing polarity of the carbon–metal bond (least → most polar):

$$\underset{1}{\text{CH}_3\text{CH}_2\text{MgCl}} \qquad \underset{2}{\text{CH}_3\text{CH}_2\text{Na}} \qquad \underset{3}{(\text{CH}_3\text{CH}_2)_3\text{Al}}$$

(a) $3 < 1 < 2$ (b) $2 < 1 < 3$ (c) $1 < 3 < 2$ (d) $2 < 3 < 1$

B-3. Which sequence of reagents would carry out the following conversion?

$$\underset{\overset{|}{\text{OH}}}{\text{CH}_3\text{CH}_2\text{CHCH}_3} \xrightarrow{?} \underset{\overset{|}{\text{D}}}{\text{CH}_3\text{CH}_2\text{CHCH}_3}$$

(a) H_2SO_4, heat; then B_2D_6; then $\text{H}_2\text{O}_2, \text{HO}^-$

(b) H_2SO_4, heat; then D_2, Pt

(c) CH_3MgBr; then D_2O

(d) HBr; then Mg; then D_2O

B-4. Arrange the following intermediates in order of decreasing basicity (strongest → weakest):

$$\underset{1}{\text{H}_2\text{C}{=}\text{CHNa}} \qquad \underset{2}{\text{CH}_3\text{CH}_2\text{Na}} \qquad \underset{3}{\text{CH}_3\text{CH}_2\text{ONa}} \qquad \underset{4}{\text{HC}{\equiv}\text{CNa}}$$

(a) $2 > 1 > 4 > 3$ (b) $4 > 1 > 2 > 3$ (c) $3 > 4 > 1 > 2$ (d) $3 > 2 > 4 > 1$

B-5. Which, if any, of the following pairs of reagents could be used to prepare 2-phenyl-2-butanol?

$$\underset{\underset{\displaystyle CH_3}{|}}{\overset{\overset{\displaystyle OH}{|}}{CH_3CH_2CC_6H_5}}$$

2-Phenyl-2-butanol

(a) CH_3CH_2MgBr + $C_6H_5CH_2\overset{\overset{\displaystyle O}{||}}{C}CH_3$

(b) CH_3CH_2MgBr + $C_6H_5CH_2\overset{\overset{\displaystyle O}{||}}{C}H$

(c) CH_3MgI + $C_6H_5CH_2\overset{\overset{\displaystyle O}{||}}{C}CH_3$

(d) C_6H_5MgCl + $CH_3\overset{\overset{\displaystyle O}{||}}{C}CH_2CH_2CH_3$

(e) None of these combinations would be effective.

B-6. Which of the following reagents would be effective for the following reaction sequence?

$$C_6H_5C\equiv CH \quad \xrightarrow[\text{3. } H_3O^+]{\substack{\text{1. ?}\\ \text{2. } H_2C=O}} \quad C_6H_5C\equiv CCH_2OH$$

(a) Sodium ethoxide

(b) Magnesium in diethyl ether

(c) Butyllithium

(d) Potassium hydroxide

B-7. What is the product of the following reaction?

+ $2CH_3MgBr$ $\xrightarrow[\text{2. } H_3O^+]{\text{1. diethyl ether}}$

(a) $\underset{\underset{\displaystyle CH_3}{|}}{HOCHCH_2CH_2CH_2}\underset{\underset{\displaystyle CH_3}{|}}{CHOH}$

(b) $HOCH_2CH_2CH_2CH_2\underset{\underset{\displaystyle CH_3}{|}}{\overset{\overset{\displaystyle CH_3}{|}}{C}OH}$

(c) $CH_3OCH_2CH_2CH_2CH_2\underset{\underset{\displaystyle OH}{|}}{CHCH_3}$

(d) $HOCH_2CH_2CH_2CH_2\underset{\underset{\displaystyle CH_3}{|}}{CHOCH_3}$

B-8. Which of the following combinations of reagents will yield a chiral product after reaction in diethyl ether followed by hydrolysis in aqueous acid?

(a) + CH_3MgBr

(b) + CH_3MgBr

(c) $CH_3CH_2\overset{\overset{\displaystyle O}{||}}{C}OCH_3$ + $2CH_3MgBr$

(d) Both (a) and (c)

B-9. Which sequence of steps describes the best synthesis of 2-phenylpropene?

(a) Benzene + 2-chloropropene, $AlCl_3$

(b) Benzene + propene, H_2SO_4

(c) 1. Benzaldehyde ($C_6H_5CH{=}O$) + CH_3CH_2MgBr, diethyl ether

    2. H_3O^+

    3. H_2SO_4, heat

(d) 1. Bromobenzene + Mg, diethyl ether
    2. Propanal ($CH_3CH_2CH{=}O$)

    3. H_3O^+

    4. H_2SO_4, heat

(e) 1. Bromobenzene + Mg, diethyl ether
    2. Acetone [$(CH_3)_2C{=}O$]

    3. H_3O^+

    4. H_2SO_4, heat

B-10. What sequence of steps represents the best synthesis of 4-heptanol $(CH_3CH_2CH_2)_2CHOH$?

(a) $CH_3CH_2CH_2MgBr$ (2 moles) + formaldehyde ($H_2C{=}O$) in diethyl ether followed by H_3O^+

(b) $CH_3CH_2CH_2MgBr$ + butanal ($CH_3CH_2CH_2CH{=}O$) in diethyl ether followed by H_3O^+

(c) $CH_3CH_2CH_2CH_2MgBr$ + acetone [$(CH_3)_2C{=}O$] in diethyl ether followed by H_3O^+

(d) $(CH_3CH_2CH_2)_2CHMgBr$ + formaldehyde ($H_2C{=}O$) in diethyl ether followed by H_3O^+

(e) $CH_3CH_2CH_2MgBr$ + ethyl acetate ($CH_3\overset{\overset{\displaystyle O}{\|}}{C}OCH_2CH_3$) in diethyl ether followed by H_3O^+

B-11. All of the following compounds react with ethylmagnesium bromide. Alcohols are formed from four of the compounds. Which one does *not* give an alcohol?

(a) benzaldehyde (C_6H_5—CH, with C=O)

(c) acetophenone (C_6H_5—COCH₃, with C=O)

(e) C_6H_5—CH₂OCCH₃ (with C=O)

(b) benzoic acid (C_6H_5—COH, with C=O)

(d) C_6H_5—CCH₃ (with C=O)

B-12. Give the major product of the following reaction:

$$(E)\text{-2-pentene} \xrightarrow{CH_2I_2,\ Zn(Cu)} \ ?$$

(a) *cis*-1-Ethyl-2-methylcyclopropane

(c) 1-Ethyl-1-methylcyclopropane

(b) *trans*-1-Ethyl-2-methylcyclopropane

(d) An equimolar mixture of products (a) and (b)

CHAPTER 15

Alcohols, Diols, and Thiols

SOLUTIONS TO TEXT PROBLEMS

15.1 The two primary alcohols, 1-butanol and 2-methyl-1-propanol, can be prepared by hydrogenation of the corresponding aldehydes.

$$\underset{\text{Butanal}}{\overset{\overset{\displaystyle O}{\overset{\displaystyle \|}{}}}{CH_3CH_2CH_2CH}} \xrightarrow{\text{H}_2,\ \text{Ni}} \underset{\text{1-Butanol}}{CH_3CH_2CH_2CH_2OH}$$

$$\underset{\text{2-Methylpropanal}}{\overset{\overset{\displaystyle O}{\overset{\displaystyle \|}{}}}{(CH_3)_2CHCH}} \xrightarrow{\text{H}_2,\ \text{Ni}} \underset{\text{2-Methyl-1-propanol}}{(CH_3)_2CHCH_2OH}$$

The secondary alcohol 2-butanol arises by hydrogenation of a ketone.

$$\underset{\text{2-Butanone}}{\overset{\overset{\displaystyle O}{\overset{\displaystyle \|}{}}}{CH_3CCH_2CH_3}} \xrightarrow{\text{H}_2,\ \text{Ni}} \underset{\text{2-Butanol}}{\underset{\overset{\displaystyle |}{OH}}{CH_3CHCH_2CH_3}}$$

Tertiary alcohols such as 2-methyl-2-propanol, $(CH_3)_3COH$, cannot be prepared by hydrogenation of a carbonyl compound.

15.2 (*b*) A deuterium atom is transferred from $NaBD_4$ to the carbonyl group of acetone.

$$CH_3\overset{\overset{\displaystyle D-\overline{B}D_3}{\curvearrowright}}{\underset{\displaystyle CH_3}{C=O}} \longrightarrow CH_3\overset{\displaystyle D}{\underset{\displaystyle CH_3}{\overset{|}{C}-O\overline{B}D_3}} \xrightarrow{3(CH_3)_2C=O} \left(CH_3\overset{\displaystyle D}{\underset{\displaystyle CH_3}{\overset{|}{C}O-}}\right)_4 \overline{B}$$

On reaction with CH_3OD, deuterium is transferred from the alcohol to the oxygen of $[(CH_3)_2CDO]_4\overline{B}$.

$$CH_3\overset{\displaystyle D}{\underset{\displaystyle CH_3}{\overset{|}{C}}}-O\overset{\curvearrowright}{\underset{\displaystyle D-OCH_3}{\overline{B}[OCD(CH_3)_2]_3}} \longrightarrow CH_3\overset{\displaystyle D}{\underset{\displaystyle CH_3}{\overset{|}{C}OD}} + \overline{B}\underset{\displaystyle OCH_3}{[OCD(CH_3)_2]_3}$$

Overall:

$$(CH_3)_2C=O \xrightarrow[CH_3OD]{NaBD_4} (CH_3)_2\overset{\displaystyle D}{\overset{|}{C}OD}$$

Acetone   2-Propanol-2-*d-O-d*

(*c*) In this case $NaBD_4$ serves as a deuterium donor to carbon, and CD_3OH is a proton (not deuterium) donor to oxygen.

$$C_6H_5\overset{\overset{\displaystyle O}{\|}}{CH} \xrightarrow[CD_3OH]{NaBD_4} C_6H_5\overset{\displaystyle D}{\overset{|}{C}HOH}$$

Benzaldehyde   Benzyl alcohol-1-*d*

(*d*) Lithium aluminum deuteride is a deuterium donor to the carbonyl carbon of formaldehyde.

$$H\overset{\overset{\displaystyle D-AlD_3}{\curvearrowright}}{\underset{\displaystyle H}{C=O}} \longrightarrow H\overset{\displaystyle D}{\underset{\displaystyle H}{\overset{|}{C}-O\overline{A}lD_3}} \xrightarrow{3H\overset{\overset{\displaystyle O}{\|}}{CH}} (DCH_2O)_4\overline{A}l$$

On hydrolysis with D_2O, the oxygen–aluminum bond is cleaved and DCH_2OD is formed.

$$\overline{A}l(OCH_2D)_4 \xrightarrow{4D_2O} 4DCH_2OD + \overline{A}l(OD)_4$$

Methanol-*d-O-d*

15.3 The acyl portion of the ester gives a primary alcohol on reduction. The alkyl group bonded to oxygen may be primary, secondary, or tertiary and gives the corresponding alcohol.

$$CH_3CH_2\overset{\overset{\displaystyle O}{\|}}{C}OCH(CH_3)_2 \xrightarrow[2.\ H_2O]{1.\ LiAlH_4} CH_3CH_2CH_2OH + HOCH(CH_3)_2$$

Isopropyl propanoate   1-Propanol   2-Propanol

15.4 (*b*) Reaction with ethylene oxide results in the addition of a —CH_2CH_2OH unit to the Grignard reagent. Cyclohexylmagnesium bromide (or chloride) is the appropriate reagent.

Cyclohexymagnesium bromide   Ethylene oxide   2-Cyclohexylethanol

15.5 Lithium aluminum hydride is the appropriate reagent for reducing carboxylic acids or esters to alcohols.

3-Methyl-1,5-pentanedioic acid   3-Methyl-1,5-pentanediol

Any alkyl group may be attached to the oxygen of the ester function. In the following example, it is a methyl group.

Dimethyl 3-methyl-1,5-pentanedioate   3-Methyl-1,5-pentanediol   Methanol

15.6 Dihydroxylation of alkenes using osmium tetraoxide is a syn addition of hydroxyl groups to the double bond. *cis*-2-Butene yields the meso diol.

cis-2-Butene   *meso*-2,3-Butanediol

trans-2-Butene yields a racemic mixture of the two enantiomeric forms of the chiral diol.

trans-2-Butene   (2*R*,3*R*)-2,3-Butanediol   (2*S*,3*S*)-2,3-Butanediol

The Fischer projection formulas of the three stereoisomers are

meso-2,3-Butanediol   (2*R*,3*R*)-2,3-Butanediol   (2*S*,3*S*)-2,3-Butanediol

15.7 Dihydroxylation using the Sharpless method is a syn addition. Therefore, *trans*-2-butene can give the following two diols.

The problem states that the reaction was enantioselective and gave the diol having the (*R*)-configuration at C-2. Therefore, the correct structure is the one shown at the right and the configuration at C-3 must also be *R*.

15.8 The first step is proton transfer to 1,5-pentanediol to form the corresponding alkyloxonium ion.

1,5-Pentanediol   Sulfuric acid   Conjugate acid of 1,5-pentanediol   Hydrogen sulfate

Rewriting the alkyloxonium ion gives

is equivalent to

The alkyloxonium ion undergoes cyclization by intramolecular nucleophilic attack of its alcohol function on the carbon that bears the leaving group.

Conjugate acid of 1,5-pentanediol   Conjugate acid of oxane   Water

Loss of a proton gives oxane.

Conjugate acid of oxane   Hydrogen sulfate   Oxane   Sulfuric acid

15.9 (*b*) The relationship of the molecular formula of the ester ($C_{10}H_{10}O_4$) to that of the starting dicarboxylic acid ($C_8H_6O_4$) indicates that the diacid reacted with 2 moles of methanol to form a diester.

Methanol   1,4-Benzenedicarboxylic acid   Dimethyl 1,4-benzenedicarboxylate

15.10 Although neither *cis-* nor *trans-4-tert-*butylcyclohexanol is a chiral molecule, the stereochemical course of their reactions with acetic anhydride becomes evident when the relative stereochemistry of the ester function is examined for each case. The cis alcohol yields the cis acetate.

cis-4-*tert*-Butylcyclohexanol   Acetic anhydride   cis-4-*tert*-Butylcyclohexyl acetate

The trans alcohol yields the trans acetate.

trans-4-*tert*-Butylcyclohexanol   Acetic anhydride   trans-4-*tert*-Butylcyclohexyl acetate

15.11 Glycerol (1,2,3-propanetriol) has three hydroxyl groups, each of which is converted to a nitrate ester in nitroglycerin.

$$
\begin{array}{ccc}
CH_2OH & & CH_2ONO_2 \\
| & \xrightarrow[\text{H}_2\text{SO}_4]{3HNO_3} & | \\
CHOH & & CHONO_2 \\
| & & | \\
CH_2OH & & CH_2ONO_2
\end{array}
$$

Glycerol   Nitroglycerin

15.12 (*b*) The substrate is a secondary alcohol and so gives a ketone on oxidation with sodium dichromate. 2-Octanone has been prepared in 92–96% yield under these reaction conditions.

$$
\underset{\underset{OH}{|}}{CH_3CH(CH_2)_5CH_3} \xrightarrow[\text{H}_2\text{SO}_4,\ \text{H}_2\text{O}]{Na_2Cr_2O_7} \overset{\overset{O}{\|}}{CH_3C(CH_2)_5CH_3}
$$

2-Octanol   2-Octanone

(*c*) The alcohol is primary, and so oxidation can produce either an aldehyde or a carboxylic acid, depending on the reaction conditions. Here the oxidation is carried out under anhydrous conditions using pyridinium chlorochromate (PCC), and the product is the corresponding aldehyde.

$$
CH_3CH_2CH_2CH_2CH_2CH_2CH_2OH \xrightarrow[\text{CH}_2\text{Cl}_2]{PCC} \overset{\overset{O}{\|}}{CH_3CH_2CH_2CH_2CH_2CH_2CH}
$$

1-Heptanol   Heptanal

15.13 (*b*) Enzymatic oxidation of CH_3CD_2OH leads to loss of one of the C-1 deuterium atoms to NAD^+. The dihydropyridine ring of the reduced form of the coenzyme will bear a single deuterium.

| 1,1-Dideuterio-ethanol | NAD^+ | 1-Deuterio-ethanal | NADD |

(*c*) The deuterium atom of CH_3CH_2OD is lost as D^+. The reduced form of the coenzyme contains no deuterium.

| Ethanol-*O*-d | NAD^+ | Ethanal | NADH |

15.14 (*b*) Oxidation of the carbon–oxygen bonds to carbonyl groups accompanies their cleavage.

| 5-Methyl-1-phenyl-2,3-hexanediol | 3-Methylbutanal | 2-Phenylethanal |

(*c*) The CH_2OH group is cleaved from the ring as formaldehyde to leave cyclopentanone.

| 1-(Hydroxymethyl)-cyclopentanol | Cyclopentanone | Formaldehyde |

15.15 The major components of "essence of skunk" are

| 3-Methyl-1-butanethiol | *trans*-2-Butene-1-thiol |

15.16 Thiols may be prepared from the corresponding alkyl halide by nucleophilic substitution using the conjugate base of H_2S. Thus, an acceptable synthesis of 1-hexanethiol from 1-hexanol would be

$$CH_3(CH_2)_4CH_2OH \xrightarrow[\text{or HBr, heat}]{PBr_3} CH_3(CH_2)_4CH_2Br \xrightarrow[\text{ethanol}]{KSH} CH_3(CH_2)_4CH_2SH$$

1-Hexanol   1-Bromohexane   1-Hexanethiol

15.17 The molecular weight of 2-methyl-2-butanol is 88. A peak in its mass spectrum at m/z 70 corresponds to loss of water from the molecular ion. The peaks at m/z 73 and m/z 59 represent stable cations corresponding to the cleavages shown in the equation.

$$\overset{+\ddot{O}H}{\underset{\underset{CH_3}{|}}{H_3C-\overset{|}{C}-CH_2CH_3}}$$

$$\overset{+\ddot{O}H}{\underset{}{\|}} \qquad\qquad \overset{+\ddot{O}H}{\underset{}{\|}}$$
$$\cdot CH_3 + CH_3\overset{}{C}CH_2CH_3 \qquad\qquad CH_3\overset{}{C}CH_3 + \cdot CH_2CH_3$$

m/z 73 m/z 59

15.18 (*a*) The appropriate alkene for the preparation of 1-butanol by a hydroboration–oxidation sequence is 1-butene. Remember, hydroboration–oxidation leads to hydration of alkenes with a regioselectivity opposite to that seen in acid-catalyzed hydration.

$$CH_3CH_2CH{=}CH_2 \xrightarrow[\text{2. } H_2O_2, \text{ HO}^-]{\text{1. } B_2H_6} CH_3CH_2CH_2CH_2OH$$

1-Butene   1-Butanol

(*b*) 1-Butanol can be prepared by reaction of a Grignard reagent with formaldehyde.

$$CH_3CH_2CH_2CH_2OH \implies CH_3CH_2\overset{-}{\ddot{C}}H_2 + \overset{O}{\overset{\|}{HCH}}$$

An appropriate Grignard reagent is propylmagnesium bromide.

$$CH_3CH_2CH_2Br \xrightarrow[\text{diethyl ether}]{Mg} CH_3CH_2CH_2MgBr$$

1-Bromopropane   Propylmagnesium bromide

$$CH_3CH_2CH_2MgBr + \overset{O}{\overset{\|}{HCH}} \xrightarrow[\text{2. } H_3O^+]{\text{1. diethyl ether}} CH_3CH_2CH_2CH_2OH$$

1-Butanol

(*c*) Alternatively, 1-butanol may be prepared by the reaction of a Grignard reagent with ethylene oxide.

$$CH_3CH_2CH_2CH_2OH \implies CH_3\overset{-}{\ddot{C}}H_2 + \underset{\underset{O}{\diagdown\diagup}}{H_2C{-}CH_2}$$

In this case, ethylmagnesium bromide would be used.

$$CH_3CH_2Br \xrightarrow[\text{diethyl ether}]{\text{Mg}} CH_3CH_2MgBr$$

Ethyl bromide   Ethylmagnesium bromide

$$CH_3CH_2MgBr \;+\; H_2C\!\!-\!\!CH_2 \xrightarrow[\text{2. } H_3O^+]{\text{1. diethyl ether}} CH_3CH_2CH_2CH_2OH$$
$$\underset{\text{O}}{\diagdown\;\diagup}$$

Ethylene oxide   1-Butanol

(*d*) Primary alcohols may be prepared by reduction of the carboxylic acid having the same number of carbons. Among the reagents we have discussed, the only one that is effective in the reduction of carboxylic acids is lithium aluminum hydride. The four-carbon carboxylic acid butanoic acid is the proper substrate.

$$\underset{\text{Butanoic acid}}{CH_3CH_2CH_2\overset{\displaystyle O}{\overset{\|}{C}}OH} \xrightarrow[\text{2. } H_2O]{\text{1. LiAlH}_4\text{, diethyl ether}} \underset{\text{1-Butanol}}{CH_3CH_2CH_2CH_2OH}$$

(*e*) Reduction of esters can be accomplished using lithium aluminum hydride. The correct methyl ester is methyl butanoate.

$$\underset{\text{Methyl butanoate}}{CH_3CH_2CH_2\overset{\displaystyle O}{\overset{\|}{C}}OCH_3} \xrightarrow[\text{2. } H_2O]{\text{1. LiAlH}_4} \underset{\text{1-Butanol}}{CH_3CH_2CH_2CH_2OH} \;+\; \underset{\text{Methanol}}{CH_3OH}$$

(*f*) A butyl ester such as butyl acetate may be reduced with lithium aluminum hydride to prepare 1-butanol.

$$\underset{\text{Butyl acetate}}{CH_3\overset{\displaystyle O}{\overset{\|}{C}}OCH_2CH_2CH_2CH_3} \xrightarrow[\text{2. } H_2O]{\text{1. LiAlH}_4} \underset{\text{1-Butanol}}{CH_3CH_2CH_2CH_2OH} \;+\; \underset{\text{Ethanol}}{CH_3CH_2OH}$$

(*g*) Because 1-butanol is a primary alcohol having four carbons, butanal must be the aldehyde that is hydrogenated. Suitable catalysts are nickel, palladium, platinum, and ruthenium.

$$\underset{\text{Butanal}}{CH_3CH_2CH_2\overset{\displaystyle O}{\overset{\|}{C}}H} \xrightarrow{\text{H}_2,\ \text{Pt}} \underset{\text{1-Butanol}}{CH_3CH_2CH_2CH_2OH}$$

(*h*) Sodium borohydride reduces aldehydes and ketones efficiently. It does not reduce carboxylic acids, and its reaction with esters is too slow to be of synthetic value.

$$\underset{\text{Butanal}}{CH_3CH_2CH_2\overset{\displaystyle O}{\overset{\|}{C}}H} \xrightarrow[\substack{\text{water, ethanol} \\ \text{or methanol}}]{\text{NaBH}_4} \underset{\text{1-Butanol}}{CH_3CH_2CH_2CH_2OH}$$

15.19 (*a*) Both (*Z*)- and (*E*)-2-butene yield 2-butanol on hydroboration–oxidation.

$$CH_3CH{=}CHCH_3 \quad \xrightarrow[\text{2. } H_2O_2,\ HO^-]{\text{1. } B_2H_6} \quad CH_3CHCH_2CH_3$$

(*Z*)- or (*E*)-2-butene   2-Butanol

(*b*) Disconnection of one of the bonds to the carbon that bears the hydroxyl group reveals a feasible route using a Grignard reagent and propanal.

$$H_3C{\overset{\xi}{-}}CHCH_2CH_3 \quad \Longrightarrow \quad :CH_3 \ + \ \overset{O}{\overset{\|}{H}C}CH_2CH_3$$

               OH   Propanal

The synthetic sequence is

$$CH_3Br \quad \xrightarrow[\text{diethyl ether}]{Mg} \quad CH_3MgBr \quad \xrightarrow[\text{2. } H_3O^+]{\text{1. } CH_3CH_2\overset{O}{\overset{\|}{C}}H} \quad CH_3CHCH_2CH_3$$

Methyl bromide   Methylmagnesium   2-Butanol
                               bromide

(*c*) Another disconnection is related to a synthetic route using a Grignard reagent and acetaldehyde.

Disconnect this bond.

$$CH_3CH{\overset{\xi}{-}}CH_2CH_3 \quad \Longrightarrow \quad CH_3\overset{O}{\overset{\|}{C}}H \ + \ CH_3\ddot{C}H_2$$

        OH   Acetaldehyde

$$CH_3CH_2Br \quad \xrightarrow[\text{diethyl ether}]{Mg} \quad CH_3CH_2MgBr \quad \xrightarrow[\text{2. } H_3O^+]{\text{1. } CH_3\overset{O}{\overset{\|}{C}}H} \quad CH_3CH_2CHCH_3$$

                                                                  OH

Ethyl bromide   Ethylmagnesium   2-Butanol
                                bromide

(*d–f*) Because 2-butanol is a secondary alcohol, it can be prepared by reduction of a ketone having the same carbon skeleton, in this case 2-butanone. All three reducing agents indicated in the equations are satisfactory.

$$CH_3\overset{O}{\overset{\|}{C}}CH_2CH_3 \quad \xrightarrow[\text{(or Pt, Ni, Ru)}]{H_2,\ Pd} \quad CH_3CHCH_2CH_3$$

                                                   OH

            2-Butanone   2-Butanol

$$CH_3\overset{O}{\overset{\|}{C}}CH_2CH_3 \quad \xrightarrow[CH_3OH]{NaBH_4} \quad CH_3CHCH_2CH_3$$

                                                   OH

            2-Butanone   2-Butanol

$$CH_3\overset{\overset{\displaystyle O}{\|}}{C}CH_2CH_3 \quad \xrightarrow[\text{2. H}_2\text{O}]{\text{1. LiAlH}_4} \quad CH_3\underset{\underset{\displaystyle OH}{|}}{C}HCH_2CH_3$$

2-Butanone   2-Butanol

15.20 (*a*) All the carbon–carbon disconnections are equivalent.

$$H_3C-\underset{\underset{\displaystyle CH_3}{|}}{\overset{\overset{\displaystyle CH_3}{|}}{C}}-OH \quad \Longrightarrow \quad {:}CH_3 \;+\; CH_3\overset{\overset{\displaystyle O}{\|}}{C}CH_3$$

Acetone

The synthesis via a Grignard reagent and acetone is

$$CH_3Br \quad \xrightarrow[\text{diethyl ether}]{\text{Mg}} \quad CH_3MgBr \quad \xrightarrow[\text{2. H}_3\text{O}^+]{\overset{\overset{\displaystyle O}{\|}}{\text{1. CH}_3\text{CCH}_3}} \quad (CH_3)_3COH$$

Methyl   Methylmagnesium   *tert*-Butyl alcohol
bromide   bromide

(*b*) An alternative route to *tert*-butyl alcohol is addition of a Grignard reagent to an ester. Esters react with *2 moles* of Grignard reagent. Thus, *tert*-butyl alcohol may be formed by reacting methyl acetate with 2 moles of methylmagnesium iodide. Methyl alcohol is formed as a byproduct of the reaction.

$$2CH_3MgI \;+\; CH_3\overset{\overset{\displaystyle O}{\|}}{C}OCH_3 \quad \xrightarrow[\text{2. H}_3\text{O}^+]{\text{1. diethyl ether}} \quad H_3C-\underset{\underset{\displaystyle CH_3}{|}}{\overset{\overset{\displaystyle CH_3}{|}}{C}}-OH \;+\; CH_3OH$$

Methylmagnesium   Methyl acetate   *tert*-Butyl alcohol   Methyl
iodide   alcohol

15.21 (*a*) All of the primary alcohols having the molecular formula $C_5H_{12}O$ may be prepared by reduction of aldehydes. The appropriate equations are

$$CH_3CH_2CH_2CH_2\overset{\overset{\displaystyle O}{\|}}{C}H \quad \xrightarrow[\text{2. H}_2\text{O}]{\text{1. LiAlH}_4\text{, diethyl ether}} \quad CH_3CH_2CH_2CH_2CH_2OH$$

Pentanal   1-Pentanol

$$CH_3CH_2\underset{\underset{\displaystyle CH_3}{|}}{C}H\overset{\overset{\displaystyle O}{\|}}{C}H \quad \xrightarrow[\text{2. H}_2\text{O}]{\text{1. LiAlH}_4\text{, diethyl ether}} \quad CH_3CH_2\underset{\underset{\displaystyle CH_3}{|}}{C}HCH_2OH$$

2-Methylbutanal   2-Methyl-1-butanol

$$(CH_3)_2CHCH_2\overset{\overset{\displaystyle O}{\|}}{C}H \quad \xrightarrow[\text{2. H}_2\text{O}]{\text{1. LiAlH}_4\text{, diethyl ether}} \quad (CH_3)_2CHCH_2CH_2OH$$

3-Methylbutanal   3-Methyl-1-butanol

$$(CH_3)_3C\overset{\overset{\displaystyle O}{\|}}{C}H \quad \xrightarrow[\text{2. H}_2\text{O}]{\text{1. LiAlH}_4\text{, diethyl ether}} \quad (CH_3)_3CCH_2OH$$

2,2-Dimethylpropanal   2,2-Dimethyl-1-propanol (*b*) The secondary alcohols having the molecular formula $C_5H_{12}O$ may be prepared by reduction of ketones.

$$\underset{\text{2-Pentanone}}{CH_3CH_2CH_2\overset{\overset{\displaystyle O}{\|}}{C}CH_3} \xrightarrow[\text{2. H}_2\text{O}]{\text{1. LiAlH}_4, \text{ diethyl ether}} \underset{\text{2-Pentanol}}{CH_3CH_2CH_2\overset{\overset{\displaystyle OH}{|}}{C}HCH_3}$$

$$\underset{\text{3-Pentanone}}{CH_3CH_2\overset{\overset{\displaystyle O}{\|}}{C}CH_2CH_3} \xrightarrow[\text{2. H}_2\text{O}]{\text{1. LiAlH}_4, \text{ diethyl ether}} \underset{\text{3-Pentanol}}{CH_3CH_2\overset{\overset{\displaystyle OH}{|}}{C}HCH_2CH_3}$$

$$\underset{\text{3-Methyl-2-butanone}}{(CH_3)_2CH\overset{\overset{\displaystyle O}{\|}}{C}CH_3} \xrightarrow[\text{2. H}_2\text{O}]{\text{1. LiAlH}_4, \text{ diethyl ether}} \underset{\text{3-Methyl-2-butanol}}{(CH_3)_2CH\overset{\overset{\displaystyle OH}{|}}{C}HCH_3}$$

(*c*) As with the reduction of aldehydes in part (*a*), reduction of carboxylic acids yields primary alcohols. For example, 1-pentanol may be prepared by reduction of pentanoic acid.

$$\underset{\text{Pentanoic acid}}{CH_3CH_2CH_2CH_2\overset{\overset{\displaystyle O}{\|}}{C}OH} \xrightarrow[\text{2. H}_2\text{O}]{\text{1. LiAlH}_4, \text{ diethyl ether}} \underset{\text{1-Pentanol}}{CH_3CH_2CH_2CH_2CH_2OH}$$

The remaining primary alcohols, 2-methyl-1-butanol, 3-methyl-1-butanol, and 2,2-dimethyl-1-propanol, may be prepared in the same way.

(*d*) As with carboxylic acids, esters may be reduced using lithium aluminum hydride to give primary alcohols. For example, 2,2-dimethyl-1-propanol may be prepared by reduction of methyl 2,2-dimethylpropanoate.

$$\underset{\substack{\text{Methyl}\\\text{2,2-dimethylpropanoate}}}{(CH_3)_3C\overset{\overset{\displaystyle O}{\|}}{C}OCH_3} \xrightarrow[\text{2. H}_2\text{O}]{\text{1. LiAlH}_4, \text{ diethyl ether}} \underset{\text{2,2-Dimethyl-1-propanol}}{(CH_3)_3CCH_2OH}$$

15.22 (*a*) The suggested synthesis

$$\underset{\text{Butane}}{CH_3CH_2CH_2CH_3} \xrightarrow[\text{light or heat}]{\text{Br}_2} \underset{\text{1-Bromobutane}}{CH_3CH_2CH_2CH_2Br} \xrightarrow{\text{KOH}} \underset{\text{1-Butanol}}{CH_3CH_2CH_2CH_2OH}$$

is a poor one because bromination of butane yields a mixture of 1-bromobutane and 2-bromobutane, 2-bromobutane being the major product.

$$\underset{\text{Butane}}{CH_3CH_2CH_2CH_3} \xrightarrow[\text{light or heat}]{\text{Br}_2} \underset{\substack{\text{1-Bromobutane}\\\text{(minor product)}}}{CH_3CH_2CH_2CH_2Br} + \underset{\substack{\text{2-Bromobutane}\\\text{(major product)}}}{CH_3\overset{\overset{\displaystyle }{|}}{C}HCH_2CH_3}\\ \underset{Br}{}$$

(b) The suggested synthesis

$$(CH_3)_3CH \xrightarrow[\text{light or heat}]{Br_2} (CH_3)_3CBr \xrightarrow{KOH} (CH_3)_3COH$$

2-Methylpropane   2-Bromo-2-methylpropane   2-Methyl-2-propanol

will fail because the reaction of 2-bromo-2-methylpropane with potassium hydroxide will proceed by elimination rather than by substitution. The first step in the process, selective bromination of 2-methylpropane to 2-bromo-2-methylpropane, is satisfactory because bromination is selective for substitution of tertiary hydrogens in the presence of secondary and primary ones.

(c) Benzyl alcohol, unlike 1-butanol and 2-methyl-2-propanol, can be prepared effectively by this method.

Toluene   Benzyl bromide   Benzyl alcohol

Free-radical bromination of toluene is selective for the benzylic position. Benzyl bromide cannot undergo elimination, and so nucleophilic substitution of bromide by hydroxide will work well.

(d) The desired transformation

Ethylbenzene   1-Bromo-1-phenylethane   1-Phenylethanol

fails because it produces more than one enantiomer. The reactant ethylbenzene is achiral and although its bromination will be highly regioselective for the benzylic position, the product will be a racemic mixture of (R)- and (S)-1-bromo-1-phenylethane. The alcohol produced by hydrolysis will also be racemic. Furthermore, the hydrolysis step will give mostly styrene ($C_6H_5CH=CH_2$) by an E2 elimination, rather than 1-phenylethanol by nucleophilic substitution.

15.23 Glucose contains five hydroxyl groups and an aldehyde functional group. Its hydrogenation will not affect the hydroxyl groups but will reduce the aldehyde to a primary alcohol.

Glucose   Sorbitol

15.24 (a) 1-Phenylethanol is a secondary alcohol and so can be prepared by the reaction of a Grignard reagent with an aldehyde. One combination is phenylmagnesium bromide and ethanal (acetaldehyde).

1-Phenylethanol   Phenylmagnesium bromide   Ethanal (acetaldehyde)

Grignard reagents—phenylmagnesium bromide in this case—are always prepared by reaction of magnesium metal and the corresponding halide. Starting with bromobenzene, a suitable synthesis is described by the sequence

$$C_6H_5Br \xrightarrow[\text{diethyl ether}]{\text{Mg}} C_6H_5MgBr \xrightarrow[\text{2. } H_3O^+]{\text{1. } CH_3\overset{\displaystyle O}{\overset{\|}{C}}H} C_6H_5\underset{\underset{\displaystyle OH}{|}}{C}HCH_3$$

Bromobenzene   Phenylmagnesium   1-Phenylethanol
                              bromide

(b) An alternative disconnection of 1-phenylethanol reveals a second route using benzaldehyde and a methyl Grignard reagent.

$$C_6H_5\underset{\underset{\displaystyle OH}{|}}{C}HCH_3 \implies C_6H_5\overset{\displaystyle O}{\overset{\|}{C}}H \quad + \quad CH_3MgI$$

1-Phenylethanol   Benzaldehyde   Methylmagnesium
                                                        iodide

Equations representing this approach are

$$CH_3I \xrightarrow[\text{diethyl ether}]{\text{Mg}} CH_3MgI \xrightarrow[\text{2. } H_3O^+]{\text{1. } C_6H_5\overset{\displaystyle O}{\overset{\|}{C}}H} C_6H_5\underset{\underset{\displaystyle OH}{|}}{C}HCH_3$$

Iodomethane   Methylmagnesium   1-Phenylethanol
                           iodide

(c) Aldehydes are, in general, obtainable by oxidation of the corresponding primary alcohol. By recognizing that benzaldehyde can be obtained by oxidation of benzyl alcohol with PCC, we write

$$C_6H_5CH_2OH \xrightarrow[\text{CH}_2\text{Cl}_2]{\text{PCC}} C_6H_5\overset{\displaystyle O}{\overset{\|}{C}}H \xrightarrow[\text{2. } H_3O^+]{\text{1. } CH_3MgI, \text{ diethyl ether}} C_6H_5\underset{\underset{\displaystyle OH}{|}}{C}HCH_3$$

Benzyl alcohol   Benzaldehyde   1-Phenylethanol

(d) The conversion of acetophenone to 1-phenylethanol is a reduction.

$$C_6H_5\overset{\displaystyle O}{\overset{\|}{C}}CH_3 \xrightarrow{\text{reducing agent}} C_6H_5\underset{\underset{\displaystyle OH}{|}}{C}HCH_3$$

Acetophenone   1-Phenylethanol

Any of a number of reducing agents could be used. These include
1. $NaBH_4$, CH_3OH
2. $LiAlH_4$ in diethyl ether, then H_2O
3. H_2 and a Pt, Pd, Ni, or Ru catalyst

(e) Benzene can be employed as the starting material in a synthesis of 1-phenylethanol. Friedel–Crafts acylation of benzene gives acetophenone, which can then be reduced as in part (d).

Benzene   Acetyl chloride   Acetophenone

Acetic anhydride (CH_3COCCH_3) can be used in place of acetyl chloride.

15.25 2-Phenylethanol is an ingredient in many perfumes, to which it imparts a rose-like fragrance. Numerous methods have been employed for its synthesis.

(a) As a primary alcohol having two more carbon atoms than bromobenzene, it can be formed by reaction of a Grignard reagent, phenylmagnesium bromide, with ethylene oxide.

The desired reaction sequence is therefore

Bromobenzene   Phenylmagnesium   2-Phenylethanol
                              bromide

(b) Hydration of styrene with a regioselectivity contrary to that of Markovnikov's rule is required. This is accomplished readily by hydroboration–oxidation.

$$C_6H_5CH=CH_2 \xrightarrow[\text{2. } H_2O_2, HO^-]{\text{1. } B_2H_6, \text{ diglyme}} C_6H_5CH_2CH_2OH$$

Styrene   2-Phenylethanol

(c) Reduction of aldehydes yields primary alcohols.

$$C_6H_5CH_2CH\overset{O}{\overset{\|}{}} \xrightarrow{\text{reducing agent}} C_6H_5CH_2CH_2OH$$

2-Phenylethanal   2-Phenylethanol

Among the reducing agents that could be (and have been) used are
1. $NaBH_4$, CH_3OH
2. $LiAlH_4$ in diethyl ether, then H_2O
3. H_2 and a Pt, Pd, Ni, or Ru catalyst

(d) Esters are readily reduced to primary alcohols with lithium aluminum hydride.

$$C_6H_5CH_2COCH_2CH_3\overset{O}{\overset{\|}{}} \xrightarrow[\text{2. } H_2O]{\text{1. } LiAlH_4, \text{ diethyl ether}} C_6H_5CH_2CH_2OH$$

Ethyl 2-phenylethanoate   2-Phenylethanol

(*e*) The only reagent that is suitable for the direct reduction of carboxylic acids to primary alcohols is lithium aluminum hydride.

$$C_6H_5CH_2\overset{\overset{\displaystyle O}{\|}}{C}OH \xrightarrow[\text{2. H}_2\text{O}]{\text{1. LiAlH}_4\text{, diethyl ether}} C_6H_5CH_2CH_2OH$$

2-Phenylethanoic acid   2-Phenylethanol

Alternatively, the carboxylic acid could be esterified with ethanol and the resulting ethyl 2-phenylethanoate reduced.

$$C_6H_5CH_2\overset{\overset{\displaystyle O}{\|}}{C}OH \;+\; CH_3CH_2OH \xrightarrow{\text{H}^+} C_6H_5CH_2\overset{\overset{\displaystyle O}{\|}}{C}OCH_2CH_3 \xrightarrow[\text{part }(d)]{\text{reduce as in}} C_6H_5CH_2CH_2OH$$

2-Phenylethanoic   Ethanol   Ethyl 2-phenylethanoate   2-Phenylethanol
acid

15.26 (*a*) Thiols may be prepared from the corresponding alkyl halide by nucleophilic substitution using the conjugate base of H_2S. The first step must therefore be a conversion of the alcohol to an alkyl bromide.

$$CH_3CH_2CH_2CH_2OH \xrightarrow[\text{or PBr}_3]{\text{HBr}} CH_3CH_2CH_2CH_2Br \xrightarrow[\text{ethanol}]{\text{KSH}} CH_3CH_2CH_2CH_2SH$$

1-Butanol   1-Bromobutane   1-Butanethiol

(*b*) To obtain 1-hexanol from alcohols having four carbons or fewer, a two-carbon chain extension must be carried out. This suggests reaction of a Grignard reagent with ethylene oxide. The retrosynthetic path for this approach is

$$CH_3CH_2CH_2CH_2\!\overset{\xi}{-}\!CH_2CH_2OH \;\Longrightarrow\; CH_3CH_2CH_2CH_2MgBr \;+\; H_2C\!\!\overset{\displaystyle O}{\overset{\diagup\diagdown}{-}}\!\!CH_2$$

The reaction sequence therefore becomes

$$CH_3CH_2CH_2CH_2Br \xrightarrow[\text{diethyl ether}]{\text{Mg}} CH_3CH_2CH_2CH_2MgBr \xrightarrow[\text{2. H}_3\text{O}^+]{\text{1. H}_2\text{C}-\text{CH}_2 \diagup\text{O}\diagdown} CH_3CH_2CH_2CH_2CH_2CH_2OH$$

1-Bromobutane   Butylmagnesium bromide   1-Hexanol
from part (*a*)

Given the constraints of the problem that we begin with alcohols, prepare ethylene oxide by the sequence

$$CH_3CH_2OH \xrightarrow[\text{heat}]{\text{H}_2\text{SO}_4} H_2C{=}CH_2 \xrightarrow{\overset{\overset{\displaystyle O}{\|}}{CH_3COOH}} H_2C\!\!\overset{\diagup\diagdown}{\underset{O}{-}}\!\!CH_2$$

Ethanol   Ethylene

(*c*) The target molecule 2-hexanol may be mentally disconnected as shown to a four-carbon unit and a two-carbon unit.

$$CH_3\underset{\underset{\displaystyle OH}{|}}{CH}\!\overset{\xi}{-}\!CH_2CH_2CH_2CH_3 \;\Longrightarrow\; CH_3\overset{\overset{\displaystyle O}{\|}}{C}H \;+\; {:}CH_2CH_2CH_2CH_3$$

The alternative disconnection to $^-\!:CH_3$ and $\overset{\overset{\displaystyle O}{\displaystyle \|}}{H}CCH_2CH_2CH_2CH_3$ reveals a plausible approach to 2-hexanol but is inconsistent with the requirement of the problem that limits starting materials to four carbons or fewer. The five-carbon aldehyde would have to be prepared first, making for a lengthy overall synthetic scheme.

An appropriate synthesis based on alcohols as starting materials is

$$CH_3CH_2OH \xrightarrow[CH_2Cl_2]{PCC} CH_3\overset{\overset{\displaystyle O}{\displaystyle \|}}{C}H$$

Ethanol   Ethanal

$$CH_3CH_2CH_2CH_2MgBr \ + \ CH_3\overset{\overset{\displaystyle O}{\displaystyle \|}}{C}H \xrightarrow[\text{2. } H_3O^+]{\text{1. diethyl ether}} CH_3\underset{\underset{\displaystyle OH}{\displaystyle |}}{C}HCH_2CH_2CH_2CH_3$$

Butylmagnesium bromide   Ethanal   2-Hexanol
from part (*b*)

(*d*) Hexanal may be obtained from 1-hexanol [prepared in part (*b*)] by oxidation in dichloromethane using pyridinium chlorochromate (PCC) or pyridinium dichromate (PDC).

$$CH_3(CH_2)_4CH_2OH \xrightarrow[CH_2Cl_2]{PCC \text{ or } PDC} CH_3(CH_2)_4\overset{\overset{\displaystyle O}{\displaystyle \|}}{C}H$$

1-Hexanol from part (*b*)   Hexanal

(*e*) Oxidation of 2-hexanol from part (*c*) yields 2-hexanone.

$$CH_3\underset{\underset{\displaystyle OH}{\displaystyle |}}{C}HCH_2CH_2CH_2CH_3 \xrightarrow[H_2SO_4, \ H_2O]{Na_2Cr_2O_7} CH_3\overset{\overset{\displaystyle O}{\displaystyle \|}}{C}CH_2CH_2CH_2CH_3$$

2-Hexanol   2-Hexanone

PCC or PDC can also be used for this transformation.

(*f*) Oxidation of 1-hexanol with chromic acid (sodium or potassium dichromate in aqueous sulfuric acid) yields hexanoic acid. Use of PDC or PCC in dichloromethane is not acceptable because those reagents yield aldehydes on reaction with primary alcohols.

$$CH_3(CH_2)_4CH_2OH \xrightarrow[H_2SO_4, \ H_2O]{K_2Cr_2O_7} CH_3(CH_2)_4CO_2H$$

1-Hexanol from part (*b*)   Hexanoic acid

(*g*) Fischer esterification of hexanoic acid with ethanol produces ethyl hexanoate.

$$CH_3(CH_2)_4\overset{\overset{\displaystyle O}{\displaystyle \|}}{C}OH \ + \ CH_3CH_2OH \xrightarrow{H^+} CH_3(CH_2)_4\overset{\overset{\displaystyle O}{\displaystyle \|}}{C}OCH_2CH_3$$

Hexanoic acid   Ethanol   Ethyl hexanoate
from part (*f*)

(*h*) Vicinal diols are normally prepared by dihydroxylation of alkenes with osmium tetraoxide and *tert*-butyl hydroperoxide.

$$(CH_3)_2C{=}CH_2 \xrightarrow[\substack{(CH_3)_3COOH,\ HO^-\\ (CH_3)_3COH}]{OsO_4} (CH_3)_2\underset{\underset{OH}{|}}{C}CH_2OH$$

2-Methylpropene   2-Methyl-1,2-propanediol

The required alkene is available by dehydration of 2-methyl-2-propanol.

$$(CH_3)_3COH \xrightarrow[\text{heat}]{H_3PO_4} (CH_3)_2C{=}CH_2$$

2-Methyl-2-propanol   2-Methylpropene

(*i*) The desired aldehyde can be prepared by oxidation of the corresponding primary alcohol with PCC or PDC.

$$(CH_3)_3CCH_2OH \xrightarrow[\text{CH}_2\text{Cl}_2]{\text{PCC or PDC}} (CH_3)_3C\overset{\overset{O}{\|}}{C}H$$

2,2-Dimethyl-1-propanol   2,2-Dimethylpropanal

The necessary alcohol is available through reaction of a *tert*-butyl Grignard reagent with formaldehyde, as shown by the disconnection

$$(CH_3)_3C{\Large\}}CH_2OH \quad \Longrightarrow \quad (CH_3)_3CMgCl + H_2C{=}O$$

$$CH_3OH \xrightarrow[\text{CH}_2\text{Cl}_2]{\text{PCC or PDC}} H\overset{\overset{O}{\|}}{C}H$$

Methanol   Formaldehyde

$$(CH_3)_3COH \xrightarrow{HCl} (CH_3)_3CCl \xrightarrow[\text{diethyl ether}]{Mg} (CH_3)_3CMgCl$$

2-Methyl-2-propanol   2-Chloro-2-methylpropane   1,1-Dimethylethyl-magnesium chloride

$$(CH_3)_3CMgCl \xrightarrow[\text{2. }H_3O^+]{\text{1. }H_2C{=}O,\ \text{diethyl ether}} (CH_3)_3CCH_2OH$$

1,1-Dimethylethyl-magnesium chloride   2,2-Dimethyl-1-propanol

15.27 (*a*) The simplest route to this primary chloride from benzene is through the corresponding alcohol. The first step is the two-carbon chain extension used in Problem 15.25*a*.

Benzene → Bromobenzene → 2-Phenylethanol

Reagents: Br_2 / $FeBr_3$; then 1. Mg, diethyl ether 2. H_2C-CH_2 (O) 3. H_3O^+

2-Phenylethanol →(SOCl₂)→ 1-Chloro-2-phenylethane

The preparation of ethylene oxide is shown in Problem 15.26*b*.

(*b*) A Friedel–Crafts acylation is the best approach to the target ketone.

Benzene + $(CH_3)_2CHCCl$ (2-Methylpropanoyl chloride) →($AlCl_3$)→ 2-Methyl-1-phenyl-1-propanone

Because acid chlorides are prepared from the corresponding carboxylic acids, we write

$(CH_3)_2CHCH_2OH$ →($K_2Cr_2O_7$ / H_2SO_4, heat)→ $(CH_3)_2CHCOH$ →($SOCl_2$)→ $(CH_3)_2CHCCl$

2-Methyl-1-propanol → 2-Methylpropanoic acid → 2-Methylpropanoyl chloride

(*c*) Wolff–Kishner or Clemmensen reduction of the ketone just prepared in part (*b*) affords isobutylbenzene.

$C_6H_5CCH(CH_3)_2$ →(H_2NNH_2, HO^-, triethylene glycol, heat or $Zn(Hg)$, HCl)→ $C_6H_5CH_2CH(CH_3)_2$

2-Methyl-1-phenyl-1-propanone → Isobutylbenzene

A less direct approach requires three steps:

$C_6H_5CCH(CH_3)_2$ →($NaBH_4$ / CH_3OH)→ $C_6H_5CHCH(CH_3)_2$ (OH) →(H_2SO_4, heat)→ $C_6H_5CH=C(CH_3)_2$ →(H_2 / Pt)→ $C_6H_5CH_2CH(CH_3)_2$

2-Methyl-1-phenyl-1-propanone → 2-Methyl-1-phenyl-1-propanol → 2-Methyl-1-phenylpropene → Isobutylbenzene

15.28 (*a*) Because 1-phenylcyclopentanol is a tertiary alcohol, a likely synthesis would involve reaction of a ketone and a Grignard reagent. Thus, a reasonable last step is treatment of cyclopentanone with phenylmagnesium bromide.

Cyclopentanone is prepared by oxidation of cyclopentanol. Any one of a number of oxidizing agents would be suitable. These include PDC or PCC in CH_2Cl_2 or chromic acid (H_2CrO_4) generated from $Na_2Cr_2O_7$ in aqueous sulfuric acid.

(*b*) Acid-catalyzed dehydration of 1-phenylcyclopentanol gives 1-phenylcyclopentene.

(*c*) Hydroboration–oxidation of 1-phenylcyclopentene gives *trans*-2-phenylcyclopentanol. The H and OH are added across the double bond opposite to Markovnikov's rule and syn to each other.

(*d*) Oxidation of *trans*-2-phenylcyclopentanol converts this secondary alcohol to the desired ketone. Any of the Cr(VI)-derived oxidizing agents mentioned in part (*a*) for oxidation of cyclopentanol to cyclopentanone is satisfactory.

(*e*) The standard procedure for preparing *cis*-1,2-diols is by dihydroxylation of alkenes with osmium tetraoxide.

(*f*) The desired compound is available either by ozonolysis of 1-phenylcyclopentene from part (*b*):

$$\text{1-Phenylcyclopentene} \quad \xrightarrow[\text{2. H}_2\text{O, Zn}]{\text{1. O}_3} \quad C_6H_5\overset{O}{\underset{\|}{C}}CH_2CH_2CH_2\overset{O}{\underset{\|}{C}}H$$

1-Phenylcyclopentene

5-Oxo-1-phenyl-1-pentanone

or by periodic acid cleavage of the diol in part (*e*):

$$\text{1-Phenyl-}cis\text{-1,2-cyclopentanediol} \quad \xrightarrow{\text{HIO}_4} \quad C_6H_5\overset{O}{\underset{\|}{C}}CH_2CH_2CH_2\overset{O}{\underset{\|}{C}}H$$

1-Phenyl-*cis*-1,2-cyclopentanediol

5-Oxo-1-phenyl-1-pentanone

(*g*) Reduction of both carbonyl groups in the product of part (*f*) gives the desired diol.

$$C_6H_5\overset{O}{\underset{\|}{C}}CH_2CH_2CH_2\overset{O}{\underset{\|}{C}}H \quad \xrightarrow[\substack{\text{1. LiAlH}_4\text{, diethyl ether} \\ \text{2. H}_2\text{O}}]{\substack{\text{H}_2\text{, Pt (or Pd, Ni, Ru)} \\ \text{or} \\ \text{NaBH}_4\text{, H}_2\text{O} \\ \text{or}}} \quad C_6H_5\overset{}{\underset{\underset{OH}{|}}{C}}HCH_2CH_2CH_2CH_2OH$$

5-Oxo-1-phenyl-1-pentanone

1-Phenyl-1,5-pentanediol

15.29 (*a*, *b*) Primary alcohols react in two different ways on being heated with acid catalysts: they can condense to form dialkyl ethers or undergo dehydration to yield alkenes. Ether formation is favored at lower temperature, and alkene formation is favored at higher temperature.

$$2CH_3CH_2CH_2OH \quad \xrightarrow[140°C]{\text{H}_2\text{SO}_4} \quad CH_3CH_2CH_2OCH_2CH_2CH_3 \quad + \quad H_2O$$

1-Propanol

Dipropyl ether

Water

$$CH_3CH_2CH_2OH \quad \xrightarrow[200°C]{\text{H}_2\text{SO}_4} \quad CH_3CH{=}CH_2 \quad + \quad H_2O$$

1-Propanol

Propene

Water

(*c*) Nitrate esters are formed by the reaction of alcohols with nitric acid in the presence of a sulfuric acid catalyst.

$$CH_3CH_2CH_2OH \quad + \quad HONO_2 \quad \xrightarrow{\text{H}_2\text{SO}_4 \text{ (cat)}} \quad CH_3CH_2CH_2ONO_2 \quad + \quad H_2O$$

1-Propanol

Nitric acid

Propyl nitrate

Water

(*d*) Pyridinium chlorochromate (PCC) oxidizes primary alcohols to aldehydes.

$$CH_3CH_2CH_2OH \quad \xrightarrow[\text{CH}_2\text{Cl}_2]{\text{PCC}} \quad CH_3CH_2\overset{O}{\underset{\|}{C}}H$$

1-Propanol

Propanal

(*e*) Potassium dichromate in aqueous sulfuric acid oxidizes primary alcohols to carboxylic acids.

$$CH_3CH_2CH_2OH \xrightarrow[\text{K}_2\text{SO}_4,\ \text{H}_2\text{O}]{\text{K}_2\text{Cr}_2\text{O}_7} CH_3CH_2\overset{\displaystyle O}{\overset{\|}{C}}OH$$

1-Propanol   Propanoic acid

(*f*) Amide ion, a strong base, abstracts a proton from 1-propanol to form ammonia and 1-propanolate ion. This is an acid–base reaction.

$$CH_3CH_2CH_2OH\ +\ NaNH_2 \longrightarrow CH_3CH_2CH_2ONa\ +\ NH_3$$

1-Propanol   Sodium amide   Sodium 1-propanolate   Ammonia

(*g*) With acetic acid and in the presence of an acid catalyst, 1-propanol is converted to its acetate ester.

$$CH_3CH_2CH_2OH\ +\ CH_3\overset{\displaystyle O}{\overset{\|}{C}}OH \xrightarrow{\text{HCl}} CH_3\overset{\displaystyle O}{\overset{\|}{C}}OCH_2CH_2CH_3\ +\ H_2O$$

1-Propanol   Acetic acid   Propyl acetate   Water

This is an equilibrium process that slightly favors products.

(*h*) Alcohols react with *p*-toluenesulfonyl chloride to give *p*-toluenesulfonates.

$$CH_3CH_2CH_2OH\ +\ H_3C\!\!-\!\!\langle\ \rangle\!\!-\!\!SO_2Cl \xrightarrow{\text{pyridine}} CH_3CH_2CH_2O\overset{\displaystyle O}{\underset{\displaystyle O}{\overset{\|}{\underset{\|}{S}}}}\!\!-\!\!\langle\ \rangle\!\!-\!\!CH_3\ +\ HCl$$

1-Propanol   *p*-Toluenesulfonyl chloride   Propyl *p*-toluenesulfonate

(*i*) Acyl chlorides convert alcohols to esters.

$$CH_3CH_2CH_2OH\ +\ CH_3O\!\!-\!\!\langle\ \rangle\!\!-\!\!\overset{\displaystyle O}{\overset{\|}{C}}Cl \xrightarrow{\text{pyridine}} CH_3CH_2CH_2O\overset{\displaystyle O}{\overset{\|}{C}}\!\!-\!\!\langle\ \rangle\!\!-\!\!OCH_3\ +\ HCl$$

1-Propanol   *p*-Methoxybenzoyl chloride   Propyl *p*-methoxybenzoate

(*j*) The reagent is benzoic anhydride. Acid anhydrides react with alcohols to give esters.

$$CH_3CH_2CH_2OH\ +\ C_6H_5\overset{\displaystyle O\ \ O}{\overset{\|\ \ \|}{C}}OCC_6H_5 \xrightarrow{\text{pyridine}} CH_3CH_2CH_2O\overset{\displaystyle O}{\overset{\|}{C}}C_6H_5\ +\ C_6H_5\overset{\displaystyle O}{\overset{\|}{C}}OH$$

1-Propanol   Benzoic anhydride   Propyl benzoate   Benzoic acid

(*k*) The reagent is succinic anhydride, a cyclic anhydride. Esterification occurs, but in this case the resulting ester and carboxylic acid functions remain part of the same molecule.

$$CH_3CH_2CH_2OH\ +\ \text{(succinic anhydride)} \xrightarrow{\text{pyridine}} CH_3CH_2CH_2O\overset{\displaystyle O}{\overset{\|}{C}}CH_2CH_2\overset{\displaystyle O}{\overset{\|}{C}}OH$$

1-Propanol   Succinic
                        anhydride   Hydrogen propyl succinate

15.30 (*a*) On being heated in the presence of sulfuric acid, tertiary alcohols undergo elimination.

$$H_3C-\underset{OH}{\overset{C_6H_5}{\bigcirc}} \quad \xrightarrow[\text{heat}]{H_2SO_4} \quad H_3C-\bigcirc=C_6H_5$$

4-Methyl-1-   4-Methyl-1-
phenylcyclohexanol   phenylcyclohexene (81%)

(*b*) The combination of reagents specified converts alkenes to vicinal diols.

$$(CH_3)_2C{=}C(CH_3)_2 \quad \xrightarrow[\text{(CH}_3)_3\text{COH, HO}^-]{(CH_3)_3COOH,\ OsO_4(cat)} \quad \underset{\underset{HO\ \ OH}{|\ \ \ |}}{(CH_3)_2C{-}C(CH_3)_2}$$

2,3-Dimethyl-2-butene   2,3-Dimethyl-2,3-butanediol
                                        (72%)

(*c*) Hydroboration–oxidation of the double bond takes place with a regioselectivity that is opposite to Markovnikov's rule. The elements of water are added in a stereospecific syn fashion.

$$\underset{\text{1-Phenylcyclobutene}}{\square{=}C_6H_5} \quad \xrightarrow[\text{2. H}_2O_2,\ HO^-]{\text{1. B}_2H_6,\ \text{diglyme}} \quad \underset{\substack{\textit{trans}\text{-2-Phenylcyclobutanol}\\(82\%)}}{\overset{HO\ \ \ \ \ H}{\square{-}C_6H_5}}$$

(*d*) Lithium aluminum hydride reduces carboxylic acids to primary alcohols, but does not reduce carbon–carbon double bonds.

$$\underset{\substack{\text{Cyclopentene-4-}\\\text{carboxylic acid}}}{\bigcirc{-}CO_2H} \quad \xrightarrow[\text{2. H}_2O]{\text{1. LiAlH}_4,\ \text{diethyl ether}} \quad \underset{\substack{\text{(3-Cyclopentenyl)-}\\\text{methanol}}}{\bigcirc{-}CH_2OH}$$

(*e*) Chromic acid oxidizes the secondary alcohol to the corresponding ketone but does not affect the triple bond.

$$\underset{\substack{OH\\\\3\text{-Octyn-2-ol}}}{CH_3CHC{\equiv}C(CH_2)_3CH_3} \quad \xrightarrow[\substack{H_2SO_4,\ H_2O\\\text{acetone}}]{H_2CrO_4} \quad \underset{3\text{-Octyn-2-one (80\%)}}{CH_3\overset{O}{\overset{||}{C}}C{\equiv}C(CH_2)_3CH_3}$$

(*f*) Lithium aluminum hydride reduces carbonyl groups efficiently but does not normally react with double bonds.

$$\underset{\text{4-Octen-2,7-dione}}{CH_3\overset{O}{\overset{||}{C}}CH_2CH{=}CHCH_2\overset{O}{\overset{||}{C}}CH_3} \quad \xrightarrow[\text{2. H}_2O]{\text{1. LiAlH}_4,\ \text{diethyl ether}} \quad \underset{\substack{OH\ \ \ \ \ \ \ \ \ \ \ \ \ \ \ OH\\4\text{-Octen-2,7-diol (75\%)}}}{CH_3CHCH_2CH{=}CHCH_2CHCH_3}$$

(*g*) Alcohols react with acyl chlorides to yield esters. The O—H bond is broken in this reaction; the C—O bond of the alcohol remains intact on ester formation.

trans-3-Methylcyclohexanol   3,5-Dinitrobenzoyl chloride   *trans*-3-Methylcyclohexyl
3,5-dinitrobenzoate (74%)

(*h*) Acid anhydrides react with alcohols to give esters. Here, too, the spatial orientation of the C—O bond remains intact.

exo-Bicyclo[2.2.1]-   Acetic anhydride   *exo*-Bicyclo[2.2.1]hept-   Acetic acid
heptan-2-ol   2-yl acetate (90%)

(*i*) The substrate is a carboxylic acid and undergoes Fischer esterification with methanol.

4-Chloro-3,5-   Methyl 4-chloro-3,5-
dinitrobenzoic acid   dinitrobenzoate (96%)

(*j*) Both ester functions are cleaved by reduction with lithium aluminum hydride. The product is a diol.

(96%)

(*k*) Treatment of the vicinal diol obtained in part (*j*) with periodic acid brings about its cleavage to two carbonyl compounds.

(74%)

15.31 Only the hydroxyl groups on C-1 and C-4 can be involved, because only these two can lead to a five-membered cyclic ether.

$$HOCH_2CHCH_2CH_2OH \xrightarrow[\text{heat}]{H^+} \text{[structure]} + H_2O$$

OH

1,2,4-Butanetriol   3-Hydroxyoxolane
($C_4H_8O_2$)

Any other combination of hydroxyl groups would lead to a strained three-membered or four-membered ring and is unfavorable under conditions of acid catalysis.

15.32 (*a*) The task of converting a ketone to an alkene requires first the reduction of the ketone to an alcohol and then dehydration. In practice the two-step transformation has been carried out in 54% yield by treating the ketone with sodium borohydride and then heating the resulting alcohol with *p*-toluenesulfonic acid.

Of course, sodium borohydride may be replaced by other suitable reducing agents, and *p*-toluenesulfonic acid is not the only acid that could be used in the dehydration step.

(*b*) This problem and the next one illustrate the value of reasoning backward. The desired product, cyclohexanol, can be prepared cleanly from cyclohexanone.

Once cyclohexanone is recognized to be a key intermediate, the synthetic pathway becomes apparent—what is needed is a method to convert the indicated starting material to cyclohexanone. The reagent ideally suited to this task is periodic acid. The synthetic sequence to be followed is therefore

1-(Hydroxymethyl)-   Cyclohexanone   Cyclohexanol
cyclohexanol

(*c*) No direct method allows a second hydroxyl group to be introduced at C-2 of 1-phenylcyclohexanol in a single step. We recognize the product as a vicinal diol and recall that such compounds are available by dihydroxylation of alkenes.

This tells us that we must first dehydrate the tertiary alcohol, then dihydroxylate the resulting alkene.

1-Phenylcyclohexanol   1-Phenylcyclohexene   1-Phenyl-*cis*-1,2-cyclohexanediol

The syn stereoselectivity of the dihydroxylation step ensures that the product will have its hydroxyl groups cis, as the problem requires.

15.33 Because the target molecule is an eight-carbon secondary alcohol and the problem restricts our choices of starting materials to alcohols of five carbons or fewer, we are led to consider building up the carbon chain by a Grignard reaction.

4-Methyl-3-heptanol

The disconnection shown leads to a three-carbon aldehyde and a five-carbon Grignard reagent. Starting with the corresponding alcohols, the following synthetic scheme seems reasonable.

First, propanal is prepared.

1-Propanol   Propanal

After converting 2-pentanol to its bromo derivative, a solution of the Grignard reagent is prepared.

2-Pentanol   2-Bromopentane   1-Methylbutylmagnesium bromide

Reaction of the Grignard reagent with the aldehyde yields the desired 4-methyl-3-heptanol.

1-Methylbutylmagnesium bromide   Propanal   4-Methyl-3-heptanol

15.34 Our target molecule is void of functionality and so requires us to focus attention on the carbon skeleton. Notice that it can be considered to arise from three ethyl groups.

$$CH_3CH_2\text{—}CH\text{—}CH_2CH_3$$
$$|$$
$$CH_3$$

3-Methylpentane

Considering the problem retrosynthetically, we can see that a key intermediate having the carbon skeleton of the desired product is 3-methyl-3-pentanol. This becomes apparent from the fact that alkanes may be prepared from alkenes, which in turn are available from alcohols. The desired alcohol may be prepared from reaction of an acetate ester with a Grignard reagent, ethylmagnesium bromide.

$$CH_3CH_2\underset{\underset{CH_3}{|}}{C}HCH_2CH_3 \implies CH_3CH_2\underset{\underset{OH}{|}}{\overset{\overset{CH_3}{|}}{C}}CH_2CH_3 \implies CH_3\overset{\overset{O}{||}}{C}OR + 2CH_3CH_2MgBr$$

The carbon skeleton can be assembled in one step by the reaction of ethylmagnesium bromide and ethyl acetate.

$$2CH_3CH_2MgBr + CH_3\overset{\overset{O}{||}}{C}OCH_2CH_3 \xrightarrow[\text{2. } H_3O^+]{\text{1. diethyl ether}} CH_3\underset{\underset{CH_2CH_3}{|}}{\overset{\overset{OH}{|}}{C}}CH_2CH_3$$

Ethylmagnesium   Ethyl acetate   3-Methyl-3-pentanol
bromide

The resulting tertiary alcohol is converted to the desired hydrocarbon by acid-catalyzed dehydration and catalytic hydrogenation of the resulting mixture of alkenes.

3-Methyl-3-pentanol   3-Methyl-2-pentene   2-Ethyl-1-butene
                            (cis + trans)

$CH_3CH(CH_2CH_3)_2$

3-Methylpentane

Because the problem requires that ethanol be the ultimate starting material, we need to show the preparation of the ethylmagnesium bromide and ethyl acetate used in constructing the carbon skeleton.

$$CH_3CH_2OH \xrightarrow{PBr_3} CH_3CH_2Br \xrightarrow[\text{diethyl ether}]{Mg} CH_3CH_2MgBr$$

Ethanol   Ethyl bromide   Ethylmagnesium
                                                       bromide

$$CH_3CH_2OH \xrightarrow[\text{H}_2\text{SO}_4, \text{H}_2\text{O, heat}]{K_2Cr_2O_7} \underset{\text{Acetic acid}}{CH_3\overset{\displaystyle O}{\overset{\|}{C}}OH}$$

Ethanol

$$\underset{\text{Acetic acid}}{CH_3\overset{\displaystyle O}{\overset{\|}{C}}OH} + CH_3CH_2OH \xrightarrow{H^+} \underset{\text{Ethyl acetate}}{CH_3\overset{\displaystyle O}{\overset{\|}{C}}OCH_2CH_3}$$

15.35 (*a*) Retrosynthetically, we can see that the cis carbon–carbon double bond is available by hydrogenation of the corresponding alkyne over the Lindlar catalyst.

$$CH_3CH_2CH{=}CHCH_2CH_2OH \quad \Longrightarrow \quad CH_3CH_2C{\equiv}CCH_2CH_2OH$$

The —CH_2CH_2OH unit can be appended to an alkynide anion by reaction with ethylene oxide.

$$CH_3CH_2C{\equiv}CCH_2CH_2OH \quad \Longrightarrow \quad CH_3CH_2C{\equiv}C\!:^- \ + \ H_2C\overset{\displaystyle}{\underset{O}{-}}CH_2$$

The alkynide anion is derived from 1-butyne by alkylation of acetylene. This analysis suggests the following synthetic sequence:

$$HC{\equiv}CH \xrightarrow[\text{2. CH}_3\text{CH}_2\text{Br}]{\text{1. NaNH}_2, \text{NH}_3} CH_3CH_2C{\equiv}CH \xrightarrow[\text{2. H}_2\text{C}-\text{CH}_2]{\text{1. NaNH}_2, \text{NH}_3} CH_3CH_2C{\equiv}CCH_2CH_2OH$$

Acetylene   1-Butyne   3-Hexyn-1-ol

Lindlar Pd
H$_2$

$$\underset{\text{H}}{\overset{CH_3CH_2}{\diagdown}}C{=}C\underset{\text{H}}{\overset{CH_2CH_2OH}{\diagup}}$$

cis-3-Hexen-1-ol

(b) The compound cited is the aldehyde derived by oxidation of the primary alcohol in part (a). Oxidize the alcohol with PDC or PCC in CH_2Cl_2.

cis-3-Hexen-1-ol   *cis*-3-Hexenal

15.36 Even though we are given the structure of the starting material, it is still better to reason backward from the target molecule rather than forward from the starting material.

The desired product contains a cyano (—CN) group. The only method we have seen so far for introducing such a function into a molecule is by nucleophilic substitution. The last step in the synthesis must therefore be

This step should work very well because the substrate is a primary benzylic halide, cannot undergo elimination, and is very reactive in S_N2 reactions. The primary benzylic halide can be prepared from the corresponding alcohol by any of a number of methods.

Suitable reagents include HBr, PBr_3, or $SOCl_2$.

Now we only need to prepare the primary alcohol from the given starting aldehyde, which is accomplished by reduction.

Reduction can be achieved by catalytic hydrogenation, with lithium aluminum hydride, or with sodium borohydride. The actual sequence of reactions as carried out is as shown.

m-Methoxy-
benzaldehyde

m-Methoxybenzyl
alcohol

m-Methoxybenzyl
bromide

m-Methoxybenzyl
cyanide

Another three-step synthesis, which is reasonable but does not involve an alcohol as an intermediate, is

m-Methoxy-
benzaldehyde   *m*-Methoxytoluene   *m*-Methoxybenzyl
bromide   *m*-Methoxybenzyl
cyanide

15.37 (*a*) Addition of hydrogen chloride to cyclopentadiene takes place by way of the most stable carbocation. In this case it is an allylic carbocation.

(Allylic carbocation;
more stable)   not   (Not allylic;
less stable)

3-Chlorocyclopentene (80–90%)
(Compound A)

Hydrolysis of 3-chlorocyclopentene gives the corresponding alcohol. Sodium bicarbonate in water is a weakly basic solvolysis medium.

Compound A   2-Cyclopenten-1-ol (88%)
(compound B)

Oxidation of compound B (a secondary alcohol) gives the ketone 2-cyclopentenone.

Compound B   2-Cyclopentenone
(60–80%) (compound C)

(*b*) Thionyl chloride converts alcohols to alkyl chlorides.

$$H_2C{=}CHCH_2CH_2\underset{\underset{\displaystyle OH}{|}}{C}HCH_3 \xrightarrow[\text{pyridine}]{SOCl_2} H_2C{=}CHCH_2CH_2\underset{\underset{\displaystyle Cl}{|}}{C}HCH_3$$

5-Hexen-2-ol

5-Chloro-1-hexene
(compound D)

Ozonolysis cleaves the carbon–carbon double bond.

$$H_2C{=}CHCH_2CH_2\underset{\underset{\displaystyle Cl}{|}}{C}HCH_3 \xrightarrow[\text{2. Zn, H}_2O]{\text{1. O}_3} \overset{\overset{\displaystyle O}{\|}}{H C}CH_2CH_2\underset{\underset{\displaystyle Cl}{|}}{C}HCH_3 \quad + \quad \overset{\overset{\displaystyle O}{\|}}{H C}H$$

Compound D

4-Chloropentanal
(compound E)

Formaldehyde

Reduction of compound E yields the corresponding alcohol.

$$\overset{\overset{\displaystyle O}{\|}}{H C}CH_2CH_2\underset{\underset{\displaystyle Cl}{|}}{C}HCH_3 \xrightarrow{NaBH_4} HOCH_2CH_2CH_2\underset{\underset{\displaystyle Cl}{|}}{C}HCH_3$$

4-Chloropentanal

4-Chloro-1-pentanol
(compound F)

(*c*) *N*-Bromosuccinimide (NBS) is a reagent designed to accomplish benzylic bromination.

1-Bromo-2-methylnaphthalene

1-Bromo-2-(bromomethyl)naphthalene
(compound G)

Hydrolysis of the benzylic bromide gives the corresponding benzylic alcohol. The bromine that is directly attached to the naphthalene ring does not react under these conditions.

1-Bromo-2-(bromomethyl)naphthalene

(1-Bromo-2-naphthyl)methanol
(compound H)

Oxidation of the primary alcohol with PCC gives the aldehyde.

(1-Bromo-2-naphthyl)methanol

1-Bromonaphthalene-2-carboxaldehyde
(compound I)

15.38 The alcohol is tertiary and benzylic and yields a relatively stable carbocation.

2-Phenyl-2-butanol   1-Methyl-1-phenylpropyl
cation

The alcohol is chiral, but the carbocation is not. Thus, irrespective of which enantiomer of 2-phenyl-2-butanol is used, the same carbocation is formed. The carbocation reacts with ethanol to give an optically inactive mixture containing equal quantities of enantiomers (a racemic mixture).

1-Methyl-1-phenylpropyl   Ethanol   2-Ethoxy-2-phenylbutane
cation   (50% R, 50% S)

15.39 The relationship between the reactant and the product requires that a methyl migration must occur and suggests a carbocation intermediate is involved. Both alcohol functions are tertiary and are equivalent. Therefore, generate a carbocation in acid solution as the first stage.

2,3-Dimethyl-2,3-butanediol   Hydronium ion

2-Hydroxy-1,1,2-trimethylpropyl
cation

The correct carbon skeleton is generated by a methyl migration. The carbocation produced is stabilized by electron release from oxygen.

2-Hydroxy-1,1,2-trimethylpropyl
cation

Loss of a proton from oxygen gives the product of the reaction.

3,3-Dimethyl-2-butanone

For more about this and related mechanisms, see the descriptive passage following Chapter 15 in the text.

15.40 The difference between the two ethers is that 1-*O*-benzylglycerol contains a vicinal diol function, but 2-*O*-benzylglycerol does not. Periodic acid will react with 1-*O*-benzylglycerol but not with 2-*O*-benzylglycerol.

$$C_6H_5CH_2OCH_2CHCH_2OH \xrightarrow{HIO_4} C_6H_5CH_2OCH_2CH + HCH$$

OH

1-*O*-Benzylglycerol   2-Benzyloxyethanal   Formaldehyde

$$HOCH_2CHCH_2OH \xrightarrow{HIO_4} \text{no reaction}$$

OCH$_2$C$_6$H$_5$

2-*O*-Benzylglycerol

15.41 The formation of an alkanethiol by reaction of an alkyl halide or alkyl *p*-toluenesulfonate with the conjugate base of H$_2$S is a substitution that occurs by the S$_N$2 pathway and will proceed with inversion of configuration. The *p*-toluenesulfonate is formed from the corresponding alcohol by a reaction that does not involve any of the bonds to the chirality center. Therefore, begin with (*S*)-2-butanol.

(*S*)-2-Butanol   (*S*)-*sec*-Butyl *p*-toluenesulfonate   (*R*)-2-Butanethiol

15.42 (*a*) Cysteine contains an —SH group and is a thiol. Oxidation of thiols gives rise to disulfides.

$$2RSH \xrightarrow{oxidize} RSSR$$

Thiol   Disulfide

Biological oxidation of cysteine gives the disulfide cystine.

$$2HSCH_2CHCO^- \xrightarrow{oxidize} {}^-OCCHCH_2S-SCH_2CHCO^-$$

$^+$NH$_3$ $^+$NH$_3$ $^+$NH$_3$

Cysteine   Cystine

(*b*) Oxidation of a thiol yields a series of acids, including a sulfinic acid and a sulfonic acid.

$$\text{RSH} \longrightarrow \text{RS}^{+}\text{-OH} \longrightarrow \text{RS}^{2+}\text{-OH}$$

      Thiol   Sulfinic acid   Sulfonic acid

Biological oxidation of cysteine can yield, in addition to the disulfide cystine, cysteine sulfinic acid and the sulfonic acid cysteic acid.

$$\text{HSCH}_2\text{CHCO}^{-} \xrightarrow{\text{oxidize}} \text{HO-}^{+}\text{SCH}_2\text{CHCO}^{-} \xrightarrow{\text{oxidize}} \text{HO}^{2+}\text{SCH}_2\text{CHCO}^{-}$$

      Cysteine   Cysteine sulfinic acid ($C_3H_7NO_4S$)   Cysteic acid ($C_3H_7NO_5S$)

15.43 The ratio of carbon to hydrogen in the molecular formula is C_nH_{2n+2} ($C_8H_{18}O_2$), and so the compound has no double bonds or rings. The compound cannot be a vicinal diol, because it does not react with periodic acid.

The NMR spectrum is rather simple as all peaks are singlets. The 12-proton singlet at δ 1.2 must correspond to four equivalent methyl groups and the four-proton singlet at δ 1.6 to two equivalent methylene groups. No nonequivalent protons can be vicinal because no splitting is observed. The two-proton singlet at δ 2.0 is due to the hydroxyl protons of the diol. The compound is 2,5-dimethyl-2,5-hexanediol.

$$\begin{array}{ccc} & \text{CH}_3 & \text{CH}_3 \\ & | & | \\ \text{CH}_3\text{C}&\text{CH}_2\text{CH}_2&\text{C}\text{CH}_3 \\ & | & | \\ & \text{OH} & \text{OH} \end{array}$$

15.44 The molecular formula of the compound ($C_8H_{10}O$) corresponds to an index of hydrogen deficiency of 4. The 4 hydrogen signal at δ 7.2 in the ^{1}H NMR spectrum suggests these unsaturations are due to a disubstituted benzene ring. That the ring is para-substituted is supported by the symmetry of the signal; it is a pair of doublets, not a quartet.

The broad band in the IR spectrum at 3300 cm^{-1} is the O—H stretching vibration of an alcohol. The presence of an alcohol is confirmed by the disappearance of the broad signal at δ 2.1 following addition of D_2O as the hydroxyl proton undergoes rapid exchange with deuterium. As the remaining signals in the ^{1}H NMR spectrum are singlets, with areas of 2H and 3H, respectively, the compound can be identified as 4-methylbenzyl alcohol.

$$\overset{\delta\,2.4}{\text{H}_3\text{C}}-\bigcirc-\overset{\delta\,4.7}{\text{CH}_2\text{OH}}$$

$$\delta\,7.2 \qquad \delta\,2.1$$

15.45 (*a*) This compound has only two different types of carbons. One type of carbon comes at low field and is most likely a carbon bonded to oxygen and three other equivalent carbons. The spectrum leads to the conclusion that this compound is *tert*-butyl alcohol.

$$
\begin{array}{c}
CH_3 \\
| \\
H_3C-C-OH \\
\uparrow \quad | \quad \searrow \delta\ 68.9 \\
\quad CH_3 \\
\delta\ 31.2
\end{array}
$$

(*b*) Four different types of carbons occur in this compound. The $C_4H_{10}O$ isomers that have four nonequivalent carbons are $CH_3CH_2CH_2CH_2OH$, $CH_3CHCH_2CH_3$, and $CH_3OCH_2CH_2CH_3$.
$$\underset{OH}{|}$$

The lowest field signal, the one at δ 69.2 from the carbon that bears the oxygen substituent, is a methine (CH). The compound is therefore 2-butanol.

$$
\begin{array}{c}
CH_3CHCH_2CH_3 \\
| \\
OH
\end{array}
$$

(*c*) This compound has two equivalent CH_3 groups, as indicated by the signal at δ 18.9. Its lowest field carbon is a CH_2, and so the group CH_2O must be present. The compound is 2-methyl-1-propanol.

$$
\begin{array}{c}
\delta\ 30.8 \\
\downarrow \\
H_3C-CH-CH_2OH \\
\uparrow \quad | \quad \uparrow \\
\quad CH_3 \\
\delta\ 18.9 \qquad \delta\ 69.4
\end{array}
$$

15.46 The compound has only three carbons, none of which is a CH_3 group. Two of the carbon signals arise from CH_2 groups; the other corresponds to a CH group. The only structure consistent with the observed data is that of 3-chloro-1,2-propanediol.

$$
\begin{array}{c}
HOCH_2-CH-CH_2Cl \\
| \\
OH
\end{array}
$$

The structure $HOCH_2CHCH_2OH$ cannot be correct. It would exhibit only two peaks in its ^{13}C NMR
$$\underset{Cl}{|}$$
spectrum, because the two terminal carbons are equivalent to each other.

15.47 The observation of a peak at *m/z* 31 in the mass spectrum of the compound suggests the presence of a primary alcohol. This fragment is most likely $H_2C=\overset{+}{O}H$. On the basis of this fact and the appearance of four different carbons in the ^{13}C NMR spectrum, the compound is 2-ethyl-1-butanol.

$$
\begin{array}{c}
\delta\ 23 \\
\downarrow \quad \delta\ 44 \\
\delta\ 11 \longrightarrow CH_3CH_2 \searrow \downarrow \\
\qquad\qquad CH-CH_2OH \\
\quad CH_3CH_2 \nearrow \quad \uparrow \\
\qquad\qquad\qquad \delta\ 65
\end{array}
$$

15.48 Given the information specified in the problem (molecular formula and method of preparation), the only possible alcohols are:

1-Phenyl-1-propanol   1-Phenyl-2-propanol   2-Phenyl-1-propanol

The H—C—O signals in both spectra are in the δ 4–5 range and integrate for 1 proton. Therefore, we can eliminate 2-phenyl-1-propanol from consideration because its CH_2O group requires two protons.

The signal for the CH_3 group in compound A appears as a doublet at δ 1.2. This is consistent with compound A being 1-phenyl-2-propanol. Likewise, the signal for its H—C—O proton is consistent with the multiplet at δ 4 because this proton is coupled to five vicinal protons.

This leaves 1-phenyl-1-propanol as compound B which is consistent with the NMR spectrum. The CH_3 signal is a triplet indicating that CH_3 is adjacent to CH_2. The same applies to the triplet for the H—C—O proton at δ 4.5.

To give 1-phenyl-2-propanol (compound A) as the product formed by the reaction of phenyllithium with 1,2-epoxypropane, phenyllithium must have attacked the CH_2 group of the epoxide.

The same is true for the formation of 1-phenyl-1-propanol (compound B) from the reaction of methyllithium with styrene oxide. Methyllithium must have attacked the CH_2 group of the epoxide.

The generalization we can draw from these experiments is that when organolithium reagents react with unsymmetrical epoxides, they attack the less crowded carbon of the ring. In this respect, the reactions resemble S_N2 processes.

ANSWERS TO INTERPRETIVE PROBLEMS

15.49 A;   **15.50** A;   **15.51** B;   **15.52** C;   **15.53** B;   **15.54** B;   **15.55** D

SELF-TEST

PART A

A-1. For each of the following reactions give the structure of the missing reactant or reagent.

(a) ? $\xrightarrow[\text{2. H}_2\text{O}]{\text{1. LiAlH}_4}$

(b) ? + $2CH_3CH_2MgBr$ $\xrightarrow[\text{2. H}_3\text{O}^+]{\text{1. diethyl ether}}$ $C_6H_5C(CH_2CH_3)_2$ + CH_3CH_2OH (OH on central carbon)

(c) $C_6H_5CH_2\overset{\underset{\displaystyle |}{CH_3}}{C}=CH_2$ $\xrightarrow{\ ?\ }$ $C_6H_5CH_2\overset{\underset{\displaystyle |}{CH_3}}{C}HCH_2OH$

(d) [structure: 1-methylcyclohexene] $\xrightarrow{\ ?\ }$ [structure: 1-methylcyclohexane-1,2-diol with CH₃ and OH on C1 and OH on C2]

(e) $C_6H_5CH_2Br$ $\xrightarrow{\ ?\ }$ $C_6H_5CH_2SH$

A-2. For the following reactions of 2-phenylethanol, $C_6H_5CH_2CH_2OH$, give the correct reagent or product(s) omitted from the equation.

(a) $C_6H_5CH_2CH_2OH$ $\xrightarrow[CH_2Cl_2]{PCC}$?

(b) $C_6H_5CH_2CH_2OH$ $\xrightarrow{\ ?\ }$ $CH_3\overset{\overset{\displaystyle O}{||}}{C}OCH_2CH_2-$ [phenyl ring]

(c) $C_6H_5CH_2CH_2OH$ (2 mol) $\xrightarrow[\text{heat}]{H^+}$ H_2O + ?

(d) $C_6H_5CH_2CH_2OH$ $\xrightarrow{\ ?\ }$ $C_6H_5CH_2CO_2H$

A-3. Write the structure of the major organic product formed in the reaction of 2-propanol with each of the following reagents:

(a) Sodium amide ($NaNH_2$)

(b) Potassium dichromate ($K_2Cr_2O_7$) in aqueous sulfuric acid, heat

(c) PDC in dichloromethane

(d) Acetic acid ($CH_3\overset{\overset{\displaystyle O}{||}}{C}OH$) in the presence of dissolved hydrogen chloride

(e) H_3C- [phenyl ring] $-SO_2Cl$ in the presence of pyridine

(f) CH_3CH_2- [phenyl ring] $-\overset{\overset{\displaystyle O}{||}}{C}Cl$ in the presence of pyridine

(g) $CH_3\overset{\overset{\displaystyle O}{||}}{C}O\overset{\overset{\displaystyle O}{||}}{C}CH_3$ in the presence of pyridine

A-4. Outline two synthetic schemes for the preparation of 3-methyl-1-butanol using different Grignard reagents.

A-5. Give the structure of the reactant, reagent, or product omitted from each of the following. Show stereochemistry where important.

(a) [cyclohexane ring with OH, H, OH, CH$_3$ substituents] $\xrightarrow{\text{HIO}_4}$?

(b) ? (a diol) $\xrightarrow[\text{heat}]{\text{H}^+}$ [tetrahydropyran ring with CH$_3$, O]

(c) ? $\xrightarrow[\text{(CH}_3)_3\text{COH, HO}^-]{\text{OsO}_4, \text{(CH}_3)_3\text{COOH}}$ 2,3-butanediol (chiral diastereomer)

A-6. Give the reagents necessary to carry out each of the following transformations:

(a) Conversion of benzyl alcohol ($C_6H_5CH_2OH$) to benzaldehyde ($C_6H_5CH{=}O$)

(b) Conversion of benzyl alcohol to benzoic acid ($C_6H_5CO_2H$)

(c) Conversion of $H_2C{=}CHCH_2CH_2CO_2H$ to $H_2C{=}CHCH_2CH_2CH_2OH$

(d) Conversion of cyclohexene to *cis*-1,2-cyclohexanediol

A-7. Provide structures for compounds A to C in the following reaction scheme:

$A(C_5H_{12}O_2)$ $\xrightarrow[\text{H}^+, \text{H}_2\text{O}]{\text{K}_2\text{Cr}_2\text{O}_7}$ $B(C_5H_8O_3)$ $\xrightarrow{\text{CH}_3\text{OH, H}^+}$ $C(C_6H_{10}O_3)$

$\Big\downarrow \text{H}^+, \text{ heat}$ $\Big\downarrow \begin{array}{l}\text{1. LiAlH}_4 \\ \text{2. H}_2\text{O}\end{array}$

[tetrahydrofuran ring with H$_3$C, O] $A + CH_3OH$

A-8. Using any necessary organic or inorganic reagents, outline a scheme for each of the following conversions.

(a) $(CH_3)_2C{=}CHCH_3$ $\xrightarrow{?}$ $(CH_3)_2CHCCH_3$ [with O double bond]

(b) [cyclopropyl]$-CH{=}O$ $\xrightarrow{?}$ [cyclopropyl]$-CCH_2CH_3$ [with O double bond]

(c) $C_6H_5CH_3$ $\xrightarrow{?}$ $C_6H_5CH_2CH_2CO_2CH_2CH_3$

PART B

B-1. Ethanethiol (CH_3CH_2SH) is a gas at room temperature, but ethanol is a liquid. The reason for this is

(*a*) The C—S—H bonds in ethanethiol are linear.

(*b*) The C—O—H bonds in ethanol are linear.

(*c*) Ethanol has a lower molecular weight.

(*d*) Ethanethiol has a higher boiling point.

(*e*) Ethanethiol is less polar.

B-2. Which of the following would yield a secondary alcohol after the indicated reaction, followed by hydrolysis if necessary?

(*a*) $LiAlH_4$ + a ketone   (*c*) 2-Butene + aqueous H_2SO_4

(*b*) CH_3CH_2MgBr + an aldehyde   (*d*) All of these

B-3. What is the major product of the following reaction?

(*a*)   (*b*)   (*c*)   (*d*)

B-4. Which of the esters shown, after reduction with $LiAlH_4$ and aqueous workup, will yield two molecules of only a single alcohol?

(*a*) $CH_3CH_2\overset{\overset{\text{O}}{\|}}{C}OCH_2CH_3$ (*c*) $C_6H_5\overset{\overset{\text{O}}{\|}}{C}OCH_2C_6H_5$

(*b*) $C_6H_5\overset{\overset{\text{O}}{\|}}{C}OC_6H_5$ (*d*) None of these

B-5. For the following reaction, select the statement that best describes the situation.

$$RCH_2OH \ + \ PCC \ [C_5H_5NH^+ \ ClCrO_3^-] \longrightarrow$$

(*a*) The alcohol is oxidized to an acid, and the Cr(VI) is reduced.

(*b*) The alcohol is oxidized to an aldehyde, and the Cr(VI) is reduced.

(c) The alcohol is reduced to an aldehyde, and the Cr(III) is oxidized.

(d) The alcohol is oxidized to a ketone, and the Cr(VI) is reduced.

B-6. What is the product from the following esterification?

$$C_6H_5CH_2\overset{O}{\overset{\|}{C}}OH \quad + \quad CH_3CH_2-{}^{18}OH \quad \xrightarrow[\text{heat}]{H^+} \quad ?$$

(a) $C_6H_5CH_2\overset{O}{\overset{\|}{C}}OCH_2CH_3$

(c) $C_6H_5CH_2\overset{{}^{18}O}{\overset{\|}{C}}-{}^{18}OCH_2CH_3$

(b) $C_6H_5CH_2\overset{O}{\overset{\|}{C}}-{}^{18}OCH_2CH_3$

(d) $CH_3CH_2\overset{{}^{18}O}{\overset{\|}{C}}OCH_2C_6H_5$

B-7. The following substance acts as a coenzyme in which of the following biological reactions?

(R = adenine dinucleotide)

(a) Alcohol oxidation

(c) Aldehyde reduction

(b) Ketone reduction

(d) None of these

B-8. Which of the following alcohols gives the best yield of dialkyl ether on being heated with a trace of sulfuric acid?

(a) 1-Pentanol

(c) Cyclopentanol

(b) 2-Pentanol

(d) 2-Methyl-2-butanol

B-9. What is the major organic product of the following sequence of reactions?

$$(CH_3)_2CHCH_2OH \xrightarrow{PBr_3} \quad \xrightarrow{Mg} \quad \xrightarrow{H_2C\overset{O}{\overset{\triangle}{-}}CH_2} \quad \xrightarrow{H_3O^+} \quad ?$$

(a) $(CH_3)_2CH\overset{OH}{\overset{|}{C}}HCH_2CH_3$

(c) $(CH_3)_2CHCH_2CH_2OH$

(b) $(CH_3)_2CHCH_2\overset{OH}{\overset{|}{C}}HCH_3$

(d) $(CH_3)_2CHCH_2CH_2CH_2OH$

B-10. What is the product of the following reaction?

(a) Only 1

(b) Only 2

(c) Only 3

(d) A 1:1 mixture of 2 and 3.

(e) A 1:1:1 mixture of 1, 2, and 3.

B-11. Which reaction is the best method for preparing (R)-2-butanol?

(a) $CH_3CH_2\overset{\displaystyle O}{\overset{\|}{C}}CH_3$ $\xrightarrow[\text{2. }H_2O]{\text{1. LiAlH}_4\text{, diethyl ether}}$

(b) $H_3C\overset{H}{\overset{}{\underset{CH_3CH_2}{C}}}-O\overset{\displaystyle O}{\overset{\|}{C}}CH_3$ $\xrightarrow[\text{2. }H_2O]{\text{1. LiAlH}_4\text{, diethyl ether}}$

(c) $CH_3CH_2\overset{\displaystyle O}{\overset{\|}{C}}H$ $\xrightarrow[\text{2. }H_3O^+]{\text{1. CH}_3\text{MgBr, diethyl ether}}$

(d) $CH_3\overset{\displaystyle O}{\overset{\|}{C}}H$ $\xrightarrow[\text{2. }H_3O^+]{\text{1. CH}_3CH_2\text{Li, diethyl ether}}$

(e) None of these would be effective.

B-12. An organic compound B is formed by the reaction of ethylmagnesium iodide (CH_3CH_2MgI) with a substance A, followed by treatment with dilute aqueous acid. Compound B does *not* react with PCC or PDC in dichloromethane. Which of the following is a possible candidate for A?

(a) $CH_3\overset{\displaystyle O}{\overset{\|}{C}}H$

(b) $H_2C=O$

(c) $H_2C\overset{\displaystyle O}{\overset{\diagup\diagdown}{}}CH_2$

(d) $CH_3CH_2\overset{\displaystyle O}{\overset{\|}{C}}CH_3$

(e) None of these

B-13. Which alcohol of molecular formula $C_5H_{12}O$ has the fewest signals in its ^{13}C NMR spectrum?

(a) 1-Pentanol

(b) 2-Pentanol

(c) 2-Methyl-2-butanol

(d) 3-Methyl-2-butanol

(e) 2,2-Dimethyl-1-propanol

B-14. Which of the following reagents would carry out the following transformation? (D = ^{2}H, the mass-2 isotope of hydrogen)

$$\text{C}_6\text{H}_5\text{-}\overset{\text{O}}{\overset{\|}{\text{C}}}\text{CH}_3 \xrightarrow{\;?\;} \text{C}_6\text{H}_5\text{-}\overset{\text{OH}}{\underset{\text{D}}{\overset{|}{\text{C}}}}\text{CH}_3$$

(*a*) NaBD$_4$ in CH$_3$OH   (*c*) LiAlH$_4$, then D$_2$O   (*e*) NaBH$_4$ in CH$_3$OD

(*b*) NaBD$_4$ in CH$_3$OD   (*d*) LiAlD$_4$, then D$_2$O

B-15. Which sequence of steps describes the best synthesis of 2-methyl-3-pentanone?

2-Methyl-3-pentanone

(*a*) 1. 1-Propanol + (CH$_3$)$_2$CHMgBr, diethyl ether

    2. H$_3$O$^+$

    3. PDC, CH$_2$Cl$_2$

(*b*) 1. 1-Propanol + Na$_2$Cr$_2$O$_7$, H$_2$SO$_4$, H$_2$O, heat

    2. SOCl$_2$

    3. (CH$_3$)$_2$CHCl, AlCl$_3$

(*c*) 1. 1-Propanol + PCC, CH$_2$Cl$_2$

    2. (CH$_3$)$_2$CHLi, diethyl ether

    3. H$_3$O$^+$

    4. Na$_2$Cr$_2$O$_7$, H$_2$SO$_4$, H$_2$O, heat

(*d*) 1. 2-Propanol + Na$_2$Cr$_2$O$_7$, H$_2$SO$_4$, H$_2$O, heat

    2. CH$_3$CH$_2$CH$_2$Li, diethyl ether

    3. H$_3$O$^+$

    4. PCC, CH$_2$Cl$_2$

CHAPTER 16

Ethers, Epoxides, and Sulfides

SOLUTIONS TO TEXT PROBLEMS

16.1 (*b*) Oxirane is the IUPAC name for ethylene oxide. A chloromethyl group ($ClCH_2$—) is attached to position 2 of the ring in 2-(chloromethyl)oxirane.

$$\underset{\text{Oxirane}}{H_2\overset{3}{C}-\overset{2}{C}H_2\ \underset{O}{\diagdown\!\diagup}}\qquad\qquad \underset{\text{2-(Chloromethyl)oxirane}}{H_2C-CHCH_2Cl\ \underset{O}{\diagdown\!\diagup}}$$

This compound is more commonly known as epichlorohydrin.

(*c*) Epoxides may be named by adding the prefix *epoxy* to the IUPAC name of a parent compound, specifying by number both atoms to which the oxygen is attached.

$$\underset{\text{1-Butene}}{CH_3CH_2CH{=}CH_2}\qquad\qquad \underset{\text{3,4-Epoxy-1-butene}}{H_2C-CHCH{=}CH_2\ \underset{O}{\diagdown\!\diagup}}$$

16.2 1,2-Epoxybutane and tetrahydrofuran both have the molecular formula C_4H_8O (that is, they are constitutional isomers) and so it is appropriate to compare their heats of combustion directly. Angle strain from the three-membered ring of 1,2-epoxybutane causes it to have more internal energy than tetrahydrofuran, and its combustion is more exothermic.

$$\underset{\substack{\text{1,2-Epoxybutane;}\\ \text{heat of combustion 2546 kJ/mol}}}{H_2C-CHCH_2CH_3\ \underset{O}{\diagdown\!\diagup}}\qquad\qquad \underset{\substack{\text{Tetrahydrofuran;}\\ \text{heat of combustion 2499 kJ/mol}}}{}$$

16.3 Ethers can form hydrogen bonds to water, alkanes cannot; therefore, the compound that is more soluble in water is the ether.

Cyclopentane
(less soluble in water)

Tetrahydrofuran
(more soluble in water)

Hydrogen bonding between
tetrahydrofuran and water

16.4 The compound is 1,4-dioxane; it has a six-membered ring and two oxygens separated by H_2C-CH_2 units.

1,4-dioxane
("6-crown-2")

16.5 Protonation of the carbon–carbon double bond leads to the more stable carbocation.

$$(CH_3)_2C{=}CH_2 \ + \ H^+ \ \longrightarrow \ (CH_3)_2\overset{+}{C}-CH_3$$

2-Methylpropene   *tert*-Butyl cation

Methanol acts as a nucleophile to capture *tert*-butyl cation.

Deprotonation of the alkyloxonium ion leads to formation of *tert*-butyl methyl ether.

$$(CH_3)_3C-\overset{+}{\underset{H}{O}}CH_3 \ + \ :\!\overset{CH_3}{\underset{H}{O}} \ \longrightarrow \ (CH_3)_3C\overset{..}{\underset{..}{O}}CH_3 \ + \ H_2\overset{+}{O}CH_3$$

tert-Butyl methyl ether

16.6 Both alkyl groups in benzyl ethyl ether are primary, thus either may come from the alkyl halide in a Williamson ether synthesis. The two routes to benzyl ethyl ether are

$$C_6H_5CH_2ONa \ + \ CH_3CH_2Br \ \longrightarrow \ C_6H_5CH_2OCH_2CH_3 \ + \ NaBr$$

Sodium benzyloxide   Bromoethane   Benzyl ethyl ether   Sodium bromide

$$C_6H_5CH_2Br \ + \ CH_3CH_2ONa \ \longrightarrow \ C_6H_5CH_2OCH_2CH_3 \ + \ NaBr$$

Benzyl bromide   Sodium ethoxide   Benzyl ethyl ether   Sodium bromide

16.7 (*b*) A primary carbon and a secondary carbon are attached to the ether oxygen. The secondary carbon can only be derived from the alkoxide, because secondary alkyl halides cannot be used in the preparation of ethers by the Williamson method. The only effective method uses an allyl halide and sodium isopropoxide.

$$(CH_3)_2CHONa \ + \ H_2C{=}CHCH_2Br \ \longrightarrow \ H_2C{=}CHCH_2OCH(CH_3)_2 \ + \ NaBr$$

Sodium isopropoxide   Allyl bromide   Allyl isopropyl ether   Sodium bromide

Elimination will be the major reaction of an isopropyl halide with an alkoxide base.

(c) Here the ether is a mixed primary–tertiary one. The best combination is the one that uses the primary alkyl halide.

$$(CH_3)_3COK \quad + \quad C_6H_5CH_2Br \quad \longrightarrow \quad (CH_3)_3COCH_2C_6H_5 \quad + \quad KBr$$

| Potassium | Benzyl bromide | | Benzyl *tert*-butyl ether | Potassium |
| *tert*-butoxide | | | | bromide |

The reaction between $(CH_3)_3CBr$ and $C_6H_5CH_2O^-$ is elimination, not substitution.

16.8
$$CH_3CH_2OCH_2CH_3 \quad + \quad 6O_2 \quad \longrightarrow \quad 4CO_2 \quad + \quad 5H_2O$$

| Diethyl ether | Oxygen | | Carbon dioxide | Water |

16.9 (b) If benzyl bromide is the only organic product from reaction of a dialkyl ether with hydrogen bromide, then both alkyl groups attached to oxygen must be benzyl.

$$C_6H_5CH_2OCH_2C_6H_5 \quad \xrightarrow[\text{heat}]{\text{HBr}} \quad 2C_6H_5CH_2Br \quad + \quad H_2O$$

| Dibenzyl ether | | Benzyl bromide | Water |

(c) Because *1 mole of a dihalide*, rather than 2 moles of a monohalide, is produced per mole of ether, the ether must be cyclic.

| Tetrahydropyran | | 1,5-Dibromopentane | Water |

16.10 As outlined in Mechanism 16.1, the first step is protonation of the ether oxygen to give a dialkyloxonium ion.

| Tetrahydrofuran | Hydrogen iodide | Dialkyloxonium ion | Iodide ion |

In the second step, nucleophilic attack of the halide ion on carbon of the dialkyloxonium ion gives 4-iodo-1-butanol.

| Iodide ion | Dialkyloxonium ion | 4-Iodo-1-butanol |

The remaining two steps of the mechanism correspond to those in which an alcohol is converted to an alkyl halide, as discussed in Chapter 4.

| 4-Iodo-1-butanol | Hydrogen iodide | | 1,4-Diiodobutane | Water |

16.11 Inasmuch as Sharpless oxidation using diethyl (2R,3R)-tartrate yields (2S,3S)-2,3-epoxy-1-hexanol from *trans*-2-hexen-1-ol, the corresponding (2S,3S) tartrate will yield the mirror image epoxide. The product is (2R,3R)-2,3-epoxy-1-hexanol.

16.12 The cis epoxide is achiral. It is a meso form containing a plane of symmetry. The trans isomer is chiral; its two mirror-image representations are not superimposable.

cis-2,3-Epoxybutane
(Plane of symmetry passes
through oxygen and midpoint
of carbon–carbon bond.)

Nonsuperimposable mirror-image
(enantiomeric) forms of
trans-2,3-epoxybutane

Neither the cis nor the trans epoxide is optically active when formed from the alkene. The cis epoxide is achiral; it cannot be optically active. The trans epoxide is capable of optical activity but is formed as an optically inactive racemic mixture because achiral starting materials are used.

16.13 (b) Azide ion $[:\!N\!\!=\!\!N\!\!=\!\!\overset{..}{N}\!:]^-$ is a good nucleophile, reacting readily with ethylene oxide to yield 2-azidoethanol.

Ethylene oxide   2-Azidoethanol

(c) Ethylene oxide is hydrolyzed to ethylene glycol in the presence of aqueous base.

Ethylene oxide   Ethylene glycol

(d) Phenyllithium reacts with ethylene oxide in a manner similar to that of a Grignard reagent.

Ethylene oxide   2-Phenylethanol

(*e*) The nucleophilic species here is the acetylenic anion $CH_3CH_2C{\equiv}C{:}^-$, which attacks a carbon atom of ethylene oxide to give 3-hexyn-1-ol.

$$H_2C{-}CH_2 \xrightarrow[NH_3]{NaC{\equiv}CCH_2CH_3} CH_3CH_2C{\equiv}CCH_2CH_2OH$$

Ethylene oxide

3-Hexyn-1-ol (48%)

16.14 Nucleophilic attack at C-2 of the starting epoxide will be faster than attack at C-1, because C-1 is more sterically hindered. Compound A, corresponding to attack at C-1, is not as likely as compound B. Compound B not only arises by methoxide ion attack at C-2 but also satisfies the stereochemical requirement that epoxide ring opening take place with inversion of configuration at the site of substitution. Compound B is correct. Compound C, although it is formed by methoxide substitution at the less crowded carbon of the epoxide, is wrong stereochemically. It requires substitution with retention of configuration, which is not the normal mode of epoxide ring opening.

16.15 Acid-catalyzed nucleophilic ring opening proceeds by attack of methanol at the more substituted carbon of the protonated epoxide. Inversion of configuration is observed at the site of attack. The correct product is compound A.

Protonated form of 1-methyl-1,2-epoxycylopentane

Compound A

The nucleophilic ring openings in both this problem and Problem 16.14 occur by inversion of configuration. Attack under basic conditions by methoxide ion, however, occurs at the *less* hindered carbon of the epoxide ring, whereas attack by methanol under acid-catalyzed conditions occurs at the *more* substituted carbon.

16.16 Begin by drawing *meso*-2,3-butanediol, recalling that a meso form is achiral. The eclipsed conformation has a plane of symmetry.

meso-2,3-Butanediol

Epoxidation followed by acid-catalyzed hydrolysis results in anti addition of hydroxyl groups to the double bond. *trans*-2-Butene is the required starting material.

trans-2-Butene   *trans*-2,3-Epoxybutane   *meso*-2,3-Butanediol

Osmium tetraoxide dihydroxylation is a method of achieving syn dihydroxylation. The necessary starting material is *cis*-2-butene.

cis-2-Butene   *meso*-2,3-Butanediol

16.17 Reaction of (*R*)-2-octanol with *p*-toluenesulfonyl chloride yields a *p*-toluenesulfonate (tosylate) having the same configuration; the chirality center is not involved in this step. Reaction of the tosylate with a nucleophile proceeds by inversion of configuration in an S_N2 process. The product has the *S* configuration.

(*R*)-2-Octanol   *p*-Toluenesulfonyl   (*R*)-1-Methylheptyl tosylate
                            chloride

(*R*)-1-Methylheptyl tosylate   Sodium   (*S*)-1-Methylheptyl phenyl sulfide
                          benzenethiolate

16.18 None of the carbons in omeprazole is a chirality center. The only chirality center is sulfur. In order of decreasing precedence, the groups attached to sulfur are: O > C=N > CH$_2$. Consider sulfur's unshared electron pair as its fourth and lowest-ranked "group." Use the structural formula given in the text and assume the unshared pair of sulfur points away from you. Therefore, oxygen points toward you as shown in the structure at the left. The order of decreasing precedence traces a clockwise path making this the (*R*)-enantiomer. Convert it to the (*S*)-enantiomer (Nexium) by reversing the orientation of oxygen.

(*R*)-Omeprazole   (*S*)-Omeprazole

16.19 As shown in the text, dodecyldimethylsulfonium iodide may be prepared by reaction of dodecyl methyl sulfide with methyl iodide. An alternative method is the reaction of dodecyl iodide with dimethyl sulfide.

$$(CH_3)_2S \ + \ CH_3(CH_2)_{10}CH_2I \longrightarrow CH_3(CH_2)_{10}CH_2\overset{+}{S}(CH_3)_2 \ I^-$$

Dimethyl   Dodecyl iodide   Dodecyldimethylsulfonium
sulfide   iodide

The reaction of a sulfide with an alkyl halide is an S_N2 process. The faster reaction will be the one that uses the less sterically hindered alkyl halide. The method presented in the text will proceed faster.

16.20 The molecular ion from *sec*-butyl ethyl ether can also fragment by cleavage of a carbon–carbon bond in its ethyl group to give an oxygen-stabilized cation of *m/z* 87.

$$H_3C-CH_2-\overset{+}{\underset{\cdot\cdot}{O}}-\underset{\underset{CH_3}{|}}{CHCH_2CH_3} \longrightarrow \ \cdot CH_3 \ + \ H_2C=\overset{+}{\underset{\cdot\cdot}{O}}-\underset{\underset{CH_3}{|}}{CHCH_2CH_3}$$

m/z 87

16.21 All the constitutionally isomeric ethers of molecular formula $C_5H_{12}O$ belong to one of two general groups: $CH_3OC_4H_9$ and $CH_3CH_2OC_3H_7$. Thus, we have

$$CH_3OCH_2CH_2CH_2CH_3 \qquad\qquad CH_3O\underset{\underset{CH_3}{|}}{CH}CH_2CH_3$$

Butyl methyl ether   *sec*-Butyl methyl ether

$$CH_3OCH_2CH(CH_3)_2 \qquad\qquad CH_3OC(CH_3)_3$$

Isobutyl methyl ether   *tert*-Butyl methyl ether

$$CH_3CH_2OCH_2CH_2CH_3 \quad \text{and} \quad CH_3CH_2OCH(CH_3)_2$$

Ethyl propyl ether   Ethyl isopropyl ether

These ethers could also have been named as "alkoxyalkanes." Thus, *sec*-butyl methyl ether would become 2-methoxybutane.

16.22 Isoflurane and enflurane are both halogenated derivatives of ethyl methyl ether.

Isoflurane: $F-\underset{\underset{F}{\overset{\overset{F}{|}}{C}}}{}\overset{}{\underset{\underset{Cl}{|}}{-CH}}-O-CHF_2$ **Enflurane:** $Cl-\underset{\underset{F}{\overset{\overset{H}{|}}{C}}}{}\overset{}{\underset{\underset{F}{\overset{\overset{F}{|}}{C}}}{-C}}-O-CHF_2$

1-Chloro-2,2,2-trifluoroethyl   2-Chloro-1,1,2-trifluoroethyl
difluoromethyl ether   difluoromethyl ether

16.23 (*a*) The parent compound is cyclopropane. It has a three-membered epoxide function, and thus a reasonable name is epoxycyclopropane. Numbers locating positions of attachment (as in "1,2-epoxycyclopropane") are not necessary, because no other structures (1,3 or 2,3) are possible here.

Epoxycyclopropane

(*b*) The longest continuous carbon chain has seven carbons, and so the compound is named as a derivative of heptane. The epoxy function bridges C-2 and C-4. Therefore

is 2,4-epoxy-2-methylheptane.

(*c*) The oxygen atom bridges the C-1 and C-4 atoms of a cyclohexane ring.

1,4-Epoxycyclohexane

(*d*) Eight carbon atoms are continuously linked and bridged by an oxygen. We name the compound as an epoxy derivative of cyclooctane.

1,5-Epoxycyclooctane

16.24 The ethers that are to be prepared are

$$CH_3OCH_2CH_2CH_3 \qquad CH_3OCH(CH_3)_2 \quad \text{and} \quad CH_3CH_2OCH_2CH_3$$

Methyl propyl ether   Isopropyl methyl ether   Diethyl ether

First examine the preparation of each ether by the Williamson method. Methyl propyl ether can be prepared in two ways:

$$CH_3ONa \; + \; CH_3CH_2CH_2Br \; \longrightarrow \; CH_3OCH_2CH_2CH_3$$

Sodium   1-Bromopropane   Methyl propyl ether
methoxide

$$CH_3Br \; + \; CH_3CH_2CH_2ONa \; \longrightarrow \; CH_3OCH_2CH_2CH_3$$

Methyl   Sodium propoxide   Methyl propyl ether
bromide

Either combination is satisfactory. The necessary reagents are prepared as shown.

$$CH_3OH \xrightarrow{\text{Na}} CH_3ONa$$

Methanol — Sodium methoxide

$$CH_3CH_2CH_2OH \xrightarrow[\text{(or HBr)}]{\text{PBr}_3} CH_3CH_2CH_2Br$$

1-Propanol — 1-Bromopropane

$$CH_3OH \xrightarrow[\text{(or HBr)}]{\text{PBr}_3} CH_3Br$$

Methanol — Methyl bromide

$$CH_3CH_2CH_2OH \xrightarrow{\text{Na}} CH_3CH_2CH_2ONa$$

1-Propanol — Sodium propoxide

Isopropyl methyl ether is best prepared by the reaction

$$CH_3Br + (CH_3)_2CHONa \longrightarrow CH_3OCH(CH_3)_2$$

Methyl bromide   Sodium isopropoxide   Isopropyl methyl ether

The reaction of sodium methoxide with isopropyl bromide will proceed mainly by elimination. Methyl bromide is prepared as shown previously; sodium isopropoxide can be prepared by adding sodium to isopropyl alcohol.

Diethyl ether may be prepared as outlined:

$$CH_3CH_2OH \xrightarrow{\text{Na}} CH_3CH_2ONa$$

Ethanol — Sodium ethoxide

$$CH_3CH_2OH \xrightarrow[\text{(or HBr)}]{\text{PBr}_3} CH_3CH_2Br$$

Ethanol — Ethyl bromide

$$CH_3CH_2ONa + CH_3CH_2Br \longrightarrow CH_3CH_2OCH_2CH_3 + NaBr$$

Sodium ethoxide   Ethyl bromide   Diethyl ether   Sodium bromide

16.25 (*a*) Secondary alkyl halides react with alkoxide bases by E2 elimination as the major pathway. The Williamson ether synthesis is not a useful reaction with secondary alkyl halides.

Sodium 2-butanolate   Bromocyclohexane   2-Butanol   Cyclohexene   Sodium bromide

(*b*) Sodium alkoxide acts as a nucleophile toward iodoethane to yield an alkyl ethyl ether.

(*R*)-2-Ethoxybutane

The ether product has the same absolute configuration as the starting alkoxide because no bonds to the chirality center are made or broken in the reaction.

(c) Vicinal halohydrins are converted to epoxides on being treated with base.

1-Bromo-2-butanol

1,2-Epoxybutane

(d) The reactants, an alkene plus a peroxy acid, are customary ones for epoxide preparation. The reaction is a stereospecific syn addition of oxygen to the double bond.

(Z)-1-Phenylpropene   Peroxybenzoic acid   cis-2-Methyl-3-phenyloxirane   Benzoic acid

(e) Azide ion is a good nucleophile and attacks the epoxide function. Substitution occurs at carbon with inversion of configuration. The product is *trans*-2-azidocyclohexanol.

1,2-Epoxycyclohexane   *trans*-2-Azidocyclohexanol (61%)

(f) Ammonia is a nucleophile capable of reacting with epoxides. It attacks the less hindered carbon of the epoxide function.

2-(o-Bromophenyl)-2-methyloxirane   1-Amino-2-(o-bromophenyl)-2-propanol

Aryl halides do not react with nucleophiles under these conditions, and so the bromine substituent on the ring is unaffected.

(g) Methoxide ion attacks the less substituted carbon of the epoxide ring with inversion of configuration.

1-Benzyl-1,2-epoxycyclohexane   1-Benzyl-*trans*-2-methoxycyclohexanol (98%)

(*h*) Under acidic conditions, substitution is favored at the carbon that can better support a positive charge. Aryl substituents stabilize carbocations, making the benzylic position the one that is attacked in an aryl substituted epoxide.

2-Phenyloxirane   2-Chloro-2-phenylethanol (71%)

(*i*) Lithium aluminum hydride reduces epoxides to alcohols, and hydride is transferred to the less substituted carbon of the epoxide ring. The alkene double bond is not reduced by $LiAlH_4$.

3,4-Epoxy-3-methyl-1-butene   2-Methyl-3-buten-2-ol

(*j*) Tosylates undergo substitution with nucleophiles such as sodium butanethiolate.

$$CH_3(CH_2)_{16}CH_2OTs \ + \ CH_3CH_2CH_2CH_2SNa \ \longrightarrow \ CH_3CH_2CH_2CH_2SCH_2(CH_2)_{16}CH_3$$

Octadecyl tosylate   Sodium butanethiolate   Butyl octadecyl sulfide

(*k*) Nucleophilic substitution proceeds with inversion of configuration.

16.26 Protonation of oxygen to form an alkyloxonium ion is followed by loss of water. The resulting carbocation has a plane of symmetry and is achiral. Capture of the carbocation by methanol yields both enantiomers of 2-methoxy-2-phenylbutane. The product is racemic.

(*R*)-(+)-2-Phenyl-2-butanol   (Achiral carbocation)

2-Methoxy-2-phenylbutane (racemic)

16.27 The best approach to this problem is to first write the equations in full stereochemical detail.

(a)

(R)-1,2-Epoxypropane   (R)-1,2-Propanediol

It now becomes clear that the arrangement of groups around the chirality center remains unchanged in going from starting materials to products. Therefore, choose conditions such that the nucleophile attacks the CH_2 group of the epoxide rather than the chirality center. Base-catalyzed hydrolysis is required; aqueous sodium hydroxide is appropriate.

The nucleophile (hydroxide ion) attacks the less hindered carbon of the epoxide ring.

(b)

(S)-1,2-Propanediol

Inversion of configuration at the chirality center is required. The nucleophile must therefore attack the chirality center, and acid-catalyzed hydrolysis should be chosen. Dilute sulfuric acid would be satisfactory.

The nucleophile (a water molecule) attacks that carbon atom of the ring that can better support a positive charge. Carbocation character develops at the transition state and is better supported by the carbon atom that is more highly substituted.

16.28 The key intermediate in the preparation of bis(2-chloroethyl) ether from ethylene is 2-chloroethanol, formed from ethylene by reaction with chlorine in water. Heating 2-chloroethanol in acid gives the desired ether.

Ethylene   2-Chloroethanol   Bis(2-chloroethyl) ether

16.29 (*a*) There is a temptation to try to do this transformation in a single step by using a reducing agent to convert the carbonyl to a methylene group. No reagent is available that reduces esters in this way! The Clemmensen and Wolff–Kishner reduction methods are suitable only for aldehydes and ketones. The best way to approach this problem is by reasoning backward. The desired product is an ether. Ethers can be prepared by the Williamson ether synthesis involving an alkyl halide and an alkoxide ion.

$$\text{C}_6\text{H}_5\text{—CH}_2\text{OCH}_3 \implies \text{C}_6\text{H}_5\text{—CH}_2\text{X} + \text{CH}_3\ddot{\text{O}}\text{:}^-$$

or

$$\text{C}_6\text{H}_5\text{—CH}_2\text{OCH}_3 \implies \text{C}_6\text{H}_5\text{—CH}_2\ddot{\text{O}}\text{:}^- + \text{CH}_3\text{X}$$

Both the alkyl halide and the alkoxide ion are prepared from alcohols. The problem then becomes one of preparing the appropriate alcohol (or alcohols) from the starting ester. This is readily done using lithium aluminum hydride.

$$\text{C}_6\text{H}_5\text{—C(=O)OCH}_3 \xrightarrow[\text{2. H}_2\text{O}]{\text{1. LiAlH}_4} \text{C}_6\text{H}_5\text{—CH}_2\text{OH} + \text{CH}_3\text{OH}$$

Methyl benzoate   Benzyl alcohol   Methanol

Then

$$\text{CH}_3\text{OH} \xrightarrow{\text{Na}} \text{CH}_3\text{ONa}$$

Methanol   Sodium methoxide

$$\text{C}_6\text{H}_5\text{—CH}_2\text{OH} \xrightarrow[\text{or PBr}_3]{\text{HBr}} \text{C}_6\text{H}_5\text{—CH}_2\text{Br}$$

Benzyl alcohol   Benzyl bromide

and

$$\text{C}_6\text{H}_5\text{—CH}_2\text{Br} + \text{NaOCH}_3 \longrightarrow \text{C}_6\text{H}_5\text{—CH}_2\text{OCH}_3 + \text{NaBr}$$

Benzyl bromide   Sodium methoxide   Benzyl methyl ether

The following sequence is also appropriate once methanol and benzyl alcohol are obtained by reduction of methyl benzoate:

$$\text{C}_6\text{H}_5\text{—CH}_2\text{OH} \xrightarrow{\text{Na}} \text{C}_6\text{H}_5\text{—CH}_2\text{ONa}$$

Benzyl alcohol   Sodium benzyloxide

$$\text{CH}_3\text{OH} \xrightarrow[\text{or HBr}]{\text{PBr}_3} \text{CH}_3\text{Br}$$

Methanol   Bromomethane

and

Sodium benzyloxide   Bromomethane   Benzyl methyl ether

(b) All the methods that we have so far discussed for the preparation of epoxides are based on alkenes as starting materials. This leads us to consider the partial retrosynthesis shown.

Target molecule   Key intermediate

The key intermediate, 1-phenylcyclohexene, is both a proper precursor to the desired epoxide and readily available from the given starting materials. A reasonable synthesis is

Preparation of the required tertiary alcohol, 1-phenylcyclohexanol, completes the synthesis.

Cyclohexanol   Cyclohexanone

Bromobenzene   1-Phenylcyclohexanol

(c) The necessary carbon skeleton can be assembled through the reaction of a Grignard reagent with 1,2-epoxypropane.

The reaction sequence is therefore

Phenylmagnesium   1,2-Epoxypropane   1-Phenyl-2-propanol
bromide
(from bromobenzene
and magnesium)

The epoxide required in the first step, 1,2-epoxypropane, is prepared as follows from isopropyl alcohol:

$$CH_3CHCH_3 \quad \xrightarrow[\text{heat}]{H_2SO_4} \quad CH_3CH{=}CH_2 \quad \xrightarrow{\overset{\overset{O}{\|}}{CH_3COOH}} \quad CH_3CH{-}CH_2$$
$$\underset{OH}{|}$$

2-Propanol   Propene   1,2-Epoxypropane
(isopropyl alcohol)

(*d*) Because the target molecule is an ether, it ultimately derives from two alcohols.

$$C_6H_5CH_2CH_2CH_2OCH_2CH_3 \quad \Longrightarrow \quad C_6H_5CH_2CH_2CH_2OH \;+\; CH_3CH_2OH$$

Our first task is to assemble 3-phenyl-1-propanol from the designated starting material benzyl alcohol. This requires formation of a primary alcohol with the original carbon chain extended by two carbons. The standard method for this transformation involves reaction of a Grignard reagent with ethylene oxide.

$$C_6H_5CH_2OH \quad \xrightarrow[\text{or HBr}]{PBr_3} \quad C_6H_5CH_2Br \quad \xrightarrow[\substack{\text{2. } H_2C{-}CH_2 \\ \quad\quad O \\ \text{3. } H_3O^+}]{\text{1. Mg, diethyl ether}} \quad C_6H_5CH_2CH_2CH_2OH$$

Benzyl alcohol   Benzyl bromide   3-Phenyl-1-propanol

After 3-phenyl-1-propanol has been prepared, its conversion to the corresponding ethyl ether can be accomplished in either of two ways:

$$C_6H_5CH_2CH_2CH_2OH \quad \xrightarrow[\text{or HBr}]{PBr_3} \quad C_6H_5CH_2CH_2CH_2Br \quad \xrightarrow[\text{ethanol}]{NaOCH_2CH_3} \quad C_6H_5CH_2CH_2CH_2OCH_2CH_3$$

3-Phenyl-1-propanol   1-Bromo-3-phenylpropane   Ethyl 3-phenylpropyl ether

or alternatively

$$C_6H_5CH_2CH_2CH_2OH \quad \xrightarrow[\text{2. } CH_3CH_2Br]{\text{1. Na}} \quad C_6H_5CH_2CH_2CH_2OCH_2CH_3$$

3-Phenyl-1-propanol   Ethyl 3-phenylpropyl ether

The reagents in each step are prepared from ethanol.

$$CH_3CH_2OH \quad \xrightarrow{Na} \quad CH_3CH_2ONa$$

Ethanol   Sodium ethoxide

$$CH_3CH_2OH \quad \xrightarrow[\text{or HBr}]{PBr_3} \quad CH_3CH_2Br$$

Ethanol   Ethyl bromide

(*e*) The target epoxide can be prepared in a single step from the corresponding alkene.

Bicyclo[2.2.2]oct-2-ene   2,3-Epoxybicyclo[2.2.2]octane

Disconnections show that this alkene is available through a Diels–Alder reaction.

The reaction of 1,3-cyclohexadiene with ethylene gives the desired substance.

1,3-Cyclohexadiene	Ethylene	Bicyclo[2.2.2]oct-2-ene

1,3-Cyclohexadiene is one of the given starting materials. Ethylene is prepared from ethanol.

$$CH_3CH_2OH \xrightarrow[\text{heat}]{H_2SO_4} H_2C=CH_2$$

Ethanol   Ethylene

(*f*) Retrosynthetic analysis reveals that the desired target molecule may be prepared by reaction of an epoxide with an ethanethiolate ion.

$$C_6H_5\underset{\underset{OH}{|}}{C}HCH_2SCH_2CH_3 \implies C_6H_5\overset{O}{\overset{/\backslash}{CH-CH_2}} + {}^-SCH_2CH_3$$

Styrene oxide may be prepared by reaction of styrene with peroxyacetic acid.

$$C_6H_5CH=CH_2 + CH_3\overset{O}{\overset{||}{C}}OOH \longrightarrow C_6H_5\overset{O}{\overset{/\backslash}{CH-CH_2}} + CH_3CO_2H$$

Styrene   Peroxyacetic acid   Styrene oxide   Acetic acid

The necessary thiolate anion is prepared from ethanol by way of the corresponding thiol.

$$CH_3CH_2OH \xrightarrow[\text{2. NaSH}]{\text{1. HBr}} CH_3CH_2SH \xrightarrow{\text{NaOH}} CH_3CH_2SNa$$

Ethanol   Ethanethiol   Sodium ethanethiolate

Reaction of styrene oxide with sodium ethanethiolate completes the synthesis.

$$C_6H_5\overset{O}{\overset{/\backslash}{CH-CH_2}} + CH_3CH_2SNa \xrightarrow{CH_3CH_2OH} C_6H_5\underset{\underset{}{}}{\overset{OH}{\overset{|}{C}}}HCH_2SCH_2CH_3$$

Styrene oxide   Sodium ethanethiolate

16.30 The simplest approach to this problem is to recognize that no bonds to the chirality center are made or broken beginning with (S)-glycidol. The bond to the chirality center projects outward from the paper in all the intermediates in the synthesis. The propranolol produced has the (S)-configuration.

(S)-Glycidol   (S)-Propranolol

16.31 Reduction of (S)-1,2-epoxybutane with lithium aluminum hydride results in hydride transfer to the less substituted carbon. The product is (R)-2-butanol.

(S)-1,2-Epoxybutane   (R)-2-Butanol

Notice that even though the chirality center is not involved in the reaction, the (S) configuration of the starting material has become (R) in the product. The reason is that the precedence order changes; the ethyl group is third highest in the epoxide but second highest in the alcohol.

Reaction of the second epoxide, (S)-1,2-epoxypropane, with methylmagnesium iodide also occurs at the less substituted carbon. The product is (S)-2-butanol.

(S)-1,2-Epoxypropane   (S)-2-Butanol

16.32 (*a*) A reasonable mechanism is one that parallels the usual one for acid-catalyzed ether formation from alcohols, modified to accommodate these particular starting materials and products. Begin with protonation of one of the oxygen atoms of ethylene glycol.

$$HOCH_2CH_2OH + H_2SO_4 \rightleftharpoons HOCH_2CH_2-\overset{+}{\underset{H}{O}}-H + HSO_4^-$$

Ethylene glycol

The protonated alcohol then reacts in the usual way with another molecule of alcohol to give an ether. (This ether is known as *diethylene glycol*.)

Diethylene glycol

Diethylene glycol then undergoes intramolecular ether formation to yield 1,4-dioxane.

$$HOCH_2CH_2OCH_2CH_2OH + H_2SO_4 \rightleftharpoons HOCH_2CH_2OCH_2CH_2\overset{+}{\underset{H}{O}}-H + HSO_4^-$$

(*b*) The substrate is a primary alkyl halide and reacts with aqueous sodium hydroxide by nucleophilic substitution.

$$ClCH_2CH_2OCH_2CH_2Cl + HO^- \xrightarrow{H_2O} HOCH_2CH_2OCH_2CH_2Cl + Cl^-$$

Bis(2-chloroethyl) ether

The product of this reaction now has an alcohol function and a primary chloride built into the same molecule. It contains the requisite functionality to undergo an intramolecular Williamson reaction.

$$HOCH_2CH_2OCH_2CH_2Cl + HO^- \rightleftharpoons {}^-OCH_2CH_2OCH_2CH_2Cl + H_2O$$

1,4-Dioxane

16.33 Organolithium reagents react readily with the carbonyl group of ketones. The intermediate alkoxide ion can then undergo an intramolecular displacement of chloride to give the observed epoxide product.

16.34 (*a*) The first step is a standard Grignard synthesis of a primary alcohol using formaldehyde. Compound A is 3-buten-1-ol.

$$H_2C=CHCH_2Br \xrightarrow[\substack{2.\ H_2C=O \\ 3.\ H_3O^+}]{1.\ Mg} H_2C=CHCH_2CH_2OH$$

Allyl bromide　　　　　　　　　　3-Buten-1-ol
(compound A)

Addition of bromine to the carbon–carbon double bond of 3-buten-1-ol takes place readily to yield the vicinal dibromide.

$$H_2C=CHCH_2CH_2OH \xrightarrow{Br_2} BrCH_2CHCH_2CH_2OH$$
$$\qquad\qquad\qquad\qquad\qquad\qquad |$$
$$\qquad\qquad\qquad\qquad\qquad\qquad Br$$

3-Buten-1-ol

3,4-Dibromo-1-butanol
(compound B)

When compound B is treated with potassium hydroxide, it loses the elements of HBr to give compound C. Because further treatment of compound C with potassium hydroxide converts it to D by a second dehydrobromination, a reasonable candidate for C is 3-bromotetrahydrofuran.

3,4-Dibromo-1-butanol
(compound B)

3-Bromotetrahydrofuran
(compound C)

Compound D

Ring closure occurs by an intramolecular Williamson reaction.

Compound B

Compound C

Dehydrohalogenation of compound C converts it to the final product, D.

The alternative series of events, in which double-bond formation precedes ring closure, is unlikely, because it requires nucleophilic attack by the alkoxide on a vinyl bromide.

(Cyclization of this intermediate does not occur.)

(*b*) Lithium aluminum hydride reduces the carboxylic acid to the corresponding primary alcohol, compound E. Treatment of the vicinal chlorohydrin with base results in formation of an epoxide, compound F.

(*S*)-2-Chloro-1-propanol
(compound E)

(*R*)-1,2-Epoxypropane
(compound F)

As actually carried out, the first step proceeded in 56–58% yield, the second step in 65–70% yield.

(c) Treatment of the vicinal chlorohydrin with base results in ring closure to form an epoxide (compound G). Recall that attack occurs on the side opposite that of the carbon–chlorine bond. Compound G undergoes ring opening on reaction with sodium methanethiolate to give compound H.

(2R,3S)-3-Chloro-2-butanol

trans-2,3-Epoxybutane
(compound G)

Compound G

Compound H

(d) Because it gives an epoxide on treatment with a peroxy acid, compound I must be an alkene; more specifically, it is 1,2-dimethylcyclopentene.

1,2-Dimethylcyclopentene
(compound I)

1,2-Epoxy-1,2-
dimethylcyclopentane
(compound K)

Compounds J and L have the same molecular formula, $C_7H_{14}O_2$, but J is a liquid and L is a crystalline solid. Their molecular formulas correspond to the addition of two OH groups to compound I. Osmium tetraoxide brings about syn dihydroxylation of an alkene; therefore compound J must be the cis diol.

1,2-Dimethylcyclopentene
(compound I)

cis-1,2-Dimethylcyclopentane-
1,2-diol (compound J)

Acid-catalyzed hydrolysis of an epoxide yields a trans diol (compound L):

trans-1,2-Dimethylcyclopentane-
1,2-diol (compound L)

16.35 Cineole contains no double or triple bonds and therefore must be bicyclic, on the basis of its molecular formula ($C_{10}H_{18}O$, index of hydrogen deficiency = 2). When cineole reacts with hydrogen chloride, one of the rings is broken and water is formed.

$$\text{Cineole} + 2\text{HCl} \longrightarrow \text{(product)} + \text{H}_2\text{O}$$

$(C_{10}H_{18}O)$ $(C_{10}H_{18}Cl_2)$

The reaction that takes place is hydrogen halide-promoted ether cleavage. In such a reaction with excess hydrogen halide, the C—O—C unit is cleaved and two carbon–halogen bonds are formed. This suggests that cineole is a cyclic ether because the product contains both newly formed carbon–halogen bonds. A reasonable structure consistent with these facts is

Cineole

16.36 Recall that *p*-toluenesulfonate (tosylate) is a good leaving group in nucleophilic substitution reactions. The nucleophile that displaces tosylate from carbon is the alkoxide ion derived from the hydroxyl group within the molecule. The product is a cyclic ether, and the nature of the union of the two rings is that they are spirocyclic.

$C_{15}H_{20}O$

16.37 Lithium aluminum deuteride ($LiAlD_4$) will transfer a deuterium from the side of the ring opposite the oxygen of the epoxide.

The *tert*-butyl group has the greatest preference for an equatorial orientation, so the deuterium will be axial in the most stable conformation of the product.

16.38 (*a*) There are three methyl-substituted thianes, two of which are chiral.

2-Methylthiane
(chiral)

3-Methylthiane
(chiral)

4-Methylthiane
(achiral)

(*b*) The locants in the name indicate the positions of the sulfur atoms in 1,4-dithiane and 1,3,5-trithiane.

1,4-Dithiane   1,3,5-Trithiane

(*c*) Disulfides possess two adjacent sulfur atoms. 1,2-Dithiane is a disulfide.

1,2-Dithiane

(*d*) Two chair conformations of the sulfoxide derived from thiane are possible; the oxygen atom may be either equatorial or axial.

16.39 Oxidation of 4-*tert*-butylthiane yields two sulfoxides that are diastereomers of each other.

4-*tert*-Butylthiane

Oxidation of both stereoisomeric sulfoxides yields the same sulfone.

16.40 (*a*) Recall from Chapter 6 (text Section 6.19) that epoxidation is a syn addition, and that substituents that are cis to each other in the alkene remain cis in the epoxide.

| (*E*)-1-(*p*-Methoxyphenyl)-propene | *m*-Chloroperoxybenzoic acid | *trans*-1,2-Epoxy-1-(*p*-methoxyphenyl)propane | *m*-Chlorobenzoic acid |

(*b*) The splitting patterns allow assignment of the protons on the epoxide ring; the proton on C-1 is a doublet, whereas the proton on C-2 is split both by the proton on C-1 and the protons of the methyl group. The higher field ring protons are those ortho to the electron-donating methoxy group.

(*c*) From Table 13.3 in the text you can see that the chemical shift range δ 55–60 in a ^{13}C NMR spectrum is typical for carbon atoms adjacent to an oxygen. The three signals in this region of the ^{13}C NMR spectrum of the epoxide arise from the two carbon atoms of the epoxide ring and the carbon of the methoxy group.

(*d*) The molecular formula of the product formed under acidic conditions $(C_{16}H_{17}O_4Cl)$ corresponds to addition of *m*-chlorobenzoic acid to the epoxide. Recall from text Section 16.13 that epoxides undergo acid-catalyzed ring opening. Because carbocation character develops at the transition state, substitution is favored at the carbon that can better support a developing positive charge. In this case, substitution occurs at the benzylic carbon. For clarity, the *p*-methoxyphenyl group has been abbreviated as Ar in the following mechanism.

16.41 (*a*) Because all the peaks in the ^{1}H NMR spectrum of this ether are singlets, none of the protons can be vicinal to any other nonequivalent proton. The only $C_5H_{12}O$ ether that satisfies this requirement is *tert*-butyl methyl ether.

$$H_3C-O-C(CH_3)_3$$

Singlet at δ 3.2   Singlet at δ 1.2

(*b*) A doublet–septet pattern is characteristic of an isopropyl group. Two isomeric $C_5H_{12}O$ ethers contain an isopropyl group: ethyl isopropyl ether and isobutyl methyl ether.

$$(CH_3)_2CHOCH_2CH_3 \qquad\qquad (CH_3)_2CHCH_2OCH_3$$

Ethyl isopropyl ether   Isobutyl methyl ether

The signal of the methine proton in isobutyl methyl ether will be split into more than a septet, however, because in addition to being split by two methyl groups, it is coupled to the two protons in the methylene group. Thus, isobutyl methyl ether does not have the correct splitting pattern to be the answer. The correct answer is ethyl isopropyl ether.

(*c*) The low-field signals are due to the protons on the carbon atoms of the C—O—C linkage. Because one gives a doublet, it must be vicinal to only one other proton. We can therefore specify the partial structure:

```
          C
          |
C-O-C-C-C
      |  |
      H  H
```
Low-field doublet

This partial structure contains all the carbon atoms in the molecule. Fill in the remaining valences with hydrogen atoms to reveal isobutyl methyl ether as the correct choice.

$$H_3C-O-CH_2-CH(CH_3)_2$$

Low-field singlet   Low-field doublet

(*d*) Here again, signals at low field arise from protons on the carbons of the C—O—C unit. One of these signals is a quartet and so corresponds to a proton on a carbon bearing a methyl group.

```
H_3C-C-O-C
      |
      H
```
Quartet

The other carbon of the C—O—C unit has a hydrogen whose signal is split into a triplet. This hydrogen must therefore be attached to a carbon that bears a methylene group.

```
H_3C-C-O-C-CH_2-
      |    |
      H    H
```
Quartet   Triplet

These data permit us to complete the structure by adding an additional carbon and the requisite number of hydrogens in such a way that the signals of the protons attached to the carbons of the ether linkage are not split further. The correct structure is ethyl propyl ether.

$$CH_3CH_2OCH_2CH_2CH_3$$

Quartet   Triplet

16.42 A good way to address this problem is to consider the dibromide derived by treatment of compound A with hydrogen bromide. The presence of an NMR signal equivalent to four protons in the aromatic region at δ 7.3 indicates that this dibromide contains a disubstituted aromatic ring. The four remaining protons appear as a sharp singlet at δ 4.7 and are most reasonably contained in two equivalent methylene groups of the type $ArCH_2Br$. Because the dibromide contains all the carbons and hydrogens of the starting material and is derived from it by treatment with hydrogen bromide, it is likely that compound A is a cyclic ether in which a CH_2OCH_2 unit spans two of the carbons of a benzene ring. This can occur only when the positions involved are ortho to each other. Therefore

Compound A

16.43 The molecular formula ($C_{10}H_{13}BrO$) indicates an index of hydrogen deficiency of 4. One of the products obtained on treatment of the compound with HBr is benzyl bromide ($C_6H_5CH_2Br$), which accounts for seven of its ten carbons and all the double bonds and rings. Thus, the compound is a benzyl ether having the formula $C_6H_5CH_2OC_3H_6Br$. The ^1H NMR spectrum includes a five-proton signal at δ 7.4 for a monosubstituted benzene ring and a two-proton singlet at δ 4.6 for the benzylic protons. This singlet appears at low field because the benzylic protons are bonded to oxygen.

$$C_6H_5CH_2OC_3H_6Br \xrightarrow[\text{heat}]{\text{HBr}} C_6H_5CH_2Br + C_3H_6Br_2$$

The six remaining protons appear as two overlapping 2-proton triplets at δ 3.6 and 3.7, along with a 2-proton pentet at δ 2.2, consistent with the unit —$OCH_2CH_2CH_2Br$. The compound is $C_6H_5CH_2OCH_2CH_2CH_2Br$.

16.44 The high index of hydrogen deficiency (5) of the unknown compound $C_9H_{10}O$ and the presence of six signals in the δ 120–140 region of the ^{13}C NMR spectrum suggests the presence of an aromatic ring. The problem states that the compound is a cyclic ether, thus the oxygen atom is contained in a second ring fused to the benzene ring. As oxidation yields 1,2-benzenedicarboxylic acid, the second ring must be attached to the benzene ring by carbon atoms.

($C_9H_{10}O$)   1,2-Benzenedicarboxylic acid

Two structures are possible with this information; however, only one of them is consistent with the presence of three CH_2 groups in the ^{13}C NMR spectrum. The compound is

ANSWERS TO INTERPRETIVE PROBLEMS

16.45 C; **16.46** B; **16.47** B; **16.48** A; **16.49** C; **16.50** D

SELF-TEST

PART A

A-1. Write the structures and give a correct name for all the isomeric ethers of molecular formula $C_4H_{10}O$.

A-2. Give the structure of the product obtained from each of the following reactions. Show stereochemistry where it is important.

A-3. Outline a scheme for the preparation of cyclohexyl ethyl ether using the Williamson method.

A-4. Outline a synthesis of 2-ethoxyethanol, $CH_3CH_2OCH_2CH_2OH$, using ethanol as the source of all the carbon atoms.

A-5. Provide the reagents necessary to complete each of the following conversions. In each case give the structure of the intermediate product.

A-6. Provide structures for compounds A and B in the following reaction scheme:

$$\text{C}_6\text{H}_5\text{—}\overset{\overset{\displaystyle O}{\|}}{\text{C}}\text{OCH}_3 \quad \xrightarrow[\text{2. H}_2\text{O}]{\text{1. LiAlH}_4} \quad \text{A (C}_7\text{H}_8\text{O)} \;+\; \text{CH}_3\text{OH}$$

$$\text{A} \quad \xrightarrow[\text{2. CH}_3\text{CH}_2\text{I}]{\text{1. Na}} \quad \text{B (C}_9\text{H}_{12}\text{O)}$$

A-7. Using any necessary organic or inorganic reagents, provide the steps to carry out the following synthetic conversion:

A-8. Give the final product, including stereochemistry, of the following reaction sequence:

$$\xrightarrow[\text{H}_2\text{O}]{\text{Br}_2} \quad \xrightarrow[\text{H}_2\text{O}]{\text{NaOH}} \quad \xrightarrow[\text{H}_2\text{O}]{\text{NH}_3} \quad ?$$

PART B

B-1. An acceptable IUPAC name of the compound shown is

$$\text{CH}_3\text{CH}_2\text{CHCH}_2\overset{\overset{\displaystyle \text{OCH}_2\text{CH}_3}{|}}{\text{CH}}\text{CH}_2\text{—C}_6\text{H}_5$$
$$\underset{\text{CH}_3}{|}$$

(*a*) 1-Benzyl-3-methylpentyl ethyl ether

(*b*) Ethyl 3-methyl-1-methylphenyl-2-hexyl ether

(*c*) Ethyl 4-methyl-1-phenyl-2-hexyl ether

(*d*) 5-Ethoxy-3-methyl-6-phenylhexane

B-2. The most effective pair of reagents for the preparation of *tert*-butyl ethyl ether is

(*a*) Potassium *tert*-butoxide and ethyl bromide

(*b*) Potassium *tert*-butoxide and ethanol

(*c*) Sodium ethoxide and *tert*-butyl bromide

(*d*) *tert*-Butyl alcohol and ethyl bromide

B-3. The best choice of reactant for the following conversion is

(*a*)   (*b*)   (*c*)   (*d*)

B-4. For which of the following ethers would the ^1H NMR spectrum consist of only two singlets?

(a) CH_3OCCH_3 with CH_3 above and CH_3 below the central carbon

(b) $CH_3OCH_2CH_2OCH_3$

(c) epoxide with CH_3 and CH_3 substituents

(d) All these

B-5. Heating a particular ether with HBr yielded a single organic product. Which of the following conclusions may be reached?

(a) The reactant was a methyl ether.

(b) The reactant was a symmetric ether.

(c) The reactant was a cyclic ether.

(d) Both (b) and (c) may be correct.

B-6. Treating anisole ($C_6H_5OCH_3$) with the following reagents will give, as the major product,

1. $(CH_3)_3CCl$, $AlCl_3$;   2. Cl_2, $FeCl_3$;   3. HBr, heat

(a) phenol with Br ortho and $C(CH_3)_3$ para

(c) anisole (OCH₃) with Br ortho and $C(CH_3)_3$ para

(e) phenol with Cl ortho and $C(CH_3)_3$ para

(b) benzene with Br and Br adjacent and $C(CH_3)_3$ para

(d) benzene with Br and Cl adjacent and $C(CH_3)_3$ para

B-7. What is the product of the following reaction?

$(CH_3)_3C$—epoxide + $NaSCH_3$ $\xrightarrow{\text{ethanol}}$?

(a) $CH_3SCH_2CHC(CH_3)_3$ with OH substituent

(b) $(CH_3)_3CCHCH_2OH$ with SCH_3 substituent

(c) $CH_3SCH_2CHC(CH_3)_3$ with OCH_2CH_3 substituent

(d) $(CH_3)_3CCH_2CHSCH_3$ with OH substituent

B-8. Identify product Z in the following reaction sequence:

$HOCH_2-C\equiv C-CH_2CH_2OH$ $\xrightarrow[\text{Lindlar Pd}]{H_2}$ X $\xrightarrow[\text{heat}]{H_2SO_4}$ Y $\xrightarrow{\overset{\overset{O}{\|}}{CH_3COOH}}$ Z

(a) tetrahydropyran ring with HO and OH substituents

(b) tetrahydropyran ring with $O\overset{\overset{O}{\|}}{C}CH_3$ substituent

(c)

(e) $CH_3\overset{\overset{\displaystyle O}{\|}}{C}OCH_2$ ... $\overset{H}{\underset{H}{C=C}}$ ... $CH_2CH_2O\overset{\overset{\displaystyle O}{\|}}{C}CH_3$

(d) $CH_3\overset{\overset{\displaystyle O}{\|}}{C}OCH_2$ ... $\overset{\underset{\displaystyle H}{}}{C=C}\underset{\displaystyle H}{}$ ... $CH_2CH_2O\overset{\overset{\displaystyle O}{\|}}{C}CH_3$

B-9. The major product of the following sequence is

$\xrightarrow[\text{light}]{Br_2} \xrightarrow{CH_3CH_2OK}$?

with o-methyl, Br substituents

(a) aromatic ring with CH3 and OCH2CH3

(d) aromatic ring with CH2OCH2CH3 and Br

(b) aromatic ring with CH2Br and OCH2CH3

(e) aromatic ring with CH2OCH2CH3 and OCH2CH3

(c) aromatic ring with OCH2CH3 and Br

B-10. Which of the following best represents the rate-determining transition state for the reaction shown?

$\text{C}_6\text{H}_5\text{—ONa} + CH_3Br \longrightarrow$

(a) $\overset{\delta-}{Br}\text{----}\overset{\overset{\displaystyle H}{\underset{\displaystyle H}{|}}}{C}\overset{H}{}\delta+$

(d) $\overset{\delta-}{Br}\text{----}\overset{\overset{\displaystyle H\ H}{}}{\underset{\displaystyle H}{C}}\text{----}\overset{\delta-}{O}\text{—C}_6\text{H}_5$

(b) $\overset{\delta+}{Na}\text{----}\overset{\delta-}{Br}\text{----}\overset{\overset{\displaystyle H\ H}{}}{C}\overset{}{\underset{\displaystyle H}{}}\delta+$

(e) $\overset{H}{\underset{H}{}}C\overset{\overset{\displaystyle \delta-}{Br}}{\underset{\displaystyle \delta-}{O}\text{—C}_6\text{H}_5}$

(c) $\overset{\overset{\displaystyle H}{H}}{\underset{\displaystyle H}{\delta+ C}}\text{----}\overset{\delta-}{O}\text{—C}_6\text{H}_5$

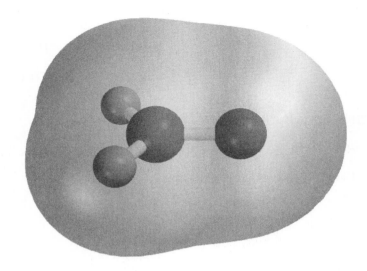

CHAPTER 17

Aldehydes and Ketones: Nucleophilic Addition to the Carbonyl Group

SOLUTIONS TO TEXT PROBLEMS

17.1 (*b*) The longest continuous chain in glutaraldehyde has five carbons and terminates in aldehyde functions at both ends. *Pentanedial* is an acceptable IUPAC name for this compound.

$$\underset{1}{\overset{O}{\overset{\|}{H}C}}\underset{2}{CH_2}\underset{3}{CH_2}\underset{4}{CH_2}\underset{5}{\overset{O}{\overset{\|}{C}H}}$$

Pentanedial (glutaraldehyde)

(*c*) The three-carbon chain has hydroxy groups on C-2 and C-3. The aldehyde function is at C-1 and is not numbered.

$$HOCH_2\underset{\underset{OH}{|}}{CH}\overset{O}{\overset{\|}{C}H}$$

2,3-Dihydroxypropanal
(glyceraldehyde)

(*d*) Vanillin can be named as a derivative of benzaldehyde. Remember to cite the remaining substituents in alphabetical order.

4-Hydroxy-3-methoxybenzaldehyde
(vanillin)

17.2 (*b*) First write the structure from the name given. Ethyl isopropyl ketone has an ethyl group and an isopropyl group bonded to a carbonyl group.

$$\underset{\underset{CH_3}{|}}{CH_3CH_2\overset{\overset{O}{\|}}{C}CHCH_3}$$

The substitutive name of ethyl isopropyl ketone is 2-methyl-3-pentanone. Its longest continuous chain has five carbons. The carbonyl carbon is C-3 irrespective of the direction in which the chain is numbered, and so we choose the direction that gives the lower number to the position that bears the methyl group. Alternatively, the ketone can be named 2-methylpentan-3-one.

(*c*) Methyl 2,2-dimethylpropyl ketone has a methyl group and a 2,2-dimethylpropyl group bonded to a carbonyl group.

$$\underset{\underset{CH_3}{|}}{CH_3\overset{\overset{O}{\|}}{C}CH_2\overset{\overset{CH_3}{|}}{C}CH_3}$$

The longest continuous chain has five carbons, and the carbonyl carbon is C-2. Thus, the substitutive name of methyl 2,2-dimethylpropyl ketone is 4,4-dimethyl-2-pentanone or 4,4-dimethylpentan-2-one.

(*d*) The structure corresponding to allyl methyl ketone is

$$CH_3\overset{\overset{O}{\|}}{C}CH_2CH{=}CH_2$$

Because the carbonyl group is given the lowest possible number in the chain, the substitutive name is 4-penten-2-one *not* 1-penten-4-one. An alternative name is pent-4-en-2-one.

17.3 The carbonyl oxygen of an aldehyde and a proton of water can form a hydrogen bond. The hydrogen bond between benzaldehyde and water can be represented as

Benzaldehyde   Water

17.4 No. Lithium aluminum hydride is the only reagent we have discussed that is capable of reducing carboxylic acids (text Section 15.3).

17.5 The target molecule, 2-butanone, contains four carbon atoms. The problem states that all of the carbons originate in acetic acid, which has two carbon atoms. This suggests the following disconnections:

$$\underset{\text{2-Butanone}}{CH_3\overset{\overset{O}{\|}}{C}CH_2CH_3} \Longrightarrow CH_3\overset{\overset{OH}{|}}{C}HCH_2CH_3 \Longrightarrow CH_3\overset{\overset{O}{\|}}{C}H \ + \ {}^-{:}CH_2CH_3$$

The necessary aldehyde (acetaldehyde) is prepared from acetic acid by reduction followed by oxidation in an anhydrous medium.

$$CH_3CO_2H \xrightarrow[\text{2. H}_2\text{O}]{\text{1. LiAlH}_4} CH_3CH_2OH \xrightarrow[\text{CH}_2\text{Cl}_2]{\text{PDC}} CH_3\overset{\overset{\displaystyle O}{\|}}{C}H$$

Acetic acid   Ethanol   Acetaldehyde

Ethylmagnesium bromide may be obtained from acetic acid by the following sequence:

$$CH_3CH_2OH \xrightarrow[\text{PBr}_3]{\text{HBr or}} CH_3CH_2Br \xrightarrow[\text{diethyl ether}]{\text{Mg}} CH_3CH_2MgBr$$

Ethanol   Ethyl bromide   Ethylmagnesium
(prepared as   bromide
previously)

The preparation of 2-butanone is completed as follows:

$$CH_3\overset{\overset{\displaystyle O}{\|}}{C}H + CH_3CH_2MgBr \xrightarrow[\text{2. H}_3\text{O}^+]{\text{1. diethyl ether}} CH_3\overset{\overset{\displaystyle OH}{|}}{C}HCH_2CH_3 \xrightarrow[\text{H}_2\text{SO}_4,\ \text{H}_2\text{O}]{\text{K}_2\text{Cr}_2\text{O}_7} CH_3\overset{\overset{\displaystyle O}{\|}}{C}CH_2CH_3$$

Acetaldehyde   Ethylmagnesium   2-Butanol   2-Butanone
                  bromide

17.6 Chloral is trichloroethanal, $Cl_3C\overset{\overset{\displaystyle O}{\|}}{C}H$. Chloral hydrate is the addition product of chloral and water.

$$Cl_3C\overset{\overset{\displaystyle OH}{|}}{\underset{\underset{\displaystyle OH}{|}}{C}}H$$

Chloral hydrate

17.7 Methacrylonitrile is formed by the dehydration of acetone cyanohydrin and thus has the structure shown.

$$CH_3\overset{\overset{\displaystyle OH}{|}}{\underset{\underset{\displaystyle CN}{|}}{C}}CH_3 \xrightarrow[\text{(−H}_2\text{O)}]{\text{H}^+,\ \text{heat}} CH_3\overset{}{\underset{\underset{\displaystyle CN}{|}}{C}}=CH_2$$

Acetone   Methacrylonitrile
cyanohydrin

17.8 The cyanohydrin and the ketone that it is derived from are revealed by the disconnections shown:

Gynocardin

17.9 The overall reaction is

$$C_6H_5\overset{\overset{\displaystyle O}{\|}}{C}H \;+\; 2CH_3CH_2OH \;\xrightleftharpoons{HCl}\; C_6H_5CH(OCH_2CH_3)_2 \;+\; H_2O$$

| Benzaldehyde | Ethanol | Benzaldehyde diethyl acetal | Water |

HCl is a strong acid and, when dissolved in ethanol, transfers a proton to ethanol to give ethyloxonium ion. Thus, we can represent the acid catalyst as the conjugate acid of ethanol.

The first three steps correspond to acid-catalyzed addition of ethanol to the carbonyl group to yield a hemiacetal.

Step 1:

Step 2:

Step 3:

Hemiacetal

Formation of the hemiacetal is followed by loss of water to give a carbocation in steps 4 and 5.

Step 4:

Step 5:

Carbocation
intermediate

The remaining two steps, describing the capture of the carbocation by ethanol to give the acetal, are given in text Mechanism 17.4.

17.10 (*b*) 1,3-Propanediol forms acetals that contain a six-membered 1,3-dioxane ring.

$$\underset{\text{Benzaldehyde}}{\text{C}_6\text{H}_5\overset{\overset{\displaystyle O}{\|}}{\text{C}}\text{H}} + \underset{\text{1,3-Propanediol}}{\text{HOCH}_2\text{CH}_2\text{CH}_2\text{OH}} \xrightarrow{\text{H}^+} \underset{\text{2-Phenyl-1,3-dioxane}}{\text{(ring structure)}} + \underset{\text{Water}}{\text{H}_2\text{O}}$$

(*c*) The cyclic acetal derived from isobutyl methyl ketone and ethylene glycol bears an isobutyl group and a methyl group at C-2 of a 1,3-dioxolane ring.

$$\underset{\text{Isobutyl methyl ketone}}{(\text{CH}_3)_2\text{CHCH}_2\overset{\overset{\displaystyle O}{\|}}{\text{C}}\text{CH}_3} + \underset{\text{Ethylene glycol}}{\text{HOCH}_2\text{CH}_2\text{OH}} \xrightarrow{\text{H}^+} \underset{\text{2-Isobutyl-2-methyl-1,3-dioxolane}}{(\text{CH}_3)_2\text{CHCH}_2 \ \text{CH}_3} + \underset{\text{Water}}{\text{H}_2\text{O}}$$

(*d*) Because the starting diol is 2,2-dimethyl-1,3-propanediol, the cyclic acetal is six-membered and bears two methyl substituents at C-5 in addition to isobutyl and methyl groups at C-2.

$$\underset{\text{Isobutyl methyl ketone}}{(\text{CH}_3)_2\text{CHCH}_2\overset{\overset{\displaystyle O}{\|}}{\text{C}}\text{CH}_3} + \underset{\substack{\text{2,2-Dimethyl-1,3-}\\\text{propanediol}}}{\text{HOCH}_2\overset{\overset{\displaystyle \text{CH}_3}{|}}{\underset{\underset{\displaystyle \text{CH}_3}{|}}{\text{C}}}\text{CH}_2\text{OH}} \xrightarrow{\text{H}^+} \underset{\substack{\text{2-Isobutyl-2,5,5-trimethyl-}\\\text{1,3-dioxane}}}{(\text{CH}_3)_2\text{CHCH}_2 \ \text{CH}_3} + \underset{\text{Water}}{\text{H}_2\text{O}}$$

17.11 The overall reaction is

$$\underset{\text{Benzaldehyde diethyl acetal}}{\text{C}_6\text{H}_5\text{CH}(\text{OCH}_2\text{CH}_3)_2} + \underset{\text{Water}}{\text{H}_2\text{O}} \underset{\text{HCl}}{\rightleftharpoons} \underset{\text{Benzaldehyde}}{\text{C}_6\text{H}_5\overset{\overset{\displaystyle O}{\|}}{\text{C}}\text{H}} + \underset{\text{Ethanol}}{2\text{CH}_3\text{CH}_2\text{OH}}$$

The mechanism of acetal hydrolysis is the reverse of acetal formation. The first four steps convert the acetal to the hemiacetal.

Step 1:

Step 2:

Step 3:

Step 4:

Hemiacetal

Step 5:

Step 6:

Step 7:

17.12 The conversion requires reduction; however, the conditions necessary ($LiAlH_4$) would also reduce the ketone carbonyl. The ketone functionality is therefore protected as the cyclic acetal.

4-Acetylbenzoic acid

Reduction of the carboxylic acid may now be carried out.

Hydrolysis to remove the protecting group completes the synthesis.

4-Acetylbenzyl alcohol

17.13 (*b*) Nucleophilic addition of butylamine to benzaldehyde gives the carbinolamine.

Benzaldehyde   Butylamine   Carbinolamine intermediate

Dehydration of the carbinolamine produces the imine.

N-Benzylidenebutylamine

(*c*) Cyclohexanone and *tert*-butylamine react according to the equation

Cyclohexanone   *tert*-Butylamine   Carbinolamine intermediate   *N*-Cyclohexylidene-*tert*-butylamine

(*d*) $C_6H_5CCH_3$ +

Acetophenone   Cyclohexylamine   Carbinolamine intermediate   *N*-(1-Phenylethylidene)-cyclohexylamine

17.14 The compound $C_5H_9NO_3$ differs from its imine precursor ($C_5H_7NO_2$) by a molecule of H_2O. Therefore, add H_2O across the double bond of the imine to give a carbinolamine as one of the answers. Recognize that the carbinolamine is in equilibrium with a noncyclic compound that contains both an aldehyde and a primary amine function to deduce the other answer.

17.15 (*b*) Pyrrolidine, a secondary amine, adds to 3-pentanone to give a carbinolamine.

3-Pentanone   Pyrrolidine   Carbinolamine
                                                              intermediate

Dehydration produces the enamine.

Carbinolamine   3-Pyrrolidino-2-pentene
intermediate

(*c*) $C_6H_5\overset{O}{\overset{\|}{C}}CH_3$ +

Acetophenone   Piperidine   Carbinolamine   1-Phenyl-1-piperidinoethene
                                      intermediate

17.16 (*b*) Here we see an example of the Wittig reaction applied to diene synthesis by use of an ylide containing a carbon–carbon double bond.

$CH_3CH_2CH_2\overset{O}{\overset{\|}{C}}H$ + $(C_6H_5)_3\overset{+}{P}-\overset{\cdot\cdot}{\overset{-}{C}}HCH=CH_2$ ⟶ $CH_3CH_2CH_2CH=CHCH=CH_2$ + $(C_6H_5)_3\overset{+}{P}-O^-$

Butanal   Allylidenetriphenylphosphorane   1,3-Heptadiene (52%)   Triphenylphosphine
                                                                                        oxide

(*c*) Methylene transfer from methylenetriphenylphosphorane is one of the most commonly used Wittig reactions.

Cyclohexyl methyl   Methylenetriphenyl-   2-Cyclohexylpropene   Triphenylphosphine
ketone   phosphorane   (66%)   oxide

17.17 A second resonance structure can be written for a phosphorus ylide with a double bond between phosphorus and carbon. As a third-row element, phosphorus can have more than 8 electrons in its valence shell.

$$(C_6H_5)_3\overset{+}{P}-\overset{-}{\ddot{C}}H_2 \longleftrightarrow (C_6H_5)_3P=CH_2$$

Methylenetriphenylphosphorane

17.18 (*b*) Two Wittig reaction routes lead to 1-pentene. One is represented retrosynthetically by the disconnection

$$CH_3CH_2CH_2CH{=}CH_2 \implies \overset{O}{\overset{||}{CH_3CH_2CH_2CH}} + (C_6H_5)_3\overset{+}{P}-\overset{-}{\ddot{C}}H_2$$

1-Pentene   Butanal   Methylenetriphenyl-phosphorane

The other route is

$$CH_3CH_2CH_2CH{=}CH_2 \implies CH_3CH_2CH_2\overset{-}{\ddot{C}}H{-}\overset{+}{P}(C_6H_5)_3 + \overset{O}{\overset{||}{HCH}}$$

1-Pentene   Butylidenetriphenylphosphorane   Formaldehyde

17.19 Ylides are prepared by the reaction of an alkyl halide with triphenylphosphine, followed by treatment with strong base. 2-Bromobutane is the alkyl halide needed in this case.

$$(C_6H_5)_3P + \overset{Br}{\overset{|}{CH_3CHCH_2CH_3}} \longrightarrow (C_6H_5)_3\overset{+}{P}-\underset{CH_3}{\overset{|}{CHCH_2CH_3}} Br^-$$

Triphenyl-phosphine   2-Bromobutane   (1-Methylpropyl)triphenyl-phosphonium bromide

$$(C_6H_5)_3\overset{+}{P}-\underset{CH_3}{\overset{|}{CHCH_2CH_3}} Br^- + \overset{O}{\overset{||}{NaCH_2SCH_3}} \longrightarrow (C_6H_5)_3\overset{+}{P}-\underset{CH_3}{\overset{|}{\overset{-}{\ddot{C}}CH_2CH_3}}$$

(1-Methylpropyl)triphenyl-phosphonium bromide   Sodiomethyl methyl sulfoxide   Ylide

17.20 The two products are stereoisomers of each other. However, they are not mirror images and thus are not enantiomers. They are diastereomers — stereoisomers that are not enantiomers. The reaction is diastereoselective.

17.21 In the first step the peroxy acid adds to the carbonyl group of the ketone. This step is a nucleophilic addition analogous to hemiacetal formation. Following protonation of the intermediate, alkyl group migration occurs at the same time as cleavage of the O—O bond of the peroxy ester. The more substituted alkyl group is the one that migrates.

17.22 (*a*) First consider all the isomeric aldehydes of molecular formula $C_5H_{10}O$.

Pentanal   3-Methylbutanal

(*S*)-2-Methylbutanal   (*R*)-2-Methylbutanal   2,2-Dimethylpropanal

There are three isomeric ketones:

2-Pentanone   3-Pentanone   3-Methyl-2-butanone

(*b*) Reduction of an aldehyde to a primary alcohol does not introduce a chirality center into the molecule. The only aldehydes that yield chiral alcohols on reduction are therefore those that already contain a chirality center.

(*S*)-2-Methylbutanal   (*S*)-2-Methyl-1-butanol

(*R*)-2-Methylbutanal   (*R*)-2-Methyl-1-butanol

Among the ketones, 2-pentanone and 3-methyl-2-butanone are reduced to chiral alcohols.

2-Pentanone   2-Pentanol (chiral but racemic)

3-Pentanone   3-Pentanol (achiral)

3-Methyl-2-butanone   3-Methyl-2-butanol (chiral but racemic)

(*c*) All the $C_5H_{10}O$ aldehydes yield chiral alcohols on reaction with methylmagnesium iodide according to the equation:

CHAPTER 17: Aldehydes and Ketones

503

None of the ketones yield chiral alcohols.

2-Pentanone → (1. CH$_3$MgI, 2. H$_3$O$^+$) → 2-Methyl-2-pentanol (achiral)

3-Pentanone → (1. CH$_3$MgI, 2. H$_3$O$^+$) → 3-Methyl-3-pentanol (achiral)

3-Methyl-2-butanone → (1. CH$_3$MgI, 2. H$_3$O$^+$) → 2,3-Dimethyl-2-butanol (achiral)

17.23 (*a*) Chloral is the trichloro derivative of ethanal (acetaldehyde).

CH$_3$CH(=O) — Ethanal

Cl—C(Cl)(Cl)—CH(=O) — Trichloroethanal (chloral)

(*b*) Pivaldehyde has two methyl groups attached to C-2 of propanal.

CH$_3$CH$_2$CH(=O) — Propanal

CH$_3$C(CH$_3$)(CH$_3$)—CH(=O) — 2,2-Dimethylpropanal (pivaldehyde)

(*c*) Acrolein has a double bond between C-2 and C-3 of a three-carbon aldehyde.

H$_2$C=CHCH(=O) — 2-Propenal (acrolein)

(*d*) Crotonaldehyde has a trans double bond between C-2 and C-3 of a four-carbon aldehyde.

(*E*)-2-Butenal (crotonaldehyde)

(e) Citral has two double bonds: one between C-2 and C-3 and the other between C-6 and C-7. The one at C-2 has the E configuration. There are methyl substituents at C-3 and C-7.

(E)-3,7-Dimethyl-2,6-octadienal
(citral)

(f) Diacetone alcohol is

4-Hydroxy-4-methyl-
2-pentanone

(g) The parent ketone is 2-cyclohexenone.

2-Cyclohexenone

Carvone has an isopropenyl group at C-5 and a methyl group at C-2.

5-Isopropenyl-2-methyl-2-
cyclohexenone (carvone)

(h) Biacetyl is 2,3-butanedione. It has a four-carbon chain that incorporates ketone carbonyls as C-2 and C-3.

2,3-Butanedione
(biacetyl)

17.24 (a) The longest chain of carbon atoms in the crocodile pheromone is numbered from the end closest to the ketone carbonyl group.

The absolute stereochemistry of the chirality centers can be determined by assigning the Cahn-Ingold-Prelog rankings to the four groups at each center.

$$\text{C-3:} \quad \overset{\overset{\textstyle O}{\|}}{C} > CH_2CH_2 > CH_2CH_3 > H \qquad \text{C-7:} \quad CH_2CH_2C_6H_5 > CH_2CH_2CH_2 > CH_2CH_3 > H$$

In order of decreasing ranking the chirality center at C-3 is S; the chirality center at C-7 is R. The systematic name for the pheromone is (3S,7R)-3,7-diethyl-9-phenylnonan-2-one.

(*b*) The stereoisomer in which both chirality centers have the S configuration (3S,7S) is a diastereomer of the (3S,7R) stereoisomer shown in the problem. The enantiomer of the (3S,7R) stereoisomer would be (3R,7S).

17.25 (*a*) Lithium aluminum hydride reduces aldehydes to primary alcohols.

$$\underset{\text{Propanal}}{CH_3CH_2\overset{\overset{\textstyle O}{\|}}{C}H} \quad \xrightarrow[\text{2. } H_2O]{\text{1. LiAlH}_4} \quad \underset{\text{1-Propanol}}{CH_3CH_2CH_2OH}$$

(*b*) Sodium borohydride reduces aldehydes to primary alcohols.

$$\underset{\text{Propanal}}{CH_3CH_2\overset{\overset{\textstyle O}{\|}}{C}H} \quad \xrightarrow[\text{CH}_3OH]{\text{NaBH}_4} \quad \underset{\text{1-Propanol}}{CH_3CH_2CH_2OH}$$

(*c*) Aldehydes can be reduced to primary alcohols by catalytic hydrogenation.

$$\underset{\text{Propanal}}{CH_3CH_2\overset{\overset{\textstyle O}{\|}}{C}H} \quad \xrightarrow[\text{Ni}]{H_2} \quad \underset{\text{1-Propanol}}{CH_3CH_2CH_2OH}$$

(*d*) Aldehydes react with Grignard reagents to form secondary alcohols.

$$\underset{\text{Propanal}}{CH_3CH_2\overset{\overset{\textstyle O}{\|}}{C}H} \quad \xrightarrow[\text{2. } H_3O^+]{\substack{\text{1. CH}_3\text{MgI,} \\ \text{diethyl ether}}} \quad \underset{\text{2-Butanol}}{CH_3CH_2\overset{\overset{\textstyle OH}{|}}{C}HCH_3}$$

(*e*) Sodium acetylide adds to the carbonyl group of propanal to give an acetylenic alcohol.

$$\underset{\text{Propanal}}{CH_3CH_2\overset{\overset{\textstyle O}{\|}}{C}H} \quad \xrightarrow[\text{2. } H_3O^+]{\substack{\text{1. HC} \equiv \text{CNa} \\ \text{liquid ammonia}}} \quad \underset{\text{1-Pentyn-3-ol}}{CH_3CH_2\overset{\overset{\textstyle OH}{|}}{C}HC \equiv CH}$$

(*f*) Alkyl- or aryllithium reagents react with aldehydes in much the same way that Grignard reagents do.

$$\underset{\text{Propanal}}{CH_3CH_2\overset{\overset{\textstyle O}{\|}}{C}H} \quad \xrightarrow[\text{2. } H_3O^+]{\substack{\text{1. C}_6H_5\text{Li,} \\ \text{diethyl ether}}} \quad \underset{\text{1-Phenyl-1-propanol}}{CH_3CH_2\underset{\underset{\textstyle OH}{|}}{C}HC_6H_5}$$

(g) Aldehydes are converted to acetals on reaction with alcohols in the presence of an acid catalyst.

$$CH_3CH_2\overset{O}{\overset{\|}{C}}H + 2CH_3OH \xrightarrow{HCl} CH_3CH_2CH(OCH_3)_2$$

     Propanal   Methanol   Propanal dimethyl acetal

(h) Cyclic acetal formation occurs when aldehydes react with ethylene glycol.

$$CH_3CH_2\overset{O}{\overset{\|}{C}}H + HOCH_2CH_2OH \xrightarrow[\text{benzene}]{p\text{-toluenesulfonic acid}}$$

     Propanal   Ethylene glycol   2-Ethyl-1,3-dioxolane

(i) Aldehydes react with primary amines to yield imines.

$$CH_3CH_2\overset{O}{\overset{\|}{C}}H + C_6H_5NH_2 \xrightarrow{-H_2O} CH_3CH_2CH=NC_6H_5$$

     Propanal   Aniline   N-Propylideneaniline

(j) Secondary amines combine with aldehydes to yield enamines.

$$CH_3CH_2\overset{O}{\overset{\|}{C}}H + (CH_3)_2NH \xrightarrow[\text{benzene}]{p\text{-toluenesulfonic acid}} CH_3CH=\overset{N(CH_3)_2}{\overset{|}{C}}H$$

     Propanal   Dimethylamine   1-(Dimethylamino)propene

(k) Oximes are formed on reaction of hydroxylamine with aldehydes.

$$CH_3CH_2\overset{O}{\overset{\|}{C}}H \xrightarrow{H_2NOH} CH_3CH_2CH=NOH$$

     Propanal   Propanal oxime

(l) Hydrazine reacts with aldehydes to form hydrazones.

$$CH_3CH_2\overset{O}{\overset{\|}{C}}H \xrightarrow{H_2NNH_2} CH_3CH_2CH=NNH_2$$

     Propanal   Propanal hydrazone

(m) Hydrazone formation is the first step in the Wolff–Kishner reduction (text Section 12.8).

$$CH_3CH_2CH=NNH_2 \xrightarrow[\text{triethylene glycol, heat}]{NaOH} CH_3CH_2CH_3 + N_2$$

     Propanal hydrazone   Propane

(n) The reaction of an aldehyde with p-nitrophenylhydrazine is analogous to that with hydrazine.

$$CH_3CH_2\overset{O}{\overset{\|}{C}}H + O_2N-\!\!\!\bigcirc\!\!\!-NHNH_2 \longrightarrow CH_3CH_2CH=NNH-\!\!\!\bigcirc\!\!\!-NO_2 + H_2O$$

     Propanal   p-Nitrophenylhydrazine   Propanal p-nitrophenylhydrazone (*o*) Semicarbazide converts aldehydes to the corresponding semicarbazone.

$$CH_3CH_2\overset{O}{\overset{||}{C}}H + H_2NNH\overset{O}{\overset{||}{C}}NH_2 \longrightarrow CH_3CH_2CH=NNH\overset{O}{\overset{||}{C}}NH_2 + H_2O$$

Propanal   Semicarbazide   Propanal semicarbazone

(*p*) Phosphorus ylides convert aldehydes to alkenes by a Wittig reaction.

$$CH_3CH_2\overset{O}{\overset{||}{C}}H + (C_6H_5)_3\overset{+}{P}-\overset{..}{\overset{-}{C}}HCH_3 \longrightarrow CH_3CH_2CH=CHCH_3 + (C_6H_5)_3\overset{+}{P}-\overset{-}{O}$$

Propanal   Ethylidenetriphenyl-phosphorane   2-Pentene   Triphenylphosphine oxide

(*q*) Acidification of solutions of sodium cyanide generates HCN, which reacts with aldehydes to form cyanohydrins.

$$CH_3CH_2\overset{O}{\overset{||}{C}}H + HCN \longrightarrow CH_3CH_2\overset{OH}{\overset{|}{C}}HCN$$

Propanal   Hydrogen cyanide   Propanal cyanohydrin

(*r*) Chromic acid oxidizes aldehydes to carboxylic acids.

$$CH_3CH_2\overset{O}{\overset{||}{C}}H \xrightarrow{H_2CrO_4} CH_3CH_2CO_2H$$

Propanal   Propanoic acid

17.26 (*a*) Lithium aluminum hydride reduces ketones to secondary alcohols.

Cyclopentanone $\xrightarrow[\text{2. } H_2O]{\text{1. LiAlH}_4}$ Cyclopentanol

(*b*) Sodium borohydride converts ketones to secondary alcohols.

Cyclopentanone $\xrightarrow[\text{CH}_3OH]{\text{NaBH}_4}$ Cyclopentanol

(*c*) Catalytic hydrogenation of ketones yields secondary alcohols.

Cyclopentanone $\xrightarrow[\text{Ni}]{H_2}$ Cyclopentanol

(*d*) Grignard reagents react with ketones to form tertiary alcohols.

Cyclopentanone   1-Methylcyclopentanol

(*e*) Addition of sodium acetylide to cyclopentanone yields a tertiary acetylenic alcohol.

Cyclopentanone   1-Ethynylcyclopentanol

(*f*) Phenyllithium adds to the carbonyl group of cyclopentanone to yield 1-phenylcyclopentanol.

1. C_6H_5Li,
     diethyl ether
2. H_3O^+

C_6H_5 OH

Cyclopentanone   1-Phenylcyclopentanol

(*g*) The equilibrium constant for acetal formation from ketones is generally unfavorable.

$+ \ 2CH_3OH$ $\underset{}{\overset{HCl}{\rightleftharpoons}}$ $K < 1$

CH_3O OCH_3

Cyclopentanone   Methanol   Cyclopentanone
                                                dimethyl acetal

(*h*) Cyclic acetal formation is favored even for ketones.

$+ \ HOCH_2CH_2OH$ $\xrightarrow[\text{benzene}]{p\text{-toluenesulfonic acid}}$

O O

Cyclopentanone   Ethylene glycol   1,4-Dioxaspiro[4.4]nonane

(*i*) Ketones react with primary amines to form imines.

$+ \ C_6H_5NH_2$ $\longrightarrow$

NC_6H_5

Cyclopentanone   Aniline   *N*-Cyclopentylideneaniline

(*j*) Dimethylamine reacts with cyclopentanone to yield an enamine.

Cyclopentanone   Dimethylamine   1-(Dimethylamino)-
                                                                    cyclopentene

(*k*) An oxime is formed when cyclopentanone is treated with hydroxylamine.

Cyclopentanone   Cyclopentanone
                                            oxime

(*l*) Hydrazine reacts with cyclopentanone to form a hydrazone.

Cyclopentanone   Cyclopentanone
                                            hydrazone

(*m*) Heating a hydrazone in base with a high-boiling alcohol as solvent converts it to an alkane.

Cyclopentanone   Cyclopentane
hydrazone

(*n*) A *p*-nitrophenylhydrazone is formed.

Cyclopentanone   *p*-Nitrophenylhydrazine   Cyclopentanone
                                                              p-nitrophenylhydrazone

(*o*) Cyclopentanone is converted to a semicarbazone on reaction with semicarbazide.

Cyclopentanone   Semicarbazide   Cyclopentanone
                                                         semicarbazone

(*p*) A Wittig reaction takes place, forming ethylidenecyclopentane.

Cyclopentanone   Ethylidenetriphenyl-   Ethylidenecyclo-   Triphenylphosphine
                              phosphorane   pentane   oxide

(*q*) Cyanohydrin formation takes place.

Cyclopentanone   Cyclopentanone
                               cyanohydrin

(*r*) Cyclopentanone is not oxidized readily with chromic acid.

17.27 (*a*) The first step in analyzing this problem is to write the structure of the starting ketone in stereochemical detail.

(*S*)-3-Phenyl-2-   (2*R*,3*S*)-3-Phenyl-   (2*S*,3*S*)-3-Phenyl-
butanone   2-butanol   2-butanol

Reduction of the ketone introduces a new chirality center, which may have either the *R* or the *S* configuration; the configuration of the original chirality center is unaffected. In practice the 2*R*,3*S* diastereomer is observed to form in greater amounts than the 2*S*,3*S* (ratio 2.5:1 for LiAlH$_4$ reduction).

(*b*) Reduction of the ketone can yield either *cis*- or *trans*-4-*tert*-butylcyclohexanol.

4-*tert*-Butylcyclo-
hexanone

trans-4-*tert*-   *cis*-4-*tert*-
Butylcyclohexanol   Butylcyclohexanol

The major product obtained on reduction with either lithium aluminum hydride or sodium borohydride is the trans alcohol (trans/cis ≈ 9:1).

(*c*) The two reduction products are the exo and endo alcohols.

| Bicyclo[2.2.1]- | *exo*-Bicyclo[2.2.1]- | *endo*-Bicyclo[2.2.1]- |
| heptan-2-one | heptan-2-ol | heptan-2-ol |

The major product is the endo alcohol (endo/exo 9:1) for reduction with $NaBH_4$ or $LiAlH_4$. The stereoselectivity observed in this reaction is due to decreased steric hindrance to attack of the hydride reagent from the exo face of the molecule, giving rise to the endo alcohol.

(*d*) The hydroxyl group may be on the same side as the double bond or on the opposite side.

| Bicyclo[2.2.1]- | *syn*-Bicyclo[2.2.1]- | *anti*-Bicyclo[2.2.1]- |
| hept-2-en-7-one | hept-2-en-7-ol | hept-2-en-7-ol |

The anti alcohol is formed in greater amounts (85:15) on reduction of the ketone with $LiAlH_4$. Steric factors governing attack of the hydride reagent again explain the major product observed.

17.28 (*a*) Aldehydes undergo nucleophilic addition faster than ketones. Steric crowding in the rate-determining step of the ketone reaction raises the energy of the transition state, giving rise to a slower rate of reaction. Thus benzaldehyde is reduced by sodium borohydride more rapidly than is acetophenone. The measured relative rates are

$$k_{rel} = \frac{C_6H_5\overset{\overset{O}{\|}}{C}H}{C_6H_5\overset{\overset{O}{\|}}{C}CH_3} = 440$$

(*b*) The presence of an electronegative substituent on the α-carbon atom causes a dramatic increase in K_{hydr}. Trichloroethanal (chloral) is almost completely converted to its geminal diol (chloral hydrate) in aqueous solution.

$$Cl_3\overset{\overset{O}{\|}}{C}CH \quad + \quad H_2O \quad \longrightarrow \quad Cl_3\underset{\underset{OH}{|}}{\overset{\overset{OH}{|}}{C}}CH$$

| Trichloroethanal | 2,2,2-Trichloro-1,1-ethanediol |
| (chloral) | (chloral hydrate) |

Electron-withdrawing groups such as Cl_3C destabilize carbonyl groups to which they are attached and make the energy change favoring the products of nucleophilic addition more favorable.

$$K_{rel} = \frac{Cl_3\overset{\overset{O}{\|}}{C}CH}{CH_3\overset{\overset{O}{\|}}{C}H} \approx 20,000$$

(*c*) Recall that the equilibrium constants for nucleophilic addition to carbonyl groups are governed by a combination of electronic effects and steric effects. Electronically there is little difference between acetone and 3,3-dimethyl-2-butanone, but sterically there is a significant difference. The cyanohydrin products are more crowded than the starting ketones, and so the bulkier the alkyl groups attached to the carbonyl, the more strained and less stable the cyanohydrin.

Ketone   Hydrogen cyanide   Cyanohydrin
[less strained for R = CH_3 than for R = $C(CH_3)_3$]

$$K_{rel} = \frac{\overset{O}{\overset{\|}{CH_3CCH_3}}}{\underset{\underset{CH_3CC(CH_3)_3}{\|}}{O}} = 40$$

(*d*) Steric effects influence the rate of nucleophilic addition to these two ketones. Carbon is on its way from sp^2 to sp^3 at the transition state, and alkyl groups are forced closer together than they are in the ketone.

Transition state

The transition state is of lower energy, and the reaction is faster, when R is smaller. Acetone (R = methyl) is reduced faster than 3,3-dimethyl-2-butanone (R = *tert*-butyl).

$$k_{rel} = \frac{\overset{O}{\overset{\|}{CH_3CCH_3}}}{\underset{\underset{CH_3CC(CH_3)_3}{\|}}{O}} = 12$$

(*e*) In this problem we examine the rate of hydrolysis of acetals to the corresponding ketone or aldehyde. The rate-determining step is carbocation formation.

Hybridization at carbon changes from sp^3 to sp^2; crowding at this carbon is relieved as the carbocation is formed. The more crowded acetal (R = CH_3) forms a carbocation faster than the less crowded one (R = H). Another factor of even greater importance is the extent of stabilization of the carbocation intermediate; the more stable carbocation (R = CH_3) is formed faster than the less stable one (R = H).

$$k_{rel} = \frac{(CH_3)_2C(OCH_2CH_3)_2}{CH_2(OCH_2CH_3)_2} = 1.8 \times 10^7$$

17.29 (*a*) The reaction as written is the reverse of cyanohydrin formation, and the principles that govern equilibria in nucleophilic addition to carbonyl groups apply in reverse order to the dissociation of cyanohydrins to aldehydes and ketones. Cyanohydrins of ketones dissociate more at equilibrium than do cyanohydrins of aldehydes. More strain due to crowding is relieved when a ketone cyanohydrin dissociates and a more stabilized carbonyl group is formed. The equilibrium constant K_{diss} is larger for

$$\underset{\substack{\text{Acetone}\\\text{cyanohydrin}}}{\overset{\overset{\text{OH}}{|}}{\underset{\underset{\text{CN}}{|}}{CH_3CCH_3}}} \quad \xrightleftharpoons{K_{diss}} \quad \underset{\text{Acetone}}{\overset{\overset{O}{||}}{CH_3CCH_3}} \quad + \quad \underset{\substack{\text{Hydrogen}\\\text{cyanide}}}{HCN}$$

than it is for

$$\underset{\substack{\text{Propanal cyanohydrin}}}{\overset{\overset{\text{OH}}{|}}{CH_3CH_2CHCN}} \quad \xrightleftharpoons{K_{diss}} \quad \underset{\text{Propanal}}{\overset{\overset{O}{||}}{CH_3CH_2CH}} \quad + \quad \underset{\substack{\text{Hydrogen}\\\text{cyanide}}}{HCN}$$

(*b*) Cyanohydrins of ketones have a more favorable equilibrium constant for dissociation than do cyanohydrins of aldehydes. Crowding is relieved to a greater extent when a ketone cyanohydrin dissociates and a more stable carbonyl group is formed. The measured dissociation constants are

$$\underset{\substack{\text{Benzaldehyde}\\\text{cyanohydrin}}}{\overset{\overset{\text{OH}}{|}}{C_6H_5CHCN}} \quad \xrightleftharpoons{} \quad \underset{\text{Benzaldehyde}}{\overset{\overset{O}{||}}{C_6H_5CH}} \quad + \quad HCN \quad K = 4.7 \times 10^{-3}$$

$$\underset{\substack{\text{Acetophenone}\\\text{cyanohydrin}}}{\overset{\overset{\text{OH}}{|}}{\underset{\underset{\text{CH}_3}{|}}{C_6H_5CCN}}} \quad \xrightleftharpoons{} \quad \underset{\text{Acetophenone}}{\overset{\overset{O}{||}}{C_6H_5CCH_3}} \quad + \quad HCN \quad K = 1.3$$

17.30 (*a*) The reaction of an aldehyde with 1,3-propanediol in the presence of *p*-toluenesulfonic acid forms a cyclic acetal.

2-Bromo-3,4,5-
trimethoxybenzaldehyde   1,3-Propanediol   2-(2′-Bromo-3′,4′,5′-
trimethoxyphenyl)-1,3-dioxane
(81%)

(b) The reagent CH_3ONH_2 is called *O*-methylhydroxylamine, and it reacts with aldehydes in a manner similar to hydroxylamine.

| 4-Hydroxy-2-methoxybenzaldehyde | *O*-Methyl-hydroxylamine | 4-Hydroxy-2-methoxybenzaldehyde *O*-methyloxime |

(c) Propanal reacts with 1,1-dimethylhydrazine to yield the corresponding hydrazone.

$$CH_3CH_2CH(O) + (CH_3)_2NNH_2 \longrightarrow CH_3CH_2CH=NN(CH_3)_2$$

| Propanal | 1,1-Dimethylhydrazine | Propanal dimethylhydrazone |

(d) Acid-catalyzed hydrolysis of the acetal gives the aldehyde in 87% yield.

4-(*p*-Methylphenyl)pentanal

(e) Hydrogen cyanide adds to carbonyl groups to form cyanohydrins.

| Acetophenone | Acetophenone cyanohydrin |

(f) The reagent is a secondary amine known as *morpholine*. Secondary amines react with ketones to give enamines.

| Acetophenone | Morpholine | Carbinolamine intermediate | 1-Morpholinostyrene (57–64%) |

(*g*) Migration of the alkyl group in a Baeyer–Villiger oxidation occurs with retention of configuration.

(*R*)-3-Methyl-3-phenyl-2-
pentanone   Peroxybenzoic acid   (*R*)-1-Methyl-1-phenylpropyl
                                                                            acetate

(*h*) 2-Mercaptoethanol reacts with a ketone in the presence of an acid catalyst to yield the sulfur
analog of an acetal.

Diisopropyl ketone   2-Mercaptoethanol   2,2-Diisopropyl-1,3-oxathiolane

17.31 Wolff–Kishner reduction converts a carbonyl group ($C=O$) to a methylene group (CH_2).

Bicyclo[4.3.0]non-   Bicyclo[4.3.0]non-3-ene
3-en-8-one   (compound A, 90%)

Treatment of the alkene with *m*-chloroperoxybenzoic acid produces an epoxide, compound B.

Bicyclo[4.3.0]non-3-ene   3,4-Epoxybicyclo[4.3.0]nonane
                                                        (compound B, 92%)

Epoxides undergo reduction with lithium aluminum hydride to form alcohols (text Section 16.12).

3,4-Epoxybicyclo[4.3.0]nonane   Bicyclo[4.3.0]nonan-3-ol
                                                        (compound C, 90%)

Chromic acid oxidizes the alcohol to a ketone.

Bicyclo[4.3.0]nonan-3-ol   Bicyclo[4.3.0]nonan-3-one
                                                        (compound D, 75%)

17.32 Hydration of formaldehyde by $H_2{}^{17}O$ produces a *gem*-diol in which the labeled and unlabeled hydroxyl groups are equivalent. When this *gem*-diol reverts to formaldehyde, loss of either of the hydroxyl groups is equally likely and leads to eventual replacement of the mass-16 isotope of oxygen by ^{17}O.

This reaction has been monitored by ^{17}O NMR spectroscopy; ^{17}O gives an NMR signal, but ^{16}O does not.

17.33 First write out the chemical equation for the reaction that takes place. Vicinal diols (1,2-diols) react with aldehydes to give cyclic acetals.

| Benzaldehyde | 1,2-Octanediol | 4-Hexyl-2-phenyl-1,3-dioxolane |

Notice that the phenyl and hexyl substituents may be either cis or trans to each other. The two products are the cis and trans stereoisomers.

cis-4-Hexyl-2-phenyl-1,3-dioxolane   *trans*-4-Hexyl-2-phenyl-1,3-dioxolane

17.34 Cyclic hemiacetals are formed by intramolecular nucleophilic addition of a hydroxyl group to a carbonyl.

Cyclic hemiacetal

The ring oxygen is derived from the hydroxyl group; the carbonyl oxygen becomes the hydroxyl oxygen of the hemiacetal.

(*a*) This compound is the cyclic hemiacetal of 5-hydroxypentanal.

Indeed, 5-hydroxypentanal seems to exist entirely as the cyclic hemiacetal. Its IR spectrum lacks absorption in the carbonyl region.

(*b*) The carbon connected to two oxygens is the one that is derived from the carbonyl group. Using retrosynthetic symbolism, disconnect the ring oxygen from this carbon.

$$HC\mathrm{CH_2CH_2CHCH{=}CHCH{=}CH_2}$$

4-Hydroxy-5,7-octadienal

The next two compounds are cyclic acetals. The original carbonyl group is identifiable as the one that bears two oxygen substituents, which originate as hydroxyl oxygens of a diol.

(*c*)

Brevicomin

$$\mathrm{CH_3CH_2CHCHCH_2CH_2CH_2CCH_3}$$

6,7-Dihydroxy-
2-nonanone

(*d*)

Talaromycin A

2,8-Di(hydroxymethyl)-1,3-
dihydroxy-5-decanone

17.35 (*a*) The *Z* stereoisomer of $CH_3CH{=}NCH_3$ has its higher ranked substituents on the same side of the double bond,

(*Z*)-*N*-Ethylidenemethylamine

The unshared electron pair of nitrogen is lower in rank than any other substituent.

(*b*) Higher ranked groups are on opposite sides of the carbon–nitrogen double bond in the *E* oxime of acetaldehyde.

(*E*)-Acetaldehyde
oxime

(*c*) (*Z*)-2-Butanone hydrazone is

(d) (E)-Acetophenone semicarbazone is

17.36 Cyclopentanone reacts with peroxybenzoic acid to form a peroxy monoester. The alkyl group that migrates is the ring itself, leading to formation of a six-membered lactone.

| Cyclopentanone | Peroxybenzoic acid | | 5-Pentanolide | Benzoic acid |

17.37 (a) The bacterial enzyme cyclohexanone monoxygenase, described in text Section 17.16, is able to catalyze a biological Baeyer–Villiger reaction. Compound A is 4-methylcyclohexanone.

4-Methylcyclohexanone (compound A)   Compound B

(b) The product of Baeyer–Villiger oxidation of 4-methylcyclohexanone with peroxyacetic acid would be the racemic cyclic ester (lactone), not the single enantiomer shown in part (a) from the enzyme-catalyzed oxidation.

17.38 (a) Nucleophilic ring opening of the epoxide occurs by attack of methoxide at the less hindered carbon.

The anion formed in this step loses a chloride ion to form the carbon–oxygen double bond of the product.

(*b*) Nucleophilic addition of methoxide ion to the aldehyde carbonyl generates an oxyanion, which can close to an epoxide by an intramolecular nucleophilic substitution reaction.

$$(CH_3)_3CCHCH \quad + \quad \overset{..}{\underset{..}{:}}OCH_3 \quad \longrightarrow \quad (CH_3)_3CCH{-}CHOCH_3 \quad \longrightarrow \quad (CH_3)_3CC{-}CHOCH_3$$

The epoxide formed in this process then undergoes nucleophilic ring opening on attack by a second methoxide ion.

$$(CH_3)_3CC{-}CHOCH_3 \quad + \quad \overset{..}{\underset{..}{:}}OCH_3 \quad \longrightarrow \quad (CH_3)_3CCH{-}CHOCH_3 \quad \overset{CH_3OH}{\longrightarrow} \quad (CH_3)_3CCHCHOCH_3$$

17.39 (*a*) The target molecule is the diethyl acetal of acetaldehyde (ethanal).

$$CH_3CH(OCH_2CH_3)_2 \quad \Longrightarrow \quad CH_3\overset{O}{\overset{||}{C}}H \quad \text{and} \quad CH_3CH_2OH$$

Acetaldehyde diethyl acetal

Acetaldehyde may be prepared by oxidation of ethanol.

$$CH_3CH_2OH \quad \overset{PCC}{\underset{CH_2Cl_2}{\longrightarrow}} \quad CH_3\overset{O}{\overset{||}{C}}H$$

Ethanol   Acetaldehyde

Reaction with ethanol in the presence of hydrogen chloride yields the desired acetal.

$$CH_3\overset{O}{\overset{||}{C}}H \quad + \quad 2CH_3CH_2OH \quad \overset{HCl}{\longrightarrow} \quad CH_3CH(OCH_2CH_3)_2$$

Acetaldehyde   Ethanol   Acetaldehyde diethyl acetal

(*b*) In this case the target molecule is a cyclic acetal of acetaldehyde.

$$\Longrightarrow \quad CH_3\overset{O}{\overset{||}{C}}H \quad \text{and} \quad HOCH_2CH_2OH$$

2-Methyl-1,3-dioxolane

Acetaldehyde has been prepared in part (*a*). Recalling that vicinal diols are available from the dihydroxylation of alkenes, 1,2-ethanediol may be prepared by the sequence

$$CH_3CH_2OH \quad \overset{H_2SO_4}{\underset{heat}{\longrightarrow}} \quad H_2C{=}CH_2 \quad \overset{OsO_4,\ (CH_3)_3COOH}{\underset{(CH_3)_3COH,\ HO^-}{\longrightarrow}} \quad HOCH_2CH_2OH$$

Ethanol   Ethylene   1,2-Ethanediol

Hydrolysis of ethylene oxide is also reasonable.

$$H_2C=CH_2 \xrightarrow{CH_3\overset{\displaystyle O}{\overset{\displaystyle \|}{C}}OOH} H_2C\overset{\displaystyle}{\underset{\displaystyle O}{\diagdown\diagup}}CH_2 \xrightarrow[HO^-]{H_2O} HOCH_2CH_2OH$$

Ethylene   Ethylene oxide   1,2-Ethanediol

Reaction of acetaldehyde with 1,2-ethanediol yields the cyclic acetal.

$$CH_3\overset{\displaystyle O}{\overset{\displaystyle \|}{C}}H \ + \ HOCH_2CH_2OH \xrightarrow{H^+}$$

Acetaldehyde   1,2-Ethanediol   2-Methyl-1,3-dioxolane

(*c*) The target molecule is, in this case, the cyclic acetal of 1,2-ethanediol and formaldehyde.

$$\Longrightarrow H\overset{\displaystyle O}{\overset{\displaystyle \|}{C}}H \ \text{ and } \ HOCH_2CH_2OH$$

1,3-Dioxolane

The preparation of 1,2-ethanediol was described in part (*b*). One method of preparing formaldehyde is by ozonolysis of ethylene.

$$CH_3CH_2OH \xrightarrow[\text{heat}]{H_2SO_4} H_2C=CH_2 \xrightarrow[\text{2. }H_2O,\ Zn]{\text{1. }O_3} 2H\overset{\displaystyle O}{\overset{\displaystyle \|}{C}}H$$

Ethanol   Ethylene   Formaldehyde

Another method is periodic acid cleavage of 1,2-ethanediol.

$$HOCH_2CH_2OH \xrightarrow{HIO_4} 2H\overset{\displaystyle O}{\overset{\displaystyle \|}{C}}H$$

1,2-Ethanediol   Formaldehyde

Cyclic acetal formation is then carried out in the usual way.

$$H\overset{\displaystyle O}{\overset{\displaystyle \|}{C}}H \ + \ HOCH_2CH_2OH \xrightarrow{H^+}$$

Formaldehyde   1,2-Ethanediol   1,3-Dioxolane

(*d*) Acetylenic alcohols are best prepared from carbonyl compounds and acetylide anions.

$$CH_3\underset{\displaystyle \underset{\displaystyle OH}{|}}{C}HC\equiv CH \Longrightarrow CH_3\overset{\displaystyle O}{\overset{\displaystyle \|}{C}}H \ + \ {}^-\!:C\equiv CH$$

3-Butyn-2-ol

Acetaldehyde is available as in part (*a*). Alkynes such as acetylene are available from the corresponding alkene by bromination followed by double dehydrobromination. Using ethylene, prepared in part (*b*), the sequence becomes

$$H_2C=CH_2 \xrightarrow{Br_2} BrCH_2CH_2Br \xrightarrow[NH_3]{NaNH_2} HC\equiv CH$$

Ethylene   1,2-Dibromoethane   Acetylene

Then

$$HC\equiv CH \xrightarrow{NaNH_2} HC\equiv CNa \xrightarrow[2.\ H_3O^+]{1.\ CH_3\overset{\displaystyle O}{\overset{\|}{C}}H} HC\equiv C\underset{\underset{\displaystyle OH}{|}}{C}HCH_3$$

Acetylene   Sodium acetylide   3-Butyn-2-ol

(*e*) The target aldehyde may be prepared from the corresponding alcohol.

$$H\overset{\displaystyle O}{\overset{\|}{C}}CH_2C\equiv CH \quad \Longrightarrow \quad HOCH_2CH_2C\equiv CH$$

3-Butynal   3-Butyn-1-ol

The best route to this alcohol is through reaction of an acetylide ion with ethylene oxide.

$$HC\equiv CNa \quad + \quad H_2\overset{\diagdown}{C}\underset{O}{-}CH_2 \xrightarrow[2.\ H_3O^+]{1.\ diethyl\ ether} HC\equiv CCH_2CH_2OH$$

Sodium acetylide   Ethylene oxide   3-Butyn-1-ol
[prepared in part (*d*)]   [prepared in part (*b*)]

Oxidation with PCC or PDC is appropriate for the final step.

$$HC\equiv CCH_2CH_2OH \xrightarrow[CH_2Cl_2]{PCC\ or\ PDC} HC\equiv CCH_2\overset{\displaystyle O}{\overset{\|}{C}}H$$

3-Butyn-1-ol   3-Butynal

(*f*) The target molecule has four carbon atoms, suggesting a route involving reaction of an ethyl Grignard reagent with ethylene oxide.

$$CH_3CH_2CH_2CH_2OH \quad \Longrightarrow \quad CH_3\overset{..}{\overset{-}{C}}H_2 \quad + \quad H_2\overset{\overset{\displaystyle O}{\diagup\diagdown}}{C}-CH_2$$

Ethylmagnesium bromide is prepared in the usual way.

$$CH_3CH_2OH \xrightarrow[PBr_3]{HBr\ or} CH_3CH_2Br \xrightarrow[diethyl\ ether]{Mg} CH_3CH_2MgBr$$

Ethanol   Bromoethane   Ethylmagnesium
                                                          bromide

Reaction of the Grignard reagent with ethylene oxide, prepared in part (*b*), completes the synthesis.

$$CH_3CH_2MgBr \ + \ H_2C{-}CH_2 \xrightarrow[\text{2. } H_3O^+]{\text{1. diethyl ether}} CH_3CH_2CH_2CH_2OH$$

Ethylmagnesium bromide   Ethylene oxide   1-Butanol

17.40 (*a*) Friedel–Crafts acylation of benzene with benzoyl chloride is a direct route to benzophenone.

Benzoyl chloride   Benzene   Benzophenone

(*b*) On analyzing the overall transformation retrosynthetically, we see that the target molecule may be prepared by a Grignard synthesis followed by oxidation of the alcohol formed.

$$C_6H_5\overset{O}{\overset{\|}{C}}C_6H_5 \Rightarrow C_6H_5\overset{OH}{\overset{|}{C}H}C_6H_5 \Rightarrow C_6H_5\overset{O}{\overset{\|}{C}}H \ + \ C_6H_5MgBr$$

In the desired synthesis, benzyl alcohol must first be oxidized to benzaldehyde.

$$C_6H_5CH_2OH \xrightarrow[CH_2Cl_2]{PDC} C_6H_5\overset{O}{\overset{\|}{C}}H$$

Benzyl alcohol   Benzaldehyde

Reaction of benzaldehyde with the Grignard reagent of bromobenzene followed by oxidation of the resulting secondary alcohol gives benzophenone.

$$C_6H_5\overset{O}{\overset{\|}{C}}H \ + \ C_6H_5MgBr \xrightarrow[\text{2. } H_3O^+]{\text{1. diethyl ether}} C_6H_5\underset{OH}{\overset{}{C}H}C_6H_5 \xrightarrow[CH_2Cl_2]{PDC} C_6H_5\overset{O}{\overset{\|}{C}}C_6H_5$$

Benzaldehyde   Phenylmagnesium bromide   Diphenylmethanol   Benzophenone

(*c*) Hydrolysis of bromodiphenylmethane yields the corresponding alcohol, which can be oxidized to benzophenone as in part (*b*).

$$C_6H_5\underset{Br}{CH}C_6H_5 \xrightarrow{H_2O} C_6H_5\underset{OH}{CH}C_6H_5 \xrightarrow{\text{oxidize}} C_6H_5\overset{O}{\overset{\|}{C}}C_6H_5$$

Bromodiphenylmethane   Diphenylmethanol   Benzophenone

(*d*) The starting material is the dimethyl acetal of benzophenone. All that is required is acid-catalyzed hydrolysis.

$$\underset{\overset{\displaystyle |}{OCH_3}}{\overset{\displaystyle OCH_3}{C_6H_5\overset{|}{\underset{|}{C}}C_6H_5}} + 2H_2O \xrightarrow{H^+} \underset{}{\overset{\displaystyle O}{C_6H_5\overset{||}{C}C_6H_5}} + 2CH_3OH$$

Dimethoxydiphenyl-   Water   Benzophenone   Methanol
methane

(*e*) Oxidative cleavage of the alkene yields benzophenone. Ozonolysis may be used.

$$(C_6H_5)_2C{=}C(C_6H_5)_2 \xrightarrow[\text{2. }H_2O,\ Zn]{\text{1. }O_3} 2(C_6H_5)_2C{=}O$$

1,1,2,2-Tetraphenylethene   Benzophenone

17.41 Each of the specified starting materials in the problem is a primary alcohol and thus terminates in a CH_2OH group. The desired carbon skeleton can be obtained by connecting the carbons of these two groups. For clarity we can abbreviate the structures in the synthesis:

$$CH_3(CH_2)_8CH_2\overset{\overset{\displaystyle O}{||}}{C}CH_2CH_2CH_2\underset{H}{\overset{H}{\underset{\diagdown}{C}}}{=}\underset{CH_2CH_2CH_2CH{=}CH_2}{\overset{\diagup}{C}} \qquad \text{becomes}$$

$$CH_3(CH_2)_8CH_2\overset{\overset{\displaystyle O}{||}}{C}CH_2CH_2CH_2R \quad \text{where R equals} \quad \underset{H}{\overset{H}{\underset{\diagdown}{C}}}{=}\underset{CH_2CH_2CH_2CH{=}CH_2}{\overset{\diagup}{C}}$$

Using retrosynthetic analysis we can see that the carbonyl group of the target molecule can be obtained by oxidation of the corresponding secondary alcohol.

$$CH_3(CH_2)_8CH_2\overset{\overset{\displaystyle O}{||}}{C}CH_2CH_2CH_2R \quad\Longrightarrow\quad CH_3(CH_2)_8CH_2\overset{\overset{\displaystyle OH}{|}}{C}HCH_2CH_2CH_2R$$

The secondary alcohol can be made by a Grignard reaction.

disconnect here

$$CH_3(CH_2)_8CH_2{\dashv}\overset{\overset{\displaystyle OH}{|}}{C}HCH_2CH_2CH_2R \quad\Longrightarrow\quad CH_3(CH_2)_8CH_2MgBr \quad \overset{\overset{\displaystyle O}{||}}{H}CCH_2CH_2CH_2R$$

The Grignard reagent and the aldehyde can both be traced back to the specified starting materials.

$$CH_3(CH_2)_8CH_2MgBr \Longrightarrow CH_3(CH_2)_8CH_2Br \Longrightarrow CH_3(CH_2)_8CH_2OH$$

$$\overset{\overset{\displaystyle O}{||}}{H}CCH_2CH_2CH_2R \Longrightarrow HOCH_2CH_2CH_2CH_2R$$

This retrosynthetic analysis suggests the following synthesis:

First, prepare the aldehyde and the Grignard reagent.

$$HOCH_2CH_2CH_2CH_2R \xrightarrow[CH_2Cl_2]{PCC} HCCH_2CH_2CH_2R$$

(E)-5,10-Undecadien-1-ol   (E)-5,10-Undecadienal

$$CH_3(CH_2)_8CH_2OH \xrightarrow[heat]{HBr} CH_3(CH_2)_8CH_2Br \xrightarrow[diethyl\ ether]{Mg} CH_3(CH_2)_8CH_2MgBr$$

1-Decanol   1-Bromodecane   Decylmagnesium bromide

Add the aldehyde to the Grignard reagent, then add aqueous acid to "work-up" the reaction. The product is a secondary alcohol having the required carbon skeleton.

$$CH_3(CH_2)_8CH_2MgBr \quad HCCH_2CH_2CH_2R \xrightarrow[2.\ H_3O]{1.\ diethyl\ ether} CH_3(CH_2)_8CH_2CHCH_2CH_2CH_2R$$

(E)-1,6-Henicosadien-11-ol

Complete the synthesis by oxidizing the secondary alcohol to the target ketone.

$$CH_3(CH_2)_8CH_2CHCH_2CH_2CH_2R \xrightarrow[CH_2Cl_2]{PCC} CH_3(CH_2)_8CH_2CCH_2CH_2CH_2$$

with the =C fragment:

$$\underset{H}{\overset{H}{C}}=\underset{CH_2CH_2CH_2CH=CH_2}{\overset{}{C}}$$

(E)-1,6-Henicosadien-11-one

17.42 The two alcohols given as starting materials contain all the carbon atoms of the desired product.

$$CH_3(CH_2)_8CH=CHCH_2CH=CHCH_2CH \xrightarrow{} CHCH=CH_2 \implies$$

$$CH_3(CH_2)_8CH=CHCH_2CH=CHCH_2CH_2OH \ \ and \ \ HOCH_2CH=CH_2$$

3,6-Hexadecadien-1-ol   Allyl alcohol

What is needed is to attach the two groups together so that the two primary alcohol carbons become doubly bonded to each other. This can be accomplished by using a Wittig reaction as the key step.

$$CH_3(CH_2)_8CH=CHCH_2CH=CHCH_2CH_2OH \xrightarrow[CH_2Cl_2]{PCC} CH_3(CH_2)_8CH=CHCH_2CH=CHCH_2CH$$

3,6-Hexadecadien-1-ol   3,6-Hexadecadienal

$$H_2C=CHCH_2OH \xrightarrow{\text{PBr}_3} H_2C=CHCH_2Br \xrightarrow{(C_6H_5)_3P} (C_6H_5)_3\overset{+}{P}CH_2CH=CH_2 \ Br^-$$

Allyl alcohol   Allyl bromide   Allyltriphenylphosphonium bromide

$$\downarrow \text{CH}_3\text{CH}_2\text{CH}_2\text{CH}_2\text{Li, THF}$$

$$(C_6H_5)_3\overset{+}{P}-\overset{..}{\overset{-}{C}}HCH=CH_2$$

Allylidenetriphenylphosphorane

$$CH_3(CH_2)_8CH=CHCH_2CH=CHCH_2\overset{\overset{\displaystyle O}{\displaystyle \|}}{C}H \ + \ (C_6H_5)_3\overset{+}{P}-\overset{..}{\overset{-}{C}}HCH=CH_2 \longrightarrow$$

3,6-Hexadecadienal   Allylidenetriphenylphosphorane

$$CH_3(CH_2)_8CH=CHCH_2CH=CHCH_2CH=CHCH=CH_2$$

1,3,6,9-Nonadecatetraene

Alternatively, allyl alcohol could be oxidized to $H_2C=CHCHO$ for subsequent reaction with the ylide derived from $CH_3(CH_2)_8CH=CHCH_2CH=CHCH_2CH_2OH$ via its bromide and triphenylphosphonium salt.

17.43 (*a*) Recalling that alkanes may be prepared by hydrogenation of the appropriate alkene, a synthesis of the desired product becomes apparent. What is needed is to convert $-C=O$ into $-C=CH_2$; a Wittig reaction is appropriate.

5,5-Dimethylcyclononanone   1,1,5-Trimethylcyclononane

The two-step procedure that was followed used a Wittig reaction to form the carbon–carbon bond, then catalytic hydrogenation of the resulting alkene.

5,5-Dimethylcyclononanone

5,5-Dimethyl-1-
methylenecyclononane (59%)   1,1,5-Trimethylcyclononane
(73%)

(*b*) In putting together the carbon skeleton of the target molecule, a methyl group has to be added to the original carbonyl carbon.

The logical way to do this is by way of a Grignard reagent.

Cyclopentyl phenyl ketone

Methylmagnesium iodide

1-Cyclopentyl-1-phenylethanol

Acid-catalyzed dehydration yields the more highly substituted alkene, the desired product, in accordance with the Zaitsev rule.

1-Cyclopentyl-1-phenylethanol

(1-Phenylethylidene)cyclopentane

(*c*) Analyzing the transformation retrosynthetically, keeping in mind the starting materials stated in the problem, we see that the carbon skeleton may be constructed in a straightforward manner.

Proceeding with the synthesis in the forward direction, reaction between the Grignard reagent of *o*-bromotoluene and 5-hexenal produces most of the desired carbon skeleton.

o-Methylphenylmagnesium bromide

5-Hexenal

1-(*o*-Methylphenyl)-5-hexen-1-ol

Oxidation of the resulting alcohol to the ketone followed by a Wittig reaction leads to the final product.

1-(o-Methylphenyl)-5-hexen-1-ol

1-(o-Methylphenyl)-5-hexen-1-one   2-(o-Methylphenyl)-1,6-heptadiene

Acid-catalyzed dehydration of the corresponding tertiary alcohol would *not* be suitable because the major elimination product would have the more highly substituted double bond.

2-(o-Methylphenyl)-6-hepten-2-ol   6-(o-Methylphenyl)-1,5-heptadiene

(*d*) Remember that terminal alkynes can serve as sources of methyl ketones by hydration.

$$CH_3\overset{O}{\overset{||}{C}}CH_2CH_2\overset{O}{\overset{||}{C}}(CH_2)_5CH_3 \implies HC\equiv CCH_2CH_2\overset{O}{\overset{||}{C}}(CH_2)_5CH_3$$

This gives us a clue as to how to proceed, because the acetylenic ketone may be prepared from the starting acetylenic alcohol.

$$HC\equiv CCH_2CH_2\overset{O}{\overset{||}{C}}(CH_2)_5CH_3 \implies HC\equiv CCH_2CH_2\overset{O}{\overset{||}{C}}H + CH_3(CH_2)_4\overset{..}{C}H_2 \implies HC\equiv CCH_2CH_2CH_2OH$$

The first synthetic step is oxidation of the primary alcohol to the aldehyde and construction of the carbon skeleton by a Grignard reaction.

$$HC\equiv CCH_2CH_2CH_2OH \xrightarrow[CH_2Cl_2]{PDC} HC\equiv CCH_2CH_2\overset{O}{\overset{||}{C}}H \xrightarrow[2.\ H_3O^+]{1.\ CH_3(CH_2)_5MgBr} HC\equiv CCH_2CH_2\underset{OH}{CH}(CH_2)_5CH_3$$

4-Pentyn-1-ol   4-Pentynal   1-Undecyn-5-ol

Oxidation of the secondary alcohol to a ketone and hydration of the terminal triple bond complete the synthesis.

$$HC\equiv CCH_2CH_2\underset{OH}{CH}(CH_2)_5CH_3 \xrightarrow[CH_2Cl_2]{PDC} HC\equiv CCH_2CH_2\overset{O}{\overset{||}{C}}(CH_2)_5CH_3 \xrightarrow[HgSO_4]{H_2O,\ H_2SO_4} CH_3\overset{O}{\overset{||}{C}}CH_2CH_2\overset{O}{\overset{||}{C}}(CH_2)_5CH_3$$

1-Undecyn-5-ol   1-Undecyn-5-one   2,5-Undecanedione (*e*) The desired product is a benzylic ether. To prepare it, the aldehyde must first be reduced to the corresponding primary alcohol. Sodium borohydride was used in the preparation described in the literature, but lithium aluminum hydride or catalytic hydrogenation would also be possible. Once the alcohol is prepared, it can be converted to its alkoxide ion and this alkoxide ion treated with methyl iodide.

Alternatively, the alcohol could be treated with hydrogen bromide or with phosphorus tribromide to give the benzylic bromide and the bromide then allowed to react with sodium methoxide.

(*f*) Two operations need to be carried out: reduction of C=O to CH$_2$, and Friedel–Crafts acylation. A successful synthesis depends on the order in which the operations are accomplished.

By reducing C=O to CH$_2$ first, a compound is formed in which both positions on the benzene ring that are available for Friedel–Crafts acylation are equivalent and are activated toward eletrophilic aromatic substitution.

If Friedel–Crafts acylation were attempted first, it would probably fail because the C=O group in the starting material deactivates the aromatic ring toward electrophilic aromatic substitution. Furthermore, even if Friedel–Crafts acylation were successful, the product of that reaction would contain two carbonyl groups and there is no reason to believe that one could be reduced to CH$_2$ and not the other.

(*g*) Retrosynthetically, the desired product is available by a Diels-Alder cycloaddition.

This suggests the following synthesis. The first reaction is the formation of the conjugated diene unit via a Wittig reaction.

17.44 Step 1 of the synthesis is formation of a cyclic acetal protecting group; the necessary reagents are ethylene glycol ($HOCH_2CH_2OH$) and p-toluenesulfonic acid, with heating in benzene. In step 2 the ester function is reduced to a primary alcohol. Lithium aluminum hydride ($LiAlH_4$) is the reagent of choice. Oxidation with PCC in CH_2Cl_2 converts the primary alcohol to an aldehyde in step 3. Wolff–Kishner reduction (N_2H_4, KOH, ethylene glycol, heat) converts the aldehyde group to a methyl group in step 4. The synthesis is completed in step 5 by hydrolysis (H_3O^+) of the acetal-protecting group.

17.45 We need to assess the extent of resonance donation to the carbonyl group by the π electrons of the aromatic rings. Such resonance for benzaldehyde may be written as

Electron-releasing groups such as methoxy at positions ortho and para to the aldehyde function increase the "single-bond character" of the aldehyde by stabilizing the dipolar resonance forms and increasing their contribution to the overall electron distribution in the molecule. Electron-withdrawing groups such as nitro decrease this single-bond character. The aldehyde with the lowest carbonyl stretching frequency is 2,4,6-trimethoxybenzaldehyde; the one with the highest is 2,4,6-trinitrobenzaldehyde. The measured values are

2,4,6-Trimethoxybenzaldehyde
(1665 cm^{-1})

Benzaldehyde
(1700 cm^{-1})

2,4,6-Trinitrobenzaldehyde
(1715 cm^{-1})

17.46 The signal in the ^{1}H NMR spectrum at δ 9.7 tells us that the compound is an aldehyde rather than a ketone. The 2H signal at δ 2.4 indicates that the group adjacent to the carbonyl is a CH_2 group. The remaining signals support the assignment of the compound as butanal.

17.47 A carbonyl group is evident from the strong IR absorption at 1710 cm^{-1}. Because all the ^{1}H NMR signals are singlets, there are no nonequivalent hydrogens in a vicinal or "three-bond" relationship. The three-proton signal at δ 2.1 and the 2-proton signal at δ 2.3 can be understood as arising from a

$$\overset{\text{O}}{\overset{\|}{\text{CH}_2\text{CCH}_3}}$$

unit. The intense 9-proton singlet at δ 1.0 is due to the three equivalent methyl groups of a $(CH_3)_3C$ unit. The compound is 4,4-dimethyl-2-pentanone.

4,4-Dimethyl-2-pentanone

17.48 The molecular formula of compounds A and B ($C_6H_{10}O_2$) indicates an index of hydrogen deficiency of 2. Because we are told the compounds are diketones, the two carbonyl groups account for all the unsaturations.

The ^{1}H NMR spectrum of compound A has only two peaks, both singlets, at δ 2.2 and 2.8. Their intensity ratio (6:4) is consistent with two equivalent methyl groups and two equivalent methylene groups. The chemical shifts are appropriate for

$$\overset{\text{O}}{\overset{\|}{\text{CH}_3\text{C}}} \quad \text{and} \quad \overset{\text{O}}{\overset{\|}{\text{CH}_2\text{C}}}$$

The simplicity of the spectrum can be understood if we are dealing with a symmetric diketone. The correct structure is

$$\overset{\text{O}\qquad\text{O}}{\overset{\|\qquad\quad\|}{\text{CH}_3\text{CCH}_2\text{CH}_2\text{CCH}_3}}$$

Equivalent methylene
groups do not split
each other.

2,5-Hexanedione (compound A)

Compound B is an isomer of compound A. The triplet–quartet pattern in the ^{1}H NMR spectrum is consistent with an ethyl group and, because the triplet is equivalent to six protons and the quartet to four, it is likely that two equivalent ethyl groups are present. The two ethyl groups account for four carbons, and because the problem stipulates that the molecule is a diketone, all the carbons are accounted for. The only $C_6H_{10}O_2$ diketone with two equivalent ethyl groups is 3,4-hexanedione.

$$CH_3CH_2\overset{\overset{\displaystyle O}{||}}{C}-\overset{\overset{\displaystyle O}{||}}{C}CH_2CH_3$$

δ 1.3 triplet   δ 2.8 quartet

3,4-Hexanedione (compound B)

17.49 From its molecular formula ($C_{11}H_{14}O$), the compound has a total of five double bonds and rings. The presence of signals in the region δ 7 to 8 suggests an aromatic ring is present, accounting for four of the elements of unsaturation. The presence of a strong peak near 1700 cm^{-1} in the IR spectrum indicates the presence of a carbonyl group, accounting for the remaining element of unsaturation. The highest field peak in the NMR spectrum is a three-proton triplet, corresponding to the methyl group of a CH_3CH_2 unit. The 2-proton signal at δ 3.0 corresponds to a CH_2 unit adjacent to the carbonyl group and, because it is a triplet, suggests the grouping $CH_2CH_2C=O$. The compound is butyl phenyl ketone (1-phenyl-1-pentanone).

$$\overset{\overset{\displaystyle O}{||}}{C}CH_2CH_2CH_2CH_3$$

δ 3.0 triplet   Multiplets at δ 1.4 and 1.7   δ 1.0 triplet

Butyl phenyl ketone

17.50 With a molecular formula of $C_7H_{14}O$, the compound has an index of hydrogen deficiency of 1. We are told that it is a ketone, so it has no rings or double bonds other than the one belonging to its C=O group. The peak at δ 211 in the ^{13}C NMR spectrum corresponds to the carbonyl carbon. Only three other signals occur in the spectrum, and so there are only three types of carbons other than the carbonyl carbon. This suggests that the compound is the symmetrical ketone 4-heptanone.

$$CH_3CH_2CH_2\overset{\overset{\displaystyle O}{||}}{C}CH_2CH_2CH_3$$

14   17   45   211   45   17   14

4-Heptanone
(all chemical shifts in ppm)

17.51 Compounds A and B are isomers and have an index of hydrogen deficiency of 5. Signals in the region 125–140 in their ^{13}C NMR spectra suggest an aromatic ring, and a peak at δ 200 indicates a carbonyl group. An aromatic ring contributes one ring and three double bonds, and a carbonyl group contributes one double bond, and so the index of hydrogen deficiency of 5 is satisfied by a benzene ring and a carbonyl group. The carbonyl group is attached directly to the benzene ring, as evidenced by the presence of a peak at m/z 105 in the mass spectra of compounds A and B.

m/z 105

Each ^{13}C NMR spectrum shows four aromatic signals, and so the rings are monosubstituted. Compound A has three unique carbons in addition to $C_6H_5C=O$ and so must be 1-phenyl-1-butanone. Compound B has only two additional signals and so must be 2-methyl-1-phenyl-1-propanone.

Compound A   Compound B

ANSWERS TO INTERPRETIVE PROBLEMS

17.52 A;   **17.53** C;   **17.54** A;   **17.55** A;   **17.56** B;   **17.57** A

SELF-TEST

PART A

A-1. Give an acceptable IUPAC name for each of the following:

(a)

(c)

(b)

(d)

A-2. Write the structural formulas for

(a) (E)-3-Hexen-2-one

(b) 3-Cyclopropyl-2,4-pentanedione

(c) 3-Ethyl-4-phenylpentanal

A-3. For each of the following reactions supply the structure of the missing reactant, reagent, or product:

(a)

(b)

(c) $(CH_3)_2CHCH$ (cyclic acetal) $\xrightarrow{H_2O,\ H^+}$? (two products)

(d) (cyclopentanone) $=O$ + ? $\longrightarrow$ (cyclopentane)$=CHCH_2CH_3$

(e) (2-methylcyclohexanone) $=O$ + $C_6H_5NHNH_2$ $\longrightarrow$?

(f) $CH_3CH_2CH_2\overset{\overset{O}{\|}}{C}H$ + $2CH_3CH_2OH$ $\xrightarrow{HCl}$?

(g) ? + ? $\longrightarrow$ $C_6H_5\overset{\overset{N(CH_3)_2}{|}}{C}=CHCH_3$

(h) $CH_3CH_2CH_2CH=O$ $\xrightarrow[H^+,\ H_2O]{Na_2Cr_2O_7}$?

A-4. Write the structures of the products, compounds A through E, of the reaction steps shown.

(a) $(C_6H_5)_3P$ + $(CH_3)_2CHCH_2Br$ $\longrightarrow$ A

A + $CH_3CH_2CH_2CH_2Li$ $\longrightarrow$ B + C_4H_{10}

B + benzaldehyde $\longrightarrow$ C + $(C_6H_5)_3\overset{+}{P}-O^-$

(b) $C_6H_5CH_2\overset{\overset{OH}{|}}{C}HCH_3$ $\xrightarrow[CH_2Cl_2]{PCC}$ D

D + $CH_3\overset{\overset{O}{\|}}{C}OOH$ $\longrightarrow$ E + $CH_3\overset{\overset{O}{\|}}{C}OH$

A-5. Give the reagents necessary to convert cyclohexanone into each of the following compounds. More than one step may be necessary.

(a) (1-methylcyclohexene)CH_3 (b) (methylenecyclohexane)CH_2 (c) (spiro dioxolane) (d) (lactone)

A-6. (a) What two organic compounds react together (in the presence of an acid catalyst) to give the compound shown, plus a molecule of water?

(*b*) Draw the structure of the open-chain form of the following cyclic acetal:

A-7. Outline reaction schemes to carry out each of the following interconversions, using any necessary organic or inorganic reagents.

(*a*) $(CH_3)_2C{=}O$ to $(CH_3)_2C{-}CHCH_3$ (with epoxide O)

(*b*)

to

A-8. Write a stepwise mechanism for the formation of $CH_3CH(OCH_3)_2$ from acetaldehyde and methanol under conditions of acid catalysis.

A-9. Suggest a structure for an unknown compound, $C_9H_{10}O$, that exhibits a strong infrared absorption at $1710\ cm^{-1}$ and has a ^1H NMR spectrum that consists of three singlets at δ 2.1 (3H), 3.7 (2H), and 7.2 (5H).

PART B

B-1. Which of the compounds shown is (are) correctly named as pentane derivatives, either as pentanals or pentanones?

(*a*) 1 only   (*b*) 2 only   (*c*) 3 only   (*d*) 1 and 3   (*e*) None of them

B-2. The compound shown is best classified as a(n)

$$(CH_3)_3CCH_2CH{=}NCH_3$$

(*a*) Carbinolamine   (*d*) Imine

(*b*) Enamine   (*e*) Oxime

(*c*) Hydrazone

B-3. When a nucleophile encounters a ketone, the site of attack is

(*a*) The carbon atom of the carbonyl

(*b*) The oxygen atom of the carbonyl

(*c*) Both the carbon and oxygen atoms, with equal probability

(*d*) No attack occurs—ketones do not react with nucleophiles.

B-4. What reagent and/or reaction conditions would you choose to bring about the following conversion?

(*a*) 1. $LiAlH_4$, 2. H_2O

(*c*) H_2O, H_2SO_4, heat

(*b*) H_2O, NaOH, heat

(*d*) PCC, CH_2Cl_2

B-5. Rank the following in order of increasing value of the equilibrium constant for hydration, K_{hyd} (smallest value first).

1	2	3

$(CH_3)_3CCC(CH_3)_3$

(*a*) $1 < 2 < 3$ (*b*) $3 < 1 < 2$ (*c*) $2 < 1 < 3$ (*d*) $2 < 3 < 1$

B-6. The structure

would be best classified as a(n)

(*a*) Acetal   (*b*) Hemiacetal   (*c*) Hydrate   (*d*) Cyanohydrin

B-7. Which of the following pairs of reactants is most effective in forming an enamine?

(*a*) CH_3CH_2CH + $[(CH_3)_2CH]_2NH$

(*c*) + $(CH_3)_3CNH_2$

(*b*) $(CH_3)_3CCH$ + $(CH_3)_2NH$

(*d*) None of these forms an enamine.

B-8. Which of the following species is an ylide?

(*a*) $(C_6H_5)_3\overset{+}{P}CH_2CH_3$ Br^-

(*c*) $(C_6H_5)_3P—CHCH_3$
$\quad\quad\quad\quad\ O—CH_2$

(*b*) $(C_6H_5)_3\overset{+}{P}\overset{..}{C}HCH_3$

(*d*) None of these

B-9. Which pair of the following compounds could serve as the reagents X and Y in the following reaction sequence?

$$X \xrightarrow{(C_6H_5)_3P} \xrightarrow{CH_3CH_2CH_2CH_2Li} \xrightarrow{Y} (CH_3)_2CHCH=C{\overset{CH_3}{\underset{CH_2CH_3}{}}}$$

$$(CH_3)_2CHCH_2Br \quad (CH_3)_2CHBr \quad CH_3CH_2\overset{CH_3}{\underset{|}{C}}HBr \quad (CH_3)_2CH\overset{O}{\overset{||}{C}}H \quad CH_3\overset{O}{\overset{||}{C}}CH_3$$

| | 1 | | 2 | | 3 | | 4 | | 5 |

	X	Y		X	Y
(a)	1	5	(d)	2	5
(b)	1	4	(e)	3	4
(c)	2	4			

B-10. The final product of the following sequence of reactions is

$$(CH_3O)_2CHCH_2CH_2CH_2Br \xrightarrow{Mg} \xrightarrow{H_2C=O} \xrightarrow[\text{heat}]{H_3O^+} ?$$

(a) $CH_3O\overset{O}{\overset{||}{C}}CH_2CH_2CH_2CH_2OH$

(b) $CH_3\overset{O}{\overset{||}{C}}CH_2CH_2CH_2CH_2OH$

(c) $H\overset{O}{\overset{||}{C}}CH_2CH_2CH_2CH_2OH$

(d) $H\overset{O}{\overset{||}{C}}CH_2CH_2CH_2\overset{O}{\overset{||}{C}}H$

(e) $(CH_3O)_2CHCH_2CH_2CH_2\overset{O}{\overset{||}{C}}H$

B-11. Which of the following sets of reagents, used in the order shown, would successfully accomplish the conversion shown?

conversion: cyclohexyl–$\overset{O}{\overset{||}{C}}CH_3 \xrightarrow{?}$ cyclohexyl–$\overset{CH_3}{\underset{OH}{\overset{|}{C}H}}CHCH_2CH_3$

(a) $CH_3CH_2CH_2MgBr$; H_3O^+; PCC, CH_2Cl_2

(b) $CH_3CH_2CH_2MgBr$; H_3O^+; H_2SO_4, heat; PCC, CH_2Cl_2

(c) $(C_6H_5)_3\overset{+}{P}-\overset{..}{C}HCH_2CH_3$; B_2H_6; H_2O_2, HO⁻

(d) $(C_6H_5)_3\overset{+}{P}-\overset{..}{C}HCH_2CH_3$; H_2SO_4, H_2O

B-12. Which of the following species is the conjugate acid of the hemiacetal formed by reaction of benzaldehyde with methanol containing a trace of acid?

(a) $C_6H_5\overset{OH}{\underset{OCH_3}{C}}H$

(c) $C_6H_5\overset{OH}{\underset{\underset{H}{+OCH_3}}{C}}H$

(e) $C_6H_5\overset{+OH}{\underset{}{C}}H$

(b) $C_6H_5CH=\overset{+}{O}CH_3$

(d) $C_6H_5\overset{OH}{\underset{+OH_2}{C}}H$

B-13. Which sequence represents the best synthesis of hexanal?

$$CH_3CH_2CH_2CH_2CH_2CH=O$$

Hexanal

(a) 1. $CH_3CH_2CH_2CH_2Br$ + $NaC\equiv CH$
    2. H_2O, H_2SO_4, $HgSO_4$

(c) 1. $CH_3CH_2CH_2CH=CH_2$ + $CH_3\overset{O}{\overset{\|}{C}}OOH$
    2. CH_3MgBr, diethyl ether
    3. H_3O^+
    4. PCC, CH_2Cl_2

(b) 1. $CH_3CH_2CH_2CH_2\overset{O}{\overset{\|}{C}}CH_3$
    2. $CH_3\overset{O}{\overset{\|}{C}}OOH$
    3. $LiAlH_4$
    4. H_2O
    5. PCC, CH_2Cl_2

(d) 1. $CH_3CH_2CH_2CH_2MgBr$ + $H_2C\overset{}{\underset{O}{-}}CH_2$
    2. H_3O^+
    3. PCC, CH_2Cl_2

B-14. The amino ketone shown undergoes a spontaneous cyclization on standing. What is the product of this intramolecular reaction?

(a)

(b)

(c)

(d)

(e)

B-15. Which of the following compounds would have a ^{1}H NMR spectrum consisting of three singlets?

$\qquad$ (a) $\underset{\underset{\displaystyle CH_3}{|}}{\overset{\overset{\displaystyle CH_3\ \ \ O}{\ |\ \ \ \|}}{CH_3CCH_2CCH_3}}$

$\qquad$ (c) $\overset{\overset{\displaystyle O\qquad\quad O}{\|\qquad\quad\|}}{HCCH_2CH_2CH_2CH}$

$\qquad$ (b) $\overset{\overset{\displaystyle O}{\|}}{CH_3CH_2CH_2CCH_2CH_2CH_3}$

$\qquad$ (d) $\overset{\overset{\displaystyle O}{\|}}{\langle\!\!\!\rangle\!\!-\!CH_2CH_2CH}$

B-16. Which of the following compounds would have the fewest number of signals in its ^{13}C NMR spectrum?

$\qquad$ (a) $\underset{\underset{\displaystyle CH_3}{|}}{\overset{\overset{\displaystyle CH_3\ \ \ O}{\ |\ \ \ \|}}{CH_3CCH_2CCH_3}}$

$\qquad$ (c) $\underset{\underset{\displaystyle CH_3\ \ CH_3}{|\qquad|}}{\overset{\overset{\displaystyle O}{\|}}{CH_3CHCCHCH_3}}$

$\qquad$ (b) $\overset{\overset{\displaystyle O}{\|}}{CH_3CH_2CH_2CCH_2CH_2CH_3}$

$\qquad$ (d) $\underset{\underset{\displaystyle CH_3}{|}}{\overset{\overset{\displaystyle O}{\|}}{CH_3CHCCH_2CH_3}}$

CHAPTER 18

Enols and Enolates

SOLUTIONS TO TEXT PROBLEMS

18.1 (*b*) Hydroxide acts as a base and removes a proton from the α carbon. Although the structure of the enolate is shown as having a trans double bond, it is actually present as a cis/trans mixture.

Hydroxide ion (base)	3-Methylbutanal (acid)	Water (conjugate acid)	Enolate (conjugate base)

(*c*) There are four equivalent α hydrogens in cyclohexanone, two of which are shown in the equation.

Hydroxide ion (base)	Cyclohexanone (acid)	Water (conjugate acid)	Enolate (conjugate base)

18.2 (*b*) Removal of a proton from 1-phenyl-1,3-butanedione occurs on the methylene group between the carbonyls.

The three most stable resonance contributors of this anion are

$$C_6H_5\overset{\overset{\ddot{O}:}{\|}}{C}CH=\overset{\overset{:\ddot{O}:^-}{|}}{C}CH_3 \longleftrightarrow C_6H_5\overset{\overset{\ddot{O}:}{\|}}{C}\overset{..}{C}H\overset{\overset{\ddot{O}:}{\|}}{C}CH_3 \longleftrightarrow C_6H_5\overset{\overset{:\ddot{O}:^-}{|}}{C}=CH\overset{\overset{\ddot{O}:}{\|}}{C}CH_3$$

(*c*) Removal of a proton from C-2 of this β-dicarbonyl compound yields the carbanion shown.

The three most stable resonance contributors of the anion are:

18.3 Methyllithium adds to the carbonyl group of aldehydes and ketones to give alcohols (text Section 14.7).

18.4 (*b*) Approaching this problem mechanistically in the same way as part (*a*), write the structure of the enolate ion from 2-methylbutanal.

$$\underset{\underset{CH_3}{|}}{CH_3CH_2\overset{\overset{O}{\|}}{C}HCH} + HO^- \rightleftharpoons \underset{\underset{CH_3}{|}}{CH_3CH_2\overset{\overset{O}{\|}}{C}CH} \longleftrightarrow \underset{\underset{CH_3}{|}}{CH_3CH_2C\overset{\overset{O^-}{|}}{=}CH}$$

2-Methylbutanal   Enolate of 2-methylbutanal

This enolate adds to the carbonyl group of the aldehyde.

$$\underset{\underset{CH_3}{|}}{CH_3CH_2\overset{\overset{O}{\|}}{C}HCH}\quad \underset{\underset{HC=O}{|}}{{}^-:\overset{\overset{CH_3}{|}}{C}CH_2CH_3} \longrightarrow \underset{\underset{CH_3}{|}}{CH_3CH_2CHCH}-\underset{\underset{HC=O}{|}}{\overset{\overset{O^-}{|}}{C}\overset{\overset{CH_3}{|}}{C}CH_2CH_3}$$

2-Methylbutanal   Enolate of
                      2-methylbutanal

A proton transfer from solvent yields the product of aldol addition.

$$\underset{\underset{HC=O}{|}}{CH_3CH_2CHCH}-\underset{\underset{CH_3}{|}}{\overset{\overset{O^-}{|}}{C}CH_2CH_3} + H_2O \longrightarrow \underset{\underset{HC=O}{|}}{CH_3CH_2CHCH}-\underset{\underset{CH_3}{|}}{\overset{\overset{HO}{|}}{C}CH_2CH_3} + HO^-$$

2-Ethyl-3-hydroxy-2,4-
dimethylhexanal

(*c*) The aldol addition product of 3-methylbutanal can be identified through the same mechanistic approach.

$$(CH_3)_2CHCH_2\overset{\overset{O}{\|}}{C}H \ + \ HO^- \ \rightleftharpoons \ (CH_3)_2CH\ddot{\overset{-}{C}}H\overset{\overset{O}{\|}}{C}H \ + \ H_2O$$

3-Methylbutanal   Enolate of
                                               3-methylbutanal

$$(CH_3)_2CHCH_2\overset{\overset{O}{\|}}{C}H \ + \ :\overset{-}{\underset{HC=O}{C}}HCH(CH_3)_2 \ \longrightarrow \ (CH_3)_2CHCH_2\overset{\overset{O^-}{|}}{C}H-\underset{HC=O}{C}HCH(CH_3)_2$$

3-Methylbutanal   Enolate of
                        3-methylbutanal

$$\Big\downarrow H_2O$$

$$(CH_3)_2CHCH_2\overset{\overset{OH}{|}}{C}H-\underset{HC=O}{C}HCH(CH_3)_2 \ + \ OH^-$$

3-Hydroxy-2-isopropyl-5-methylhexanal

18.5 Dehydration of the aldol addition product involves loss of a proton from the α-carbon atom and hydroxide from the β-carbon atom.

$$R_2\overset{\overset{OH}{|}}{C}-\underset{\underset{\overset{|}{:\ddot{O}H}}{H}}{C}H\overset{\overset{O}{\|}}{C}H \ \xrightarrow{\ heat\ } \ R_2C=CH\overset{\overset{O}{\|}}{C}H \ + \ H_2O \ + \ HO^-$$

(*b*) The product of aldol addition of 2-methylbutanal has no α hydrogens. It cannot dehydrate to an aldol condensation product.

$$2CH_3CH_2\underset{CH_3}{\overset{\overset{O}{\|}}{C}H}CH \ \xrightarrow[\ ]{HO^-} \ CH_3CH_2\underset{CH_3}{C}HCH-\overset{\overset{HO}{|}\ \overset{\alpha}{\searrow}\ \overset{CH_3}{|}}{\underset{HC=O}{C}}CH_2CH_3$$

2-Methylbutanal   (No protons on α-carbon atom)

(*c*) Aldol condensation is possible with 3-methylbutanal.

$$2(CH_3)_2CHCH_2\overset{\overset{O}{\|}}{C}H \ \xrightarrow[\ ]{HO^-} \ (CH_3)_2CHCH_2\underset{HC=O}{\overset{\overset{HO}{|}}{C}}HCHCH(CH_3)_2 \ \xrightarrow{-H_2O} \ (CH_3)_2CHCH_2CH=\underset{HC=O}{C}CH(CH_3)_2$$

3-Methylbutanal   Aldol addition product   2-Isopropyl-5-methyl-2-hexenal

18.6 The carbon skeleton of 2-ethyl-1-hexanol is the same as that of the aldol condensation product derived from butanal. Hydrogenation of this compound under conditions in which both the carbon–carbon double bond and the carbonyl group are reduced gives 2-ethyl-1-hexanol.

$$\underset{\text{Butanal}}{CH_3CH_2CH_2\overset{\displaystyle O}{\overset{\|}{C}}H} \quad \xrightarrow[\text{heat}]{NaOH, H_2O} \quad \underset{\text{2-Ethyl-2-hexenal}}{CH_3CH_2CH_2CH=\overset{\displaystyle O}{\overset{\|}{C}}\underset{\underset{\displaystyle CH_2CH_3}{|}}{C}H} \quad \xrightarrow{H_2, Ni} \quad \underset{\text{2-Ethyl-1-hexanol}}{CH_3CH_2CH_2CH_2\underset{\underset{\displaystyle CH_2CH_3}{|}}{C}HCH_2OH}$$

18.7 Acetaldehyde can lose an α hydrogen to form an enolate.

$$CH_3\overset{\displaystyle O}{\overset{\|}{C}}H \;+\; HO^- \;\; \rightleftharpoons \;\; H_2C=\overset{\displaystyle O^-}{\overset{|}{C}}H \;+\; H_2O$$

The enolate can add to either a second molecule of acetaldehyde (to give 3-hydroxybutanal) or to a molecule of propanal (to give 3-hydroxypentanal).

$$\xrightarrow{H_2O} \quad \underset{\text{3-Hydroxybutanal}}{CH_3\overset{\overset{\displaystyle OH}{|}}{C}HCH_2\overset{\overset{\displaystyle O}{\|}}{C}H}$$

$$\xrightarrow{H_2O} \quad \underset{\text{3-Hydroxypentanal}}{CH_3CH_2\overset{\overset{\displaystyle OH}{|}}{C}HCH_2\overset{\overset{\displaystyle O}{\|}}{C}H}$$

In a similar manner the enolate of propanal can add to either a second molecule of propanal (to give 3-hydroxy-2-methylpentanal) or to a molecule of acetaldehyde (to give 3-hydroxy-2-methylbutanal).

$$\xrightarrow{H_2O} \quad \underset{\text{3-Hydroxy-2-methylpentanal}}{CH_3CH_2\overset{\overset{\displaystyle OH}{|}}{C}H\underset{\underset{\displaystyle CH_3}{|}}{C}H\overset{\overset{\displaystyle O}{\|}}{C}H}$$

$$\xrightarrow{H_2O} \quad \underset{\text{3-Hydroxy-2-methylbutanal}}{CH_3\overset{\overset{\displaystyle OH}{|}}{C}H\underset{\underset{\displaystyle CH_3}{|}}{C}H\overset{\overset{\displaystyle O}{\|}}{C}H}$$

18.8 (*b*) The only enolate that can be formed from *tert*-butyl methyl ketone arises by proton abstraction from the methyl group attached to the carbonyl.

$$(CH_3)_3CCCH_3 \;+\; HO^- \;\rightleftharpoons\; (CH_3)_3C\overset{O}{\overset{||}{C}}\ddot{C}H_2$$

tert-Butyl methyl
ketone

Enolate of *tert*-butyl
methyl ketone

This enolate adds to the carbonyl group of benzaldehyde to give the mixed aldol addition product, which then dehydrates under the reaction conditions.

$$C_6H_5CH \;+\; \overset{-}{:}CH_2CC(CH_3)_3 \longrightarrow C_6H_5\overset{O^-}{\underset{}{C}}HCH_2CC(CH_3)_3 \xrightarrow{H_2O} C_6H_5\overset{OH}{\underset{}{C}}HCH_2CC(CH_3)_3$$

Benzaldehyde

Enolate of *tert*-butyl
methyl ketone

Product of mixed aldol addition

$$\downarrow -H_2O$$

$$C_6H_5CH{=}CHCC(CH_3)_3$$

4,4-Dimethyl-1-phenyl-1-penten-3-one
(product of mixed aldol condensation)

(*c*) The enolate of cyclohexanone adds to benzaldehyde. Dehydration of the mixed aldol addition product takes place under the reaction conditions to give the following mixed aldol condensation product.

Cyclohexanone   Benzaldehyde

Benzylidenecyclohexanone

18.9 Enolate formation can occur at either of the two nonequivalent α carbons of 2-methylcyclohexanone. Two constitutionally isomeric enolates are formed and these give different products.

2-Methylcyclohexanone

2-Benzyl-6-
methylcyclohexanone

2-Benzyl-2-
methylcyclohexanone

18.10 (*b*) Acetophenone can enolize only in the direction of the methyl group.

| Acetophenone | Enol form of acetophenone |

(*c*) Enolization of 2-methylcyclohexanone can take place in two different directions.

2-Methylcyclohex-1-enol   2-Methylcyclohexanone   6-Methylcyclohex-1-enol
(enol form)   (enol form)

18.11 Intramolecular hydrogen bonding is not possible in the enol form of 1,3-cyclohexanedione because the carbonyl groups are held apart by the six-membered ring.

1,3-Cyclohexanedione   Enol form; intramolecular
                                                    hydrogen bonding not possible

Because cyclohexanedione exists predominantly in the enol form even though intramolecular hydrogen bonding is not possible, conjugation of the enol double bond must be the more important stabilizing effect.

18.12 (*b*) Enolization of the central methylene group can involve either of the two carbonyl groups.

Enol form   1-Phenyl-1,3-butanedione   Enol form

18.13 As shown in the general equation and the examples in the text, halogen substitution is specific for the α-carbon atom. The ketone 2-butanone has two nonequivalent α carbons, and so substitution is possible at both positions. Both 1-chloro-2-butanone and 3-chloro-2-butanone are formed in the reaction.

2-Butanone   Chlorine   1-Chloro-2-butanone   3-Chloro-2-butanone

18.14 The carbon–carbon double bond of the enol always involves the original carbonyl carbon and the α-carbon atom. 2-Butanone can form two different enols, each of which yields a different α-chloro ketone.

$$CH_3\overset{\displaystyle O}{\overset{\|}{C}}CH_2CH_3 \quad \underset{}{\overset{slow}{\rightleftharpoons}} \quad H_2C=\overset{\displaystyle OH}{\underset{}{\overset{|}{C}}}CH_2CH_3 \quad \xrightarrow[fast]{Cl_2} \quad ClCH_2\overset{\displaystyle O}{\overset{\|}{C}}CH_2CH_3$$

2-Butanone   1-Buten-2-ol (enol)   1-Chloro-2-butanone

$$CH_3\overset{\displaystyle O}{\overset{\|}{C}}CH_2CH_3 \quad \underset{}{\overset{slow}{\rightleftharpoons}} \quad CH_3\overset{\displaystyle OH}{\underset{}{\overset{|}{C}}}=CHCH_3 \quad \xrightarrow[fast]{Cl_2} \quad CH_3\overset{\displaystyle O}{\overset{\|}{C}}\underset{\displaystyle Cl}{\underset{|}{C}}HCH_3$$

2-Butanone   2-Buten-2-ol (enol)   3-Chloro-2-butanone

18.15 The electron-rich carbon–carbon double bond of each enol attacks chlorine.

$$H_2C=\overset{:\ddot{O}H}{\underset{}{\overset{|}{C}}}CH_2CH_3 \quad \longrightarrow \quad ClCH_2-\overset{\overset{+}{\ddot{O}H}}{\overset{\|}{C}}CH_2CH_3 \ + \ Cl^-$$

$$Cl{-}Cl$$

$$CH_3\overset{:\ddot{O}H}{\underset{}{\overset{|}{C}}}=CHCH_3 \quad \longrightarrow \quad CH_3\overset{\overset{+}{\ddot{O}H}}{\overset{\|}{C}}-\underset{\displaystyle Cl}{\underset{|}{C}}HCH_3 \ + \ Cl^-$$

$$Cl{-}Cl$$

18.16 Each of the five α hydrogens has been replaced by deuterium by base-catalyzed enolization. Only the OCH$_3$ hydrogens and the hydrogens on the aromatic ring are observed in the ^{1}H NMR spectrum at δ 3.9 and δ 6.7–6.9, respectively.

18.17 α-Chlorination of (R)-sec-butyl phenyl ketone in acetic acid proceeds via the enol. The enol is achiral and yields equal amounts of (R)- and (S)-2-chloro-2-methyl-1-phenyl-1-butanone. The product is chiral. It is formed as a racemic mixture, however, and this mixture is not optically active.

(R)-sec-Butyl phenyl ketone   Enol (achiral)   2-Chloro-2-methyl-1-phenyl-1-butanone (50% R; 50% S)

18.18 Mesityl oxide is an α,β-unsaturated ketone. Traces of acids or bases can catalyze its isomerization so that some of the less stable β,γ-unsaturated isomer is present.

Mesityl oxide; 4-methyl-
3-penten-2-one
(more stable)

4-Methyl-4-penten-2-one
(less stable)

18.19 The relationship between the molecular formula of acrolein (C_3H_4O) and the product ($C_3H_5N_3O$) corresponds to the addition of HN_3 to acrolein. Because propanal ($CH_3CH_2CH=O$) does not react under these conditions, the carbon–carbon, not the carbon–oxygen, double bond of acrolein is the reactive site. Conjugate addition is the reaction that occurs.

Acrolein

3-Azidopropanal

18.20 The enolate of dibenzyl ketone adds to methyl vinyl ketone in the conjugate addition step.

Dibenzyl ketone   Methyl vinyl ketone   1,3-Diphenyl-2,6-heptanedione

via

The intramolecular aldol condensation that gives the observed product is

1,3-Diphenyl-2,6-heptanedione

3-Methyl-2,6-diphenyl-2-
cyclohexenone

18.21 A second solution to the synthesis of 4-methyl-2-octanone by conjugate addition of a lithium dialkylcuprate reagent to an α,β-unsaturated ketone is revealed by the disconnection shown:

Disconnect this bond.

According to this disconnection, the methyl group is derived from lithium dimethylcuprate.

$$CH_3CH_2CH_2CH_2CH=CHCCH_3 \;+\; LiCu(CH_3)_2 \longrightarrow CH_3CH_2CH_2CH_2CHCH_2CCH_3$$

3-Octen-2-one   Lithium dimethylcuprate   4-Methyl-2-octanone

18.22 (*a*) In addition to the double bond of the carbonyl group, there must be a double bond elsewhere in the molecule to satisfy the molecular formula C_4H_6O (the problem states that the compounds are noncyclic). There are a total of five isomers:

$$H_2C=CHCH_2CH$$

3-Butenal   (*E*)-2-Butenal   (*Z*)-2-Butenal

$$H_2C=CCH \qquad H_2C=CHCCH_3$$
$$CH_3$$

2-Methylpropenal   3-Buten-2-one (methyl vinyl ketone)

(*b*) The *E* and *Z* isomers of 2-butenal are stereoisomers.

(*c*) None of the C_4H_6O aldehydes and ketones is chiral.

(*d*) The α,β-unsaturated aldehydes are (*E*)- and (*Z*)-$CH_3CH=CHCHO$; and $H_2C=CCHO$.
$$CH_3$$

There is one α,β-unsaturated ketone in the group: $H_2C=CHCCH_3$.

(*e*) The *E* and *Z* isomers of 2-butenal are formed by the aldol condensation of acetaldehyde.

18.23 The main flavor component of the hazelnut has the structure shown.

(2*E*,5*S*)-5-Methyl-2-hepten-4-one

18.24 The characteristic reaction of an alcohol on being heated with $KHSO_4$ is acid-catalyzed dehydration. Secondary alcohols dehydrate faster than primary alcohols, and so a reasonable first step is

$$HOCH_2CHCH_2OH \xrightarrow[\text{heat}]{KHSO_4} HOCH_2CH=CHOH$$
$$OH$$

1,2,3-Propanetriol   Propene-1,3-diol

The product of this dehydration is an enol, which tautomerizes to an aldehyde. The aldehyde then undergoes dehydration to form acrolein.

$$HOCH_2CH=CHOH \longrightarrow HOCH_2CH_2\overset{\overset{\displaystyle O}{\|}}{C}H \xrightarrow[\substack{\text{heat} \\ (-H_2O)}]{KHSO_4} H_2C=CH\overset{\overset{\displaystyle O}{\|}}{C}H$$

Propene-1,3-diol · 3-Hydroxypropanal · Acrolein

18.25 (a) 2-Methylpropanal has the greater enol content.

$$(CH_3)_2CH\overset{\overset{\displaystyle O}{\|}}{C}H \rightleftharpoons (CH_3)_2C=\overset{\overset{\displaystyle OH}{|}}{C}H$$

2-Methylpropanal · Enol form

Although the enol content of 2-methylpropanal is quite small, the compound is nevertheless capable of enolization, whereas the other compound, 2,2-dimethylpropanal, cannot enolize—it has no α hydrogens.

$$CH_3\underset{\underset{\displaystyle CH_3}{|}}{\overset{\overset{\displaystyle CH_3}{|}}{C}}-C\overset{\diagup O}{\underset{\diagdown H}{}}$$

(Enolization is impossible.)

(b) Benzophenone has no α hydrogens; it cannot form an enol.

(Enolization is impossible.)

Dibenzyl ketone enolizes slightly to form a small amount of enol.

$$C_6H_5CH_2\overset{\overset{\displaystyle O}{\|}}{C}CH_2C_6H_5 \rightleftharpoons C_6H_5CH=\overset{\overset{\displaystyle OH}{|}}{C}CH_2C_6H_5$$

Dibenzyl ketone · Enol form

(c) Here we are comparing a simple ketone, dibenzyl ketone, with a β-diketone. The β-diketone enolizes to a much greater extent than the simple ketone because its enol form is stabilized by conjugation of the double bond with the remaining carbonyl group and by intramolecular hydrogen bonding.

$$C_6H_5\overset{\overset{\displaystyle O}{\|}}{C}CH_2\overset{\overset{\displaystyle O}{\|}}{C}C_6H_5 \rightleftharpoons$$

1,3-Diphenyl-1,3-propanedione · Enol form

(d) The enol content of cyclohexanone is quite small, whereas the enol form of 2,4-cyclohexadienone is the aromatic compound phenol, and therefore enolization is essentially complete.

Keto form · Enol form (aromatic; much more stable)

(*e*) A small amount of enol is in equilibrium with cyclopentanone.

Cyclopentanone   Enol form

Cyclopentadienone does not form a stable enol. Enolization would lead to a highly strained allene-type compound.

(Not stable;
highly strained)

(*f*) The β-diketone is more extensively enolized.

1,3-Cyclohexanedione   Enol form (double bond
                                        conjugated with carbonyl group)

The double bond of the enol form of 1,4-cyclohexanedione is not conjugated with the carbonyl group. Its enol content is expected to be similar to that of cyclohexanone.

1,3-Cyclohexanedione   Enol form (not particularly stable;
                                        double bond and carbonyl group
                                        not conjugated)

18.26 (*a*) Chlorination of 3-phenylpropanal under conditions of acid catalysis occurs via the enol form and yields the α-chloro derivative.

$$C_6H_5CH_2CH_2\overset{\overset{\displaystyle O}{\|}}{C}H \ + \ Cl_2 \ \xrightarrow{\text{acetic acid}} \ C_6H_5CH_2\underset{\underset{\displaystyle Cl}{|}}{\overset{\overset{\displaystyle O}{\|}}{C}}HCH \ + \ HCl$$

3-Phenylpropanal   2-Chloro-3-
                                        phenylpropanal

(*b*) Aldehydes undergo aldol addition on treatment with base.

$$2C_6H_5CH_2CH_2\overset{\overset{\displaystyle O}{\|}}{C}H \xrightarrow[\text{ethanol, 10°C}]{\text{NaOH}} C_6H_5CH_2CH_2\underset{\underset{\displaystyle OH}{|}}{C}H\underset{}{C}H\overset{\overset{\displaystyle HC=O}{|}}{}CH_2C_6H_5$$

3-Phenylpropanal   2-Benzyl-3-hydroxy-5-phenylpentanal

(*c*) Dehydration of the aldol addition product occurs when the reaction is carried out at elevated temperature.

$$2C_6H_5CH_2CH_2\overset{\overset{\displaystyle O}{\|}}{C}H \xrightarrow[\text{ethanol, 70°C}]{\text{NaOH}} C_6H_5CH_2CH_2CH=\overset{\overset{\displaystyle HC=O}{|}}{C}CH_2C_6H_5$$

3-Phenylpropanal   2-Benzyl-5-phenyl-2-pentenal

(*d*) Lithium aluminum hydride reduces the aldehyde function to the corresponding primary alcohol.

$$C_6H_5CH_2CH_2CH=\overset{\overset{\displaystyle HC=O}{|}}{C}CH_2C_6H_5 \xrightarrow[\text{2. H}_2\text{O}]{\text{1. LiAlH}_4} C_6H_5CH_2CH_2CH=\overset{\overset{\displaystyle CH_2OH}{|}}{C}CH_2C_6H_5$$

2-Benzyl-5-phenyl-2-pentenal   2-Benzyl-5-phenyl-2-penten-1-ol

(*e*) A characteristic reaction of α,β-unsaturated carbonyl compounds is their tendency to undergo conjugate addition on treatment with weakly basic nucleophiles.

$$C_6H_5CH_2CH_2CH=\overset{\overset{\displaystyle HC=O}{|}}{C}CH_2C_6H_5 \xrightarrow[\text{H}^+]{\text{NaCN}} C_6H_5CH_2CH_2\underset{\underset{\displaystyle CN}{|}}{C}H\overset{\overset{\displaystyle HC=O}{|}}{C}HCH_2C_6H_5$$

2-Benzyl-5-phenyl-2-pentenal   2-Benzyl-3-cyano-5-phenylpentanal

18.27 (*a*) Ketones undergo α halogenation by way of their enol form.

1-(*o*-Chlorophenyl)-1-propanone   2-Chloro-1-(*o*-chlorophenyl)-1-propanone

(*b*) The combination of $C_6H_5CH_2SH$ and NaOH yields $C_6H_5CH_2S^-$ (as its sodium salt), which is a weakly basic nucleophile and adds to α,β-unsaturated ketones by conjugate addition.

2-Isopropylidene-5-methylcyclohexanone   2-(1-Benzylthio-1-methylethyl)-5-methylcyclohexanone (89–90%)

(*c*) Bromination occurs at the carbon atom that is α to the carbonyl group.

2,2-Diphenylcyclopentanone

5-Bromo-2,2-
diphenylcyclopentanone (76%)

(*d*) The reaction is a mixed aldol condensation. The enolate of 2,2-diphenylcyclohexanone reacts with *p*-chlorobenzaldehyde. Elimination of the aldol addition product occurs readily to yield the α,β-unsaturated ketone as the isolated product.

p-Chlorobenzaldehyde   2,2-Diphenylcyclohexanone

(Not isolated)

$-H_2O$

6-(*p*-Chlorobenzylidene)-2,2-
diphenylcyclohexanone (84%)

(*e*) The aldehyde given as the starting material is called *furfural* and is based on a furan unit as an aromatic ring. Furfural cannot form an enolate. It reacts with the enolate of acetone in a manner much as benzaldehyde would.

Furfural   Acetone   (Not isolated)   4-Furyl-3-buten-2-one
(60–66%)

(*f*) Lithium dialkylcuprates transfer an alkyl group to the β-carbon atom of α,β-unsaturated ketones.

2,4,4-Trimethyl-2-
cyclohexenone

2,3,4,4-Tetramethylcyclohexanone

A mixture of stereoisomers was obtained in 67% yield in this reaction.

(*g*) Two nonequivalent α-carbon atoms occur in the starting ketone. Although enolate formation is possible at either position, only reaction at the methylene carbon leads to an intermediate that can undergo dehydration.

Observed product
(75% yield)

Reaction at the other α position gives an intermediate that cannot dehydrate.

(Cannot dehydrate;
reverts to starting materials)

(*h*) β-Diketones readily undergo alkylation by primary halides at the most acidic position, on the carbon between the carbonyls.

1,3-Cyclohexanedione　　　　Allyl bromide　　　　　　　　2-Allyl-1,3-cyclohexanedione
(75%)

18.28 (*a*) Conversion of 3-pentanone to 2-bromo-3-pentanone is best accomplished by acid-catalyzed bromination via the enol. Bromine in acetic acid is the customary reagent for this transformation.

$$CH_3CH_2\overset{\overset{\displaystyle O}{\|}}{C}CH_2CH_3 \xrightarrow[\text{acetic acid}]{Br_2} CH_3\overset{\overset{\displaystyle O}{\|}}{\underset{\underset{\displaystyle Br}{|}}{CH}}CCH_2CH_3$$

3-Pentanone　　　　　　　　　　　　2-Bromo-3-pentanone

(*b*) Once 2-bromo-3-pentanone has been prepared, its dehydrohalogenation by base converts it to the desired α,β-unsaturated ketone 1-penten-3-one.

$$CH_3\overset{\overset{\displaystyle O}{\|}}{\underset{\underset{\displaystyle Br}{|}}{CH}}CCH_2CH_3 \xrightarrow{KOC(CH_3)_3} H_2C{=}CH\overset{\overset{\displaystyle O}{\|}}{C}CH_2CH_3$$

2-Bromo-3-pentanone　　　　　　　　　　　1-Penten-3-one

Potassium *tert*-butoxide is a good base for bringing about elimination reactions of secondary alkyl halides; suitable solvents include *tert*-butyl alcohol and dimethyl sulfoxide.

(*c*) Reduction of the carbonyl group of 1-penten-3-one converts it to the desired alcohol.

$$H_2C=CHCCH_2CH_3 \quad \xrightarrow[\text{2. H}_2\text{O}]{\text{1. LiAlH}_4, \text{ diethyl ether}} \quad H_2C=CHCHCH_2CH_3$$

1-Penten-3-one   1-Penten-3-ol

Catalytic hydrogenation would not be suitable for this reaction because reduction of the double bond would accompany carbonyl reduction.

(*d*) Conversion of 3-pentanone to 3-hexanone requires addition of a methyl group to the β-carbon atom.

$$CH_3CH_2CH_2CCH_2CH_3 \quad \Longrightarrow \quad {}^{-}:CH_3 \; + \; H_2\overset{\beta}{C}=CHCCH_2CH_3$$

The best way to add an alkyl group to the β carbon of a ketone is via conjugate addition of a dialkylcuprate reagent to an α,β-unsaturated ketone.

$$H_2C=CHCCH_2CH_3 \quad \xrightarrow[\text{2. H}_2\text{O}]{\text{1. LiCu(CH}_3)_2} \quad CH_3CH_2CH_2CCH_2CH_3$$

1-Penten-3-one   3-Hexanone
[prepared as described in part (*b*)]

(*e*) The compound to be prepared is the mixed aldol condensation product of 3-pentanone and benzaldehyde.

$$\begin{array}{c} CH_3CCCH_2CH_3 \\ | \\ C_6H_5CH \end{array} \quad \Longrightarrow \quad CH_3\overset{..}{C}HCCH_2CH_3 \; + \; C_6H_5CH$$

2-Methyl-1-phenyl-
1-penten-3-one

The desired reaction sequence is

$$CH_3CH_2CCH_2CH_3 \xrightarrow{HO^-} CH_3\overset{..}{C}HCCH_2CH_3 \xrightarrow{C_6H_5CH} \begin{array}{c} CH_3CHCCH_2CH_3 \\ | \\ C_6H_5CHOH \end{array} \xrightarrow{-H_2O} \begin{array}{c} CH_3CCCH_2CH_3 \\ | \\ C_6H_5CH \end{array}$$

3-Pentanone   Enolate of 3-pentanone   Aldol addition product   2-Methyl-1-phenyl-
                                                       (not isolated; dehydration   1-penten-3-one
                                                       occurs under conditions
                                                       of its formation)

18.29 (*a*) The first step is an α halogenation of a ketone. This is customarily accomplished under conditions of acid catalysis.

$$(CH_3)_3CCCH_3 \quad \xrightarrow[\text{H}^+]{\text{Br}_2} \quad (CH_3)_3CCCH_2Br$$

3,3-Dimethyl-2-butanone   1-Bromo-3,3-dimethyl-2-butanone (58%)

In the second step the carbonyl group of the α-bromo ketone is reduced to a secondary alcohol. As actually carried out, sodium borohydride in water was used to achieve this transformation.

$$(CH_3)_3CCCH_2Br \xrightarrow[\text{H}_2\text{O}]{\text{NaBH}_4} (CH_3)_3CCHCH_2Br$$

1-Bromo-3,3-dimethyl-2-butanone   1-Bromo-3,3-dimethyl-2-butanol (54%)

The third step is conversion of a vicinal bromohydrin to an epoxide in aqueous base.

$$(CH_3)_3CCHCH_2Br \xrightarrow[\text{H}_2\text{O}]{\text{KOH}} (CH_3)_3CC{-}CH_2$$

1-Bromo-3,3-dimethyl-2-butanol   2-*tert*-Butyloxirane (68%)

(*b*) The overall yield is the product of the yields of the individual steps.

$$\text{Yield} = 100(0.58 \times 0.54 \times 0.68)$$

$$= 21\%$$

18.30 The product is a sulfide (thioether). Retrosynthetic analysis reveals a pathway that begins with benzene and acetic anhydride.

The desired synthesis can be accomplished with the following series of reactions:

Benzene   Acetic anhydride   Acetophenone   Bromomethyl phenyl ketone

The synthesis is completed by reacting bromomethyl phenyl ketone with 1-propanethiolate anion.

1-Propanethiol   1-Propanethiolate   Phenyl (1-thiopropyl)methyl ketone

18.31 All these problems begin in the same way, with exchange of all the α protons for deuterium (text Section 18.10).

Cyclopentanone   Cyclopentanone-2,2,5,5-d$_4$

Once the tetradeuterated cyclopentanone has been prepared, functional group transformations are employed to convert it to the desired products.

(*a*) Reduction of the carbonyl group can be achieved by using any of the customary reagents.

Cyclopentanone-2,2,5,5-d$_4$ Cyclopentanol-2,2,5,5-d$_4$

(*b*) Acid-catalyzed dehydration of the alcohol prepared in part (*a*) yields the desired alkene.

Cyclopentanol-2,2,5,5-d$_4$ Cyclopentene-1,3,3-d$_3$

(*c*) Catalytic hydrogenation of the alkene in part (*b*) yields cyclopentane-1,1,3-d$_3$.

Cyclopentene-1,3,3-d$_3$ Hydrogen   Cyclopentane-1,1,3-d$_3$

(*d*) Carbonyl reduction of the tetradeuterated ketone under Wolff–Kishner conditions furnishes the desired product.

Cyclopentanone-2,2,5,5-d$_4$ Cyclopentane-1,1,3,3-d$_4$

Alternatively, Clemmensen reduction conditions (Zn(Hg), HCl) could be used.

18.32 (*a*) Hydroformylation converts alkenes to aldehydes having one more carbon atom by reaction with carbon monoxide and hydrogen in the presence of a cobalt octacarbonyl catalyst.

$$CH_3CH{=}CH_2 \;+\; CO \;+\; H_2 \xrightarrow{Co_2(CO)_8} CH_3CH_2CH_2\overset{\displaystyle O}{\overset{\|}{C}}H$$

Propene   Carbon   Hydrogen   Butanal
           monoxide

(*b*) Aldol condensation of acetaldehyde to 2-butenal, followed by catalytic hydrogenation of the carbon–carbon double bond, gives butanal.

$$2CH_3\overset{\overset{\displaystyle O}{\|}}{C}H \xrightarrow[\text{heat}]{\text{NaOH}} CH_3CH=CH\overset{\overset{\displaystyle O}{\|}}{C}H \xrightarrow[\text{Ni}]{H_2} CH_3CH_2CH_2\overset{\overset{\displaystyle O}{\|}}{C}H$$

Acetaldehyde   2-Butenal   Butanal

18.33 (*a*) The first conversion is the α halogenation of an aldehyde. As described in text Section 18.7, this particular conversion has been achieved in 80% yield simply by treatment with bromine in chloroform.

Cyclohexane-   1-Bromocyclohexane-
carbaldehyde   carbaldehyde

Dehydrohalogenation of this compound can be accomplished under E2 conditions by treatment with base. Sodium methoxide in methanol would be appropriate, for example, although almost any alkoxide could be employed to dehydrohalogenate this tertiary bromide.

1-Bromocyclohexane-   Cyclohexene-1-
carbaldehyde   .carbaldehyde

As the reaction was actually carried out, the bromide was heated with the weak base *N*, *N*-diethylaniline to effect dehydrobromination in 71% yield.

(*b*) Cleavage of vicinal diols to carbonyl compounds can be achieved by using periodic acid (HIO_4) (text Section 15.12).

trans-1,2-Cyclohexanediol   1,6-Hexanedial

The conversion of this dialdehyde to cyclopentene-1-carbaldehyde is an intramolecular aldol condensation and is achieved by treatment with potassium hydroxide.

Cyclopentene-1-
carbaldehyde

As the reaction was actually carried out, cyclopentene-1-carbaldehyde was obtained in 58% yield from *trans*-1,2-cyclohexanediol by this method.

(*c*) The first transformation requires an oxidative cleavage of a carbon–carbon double bond. Treatment with ozone followed by hydrolysis in the presence of zinc is indicated.

4-Isopropyl-1-
methylcyclohexene

3-Isopropyl-6-
oxoheptanal

Cyclization of the resulting keto aldehyde is an intramolecular aldol condensation. Base is required.

(*d*) The first step in this synthesis is the hydration of the alkene function to an alcohol. Notice that this hydration must take place with a regioselectivity opposite to that of Markovnikov's rule and therefore requires a hydroboration–oxidation sequence.

$$(CH_3)_2C{=}CHCH_2CH_2\overset{\overset{O}{\|}}{C}CH_3 \quad \xrightarrow[\text{2. } H_2O_2, HO^-]{\text{1. } B_2H_6, THF} \quad (CH_3)_2CHCHCH_2CH_2\overset{\overset{O}{\|}}{C}CH_3$$
$$\hspace{8.5cm}\overset{|}{OH}$$

6-Methyl-5-hepten-2-one

5-Hydroxy-6-methyl-2-heptanone

Conversion of the secondary alcohol function to a carbonyl group can be achieved with any of a number of oxidizing agents.

$$(CH_3)CHCHCH_2CH_2\overset{\overset{O}{\|}}{C}CH_3 \quad \xrightarrow{H_2CrO_4} \quad (CH_3)_2CH\overset{\overset{O}{\|}}{C}CH_2CH_2\overset{\overset{O}{\|}}{C}CH_3$$
$$\hspace{1.2cm}\overset{|}{OH}$$

5-Hydroxy-6-methyl-2-heptanone

6-Methyl-2,5-heptanedione

Cyclization of the dione to the final product is a base-catalyzed intramolecular aldol condensation and was accomplished in 71% yield by treatment of the dione with a 2% solution of sodium hydroxide in aqueous ethanol.

$$(CH_3)_2CH\overset{\overset{O}{\|}}{C}CH_2CH_2\overset{\overset{O}{\|}}{C}CH_3 \quad \xrightarrow{HO^-} \quad (CH_3)_2CH\overset{\overset{O}{\|}}{C}CH_2CH_2\overset{\overset{O^-}{|}}{C}{=}CH_2$$

18.34 Intramolecular aldol condensations occur best when a five- or six-membered ring is formed. Carbon–carbon bond formation therefore involves the aldehyde and the methyl group attached to the ketone carbonyl.

$$CH_3CCH_2CCH=O \quad \xrightarrow{\text{KOH}} \quad \cdots \quad \longrightarrow \quad \cdots \quad \longrightarrow \quad \cdots$$

2,2-Dimethyl-4-
oxopentanal

4,4-Dimethyl-2-
cyclopentenone (63%)

18.35 (*a*) By realizing that the primary alcohol function of the target molecule can be introduced by reduction of an aldehyde, it can be seen that the required carbon skeleton is the same as that of the aldol addition product of 2-methylpropanal.

$$(CH_3)_2CHCHCCH_2OH \quad \Longrightarrow \quad (CH_3)_2CHCH{-}CC \quad \Longrightarrow \quad 2(CH_3)_2CHCH$$

The synthetic sequence is

$$(CH_3)_2CHCH \quad \xrightarrow[\text{ethanol}]{\text{NaOH}} \quad (CH_3)_2CHCHCC \quad \xrightarrow[\text{CH}_3\text{OH}]{\text{NaBH}_4} \quad (CH_3)_2CHCHCCH_2OH$$

2-Methylpropanal

3-Hydroxy-2,2,4-
trimethylpentanal

2,2,4-Trimethyl-1,3-pentanediol

The starting aldehyde is prepared by oxidation of 2-methyl-1-propanol.

$$(CH_3)_2CHCH_2OH \quad \xrightarrow[\text{CH}_2\text{Cl}_2]{\text{PCC}} \quad (CH_3)_2CHCH$$

2-Methyl-1-propanol

2-Methylpropanal

(*b*) Retrosynthetic analysis of the desired product shows that the carbon skeleton can be constructed by a mixed aldol condensation between benzaldehyde and propanal.

$$C_6H_5CH{=}CCH_2OH \quad \Longrightarrow \quad C_6H_5CH{=}CCH \quad \Longrightarrow \quad C_6H_5CH \;+\; CH_3CH_2CH$$

The reaction scheme therefore becomes

$$C_6H_5CH \;+\; CH_3CH_2CH \quad \xrightarrow{\text{HO}^-} \quad C_6H_5CH{=}CCH$$

Benzaldehyde   Propanal

2-Methyl-3-phenyl-2-
propenal

Reduction of the aldehyde to the corresponding primary alcohol gives the desired compound.

$$C_6H_5CH=\overset{\overset{\displaystyle O}{\|}}{\underset{\underset{\displaystyle CH_3}{|}}{C}}CH \quad \xrightarrow[\text{or NaBH}_4\text{, CH}_3\text{OH}]{\text{LiAlH}_4\text{, then H}_2\text{O}} \quad C_6H_5CH=\underset{\underset{\displaystyle CH_3}{|}}{C}CH_2OH$$

2-Methyl-3-phenyl-2-
propenal

2-Methyl-3-phenyl-2-
propen-1-ol

The starting materials for the mixed aldol condensation (benzaldehyde and propanal) are prepared by oxidation of benzyl alcohol and 1-propanol, respectively.

$$C_6H_5CH_2OH \quad \xrightarrow[\text{CH}_2\text{Cl}_2]{\text{PCC}} \quad C_6H_5\overset{\overset{\displaystyle O}{\|}}{C}H$$

Benzyl alcohol

Benzaldehyde

$$CH_3CH_2CH_2OH \quad \xrightarrow[\text{CH}_2\text{Cl}_2]{\text{PCC}} \quad CH_3CH_2\overset{\overset{\displaystyle O}{\|}}{C}H$$

1-Propanol

Propanal

(*c*) The cyclohexene ring in this case can be assembled by a Diels–Alder reaction.

1,3-Butadiene is one of the given starting materials; the α,β-unsaturated ketone is the mixed aldol condensation product of 4-methylbenzaldehyde and acetophenone.

The complete synthetic sequence is

4-Methylbenzyl alcohol

4-Methylbenzaldehyde

4-Methylbenzaldehyde   Acetophenone

trans-4-Benzoyl-5-
(4-methylphenyl)cyclohexene

α,β-Unsaturated ketones are good dienophiles in Diels–Alder reactions.

18.36 The carbon atom flanked by two carbonyl groups is the one involved in the enolization of terreic acid.

Terreic acid   Enol A   Enol B

Of these two structures, enol A, with its double bond conjugated to two carbonyl groups, is more stable than enol B, in which the double bond is conjugated to only one carbonyl.

18.37 (*a*) Recall that aldehydes and ketones are in equilibrium with their *hydrates* in aqueous solution (text Section 17.6). Thus, the principal substance present when $(C_6H_5)_2CHCH{=}O$ is dissolved in aqueous acid is $(C_6H_5)_2CHCH(OH)_2$ (81%).

(*b*) The problem states that the major species present in aqueous base is *not* $(C_6H_5)_2CHCH{=}O$, its enol, or its hydrate. The most reasonable species is the *enolate ion:*

$$(C_6H_5)_2CHCH{=}O \xrightarrow{-H^+} (C_6H_5)_2\overset{-}{C}{-}CH{=}\overset{..}{\underset{..}{O}}{:} \longleftrightarrow (C_6H_5)_2C{=}CH{-}\overset{..}{\underset{..}{O}}{:}^-$$

18.38 (*a*) At first glance this transformation seems to be an internal oxidation–reduction reaction. An aldehyde function is reduced to a primary alcohol, and a secondary alcohol is oxidized to a ketone.

Compound A   Compound B

Once one realizes that enolization can occur, however, a simpler explanation, involving only proton-transfer reactions, emerges.

Compound A   Enol form of compound A

The enol form of compound A is an enediol; it is at the same time the enol form of compound B. The enediol can revert to compound A or to compound B.

At equilibrium, compound B predominates because it is more stable than A. A ketone carbonyl is more stabilized than an aldehyde, and the carbonyl in B is conjugated with the benzene ring.

(*b*) The isolated product is the double hemiacetal formed between two molecules of compound A.

Compound C

18.39 (*a*) The only chirality center in piperitone is adjacent to a carbonyl group. Base-catalyzed enolization causes this carbon to lose its stereochemical integrity.

(–)-Piperitone   Enolate of   Enol of piperitone
                                  piperitone

Both the enolate and enol of piperitone are achiral and revert to a racemic mixture of piperitones.

(*b*) The enol formed from menthone can revert to either menthone or isomenthone.

Menthone   Enol form   Isomenthone

Only the stereochemistry at the α-carbon atom is affected by enolization. The other chirality center in menthone (the one bearing the methyl group) is not affected.

18.40 In all parts of this problem the bonding change that takes place is described by the general equation

$$HX-N=Z \quad \Longleftrightarrow \quad X=N-ZH$$

(*a*) The compound given is nitrosoethane. Nitrosoalkanes are less stable than their oxime isomers formed by proton transfer.

$$CH_3\overset{\displaystyle |}{\underset{\displaystyle H}{C}H}-N=O \quad \Longleftrightarrow \quad CH_3CH=N-OH$$

Nitrosoethane   Acetaldehyde oxime
(less stable)   (more stable)

(*b*) You may recognize this compound as an enamine. It is slightly different, however, from the enamines we discussed earlier (text Section 17.11) in that nitrogen bears a hydrogen. Stable enamines are compounds of the type

where neither R group is hydrogen; both R's must be alkyl or aryl. Enamines that bear a hydrogen substituent are converted to imines in a proton-transfer equilibrium.

$$(CH_3)_2C=CH-\overset{\displaystyle |}{\underset{\displaystyle H}{N}}CH_3 \quad \Longleftrightarrow \quad (CH_3)_2CH-CH=NCH_3$$

Enamine (less stable)   Imine (more stable)

(c) The compound given is known as a nitronic acid; its more stable tautomeric form is a nitroalkane.

Nitronic acid   Nitroalkane

(d) The six-membered ring is aromatic in the tautomeric form derived from the compound given.

(e) This compound is called isourea. Urea has a carbon–oxygen double bond and is more stable.

Isourea (less stable)   Urea (more stable)

18.41 (a) This reaction is an intramolecular alkylation of a ketone. Although alkylation of a ketone with a separate alkyl halide molecule is usually difficult, *intramolecular* alkylation reactions can be carried out effectively. The enolate formed by proton abstraction from the α-carbon atom carries out a nucleophilic attack on the carbon that bears the leaving group.

(b) The starting material, known as *citral*, is converted to the two products by a reversal of an aldol condensation. The first step is conjugate addition of hydroxide.

The product of this conjugate addition is a β-hydroxy aldehyde. It undergoes base-catalyzed cleavage to the observed products.

(c) The product is formed by an intramolecular aldol condensation.

(d) In this problem stereochemical isomerization involving a proton attached to the α-carbon atom of a ketone takes place. Enolization of the ketone yields an intermediate in which the stereochemical integrity of the α carbon is lost. Reversion to ketone eventually leads to the formation of the more stable stereoisomer at equilibrium.

Less stable ketone;   Enol   More stable ketone;
starting material   preferred at equilibriun

The rate of enolization is increased by heating or by base catalysis. The cis ring fusion in the product is more stable than the trans because there are not enough atoms in the six-membered ring to span *trans*-1,2 positions in the four-membered ring without excessive strain.

(*e*) Working backward from the product, we can see that the transformation involves two aldol condensations: one intermolecular and the other intramolecular.

The first reaction is a mixed aldol condensation between the enolate of dibenzyl ketone and one of the carbonyl groups of the dione.

Dibenzyl ketone   Benzil

This is followed by an intramolecular aldol condensation.

2,3,4,5-Tetraphenyl-
cyclopentadienone

(*f*) This is a fairly difficult problem because it is not obvious at the outset which of the two possible enolates of benzyl ethyl ketone undergoes conjugate addition to the α,β-unsaturated ketone. A good idea here is to work backward from the final product—in effect, do a retrosynthetic analysis. The first step is to recognize that the enone arises by dehydration of a β-hydroxy ketone.

Now, mentally disconnect the bond between the α-carbon atom and the carbon that bears the hydroxyl group to reveal the intermediate that undergoes intramolecular aldol condensation.

Disconnect
this bond.

The β-hydroxy ketone is the intermediate formed in the intramolecular aldol addition step, and the diketone that leads to it is the intermediate that is formed in the conjugate addition step. The relationship of the starting materials to the intermediates and product is now more evident.

Intermediate formed in conjugate addition step

18.42 (*a*) The reduced C=O stretching frequency of α,β-unsaturated ketones is consistent with an enhanced degree of single bond character as compared with simple dialkyl ketones.

Resonance is more important in α,β-unsaturated ketones. Conjugation of the carbonyl group with the carbon–carbon double bond increases opportunities for electron delocalization.

(*b*) Even more single-bond character is indicated in the carbonyl group of cyclopropenone than in that of typical α,β-unsaturated ketones. The dipolar resonance form contributes substantially to the electron distribution because of the aromatic character of the three-membered ring. Recall that cyclopropenyl cation satisfies the $4n + 2$ rule for aromaticity (text Section 11.22).

Equivalent to an oxyanion-substituted cyclopropenyl cation

(*c*) The dipolar resonance form is a more important contributor to the electron distribution in diphenylcyclopropenone than in benzophenone.

is more pronounced than

The dipolar resonance form of diphenylcyclopropenone has aromatic character. Its stability leads to increased charge separation and a larger dipole moment.

(*d*) Decreased electron density at the β-carbon atom of an α,β-unsaturated ketone is responsible for its decreased shielding. The decreased electron density arises from the polarization of its π electrons as represented by a significant contribution of the dipolar resonance form.

$$H_2C=CH-\overset{\overset{\displaystyle O}{\|}}{C}R \quad\longleftrightarrow\quad H_2\overset{+}{C}-CH=\overset{\overset{\displaystyle O^-}{\|}}{C}R$$

18.43 Bromination can occur at either of the two α-carbon atoms.

$$CH_3\overset{\overset{\displaystyle O}{\|}}{C}CH(CH_3)_2 \xrightarrow{\;Br_2\;} BrCH_2\overset{\overset{\displaystyle O}{\|}}{C}CH(CH_3)_2 \;+\; CH_3\overset{\overset{\displaystyle O}{\|}}{C}\underset{\underset{\displaystyle Br}{|}}{C}(CH_3)_2$$

3-Methyl-2-butanone   1-Bromo-3-methyl-   3-Bromo-3-methyl-
                                    2-butanone   2-butanone

The ^{1}H NMR spectrum of the major product, compound A, is consistent with the structure of 1-bromo-3-methyl-2-butanone. The minor product B is identified as 3-bromo-3-methyl-2-butanone on the basis of its NMR spectrum.

Compound A — δ 1.2 (doublet), δ 4.1 (singlet), δ 3.0 (septet)

Compound B — δ 1.9 (singlet), δ 2.5 (singlet)

18.44 Three dibromination products are possible from α halogenation of 2-butanone.

$$Br_2CHCCH_2CH_3 \qquad BrCH_2\underset{\underset{\displaystyle Br}{|}}{\overset{\overset{\displaystyle O}{\|}}{C}}CHCH_3 \qquad CH_3\overset{\overset{\displaystyle O}{\|}}{C}-\underset{\underset{\displaystyle Br}{|}}{\overset{\overset{\displaystyle Br}{|}}{C}}CH_3$$

1,1-Dibromo-2-butanone   1,3-Dibromo-2-butanone   3,3-Dibromo-2-butanone

The product is *1,3-dibromo-2-butanone*, on the basis of its observed ^{1}H NMR spectrum, which showed two signals at low field. One is a two-proton singlet at δ 4.6 assignable to CH_2Br and the other a one-proton quartet at δ 5.2 assignable to CHBr.

ANSWERS TO INTERPRETIVE PROBLEMS

18.45 A;   **18.46** A;   **18.47** C;   **18.48** C;   **18.49** C

SELF-TEST

PART A

A-1. Write the correct structure(s) for each of the following:

(*a*) The two enol forms of 2-butanone

(*b*) The most stable enolate derived from reaction of 1,3-cyclohexanedione with sodium methoxide

(*c*) The carbonyl form of the following enol

A-2. Give the correct structures for compounds A and B in the following reaction schemes:

$$(a) \quad 2C_6H_5CH_2\overset{\overset{\displaystyle O}{\|}}{C}H \quad \xrightarrow[\text{2. heat } (-H_2O)]{\text{1. } HO^-} \quad A$$

$$(b) \quad CH_3CH_2CH{=}CH\overset{\overset{\displaystyle O}{\|}}{C}CH_2CH_3 \; + \; LiCu(CH_2CH_3)_2 \quad \xrightarrow[\text{2. } H_2O]{\text{1. diethyl ether}} \quad B$$

A-3. Write the structures of all the possible aldol addition products that may be obtained by reaction of a mixture of propanal and 2-methylpropanal with base.

$$\underset{\text{Propanal}}{CH_3CH_2\overset{\overset{\displaystyle O}{\|}}{C}H} \qquad\qquad \underset{\text{2-Methylpropanal}}{CH_3\overset{\overset{\displaystyle CH_3}{|}}{C}H\overset{\underset{\displaystyle \|}{\underset{\displaystyle O}{}}}{C}H}$$

A-4. Using any necessary organic or inorganic reagents, outline a synthesis of 1,3-butanediol from ethanol as the only source of carbons.

A-5. Outline a series of reaction steps that will allow the preparation of compound B from 1,3-cyclopentanedione, compound A.

A   B

A-6. Give the structure of the product formed in each of the following reactions:

$$(a) \quad CH_3CH_2CH_2CH_2\overset{\overset{\displaystyle O}{\|}}{C}H \quad \xrightarrow[\text{acetic acid}]{Br_2}$$

$$(b) \quad 2CH_3CH_2CH_2CH_2\overset{\overset{\displaystyle O}{\|}}{C}H \quad \xrightarrow[\text{5°C}]{NaOH}$$

$$(c) \qquad \xrightarrow[\text{(}-H_2O\text{)}]{NaOH, \text{ heat}}$$

$$(d) \qquad \xrightarrow[\text{ethanol}]{NaSCH_3}$$

A-7. Write out the mechanism, using curved arrows to show electron movement, of the following aldol addition reaction.

$$\overset{\overset{\displaystyle O}{\|}}{H}CH \;+\; CH_3CH_2\overset{\overset{\displaystyle O}{\|}}{C}H \quad \xrightarrow[\text{5°C}]{NaOH,\ H_2O}$$

A-8. Identify the two starting materials needed to make the following compound by a mixed aldol condensation.

$$? \xrightarrow[\text{heat}]{\text{NaOH, H}_2\text{O}} \text{H}_2\text{O} + \text{C}_6\text{H}_5-\text{CH}=\overset{\displaystyle O}{\overset{\displaystyle \|}{\text{C}}}\overset{\displaystyle}{\underset{\displaystyle \text{CH}_3}{\text{C}}}\text{CH}_2\text{CH}_3$$

PART B

B-1. When enolate A is compared with enolate B

A   B

which of the following statements is true?

(a) A is more stable than B.

(b) B is more stable than A.

(c) A and B have the same stability.

(d) No comparison of stability can be made.

B-2. Which structure is the most stable?

(a)   (b)   (c)   (d)   (e)

B-3. Which one of the following molecules contains deuterium (D = ^{2}H) after reaction with NaOD in D_2O?

(a) $C_6H_5\overset{\displaystyle O}{\overset{\displaystyle \|}{C}}H$

(b) $C_6H_5CH_2\overset{\displaystyle O}{\overset{\displaystyle \|}{C}}H$

(c) $C_6H_5\overset{\displaystyle O}{\overset{\displaystyle \|}{C}}C(CH_3)_3$

(d) $(CH_3)_3C\overset{\displaystyle O}{\overset{\displaystyle \|}{C}}C(CH_3)_3$

B-4. Which of the following RX compounds is the best alkylating agent in the reaction shown?

(a) $(CH_3)_3CBr$

(b) $(CH_3)_2CHCH_2OH$

(c) C_6H_5Br

(d) $(CH_3)_2CHCH_2Br$

B-5. Which of the following pairs of aldehydes gives a single product in a mixed aldol condensation?

(a) $\underset{\text{O}}{\overset{\text{O}}{\text{C}_6\text{H}_5\text{CH}_2\overset{\parallel}{\text{C}}\text{H}}}$ + $\text{C}_6\text{H}_5\overset{\overset{\text{O}}{\parallel}}{\text{C}}\text{H}$

(c) $\text{C}_6\text{H}_5\overset{\overset{\text{O}}{\parallel}}{\text{C}}\text{H}$ + $\text{H}_2\text{C}{=}\text{O}$

(b) $\text{C}_6\text{H}_5\overset{\overset{\text{O}}{\parallel}}{\text{C}}\text{H}$ + $(\text{CH}_3)_3\text{C}\overset{\overset{\text{O}}{\parallel}}{\text{C}}\text{H}$

(d) $\text{CH}_3\overset{\overset{\text{O}}{\parallel}}{\text{C}}\text{H}$ + $(\text{CH}_3)_2\text{CH}\overset{\overset{\text{O}}{\parallel}}{\text{C}}\text{H}$

B-6. What is the principal product of the following reaction?

$$\underset{\text{ethanol, water, heat}}{\overset{\text{NaOH}}{\longrightarrow}} \quad \text{H}_2\text{O} \;+\; ?$$

(a)

(c)

(b)

(d)

B-7. Which of the following forms an enol to the greatest extent?

(a) $\text{CH}_3\text{CH}_2\overset{\overset{\text{O}}{\parallel}}{\text{C}}\text{H}$

(c) $\text{CH}_3\overset{\overset{\text{O}}{\parallel}}{\text{C}}\text{CH}_2\overset{\overset{\text{O}}{\parallel}}{\text{C}}\text{H}$

(b) $\text{CH}_3\overset{\overset{\text{O}}{\parallel}}{\text{C}}\text{CH}_2\text{CH}_2\overset{\overset{\text{O}}{\parallel}}{\text{C}}\text{CH}_3$

(d) $\text{CH}_3\overset{\overset{\text{O O}}{\parallel\parallel}}{\text{C}}\text{C}\text{CH}_2\text{CH}_3$

B-8. Which of the following species is (are) *not* intermediates in the aldol condensation of acetaldehyde (ethanal) in aqueous base?

$$\underset{1}{\overset{\overset{\text{O}^-}{|}}{\text{H}_2\text{C}{=}\text{CH}}} \qquad \underset{2}{\overset{\overset{\text{O}^-}{|}}{\text{CH}_3\text{CH}\text{CH}_2\overset{\overset{\text{O}}{\parallel}}{\text{C}}\text{H}}} \qquad \underset{3}{\overset{\overset{{}^+\text{OH}}{\parallel}}{\text{CH}_3\text{CH}}} \qquad \underset{4}{\overset{\overset{\text{O}^-}{|}}{\text{CH}_3\text{CH}}\overset{\overset{\text{O}^-}{|}}{\text{CH}{=}\text{CH}}}$$

(a) 1 and 2   (b) 3 only   (c) 4 only   (d) 2 and 3   (e) 3 and 4

B-9. The compound shown in the box undergoes racemization on reaction with aqueous acid. Which of the following structures best represents the intermediate responsible for this process?

(a)   (b)   (c)   (d)   (e)

B-10. Which one of the following compounds is the best candidate for being prepared by an efficient mixed aldol addition reaction?

(a)

(c)

(e)

(b)

(d)

B-11. Which one of the following undergoes 1,4-addition with CH_3SK (in ethanol)?

(a)   (b)   (c)   (d)

B-12. Benzalacetone is the mixed aldol condensation product formed between benzaldehyde ($C_6H_5CH=O$) and acetone [$(CH_3)_2C=O$]. What is its structure?

(a) $C_6H_5CH=CHCCH_3$ (with O)

(c) $C_6H_5CH=C(CH_3)_2$

(b) $C_6H_5CCH=CHCH_3$ (with O)

(d) $C_6H_5CH_2CCH=CH_2$ (with O)

CHAPTER 19

Carboxylic Acids

SOLUTIONS TO TEXT PROBLEMS

19.1 (*b*) The four carbon atoms of crotonic acid form a continuous chain. Because there is a double bond between C-2 and C-3, crotonic acid is one of the stereoisomers of 2-butenoic acid. The stereochemistry of the double bond is *E*.

$$H_3C \quad \quad H$$
$$C{=}C$$
$$H \quad \quad CO_2H$$

(*E*)-2-Butenoic acid or
(*E*)-But-2-enoic acid
(crotonic acid)

(*c*) Oxalic acid is a dicarboxylic acid that contains two carbons. It is *ethanedioic* acid.

$$HO_2CCO_2H$$

Ethanedioic acid
(oxalic acid)

(*d*) The name given to $C_6H_5CO_2H$ is benzoic acid. Because it has a methyl group at the para position, the compound shown is *p-methylbenzoic acid*, or *4-methylbenzoic acid*.

$$H_3C{-}\bigcirc{-}CO_2H$$

p-Methylbenzoic acid or
4-methylbenzoic acid
(*p*-toluic acid)

19.2 (*b*) As in part (*a*) of this problem we can calculate the conjugate base/acid ratio using the Henderson-Hasselbalch relationship.

$$\frac{[\text{conjugate base}]}{[\text{acid}]} = 10^{(\text{pH}-\text{p}Ka)}$$

$$\frac{[\text{lactate}]}{[\text{lactic acid}]} = 10^{(2.5-3.9)} = 10^{-1.4} = 0.04$$

At pH = 2.5 there is 25 times as much lactic acid present as lactate ion.

19.3 Recall from Chapter 1 (text Section 1.16) that an acid–base equilibrium favors formation of the weaker acid and base. Also remember that the weaker acid forms the stronger conjugate base, and vice versa.

(*b*) The acid–base reaction between acetic acid and *tert*-butoxide ion is represented by the equation

$$CH_3CO_2H \ + \ (CH_3)_3CO^- \ \rightleftharpoons \ CH_3CO_2^- \ + \ (CH_3)_3COH$$

Acetic acid (stronger acid)	*tert*-Butoxide (stronger base)	Acetate ion (weaker base)	*tert*-Butyl alcohol (weaker acid)

Alcohols ($\text{p}K_a = 16-18$) are weaker acids than acetic acid ($\text{p}K_a = 4.7$); the equilibrium lies to the right. The equilibrium constant K is $10^{(18-4.7)} = 10^{13.3}$.

(*c*) Bromide ion is the conjugate base of hydrogen bromide, a strong acid.

$$CH_3CO_2H \ + \ Br^- \ \rightleftharpoons \ CH_3CO_2^- \ + \ HBr$$

Acetic acid (weaker acid)	Bromide ion (weaker base)	Acetate ion (stronger base)	Hydrogen bromide (stronger acid)

In this case, the position of equilibrium favors the starting materials, because acetic acid is a much weaker acid than hydrogen bromide. The equilibrium constant K is $10^{(-5.8-4.7)} = 10^{-10.5}$.

(*d*) Acetylide ion is a rather strong base, and acetylene ($\text{p}K_a = 26$) is a much weaker acid than acetic acid. The position of equilibrium favors the formation of products. The equilibrium constant K is $10^{(26-4.7)} = 10^{21.3}$.

$$CH_3CO_2H \ + \ HC{\equiv}C{:}^- \ \rightleftharpoons \ CH_3CO_2^- \ + \ HC{\equiv}CH$$

Acetic acid (stronger acid)	Acetylide ion (stronger base)	Acetate ion (weaker base)	Acetylene (weaker acid)

(*e*) Nitrate ion is a very weak base; it is the conjugate base of nitric acid. The position of equilibrium lies far to the left. The equilibrium constant K is $10^{(-1.4-4.7)} = 10^{-6.1}$.

$$CH_3CO_2H \ + \ NO_3^- \ \rightleftharpoons \ CH_3CO_2^- \ + \ HNO_3$$

Acetic acid (weaker acid)	Nitrate ion (weaker base)	Acetate ion (stronger base)	Nitric acid (stronger acid)

(*f*) Amide ion is a very strong base; it is the conjugate base of ammonia ($\text{p}K_a = 36$). The position of equilibrium lies far to the right. The equilibrium constant K is $10^{(36-4.7)} = 10^{31.3}$.

$$CH_3CO_2H \ + \ H_2N^- \ \rightleftharpoons \ CH_2CO_2^- \ + \ NH_3$$

Acetic acid (stronger acid)	Amide ion (stronger base)	Acetate ion (weaker base)	Ammonia (weaker acid)

19.4 (*b*) Propanoic acid is similar to acetic acid in its acidity. A hydroxyl group at C-2 is electron-withdrawing and stabilizes the carboxylate ion of lactic acid by a combination of inductive and field effects.

$$CH_3\underset{\uparrow OH}{\overset{+}{C}H}-C\overset{\displaystyle O}{\underset{\displaystyle O^-}{\diagdown}}$$

Hydroxyl group stabilizes
negative charge by attracting electrons.

Lactic acid is more acidic than propanoic acid. The measured pK_a's are

$$CH_3\underset{OH}{\overset{|}{C}H}CO_2H \qquad\qquad CH_3CH_2CO_2H$$

Lactic acid   Propanoic acid
$pK_a = 3.8$ $pK_a = 4.9$

(*c*) A carbonyl group is more strongly electron-withdrawing than a carbon–carbon double bond. Pyruvic acid is a stronger acid than acrylic acid.

$$CH_3\overset{\displaystyle O}{\overset{\displaystyle \|}{C}}CO_2H \qquad\qquad H_2C{=}CHCO_2H$$

Pyruvic acid   Acrylic acid
$pK_a = 3.3$ $pK_a = 4.3$

(*d*) Viewing the two compounds as substituted derivatives of acetic acid, RCH_2CO_2H, we judge $CH_3SO_2{-}$ to be strongly electron-withdrawing and acid-strengthening, whereas an ethyl group has only a small effect.

$$CH_3\overset{\displaystyle O}{\underset{\displaystyle O}{\overset{\displaystyle \|}{\underset{\displaystyle \|}{S}}}}CH_2CO_2H \qquad\qquad CH_3CH_2CH_2CO_2H$$

Methanesulfonylacetic acid   Butanoic acid
$pK_a = 2.4$ $pK_a = 4.7$

19.5 The compound can only be a carboxylic acid; no other class containing only carbon, hydrogen, and oxygen is more acidic. A reasonable choice is $HC{\equiv}CCO_2H$; C-2 is *sp*-hybridized and therefore electron-withdrawing and acid-strengthening. This is borne out by its measured pK_a of 1.8.

19.6 For carbonic acid, the "true K_1" is given by

$$\text{True } K_1 = \frac{[H^+][HCO_3^-]}{[H_2CO_3]}$$

The "observed K" is given by the expression

$$4.3\times10^{-7} = \frac{[H^+][HCO_3^-]}{[CO_2]}$$

which can be rearranged to

$$[H^+][HCO_3^-] = (4.3\times10^{-7})[CO_2]$$

and therefore

$$\text{True } K_1 = \frac{(4.3 \times 10^{-7})[CO_2]}{[H_2CO_3]}$$

$$= \frac{(4.3 \times 10^{-7})(99.7)}{0.3}$$

$$= 1.4 \times 10^{-4}$$

Thus, when corrected for the small degree to which carbon dioxide is hydrated, it can be seen that carbonic acid is actually a stronger acid than acetic acid. Carboxylic acids dissolve in sodium bicarbonate solution because the equilibrium that leads to carbon dioxide formation is favorable, not because carboxylic acids are stronger acids than carbonic acid.

19.7 (*b*) 2-Chloroethanol has been converted to 3-hydroxypropanoic acid by way of the corresponding nitrile.

$$HOCH_2CH_2Cl \xrightarrow[\text{H}_2\text{O}]{\text{NaCN}} HOCH_2CH_2CN \xrightarrow[\text{heat}]{\text{H}_3\text{O}^+} HOCH_2CH_2CO_2H$$

| 2-Chloroethanol | | 2-Cyanoethanol | | 3-Hydroxypropanoic acid |

The presence of the hydroxyl group in 2-chloroethanol precludes the preparation of a Grignard reagent from this material, and so any attempt at the preparation of 3-hydroxypropanoic acid via the Grignard reagent of 2-chloroethanol is certain to fail.

(*c*) Grignard reagents can be prepared from tertiary halides and react in the expected manner with carbon dioxide. The procedure shown is entirely satisfactory.

$$(CH_3)_3CCl \xrightarrow[\text{diethyl ether}]{\text{Mg}} (CH_3)_3CMgCl \xrightarrow[\text{2. H}_3\text{O}^+]{\text{1. CO}_2} (CH_3)_3CCO_2H$$

| *tert*-Butyl chloride | | *tert*-Butylmagnesium chloride | | 2,2-Dimethylpropanoic acid (61–70%) |

Preparation by way of the nitrile will not be feasible. Rather than react with sodium cyanide by substitution, *tert*-butyl chloride will undergo elimination exclusively. The S_N2 reaction with cyanide ion is limited to primary and secondary alkyl halides.

19.8 Incorporation of ^{18}O into benzoic acid proceeds by a mechanism analogous to that of esterification. The nucleophile that adds to the protonated form of benzoic acid is ^{18}O-enriched water (the ^{18}O atom is represented by the shaded letter Ø in the following equations).

Benzoic acid

Tetrahedral intermediate

The three hydroxyl groups of the tetrahedral intermediate are equivalent except that one of them is labeled with ^{18}O. Any one of these three hydroxyl groups may be lost in the dehydration step; when the hydroxyl group that is lost is unlabeled, an ^{18}O label is retained in the benzoic acid.

Tetrahedral intermediate

^{18}O-enriched benzoic acid

19.9 (*b*) The 16-membered ring of 15-pentadecanolide is formed from 15-hydroxypentadecanoic acid.

15-Pentadecanolide   15-Hydroxypentadecanoic acid

(*c*) Vernolepin has two lactone rings, which can be related to two hydroxy acid combinations.

Be sure to keep the relative stereochemistry unchanged. Remember, the carbon–oxygen bond of an alcohol remains intact when the alcohol reacts with a carboxylic acid to give an ester.

19.10 Alkyl chlorides and bromides undergo nucleophilic substitution when treated with sodium iodide in acetone (text Section 8.1). A reasonable approach is to brominate octadecanoic acid at its α-carbon atom, then replace the bromine substituent with iodine by nucleophilic substitution.

Octadecanoic acid   2-Bromooctadecanoic acid   2-Iodooctadecanoic acid

19.11 (*b*) The starting material is a derivative of malonic acid. It undergoes efficient thermal decarboxylation in the manner shown.

2-Heptylmalonic acid   Carbon dioxide

$$CH_3(CH_2)_6CH_2\overset{\overset{\textstyle O}{\|}}{C}OH$$

Nonanoic acid

(*c*) The phenyl and methyl substituents attached to C-2 of malonic acid play no role in the decarboxylation process.

2-Phenylpropanoic acid

19.12 (*b*) The thermal decarboxylation of β-keto acids resembles that of substituted malonic acids. The structure of 2,2-dimethylacetoacetic acid and the equation representing its decarboxylation were given in the text. The overall process involves the bonding changes shown.

| 2,2-Dimethylacetoacetic acid | Enol form of 3-methyl-2-butanone | 3-Methyl-2-butanone |

19.13 (*a*) Lactic acid (2-hydroxypropanoic acid) is a three-carbon carboxylic acid that bears a hydroxyl group at C-2.

$$\overset{3}{C}H_3\overset{2}{C}H\overset{1}{C}O_2H$$
$$|$$
$$OH$$

2-Hydroxypropanoic acid

(*b*) The parent name *ethanoic acid* tells us that the chain that includes the carboxylic acid function contains only two carbons. A hydroxyl group and a phenyl substituent are present at C-2.

2-Hydroxy-2-phenylethanoic acid
(mandelic acid)

(*c*) The parent alkane is *tetradecane,* which has an unbranched chain of 14 carbons. The terminal methyl group is transformed to a carboxyl function in tetradecanoic acid.

$$\overset{O}{\overset{||}{CH_3(CH_2)_{12}COH}}$$

Tetradecanoic acid
(myristic acid)

(*d*) Undecane is the unbranched alkane with 11 carbon atoms, undecanoic acid is the corresponding carboxylic acid, and *undecenoic acid* is an 11-carbon carboxylic acid that contains a double bond.

Because the carbon chain is numbered beginning with the carboxyl group, 10-undecenoic acid has its double bond at the opposite end of the chain from the carboxyl group.

$$H_2C{=}CH(CH_2)_8CO_2H$$

10-Undecenoic acid
(undecylenic acid)

(*e*) Mevalonic acid has a five-carbon chain with hydroxyl groups at C-3 and C-5, along with a methyl group at C-3.

$$\underset{\underset{OH}{|}}{HOCH_2CH_2\overset{\overset{CH_3}{|}}{C}CH_2CO_2H}$$

3,5-Dihydroxy-3-methylpentanoic acid
(mevalonic acid)

(*f*) The constitution represented by the systematic name 2-methyl-2-butenoic acid gives rise to two stereoisomers.

$$CH_3CH{=}\underset{\underset{CH_3}{|}}{C}CO_2H$$

2-Methyl-2-butenoic acid

Tiglic acid is the *E* isomer, and the *Z* isomer is known as *angelic acid*. The higher ranked substituents, methyl and carboxyl, are placed on opposite sides of the double bond in tiglic acid and on the same side in angelic acid.

(*E*)-2-Methyl-2-butenoic　　　(*Z*)-2-Methyl-2-butenoic
acid (tiglic acid)　　　　　　acid (angelic acid)

(*g*) Butanedioic acid is a four-carbon chain in which both terminal carbons are carboxylic acid groups. Malic acid has a hydroxyl group at C-2.

$$\underset{\underset{OH}{|}}{HO_2CCHCH_2CO_2H}$$

2-Hydroxybutanedioic acid
(malic acid)

(*h*) Each of the carbon atoms of propane bears a carboxyl group as a substituent in 1,2,3-propanetricarboxylic acid. In citric acid C-2 also bears a hydroxyl group.

$$\underset{\underset{OH}{|}}{HO_2CCH_2\overset{\overset{CO_2H}{|}}{C}CH_2CO_2H}$$

2-Hydroxy-1,2,3-propanetricarboxylic acid
(citric acid)

(*i*) There is an aryl substituent at C-2 of propanoic acid in ibuprofen. This aryl substituent is a benzene ring bearing an isobutyl group at the para position.

$$CH_3CHCO_2H$$

$$CH_2CH(CH_3)_2$$

2-(*p*-Isobutylphenyl)-
propanoic acid

(*j*) Benzenecarboxylic acid is the systematic name for benzoic acid. *Salicylic acid* is a derivative of benzoic acid bearing a hydroxyl group at the position ortho to the carboxyl.

OH

CO_2H

o-Hydroxybenzenecarboxylic acid
(salicylic acid)

19.14 (*a*) The carboxylic acid contains a linear chain of eight carbon atoms. The parent alkane is *octane,* and so the systematic name of $CH_3(CH_2)_6CO_2H$ is *octanoic acid.*

(*b*) The compound shown is the potassium salt of octanoic acid. It is *potassium octanoate.*

(*c*) The presence of a double bond in $H_2C=CH(CH_2)_5CO_2H$ is indicated by the ending *-enoic acid.* Numbering of the chain begins with the carboxylic acid, and so the double bond is between C-7 and C-8. The compound is named 7-*octenoic acid* or oct-7-enoic acid.

(*d*) Stereochemistry is systematically described by the *E–Z* notation. Here, the double bond between C-6 and C-7 in octenoic acid has the *Z* configuration; the higher ranked substituents are on the same side.

$$H_3C \qquad (CH_2)_4CO_2H$$
$$C=C$$
$$H \qquad H$$

(*Z*)-6-Octenoic acid or
(*Z*)-oct-6-enoic acid

(*e*) A dicarboxylic acid is named as a *dioic acid.* The carboxyl functions are the terminal carbons of an eight-carbon chain; $HO_2C(CH_2)_6CO_2H$ is *octanedioic acid.* It is not necessary to identify the carboxylic acid locations by number because they can only be at the ends of the chain when the *-dioic acid* name is used.

(*f*) Pick the longest continuous chain that includes both carboxyl groups and name the compound as a *-dioic acid.* This chain contains only three carbons and bears a pentyl group as a substituent at C-2. It is not necessary to specify the position of the pentyl group, because it can only be attached to C-2.

$$CH_3(CH_2)_4\overset{2}{C}H\overset{1}{C}O_2H$$
$$\underset{3}{C}O_2H$$

Pentylpropanedioic acid

Malonic acid is an acceptable synonym for propanedioic acid; this compound may also be named *pentylmalonic acid.*

(*g*) A carboxylic acid function is attached as a substituent on a seven-membered ring. The compound is *cycloheptanecarboxylic acid.*

(*h*) The aromatic ring is named as a substituent attached to the eight-carbon carboxylic acid. Numbering of the chain begins with the carboxyl group.

6-Phenyloctanoic acid

19.15 (*a*) Carboxylic acids are the most acidic class of organic compounds containing only the elements C, H, and O. The order of decreasing acidity is

		pK_a
Acetic acid	CH_3CO_2H	4.7
Ethanol	CH_3CH_2OH	16
Ethane	CH_3CH_3	≈62

(*b*) Here again, the carboxylic acid is the strongest acid and the hydrocarbon the weakest:

		pK_a
Benzoic acid	$C_6H_5CO_2H$	4.2
Benzyl alcohol	$C_6H_5CH_2OH$	16–18
Benzene	C_6H_6	≈43

(*c*) Propanedioic acid is a stronger acid than propanoic acid because the electron-withdrawing effect of one carboxyl group enhances the ionization of the other. Propanedial is a 1,3-dicarbonyl compound that yields a stabilized enolate; it is more acidic than 1,3-propanediol.

		pK_a
Propanedioic acid	$HO_2CCH_2CO_2H$	2.9
Propanoic acid	$CH_3CH_2CO_2H$	4.9
Propanedial	$O{=}CHCH_2CH{=}O$	≈9
1,3-Propanediol	$HOCH_2CH_2CH_2OH$	≈16

(*d*) Trifluoromethanesulfonic acid is by far the strongest acid in the group. It is structurally related to sulfuric acid, but its three fluorine substituents make it much stronger. Fluorine substituents increase the acidity of carboxylic acids and alcohols relative to their nonfluorinated analogs, but not enough to make fluorinated alcohols as acidic as carboxylic acids.

		pK_a
Trifluoromethanesulfonic acid	CF_3SO_2OH	−6
Trifluoroacetic acid	CF_3CO_2H	0.2
Acetic acid	CH_3CO_2H	4.7
2,2,2-Trifluoroethanol	CF_3CH_2OH	12.4
Ethanol	CH_3CH_2OH	≈16

(*e*) The order of decreasing acidity is carboxylic acid > β-diketone > ketone > hydrocarbon.

		pK_a
Cyclopentanecarboxylic acid	—CO_2H	5.0
2,4-Pentanedione	$CH_3CCH_2CCH_3$ (with two C=O)	9
Cyclopentanone	=O	20
Cyclopentene		45

19.16 (*a*) A trifluoromethyl group is strongly electron-withdrawing and acid-strengthening. Its ability to attract electrons from the carboxylate ion decreases as its distance down the chain increases. 3,3,3-Trifluoropropanoic acid is a stronger acid than 4,4,4-trifluorobutanoic acid.

$$CF_3CH_2CO_2H \qquad\qquad CF_3CH_2CH_2CO_2H$$

3,3,3-Trifluoropropanoic acid   4,4,4-Trifluorobutanoic acid
pK_a = 3.0   pK_a = 4.2

(*b*) The carbon that bears the carboxyl group in 2-butynoic acid is *sp*-hybridized and is, therefore, more electron-withdrawing than the *sp*³-hybridized α carbon of butanoic acid. The anion of 2-butynoic acid is therefore stabilized better than the anion of butanoic acid, and 2-butynoic acid is a stronger acid.

$$CH_3C \equiv CCO_2H \qquad CH_3CH_2CH_2CO_2H$$

2-Butynoic acid
pK_a = 2.6

Butanoic acid
pK_a = 4.8

(*c*) Cyclohexanecarboxylic acid is a typical aliphatic carboxylic acid and is expected to be similar to acetic acid in acidity. The greater electronegativity of the *sp*²-hybridized carbon attached to the carboxyl group in benzoic acid stabilizes benzoate anion better than the corresponding *sp*³-hybridized carbon stabilizes cyclohexanecarboxylate. Benzoic acid is a stronger acid.

Benzoic acid
pK_a = 4.2

Cyclohexanecarboxylic acid
pK_a = 4.9

(*d*) Its five fluorine substituents make the pentafluorophenyl group more electron-withdrawing than an unsubstituted phenyl group. Thus, pentafluorobenzoic acid is a stronger acid than benzoic acid.

Pentafluorobenzoic acid
pK_a = 3.4

Benzoic acid
pK_a = 4.2

(*e*) The pentafluorophenyl substituent is electron-withdrawing and increases the acidity of a carboxyl group to which it is attached. Its electron-withdrawing effect decreases with distance. Pentafluorobenzoic acid is a stronger acid than *p*-(pentafluorophenyl)benzoic acid.

Pentafluorobenzoic acid
pK_a = 3.4

p-(Pentafluorophenyl)benzoic acid
(pK_a not measured in water; comparable
with benzoic acid in acidity)

(*f*) The oxygen of the ring exercises an acidifying effect on the carboxyl group. This effect is largest when the oxygen is attached directly to the carbon that bears the carboxyl group. Furan-2-carboxylic acid is thus a stronger acid than furan-3-carboxylic acid.

Furan-2-carboxylic acid
pK_a = 3.2

Furan-3-carboxylic acid
pK_a = 3.9

(*g*) Furan-2-carboxylic acid has an oxygen attached to the carbon that bears the carboxyl group, whereas pyrrole-2-carboxylic acid has a nitrogen in that position. Oxygen is more electronegative than nitrogen and so stabilizes the carboxylate anion better. Furan-2-carboxylic acid is a stronger acid than pyrrole-2-carboxylic acid.

Furan-2-carboxylic acid
$pK_a = 3.2$ Pyrrole-2-carboxylic acid
$pK_a = 4.4$

19.17 (*a*) The conversion of 1-butanol to butanoic acid is simply the oxidation of a primary alcohol to a carboxylic acid. Chromic acid is a suitable oxidizing agent.

$$CH_3CH_2CH_2CH_2OH \xrightarrow{\ H_2CrO_4\ } CH_3CH_2CH_2CO_2H$$

1-Butanol   Butanoic acid

(*b*) Aldehydes may be oxidized to carboxylic acids by any of the oxidizing agents that convert primary alcohols to carboxylic acids.

$$CH_3CH_2CH_2\overset{\overset{\displaystyle O}{\|}}{C}H \xrightarrow[\text{H}_2\text{SO}_4,\ \text{H}_2\text{O}]{\ K_2Cr_2O_7\ } CH_3CH_2CH_2\overset{\overset{\displaystyle O}{\|}}{C}OH$$

Butanal   Butanoic acid

(*c*) The starting material has the same number of carbon atoms as does butanoic acid, and so all that is required is a series of functional group transformations. Carboxylic acids may be obtained by oxidation of the corresponding primary alcohol. The alcohol is available from the designated starting material, 1-butene.

$$CH_3CH_2CH_2CO_2H \quad \Longrightarrow \quad CH_3CH_2CH_2CH_2OH \quad \Longrightarrow \quad CH_3CH_2CH{=}CH_2$$

Hydroboration–oxidation of 1-butene yields 1-butanol, which can then be oxidized to butanoic acid as in part (*a*).

$$CH_3CH_2CH{=}CH_2 \xrightarrow[\text{2. H}_2\text{O}_2,\ \text{HO}^-]{\text{1. B}_2\text{H}_6} CH_3CH_2CH_2CH_2OH \xrightarrow{\ H_2CrO_4\ } CH_3CH_2CH_2CO_2H$$

1-Butene   1-Butanol   Butanoic acid

(*d*) Converting 1-propanol to butanoic acid requires the carbon chain to be extended by one atom. Both methods for achieving this conversion, carboxylation of a Grignard reagent and formation and hydrolysis of a nitrile, begin with alkyl halides. Alkyl halides in turn are prepared from alcohols.

Either of the two following procedures is satisfactory:

$$CH_3CH_2CH_2OH \xrightarrow[\text{or HBr}]{PBr_3} CH_3CH_2CH_2Br \xrightarrow[\text{diethyl ether}]{Mg} CH_3CH_2CH_2MgBr \xrightarrow[\text{2. } H_3O^+]{\text{1. } CO_2} CH_3CH_2CH_2CO_2H$$

1-Propanol   1-Bromopropane   Butanoic acid

$$CH_3CH_2CH_2OH \xrightarrow[\text{or HBr}]{PBr_3} CH_3CH_2CH_2Br \xrightarrow[\text{DMSO}]{KCN} CH_3CH_2CH_2CN \xrightarrow[\text{heat}]{H_2O, HCl} CH_3CH_2CH_2CO_2H$$

1-Propanol   1-Bromopropane   Butanenitrile   Butanoic acid

(*e*) Dehydration of 2-propanol to propene followed by free-radical addition of hydrogen bromide affords 1-bromopropane.

$$CH_3CHCH_3 \xrightarrow[\text{heat}]{H_2SO_4} CH_3CH{=}CH_2 \xrightarrow[\text{peroxides}]{HBr} CH_3CH_2CH_2Br$$
   |
   OH

2-Propanol   Propene   1-Bromopropane

Once 1-bromopropane has been prepared it is converted to butanoic acid as in part (*d*).

(*f*) The carbon skeleton of butanoic acid may be assembled by an aldol condensation of acetaldehyde.

$$CH_3CH_2CH_2CO_2H \Rightarrow CH_3CH{\Leftarrow}CHCH\overset{O}{\overset{\|}{}} \Rightarrow 2CH_3CH\overset{O}{\overset{\|}{}}$$

$$2CH_3CH\overset{O}{\overset{\|}{}} \xrightarrow[\text{heat}]{KOH, \text{ ethanol}} CH_3CH{=}CHCH\overset{O}{\overset{\|}{}}$$

Acetaldehyde   2-Butenal

Oxidation of the aldehyde followed by hydrogenation of the double bond yields butanoic acid.

$$CH_3CH{=}CHCH\overset{O}{\overset{\|}{}} \xrightarrow{H_2CrO_4} CH_3CH{=}CHCO_2H \xrightarrow[\text{Pt}]{H_2} CH_3CH_2CH_2CO_2H$$

2-Butenal   2-Butenoic acid   Butanoic acid

(*g*) Ethylmalonic acid belongs to the class of substituted malonic acids that undergo ready thermal decarboxylation. Decarboxylation yields butanoic acid.

$$CH_3CH_2CHCO_2H \xrightarrow{heat} CH_3CH_2CH_2CO_2H + CO_2$$
        |
        CO_2H

Ethylmalonic acid   Butanoic acid   Carbon dioxide

19.18 (*a*) The Friedel–Crafts alkylation of benzene by methyl chloride can be used to prepare ^{14}C-labeled toluene ($C^* = {}^{14}C$). Once prepared, toluene could be oxidized to benzoic acid.

$$\text{Benzene} + \overset{*}{C}H_3Cl \xrightarrow{AlCl_3} \text{(toluene)}{-}\overset{*}{C}H_3 \xrightarrow[\text{H}_2\text{O, heat}]{K_2Cr_2O_7, H_2SO_4} \text{(benzoic acid)}{-}\overset{*}{C}O_2H$$

Benzene   Methyl chloride   Toluene   Benzoic acid

(b) Formaldehyde can serve as a one-carbon source if it is attacked by the Grignard reagent derived from bromobenzene.

Benzene   Bromobenzene   Phenylmagnesium bromide   Benzyl alcohol

This sequence yields ^{14}C-labeled benzyl alcohol, which can be oxidized to ^{14}C-labeled benzoic acid.

Benzyl alcohol   Benzoic acid

(c) A direct route to ^{14}C-labeled benzoic acid utilizes a Grignard synthesis employing ^{14}C-labeled carbon dioxide.

Benzene   Bromobenzene   Phenylmagnesium bromide   Benzoic acid

19.19 (a) An acid–base reaction takes place when pentanoic acid is combined with sodium hydroxide.

$$CH_3CH_2CH_2CH_2CO_2H \ + \ NaOH \ \longrightarrow \ CH_3CH_2CH_2CH_2CO_2Na \ + \ H_2O$$

Pentanoic acid   Sodium hydroxide   Sodium pentanoate   Water

(b) Carboxylic acids react with sodium bicarbonate to give carbonic acid, which dissociates to carbon dioxide and water, so that the actual reaction that takes place is

$$CH_3CH_2CH_2CH_2CO_2H \ + \ NaHCO_3 \ \longrightarrow \ CH_3CH_2CH_2CH_2CO_2Na \ + \ CO_2 \ + \ H_2O$$

Pentanoic acid   Sodium bicarbonate   Sodium pentanoate   Carbon dioxide   Water

(c) Thionyl chloride is a reagent that converts carboxylic acids to the corresponding acyl chlorides.

$$CH_3CH_2CH_2CH_2CO_2H \ + \ SOCl_2 \ \longrightarrow \ CH_3CH_2CH_2CH_2\overset{\displaystyle O}{\overset{\|}{C}}Cl \ + \ SO_2 \ + \ HCl$$

Pentanoic acid   Thionyl chloride   Pentanoyl chloride   Sulfur dioxide   Hydrogen chloride

(d) Phosphorus tribromide is used to convert carboxylic acids to their acyl bromides.

$$3CH_3CH_2CH_2CH_2CO_2H \ + \ PBr_3 \ \longrightarrow \ 3CH_3CH_2CH_2CH_2\overset{\displaystyle O}{\overset{\|}{C}}Br \ + \ H_3PO_3$$

Pentanoic acid   Phosphorus tribromide   Pentanoyl bromide   Phosphorous acid

(e) Carboxylic acids react with alcohols in the presence of acid catalysts to give esters.

$$CH_3CH_2CH_2CH_2CO_2H \ + \ C_6H_5CH_2OH \ \underset{\longleftarrow}{\overset{H_2SO_4}{\longrightarrow}} \ CH_3CH_2CH_2CH_2\overset{\displaystyle O}{\overset{\|}{C}}OCH_2C_6H_5 \ + \ H_2O$$

Pentanoic acid   Benzyl alcohol   Benzyl pentanoate   Water

(f) Chlorine is introduced at the α-carbon atom of a carboxylic acid. The reaction is catalyzed by a small amount of phosphorus or a phosphorus trihalide and is called the Hell–Volhard–Zelinsky reaction.

$$\text{CH}_3\text{CH}_2\text{CH}_2\text{CH}_2\text{CO}_2\text{H} \;+\; \text{Cl}_2 \;\xrightarrow[\text{(catalyst)}]{\text{PBr}_3}\; \underset{\text{Cl}}{\text{CH}_3\text{CH}_2\text{CH}_2\text{CHCO}_2\text{H}} \;+\; \text{HCl}$$

| Pentanoic acid | Chlorine | 2-Chloropentanoic acid | Hydrogen chloride |

The α-halo substituent is derived from the halogen used, not from the phosphorus trihalide.

(g) In the case, bromine is introduced at the α carbon.

$$\text{CH}_3\text{CH}_2\text{CH}_2\text{CH}_2\text{CO}_2\text{H} \;+\; \text{Br}_2 \;\xrightarrow[\text{(catalyst)}]{\text{PCl}_3}\; \underset{\text{Br}}{\text{CH}_3\text{CH}_2\text{CH}_2\text{CHCO}_2\text{H}} \;+\; \text{HBr}$$

| Pentanoic acid | Bromine | 2-Bromopentanoic acid | Hydrogen bromide |

(h) α-Halo carboxylic acids are reactive substrates in nucleophilic substitution. Iodide acts as a nucleophile to displace bromide from 2-bromopentanoic acid.

$$\underset{\text{Br}}{\text{CH}_3\text{CH}_2\text{CH}_2\text{CHCO}_2\text{H}} \;+\; \text{NaI} \;\xrightarrow{\text{acetone}}\; \underset{\text{I}}{\text{CH}_3\text{CH}_2\text{CH}_2\text{CHCO}_2\text{H}} \;+\; \text{NaBr}$$

| 2-Bromopentanoic acid | Sodium iodide | 2-Iodopentanoic acid | Sodium bromide |

(i) Aqueous ammonia converts α-halo acids to α-amino acids.

$$\underset{\text{Br}}{\text{CH}_3\text{CH}_2\text{CH}_2\text{CHCO}_2\text{H}} \;+\; 2\text{NH}_3 \;\longrightarrow\; \underset{\text{NH}_2}{\text{CH}_3\text{CH}_2\text{CH}_2\text{CHCO}_2\text{H}} \;+\; \text{NH}_4\text{Br}$$

| 2-Bromopentanoic acid | Ammonia | 2-Aminopentanoic acid | Ammonium bromide |

(j) Lithium aluminum hydride is a powerful reducing agent and reduces carboxylic acids to primary alcohols.

$$\text{CH}_3\text{CH}_2\text{CH}_2\text{CH}_2\text{CO}_2\text{H} \;\xrightarrow[\text{2. H}_2\text{O}]{\text{1. LiAlH}_4}\; \text{CH}_3\text{CH}_2\text{CH}_2\text{CH}_2\text{CH}_2\text{OH}$$

| Pentanoic acid | 1-Pentanol |

(k) Phenylmagnesium bromide acts as a base to abstract the carboxylic acid proton because Grignard reagents are not compatible with carboxylic acids. Proton transfer converts the Grignard reagent to the corresponding hydrocarbon.

$$\text{CH}_3\text{CH}_2\text{CH}_2\text{CH}_2\text{CO}_2\text{H} \;+\; \text{C}_6\text{H}_5\text{MgBr} \;\longrightarrow\; \text{CH}_3\text{CH}_2\text{CH}_2\text{CH}_2\text{CO}_2\text{MgBr} \;+\; \text{C}_6\text{H}_6$$

| Pentanoic acid | Phenylmagnesium bromide | Bromomagnesium pentanoate | Benzene |

19.20 (a) Conversion of butanoic acid to 1-butanol is a reduction and requires lithium aluminum hydride as the reducing agent.

$$\overset{\overset{\displaystyle O}{\|}}{\text{CH}_3\text{CH}_2\text{CH}_2\text{COH}} \;\xrightarrow[\text{2. H}_2\text{O}]{\text{1. LiAlH}_4}\; \text{CH}_3\text{CH}_2\text{CH}_2\text{CH}_2\text{OH}$$

| Butanoic acid | 1-Butanol |

(*b*) Carboxylic acids cannot be reduced directly to aldehydes. The following two-step procedure may be used:

$$
\underset{\text{Butanoic acid}}{CH_3CH_2CH_2\overset{\overset{O}{\|}}{C}OH} \quad \xrightarrow[\text{2. H}_2\text{O}]{\text{1. LiAlH}_4} \quad \underset{\text{1-Butanol}}{CH_3CH_2CH_2CH_2OH} \quad \xrightarrow[\text{CH}_2\text{Cl}_2]{\text{PCC}} \quad \underset{\text{Butanal}}{CH_3CH_2CH_2\overset{\overset{O}{\|}}{C}H}
$$

(*c*) Remember that alkyl halides are usually prepared from alcohols. 1-Butanol is therefore needed in order to prepare 1-chlorobutane.

$$
\underset{\text{1-Butanol [from part (}a\text{)]}}{CH_3CH_2CH_2CH_2OH} \quad \xrightarrow{\text{SOCl}_2} \quad \underset{\text{1-Chlorobutane}}{CH_3CH_2CH_2CH_2Cl}
$$

(*d*) Carboxylic acids are converted to their corresponding acyl chlorides with thionyl chloride.

$$
\underset{\text{Butanoic acid}}{CH_3CH_2CH_2\overset{\overset{O}{\|}}{C}OH} \quad \xrightarrow{\text{SOCl}_2} \quad \underset{\text{Butanoyl chloride}}{CH_3CH_2CH_2\overset{\overset{O}{\|}}{C}Cl}
$$

(*e*) Butanoyl chloride, prepared in part (*d*), can be used to acylate benzene in a Friedel–Crafts reaction.

Benzene + Butanoyl chloride $CH_3CH_2CH_2\overset{\overset{O}{\|}}{C}Cl$ $\xrightarrow{\text{AlCl}_3}$ Phenyl propyl ketone $\overset{\overset{O}{\|}}{C}CH_2CH_2CH_3$

(*f*) The preparation of 4-octanone using compounds derived from butanoic acid may be seen by using disconnections in a retrosynthetic analysis.

$$
CH_3CH_2CH_2\overset{\overset{O}{\|}}{C}CH_2CH_2CH_2CH_3 \quad \Longrightarrow \quad CH_3CH_2CH_2\underset{\overset{\displaystyle OH}{\big|}}{C}H\text{---}CH_2CH_2CH_2CH_3
$$

$$
CH_3CH_2CH_2\overset{\overset{O}{\|}}{C}H \quad + \quad CH_3CH_2CH_2CH_2MgBr
$$

The reaction scheme which may be used is

$$
\underset{\text{1-Chlorobutane [from part (}c\text{)]}}{CH_3CH_2CH_2CH_2Cl} \quad \xrightarrow{\text{Mg}} \quad \underset{\text{Butylmagnesium chloride}}{CH_3CH_2CH_2CH_2MgCl}
$$

$$CH_3CH_2CH_2CH_2MgCl \ + \ CH_3CH_2CH_2\overset{\overset{\displaystyle O}{\|}}{C}H \quad \xrightarrow[\text{2. } H_3O^+]{\text{1. diethyl ether}} \quad CH_3CH_2CH_2\underset{\underset{\displaystyle OH}{|}}{C}HCH_2CH_2CH_2CH_3$$

Butylmagnesium chloride   Butanal [from part (*b*)]   4-Octanol

$$\Big\downarrow H_2CrO_4$$

$$CH_3CH_2CH_2\overset{\overset{\displaystyle O}{\|}}{C}CH_2CH_2CH_2CH_3$$

4-Octanone

(*g*) Carboxylic acids are halogenated at their α-carbon atom by the Hell–Volhard–Zelinsky reaction.

$$CH_3CH_2CH_2CO_2H \quad \xrightarrow[Br_2]{P} \quad CH_3CH_2\underset{\underset{\displaystyle Br}{|}}{C}HCO_2H$$

Butanoic acid   2-Bromobutanoic acid

A catalytic amount of PCl_3 may be used in place of phosphorus in the reaction.

(*h*) Dehydrohalogenation of 2-bromobutanoic acid gives 2-butenoic acid.

$$CH_3CH_2\underset{\underset{\displaystyle Br}{|}}{C}HCO_2H \quad \xrightarrow[\substack{DMSO \\ \text{2. } H^+}]{\text{1. KOC(CH}_3)_3} \quad CH_3CH{=}CHCO_2H$$

2-Bromobutanoic acid   2-Butenoic acid

19.21 (*a*) The compound to be prepared is *glycine*, an α-amino acid. The amino functional group can be introduced by a nucleophilic substitution reaction on an α-halo acid, which is available by way of the Hell–Volhard–Zelinsky reaction.

$$CH_3CO_2H \quad \xrightarrow[P]{Br_2} \quad BrCH_2CO_2H \quad \xrightarrow{NH_3,\ H_2O} \quad H_2NCH_2CO_2H$$

Acetic acid   Bromoacetic acid   Aminoacetic acid
(glycine)

(*b*) Phenoxyacetic acid is used as a fungicide. It can be prepared by a nucleophilic substitution using sodium phenoxide and bromoacetic acid.

$$BrCH_2CO_2H \quad \xrightarrow[\text{2. } H^+]{\text{1. } C_6H_5ONa} \quad C_6H_5OCH_2CO_2H \ + \ NaCl$$

Bromoacetic acid   Phenoxyacetic acid
[from part (*a*)]

(*c*) Cyanide ion is a good nucleophile and will displace bromide from bromoacetic acid.

$$BrCH_2CO_2H \quad \xrightarrow[Na_2CO_3,\ H_2O]{NaCN} \quad NCCH_2CO_2Na \quad \xrightarrow{H^+} \quad NCCH_2CO_2H$$

Bromoacetic acid   Sodium cyanoacetate   Cyanoacetic acid
[from part (*a*)]

(*d*) Cyanoacetic acid, prepared as in part (*c*), serves as a convenient precursor to malonic acid. Hydrolysis of the nitrile substituent converts it to a carboxyl group.

$$NCCH_2CO_2H \quad \xrightarrow[\text{heat}]{H_2O,\ H^+} \quad HO_2CCH_2CO_2H$$

Cyanoacetic acid   Malonic acid

(*e*) Iodoacetic acid is not prepared directly from acetic acid but is derived by nucleophilic substitution of iodide on bromoacetic acid.

$$BrCH_2CO_2H \xrightarrow[\text{acetone}]{\text{NaI}} ICH_2CO_2H$$

Bromoacetic acid
[from part (*a*)]

Iodoacetic acid

(*f*) Two transformations need to be accomplished, α bromination and esterification. The correct sequence is bromination followed by esterification.

$$CH_3CO_2H \xrightarrow[P]{Br_2} BrCH_2CO_2H \xrightarrow[H^+]{CH_3CH_2OH} BrCH_2CO_2CH_2CH_3$$

Acetic acid

Bromoacetic acid

Ethyl bromoacetate

Reversing the order of steps is not appropriate. It must be the carboxylic acid that is subjected to halogenation because the Hell–Volhard–Zelinsky reaction is a reaction of carboxylic acids, not esters.

(*g*) The compound shown is an ylide. It can be prepared from ethyl bromoacetate as shown.

$$BrCH_2CO_2CH_2CH_3 \quad + \quad (C_6H_5)_3P \longrightarrow (C_6H_5)_3\overset{+}{P}CH_2CO_2CH_2CH_3 \ Br^-$$

Ethyl bromoacetate
[from part (*f*)]

Triphenyl-
phosphine

$$\downarrow NaOCH_2CH_3$$

$$(C_6H_5)_3\overset{+}{P}-\overset{..}{\underset{}{C}}HCO_2CH_2CH_3$$

Ylide

The first step is a nucleophilic substitution of bromide by triphenylphosphine. Treatment of the derived triphenylphosphonium salt with base removes the relatively acidic α proton, forming the ylide. (For a review of ylide formation, refer to text Section 17.12.)

(*h*) Reaction of the ylide formed in part (*g*) with benzaldehyde gives the desired alkene by a Wittig reaction.

$$(C_6H_5)_3\overset{+}{P}-\overset{..}{C}HCO_2CH_2CH_3 \ + \ C_6H_5\overset{\overset{O}{\|}}{C}H \longrightarrow C_6H_5CH=CHCO_2CH_2CH_3 \ + \ (C_6H_5)_3\overset{+}{P}-\overset{-}{O}$$

Ylide from part (*g*)

Benzaldehyde

Ethyl cinnamate

Triphenylphosphine
oxide

19.22 (*a*) Carboxylic acids are converted to ethyl esters when they are allowed to stand in ethanol in the presence of an acid catalyst.

$$\underset{\substack{H}}{\overset{\substack{H_3C}}{}}C=C\underset{\substack{CO_2H}}{\overset{\substack{CH_3}}{}} \ + \ CH_3CH_2OH \xrightarrow{H_2SO_4} \underset{\substack{H}}{\overset{\substack{H_3C}}{}}C=C\underset{\substack{COCH_2CH_3 \\ \| \\ O}}{\overset{\substack{CH_3}}{}} \ + \ H_2O$$

(*E*)-2-Methyl-2-butenoic acid

Ethanol

Ethyl (*E*)-2-methyl-2-butenoate
(74–80%)

Water

(*b*) Lithium aluminum hydride, LiAlH$_4$, reduces carboxylic acids to primary alcohols. When LiAlD$_4$ is used, deuterium is transferred to the carbonyl carbon.

Cyclopropanecarboxylic   1-Cyclopropyl-1,1-
acid   dideuteriomethanol
                      (75%)

Notice that deuterium is bonded only to carbon. The hydroxyl proton is derived from water, not from the reducing agent.

(*c*) In the presence of a catalytic amount of phosphorus, bromine reacts with carboxylic acids to yield the corresponding α-bromo derivative.

Cyclohexanecarboxylic   1-Bromocyclohexanecarboxylic
acid   acid (96%)

(*d*) Alkyl fluorides are not readily converted to Grignard reagents, and so it is the bromine substituent that is attacked by magnesium.

m-Bromo(trifluoromethyl)-   *m*-(Trifluoromethyl)-
benzene   benzoic acid

(*e*) Cyano substituents are hydrolyzed to carboxyl groups in the presence of acid catalysts.

m-Chlorobenzyl   *m*-Chlorophenylacetic
cyanide   acid (61%)

(*f*) The carboxylic acid function plays no part in this reaction; free-radical addition of hydrogen bromide to the carbon–carbon double bond occurs.

$$H_2C=CH(CH_2)_8CO_2H \xrightarrow[\text{benzoyl peroxide}]{\text{HBr}} BrCH_2CH_2(CH_2)_8CO_2H$$

10-Undecenoic acid   11-Bromoundecanoic acid
                                         (66–70%)

Recall that hydrogen bromide adds to alkenes in the presence of peroxides with a regioselectivity opposite to that of Markovnikov's rule.

19.23 The series of reactions given in this problem were used to convert the aldehyde starting material to a primary alcohol having one more carbon atom.

Reaction 1: The starting material is an aldehyde and is reduced by lithium aluminum hydride to the corresponding primary alcohol.

Reaction 2: Thionyl chloride converts alcohols to alkyl chlorides.

Reaction 3: The reactant in this step is a primary, benzylic chloride. It undergoes an S_N2 reaction with sodium cyanide to give a nitrile.

Reaction 4: The nitrile is converted to a carboxylic acid (as its potassium salt) on reaction with potassium hydroxide. Acidification of the reaction mixture gives the carboxylic acid.

Reaction 5: Lithium aluminum hydride reduces carboxylic acids to primary alcohols.

19.24 (*a*) The desired product and the starting material have the same carbon skeleton, and so all that is required is a series of functional group transformations. Recall that, as seen in Problem 19.17, a carboxylic acid may be prepared by oxidation of the corresponding primary alcohol. The needed alcohol is available from the appropriate alkene.

(*b*) The target molecule contains one more carbon than the starting material, and so a carbon–carbon bond-forming step is indicated. Two approaches are reasonable; one proceeds by way of nitrile formation and hydrolysis, the other by carboxylation of a Grignard reagent. In either case the key intermediate is 1-bromo-2-methylpropane.

The desired alkyl bromide may be prepared by free-radical addition of hydrogen bromide to 2-methylpropene.

Another route to the alkyl bromide utilizes the alcohol prepared in part (*a*).

$$(CH_3)_2CHCH_2OH \xrightarrow[\text{or HBr}]{PBr_3} (CH_3)_2CHCH_2Br$$

2-Methyl-1-propanol   1-Bromo-2-
                                         methylpropane

Conversion of the alkyl bromide to the carboxylic acid is then carried out as follows:

$$(CH_3)_2CHCH_2Br \xrightarrow{KCN} (CH_3)_2CHCH_2CN \xrightarrow[\text{heat}]{H_2O, H^+} (CH_3)_2CHCH_2CO_2H$$

3-Methylbutanoic acid

1-Bromo-2-methylpropane $\xrightarrow[\text{diethyl ether}]{Mg} (CH_3)_2CHCH_2MgBr \xrightarrow[\text{2. } H_3O^+]{\text{1. } CO_2} (CH_3)_2CHCH_2CO_2H$

3-Methylbutanoic acid

(*c*) Examining the target molecule reveals that it contains two more carbon atoms than the indicated starting material, suggesting use of ethylene oxide in a two-carbon chain-extension process.

$$(CH_3)_3CCH_2CO_2H \implies (CH_3)_3C \overset{\S}{\longmapsto} CH_2CH_2OH \implies (CH_3)_3CMgX \ + \ H_2C\overset{O}{\underset{}{\frown}}CH_2$$

This suggests the following sequence of steps:

$$(CH_3)_3COH \xrightarrow{HBr} (CH_3)_3CBr \xrightarrow{Mg} (CH_3)_3CMgBr \xrightarrow[\text{2. } H_3O^+]{\text{1. } H_2C-CH_2 (O)} (CH_3)_3CCH_2CH_2OH$$

tert-Butyl   2-Bromo-2-   *tert*-Butylmagnesium   3,3-Dimethyl-1-butanol
alcohol   methylpropane   bromide

$$\bigg\downarrow \begin{array}{l} K_2Cr_2O_7, H^+ \\ H_2O \end{array}$$

$$(CH_3)_3CCH_2CO_2H$$

3,3-Dimethylbutanoic
acid

(*d*) This synthesis requires extending a carbon chain by two carbon atoms. One way to form dicarboxylic acids is by hydrolysis of dinitriles.

$$HO_2C(CH_2)_5CO_2H \implies NC(CH_2)_5CN \implies Br(CH_2)_5Br$$

This suggests the following sequence of steps:

$$HO_2C(CH_2)_3CO_2H \xrightarrow[\text{2. } H_2O]{\text{1. LiAlH}_4} HOCH_2(CH_2)_3CH_2OH \xrightarrow[PBr_3]{HBr \text{ or}} BrCH_2(CH_2)_3CH_2Br$$

Pentanedioic acid   1,5-Pentanediol   1,5-Dibromopentane

$$\bigg\downarrow KCN$$

$$HO_2CCH_2(CH_2)_3CH_2CO_2H \xleftarrow[\text{heat}]{H_2O, H^+} NCCH_2(CH_2)_3CH_2CN$$

Heptanedioic acid   1,5-Dicyanopentane (*e*) The desired alcohol cannot be prepared directly from the nitrile. It is available, however, by lithium aluminum hydride reduction of the carboxylic acid obtained by hydrolysis of the nitrile.

$$CH_3\underset{\underset{\displaystyle C_6H_5}{|}}{C}HCH_2CN \xrightarrow[\text{heat}]{H_2O,\ H^+} CH_3\underset{\underset{\displaystyle C_6H_5}{|}}{C}HCH_2\overset{\displaystyle \overset{O}{\|}}{C}OH \xrightarrow[\text{2. } H_2O]{\text{1. } LiAlH_4} CH_3\underset{\underset{\displaystyle C_6H_5}{|}}{C}HCH_2CH_2OH$$

3-Phenylbutanenitrile　　　　3-Phenylbutanoic acid　　　　3-Phenyl-1-butanol

(*f*) In spite of the structural similarity between the starting material and the desired product, a one-step transformation cannot be achieved.

Cyclopentyl　　　　1-Bromocyclopentanecarboxylic acid
bromide

Instead, recall that α-bromo acids are prepared from carboxylic acids by the Hell–Volhard–Zelinsky reaction:

Cyclopentane-　　　　1-Bromocyclopentanecarboxylic acid
carboxylic acid

The problem now simplifies to one of preparing cyclopentanecarboxylic acid from cyclopentyl bromide. Two routes are possible:

Cyclopentyl　　　　Cyclopentyl　　　　Cyclopentanecarboxylic
bromide　　　　cyanide　　　　acid

Cyclopentyl　　　　Cyclopentylmagnesium　　　　Cyclopentanecarboxylic
bromide　　　　bromide　　　　acid

The Grignard route is slightly better; it is a "one-pot" transformation. Converting the secondary bromide to a nitrile requires two separate operations.

(*g*) In this case the halogen substituent is present at the β carbon rather than the α-carbon atom of the carboxylic acid. The starting material, a β-chloro unsaturated acid, can lead to the desired carbon skeleton by a Diels–Alder reaction.

1,3-Butadiene	(E)-3-Chloropropenoic acid	trans-2-Chloro-4-cyclohexenecarboxylic acid

The required trans stereochemistry is a consequence of the stereospecificity of the Diels–Alder reaction.

Hydrogenation of the double bond of the Diels–Alder adduct gives the required product.

trans-2-Chloro-4-cyclohexenecarboxylic acid   trans-2-Chlorocyclohexanecarboxylic acid

(h) The target molecule can be related to the starting material by the retrosynthesis

2,4-Dimethylbenzoic acid   m-Xylene

Bromine can be introduced by electrophilic substitution in the activated aromatic ring of m-xylene.

m-Xylene   1-Bromo-2,4-dimethylbenzene

The aryl bromide cannot be converted to a carboxylic acid by way of the corresponding nitrile, because aryl bromides are not reactive toward nucleophilic substitution. The Grignard route is necessary.

1-Bromo-2,4-dimethylbenzene → (1. Mg, diethyl ether 2. CO_2 3. H_3O^+) → 2,4-Dimethylbenzoic acid

(*i*) The relationship of the target molecule to the starting material

4-Chloro-3-nitrobenzoic acid → *p*-Chlorotoluene

requires that there be two synthetic operations: oxidation of the methyl group and nitration of the ring. The orientation of the nitro group requires that nitration must follow oxidation of the methyl group of the starting material.

p-Chlorotoluene → ($K_2Cr_2O_7$, H_2SO_4, H_2O) → *p*-Chlorobenzoic acid

Nitration of *p*-chlorobenzoic acid gives the desired product, because the directing effects of the chlorine (ortho, para) and the carboxyl (meta) groups reinforce each other.

p-Chlorobenzoic acid → (HNO_3, H_2SO_4) → 4-Chloro-3-nitrobenzoic acid

(*j*) The desired synthetic route becomes apparent when it is recognized that the Z alkene stereoisomer may be obtained from an alkyne, which, in turn, is available by carboxylation of the anion derived from the starting material.

$CH_3C{\equiv}CCO_2H$ → $CH_3C{\equiv}C{:}^-$ + CO_2

The desired reaction sequence is

$$CH_3C{\equiv}CH \xrightarrow[NH_3]{NaNH_2} CH_3C{\equiv}CNa \xrightarrow[\text{2. } H_3O^+]{\text{1. } CO_2} CH_3C{\equiv}CCO_2H$$

Propyne   Propynylsodium   2-Butynoic acid

Hydrogenation of the carbon–carbon triple bond of 2-butynoic acid over the Lindlar catalyst converts this compound to the Z isomer of 2-butenoic acid.

2-Butynoic acid   (Z)-2-Butenoic acid

19.25 (*a*) Only the cis stereoisomer of 4-hydroxycyclohexanecarboxylic acid is capable of forming a lactone. The most stable conformation of the starting hydroxy acid is a chair conformation; however, in the lactone, the cyclohexane ring adopts a boat conformation.

cis-4-Hydroxycyclohexane-
carboxylic acid
(chair conformation)

cis-4-Hydroxycyclohexane-
carboxylic acid
(boat conformation)

Lactone

(*b*) As in part (*a*), lactone formation is possible only when the hydroxyl and carboxyl groups are cis.

cis-3-Hydroxycyclo-
hexanecarboxylic acid

Lactone

Although the most stable conformation of *cis*-3-hydroxycyclohexanecarboxylic acid has both substituents equatorial and is unable to close to a lactone, the diaxial orientation is accessible and is capable of lactone formation.

Neither conformation of *trans*-3-hydroxycyclohexanecarboxylic acid has the substituents close enough to each other to form an unstrained lactone.

trans-3-Hydroxycyclohexanecarboxylic acid: lactone formation impossible

19.26 Dicarboxylic acids in which both carboxyl groups are attached to the same carbon undergo ready thermal decarboxylation to produce the enol form of an acid.

Compound A

This enol yields a mixture of *cis*- and *trans*-3-chlorocyclobutanecarboxylic acid. The two products are stereoisomers.

cis-3-Chlorocyclobutane-
carboxylic acid

trans-3-Chlorocyclobutane-
carboxylic acid

19.27 Examination of the molecular formula $C_{14}H_{26}O_2$ reveals that the compound has an index of hydrogen deficiency of 2. Because we are told that the compound is a carboxylic acid, one of these elements of unsaturation must be a carbon–oxygen double bond. The other must be a carbon–carbon double bond because the compound undergoes cleavage on ozonolysis. Examining the products of ozonolysis serves to locate the position of the double bond.

Cleavage by ozone
occurs here.

Nonanal

5-Oxopentanoic acid

The starting acid must be 5-tetradecenoic acid. The stereochemistry of the double bond is not revealed by these experiments.

19.28 Hydrogenation of the starting material is expected to result in reduction of the ketone carbonyl while leaving the carboxyl group unaffected. Because the isolated product lacks a carboxyl group, however, that group must react in some way. The most reasonable reaction is intramolecular esterification to form a γ-lactone.

Levulinic acid

4-Hydroxypentanoic
acid (not isolated)

4-Pentanolide
$(C_5H_8O_2)$

19.29 Compound A is a cyclic acetal and undergoes hydrolysis in aqueous acid to produce acetaldehyde, along with a dihydroxy carboxylic acid.

Compound A   3,5-Dihydroxy-3-methylpentanoic   Acetaldehyde
                              acid

The dihydroxy acid that is formed in this step cyclizes to the δ-lactone mevalonolactone.

3,5-Dihydroxy-3-
methylpentanoic acid

Mevalonolactone

19.30 Compound A is a δ-lactone. To determine its precursor, disconnect the ester linkage to a hydroxy acid.

Compound A

The precursor has the same carbon skeleton as the designated starting material. All that is necessary is to hydrogenate the double bond of the alkynoic acid to the cis alkene. This can be done by using the Lindlar catalyst. Cyclization of the hydroxy acid to the lactone is spontaneous.

$$CH_3CHCH_2C{\equiv}CCO_2H \quad \xrightarrow[\text{Lindlar Pd}]{H_2}$$
OH

5-Hydroxy-2-hexynoic
acid

(Not isolated)

Compound A

19.31 Hydration of the double bond can occur in two different directions:

(*a*) The achiral isomer is citric acid.

$$\underset{\text{OH}}{HO_2CCH_2CCH_2CO_2H}$$
$$\overset{CO_2H}{|}$$

Citric acid has no chirality centers.

(*b*) The other isomer, isocitric acid, has two chirality centers (marked with an asterisk*). Isocitric acid has the constitution

$$CO_2H$$
$$|$$
$$HO_2C\overset{*}{C}HCHCH_2CO_2H$$
$$|\ \ \ \ \ \ |^{*}$$
$$OH$$

Isocitric acid

With two chirality centers, there are 2^2, or four, stereoisomers represented by this constitution. The one that is actually formed in this enzyme-catalyzed reaction is the 2*R*,3*S* isomer.

19.32 Carboxylic acid protons give signals in the range δ 10–12. A signal in this region suggests the presence of a carboxyl group but tells little about its environment. Thus, in assigning structures to compounds A, B, and C, the most useful data are the chemical shifts of the protons other than the carboxyl protons. Compare the three structures:

$$\overset{\displaystyle O}{\overset{\displaystyle \|}{HCOH}}$$

$$\overset{\displaystyle H}{\underset{\displaystyle HO_2C}{}}C=C\overset{\displaystyle H}{\underset{\displaystyle CO_2H}{}}$$

$$HO_2CCH_2CO_2H$$

Formic acid   Maleic acid   Malonic acid

The proton that is diagnostic of structure in formic acid is bonded to a carbonyl group; it is an aldehyde proton. Typical chemical shifts of aldehyde protons are δ 8–10, and therefore formic acid is compound C.

$$\overset{\displaystyle O}{\overset{\displaystyle \|}{H-C-O-H}}$$

δ 8.0   δ 11.4

Compound C

The critical signal in maleic acid is that of the vinyl protons, which normally is found in the range δ 5–7. Maleic acid is compound B.

$$\overset{\displaystyle H}{\underset{\displaystyle HO_2C}{}}C=C\overset{\displaystyle H}{\underset{\displaystyle CO_2H}{}}$$

δ 6.3

δ 12.4

Compound B

Compound A is malonic acid. Here we have a methylene group bearing two carbonyl substituents. These methylene protons are more shielded than the aldehyde proton of formic acid or the vinyl protons of maleic acid.

$$HO_2CCH_2CO_2H$$

δ 3.2   δ 12.1

Compound A

19.33 Compounds A and B both exhibit ^1H NMR absorptions in the region δ 10–12 characteristic of carboxylic acids. The formula $C_4H_8O_3$ suggests an index of hydrogen deficiency of 1, accounted for by the carbonyl of the carboxyl group. Compound A has the triplet–quartet splitting indicative of an ethyl group, and compound B has two triplets, suggesting $—CH_2CH_2—$.

$$3.6\,(q) \quad 11.1\,(s) \qquad\qquad 3.7\,(t) \quad 11.3\,(s)$$

$$CH_3CH_2OCH_2COH \qquad\qquad CH_3OCH_2CH_2COH$$

$$1.3\,(t) \qquad 4.1\,(s) \qquad\qquad 3.4\,(s) \qquad 2.6\,(t)$$

Compound A   Compound B

19.34 (*a*) The formula of compound A ($C_3H_5ClO_2$) has an index of hydrogen deficiency of 1—the carboxyl group. Only two structures are possible:

$$ClCH_2CH_2C\overset{O}{\underset{OH}{}} \quad \text{and} \quad CH_3CHC\overset{O}{\underset{OH}{}}$$
$$Cl$$

3-Chloropropanoic acid   2-Chloropropanoic acid

Compound A is determined to be 3-chloropropanoic acid on the basis of its ^1H NMR spectrum, which shows two triplets at δ 2.9 and δ 3.8.

$$ClCH_2—CH_2—C\overset{O}{\underset{O—H}{}}$$

Triplet 3.8   Triplet 2.9   11.9   Compound A

Compound A cannot be 2-chloropropanoic acid, because that compound's ^1H NMR spectrum would show a three-proton doublet for the methyl group and a one-proton quartet for the methine proton.

(*b*) The formula of compound B ($C_9H_9NO_4$) corresponds to an index of hydrogen deficiency of 6. The presence of an aromatic ring, as evidenced by the ^1H NMR absorptions at δ 7.5 and 8.2, accounts for four of the unsaturations. The appearance of the aromatic protons as a pair of doublets with a total area of 4 suggests a para-disubstituted ring.

$$X—\langle\text{ring}\rangle—Y$$

That compound B is a carboxylic acid is evidenced by the singlet (area = 1) at δ 12.1. The remaining ^1H NMR signals [a quartet at δ 3.9 (1H) and a doublet at δ 1.6 (3H)] suggest the fragment $CH—CH_3$. All that remains of the molecular formula is $—NO_2$. Combining this information identifies compound B as 2-(4-nitrophenyl)propanoic acid.

$$CH_3$$
$$O_2N—\langle\text{ring}\rangle—CHCO_2H$$

2-(4-Nitrophenyl)propanoic acid
(compound B)

ANSWERS TO INTERPRETIVE PROBLEMS

19.35 C; **19.36** B; **19.37** C; **19.38** D; **19.39** B

SELF-TEST

PART A

A-1. Provide an acceptable IUPAC name for each of the following:

(a) $\underset{\underset{CH_3}{|}}{C_6H_5CH}\underset{\underset{CH_3}{|}}{CHCH_2CH_2CO_2H}$

(b) (cyclohexyl)—CO_2H

(c) $\underset{\underset{CH_2CH_3}{|}}{CH_3}\underset{\underset{}{|}}{\overset{\overset{Br}{|}}{C}HCHCO_2H}$

A-2. Both of the following compounds may be converted into 4-phenylbutanoic acid by one or more reaction steps. Give the reagents and conditions necessary to carry out these conversions.

$$C_6H_5CH_2CH_2CH(CO_2H)_2 \qquad C_6H_5CH_2CH_2CH_2Br$$

(Two methods)

A-3. The species whose structure is shown is an intermediate in an esterification reaction. Write the complete, balanced equation for this process.

$$\underset{\underset{OH}{|}}{C_6H_5CH_2}\overset{\overset{OH}{|}}{C}OCH_2CH_3$$

A-4. Give the correct structures for compounds A through C in the following reactions:

(a) $(CH_3)_2CHCH_2CH_2CO_2H \xrightarrow[P]{Br_2} A \xrightarrow[acetone]{KI} B$

(b) $C \xrightarrow{heat} C_6H_5\overset{\overset{O}{||}}{C}CH(CH_3)_2 + CO_2$

A-5. Give the missing reagent(s) and the missing compound in each of the following:

(a) (phenyl)—Br $\xrightarrow{?}$? $\xrightarrow[2.\ H_3O^+]{1.\ CO_2}$ (phenyl)—CO_2H

(b) $CH_3CH_2CH_2\overset{\overset{O}{||}}{C}OH \xrightarrow{?}$? $\xrightarrow[CH_2Cl_2]{PCC} CH_3CH_2CH_2\overset{\overset{O}{||}}{C}H$

(c) $CH_3CH_2CH_2\overset{\overset{O}{||}}{C}OH \xrightarrow{?}$? $\xrightarrow{NaSCH_3} \underset{\underset{SCH_3}{|}}{CH_3CH_2}\overset{\overset{O}{||}}{C}HCOH$

(d) $CH_3CH_2\overset{\overset{O}{||}}{C}H \xrightarrow[HCl]{NaCN}$? $\xrightarrow{?} \underset{\underset{OH}{|}}{CH_3CH_2}\overset{\overset{O}{||}}{C}HCOH$

A-6. Identify the carboxylic acid ($C_4H_7BrO_2$) having the ^1H NMR spectrum consisting of

δ 1.1, 3H (triplet)
δ 2.0, 2H (pentet)
δ 4.2, 1H (triplet)
δ 12.1, 1H (singlet)

A-7. Draw the structure of the tetrahedral intermediate in the esterification of formic acid with 1-butanol.

A-8. Write a mechanism for the esterification reaction shown.

$$CH_3\overset{\overset{O}{\|}}{C}OH + CH_3OH \xrightarrow{H^+} CH_3CO_2CH_3 + H_2O$$

PART B

B-1. Which of the following is a correct IUPAC name for the compound shown?

$$\underset{(CH_3CH_2)_2\overset{\overset{CO_2H}{|}}{C}CH_2CH(CH_2CH_3)_2}{}$$

(*a*) 1,1,3-Triethylhexanoic acid

(*c*) 3,5-Diethyl-3-heptylcarboxylic acid

(*b*) 2,2,4-Triethylhexanoic acid

(*d*) 3,5,5-Triethyl-6-hexanoic acid

B-2. Rank the following substances in order of decreasing acid strength (strongest → weakest):

$$CH_3CH_2CH_2CO_2H \qquad CH_3CH{=}CHCO_2H \qquad CH_3CH_2CH_2CH_2OH \qquad CH_3C{\equiv}CCO_2H$$

1   2   3   4

(*a*) 4 > 2 > 1 > 3   (*b*) 1 > 2 > 4 > 3   (*c*) 3 > 1 > 2 > 4   (*d*) 2 > 4 > 1 > 3

B-3. Which of the following compounds will undergo decarboxylation on heating?

1   2   3   4

(*a*) 2 and 3   (*b*) 3 and 4   (*c*) 3 only   (*d*) 1 and 4

B-4. Which of the following is *least* likely to form a lactone?

(*a*) $CH_3\overset{\overset{OH}{|}}{C}HCH_2CH_2CO_2H$

B-5. Compare the two methods shown for the preparation of carboxylic acids:

Method 1:

$$RBr \xrightarrow[\text{diethyl ether}]{Mg} RMgBr \xrightarrow[\text{2. } H_3O^+]{\text{1. } CO_2} RCO_2H$$

Method 2:

$$RBr \xrightarrow{\text{NaCN}} RCN \xrightarrow[\text{heat}]{H_2O, HCl} RCO_2H$$

Which one of the following statements correctly describes this conversion?

(a) Both method 1 and method 2 are appropriate for carrying out this conversion.

(b) Neither method 1 nor method 2 is appropriate for carrying out this conversion.

(c) Method 1 will work well, but method 2 is not appropriate.

(d) Method 2 will work well, but method 1 is not appropriate.

B-6. Which one of the following is *not* an intermediate in the generally accepted mechanism for the reaction shown?

B-7. Identify compound C in the following sequence:

B-8. What is the final product (B) of this sequence?

(a) 2-methyl benzoic acid with CH₃ and CO₂H + para

(c) benzoic acid with CO₂H

(e) None of these

(b) 3-methyl benzene with CH₃ and CO₂H

(d) phenylacetic acid CH₂CO₂H

B-9. Which one of the following undergoes decarboxylation (loses carbon dioxide) most readily on being heated?

(a)

(c)

(b)

(d)

(e)

B-10. Which of the compounds in the previous problem yields a δ-lactone after being reduced with sodium borohydride?

B-11. What is compound Z?

$$CH_3CH_2CH_2Br \xrightarrow{NaCN} X \xrightarrow[\text{heat}]{H_3O^+} Y \xrightarrow[\text{H}^+]{CH_3CH_2OH} Z$$

(a) $CH_3CH{=}CHCOH$ (with =O)

(c) $CH_3CH_2CHC{\equiv}N$ with OCH_2CH_3

(b) $CH_3CH_2CH_2CH(OCH_2CH_3)_2$

(d) $CH_3CH_2CH_2CH{=}NOCH_2CH_3$

(e) $CH_3CH_2CH_2COCH_2CH_3$ (with =O)

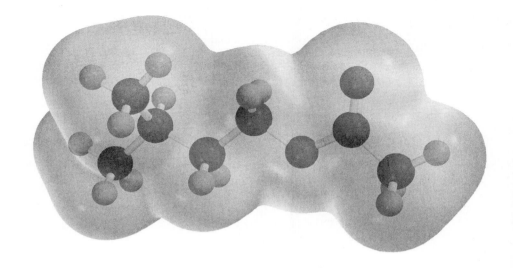

CHAPTER 20

Carboxylic Acid Derivatives: Nucleophilic Acyl Substitution

SOLUTIONS TO TEXT PROBLEMS

20.1 (*b*) Acid anhydrides bear two acyl groups on oxygen, as in $\overset{\overset{O}{\parallel}\;\overset{O}{\parallel}}{RCOCR}$. They are named as derivatives of carboxylic acids.

$$\underset{\underset{C_6H_5}{|}}{CH_3CH_2\overset{\overset{O}{\parallel}}{C}HCOH}$$

2-Phenylbutanoic acid

$$\underset{\underset{C_6H_5}{|}\;\;\;\;\underset{C_6H_5}{|}}{CH_3CH_2\overset{\overset{O}{\parallel}\;\;\overset{O}{\parallel}}{C}HCOCCHCH_2CH_3}$$

2-Phenylbutanoic anhydride

(*c*) Butyl 2-phenylbutanoate is the butyl ester of 2-phenylbutanoic acid.

$$\underset{\underset{C_6H_5}{|}}{CH_3CH_2\overset{\overset{O}{\parallel}}{C}HCOCH_2CH_2CH_2CH_3}$$

Butyl 2-phenylbutanoate

(*d*) In 2-phenylbutyl butanoate the 2-phenylbutyl group is an alkyl group bonded to oxygen of the ester. It is not involved in the acyl group of the molecule.

$$\underset{\underset{C_6H_5}{|}}{CH_3CH_2CH_2\overset{\overset{O}{\parallel}}{C}OCH_2CHCH_2CH_3}$$

2-Phenylbutyl butanoate

(*e*) The ending -*amide* reveals this to be a compound of the type $\overset{\overset{\textstyle O}{\|}}{RCNH_2}$.

$$\overset{\overset{\textstyle O}{\|}}{CH_3CH_2\underset{\underset{\textstyle C_6H_5}{|}}{CH}CNH_2}$$

2-Phenylbutanamide

(*f*) This compound differs from 2-phenylbutanamide in part (*e*) only in that it bears an ethyl substituent on nitrogen.

$$\overset{\overset{\textstyle O}{\|}}{CH_3CH_2\underset{\underset{\textstyle C_6H_5}{|}}{CH}CNHCH_2CH_3}$$

N-Ethyl-2-phenylbutanamide

(*g*) The -*nitrile* ending signifies a compound of the type $RC{\equiv}N$ containing the same number of carbons as the alkane RCH_3. Alternatively the compound may be named as an alkyl cyanide.

$$CH_3CH_2\underset{\underset{\textstyle C_6H_5}{|}}{CH}C{\equiv}N$$

2-Phenylbutanenitrile or
1-phenylpropyl cyanide

20.2 The methyl groups in *N*,*N*-dimethylformamide are nonequivalent; one is cis to oxygen, the other is trans. The two methyl groups have different chemical shifts. The third signal is due to the carbonyl carbon.

Rotation about the carbon–nitrogen bond is required to average the environments of the two methyl groups, but this rotation is relatively slow in amides as the result of the double-bond character imparted to the carbon–nitrogen bond, as shown by these two resonance contributors.

20.3 (*b*) Benzoyl chloride reacts with benzoic acid to give benzoic anhydride.

$$\overset{\overset{\textstyle O}{\|}}{C_6H_5CCl} + \overset{\overset{\textstyle O}{\|}}{C_6H_5COH} \longrightarrow \overset{\overset{\textstyle O\ \ O}{\|\ \ \|}}{C_6H_5COCC_6H_5} + HCl$$

Benzoyl
chloride

Benzoic acid

Benzoic anhydride

Hydrogen
chloride

(*c*) Acyl chlorides react with alcohols to form esters. The organic product is the ethyl ester of benzoic acid, ethyl benzoate.

$$
\underset{\substack{\text{Benzoyl} \\ \text{chloride}}}{C_6H_5\overset{\overset{\displaystyle O}{\|}}{C}Cl} \;+\; \underset{\text{Ethanol}}{CH_3CH_2OH} \;\longrightarrow\; \underset{\text{Ethyl benzoate}}{C_6H_5\overset{\overset{\displaystyle O}{\|}}{C}OCH_2CH_3} \;+\; \underset{\substack{\text{Hydrogen} \\ \text{chloride}}}{HCl}
$$

(*d*) Acyl transfer from benzoyl chloride to the nitrogen of methylamine yields the amide *N*-methylbenzamide.

$$
\underset{\substack{\text{Benzoyl} \\ \text{chloride}}}{C_6H_5\overset{\overset{\displaystyle O}{\|}}{C}Cl} \;+\; \underset{\text{Methylamine}}{2CH_3NH_2} \;\longrightarrow\; \underset{\textit{N}\text{-Methylbenzamide}}{C_6H_5\overset{\overset{\displaystyle O}{\|}}{C}NHCH_3} \;+\; CH_3\overset{+}{N}H_3\ Cl^-
$$

(*e*) As in part (*d*), an amide is formed. In this case, the amide has two methyl groups on nitrogen.

$$
\underset{\substack{\text{Benzoyl} \\ \text{chloride}}}{C_6H_5\overset{\overset{\displaystyle O}{\|}}{C}Cl} \;+\; \underset{\text{Dimethylamine}}{2(CH_3)_2NH} \;\longrightarrow\; \underset{\textit{N,N}\text{-Dimethylbenzamide}}{C_6H_5\overset{\overset{\displaystyle O}{\|}}{C}N(CH_3)_2} \;+\; (CH_3)_2\overset{+}{N}H_2\ Cl^-
$$

(*f*) Acyl chlorides undergo hydrolysis on reaction with water. The product is a carboxylic acid.

$$
\underset{\substack{\text{Benzoyl} \\ \text{chloride}}}{C_6H_5\overset{\overset{\displaystyle O}{\|}}{C}Cl} \;+\; \underset{\text{Water}}{H_2O} \;\longrightarrow\; \underset{\text{Benzoic acid}}{C_6H_5\overset{\overset{\displaystyle O}{\|}}{C}OH} \;+\; \underset{\substack{\text{Hydrogen} \\ \text{chloride}}}{HCl}
$$

20.4 (*b*) Nucleophilic addition of benzoic acid to benzoyl chloride gives the tetrahedral intermediate shown.

$$
\underset{\substack{\text{Benzoyl} \\ \text{chloride}}}{C_6H_5\overset{\overset{\displaystyle O}{\|}}{C}Cl} \;+\; \underset{\text{Benzoic acid}}{C_6H_5\overset{\overset{\displaystyle O}{\|}}{C}OH} \;\longrightarrow\; \underset{\text{Tetrahedral intermediate}}{C_6H_5\underset{\underset{\displaystyle Cl}{|}}{\overset{\overset{\displaystyle HO}{|}}{C}}O\overset{\overset{\displaystyle O}{\|}}{C}C_6H_5}
$$

Dissociation of the tetrahedral intermediate occurs by loss of chloride and of the proton on the oxygen.

$$
\underset{\text{Tetrahedral intermediate}}{C_6H_5\underset{\underset{\displaystyle Cl}{|}}{\overset{\overset{\displaystyle O}{|}}{C}}O\overset{\overset{\displaystyle O}{\|}}{C}C_6H_5} \;\longrightarrow\; \underset{\text{Benzoic anhydride}}{C_6H_5\overset{\overset{\displaystyle O}{\|}}{C}O\overset{\overset{\displaystyle O}{\|}}{C}C_6H_5} \;+\; \overset{+}{B}-H \;+\; Cl^-
$$

B: H

(*c*) Ethanol is the nucleophile that adds to the carbonyl group of benzoyl chloride to form the tetrahedral intermediate.

$$
\underset{\substack{\text{Benzoyl}\\\text{chloride}}}{C_6H_5\overset{\overset{\displaystyle O}{\|}}{C}Cl} \quad + \quad \underset{\text{Ethanol}}{CH_3CH_2OH} \quad \longrightarrow \quad \underset{\text{Tetrahedral intermediate}}{C_6H_5\underset{\underset{\displaystyle Cl}{|}}{\overset{\overset{\displaystyle OH}{|}}{C}}OCH_2CH_3}
$$

In analogy with parts (*a*) and (*b*) of this problem, a proton is lost from the hydroxyl group along with chloride to restore the carbon–oxygen double bond.

$$
\underset{\text{Tetrahedral intermediate}}{\text{B:} \overset{\displaystyle H}{\underset{\underset{\displaystyle Cl}{|}}{\underset{\displaystyle \underset{\displaystyle |}{O}}{\;}}}C_6H_5COCH_2CH_3} \quad \longrightarrow \quad \underset{\text{Ethyl benzoate}}{C_6H_5\overset{\overset{\displaystyle O}{\|}}{C}OCH_2CH_3} \quad + \quad \overset{+}{B}{-}H \quad + \quad Cl^-
$$

(*d*) The tetrahedral intermediate formed from benzoyl chloride and methylamine has a carbon–nitrogen bond.

$$
\underset{\substack{\text{Benzoyl}\\\text{chloride}}}{C_6H_5\overset{\overset{\displaystyle O}{\|}}{C}Cl} \quad + \quad \underset{\text{Methylamine}}{CH_3NH_2} \quad \longrightarrow \quad \underset{\text{Tetrahedral intermediate}}{C_6H_5\underset{\underset{\displaystyle Cl}{|}}{\overset{\overset{\displaystyle OH}{|}}{C}}NHCH_3}
$$

The dissociation of the tetrahedral intermediate may be shown as

$$
\underset{\text{Tetrahedral intermediate}}{\text{B:} \overset{\displaystyle H}{\underset{\underset{\displaystyle Cl}{|}}{\underset{\displaystyle \underset{\displaystyle |}{O}}{\;}}}C_6H_5CNHCH_3} \quad \longrightarrow \quad \underset{\text{N-Methylbenzamide}}{C_6H_5\overset{\overset{\displaystyle O}{\|}}{C}NHCH_3} \quad + \quad \overset{+}{B}{-}H \quad + \quad Cl^-
$$

More specifically, a second methylamine molecule is the Brønsted base that abstracts a proton from oxygen.

$$
\underset{}{CH_3\overset{\displaystyle ..}{N}H_2}
$$
$$
\underset{}{\overset{\displaystyle H}{\underset{\underset{\displaystyle Cl}{|}}{\underset{\displaystyle \underset{\displaystyle |}{O}}{\;}}}C_6H_5CNHCH_3} \quad \longrightarrow \quad \underset{\textit{N}\text{-Methylbenzamide}}{C_6H_5\overset{\overset{\displaystyle O}{\|}}{C}NHCH_3} \quad + \quad \underset{\substack{\text{Methylammonium}\\\text{chloride}}}{CH_3\overset{+}{N}H_3\ Cl^-}
$$

(*e*) The intermediates in the reaction of benzoyl chloride with dimethylamine are similar to those in part (*d*). The methyl substituents on nitrogen are not directly involved in the reaction.

$$C_6H_5\overset{\displaystyle O}{\overset{\|}{C}}Cl \quad + \quad (CH_3)_2NH \quad \longrightarrow \quad C_6H_5\overset{\displaystyle OH}{\underset{\displaystyle Cl}{\overset{|}{\underset{|}{C}}}}N(CH_3)_2$$

| Benzoyl chloride | Dimethylamine | Tetrahedral intermediate |

Then

$$(CH_3)_2\overset{..}{N}H$$

$$C_6H_5\overset{\displaystyle O}{\underset{\displaystyle Cl}{\overset{\|}{\underset{|}{C}}}}N(CH_3)_2 \quad \longrightarrow \quad C_6H_5\overset{\displaystyle O}{\overset{\|}{C}}N(CH_3)_2 \quad + \quad (CH_3)_2\overset{+}{N}H_2\ Cl^-$$

| | | *N*,*N*-Dimethylbenzamide | Dimethylammonium chloride |

(*f*) Water attacks the carbonyl group of benzoyl chloride to form the tetrahedral intermediate.

$$C_6H_5\overset{\displaystyle O}{\overset{\|}{C}}Cl \quad + \quad H_2O \quad \longrightarrow \quad C_6H_5\overset{\displaystyle OH}{\underset{\displaystyle OH}{\overset{|}{\underset{|}{C}}}}Cl$$

| Benzoyl chloride | Water | Tetrahedral intermediate |

Dissociation of the tetrahedral intermediate occurs by loss of chloride and the proton on oxygen.

$$B\colon \quad H$$

$$C_6H_5\overset{\displaystyle }{\underset{\displaystyle OH}{\overset{|}{\underset{|}{C}}}}Cl \quad \longrightarrow \quad C_6H_5\overset{\displaystyle O}{\overset{\|}{C}}OH \quad + \quad \overset{+}{B}-H \quad + \quad Cl^-$$

| Tetrahedral intermediate | | Benzoic acid |

20.5 (*b*) Acyl transfer from an acid anhydride to ammonia yields an amide. The organic products are acetamide and ammonium acetate.

$$CH_3\overset{\displaystyle O}{\overset{\|}{C}}O\overset{\displaystyle O}{\overset{\|}{C}}CH_3 \quad + \quad 2NH_3 \quad \longrightarrow \quad CH_3\overset{\displaystyle O}{\overset{\|}{C}}NH_2 \quad + \quad CH_3\overset{\displaystyle O}{\overset{\|}{C}}O^-\ \overset{+}{N}H_4$$

| Acetic anhydride | Ammonia | Acetamide | Ammonium acetate |

(*c*) The reaction of phthalic anhydride with dimethylamine is analogous to that of part (*b*). The organic products are an amide and the carboxylate salt of an amine.

| Phthalic anhydride | Dimethylamine | | Product is an amine salt and contains an amide function. |

$$+ \quad 2(CH_3)_2NH \quad \longrightarrow$$

$$H_2\overset{+}{N}(CH_3)_2$$

In this case both the amide function and the ammonium carboxylate salt are incorporated into the same molecule.

(*d*) The disodium salt of phthalic acid is the product of hydrolysis of phthalic acid in excess sodium hydroxide.

| Phthalic anhydride | Sodium hydroxide | Sodium phthalate | Water |

20.6 (*b*) The tetrahedral intermediate is formed by nucleophilic addition of ammonia to one of the carbonyl groups of acetic anhydride.

Tetrahedral intermediate

(*c*) Dimethylamine is the nucleophile; it adds to one of the two equivalent carbonyl groups of phthalic anhydride.

| Phthalic anhydride | Dimethylamine | Tetrahedral intermediate |

(*d*) Hydroxide acts as a nucleophile to form the tetrahedral intermediate.

Phthalic anhydride

Tetrahedral intermediate

20.7 The starting material contains three acetate ester functions. All three undergo hydrolysis in aqueous sulfuric acid.

1,2,5-Pentanetriol ($C_5H_{12}O_3$)   Acetic acid

The product is 1,2,5-pentanetriol. Also formed in the hydrolysis of the starting triacetate are three molecules of acetic acid.

20.8 **Step 1: Protonation of the carbonyl oxygen**

Ethyl benzoate   Hydronium ion   Protonated form of ester   Water

Step 2: Nucleophilic addition of water

Water   Protonated form of ester   Oxonium ion

Step 3: Deprotonation of oxonium ion to give neutral form of tetrahedral intermediate

Tetrahedral intermediate   Hydronium ion

Step 4: Protonation of ethoxy oxygen

Tetrahedral intermediate   Hydronium ion   Oxonium ion   Water

Step 5: Dissociation of protonated form of tetrahedral intermediate

This step yields ethyl alcohol and the protonated form of benzoic acid.

Oxonium ion   Protonated form   Ethyl alcohol
                               of benzoic acid

Step 6: Deprotonation of protonated form of benzoic acid

Protonated form   Water   Benzoic acid   Hydronium
of benzoic acid   ion

20.9 To determine which oxygen of 4-butanolide becomes labeled with ^{18}O, trace the path of ^{18}O-labeled water ($\oslash = {}^{18}O$) as it undergoes nucleophilic addition to the carbonyl group to form the tetrahedral intermediate.

4-Butanolide ^{18}O-labeled   Tetrahedral
                      water   intermediate

The tetrahedral intermediate can revert to unlabeled 4-butanolide by loss of ^{18}O-labeled water. Alternatively it can lose ordinary water to give ^{18}O-labeled lactone.

Tetrahedral ^{18}O-labeled   Water
intermediate   4-butanolide

The carbonyl oxygen is the one that is isotopically labeled in the ^{18}O-enriched 4-butanolide.

20.10 On the basis of trimyristin's molecular formula $C_{45}H_{86}O_6$ and of the fact that its hydrolysis gives only glycerol and tetradecanoic acid $CH_3(CH_2)_{12}CO_2H$, it must have the structure shown.

Trimyristin
($C_{45}H_{86}O_6$)

20.11 Because ester hydrolysis in base proceeds by acyl–oxygen cleavage, the ^{18}O label becomes incorporated into acetate ion ($\oslash = {}^{18}O$).

$$CH_3CH_2CH_2CH_2CH_2O\!\!-\!\!\overset{\overset{\textstyle O}{\|}}{C}CH_3 + {}^-\!\!:\!\!\oslash H \longrightarrow CH_3CH_2CH_2CH_2CH_2OH + \overset{O}{\underset{\oslash}{\diagdown}}\!\!CCH_3$$

Pentyl acetate	Hydroxide ion	1-Pentanol	Acetate ion

20.12 Step 1: Nucleophilic addition of hydroxide ion to the carbonyl group

$$HO\!:\!{}^- + C_6H_5C\overset{\overset{\textstyle :\!\ddot{O}:}{}}{\underset{\ddot{O}CH_2CH_3}{\diagup}} \rightleftharpoons C_6H_5\overset{\overset{\textstyle {}^-\!:\!\ddot{O}:}{}}{\underset{:\!\ddot{O}H}{C}}\!\!-\!\!\ddot{O}CH_2CH_3$$

Hydroxide ion	Ethyl benzoate	Anionic form of tetrahedral intermediate

Step 2: Proton transfer from water to give neutral form of tetrahedral intermediate

$$C_6H_5\overset{\overset{\textstyle :\ddot{O}:{}^-}{}}{\underset{:\ddot{O}H}{C}}\!\!-\!\!\ddot{O}CH_2CH_3 + H\!\!-\!\!\ddot{O}H \rightleftharpoons C_6H_5\overset{\overset{\textstyle :\ddot{O}H}{}}{\underset{:\ddot{O}H}{C}}\!\!-\!\!\ddot{O}CH_2CH_3 + {}^-\!:\!\ddot{O}H$$

Anionic form of tetrahedral intermediate	Water	Tetrahedral intermediate	Hydroxide ion

Step 3: Dissociation of tetrahedral intermediate

$$HO\!:\!{}^- + C_6H_5\overset{\overset{\textstyle H\!\!-\!\!\ddot{O}:}{}}{\underset{:\ddot{O}H}{C}}\!\!-\!\!\ddot{O}CH_2CH_3 \rightleftharpoons H\ddot{O}H + C_6H_5\overset{\overset{\textstyle :\ddot{O}:}{}}{\underset{\ddot{O}H}{C}} + {}^-\!:\!\ddot{O}CH_2CH_3$$

Hydroxide ion	Tetrahedral intermediate	Water	Benzoic acid	Ethoxide ion

Step 4: Proton transfer processes

$$C_6H_5\overset{\overset{\textstyle O}{\diagup}}{\underset{\ddot{O}\!\!-\!\!H}{C}} + {}^-\!:\!\ddot{O}H \longrightarrow C_6H_5\overset{\overset{\textstyle :\ddot{O}:}{\diagup}}{\underset{\ddot{O}:{}^-}{C}} + H\ddot{O}H$$

Benzoic acid	Hydroxide ion	Benzoate ion	Water

$$CH_3CH_2\ddot{O}:{}^- + H\!\!-\!\!\ddot{O}H \rightleftharpoons CH_3CH_2\ddot{O}\!\!-\!\!H + {}^-\!:\!\ddot{O}H$$

Ethoxide ion	Water	Ethanol	Hydroxide ion

20.13 The starting material is a lactone, a cyclic ester. The ester function is converted to an amide by nucleophilic acyl substitution.

Methylamine   4-Pentanolide   4-Hydroxy-*N*-methylpentanamide

20.14 The difference in the compounds lies in the number of N—H bonds in each.

Propanamide	*N*-methylacetamide	*N,N*-Dimethylformamide
2 N—H bonds	1 N—H bond	0 N—H bonds

The compound with the greatest number of N—H bonds will engage in the most intermolecular hydrogen bonds and have the highest boiling point. Propanamide will have the highest boiling point; *N,N*-dimethylformamide will have the lowest. The actual boiling points are: propanamide: 222°C; *N*-methylacetamide: 206°C; *N,N*-dimethylformamide: 153°C.

20.15 Recall (from text Section 20.2) that amide nitrogens are planar. Thus structure B is a planar monocyclic array of sp^2-hybridized atoms. Structure B also has six π electrons in the ring, two from each nitrogen and two from the double bond, and thus satisfies Hückel's rule.

A resonance structure of B can be drawn with a ring analogous to benzene:

20.16 (*b*) Acetic anhydride is the acid anhydride that must be used; it transfers an acetyl group to suitable nucleophiles. The nucleophile in this case is methylamine.

| Acetic anhydride | Methylamine | *N*-Methylacetamide | Methylammonium acetate |

(*c*) The acyl group is $\overset{\overset{\displaystyle O}{\|}}{HC}$— . Because the problem specifies that the acyl transfer agent is a methyl ester, methyl formate is one of the starting materials.

$$\underset{\substack{\text{Methyl} \\ \text{formate}}}{\overset{\overset{\displaystyle O}{\|}}{HCOCH_3}} \quad + \quad \underset{\text{Dimethylamine}}{HN(CH_3)_2} \quad \longrightarrow \quad \underset{\textit{N,N}\text{-Dimethylformamide}}{\overset{\overset{\displaystyle O}{\|}}{HCN(CH_3)_2}} \quad + \quad \underset{\substack{\text{Methyl} \\ \text{alcohol}}}{CH_3OH}$$

20.17 Step 1: Protonation of the carbonyl oxygen

Acetanilide   Hydronium ion   Protonated form   Water
                                                              of amide

Step 2: Nucleophilic addition of water

Water   Protonated form   Oxonium ion
             of amide

Step 3: Deprotonation of oxonium ion to give neutral form of tetrahedral intermediate

Oxonium ion   Water   Tetrahedral   Hydronium ion
                                              intermediate

Step 4: Protonation of amino group of tetrahedral intermediate

Tetrahedral   Hydronium ion   *N*-Protonated form of   Water
intermediate   tetrahedral intermediate

Step 5: Dissociation of *N*-protonated form of tetrahedral intermediate

$$
\underset{\substack{\text{N-Protonated form of} \\ \text{tetrahedral intermediate}}}{CH_3\overset{\overset{\displaystyle :\ddot{O}H}{|}}{\underset{\underset{\displaystyle :\ddot{O}H}{|}}{C}}-\overset{\overset{\displaystyle H}{|}}{\underset{\underset{\displaystyle H}{|}}{\overset{+}{N}}C_6H_5} \;\;\rightleftharpoons\;\; \underset{\substack{\text{Protonated form} \\ \text{of acetic acid}}}{CH_3C\overset{\displaystyle \overset{+}{\ddot{O}}H}{\underset{\displaystyle \ddot{O}H}{\diagup\kern-0.6em\diagdown}}} \;+\; \underset{\text{Aniline}}{H_2\ddot{N}C_6H_5}
$$

Step 6: Proton-transfer processes

$$
\underset{\text{Hydronium ion}}{\overset{\overset{\displaystyle H}{|}}{\underset{\underset{\displaystyle H}{|}}{\overset{+}{\ddot{O}}}}-H} \;+\; \underset{\text{Aniline}}{H_2\ddot{N}C_6H_5} \;\;\rightleftharpoons\;\; \underset{\text{Water}}{\overset{\overset{\displaystyle H}{|}}{\underset{\underset{\displaystyle H}{|}}{\ddot{O}}}:} \;+\; \underset{\text{Anilinium ion}}{H_3\overset{+}{N}C_6H_5}
$$

$$
\underset{\substack{\text{Protonated form} \\ \text{of acetic acid}}}{CH_3C\overset{\displaystyle \overset{+}{\ddot{O}}-H}{\underset{\displaystyle \ddot{O}H}{\diagup\kern-0.6em\diagdown}}} \;+\; \underset{\text{Water}}{:\overset{\overset{\displaystyle H}{|}}{\underset{\underset{\displaystyle H}{|}}{\ddot{O}}}} \;\;\rightleftharpoons\;\; \underset{\text{Acetic acid}}{CH_3C\overset{\displaystyle \ddot{O}:}{\underset{\displaystyle \ddot{O}H}{\diagup\kern-0.6em\diagdown}}} \;+\; \underset{\text{Hydronium ion}}{H-\overset{\overset{\displaystyle H}{|}}{\underset{\underset{\displaystyle H}{|}}{\overset{+}{\ddot{O}}}}}
$$

20.18 Step 1: Nucleophilic addition of hydroxide ion to the carbonyl group

$$
\underset{\substack{\text{Hydroxide} \\ \text{ion}}}{H\ddot{O}:^-} \;+\; \underset{\text{N,N-Dimethylformamide}}{H\overset{\overset{\displaystyle :\ddot{O}:}{\|}}{C}\ddot{N}(CH_3)_2} \;\;\longrightarrow\;\; \underset{\substack{\text{Anionic form of tetrahedral} \\ \text{intermediate}}}{H\overset{\overset{\displaystyle :\ddot{O}:^-}{|}}{\underset{\underset{\displaystyle :\ddot{O}H}{|}}{C}}-\ddot{N}(CH_3)_2}
$$

Step 2: Proton transfer to give neutral form of tetrahedral intermediate

$$
\underset{\substack{\text{Anionic form} \\ \text{of tetrahedral} \\ \text{intermediate}}}{H\overset{\overset{\displaystyle :\ddot{O}:^-}{|}}{\underset{\underset{\displaystyle :\ddot{O}H}{|}}{C}}-\ddot{N}(CH_3)_2} \;+\; \underset{\text{Water}}{H-\ddot{O}H} \;\;\longrightarrow\;\; \underset{\substack{\text{Tetrahedral} \\ \text{intermediate}}}{H\overset{\overset{\displaystyle :\ddot{O}H}{|}}{\underset{\underset{\displaystyle :\ddot{O}H}{|}}{C}}-\ddot{N}(CH_3)_2} \;+\; \underset{\substack{\text{Hydroxide} \\ \text{ion}}}{^-:\ddot{O}H}
$$

Step 3: Proton transfer from water to nitrogen of tetrahedral intermediate

$$
\underset{\substack{\text{Tetrahedral} \\ \text{intermediate}}}{H\overset{\overset{\displaystyle :\ddot{O}H}{|}}{\underset{\underset{\displaystyle :\ddot{O}H}{|}}{C}}-\ddot{N}(CH_3)_2} \;+\; \underset{\text{Water}}{H-\ddot{O}H} \;\;\rightleftharpoons\;\; \underset{\substack{\text{N-Protonated form of} \\ \text{tetrahedral intermediate}}}{H\overset{\overset{\displaystyle :\ddot{O}H}{|}}{\underset{\underset{\displaystyle :\ddot{O}H}{|}}{C}}-\overset{+}{N}H(CH_3)_2} \;+\; \underset{\substack{\text{Hydroxide} \\ \text{ion}}}{^-:\ddot{O}H}
$$

Step 4: Dissociation of *N*-protonated form of tetrahedral intermediate

| Hydroxide ion | *N*-Protonated form of tetrahedral intermediate | | Water | Formic acid | Dimethylamine |

Step 5: Irreversible formation of formate ion

| Formic acid | Hydroxide ion | | Formate ion | Water |

20.19 Both parts of this problem deal with hydrolysis reactions. Penicillin G has two amide bonds capable of being cleaved by hydrolysis. Bond *a* is cleaved in part (*a*) and bond *b* is cleaved in part (*b*).

(*a*) The molecular formulas of penicillin G ($C_{16}H_{18}N_2O_4S$) and the product of β-lactamase-catalyzed hydrolysis ($C_{16}H_{20}N_2O_5S$) differ by H_2O. Therefore, all the atoms in penicillin G are retained on hydrolysis which is consistent only with cleaving the β-lactam ring. Therefore, the product is:

(*b*) The product of *penicillin acyl transferase*-catalyzed hydrolysis has a molecular formula ($C_8H_{12}N_2O_3S$) with 8 fewer carbons than penicillin G. Therefore, the acyl group attached to the nitrogen of the side chain is cleaved from the molecule. The product is:

20.20 (*a*) Ethanenitrile has the same number of carbon atoms as ethyl alcohol. This suggests a reaction scheme proceeding via an amide.

$$CH_3CH_2OH \longrightarrow CH_3\overset{O}{\overset{||}{C}}NH_2 \xrightarrow{P_4O_{10}} CH_3C\equiv N$$

Ethyl alcohol · · · · · · Acetamide · · · · · · Ethanenitrile

The necessary amide is prepared from ethanol.

$$CH_3CH_2OH \xrightarrow[H_2SO_4,\ heat]{Na_2Cr_2O_7,\ H_2O} CH_3\overset{O}{\overset{||}{C}}OH \xrightarrow[2.\ NH_3]{1.\ SOCl_2} CH_3\overset{O}{\overset{||}{C}}NH_2$$

Ethyl alcohol · · · · · · Acetic acid · · · · · · Acetamide

(*b*) Propanenitrile may be prepared from ethyl alcohol by way of a nucleophilic substitution reaction of the corresponding bromide.

$$CH_3CH_2OH \xrightarrow[or\ HBr]{PBr_3} CH_3CH_2Br \xrightarrow{NaCN} CH_3CH_2CN$$

Ethyl alcohol · · · · · · Ethyl bromide · · · · · · Propanenitrile

20.21 Step 1: Protonation of the nitrile

Step 2: Nucleophilic addition of water

Step 3: Deprotonation of imino acid

Steps 4 and 5: Proton transfers to give an amide

20.22 Ketones may be prepared by the reaction of nitriles with Grignard reagents. Nucleophilic addition of a Grignard reagent to a nitrile produces an imine. The imine is not normally isolated, however, but is hydrolyzed to the corresponding ketone. Ethyl phenyl ketone may be prepared by the reaction of propanenitrile with a phenyl Grignard reagent such as phenylmagnesium bromide, followed by hydrolysis of the imine.

$$CH_3CH_2C{\equiv}N \;+\; C_6H_5MgBr \xrightarrow{\text{diethyl ether}} \left[\underset{C_6H_5CCH_2CH_3}{\overset{NH}{\|}} \right] \xrightarrow[\text{heat}]{H_2O,\,H^+} \underset{C_6H_5CCH_2CH_3}{\overset{O}{\|}}$$

Propanenitrile Phenylmagnesium bromide Imine (not isolated) Ethyl phenyl ketone

20.23 (*a*) The halogen that is attached to the carbonyl group is identified in the name as a separate word following the name of the acyl group.

m-Chlorobenzoyl chloride

(*b*) Trifluoroacetic anhydride is the anhydride of trifluoroacetic acid. Notice that it contains six fluorines.

$$\underset{CF_3COCCF_3}{\overset{O\;\;O}{\|\;\;\|}}$$

Trifluoroacetic anhydride

(*c*) This compound is the cyclic anhydride of *cis*-1,2-cyclopropanedicarboxylic acid.

cis-1,2-Cyclopropanedicarboxylic acid   *cis*-1,2-Cyclopropanedicarboxylic anhydride

(*d*) Ethyl cycloheptanecarboxylate is the ethyl ester of cycloheptanecarboxylic acid.

Ethyl cycloheptanecarboxylate

(*e*) 1-Phenylethyl acetate is the ester of 1-phenylethanol and acetic acid.

1-Phenylethyl acetate

(*f*) 2-Phenylethyl acetate is the ester of 2-phenylethanol and acetic acid.

$$CH_3\overset{\displaystyle O}{\overset{\|}{C}}OCH_2CH_2-\text{⟨⟩}$$

2-Phenylethyl acetate

(*g*) The parent compound in this case is benzamide. *p*-Ethylbenzamide has an ethyl substituent at the ring position para to the carbonyl group.

$$CH_3CH_2-\text{⟨⟩}-\overset{\displaystyle O}{\overset{\|}{C}}NH_2$$

p-Ethylbenzamide

(*h*) The parent compound is benzamide. In *N*-ethylbenzamide the ethyl substituent is bonded to nitrogen.

$$\text{⟨⟩}-\overset{\displaystyle O}{\overset{\|}{C}}NHCH_2CH_3$$

N-Ethylbenzamide

(*i*) Nitriles are named by adding the suffix *-nitrile* to the name of the alkane having the same number of carbons. Numbering begins at the nitrile carbon.

$$\overset{6}{C}H_3\overset{5}{C}H_2\overset{4}{C}H_2\overset{3}{C}H_2\overset{2}{C}H\overset{1}{C}\equiv N$$
$$|$$
$$CH_3$$

2-Methylhexanenitrile

20.24 (*a*) This compound, with a chlorine attached to its carbonyl group, is named as an acyl chloride. It is 3-chlorobutanoyl chloride.

$$CH_3CHCH_2\overset{\displaystyle O}{\overset{\|}{C}}Cl$$
$$|$$
$$Cl$$

3-Chlorobutanoyl
chloride

(*b*) The group attached to oxygen, in this case *benzyl,* is identified first in the name of the ester. This compound is the benzyl ester of acetic acid.

$$CH_3\overset{\displaystyle O}{\overset{\|}{C}}OCH_2-\text{⟨⟩}$$

Benzyl acetate

(*c*) The group attached to oxygen is methyl; this compound is the methyl ester of phenylacetic acid.

$$CH_3O\overset{\displaystyle O}{\overset{\|}{C}}CH_2-\text{⟨⟩}$$

Methyl phenylacetate

(d) This compound contains the functional group $-\overset{\overset{\text{O}}{\|}}{\text{C}}\overset{\overset{\text{O}}{\|}}{\text{O}}\overset{\|}{\text{C}}-$ and thus is an anhydride of a carboxylic acid. We name the acid, in this case 3-chloropropanoic acid, drop the *acid* part of the name, and replace it by *anhydride*.

$$\text{ClCH}_2\text{CH}_2\overset{\overset{\text{O}}{\|}}{\text{C}}\text{O}\overset{\overset{\text{O}}{\|}}{\text{C}}\text{CH}_2\text{CH}_2\text{Cl}$$

3-Chloropropanoic anhydride

(e) This compound is a cyclic anhydride, whose parent acid is 3,3-dimethylpentanedioic acid.

3,3-Dimethylpentanedioic
anhydride

(f) Nitriles are named by adding *-nitrile* to the name of the alkane having the same number of carbons. Remember to count the carbon of the C≡N group.

$$\underset{\underset{\text{CH}_3}{|}}{\text{CH}_3\text{CHCH}_2\text{CH}_2\text{C}}\equiv\text{N}$$

4-Methylpentanenitrile

(g) This compound is an amide. We name the corresponding acid and then replace the *-oic acid* suffix by *-amide*.

$$\underset{\underset{\text{CH}_3}{|}}{\text{CH}_3\text{CHCH}_2\text{CH}_2}\overset{\overset{\text{O}}{\|}}{\text{C}}\text{NH}_2$$

4-Methylpentanamide

(h) This compound is the *N*-methyl derivative of 4-methylpentanamide.

$$\underset{\underset{\text{CH}_3}{|}}{\text{CH}_3\text{CHCH}_2\text{CH}_2}\overset{\overset{\text{O}}{\|}}{\text{C}}\text{NHCH}_3$$

N,4-Dimethylpentanamide

(i) The amide nitrogen bears two methyl groups. We designate this as an *N,N*-dimethyl amide.

$$\underset{\underset{\text{CH}_3}{|}}{\text{CH}_3\text{CHCH}_2\text{CH}_2}\overset{\overset{\text{O}}{\|}}{\text{C}}\text{N(CH}_3)_2$$

N,N, 4-Trimethylpentanamide

20.25 (*a*) Acetyl chloride acts as an acyl transfer agent to the aromatic ring of bromobenzene. The reaction is a Friedel–Crafts acylation. Bromine is an ortho, para-directing substituent.

Bromobenzene   Acetyl   *o*-Bromoacetophenone   *p*-Bromoacetophenone
chloride

(*b*) Sodium propanoate acts as a nucleophile toward propanoyl chloride. The product is propanoic anhydride.

Propanoate   Propanoyl chloride   Propanoic anhydride
anion

(*c*) Acyl chlorides convert alcohols to esters.

Butanoyl chloride   Benzyl alcohol   Benzyl butanoate

(*d*) Acyl chlorides react with ammonia to yield amides.

p-Chlorobenzoyl chloride   Ammonia   *p*-Chlorobenzamide

(*e*) The starting material is a cyclic anhydride. Acid anhydrides react with water to yield two carboxylic acid functions; when the anhydride is cyclic, a dicarboxylic acid results.

Succinic anhydride   Water   Succinic acid

(*f*) In dilute sodium hydroxide the anhydride is converted to the disodium salt of the diacid.

Succinic anhydride   Sodium   Sodium succinate
                  hydroxide

(*g*) One of the carbonyl groups of the cyclic anhydride is converted to an amide function on reaction with ammonia. The other, the one that would become a carboxylic acid group, is converted to an ammonium carboxylate salt.

Succinic anhydride	Ammonia	Ammonium succinamate

(*h*) Acid anhydrides are used as acylating agents in Friedel–Crafts reactions.

Succinic anhydride	Benzene	3-Benzoylpropanoic acid

(*i*) The reactant is maleic anhydride; it is a good dienophile in Diels–Alder reactions.

1,3-Pentadiene	Maleic anhydride	3-Methylcyclohexene-4,5-dicarboxylic anhydride

(*j*) Acid anhydrides react with alcohols to give an ester and a carboxylic acid.

Acetic anhydride	3-Pentanol	1-Ethylpropyl acetate	Acetic acid

(*k*) The starting material is a cyclic ester, a lactone. Esters undergo saponification in aqueous base to give an alcohol and a carboxylate salt.

4-Butanolide	Sodium hydroxide	Sodium 4-hydroxybutanoate

(*l*) Ammonia reacts with esters to give an amide and an alcohol.

4-Butanolide	Ammonia	4-Hydroxybutanamide

(*m*) Lithium aluminum hydride reduces esters to two alcohols; the one derived from the acyl group is a primary alcohol. Reduction of a cyclic ester gives a diol.

$$\xrightarrow[\text{2. H}_2\text{O}]{\text{1. LiAlH}_4} \quad HOCH_2CH_2CH_2CH_2OH$$

4-Butanolide   1,4-Butanediol

(*n*) Grignard reagents react with esters to give tertiary alcohols.

$$\xrightarrow[\text{2. H}_3\text{O}^+]{\text{1. 2CH}_3\text{MgBr}} \quad \underset{\underset{CH_3}{|}}{\overset{\overset{OH}{|}}{HOCH_2CH_2CH_2CCH_3}}$$

4-Butanolide   4-Methyl-1,4-pentanediol

(*o*) In this reaction methylamine acts as a nucleophile toward the carbonyl group of the ester. The product is an amide.

$$CH_3\overset{..}{N}H_2 \; + \; \underset{}{C_6H_5CH_2\overset{\overset{O}{||}}{C}OCH_2CH_3} \quad \longrightarrow \quad C_6H_5CH_2\overset{\overset{O}{||}}{C}NHCH_3 \; + \; CH_3CH_2OH$$

Methylamine   Ethyl phenylacetate   *N*-Methylphenylacetamide   Ethyl alcohol

(*p*) The starting material is a lactam, a cyclic amide. Amides are hydrolyzed in base to amines and carboxylate salts.

$$+ \quad NaOH \quad \xrightarrow{\text{H}_2\text{O}} \quad CH_3NHCH_2CH_2CH_2\overset{\overset{O}{||}}{C}O^- \; Na^+$$

N-Methylpyrrolidone   Sodium   Sodium 4-(methylamino)butanoate
                        hydroxide

(*q*) In acid solution amides yield carboxylic acids and ammonium salts.

$$+ \quad H_3O^+ \quad \longrightarrow \quad CH_3\overset{\overset{H}{\overset{+|}{}}}{\underset{\underset{H}{|}}{N}}CH_2CH_2CH_2\overset{\overset{O}{||}}{C}OH$$

N-Methylpyrrolidone   Hydronium   4-(Methylammonio)butanoic
                    ion   acid

(*r*) The starting material is a cyclic imide. Both its amide bonds are cleaved by nucleophilic attack by hydroxide ion.

$$+ \quad 2NaOH \quad \xrightarrow{\text{H}_2\text{O}} \quad Na^+ \; {}^-O\overset{\overset{O}{||}}{C}CH_2CH_2\overset{\overset{O}{||}}{C}O^- \; Na^+ \; + \; CH_3NH_2$$

N-Methylsuccinimide   Sodium   Sodium succinate   Methylamine
                      hydroxide

(s) In acid the imide undergoes cleavage to give a dicarboxylic acid and the conjugate acid of methylamine.

N-Methylsuccinimide		Water	Hydrogen chloride		Succinic acid	Methylammonium chloride

(t) Acetanilide is hydrolyzed in acid to acetic acid and the conjugate acid of aniline.

$$C_6H_5NHCCH_3 + H_2O + HCl \longrightarrow C_6H_5\overset{+}{N}H_3\ Cl^- + CH_3COH$$

Acetanilide	Water	Hydrogen chloride	Anilinium chloride	Acetic acid

(u) This is another example of amide hydrolysis.

$$C_6H_5CNHCH_3 + H_2O + H_2SO_4 \longrightarrow C_6H_5COH + CH_3\overset{+}{N}H_3\ HSO_4^-$$

N-Methylbenzamide	Water	Sulfuric acid	Benzoic acid	Methylammonium hydrogen sulfate

(v) One way to prepare nitriles is by dehydration of amides.

Cyclopentanecarboxamide   Cyclopentyl cyanide

(w) Nitriles are hydrolyzed to carboxylic acids in acidic media.

$$(CH_3)_2CHCH_2C \equiv N \xrightarrow[\text{heat}]{\text{HCl, H}_2\text{O}} (CH_3)_2CHCH_2COH$$

3-Methylbutanenitrile   3-Methylbutanoic acid

(x) Nitriles are hydrolyzed in aqueous base to salts of carboxylic acids.

p-Methoxybenzonitrile   Sodium p-methoxybenzoate   Ammonia

(y) Grignard reagents react with nitriles to yield ketones after addition of aqueous acid.

$$CH_3CH_2C \equiv N \xrightarrow[\text{2. H}_3\text{O}^+]{\text{1. CH}_3\text{MgBr}} CH_3CH_2CCH_3$$

Propanenitrile   2-Butanone

20.26 The reactant is a cyclic diester of carbonic acid. Hydrolysis gives a diol as the organic product plus carbonic acid. Carbonic acid spontaneously dissociates to carbon dioxide and water

5-Ethyl-5-methyl-1,3-dioxan-2-one   Carbonic acid   2-Ethyl-2-methyl-1,3-propanediol

20.27 (*a*) Acetyl chloride is prepared by reaction of acetic acid with thionyl chloride. The first task then is to prepare acetic acid by oxidation of ethanol.

Ethanol   Acetic acid   Acetyl chloride

(*b*) Acetic acid and acetyl chloride, available from part (*a*), can be combined to form acetic anhydride.

Acetic acid   Acetyl chloride   Acetic anhydride   Hydrogen chloride

(*c*) Ethanol can be converted to ethyl acetate by reaction with acetic acid, acetyl chloride, or acetic anhydride from parts (*a*) and (*b*).

Ethanol   Acetic acid   Ethyl acetate   Water

or

Ethanol   Acetyl chloride   Ethyl acetate

or

Ethanol   Acetic anhydride   Ethyl acetate

(*d*) Ethyl bromoacetate is the ethyl ester of bromoacetic acid; thus the first task is to prepare the acid. We use the acetic acid prepared in part (*a*), converting it to bromoacetic acid by the Hell–Volhard–Zelinsky reaction.

Acetic acid   Bromoacetic acid   Ethyl bromoacetate

Alternatively, bromoacetic acid could be converted to the corresponding acyl chloride, then treated with ethanol. It would be incorrect to try to brominate ethyl acetate; the Hell–Volhard–Zelinsky method requires an acid as starting material, not an ester.

(e) The alcohol $BrCH_2CH_2OH$, needed in order to prepare 2-bromoethyl acetate, is prepared from ethanol by way of ethylene.

$$CH_3CH_2OH \xrightarrow[\text{heat}]{H_2SO_4} H_2C=CH_2 \xrightarrow[H_2O]{Br_2} BrCH_2CH_2OH$$

Ethanol   Ethylene   2-Bromoethanol

Then

$$BrCH_2CH_2OH \xrightarrow[\substack{\text{or} \\ CH_3CCl \\ \text{O}}]{\substack{\text{O O} \\ CH_3COCCH_3}} CH_3COCH_2CH_2Br$$

2-Bromoethanol   2-Bromoethyl acetate

(f) Ethyl cyanoacetate may be prepared from the ethyl bromoacetate obtained in part (d). The bromide may be displaced by cyanide in a nucleophilic substitution reaction.

$$BrCH_2COCH_2CH_3 \xrightarrow[S_N2]{NaCN} N{\equiv}CCH_2COCH_2CH_3$$

Ethyl bromoacetate   Ethyl cyanoacetate

(g) Reaction of acetyl chloride, prepared in part (a), or acetic anhydride, from part (b), with ammonia gives acetamide.

$$CH_3CCl \quad \text{or} \quad CH_3COCCH_3 \xrightarrow{NH_3} CH_3CNH_2$$

Acetyl   Acetic   Acetamide
chloride   anhydride

(h) The desired hydroxy acid is available from hydrolysis of the corresponding cyanohydrin, which may be prepared by reaction of the appropriate aldehyde with cyanide ion.

$$CH_3CHCOH \Longrightarrow CH_3CHC{\equiv}N \Longrightarrow CH_3CH$$
$$\quad\;\;| \qquad\qquad\qquad\qquad | $$
$$\quad\;\;OH \qquad\qquad\qquad\; OH$$

In this synthesis the cyanohydrin is prepared from ethanol by way of acetaldehyde.

$$CH_3CH_2OH \xrightarrow[CH_2Cl_2]{PCC} CH_3CH \xrightarrow[H^+]{KCN} CH_3CHC{\equiv}N$$
$$\qquad\qquad\qquad\qquad\qquad\qquad\qquad\qquad\qquad\qquad | $$
$$\qquad\qquad\qquad\qquad\qquad\qquad\qquad\qquad\qquad\; OH$$

Ethanol   Acetaldehyde   2-Hydroxypropanenitrile

$$CH_3CHC{\equiv}N \xrightarrow[\substack{\text{or} \\ 1.\, HO^-,\, H_2O,\, \text{heat} \\ 2.\, H^+}]{H_2O,\, H^+,\, \text{heat}} CH_3CHCOH$$
$$\quad\;\;| \qquad\qquad\qquad\qquad\qquad\qquad\qquad\qquad | $$
$$\quad\;\;OH \qquad\qquad\qquad\qquad\qquad\qquad\qquad OH$$

2-Hydroxypropanenitrile   2-Hydroxypropanoic
acid

20.28 (*a*) Benzoyl chloride is made from benzoic acid. Oxidize toluene to benzoic acid, and then treat with thionyl chloride.

$$C_6H_5CH_3 \xrightarrow[\text{H}_2\text{O, heat}]{\text{K}_2\text{Cr}_2\text{O}_7,\ \text{H}_2\text{SO}_4} \underset{\text{Benzoic acid}}{C_6H_5\overset{\displaystyle O}{\overset{\|}{C}}OH} \xrightarrow{\text{SOCl}_2} \underset{\substack{\text{Benzoyl}\\\text{chloride}}}{C_6H_5\overset{\displaystyle O}{\overset{\|}{C}}Cl}$$

Toluene

(*b*) Benzoyl chloride and benzoic acid, both prepared from toluene in part (*a*), react with each other to give benzoic anhydride.

$$\underset{\text{Benzoic acid}}{C_6H_5\overset{\displaystyle O}{\overset{\|}{C}}OH} \ + \ \underset{\substack{\text{Benzoyl}\\\text{chloride}}}{C_6H_5\overset{\displaystyle O}{\overset{\|}{C}}Cl} \ \longrightarrow \ \underset{\text{Benzoic anhydride}}{C_6H_5\overset{\displaystyle O}{\overset{\|}{C}}O\overset{\displaystyle O}{\overset{\|}{C}}C_6H_5}$$

(*c*) Benzoic acid, benzoyl chloride, and benzoic anhydride were prepared in parts (*a*) and (*b*) of this problem. Any of them could be converted to benzyl benzoate on reaction with benzyl alcohol. Thus the synthesis of benzyl benzoate requires the preparation of benzyl alcohol from toluene. This is effected by a nucleophilic substitution reaction of benzyl bromide, in turn prepared by halogenation of toluene.

$$\underset{\text{Toluene}}{C_6H_5CH_3} \xrightarrow[\text{or Br}_2,\text{ light}]{\textit{N}\text{-bromosuccinimide (NBS)}} \underset{\text{Benzyl bromide}}{C_6H_5CH_2Br} \xrightarrow[\text{HO}^-]{\text{H}_2\text{O}} \underset{\text{Benzyl alcohol}}{C_6H_5CH_2OH}$$

Alternatively, recall that primary alcohols may be obtained by reduction of the corresponding carboxylic acid.

$$\underset{\text{Benzoic acid}}{C_6H_5\overset{\displaystyle O}{\overset{\|}{C}}OH} \xrightarrow[\text{2. H}_2\text{O}]{\text{1. LiAlH}_4} \underset{\text{Benzyl alcohol}}{C_6H_5CH_2OH}$$

Then

$$\underset{\substack{\text{Benzoyl}\\\text{chloride}}}{C_6H_5\overset{\displaystyle O}{\overset{\|}{C}}Cl} \ + \ \underset{\text{Benzyl alcohol}}{C_6H_5CH_2OH} \xrightarrow{\text{pyridine}} \underset{\text{Benzyl benzoate}}{C_6H_5\overset{\displaystyle O}{\overset{\|}{C}}OCH_2C_6H_5}$$

(*d*) Benzamide is prepared by reaction of ammonia with either benzoyl chloride from part (*a*) or benzoic anhydride from part (*b*).

$$\underset{\substack{\text{Benzoyl}\\\text{chloride}}}{C_6H_5\overset{\displaystyle O}{\overset{\|}{C}}Cl} \quad \text{or} \quad \underset{\text{Benzoic anhydride}}{C_6H_5\overset{\displaystyle O}{\overset{\|}{C}}O\overset{\displaystyle O}{\overset{\|}{C}}C_6H_5} \xrightarrow{\text{NH}_3} \underset{\text{Benzamide}}{C_6H_5\overset{\displaystyle O}{\overset{\|}{C}}NH_2}$$

(*e*) Benzonitrile may be prepared by dehydration of benzamide.

$$\underset{\text{Benzamide}}{C_6H_5\overset{\displaystyle O}{\overset{\|}{C}}NH_2} \xrightarrow[\text{heat}]{\text{P}_4\text{O}_{10}} \underset{\text{Benzonitrile}}{C_6H_5C\equiv N}$$

(*f*) Benzyl cyanide is the product of nucleophilic substitution by cyanide ion on benzyl bromide or benzyl chloride. The benzyl halides are prepared by free-radical halogenation of the toluene side chain.

$$C_6H_5CH_3 \xrightarrow[\text{light or heat}]{Cl_2} C_6H_5CH_2Cl \xrightarrow{NaCN} C_6H_5CH_2C{\equiv}N$$

Toluene   Benzyl chloride   Benzyl cyanide

or

$$C_6H_5CH_3 \xrightarrow[\substack{\text{or } Br_2, \\ \text{light}}]{NBS} C_6H_5CH_2Br \xrightarrow{NaCN} C_6H_5CH_2C{\equiv}N$$

Toluene   Benzyl bromide   Benzyl cyanide

(*g*) Hydrolysis of benzyl cyanide yields phenylacetic acid.

$$C_6H_5CH_2C{\equiv}N \xrightarrow[\substack{or \\ \text{1. NaOH, heat} \\ \text{2. H}^+}]{H_2O, H^+, \text{heat}} C_6H_5CH_2\overset{\displaystyle O}{\overset{\displaystyle \|}{C}}OH$$

Benzyl cyanide   Phenylacetic acid

Alternatively, the Grignard reagent derived from benzyl bromide may be carboxylated.

$$C_6H_5CH_2Br \xrightarrow[\text{diethyl ether}]{Mg} C_6H_5CH_2MgBr \xrightarrow[\text{2. H}_3O^+]{1. CO_2} C_6H_5CH_2\overset{\displaystyle O}{\overset{\displaystyle \|}{C}}OH$$

Benzyl bromide   Benzylmagnesium   Phenylacetic acid
                              bromide

(*h*) The first goal is to synthesize *p*-nitrobenzoic acid because this may be readily converted to the desired acyl chloride. First convert toluene to *p*-nitrotoluene; then oxidize. Nitration must precede oxidation of the side chain in order to achieve the desired para orientation.

Toluene   *p*-Nitrotoluene   *p*-Nitrobenzoic acid
                     (separate from ortho isomer)

Treatment of *p*-nitrobenzoic acid with thionyl chloride yields *p*-nitrobenzoyl chloride.

p-Nitrobenzoic acid   *p*-Nitrobenzoyl chloride

(*i*) To achieve the correct orientation in *m*-nitrobenzoyl chloride, oxidation of the methyl group must precede nitration.

Toluene   Benzoic acid   *m*-Nitrobenzoic acid

Once *m*-nitrobenzoic acid has been prepared, it may be converted to the corresponding acyl chloride.

m-Nitrobenzoic acid   *m*-Nitrobenzoyl chloride

20.29 The problem specifies that $CH_3CH_2\overset{O}{\overset{||}{C}}OCH_2CH_3$ is to be prepared from ^{18}O-labeled ethyl alcohol ($\varnothing = {}^{18}O$).

Propanoyl chloride   Ethyl alcohol   Ethyl propanoate

Thus, we need to prepare ^{18}O-labeled ethyl alcohol from the other designated starting materials, acetaldehyde and ^{18}O-enriched water. First, replace the oxygen of acetaldehyde with ^{18}O by the hydration–dehydration equilibrium in the presence of ^{18}O-enriched water.

Acetaldehyde ^{18}O-enriched water   Hydrate of acetaldehyde ^{18}O-enriched acetaldehyde   Water

Once ^{18}O-enriched acetaldehyde has been obtained, it can be reduced to ^{18}O-enriched ethanol.

20.30 (*a*) The rate-determining step in basic ester hydrolysis is nucleophilic addition of hydroxide ion to the carbonyl group. The intermediate formed in this step is negatively charged.

Ethyl acetate   Hydroxide ion   Rate-determining intermediate

The electron-withdrawing effect of a CF_3 group stabilizes the intermediate formed in the rate-determining step of ethyl trifluoroacetate saponification.

$$CF_3\overset{\overset{\displaystyle O}{\|}}{C}OCH_2CH_3 \quad + \quad HO^- \quad \longrightarrow \quad CF_3\overset{\overset{\displaystyle O^-}{|}}{\underset{\underset{\displaystyle OH}{|}}{C}}OCH_2CH_3$$

Ethyl trifluoroacetate   Hydroxide ion   Rate-determining intermediate

Because the intermediate is more stable, it is formed faster than the one from ethyl acetate.

(b) Crowding is increased as the transition state for nucleophilic addition to the carbonyl group is approached. The carbonyl carbon undergoes a change in hybridization from sp^2 to sp^3.

$$CH_3\overset{\overset{\displaystyle CH_3}{|}}{\underset{\underset{\displaystyle CH_3}{|}}{C}}\overset{\overset{\displaystyle O}{\|}}{C}OCH_2CH_3 \quad + \quad {}^-OH \quad \longrightarrow \quad CH_3\overset{\overset{\displaystyle CH_3}{|}}{\underset{\underset{\displaystyle CH_3}{|}}{C}}\overset{\overset{\displaystyle O^-}{|}}{\underset{\underset{\displaystyle OH}{|}}{C}}OCH_2CH_3$$

Ethyl 2,2-dimethylpropanoate   Hydroxide ion   Rate-determining intermediate; crowded

The *tert*-butyl group of ethyl 2,2-dimethylpropanoate causes more crowding than the methyl group of ethyl acetate; the tetrahedral intermediate is less stable and is formed more slowly.

(c) We see here another example of a steric effect of a *tert*-butyl group. The tetrahedral intermediate formed when hydroxide ion adds to the carbonyl group of *tert*-butyl acetate is more crowded and less stable than the tetrahedral intermediate formed from methyl acetate.

$$CH_3\overset{\overset{\displaystyle O}{\|}}{C}\overset{\overset{\displaystyle CH_3}{|}}{\underset{\underset{\displaystyle CH_3}{|}}{C}}CH_3 \quad + \quad HO^- \quad \longrightarrow \quad CH_3\overset{\overset{\displaystyle O^-}{|}}{\underset{\underset{\displaystyle HO}{|}}{C}}\overset{\overset{\displaystyle CH_3}{|}}{\underset{\underset{\displaystyle CH_3}{|}}{C}}CH_3$$

tert-Butyl acetate   Hydroxide ion   Rate-determining intermediate; more crowded

$$CH_3\overset{\overset{\displaystyle O}{\|}}{C}OCH_3 \quad + \quad HO^- \quad \longrightarrow \quad CH_3\overset{\overset{\displaystyle O^-}{|}}{\underset{\underset{\displaystyle HO}{|}}{C}}OCH_3$$

Methyl acetate   Hydroxide ion   Rate-determining intermediate; less crowded than intermediate from *tert*-butyl acetate

(d) Here, as in part (a), we have an electron-withdrawing substituent increasing the rate of ester saponification. It does so by stabilizing the negatively charged intermediate formed in the rate-determining step.

Rate-determining intermediate from methyl *m*-nitrobenzoate   more stable than   Rate-determining intermediate from methyl benzoate (*e*) Addition of hydroxide to 4-butanolide introduces torsional strain in the intermediate because of eclipsed bonds. The corresponding intermediate from 5-butanolide is more stable because the bonds are staggered in a six-membered ring.

Less stable;   More stable;
formed more slowly   formed faster

(*f*) Steric crowding increases more when hydroxide adds to the axial carbonyl group.

Cis diastereomer: greater increase in crowding   Trans diastereomer: smaller increase in crowding
when carbon changes from sp^2 to sp^3;   when carbon changes from sp^2 to sp^3;
formed more slowly   formed more rapidly

20.31 Compound A is the *p*-toluenesulfonate (tosylate) of *trans*-4-*tert*-butylcyclohexanol. The oxygen atom of the alcohol attacks the sulfur of *p*-toluenesulfonyl chloride, and so the reaction proceeds with retention of configuration.

trans-4-*tert*-Butylcyclohexanol   *p*-Toluenesulfonyl chloride   *trans*-4-*tert*-Butylcyclohexyl
p-toluenesulfonate (compound A)

The second step is a nucleophilic substitution in which benzoate ion displaces *p*-toluenesulfonate with inversion of configuration.

Benzoate ion   *trans*-4-*tert*-Butylcyclohexyl   *cis*-4-*tert*-Butylcyclohexyl
p-toluenesulfonate (compound A)   benzoate (compound B)

Saponification (basic hydrolysis) of *cis*-4-*tert*-butylcyclohexyl benzoate in step 3 proceeds with acyl–oxygen cleavage to give *cis*-4-*tert*-butylcyclohexanol.

20.32 Reaction of ethyl trifluoroacetate with ammonia yields the corresponding amide, compound A. Compound A undergoes dehydration on heating with P_4O_{10} to give trifluoroacetonitrile, compound B. Grignard reagents react with nitriles to form ketones. *tert*-Butyl trifluoromethyl ketone is formed from trifluoroacetonitrile by treatment with *tert*-butylmagnesium chloride followed by aqueous hydrolysis.

$$CF_3\overset{\overset{\displaystyle O}{\|}}{C}OCH_2CH_3 \xrightarrow{\ NH_3\ } CF_3\overset{\overset{\displaystyle O}{\|}}{C}NH_2 \xrightarrow[\text{heat}]{\ P_4O_{10}\ } CF_3C\equiv N$$

Ethyl trifluoroacetate   Trifluoroacetamide (compound A)   Trifluoroacetonitrile (compound B)

$$CF_3C\equiv N \ + \ (CH_3)_3CMgCl \xrightarrow[\text{2. }H_3O^+]{\text{1. diethyl ether}} CF_3\overset{\overset{\displaystyle O}{\|}}{C}C(CH_3)_3$$

Compound B   *tert*-Butylmagnesium chloride   *tert*-Butyl trifluoromethyl ketone

20.33 The first step is acid hydrolysis of an acetal protecting group.

Step 1:

$$\text{Compound A} \xrightarrow[\text{heat}]{H_2O,\ H^+} HO\overset{\overset{\displaystyle O}{\|}}{C}(CH_2)_5\underset{\underset{\displaystyle HO}{|}}{CH}-\underset{\underset{\displaystyle OH}{|}}{CH}(CH_2)_7CH_2OH$$

Compound B ($C_{16}H_{32}O_5$)

All three alcohol functions are converted to bromide by reaction with hydrogen bromide in step 2.

Step 2:

$$\text{Compound B} \xrightarrow{\ HBr\ } HO\overset{\overset{\displaystyle O}{\|}}{C}(CH_2)_5\underset{\underset{\displaystyle Br}{|}}{CH}-\underset{\underset{\displaystyle Br}{|}}{CH}(CH_2)_7CH_2Br$$

Compound C ($C_{16}H_{29}Br_3O_2$)

Reaction with ethanol in the presence of an acid catalyst converts the carboxylic acid to its ethyl ester in step 3.

Step 3:

$$\text{Compound C} \xrightarrow[H_2SO_4]{\text{ethanol}} CH_3CH_2O\overset{\overset{\displaystyle O}{\|}}{C}(CH_2)_5\underset{\underset{\displaystyle Br}{|}}{CH}-\underset{\underset{\displaystyle Br}{|}}{CH}(CH_2)_7CH_2Br$$

Compound D ($C_{18}H_{33}Br_3O_2$)

The problem hint points out that zinc converts vicinal dibromides to alkenes. Of the three bromine substituents in compound D, two of them are vicinal. Step 4 is a dehalogenation reaction.

Step 4:

$$\text{Compound D} \xrightarrow[\text{ethanol}]{\text{Zn}} \text{CH}_3\text{CH}_2\text{OC(CH}_2)_5\text{CH}=\text{CH(CH}_2)_7\text{CH}_2\text{Br}$$

Compound E
($C_{18}H_{33}BrO_2$)

Step 5 is a nucleophilic substitution of the S_N2 type. Acetate ion is the nucleophile and displaces bromide from the primary carbon.

Step 5:

$$\text{Compound E} \xrightarrow[\text{CH}_3\text{CO}_2\text{H}]{\text{NaOCCH}_3} \text{CH}_3\text{CH}_2\text{OC(CH}_2)_5\text{CH}=\text{CH(CH}_2)_7\text{CH}_2\text{OCCH}_3$$

Compound F
($C_{20}H_{36}O_4$)

Step 6 is ester hydrolysis. It yields a 16-carbon chain having a carboxylic acid function at one end and an alcohol at the other.

Step 6:

$$\text{Compound F} \xrightarrow[\text{2. H}^+]{\text{1. KOH, ethanol}} \text{HOC(CH}_2)_5\text{CH}=\text{CH(CH}_2)_7\text{CH}_2\text{OH}$$

Compound G
($C_{16}H_{30}O_3$)

In step 7, compound G cyclizes to ambrettolide on heating.

Step 7:

Compound G   Ambrettolide

20.34 (*a*) This step requires the oxidation of a primary alcohol to an aldehyde. As reported in the literature, pyridinium dichromate in dichloromethane was used to give the desired aldehyde in 84% yield.

$$\text{HOCH}_2\text{CH}=\text{CH(CH}_2)_7\text{CO}_2\text{CH}_3 \xrightarrow[\text{CH}_2\text{Cl}_2]{\text{PDC}} \text{HCCH}=\text{CH(CH}_2)_7\text{CO}_2\text{CH}_3$$

Compound A (*E* isomer)   Compound B

(b) Conversion of $-\overset{\overset{\textstyle O}{\|}}{C}H$ to $-CH=CH_2$ is a typical case in which a Wittig reaction is appropriate.

$$\overset{\overset{\textstyle O}{\|}}{H C}CH=CH(CH_2)_7CO_2CH_3 \quad \xrightarrow{(C_6H_5)_3\overset{+}{P}-\overset{\cdot\cdot}{C}H_2} \quad H_2C=CHCH=CH(CH_2)_7CO_2CH_3$$

Compound B

Compound C
(observed yield, 53%)

(c) Lithium aluminum hydride was used to reduce the ester to a primary alcohol in 81% yield.

$$H_2C=CHCH=CH(CH_2)_7CO_2CH_3 \quad \xrightarrow[\text{2. H}_2\text{O}]{\text{1. LiAlH}_4} \quad H_2C=CHCH=CH(CH_2)_7CH_2OH$$

Compound C

Compound D

(d) The desired sex pheromone is the acetate ester of compound D. Compound D was treated with acetic anhydride to give the acetate ester in 99% yield.

$$H_2C=CHCH=CH(CH_2)_7CH_2OH \quad \xrightarrow[\text{pyridine}]{\overset{\overset{\textstyle O\ \ O}{\| \ \ \|}}{CH_3COCCH_3}} \quad H_2C=CHCH=CH(CH_2)_7CH_2O\overset{\overset{\textstyle O}{\|}}{C}CH_3$$

Compound D

(E)-9,11-Dodecadien-1-yl acetate

Acetyl chloride could have been used in this step instead of acetic anhydride.

20.35 (a) The reaction given in the problem is between a lactone (cyclic ester) and a difunctional Grignard reagent. Esters usually react with 2 moles of a Grignard reagent; in this instance both Grignard functions of the reagent attack the lactone. The second attack is intramolecular, giving rise to the cyclopentanol ring of the product.

4-Butanolide

Protonation of both oxygens completes the reaction.

1-(3-Hydroxypropyl)-
cyclopentanol (88%)

(*b*) An intramolecular acyl transfer process takes place in this reaction. The amine group in the thiolactone starting material replaces sulfur on the acyl group to form a lactam (cyclic amide).

| Thiolactone | Tetrahedral intermediate | Lactam |

20.36 (*a*) Acyl chlorides react with alcohols to form esters.

p-Methoxybenzoyl chloride   Benzoin

Benzoin *p*-methoxybenzoate
(compound A; 95%)

(*b*) Of the two carbonyl groups in the starting material, the ketone carbonyl is more reactive than the ester. (The ester carbonyl is stabilized by electron release from oxygen.)

$$CH_3CCH_2CH_2COCH_2CH_3 \xrightarrow[\text{(1 equiv)}]{CH_3MgI} CH_3CCH_2CH_2COCH_2CH_3$$

Compound B has the molecular formula $C_6H_{10}O_2$. The initial product forms a cyclic ester (lactone), with elimination of ethoxide ion.

Compound B

(*c*) The primary amine can react with both acyl chloride groups of the starting material to give compound C.

Compound C

20.37 Compound A is an ester but has within it an amine function. Acyl transfer from oxygen to nitrogen converts the ester to a more stable amide, compound B.

Compound A
(Ar = p-nitrophenyl)

Tetrahedral
intermediate

Compound B
(Ar = p-nitrophenyl)

The tetrahedral intermediate is the key intermediate in the reaction.

20.38 (*a*) The rearrangement in this problem is an acyl transfer from nitrogen to oxygen.

Compound A
(Ar = p-nitrophenyl)

Tetrahedral
intermediate

Compound B
(Ar = p-nitrophenyl)

This rearrangement takes place in the indicated direction because it is carried out in acid solution. The amino group is protonated in acid and is no longer nucleophilic.

(*b*) The trans stereoisomer of compound A does not undergo rearrangement because when the oxygen and nitrogen atoms on the five-membered ring are trans, the necessary tetrahedral intermediate cannot form.

20.39 The ester functions of a polymer such as poly(vinyl acetate) are just like ester functions of simple molecules; they can be cleaved by hydrolysis under either acidic or basic conditions. To prepare poly(vinyl alcohol), therefore, polymerize vinyl acetate to poly(vinyl acetate), and then cleave the ester groups by hydrolysis.

Vinyl acetate

Poly(vinyl acetate)

Poly(vinyl alcohol)

20.40 The compound contains nitrogen and exhibits a prominent peak in the IR spectrum at 2270 cm^{-1}; it is likely to be a nitrile. Its molecular weight of 83 is consistent with the molecular formula C_5H_9N. The presence of four signals in the δ 10 to 30 region of the ^{13}C NMR spectrum suggests an unbranched carbon skeleton. This is confirmed by the presence of two triplets in the ^1H NMR spectrum at δ 1.0 (CH_3 coupled with adjacent CH_2) and at δ 2.3 (CH_2CN coupled with adjacent CH_2). The compound is pentanenitrile.

$$CH_3CH_2CH_2CH_2C{\equiv}N$$

Pentanenitrile

20.41 The compound has the characteristic triplet–quartet pattern of an ethyl group in its ^{1}H NMR spectrum. Because these signals correspond to ten protons, there must be two equivalent ethyl groups in the molecule. The methylene quartet appears at relatively low field (δ 4.1), which is consistent with ethyl groups bonded to oxygen, as in —OCH_2CH_3. There is a peak at 1730 cm^{-1} in the IR spectrum, suggesting that these ethoxy groups reside in ester functions. The molecular formula $C_8H_{14}O_4$ reveals that if two ester groups are present, there can be no rings or double bonds. The remaining four hydrogens are equivalent in the ^{1}H NMR spectrum, and so two equivalent CH_2 groups are present. The compound is the diethyl ester of succinic acid.

$$\underset{\text{Diethyl succinate}}{CH_3CH_2O\overset{\overset{\displaystyle O}{\|}}{C}CH_2CH_2\overset{\overset{\displaystyle O}{\|}}{C}OCH_2CH_3}$$

20.42 The compound ($C_4H_6O_2$) has an index of hydrogen deficiency of 2. With two oxygen atoms and a peak in the infrared at 1760 cm^{-1}, it is likely that one of the elements of unsaturation is the carbon–oxygen double bond of an ester. The ^{1}H NMR spectrum contains a three-proton singlet at δ 2.1, which is consistent with a $CH_3\overset{\overset{\displaystyle O}{\|}}{C}$ unit. It is likely that the compound is an acetate ester.

The ^{13}C NMR spectrum reveals that the four carbon atoms of the molecule are contained in one each of the fragments CH_3, CH_2, and CH, along with the carbonyl carbon. In addition to the two carbons of the acetate group, the remaining two carbons are the CH_2 and CH carbons of a vinyl group, $CH=CH_2$. The compound is vinyl acetate.

$$
\begin{array}{c}
\delta\ 20.2 \\
\downarrow \\
CH_3C\overset{\overset{\textstyle \delta\ 96.8}{\downarrow}}{\underset{OCH=CH_2}{\diagup O}} \\
\delta\ 167.6 \quad \uparrow \\
\delta\ 141.8
\end{array}
$$

Each vinyl proton is coupled to two other vinyl protons; each appears as a doublet of doublets in the ^{1}H NMR spectrum.

ANSWERS TO INTERPRETIVE PROBLEMS

20.43 C; **20.44** D; **20.45** B; **20.46** A; **20.47** A; **20.48** B; **20.49** C; **20.50** D

SELF-TEST

PART A

A-1. Give a correct IUPAC name for each of the following acid derivatives:

(a) $CH_3CH_2CH_2O\overset{\overset{\displaystyle O}{\|}}{C}CH_2CH_2CH_3$

(c) $(CH_3)_2CHCH_2CH_2\overset{\overset{\displaystyle O}{\|}}{C}Cl$

(b) $C_6H_5\overset{\overset{\displaystyle O}{\|}}{C}NHCH_3$

A-2. Provide the correct structure of

   (*a*) Benzoic anhydride

   (*b*) *N*-(1-Methylpropyl)acetamide

   (*c*) Phenyl benzoate

A-3. What reagents are needed to carry out each of the following conversions?

   (*a*) $C_6H_5CH_2CO_2H$ $\xrightarrow{?}$ $C_6H_5CH_2\overset{\displaystyle O}{\overset{\|}{C}}Cl$

   (*b*) $(CH_3)_2CHCH_2NH_2$ $\xrightarrow{?}$ $C_6H_5\overset{\displaystyle O}{\overset{\|}{C}}NHCH_2CH(CH_3)_2$ + CH_3OH

A-4. Write the structure of the product of each of the following reactions:

   (*a*) Cyclohexyl acetate $\xrightarrow[\text{2. H}^+]{\text{1. NaOH, H}_2\text{O}}$? (two products)

   (*b*) Cyclopentanol + benzoyl chloride $\xrightarrow{\text{pyridine}}$?

   (*c*) + CH_3CH_2OH $\xrightarrow{\text{H}^+ \text{(cat)}}$?

   (*d*) Ethyl propanoate + dimethylamine $\longrightarrow$? (two products)

   (*e*) H_3C——$\overset{\displaystyle O}{\overset{\|}{C}}NHCH_3$ $\xrightarrow[\text{heat}]{\text{H}_2\text{O, H}_2\text{SO}_4}$? (two products)

A-5. The following reaction occurs when the reactant is allowed to stand in pentane. Write the structure of the key intermediate in this process.

$$C_6H_5\overset{\displaystyle O}{\overset{\|}{C}}OCH_2CH_2NHCH_3 \longrightarrow \underset{\underset{C_6H_5\overset{}{C}=O}{|}}{CH_3NCH_2CH_2OH}$$

A-6. Give the correct structures, clearly showing stereochemistry, of each compound, A through D, in the following sequence of reactions:

$\xrightarrow{\text{SOCl}_2}$ A $\xrightarrow{\text{NH}_3}$ B $\xrightarrow[\text{heat}]{\text{P}_4\text{O}_{10}}$ C $(C_8H_{13}N)$

A-7. Write the structure of the neutral form of the tetrahedral intermediate in the

   (*a*) Acid-catalyzed hydrolysis of methyl acetate

   (*b*) Reaction of ammonia with acetic anhydride

A-8. Write the steps necessary to prepare H_3C——$\overset{\displaystyle O}{\overset{\|}{C}}NH_2$ from H_3C——Br.

A-9. Outline a synthesis of benzyl benzoate using toluene as the source of all the carbon atoms.

Benzyl benzoate

A-10. The IR spectrum of a compound (C_3H_6ClNO) has an intense peak at 1680 cm^{-1}. Its ^{1}H NMR spectrum consists of a doublet (3H, δ 1.5), a quartet (1H, δ 4.1), and a broad singlet (2H, δ 6.5). What is the structure of the compound? How would you prepare it from propanoic acid?

PART B

B-1. What are the products of the most favorable mode of decomposition of the intermediate species shown?

$$C_6H_5 - \underset{\underset{Cl}{|}}{\overset{\overset{OH}{|}}{C}} - OH$$

(a) Benzoic acid and HCl   (c) Both (a) and (b) equally likely

(b) Benzoyl chloride and H_2O (d) Neither (a) nor (b) is likely

B-2. What is the correct IUPAC name for the compound shown?

(a) 3-Bromopropyl 2-chloro-4-butylbutanoate

(b) 2-Chloro-4-phenylbutyl 3-bromopropanoate

(c) 3-Chloro-1-phenylbutyl 1-bromopropanoate

(d) 3-Chloro-1-phenylbutyl 3-bromopropanoate

(e) 7-Bromo-3-chloro-1-phenylbutyl propanoate

B-3. Rank the following in order of increasing reactivity (least ⟶ most) toward acid hydrolysis:

$$\underset{1}{CH_3\overset{\overset{O}{||}}{C}OCH_2CH_3} \qquad \underset{2}{CH_3\overset{\overset{O}{||}}{C}Cl} \qquad \underset{3}{CH_3\overset{\overset{O}{||}}{C}NHCH_3}$$

(a) 1 < 2 < 3   (b) 3 < 1 < 2   (c) 1 < 3 < 2   (d) 2 < 1 < 3

B-4. The structure of *N*-propylacetamide is

(a) $CH_3CH_2\overset{\overset{O}{||}}{C}NHCH_3$ (c) $CH_3\overset{\overset{O}{||}}{C}NHCH_2CH_2CH_3$

(b) $CH_3\overset{\overset{O}{||}}{C}N(CH_2CH_2CH_3)_2$ (d) $CH_3CH=NCH_2CH_2CH_3$

B-5. Choose the response that matches the correct functional group classification with the following group of structural formulas.

(a)	Anhydride	Lactam	Lactone
(b)	Lactam	Imide	Lactone
(c)	Imide	Lactone	Anhydride
(d)	Imide	Lactam	Lactone

B-6. Choose the best sequence of reactions for the transformation given. Semicolons indicate separate reaction steps to be used in the order shown.

$$H_3C-\!\!\!\bigcirc\!\!\!-CO_2CH_3 \xrightarrow{\;?\;} H_3C-\!\!\!\bigcirc\!\!\!-CH_2\overset{\overset{\displaystyle O}{\|}}{C}NHCH_3$$

(a) H_3O^+; $SOCl_2$; CH_3NH_2

(b) HO^-/H_2O; PBr_3; Mg; CO_2; H_3O^+; $SOCl_2$; CH_3NH_2

(c) $LiAlH_4$; H_2O; HBr; Mg; CO_2; H_3O^+; $SOCl_2$; CH_3NH_2

(d) None of these would yield the desired product.

B-7. A key step in the hydrolysis of acetamide in aqueous acid proceeds by nucleophilic addition of

(a) H_3O^+ to $CH_3\overset{\overset{\displaystyle O}{\|}}{C}NH_2$

(b) H_3O^+ to $CH_3\overset{\overset{\displaystyle {}^+OH}{\|}}{C}NH_2$

(c) HO^- to $CH_3\overset{\overset{\displaystyle O}{\|}}{C}NH_2$

(d) H_2O to $CH_3\overset{\overset{\displaystyle {}^+OH}{\|}}{C}NH_2$

(e) HO^- to $CH_3\overset{\overset{\displaystyle {}^+OH}{\|}}{C}NH_2$

B-8. Which reaction is *not* possible for acetic anhydride?

(a) $(CH_3\overset{\overset{\displaystyle O}{\|}}{C})_2O$ + $2HN(CH_3)_2$ $\longrightarrow$ $CH_3\overset{\overset{\displaystyle O}{\|}}{C}N(CH_3)_2$ + $CH_3CO_2^-\ H_2\overset{+}{N}(CH_3)_2$

(b) $(CH_3\overset{\overset{\displaystyle O}{\|}}{C})_2O$ + CH_3CH_2OH $\longrightarrow$ $CH_3\overset{\overset{\displaystyle O}{\|}}{C}OCH_2CH_3$ + CH_3CO_2H

(c) $(CH_3\overset{\overset{\displaystyle O}{\|}}{C})_2O$ + C_6H_6 $\xrightarrow{\text{AlCl}_3}$ $CH_3\overset{\overset{\displaystyle O}{\|}}{C}C_6H_5$ + CH_3CO_2H

(d) $(CH_3\overset{\overset{\displaystyle O}{\|}}{C})_2O$ + $NaCl$ $\longrightarrow$ $CH_3\overset{\overset{\displaystyle O}{\|}}{C}Cl$ + $CH_3CO_2^-\ Na^+$

B-9. All but one of the following compounds react with aniline to give acetanilide. Which one does *not*?

$$\text{(Aniline)} \longrightarrow \text{(Acetanilide)}$$

Aniline   Acetanilide

(a) $CH_3\overset{O}{\overset{||}{C}}Cl$

(b) $CH_3\overset{O}{\overset{||}{C}}H$

(c) $H_3C\overset{O}{\overset{||}{C}}-O-\overset{O}{\overset{||}{C}}CH_3$

(d) $H_3C\overset{O}{\overset{||}{C}}-O-CH_2CH_3$ (structure with CH_3 chain)

(e) (phenyl)$-O\overset{O}{\overset{||}{C}}CH_3$

B-10. Identify product Z in the following reaction sequence:

$$H_2C=CHCH_2Br \xrightarrow{\text{NaCN}} Y \xrightarrow[\substack{\text{2. } H_3O^+}]{\substack{\text{1. } C_6H_5MgBr, \\ \text{diethyl ether}}} Z$$

(a) $H_2C=CHCH_2\overset{O}{\overset{||}{C}}C_6H_5$

(b) $H_2C=CHCH_2\overset{OH}{\overset{|}{C}}HC_6H_5$

(c) $H_2C=CHCH_2\overset{O}{\overset{||}{C}}NHC_6H_5$

(d) $H_2C=CHCH_2NH\overset{O}{\overset{||}{C}}C_6H_5$

(e) $H_2C=CHCH_2\overset{NH_2}{\overset{|}{C}}HC_6H_5$

B-11. Which of the following best describes the nucleophilic addition step in the acid-catalyzed hydrolysis of acetonitrile (CH_3CN)?

(a) $H_3C-C\equiv N\mathbf{:}$ ... $\overset{..}{\underset{H}{O}}\mathbf{:}^-$

(b) $H_3C-C\equiv N\mathbf{:}$ ... $\overset{..}{\underset{H\quad H}{O}}$

(c) $H_3C-C\equiv \overset{+}{N}H$ ... $\overset{..}{\underset{H}{O}}\mathbf{:}^-$

(d) $H_3C-C\equiv \overset{+}{N}H$ ... $\overset{..}{\underset{H\quad H}{O}}$

(e) $H_3C-C\equiv N\mathbf{:}$ ... $\overset{..}{\underset{H\quad H}{O}}$

B-12. Saponification (basic hydrolysis) of $C_6H_5\overset{}{\underset{O}{C}}\!⌀CH_3$ will yield: [⌀ = mass-18 isotope of oxygen]

(a) $C_6H_5\overset{}{\underset{⌀}{C}}O^- + H⌀CH_3$

(b) $C_6H_5\overset{}{\underset{O}{C}}O^- + H⌀CH_3$

(c) $C_6H_5\overset{}{\underset{⌀}{C}}O^- + H⌀CH_3$

(d) $C_6H_5\overset{}{\underset{O}{C}}O^- + H⌀CH_3$

(e) $C_6H_5\overset{OH}{\overset{|}{C}}O^-$ with $⌀CH_3$

B-13. An unknown compound, $C_9H_{10}O_2$, did not dissolve in aqueous NaOH. The IR spectrum exhibited strong absorption at 1730 cm^{-1}. The ^{1}H NMR spectrum had signals at δ 7.2 (multiplet), 4.1 (quartet), and 1.3 (triplet). Which of the following is most likely the unknown?

(a)   (b)   (c)   (d)   (e)

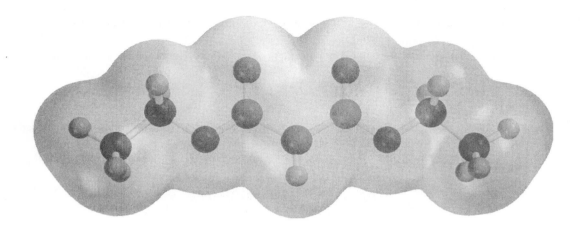

CHAPTER 21

Ester Enolates

SOLUTIONS TO TEXT PROBLEMS

21.1 Ethyl benzoate cannot undergo the Claisen condensation, because it has no protons on its α-carbon atom and so cannot form an enolate. Ethyl pentanoate and ethyl phenylacetate can undergo the Claisen condensation.

$$2CH_3CH_2CH_2CH_2\overset{\overset{\displaystyle O}{\|}}{C}OCH_2CH_3 \xrightarrow[\text{2. H}_3O^+]{\text{1. NaOCH}_2\text{CH}_3} CH_3CH_2CH_2CH_2\overset{\overset{\displaystyle O}{\|}}{C}\underset{\underset{\displaystyle CH_2CH_2CH_3}{|}}{CH}\overset{\overset{\displaystyle O}{\|}}{C}OCH_2CH_3$$

Ethyl pentanoate   Ethyl 3-oxo-2-propylheptanoate

$$2C_6H_5CH_2\overset{\overset{\displaystyle O}{\|}}{C}OCH_2CH_3 \xrightarrow[\text{2. H}_3O^+]{\text{1. NaOCH}_2\text{CH}_3} C_6H_5CH_2\overset{\overset{\displaystyle O}{\|}}{C}\underset{\underset{\displaystyle C_6H_5}{|}}{CH}\overset{\overset{\displaystyle O}{\|}}{C}OCH_2CH_3$$

Ethyl phenylacetate   Ethyl 3-oxo-2,4-diphenylbutanoate

21.2 (*b*) The enolate formed by proton abstraction from the α-carbon atom of diethyl 4-methylheptanedioate cyclizes to form a six-membered β-keto ester.

$$CH_3CH_2O\overset{\overset{\displaystyle O}{\|}}{C}CH_2CH_2\underset{\underset{\displaystyle CH_3}{|}}{CH}CH_2CH_2\overset{\overset{\displaystyle O}{\|}}{C}OCH_2CH_3 \xrightarrow{\text{NaOCH}_2\text{CH}_3}$$

Diethyl 4-methylheptanedioate

Ethyl (5-methyl-2-oxocyclohexane)-
carboxylate

(*c*) The two α carbons of this diester are not equivalent. Cyclization by attack of the enolate at C-2 gives

$$CH_3CH_2OCCHCH_2CH_2CH_2COCH_2CH_3 \xrightarrow{NaOCH_2CH_3}$$

Site of
carbanion

Enolate attacks
this carbon

Ethyl (1-methyl-2-
oxocyclopentane)-
carboxylate

This β-keto ester cannot form a stable enolate by deprotonation. It is present in only small amounts at equilibrium. The major product is formed by way of the other enolate.

$$CH_3CH_2OCCHCH_2CH_2CH_2COCH_2CH_3 \xrightarrow{NaOCH_2CH_3}$$

Enolate
attacks
this carbon

Site of
carbanion

Ethyl (3-methyl-2-
oxocyclopentane)-
carboxylate

This β-keto ester is converted to a stable enolate on deprotonation, causing the equilibrium to shift in its favor.

21.3 (*b*) Both carbonyl groups of diethyl oxalate are equivalent. The enolate of ethyl phenylacetate attacks one of them.

$$C_6H_5\overset{-}{\underset{\cdot\cdot}{C}}HCOCH_2CH_3 + CH_3CH_2O-CCOCH_2CH_3 \longrightarrow$$

$$C_6H_5CHCCOCH_2CH_3$$

$$\overset{O}{\underset{}{\underset{}{C}}}COCH_2CH_3$$

Diethyl 2-oxo-3-
phenylbutanedioate

(*c*) The enolate of ethyl phenylacetate attacks the carbonyl group of ethyl formate.

$$C_6H_5\overset{-}{\underset{\cdot\cdot}{C}}HCOCH_2CH_3 + HC-OCH_2CH_3 \longrightarrow$$

$$C_6H_5CHCH$$

$$\overset{O}{\underset{}{C}}COCH_2CH_3$$

Ethyl 3-oxo-2-
phenylpropanoate

21.4   For a five-membered ring to be formed, C-5 must be the carbanionic site that attacks the ester carbonyl.

Enolate of ethyl 4-oxohexanoate	Anionic form of tetrahedral intermediate	2-Methyl-1,3-cyclopentanedione

21.5   The desired ketone, cyclopentanone, is derived from the corresponding β-keto ester. This key intermediate is obtained from a Dieckmann cyclization of the starting material, diethyl hexanedioate.

First treat the diester with sodium ethoxide to effect the Dieckmann cyclization.

Diethyl hexanedioate	Ethyl (2-oxocyclopentane)-carboxylate

Next convert the β-keto ester to the desired product by saponification and decarboxylation.

Ethyl (2-oxocyclopentane)-carboxylate	Cyclopentanone

21.6   (*b*) Write a structural formula for the desired product; then disconnect a bond to the α-carbon atom.

4-Phenyl-2-butanone	Required alkyl halide	Derived from ethyl acetoacetate

Therefore

Benzyl bromide	Ethyl acetoacetate	4-Phenyl-2-butanone

(*c*) The disconnection approach to retrosynthetic analysis reveals that the preparation of 5-hexen-2-one by the acetoacetic ester synthesis requires an allylic halide.

$$H_2C=CHCH_2 \overset{\xi}{-} CH_2\overset{O}{\overset{||}{C}}CH_3 \quad \Longrightarrow \quad H_2C=CHCH_2 \underset{X}{-} \quad + \quad :\overset{..}{C}H_2\overset{O}{\overset{||}{C}}CH_3$$

5-Hexen-2-one

Required alkyl halide

Derived from ethyl acetoacetate

$$H_2C=CHCH_2Br \quad + \quad CH_3\overset{O}{\overset{||}{C}}CH_2\overset{O}{\overset{||}{C}}OCH_2CH_3 \quad \xrightarrow[\substack{2.\ HO^-,\ H_2O \\ 3.\ H_3O^+ \\ 4.\ heat}]{1.\ NaOCH_2CH_3} \quad H_2C=CHCH_2CH_2\overset{O}{\overset{||}{C}}CH_3$$

Allyl bromide

Ethyl acetoacetate

5-Hexen-2-one

21.7 (*b*) Nonanoic acid has a $CH_3(CH_2)_5CH_2$— unit attached to the $\overset{O}{\overset{||}{C}}H_2\overset{O}{\overset{||}{C}}OH$ synthon.

$$CH_3(CH_2)_5CH_2 \overset{\xi}{-} CH_2\overset{O}{\overset{||}{C}}OH \quad \Longrightarrow \quad CH_3(CH_2)_5CH_2X \quad + \quad :\overset{..}{C}H_2\overset{O}{\overset{||}{C}}OH$$

Nonanoic acid

Required alkyl halide

Derived from diethyl malonate

Therefore the anion of diethyl malonate is alkylated with a 1-haloheptane.

$$CH_3(CH_2)_5CH_2Br \quad + \quad CH_2(COOCH_2CH_3)_2 \quad \xrightarrow[\text{ethanol}]{NaOCH_2CH_3} \quad CH_3(CH_2)_5CH_2CH(COOCH_2CH_3)_2$$

1-Bromoheptane

Diethyl malonate

Diethyl 2-heptylmalonate

$$\Big\downarrow \substack{1.\ HO^-,\ H_2O \\ 2.\ H_3O^+ \\ 3.\ heat}$$

$$CH_3(CH_2)_5CH_2CH_2CO_2H$$

Nonanoic acid

(*c*) Disconnection of the target molecule adjacent to the α carbon reveals the alkyl halide needed to react with the enolate derived from diethyl malonate.

$$CH_3CH_2\underset{CH_3}{\overset{}{C}}HCH_2 \overset{\xi}{-} CH_2\overset{O}{\overset{||}{C}}OH \quad \Longrightarrow \quad CH_3CH_2\underset{CH_3}{\overset{}{C}}HCH_2X \quad + \quad :\overset{..}{C}H_2\overset{O}{\overset{||}{C}}OH$$

4-Methylhexanoic acid

Required alkyl halide

Derived from diethyl malonate

The necessary alkyl halide in this synthesis is 1-bromo-2-methylbutane.

$$CH_3CH_2\underset{\overset{|}{CH_3}}{C}HCH_2Br + CH_2(COOCH_2CH_3)_2 \xrightarrow[\text{ethanol}]{NaOCH_2CH_3} CH_3CH_2\underset{\overset{|}{CH_3}}{C}HCH_2CH(COOCH_2CH_3)_2$$

1-Bromo-2-methylbutane   Diethyl malonate   Diethyl 2-(2-methylbutyl)malonate

1. HO⁻, H_2O
2. H_3O^+
3. heat

$$CH_3CH_2\underset{\overset{|}{CH_3}}{C}HCH_2CH_2\overset{\overset{\displaystyle O}{\|}}{C}OH$$

4-Methylhexanoic acid

(*d*) Once again disconnection reveals the necessary halide, which is treated with diethyl malonate.

$$C_6H_5CH_2\text{⌇}CH_2\overset{\overset{\displaystyle O}{\|}}{C}OH \quad \Longrightarrow \quad C_6H_5CH_2X + \text{ :}\bar{C}H_2\overset{\overset{\displaystyle O}{\|}}{C}OH$$

3-Phenylpropanoic   Required   Derived from
acid   halide   diethyl malonate

Alkylation of diethyl malonate with benzyl bromide is the first step in the preparation of 3-phenylpropanoic acid.

$$C_6H_5CH_2Br + CH_2(COOCH_2CH_3)_2 \xrightarrow[\text{ethanol}]{NaOCH_2CH_3} C_6H_5CH_2CH(COOCH_2CH_3)_2 \xrightarrow[\substack{\text{2. } H_3O^+ \\ \text{3. heat}}]{\text{1. HO}^-,\ H_2O} C_6H_5CH_2CH_2\overset{\overset{\displaystyle O}{\|}}{C}OH$$

Benzyl bromide   Diethyl malonate   Diethyl 2-benzylmalonate   3-Phenylpropanoic acid

21.8 Retrosynthetic analysis of the formation of 3-methyl-2-butanone is carried out in the same way as for other ketones.

$$CH_3\overset{\overset{\displaystyle O}{\|}}{C}\underset{\overset{|}{CH_3}}{C}H\text{⌇}CH_3 \quad \Longrightarrow \quad CH_3\overset{\overset{\displaystyle O}{\|}}{C}\ddot{\bar{C}}H + 2CH_3X$$

3-Methyl-2-butanone   Derived from
(two disconnections as shown)   ethyl acetoacetate

The two alkylation steps are carried out sequentially.

$$CH_3\overset{\overset{\displaystyle O}{\|}}{C}CH_2\overset{\overset{\displaystyle O}{\|}}{C}OCH_2CH_3 \xrightarrow[\text{CH}_3\text{Br}]{NaOCH_2CH_3} CH_3\overset{\overset{\displaystyle O}{\|}}{C}\underset{\overset{|}{CH_3}}{C}H\overset{\overset{\displaystyle O}{\|}}{C}OCH_2CH_3 \xrightarrow[\text{CH}_3\text{Br}]{NaOCH_2CH_3} CH_3\overset{\overset{\displaystyle O}{\|}}{C}\underset{\underset{H_3C \quad CH_3}{}}{C}\overset{\overset{\displaystyle O}{\|}}{C}OCH_2CH_3 \xrightarrow[\substack{\text{2. } H_3O^+ \\ \text{3. heat}}]{\text{1. HO}^-,\ H_2O} CH_3\overset{\overset{\displaystyle O}{\|}}{C}CH(CH_3)_2$$

Ethyl acetoacetate   Ethyl   Ethyl   3-Methyl-2-butanone
                      2-methyl-3-oxobutanoate   2,2-dimethyl-3-oxobutanoate

21.9 Alkylation of ethyl acetoacetate with 1,4-dibromobutane gives a product that can cyclize to a five-membered ring.

Ethyl acetoacetate   1,4-Dibromobutane

Ethyl 1-acetylcyclopentane-
carboxylate

Saponification followed by decarboxylation gives cyclopentyl methyl ketone.

Ethyl 1-acetylcyclopentane-
carboxylate

Cyclopentyl methyl
ketone

21.10 Like diethyl malonate, ethyl acetoacetate undergoes Michael addition to an α,β-unsaturated ketone.

Basic ester hydrolysis followed by acidification and decarboxylation gives the diketone 3-(2-oxopropyl)cycloheptanone as the major product of the reaction sequence.

3-(2-Oxopropyl)-
cycloheptanone (52%)

21.11 (*b*) The α-carbon atom of the ester bears a phenyl substituent and a methyl group. Only the methyl group can be attached to the α carbon by nucleophilic substitution. Therefore generate the enolate of methyl phenylacetate with lithium diisopropylamide (LDA) in tetrahydrofuran (THF) and then alkylate with methyl iodide.

$$C_6H_5CH_2CO_2CH_3 \quad \xrightarrow[\text{THF}]{\text{LDA}} \quad C_6H_5CH=C\underset{\displaystyle OCH_3}{\overset{\displaystyle OLi}{\diagup}} \quad \xrightarrow{CH_3I} \quad C_6H_5\underset{\displaystyle CH_3}{\overset{\displaystyle |}{C}}HCO_2CH_3$$

| Methyl phenylacetate | Enolate of methyl phenylacetate | Methyl 2-phenylpropanoate |

(*c*) The desired product corresponds to an aldol addition product.

Therefore convert cyclohexanone to its enolate and then treat with benzaldehyde.

| Cyclohexanone | 1-(2-Oxocyclohexyl)-1-phenylmethanol |

(*d*) This product corresponds to the addition of the enolate of *tert*-butyl acetate to cyclohexanone.

Generate the enolate of *tert*-butyl acetate with lithium diisopropylamide; then add cyclohexanone.

$$CH_3CO_2C(CH_3)_3 \quad \xrightarrow[\substack{\text{2. cyclohexanone} \\ \text{3. } H_3O^+}]{\text{1. LDA, THF}} \quad$$

| *tert*-Butyl acetate | *tert*-Butyl (1-hydroxycyclohexyl)acetate |

21.12 To undergo a Claisen condensation, an ester must have at least two protons on the α carbon:

$$2R\overset{\alpha}{C}H_2\overset{\displaystyle O}{\overset{\displaystyle \|}{C}}OCH_2CH_3 + NaOCH_2CH_3 \longrightarrow \left[RCH_2\overset{\displaystyle O}{\overset{\displaystyle \|}{C}}\underset{\displaystyle R}{\overset{\displaystyle |}{C}}\overset{\displaystyle O}{\overset{\displaystyle \|}{C}}OCH_2CH_3 \right] Na^+ + 2CH_3CH_2OH$$

The equilibrium constant for condensation is unfavorable unless the β-keto ester can be deprotonated to form a stable anion.

(*a*) Among the esters given, ethyl pentanoate and ethyl 3-methylbutanoate undergo the Claisen condensation.

$$\underset{\text{Ethyl pentanoate}}{CH_3CH_2CH_2CH_2\overset{\overset{\displaystyle O}{\|}}{C}OCH_2CH_3} \quad \xrightarrow[\text{2. }H_3O^+]{\text{1. }NaOCH_2CH_3} \quad \underset{\text{Ethyl 3-oxo-2-propylheptanoate}}{CH_3CH_2CH_2CH_2\overset{\overset{\displaystyle O}{\|}}{C}\underset{\underset{\displaystyle CH_2CH_2CH_3}{|}}{CH}\overset{\overset{\displaystyle O}{\|}}{C}OCH_2CH_3}$$

$$\underset{\text{Ethyl 3-methylbutanoate}}{(CH_3)_2CHCH_2\overset{\overset{\displaystyle O}{\|}}{C}OCH_2CH_3} \quad \xrightarrow[\text{2. }H_3O^+]{\text{1. }NaOCH_2CH_3} \quad \underset{\substack{\text{Ethyl 2-isopropyl-5-methyl-} \\ \text{3-oxohexanoate}}}{(CH_3)_2CHCH_2\overset{\overset{\displaystyle O}{\|}}{C}\underset{\underset{\displaystyle CH(CH_3)_2}{|}}{CH}\overset{\overset{\displaystyle O}{\|}}{C}OCH_2CH_3}$$

(*b*) The Claisen condensation product of ethyl 2-methylbutanoate cannot be deprotonated; the equilibrium constant for its formation is less than 1.

$$\underset{\text{Ethyl 2-methylbutanoate}}{CH_3CH_2\underset{\underset{\displaystyle CH_3}{|}}{CH}\overset{\overset{\displaystyle O}{\|}}{C}OCH_2CH_3} \quad \underset{K<1}{\overset{NaOCH_2CH_3}{\rightleftharpoons}} \quad \underset{\substack{\text{No protons on }\alpha\text{-carbon atom; cannot} \\ \text{form stabilized enolate by deprotonation}}}{CH_3CH_2\underset{\underset{\displaystyle CH_3}{|}}{CH}\overset{\overset{\displaystyle OCH_3}{\overset{\displaystyle \|\;|}{}}}{C}\underset{\underset{\displaystyle CH_2CH_3}{|}}{C}OOCH_2CH_3}$$

(*c*) Ethyl 2,2-dimethylpropanoate has no protons on its α carbon; it cannot form the ester enolate required in the first step of the Claisen condensation.

$$\underset{\substack{\text{Ethyl 2,2-} \\ \text{dimethylpropanoate}}}{CH_3\overset{\overset{\displaystyle H_3CO}{\overset{\displaystyle |\;\|}{}}}{C}\underset{\underset{\displaystyle CH_3}{|}}{}COCH_2CH_3} \;+\; {}^-OCH_2CH_3 \quad \longrightarrow \quad \text{no reaction}$$

21.13 (*a*) The Claisen condensation of ethyl phenylacetate is given by the equation

$$\underset{\text{Ethyl phenylacetate}}{C_6H_5CH_2\overset{\overset{\displaystyle O}{\|}}{C}OCH_2CH_3} \quad \xrightarrow[\text{2. }H_3O^+]{\text{1. }NaOCH_2CH_3} \quad \underset{\text{Ethyl 3-oxo-2,4-diphenylbutanoate}}{C_6H_5CH_2\overset{\overset{\displaystyle O}{\|}}{C}\underset{\underset{\displaystyle C_6H_5}{|}}{CH}\overset{\overset{\displaystyle O}{\|}}{C}OCH_2CH_3}$$

(*b*) Saponification and decarboxylation of this β-keto ester gives dibenzyl ketone.

$$\underset{\substack{\text{Ethyl 3-oxo-2,4-} \\ \text{diphenylbutanoate}}}{C_6H_5CH_2\overset{\overset{\displaystyle O}{\|}}{C}\underset{\underset{\displaystyle C_6H_5}{|}}{CH}\overset{\overset{\displaystyle O}{\|}}{C}OCH_2CH_3} \quad \xrightarrow[\substack{\text{2. }H_3O^+ \\ \text{3. heat}}]{\text{1. }HO^-,\,H_2O} \quad \underset{\text{Dibenzyl ketone}}{C_6H_5CH_2\overset{\overset{\displaystyle O}{\|}}{C}CH_2C_6H_5}$$

(*c*) This process illustrates the alkylation of a β-keto ester with subsequent saponification and decarboxylation.

Ethyl 3-oxo-2,4-diphenylbutanoate → (NaOCH$_2$CH$_3$ / H$_2$C=CHCH$_2$Br) → $C_6H_5CH_2CCCOOCH_2CH_3$ → (1. HO⁻, H$_2$O; 2. H$_3$O⁺; 3. heat) → 1,3-Diphenyl-5-hexen-2-one

(*d*) The enolate ion of ethyl phenylacetate attacks the carbonyl carbon of ethyl benzoate.

Ethyl 3-oxo-2,3-diphenylpropanoate

(*e*) Saponification and decarboxylation yield benzyl phenyl ketone.

Ethyl 3-oxo-2,3-diphenylpropanoate → (1. HO⁻, H$_2$O; 2. H$_3$O⁺; 3. heat) → Benzyl phenyl ketone

(*f*) This sequence is analogous to that of part (*c*).

1,2-Diphenyl-4-penten-1-one

21.14 (*a*) The Dieckmann cyclization is the intramolecular version of the Claisen condensation. It employs a diester as starting material.

$CH_3CH_2OC(CH_2)_5COCH_2CH_3$ → (1. NaOCH$_2$CH$_3$; 2. H$_3$O⁺) → Ethyl (2-oxocyclohexane)-carboxylate

Diethyl heptanedioate

(*b*) Acylation of cyclohexanone with diethyl carbonate yields the same β-keto ester formed in part (*a*).

Cyclohexanone + $CH_3CH_2OCOCH_2CH_3$ → (1. NaOCH$_2$CH$_3$; 2. H$_3$O⁺) → Ethyl (2-oxocyclohexane)-carboxylate

Diethyl carbonate

(*c*) The two most stable enol forms are those that involve the proton on the carbon flanked by the two carbonyl groups.

(*d*) Deprotonation of the β-keto ester involves the acidic proton at the carbon flanked by the two carbonyl groups

(*e*) The methyl group is introduced by alkylation of the β-keto ester. Saponification and decarboxylation complete the synthesis.

Ethyl (2-oxocyclohexane)-carboxylate

Ethyl (1-methyl-2-oxocyclohexane)-carboxylate

2-Methylcyclohexanone

(*f*) The enolate ion of the β-keto ester [see part (*d*)] undergoes Michael addition to the carbon–carbon double bond of acrolein.

Ethyl (2-oxocyclohexane)-carboxylate

Acrolein

Michael adduct

This reaction has been reported in the chemical literature and proceeds in 65–75% yield.

21.15 (*a*) Ethyl acetoacetate is converted to its enolate ion with sodium ethoxide; this anion then acts as a nucleophile toward 1-bromopentane.

$$CH_3CCH_2COCH_2CH_3 \ + \ CH_3CH_2CH_2CH_2CH_2Br \ \xrightarrow{\ NaOCH_2CH_3\ } \ CH_3CCHCOCH_2CH_3$$

Ethyl acetoacetate

1-Bromopentane

Ethyl 2-acetylheptanoate

(b) Saponification and decarboxylation of the product in part (a) yields 2-octanone.

$$\underset{\substack{\text{Ethyl}\\\text{2-acetylheptanoate}}}{\overset{\displaystyle O \quad O}{\underset{\underset{\text{CH}_2\text{CH}_2\text{CH}_2\text{CH}_2\text{CH}_3}{|}}{\text{CH}_3\overset{\|}{\text{C}}\text{CH}\overset{\|}{\text{C}}\text{OCH}_2\text{CH}_3}}} \quad \xrightarrow[\substack{\text{2. H}_3\text{O}^+\\\text{3. heat}}]{\text{1. HO}^-,\ \text{H}_2\text{O}} \quad \underset{\substack{\\\text{2-Octanone}}}{\overset{\displaystyle O}{\text{CH}_3\overset{\|}{\text{C}}\text{CH}_2\text{CH}_2\text{CH}_2\text{CH}_2\text{CH}_2\text{CH}_3}}$$

(c) The product derived from the reaction in part (a) can be alkylated again:

$$\underset{\substack{\text{Ethyl}\\\text{2-acetylheptanoate}}}{\overset{\displaystyle O \quad O}{\underset{\underset{\text{CH}_2\text{CH}_2\text{CH}_2\text{CH}_2\text{CH}_3}{|}}{\text{CH}_3\overset{\|}{\text{C}}\text{CH}\overset{\|}{\text{C}}\text{OCH}_2\text{CH}_3}}} + \text{CH}_3\text{I} \xrightarrow{\text{NaOCH}_2\text{CH}_3} \underset{\substack{\text{Ethyl}\\\text{2-acetyl-2-methylheptanoate}}}{\overset{\displaystyle \quad\ OCH_3}{\underset{\underset{\text{CH}_2\text{CH}_2\text{CH}_2\text{CH}_2\text{CH}_3}{|}}{\text{CH}_3\overset{\|\ |}{\text{C}}\text{C}\text{COOCH}_2\text{CH}_3}}}$$

(d) The dialkylated derivative of acetoacetic ester formed in part (c) can be converted to a ketone by saponification and decarboxylation.

$$\underset{\substack{\text{Ethyl}\\\text{2-acetyl-2-methylheptanoate}}}{\overset{\displaystyle \quad\ OCH_3}{\underset{\underset{\text{CH}_2\text{CH}_2\text{CH}_2\text{CH}_2\text{CH}_3}{|}}{\text{CH}_3\overset{\|\ |}{\text{C}}\text{C}\text{COOCH}_2\text{CH}_3}}} \quad \xrightarrow[\substack{\text{2. H}_3\text{O}^+\\\text{3. heat}}]{\text{1. HO}^-,\ \text{H}_2\text{O}} \quad \underset{\substack{\\\text{3-Methyl-2-octanone}}}{\overset{\displaystyle O}{\underset{\underset{\text{CH}_2\text{CH}_2\text{CH}_2\text{CH}_3}{|}}{\text{CH}_3\overset{\|}{\text{C}}\text{CHCH}_3}}}$$

(e) The anion of ethyl acetoacetate acts as a nucleophile toward 1-bromo-3-chloropropane. Bromide is a better leaving group than chloride and is displaced preferentially.

$$\underset{\text{Ethyl acetoacetate}}{\overset{\displaystyle O \quad O}{\text{CH}_3\overset{\|}{\text{C}}\text{CH}_2\overset{\|}{\text{C}}\text{CH}_2\text{CH}_3}} + \underset{\substack{\text{1-Bromo-3-}\\\text{chloropropane}}}{\text{BrCH}_2\text{CH}_2\text{CH}_2\text{Cl}} \xrightarrow{\text{NaOCH}_2\text{CH}_3} \underset{\substack{\text{Ethyl}\\\text{2-acetyl-5-chloropentanoate}}}{\overset{\displaystyle O \quad O}{\underset{\underset{\text{CH}_2\text{CH}_2\text{CH}_2\text{Cl}}{|}}{\text{CH}_3\overset{\|}{\text{C}}\text{CH}\overset{\|}{\text{C}}\text{OCH}_2\text{CH}_3}}}$$

(f) Treatment of the product of part (e) with sodium ethoxide gives an enolate ion that cyclizes by intramolecular nucleophilic substitution of chloride.

$$\underset{\substack{\text{Ethyl}\\\text{2-acetyl-5-chloropentanoate}}}{\overset{\displaystyle O \quad O}{\underset{\underset{\text{CH}_2\text{CH}_2\text{CH}_2\text{Cl}}{|}}{\text{CH}_3\overset{\|}{\text{C}}\text{CH}\overset{\|}{\text{C}}\text{OCH}_2\text{CH}_3}}} \xrightarrow{\text{NaOCH}_2\text{CH}_3} \underset{\substack{\text{Ethyl}\\\text{1-acetylcyclobutanecarboxylate}}}{\overset{\displaystyle O \quad O}{\text{CH}_3\text{C}\quad\text{COCH}_2\text{CH}_3}}$$

(g) Cyclobutyl methyl ketone is formed by saponification and decarboxylation of the product in part (f).

$$\text{CH}_3\text{C}(\!=\!\text{O})\text{—C(COCH}_2\text{CH}_3)(\text{cyclobutane}) \xrightarrow[\substack{\text{2. H}_3\text{O}^+ \\ \text{3. heat}}]{\text{1. HO}^-,\ \text{H}_2\text{O}} \text{CH}_3\text{C}(\!=\!\text{O})\text{—(cyclobutane)}$$

Ethyl
1-acetylcyclobutanecarboxylate

Cyclobutyl methyl
ketone

(h) Ethyl acetoacetate undergoes Michael addition to phenyl vinyl ketone in the presence of base.

$$\text{CH}_3\text{CCH}_2\text{COCH}_2\text{CH}_3 + \text{H}_2\text{C}=\text{CHCC}_6\text{H}_5 \xrightarrow[\text{ethanol}]{\text{NaOCH}_2\text{CH}_3} \text{CH}_3\text{CCHCOCH}_2\text{CH}_3$$
$$\text{CH}_2\text{CH}_2\text{CC}_6\text{H}_5$$
$$\text{O}$$

Ethyl acetoacetate

Phenyl vinyl ketone

Ethyl 2-acetyl-5-
oxo-5-phenylpentanoate

(i) A diketone results from saponification and decarboxylation of the Michael adduct.

$$\text{CH}_3\text{CCHCOCH}_2\text{CH}_3 \xrightarrow[\substack{\text{2. H}_3\text{O}^+ \\ \text{3. heat}}]{\text{1. HO}^-,\ \text{H}_2\text{O}} \text{CH}_3\text{CCH}_2\text{CH}_2\text{CH}_2\text{CC}_6\text{H}_5$$
$$\text{CH}_2\text{CH}_2\text{CC}_6\text{H}_5$$
$$\text{O}$$

Ethyl 2-acetyl-5-
oxo-5-phenylpentanoate

1-Phenyl-1,5-hexanedione

21.16 Diethyl malonate reacts with the reagents given in the preceding problem in a manner analogous to that of ethyl acetoacetate.

(a)

$$\text{CH}_2(\text{COOCH}_2\text{CH}_3)_2 + \text{CH}_3\text{CH}_2\text{CH}_2\text{CH}_2\text{CH}_2\text{Br} \xrightarrow{\text{NaOCH}_2\text{CH}_3} \text{CH}_3\text{CH}_2\text{CH}_2\text{CH}_2\text{CH}_2\text{CH}(\text{COOCH}_2\text{CH}_3)_2$$

Diethyl malonate

1-Bromopentane

Diethyl 1,1-hexanedicarboxylate

(b) $\text{CH}_3\text{CH}_2\text{CH}_2\text{CH}_2\text{CH}_2\text{CH}(\text{COOCH}_2\text{CH}_3)_2 \xrightarrow[\substack{\text{2. H}_3\text{O}^+ \\ \text{3. heat}}]{\text{1. HO}^-,\ \text{H}_2\text{O}} \text{CH}_3\text{CH}_2\text{CH}_2\text{CH}_2\text{CH}_2\text{CH}_2\text{COH}$

Diethyl 1,1-hexanedicarboxylate

Heptanoic acid

(c) $\text{CH}_3\text{CH}_2\text{CH}_2\text{CH}_2\text{CH}_2\text{CH}(\text{COOCH}_2\text{CH}_3)_2 \xrightarrow[\text{NaOCH}_2\text{CH}_3]{\text{CH}_3\text{I}} \text{CH}_3\text{CH}_2\text{CH}_2\text{CH}_2\text{CH}_2\text{C}(\text{COOCH}_2\text{CH}_3)_2$
$$\text{CH}_3$$

Diethyl 1,1-hexanedicarboxylate

Diethyl 2,2-heptanedicarboxylate

(d) $\text{CH}_3\text{CH}_2\text{CH}_2\text{CH}_2\text{CH}_2\text{C}(\text{COOCH}_2\text{CH}_3)_2 \xrightarrow[\substack{\text{2. H}_3\text{O}^+ \\ \text{3. heat}}]{\text{1. HO}^-,\ \text{H}_2\text{O}} \text{CH}_3\text{CH}_2\text{CH}_2\text{CH}_2\text{CH}_2\text{CHCOH}$
$$\text{CH}_3 \qquad\qquad\qquad\qquad\qquad\qquad \text{CH}_3$$

Diethyl 2,2-heptanedicarboxylate

2-Methylheptanoic acid

(e) $CH_2(COOCH_2CH_3)_2$ + $BrCH_2CH_2CH_2Cl$ $\xrightarrow{\text{NaOCH}_2\text{CH}_3}$ $ClCH_2CH_2CH_2CH(COOCH_2CH_3)_2$

Diethyl malonate   1-Bromo-3-chloropropane   Diethyl 4-chloro-1,1-butanedicarboxylate

(f) $ClCH_2CH_2CH_2CH(COOCH_2CH_3)_2$ $\xrightarrow{\text{NaOCH}_2\text{CH}_3}$

Diethyl 4-chloro-1,1-butanedicarboxylate

Diethyl
cyclobutane-1,1-dicarboxylate

(g)

$\xrightarrow[\substack{\text{2. H}_3\text{O}^+ \\ \text{3. heat}}]{\text{1. HO}^-, \text{H}_2\text{O}}$

Diethyl
cyclobutane-1,1-dicarboxylate

Cyclobutanecarboxylic
acid

(h) $CH_2(COOCH_2CH_3)_2$ + $C_6H_5\overset{O}{\overset{||}{C}}CH=CH_2$ $\xrightarrow[\text{CH}_3\text{CH}_2\text{OH}]{\text{NaOCH}_2\text{CH}_3}$ $C_6H_5\overset{O}{\overset{||}{C}}CH_2CH_2CH(COOCH_2CH_3)_2$

Diethyl malonate   Phenyl vinyl ketone   Diethyl 4-oxo-4-phenylbutane-
1,1-dicarboxylate

(i) $C_6H_5\overset{O}{\overset{||}{C}}CH_2CH_2CH(COOCH_2CH_3)_2$ $\xrightarrow[\substack{\text{2. H}_3\text{O}^+ \\ \text{3. heat}}]{\text{1. HO}^-, \text{H}_2\text{O}}$ $C_6H_5\overset{O}{\overset{||}{C}}CH_2CH_2CH_2\overset{O}{\overset{||}{C}}OH$

Diethyl 4-oxo-4-phenylbutane-
1,1-dicarboxylate

5-Oxo-5-phenylpentanoic acid

21.17 (a) Both carbonyl groups of diethyl malonate are equivalent, and so enolization can occur in either direction.

Diethyl malonate

(*b*) Ethyl acetoacetate can give three constitutionally isomeric enols:

Least stable enol; double bond not conjugated with carbonyl group

Ethyl acetoacetate

Enol stable but lacking ester resonance

Most stable enol; double bond conjugated with carbonyl group; ester carbonyl stabilized by resonance

(*c*) Bromine reacts with diethyl malonate and ethyl acetoacetate by way of the corresponding enols:

$CH_2(COOCH_2CH_3)_2$

Diethyl malonate

Diethyl bromomalonate

Ethyl acetoacetate

Ethyl α-bromoacetoacetate

21.18 (*a*) Recall that Grignard reagents are destroyed by reaction with proton donors. Ethyl acetoacetate is a stronger acid than water; it transfers a proton to a Grignard reagent.

Ethyl acetoacetate

Methylmagnesium iodide

Methane

Iodomagnesium salt of ethyl acetoacetate

(*b*) Adding D_2O and DCl to the reaction mixture leads to D^+ transfer to the α-carbon atom of ethyl acetoacetate.

Iodomagnesium salt of ethyl acetoacetate

Deuterium oxide

Ethyl α-deuterioacetoacetate

21.19 (*a*) Ethyl octanoate undergoes a Claisen condensation to form a β-keto ester on being treated with sodium ethoxide.

$$CH_3(CH_2)_5CH_2\overset{\overset{\displaystyle O}{\|}}{C}OCH_2CH_3 \xrightarrow[\text{2. H}_3\text{O}^+]{\text{1. NaOCH}_2\text{CH}_3} CH_3(CH_2)_5CH_2\overset{\overset{\displaystyle O}{\|}}{C}\underset{\underset{\displaystyle (CH_2)_5CH_3}{|}}{C}H\overset{\overset{\displaystyle O}{\|}}{C}OCH_2CH_3$$

Ethyl octanoate   Ethyl 2-hexyl-3-oxodecanoate

(*b*) Saponification and decarboxylation of the β-keto ester yields a ketone.

$$CH_3(CH_2)_5CH_2\overset{\overset{\displaystyle O}{\|}}{C}\underset{\underset{\displaystyle (CH_2)_5CH_3}{|}}{C}H\overset{\overset{\displaystyle O}{\|}}{C}OCH_2CH_3 \xrightarrow[\substack{\text{2. H}_3\text{O}^+ \\ \text{3. heat}}]{\text{1. NaOH, H}_2\text{O}} CH_3(CH_2)_5CH_2\overset{\overset{\displaystyle O}{\|}}{C}CH_2(CH_2)_5CH_3$$

Ethyl 2-hexyl-3-oxodecanoate   8-Pentadecanone

(*c*) On treatment with base, ethyl acetoacetate is converted to its enolate, which reacts as a nucleophile toward 1-bromobutane.

$$CH_3\overset{\overset{\displaystyle O}{\|}}{C}CH_2\overset{\overset{\displaystyle O}{\|}}{C}OCH_2CH_3 + CH_3CH_2CH_2CH_2Br \xrightarrow[\text{ethanol}]{\text{NaOCH}_2\text{CH}_3} CH_3\overset{\overset{\displaystyle O}{\|}}{C}\underset{\underset{\displaystyle CH_2CH_2CH_2CH_3}{|}}{C}H\overset{\overset{\displaystyle O}{\|}}{C}OCH_2CH_3$$

Ethyl acetoacetate   1-Bromobutane   Ethyl 2-acetylhexanoate

(*d*) Alkylation of ethyl acetoacetate, followed by saponification and decarboxylation, gives a ketone. These steps constitute the acetoacetic ester synthesis.

$$CH_3\overset{\overset{\displaystyle O}{\|}}{C}\underset{\underset{\displaystyle CH_2CH_2CH_2CH_3}{|}}{C}H\overset{\overset{\displaystyle O}{\|}}{C}OCH_2CH_3 \xrightarrow[\substack{\text{2. H}_3\text{O}^+ \\ \text{3. heat}}]{\text{1. NaOH, H}_2\text{O}} CH_3\overset{\overset{\displaystyle O}{\|}}{C}CH_2CH_2CH_2CH_2CH_3$$

Ethyl 2-acetylhexanoate   2-Heptanone

(*e*) An alkylated derivative of ethyl acetoacetate is capable of being alkylated a second time.

$$CH_3\overset{\overset{\displaystyle O}{\|}}{C}\underset{\underset{\displaystyle CH_2CH_2CH_2CH_3}{|}}{C}H\overset{\overset{\displaystyle O}{\|}}{C}OCH_2CH_3 + CH_3CH_2CH_2CH_2I \xrightarrow[\text{ethanol}]{\text{NaOCH}_2\text{CH}_3} CH_3\overset{\overset{\displaystyle O}{\|}}{C}\underset{\underset{\displaystyle COOCH_2CH_3}{|}}{C}(CH_2CH_2CH_2CH_3)_2$$

Ethyl 2-acetylhexanoate   1-Iodobutane   Ethyl 2-acetyl-2-butylhexanoate

(*f*) The dialkylated derivative of acetoacetic ester formed in part (*e*) is converted to a ketone by saponification and decarboxylation.

$$CH_3\overset{\overset{\displaystyle O}{\|}}{C}\underset{\underset{\displaystyle COOCH_2CH_3}{|}}{C}(CH_2CH_2CH_2CH_3)_2 \xrightarrow[\substack{\text{2. H}_3\text{O}^+ \\ \text{3. heat}}]{\text{1. NaOH}} CH_3\overset{\overset{\displaystyle O}{\|}}{C}CH(CH_2CH_2CH_2CH_3)_2$$

Ethyl 2-acetyl-2-butylhexanoate   3-Butyl-2-heptanone

(g) The enolate of acetophenone attacks the carbonyl group of diethyl carbonate.

$$C_6H_5\overset{\overset{\displaystyle O}{\|}}{C}CH_3 \ + \ CH_3CH_2O\overset{\overset{\displaystyle O}{\|}}{C}OCH_2CH_3 \quad \xrightarrow[\text{2. H}_3\text{O}^+]{\text{1. NaOCH}_2\text{CH}_3} \quad C_6H_5\overset{\overset{\displaystyle O}{\|}}{C}CH_2\overset{\overset{\displaystyle O}{\|}}{C}OCH_2CH_3$$

Acetophenone   Diethyl carbonate   3-Oxo-3-phenylpropanoate

(h) Diethyl oxalate acts as an acylating agent toward the enolate of acetone.

$$CH_3\overset{\overset{\displaystyle O}{\|}}{C}CH_3 \ + \ CH_3CH_2O\overset{\overset{\displaystyle O\ O}{\|\ \|}}{C\ C}OCH_2CH_3 \quad \xrightarrow[\text{2. H}_3\text{O}^+]{\text{1. NaOCH}_2\text{CH}_3} \quad CH_3\overset{\overset{\displaystyle O}{\|}}{C}CH_2\overset{\overset{\displaystyle O\ O}{\|\ \|}}{C\ C}OCH_2CH_3$$

Acetone   Diethyl oxalate   Ethyl 2,4-dioxopentanoate

(i) The first stage of the malonic ester synthesis is the alkylation of diethyl malonate with an alkyl halide.

$$CH_2(COOCH_2CH_3)_2 \ + \ \underset{\underset{\displaystyle CH_3}{|}}{BrCH_2CHCH_2CH_3} \quad \xrightarrow[\text{ethanol}]{\text{NaOCH}_2\text{CH}_3} \quad \underset{\underset{\displaystyle CH_3}{|}}{CH_3CH_2CHCH_2CH(COOCH_2CH_3)_2}$$

Diethyl malonate   1-Bromo-2-methylbutane   Diethyl 3-methylpentane-1,1-dicarboxylate

(j) Alkylation of diethyl malonate is followed by saponification and decarboxylation to give a carboxylic acid.

$$\underset{\underset{\displaystyle CH_3}{|}}{CH_3CH_2CHCH_2CH(COOCH_2CH_3)_2} \quad \xrightarrow[\substack{\text{2. H}_3\text{O}^+ \\ \text{3. heat}}]{\text{1. NaOH, H}_2\text{O}} \quad \underset{\underset{\displaystyle CH_3}{|}}{CH_3CH_2CHCH_2CH_2\overset{\overset{\displaystyle O}{\|}}{C}OH}$$

Diethyl 3-methylpentane-1,1-dicarboxylate   4-Methylhexanoic acid

(k) The anion of diethyl malonate undergoes Michael addition to 6-methyl-2-cyclohexenone.

$$CH_2(COOCH_2CH_3)_2 \ + \quad \xrightarrow[\text{ethanol}]{\text{NaOCH}_2\text{CH}_3}$$

Diethyl malonate   6-Methyl-2-cyclohexenone   Diethyl 2-(4-methyl-3-oxocyclohexyl)malonate

(l) Acid hydrolysis converts the diester in part (k) to a malonic acid derivative, which then undergoes decarboxylation.

$$\xrightarrow[\text{heat}]{\text{H}_2\text{O, HCl}}$$

Diethyl 2-(4-methyl-3-oxocyclohexyl)malonate   (4-Methyl-3-oxocyclohexyl)acetic acid (*m*) Lithium diisopropylamide (LDA) is used to convert esters quantitatively to their enolate ions. In this reaction the enolate of *tert*-butyl acetate adds to benzaldehyde.

| *tert*-Butyl acetate | Lithium enolate of *tert*-butyl acetate | *tert*-Butyl 3-hydroxy-3-phenylpropanoate |

21.20 (*a*) Both ester functions in this molecule are β to a ketone carbonyl. Hydrolysis is followed by decarboxylation.

Diethyl
3-ethylcyclopentanone-2,5-dicarboxylate

3-Ethylcyclopentanone
(C$_7$H$_{12}$O)

(*b*) Examine each carbon that is α to an ester function to see if it can lead to a five-, six-, or seven-membered cyclic β-keto ester by a Dieckmann cyclization.

Cyclization to a five-membered ring possible, but β-keto ester cannot be deprotonated to give a stable anion.

Cyclization not likely; resulting ring is four-membered and highly strained.

Cyclization gives a five-membered ring; β-keto ester deprotonated under reaction conditions; this is the observed product (C$_{12}$H$_{18}$O$_5$).

(*c*) Both ester functions undergo hydrolysis in acid, but decarboxylation occurs only at the carboxyl group that is β to the ketone carbonyl.

Diethyl 2-methylcyclopentanone-
3,5-dicarboxylate

2-Methylcyclopentanone-3-
carboxylic acid ($C_7H_{10}O_3$)

(*d*) A Dieckmann cyclization occurs, giving a five-membered ring fused to the original three-membered ring.

Diethyl *cis*-1,2-
cyclopropanediacetate

Ethyl bicyclo[3.1.0]-
hexan-3-one-2-carboxylate
($C_9H_{12}O_3$, 79%)

(*e*) Saponification and decarboxylation convert the β-keto ester to a ketone.

Ethyl bicyclo[3.1.0]-hexan-3-
one-2-carboxylate

Bicyclo[3.1.0]hexan-3-one
(C_6H_8O, 43%)

21.21 The heart of the preparation of capsaicin is a malonic ester synthesis. The first step is bromination of the primary alcohol by phosphorous tribromide. The resulting primary alkyl bromide is used to alkylate the sodium salt of diethyl malonate. A substituted malonic acid derivative is obtained following basic hydrolysis of the ester groups.

$C_8H_{15}Br$

$C_{11}H_{18}O_4$

Malonic acid derivatives undergo decarboxylation on heating.

$C_{10}H_{18}O_2$

Formation of the amide completes the synthesis of capsaicin.

Capsaicin ($C_{18}H_{27}NO_3$)

21.22 (*a*) First write out the structure of 4-phenyl-2-butanone and identify the synthon that is derived from ethyl acetoacetate.

Therefore carry out the acetoacetic ester synthesis using a benzyl halide as the alkylating agent.

$$C_6H_5CH_2OH \xrightarrow[\text{or PBr}_3]{\text{HBr}} C_6H_5CH_2Br$$

Benzyl alcohol　　　　　Benzyl bromide

Ethyl acetoacetate　　Benzyl bromide　　Ethyl 2-benzyl-3-oxobutanoate　　4-Phenyl-2-butanone

(*b*) Identify the synthon in 3-phenylpropanoic acid that is derived from malonic ester by disconnecting the molecule at its α-carbon atom.

Here, as in part (*a*), a benzyl halide is the required alkylating agent.

$$CH_2(COOCH_2CH_3)_2 + C_6H_5CH_2Br \xrightarrow[\text{ethanol}]{\text{NaOCH}_2CH_3} C_6H_5CH_2CH(COOCH_2CH_3)_2 \xrightarrow[\substack{\text{2. H}_3O^+ \\ \text{3. heat}}]{\text{1. HO}^-, \text{H}_2O} C_6H_5CH_2CH_2COOH$$

Diethyl malonate　　Benzyl bromide　　Diethyl benzylmalonate　　3-Phenylpropanoic acid

(c) In this synthesis the desired 1,3-diol function can be derived by reduction of a malonic ester derivative. First propene must be converted to an allyl halide for use as an alkylating agent.

$$H_2C=CHCH_2CH(CH_2OH)_2 \implies H_2C=CHCH_2X \ + \ :\bar{C}H(COOCH_2CH_3)_2$$

$$H_2C=CHCH_3 \xrightarrow[\text{heat}]{Cl_2} H_2C=CHCH_2Cl$$

Propene   Allyl chloride

$$CH_2(COOCH_2CH_3)_2 \ + \ H_2C=CHCH_2Cl \xrightarrow[\text{ethanol}]{NaOCH_2CH_3} H_2C=CHCH_2CH(COOCH_2CH_3)_2$$

Diethyl malonate   Allyl chloride   Diethyl 2-allylmalonate

1. LiAlH$_4$
2. H$_2$O

$$H_2C=CHCH_2CH(CH_2OH)_2$$

2-Allyl-1,3-propanediol

(d) The desired primary alcohol may be prepared by reduction of the corresponding carboxylic acid, which in turn is available from the malonic ester synthesis using allyl chloride, including saponification and decarboxylation of the diester [prepared in part (c)].

$$H_2C=CHCH_2CH_2CH_2OH \implies H_2C=CHCH_2CH_2CO_2H \implies H_2C=CHCH_2CH(CO_2CH_2CH_3)_2$$

4-Penten-1-ol

The correct sequence of reactions is

$$H_2C=CHCH_2CH(COOCH_2CH_3)_2 \xrightarrow[\substack{2.\ H_3O^+ \\ 3.\ heat}]{1.\ HO^-,\ H_2O} H_2C=CHCH_2CH_2COOH \xrightarrow[2.\ H_2O]{1.\ LiAlH_4} H_2C=CHCH_2CH_2CH_2OH$$

Diethyl 2-allylmalonate
[prepared as in part (c)]   4-Pentenoic acid   4-Penten-1-ol

(e) The desired product is an alcohol. It can be prepared by reduction of a ketone, which in turn can be prepared by the acetoacetic ester synthesis.

$$\underset{\substack{| \\ OH}}{H_2C=CHCH_2CH_2CHCH_3} \implies \underset{\substack{\| \\ O}}{H_2C=CHCH_2CH_2CCH_3} \implies H_2C=CHCH_2X \ + \ \underset{\substack{\| \\ O}}{:\bar{C}H_2CCH_3}$$

Therefore

$$\underset{\substack{\| \quad \| \\ O \quad O}}{CH_3CCH_2COCH_2CH_3} \ + \ H_2C=CHCH_2Cl \xrightarrow[\text{ethanol}]{NaOCH_2CH_3} \underset{\substack{| \\ CH_2CH=CH_2}}{\underset{\substack{\| \quad \| \\ O \quad O}}{CH_3CCHCOCH_2CH_3}} \xrightarrow[\substack{2.\ H_3O^+ \\ 3.\ heat}]{1.\ HO^-,\ H_2O} \underset{\substack{\| \\ O}}{CH_3CCH_2CH_2CH=CH_2}$$

Ethyl acetoacetate   Allyl chloride

NaBH$_4$
CH$_3$OH

$$\underset{\substack{| \\ OH}}{CH_3CHCH_2CH_2CH=CH_2}$$

5-Hexen-2-ol

(*f*) Cyclopropanecarboxylic acid may be prepared by a malonic ester synthesis, as retrosynthetic analysis shows.

The desired reaction sequence is

Dimethyl malonate   1,2-Dibromoethane   Cyclopropanecarboxylic acid

(*g*) Treatment of the diester formed in part (*f*) with ammonia gives a diamide.

Diethyl cyclopropane-1,1-dicarboxylate [prepared as in part (*f*)]   Cyclopropane-1,1-dicarboxamide

(*h*) We need to extend the carbon chain of the starting material by *four* carbons. One way to accomplish this is by way of a malonic ester synthesis at each end of the chain.

Octanedioic acid   1,8-Dibromooctane

$2CH_2(COOCH_2CH_3)_2$ + $Br(CH_2)_8Br$ $\xrightarrow{NaOCH_2CH_3}$ $(CH_3CH_2OOC)_2CH(CH_2)_8CH(COOCH_2CH_3)_2$

Diethyl malonate   1,8-Dibromooctane

1. HO⁻, H₂O
2. H₃O⁺
3. heat

$HOCCH_2(CH_2)_8CH_2COH$

Dodecanedioic acid

21.23 The problem states that diphenadione is prepared from 1,1-diphenylacetone and dimethyl 1,2-benzenedicarboxylate. Therefore, disconnect the molecule in a way that reveals the two reactants.

Diphenadione

By treating dimethyl 1,2-benzenedicarboxylate and 1,1-diphenylacetone with base, two successive acylations of a ketone enolate occur; the first is intermolecular, the second intramolecular.

Dimethyl 1,2-benzenedicarboxylate   1,1-Diphenylacetone   β-Diketone; not isolated

Diphenadione

21.24 The first step is the conversion of the thioester to its enolate. LDA is a strong base and removes a proton from the α carbon.

The enolate adds to the carbonyl group of cyclohexanone and the species formed in this step undergoes intramolecular displacement of thiophenoxide from thioester carbonyl.

21.25 Styrene oxide will be attacked by the anion of diethyl malonate at its less hindered ring position.

The product is 4-phenylbutanolide. It has been prepared in 72% yield by this procedure.

21.26 The first task is to convert acetic acid to ethyl chloroacetate.

Chlorination must precede esterification, because the Hell–Volhard–Zelinsky reaction requires a carboxylic acid, not an ester, as the starting material. The remaining step is a nucleophilic substitution.

21.27 The compound given in the problem contains three functionalities that can undergo acid-catalyzed hydrolysis: an acetal and two equivalent ester groups. Hydrolysis yields 3-oxo-1,1-cyclobutane-dicarboxylic acid and 2 moles each of methanol and 2-propanol. The hydrolysis product is a malonic acid derivative that decarboxylates on heating. The final product of the reaction is 3-oxocyclobutanecarboxylic acid ($C_5H_6O_3$).

Diisopropyl 3,3-dimethoxycyclobutane-1,1-dicarboxylate → 3-Oxo-1,1-cyclobutanedicarboxylic acid + Methanol + 2-Propanol

3-Oxocyclobutanecarboxylic acid + Carbon dioxide

21.28 One way to do this problem is to work backward. The molecular formula of the final product is $C_6H_{10}O_4$, which differs from compound C ($C_7H_{10}O_6$) by one carbon and two oxygens (CO_2). Thus, the conversion of compound C to the final product is a decarboxylation.

$$\text{Compound C } (C_7H_{10}O_6) \xrightarrow{\text{heat}} CH_3CH(CH_2CO_2H)_2 + CO_2$$

The most likely structure for Compound C is therefore:

$$\begin{array}{c} CH_3CHCH_2CO_2H \\ | \\ CH \\ HO_2C \quad CO_2H \end{array}$$

Inasmuch as compound C is formed from diethyl malonate, and the steps leading to C are the customary ones used to convert ethyl esters to carboxylic acids, Compound B must be the triethyl ester of C.

$$\begin{array}{c} CH_3CHCH_2CO_2CH_2CH_3 \\ | \\ CH \\ CH_3CH_2O_2C \quad CO_2CH_2CH_3 \end{array}$$

The structure of Compound B corresponds to that formed by conjugate addition (Michael addition) of diethyl malonate to an α,β-unsaturated ester. This gives for the overall sequence:

$$CH_3CH{=}CHCOCH_2CH_3 + H_2C(CO_2CH_2CH_3)_2 \xrightarrow[\text{ethanol}]{NaOCH_2CH_3}$$

Compound A

$$\begin{array}{c} CH_3CHCH_2CO_2CH_2CH_3 \\ | \\ CH \\ CH_3CH_2O_2C \quad CO_2CH_2CH_3 \end{array}$$

Compound B

1. NaOH, H_2O
2. H_3O^+
3. heat

$$CH_3CHCH_2CO_2H \xleftarrow{\text{heat}} \begin{array}{c} CH_3CHCH_2CO_2H \\ | \\ CH \\ HO_2C \quad CO_2H \end{array}$$
$$| \\ CH_2CO_2H$$

Compound C

ANSWERS TO INTERPRETIVE PROBLEMS

21.29 C;   **21.30** D;   **21.31** C;   **21.32** A;   **21.33** C;   **21.34** A;   **21.35** B

SELF-TEST

PART A

A-1. Give the structure of the reactant, reagent, or product omitted from each of the following:

(a) $CH_3CH_2CH_2\overset{\overset{\displaystyle O}{\|}}{C}OCH_2CH_3$ $\xrightarrow[\text{2. } H_3O^+]{\text{1. } NaOCH_2CH_3}$?

(b) $H\overset{\overset{\displaystyle O}{\|}}{C}OCH_2CH_3$ + ? $\xrightarrow[\text{2. } H_3O^+]{\text{1. } NaOCH_2CH_3}$ $C_6H_5\underset{\underset{\displaystyle O}{\overset{\displaystyle \|}{C}\diagdown H}}{\overset{\displaystyle C}{|}}H\overset{\overset{\displaystyle O}{\|}}{C}OCH_2CH_3$

(c) [cyclobutane ring with Cl substituent and two $\overset{\overset{\displaystyle O}{\|}}{C}OCH_2CH_3$ groups] $\xrightarrow[\substack{\text{2. } H_3O^+ \\ \text{3. heat}}]{\text{1. } HO^-, H_2O}$? (two isomeric products; $C_5H_7ClO_2$)

(d) $CH_2(COOCH_2CH_3)_2$ + $H_2C{=}CH\overset{\overset{\displaystyle O}{\|}}{C}OCH_2CH_3$ $\xrightarrow[\text{ethanol}]{NaOCH_2CH_3}$?

(e) $CH_3\overset{\overset{\displaystyle O}{\|}}{C}CH_2\overset{\overset{\displaystyle O}{\|}}{C}OCH_2CH_3$ $\xrightarrow[\text{2. } C_6H_5CH_2Br]{\text{1. } NaOCH_2CH_3}$?

(f) Product of part (e) $\xrightarrow{?}$ $C_6H_5CH_2CH_2\overset{\overset{\displaystyle O}{\|}}{C}CH_3$

(g) $CH_3\overset{\overset{\displaystyle O}{\|}}{C}\underset{\underset{\displaystyle CO_2H}{|}}{C}HCH_2CO_2H$ $\xrightarrow{\text{heat}}$ CO_2 + ?

A-2. Provide the correct structures of compounds A through E in the following reaction sequences:

(a) A $\xrightarrow[\text{2. } H_3O^+]{\text{1. } NaOCH_2CH_3}$ [cyclopentanone ring with $\overset{\overset{\displaystyle O}{\|}}{C}OCH_2CH_3$ group] $\xrightarrow[\text{2. } CH_3CH_2I]{\text{1. } NaOCH_2CH_3}$ B $\xrightarrow[\substack{\text{2. } H_3O^+ \\ \text{3. heat}}]{\text{1. } HO^-, H_2O}$ C

(b) $CH_3CH_2CH_2\overset{\overset{\displaystyle O}{\|}}{C}OCH_2CH_3$ $\xrightarrow[\text{2. } H_3O^+]{\text{1. } NaOCH_2CH_3}$ D $\xrightarrow[\substack{\text{2. } H_3O^+ \\ \text{3. heat}}]{\text{1. } HO^-, H_2O}$ E + CO_2

A-3. Give a series of steps that will enable preparation of each of the following compounds from the starting material(s) given and any other necessary reagents:

(a) $CH_3\overset{\overset{\displaystyle O}{\|}}{C}CH_2CH_2\overset{\overset{\displaystyle O}{\|}}{C}OH$ from ethyl acetoacetate

(b) $C_6H_5\overset{\overset{\displaystyle O}{\|}}{C}CH_2CH_2CH_2\overset{\overset{\displaystyle O}{\|}}{C}CH_3$ from $C_6H_5\overset{\overset{\displaystyle O}{\|}}{C}CH_3$ and diethyl carbonate

A-4. Write a stepwise mechanism for the reaction of ethyl propanoate with sodium ethoxide in ethanol.

A-5. Ethyl 2-methylbutanoate does not undergo a Claisen condensation, whereas ethyl 3-methylbutanoate does. Provide a mechanistic explanation for this observation.

PART B

B-1. Which of the following compounds is the strongest acid?

(a) $HCO_2CH_2CH_3$

(c) $CH_3CH_2O_2CCH_2CH_2CO_2CH_2CH_3$

(b) $CH_3CH_2O_2CCH_2CO_2CH_2CH_3$

(d) $CH_3CO_2CH_2CH_3$

B-2. Which of the following will yield a ketone and carbon dioxide following saponification, acidification, and heating?

(a) CH_3CH_2CHCH_2CCH_3 (with =O on the C, and COCH_2CH_3 with O double bond on the branch)

(c) CH_3CH_2CHCCH_3 (=O, and CCH_2CH_3 with O= on branch)

(b) CH_3CH_2CHCOCH_2CH_3 (=O, and COCH_2CH_3 with O= on branch)

(d) CH_3CH_2CHCCH_3 (=O, and COCH_2CH_3 with O= on branch)

B-3. Which of the following keto esters is *not* likely to have been prepared by a Claisen condensation?

(a) CH_3CH_2CCHCOCH_2CH_3 (two =O groups, CH_3 branch)

(c) (CH_3)_2CHCC(CH_3)_2 (=O, and COCH_2CH_3 with O= branch)

(b) C_6H_5CCHCOCH_2CH_3 (two =O groups, CH_3 branch)

(d) (CH_3)_2CHCH_2CCHCOCH_2CH_3 (two =O groups, CH(CH_3)_2 branch)

B-4. Dieckmann cyclization of $CH_3CH_2OC(CH_2)_5COCH_2CH_3$ (with two C=O groups) will yield

(a)

(c)

(b)

(d)

B-5. What is the final product of this sequence?

$$\xrightarrow[\text{CH}_3\text{CH}_2\text{OH}]{\text{NaOCH}_2\text{CH}_3} \xrightarrow[\begin{array}{l}\text{1. HO}^-, \text{H}_2\text{O} \\ \text{2. H}_3\text{O}^+ \\ \text{3. heat}\end{array}]{}$$

(a)

(c)

(b)

(d)

B-6. What is the final product of the following sequence of reactions?

$$\text{CH}_2(\text{CO}_2\text{CH}_2\text{CH}_3)_2 \ + \ (\text{CH}_3)_2\text{C}{=}\text{CHCCH}_3 \xrightarrow[\text{ethanol}]{\text{NaOCH}_2\text{CH}_3} \xrightarrow[\begin{array}{l}\text{1. KOH} \\ \text{2. H}_3\text{O}^+ \\ \text{3. heat}\end{array}]{}$$

(a)

(c)

(b)

(d)

B-7. Which of the following would be a suitable candidate for preparation by a mixed Claisen condensation?

(a) $\text{CH}_3\text{CH}_2\overset{\text{O}}{\overset{\|}{\text{C}}}\text{CH}_2\overset{\text{O}}{\overset{\|}{\text{C}}}\text{OCH}_2\text{CH}_3$

(c) $\text{C}_6\text{H}_5\text{CH}_2\overset{\text{O}}{\overset{\|}{\text{C}}}\text{CH}_2\overset{\text{O}}{\overset{\|}{\text{C}}}\text{OCH}_2\text{CH}_3$

(b) $\text{C}_6\text{H}_5\overset{\text{O}}{\overset{\|}{\text{C}}}\text{CH}_2\overset{\text{O}}{\overset{\|}{\text{C}}}\text{OCH}_2\text{CH}_3$

(d) $\text{C}_6\text{H}_5\overset{\text{O}}{\overset{\|}{\text{C}}}{-}\overset{\text{CH}_3}{\underset{\text{CH}_3}{\overset{|}{\underset{|}{\text{C}}}}}{-}\overset{\text{O}}{\overset{\|}{\text{C}}}\text{OCH}_2\text{CH}_3$

B-8. What is the major product of the following reaction?

$$\text{CH}_3\text{CH}_2\overset{\text{O}}{\overset{\|}{\text{C}}}\text{OCH}_3 \ + \ \text{H}\overset{\text{O}}{\overset{\|}{\text{C}}}\text{OCH}_3 \xrightarrow[\text{2. H}_3\text{O}^+]{\text{1. NaOCH}_2\text{CH}_3} \ ?$$

(a) $\text{CH}_3\text{O}\overset{\text{O}}{\overset{\|}{\text{C}}}\overset{\text{O}}{\underset{\text{CH}_3}{\overset{\|}{\underset{|}{\text{CH}}}}}\overset{\text{O}}{\overset{\|}{\text{C}}}\text{OCH}_3$

(c) $\text{HOCH}_2\overset{\text{O}}{\underset{\text{CH}_3}{\overset{\|}{\underset{|}{\text{CH}}}}}\overset{\text{O}}{\overset{\|}{\text{C}}}\text{OCH}_3$

(e) $\text{CH}_3\text{O}\overset{\text{O}}{\overset{\|}{\text{C}}}\overset{}{\underset{\text{CH}_3}{\overset{}{\underset{|}{\text{CH}}}}}\text{COCH}_3$

(b) $\text{H}\overset{\text{O}}{\overset{\|}{\text{C}}}\text{O}\overset{\text{O}}{\underset{\text{CH}_3}{\overset{\|}{\underset{|}{\text{CH}}}}}\overset{\text{O}}{\overset{\|}{\text{C}}}\text{OCH}_3$

(d) $\text{H}\overset{\text{O}}{\overset{\|}{\text{C}}}\overset{\text{O}}{\underset{\text{CH}_3}{\overset{\|}{\underset{|}{\text{CH}}}}}\overset{\text{O}}{\overset{\|}{\text{C}}}\text{OCH}_3$

CHAPTER 22

Amines

SOLUTIONS TO TEXT PROBLEMS

22.1 (*b*) The amino and phenyl groups are both attached to C-1 of an ethyl group.

$$C_6H_5CHCH_3$$
$$|$$
$$NH_2$$

1-Phenylethylamine, or
1-phenylethanamine

(*c*) $$H_2C{=}CHCH_2NH_2$$

Allylamine, or
2-propen-1-amine

22.2 *N,N*-Dimethylcycloheptylamine may also be named as a dimethyl derivative of cycloheptanamine.

⬡—N(CH₃)₂

N,N-Dimethylcycloheptanamine

22.3 Three substituents are attached to the nitrogen atom; the amine is tertiary. In alphabetical order, the substituents present on the aniline nucleus are ethyl, isopropyl, and methyl. Their positions are specified as *N*-ethyl, 4-isopropyl, and *N*-methyl.

$$(CH_3)_2CH{-}\bigcirc{-}N{-}CH_2CH_3$$
with CH₃ on N

N-Ethyl-4-isopropyl-*N*-methylaniline

22.4   The electron-donating amino group and the electron-withdrawing nitro group are directly conjugated in *p*-nitroaniline. The planar geometry of *p*-nitroaniline suggests that the delocalized resonance form shown is a major contributor to the structure of the compound.

22.5   The Henderson–Hasselbalch equation described in text Section 19.4 can be applied to bases such as amines, as well as carboxylic acids. The ratio $[CH_3NH_3^+]/[CH_3NH_2]$ is given by

$$\frac{[CH_3NH_3^+]}{[CH_3NH_2]} = 10^{(pK_a - pH)}$$

The pK_a of methylammonium ion is given in the text as 10.7. Therefore

$$\frac{[CH_3NH_3^+]}{[CH_3NH_2]} = 10^{(10.7 - 7)} = 10^{3.7} = 5 \times 10^3$$

22.6   Tetrahydroisoquinoline is an alkylamine because the nitrogen atom is attached only to sp^3-hybridized carbons. Tetrahydroquinoline, on the other hand, is an arylamine because nitrogen is directly attached to the aromatic ring. Arylamines are less basic than alkylamines because the nitrogen unshared-electron pair is delocalized into the π system of the aromatic ring. Thus, tetrahydroisoquinoline is the stronger base. Recalling that the stronger base will have the weaker conjugate acid, the pK_a of the conjugate acid of tetrahydroisoquinoline is smaller than that of tetrahydroquinoline.

Tetrahydroisoquinoline
(an alkylamine): more basic

Tetrahydroquinoline
(an arylamine): less basic

pK_a (of conjugate acid):   9.4   4.0

22.7   (*b*)   An acetyl group attached directly to nitrogen as in acetanilide delocalizes the nitrogen lone pair into the carbonyl group. Amides are weaker bases than amines.

   (*c*)   An acetyl group in a position para to an amine function is conjugated to it and delocalizes the nitrogen lone pair.

22.8 An aqueous solution of imidazolium chloride would be acidic. With a pK_a of 7, imidazolium chloride is a stronger acid than water ($pK_a = 15.7$), so when dissolved in water imidazolium chloride will produce a solution with a hydrogen ion concentration greater than water, in other words, with a pH less than 7 (an acidic solution).

An aqueous solution of imidazole would be basic. Imidazole is more basic than water; the pK_a of its conjugate acid is 7, whereas the pK_a of H_3O^+ (the conjugate acid of water) is –1.7.

A solution containing equal quantities of imidazole and its conjugate acid would be neutral, with a pH = 7. Recall the Henderson–Hasselbalch equation from text Section 19.4:

$$pH = pK_a + \log \frac{[\text{conjugate base}]}{[\text{conjugate acid}]}$$

When the concentration of imidazole (the conjugate base) and imidazolium ion (the conjugate acid) are equal, the pH of the solution equals pK_a because log (1) = 0.

22.9 Nicotine contains two different nitrogen atoms. The more basic of the two is the nitrogen on the pyrrolidine ring, and is the nitrogen that becomes protonated in the conjugate acid of nicotine.

The nitrogen that becomes protonated is similar to that in a piperidine ring as mentioned in the text, and the pK_a is expected to be about 11.

22.10 The reaction that leads to allylamine is nucleophilic substitution by ammonia on allyl chloride.

$$H_2C=CHCH_2Cl + 2NH_3 \longrightarrow H_2C=CHCH_2NH_2 + NH_4Cl$$

| Allyl chloride | Ammonia | | Allylamine | Ammonium chloride |

Allyl chloride is prepared by free-radical chlorination of propene.

$$H_2C=CHCH_3 + Cl_2 \xrightarrow{400°C} H_2C=CHCH_2Cl + HCl$$

| Propene | Chlorine | | Allyl chloride | Hydrogen chloride |

22.11 (*b*) Isobutylamine is $(CH_3)_2CHCH_2NH_2$. It is a primary amine of the type RCH_2NH_2 and can be prepared from a primary alkyl halide by the Gabriel synthesis.

$(CH_3)_2CHCH_2Br$ + *N*-Potassiophthalimide ⟶ *N*-Isobutylphthalimide

Isobutyl bromide   *N*-Potassiophthalimide   *N*-Isobutylphthalimide

H_2NNH_2

$(CH_3)_2CHCH_2NH_2$ +

Isobutylamine   Phthalhydrazide

(*c*) Although *tert*-butylamine $(CH_3)_3CNH_2$ is a primary amine, it cannot be prepared by the Gabriel method, because it would require an S_N2 reaction on a tertiary alkyl halide in the first step. Elimination occurs instead.

$(CH_3)_2CBr$ + *N*-Potassiophthalimide ⟶ $(CH_3)_2C=CH_2$ + Phthalimide + KBr

tert-Butyl   *N*-Potassiophthalimide   2-Methylpropene   Phthalimide   Potassium
bromide   bromide

(*d*) The preparation of 2-phenylethylamine by the Gabriel synthesis has been described in the chemical literature.

2-Phenylethyl bromide   N-Potassiophthalimide   N-(2-Phenylethyl)phthalimide

$C_6H_5CH_2CH_2NH_2$ +

2-Phenylethylamine   Phthalhydrazide

(*e*) The Gabriel synthesis leads to primary amines; *N*-methylbenzylamine is a secondary amine and cannot be prepared by this method.

N-Methylbenzylamine
(two carbon substituents on
nitrogen; a secondary amine)

(*f*) Aniline cannot be prepared by the Gabriel method. Aryl halides do not undergo nucleophilic substitution under these conditions.

Bromobenzene   N-Potassiophthalimide

22.12 For each part of this problem, keep in mind that aromatic amines are derived by reduction of the corresponding aromatic nitro compound. Each synthesis should be approached from the standpoint of how best to prepare the necessary nitroaromatic compound.

$Ar—NH_2 \Longrightarrow Ar—NO_2 \Longrightarrow Ar—H$

(Ar = substituted aromatic ring)

(*b*) The para isomer of isopropylaniline may be prepared by a procedure analogous to that used for its ortho isomer in part (*a*).

Benzene   Isopropylbenzene   *o*-Isopropylnitrobenzene   *p*-Isopropylnitrobenzene

After separating the ortho, para mixture by distillation, the nitro group of *p*-isopropylnitrobenzene is reduced to yield the desired *p*-isopropylaniline.

(*c*) The target compound is the reduction product of 1-isopropyl-2,4-dinitrobenzene.

1-Isopropyl-2,4-
dinitrobenzene   4-Isopropyl-1,3-
benzenediamine

This reduction is carried out in the same way as reduction of an arene that contains only a single nitro group. In this case hydrogenation over a nickel catalyst gave the desired product in 90% yield. The starting dinitro compound is prepared by nitration of isopropylbenzene.

Isopropylbenzene   1-Isopropyl-2,4-
dinitrobenzene
(43%)

(*d*) The conversion of *p*-chloronitrobenzene to *p*-chloroaniline was cited as an example in the text to illustrate reduction of aromatic nitro compounds to arylamines. *p*-Chloronitrobenzene is prepared by nitration of chlorobenzene.

Benzene · Chlorobenzene · o-Chloronitrobenzene + p-Chloronitrobenzene

The para isomer accounts for 69% of the product in this reaction (30% is ortho, 1% meta). Separation of p-chloronitrobenzene and its reduction completes the synthesis.

p-Chloronitrobenzene · p-Chloroaniline

Chlorination of nitrobenzene would not be a suitable route to the required intermediate, because it would produce mainly m-chloronitrobenzene.

(e) The synthesis of m-aminoacetophenone may be carried out by the scheme shown:

Benzene · Acetophenone · m-Nitroacetophenone · m-Aminoacetophenone

The acetyl group is attached to the ring by Friedel–Crafts acylation. It is a meta director, and its nitration gives the proper orientation of substituents. The order of the first two steps cannot be reversed, because Friedel–Crafts acylation of nitrobenzene is not possible (text Section 12.16). Once prepared, m-nitroacetophenone can be reduced to m-nitroaniline by any of a number of reagents. Indeed, all three reducing combinations described in the text have been employed for this transformation.

	Reducing agent	Yield (%)
m-Nitroacetophenone	H_2, Pt	94
↓	Fe, HCl	84
m-Aminoacetophenone	Sn, HCl	82

22.13 (b) Dibenzylamine is a secondary amine and can be prepared by reductive amination of benzaldehyde with benzylamine.

$$C_6H_5\overset{\overset{\displaystyle O}{\|}}{C}H \quad + \quad C_6H_5CH_2NH_2 \quad \xrightarrow{H_2, \ Ni} \quad C_6H_5CH_2NHCH_2C_6H_5$$

Benzaldehyde · Benzylamine · Dibenzylamine

(*c*) *N,N*-Dimethylbenzylamine is a tertiary amine. Its preparation from benzaldehyde requires dimethylamine, a secondary amine.

$$C_6H_5\overset{\overset{\displaystyle O}{\|}}{C}H \;+\; (CH_3)_2NH \;\;\xrightarrow{\;H_2,\,Ni\;}\;\; C_6H_5CH_2N(CH_3)_2$$

Benzaldehyde   Dimethylamine   *N,N*-Dimethylbenzylamine

(*d*) The preparation of *N*-butylpiperidine by reductive amination is described in text Section 22.10. An analogous procedure is used to prepare *N*-benzylpiperidine.

$$C_6H_5\overset{\overset{\displaystyle O}{\|}}{C}H \;+\; \underset{\underset{\displaystyle H}{N}}{\bigcirc} \;\;\xrightarrow{\;H_2,\,Ni\;}\;\; C_6H_5CH_2\!-\!N\!\!-\!\bigcirc$$

Benzaldehyde   Piperidine   *N*-Benzylpiperidine

22.14 (*b*) First identify the available β hydrogens. Elimination must involve a proton from the carbon atom adjacent to the one that bears the nitrogen.

$$\underset{\text{A methylene group}}{(CH_3)_3C\overset{\beta}{C}H_2}\!-\!\underset{\underset{\displaystyle +N(CH_3)_3}{|}}{\overset{\overset{\displaystyle \overset{\beta}{C}H_3}{|}}{C}}\!-\!\overset{\beta}{C}H_3 \quad \text{Two equivalent methyl groups}$$

It is a proton from one of the methyl groups, rather than from the more sterically hindered methylene, that is lost on elimination.

$$(CH_3)_3CCH_2\!-\!\underset{\underset{\displaystyle +}{\overset{\displaystyle |}{\underset{N(CH_3)_3}{}}}}{\overset{\overset{\displaystyle CH_3}{|}}{C}}\!-\!CH_2\!-\!H \;\; \ddot{\underset{\cdot\cdot}{O}}H \;\longrightarrow\; (CH_3)_3CCH_2C\!=\!CH_2 \;+\; (CH_3)_3N\!:\!$$

(1,1,3,3-Tetramethylbutyl)-   2,4,4-Trimethyl-1-pentene   Trimethylamine
trimethylammonium   (only alkene formed,
hydroxide   70% isolated yield)

(*c*) The base may abstract a proton from either of two β carbons. Deprotonation of the β methyl carbon yields ethylene.

$$H\ddot{\underset{\cdot\cdot}{O}}\!:^{-}\;\;H\!-\!CH_2\!-\!CH_2\!\overset{\overset{\displaystyle CH_3}{|}}{\underset{\underset{\displaystyle CH_3}{|}}{\overset{+}{N}}}CH_2CH_2CH_2CH_3 \;\;\xrightarrow[(-H_2O)]{heat}\;\; H_2C\!=\!CH_2 \;+\; (CH_3)_2\ddot{N}CH_2CH_2CH_2CH_3$$

N-Ethyl-*N,N*-dimethylbutylammonium hydroxide   Ethylene   *N,N*-Dimethylbutylamine

Deprotonation of the β methylene carbon yields 1-butene.

$$CH_3CH_2\!\overset{+}{\underset{\underset{\displaystyle CH_3}{|}}{\overset{\overset{\displaystyle CH_3}{|}}{N}}}\!-\!CH_2\!\!-\!\underset{\underset{\displaystyle \underset{\displaystyle \ddot{\underset{\cdot\cdot}{O}}H}{H}}{}}{C}HCH_2CH_3 \;\;\xrightarrow[(-H_2O)]{heat}\;\; CH_3CH_2\ddot{N}(CH_3)_2 \;+\; H_2C\!=\!CHCH_2CH_3$$

N-Ethyl-*N,N*-dimethylbutylammonium hydroxide   *N,N*-Dimethylethylamine   1-Butene

The preferred order of proton removal in Hofmann elimination reactions is β CH_3 > β CH_2 > β CH. Ethylene is the major alkene formed, the observed ratio of ethylene to 1-butene being 98 : 2.

22.15 (*b*) The pattern of substituents in 2,4-dinitroaniline suggests that they can be introduced by dinitration. Because nitration of aniline itself is not practical, the amino group must be protected by conversion to its *N*-acetyl derivative.

| Aniline | Acetanilide | 2,4-Dinitroacetanilide |

Hydrolysis of the amide bond in 2,4-dinitroacetanilide furnishes the desired 2,4-dinitroaniline.

| 2,4-Dinitroacetanilide | 2,4-Dinitroaniline |

(*c*) Retrosynthetically, *p*-aminoacetanilide may be derived from *p*-nitroacetanilide.

| *p*-Aminoacetanilide | *p*-Nitroacetanilide |

This suggests the sequence

| Aniline | Acetanilide | *p*-Nitroacetanilide (separate from ortho isomer) |

1. Fe, HCl; 2. HO⁻ or
1. Sn, HCl; 2. HO⁻ or
H₂, Pt

p-Aminoacetanilide

22.16 The principal resonance contributors of *N*-nitrosodimethylamine are

All atoms (except hydrogen) have octets of electrons in each of these structures. Other resonance contributors are less stable because they do not have a full complement of electrons around each atom.

22.17 Deamination of 1,1-dimethylpropylamine gives products that result from 1,1-dimethylpropyl cation. Because 2,2-dimethylpropylamine gives the same products, it is likely that 1,1-dimethylpropyl cation is formed from 2,2-dimethylpropylamine by way of its diazonium ion. A carbocation rearrangement is indicated.

| 2,2-Dimethylpropylamine | 2,2-Dimethylpropyldiazonium ion | 1,1-Dimethylpropyl cation |

Once formed, 1,1-dimethylpropyl cation loses a proton to form an alkene or is captured by water to give an alcohol.

2-Methyl-1-butene   2-Methyl-2-butene

2-Methyl-2-butanol

22.18 Phenols may be prepared by diazotization of the corresponding aniline derivative. The problem simplifies itself, therefore, to the preparation of *m*-bromoaniline. Recognizing that arylamines are ultimately derived from nitroarenes, we derive the retrosynthetic sequence of intermediates:

| *m*-Bromophenol | *m*-Bromoaniline | *m*-Bromonitrobenzene | Nitrobenzene |

The desired reaction sequence is straightforward, using reactions discussed previously in the text.

22.19 The key to this problem is to recognize that the iodine substituent in *m*-bromoiodobenzene is derived from an arylamine by diazotization.

m-Bromoiodobenzene   *m*-Bromoaniline

The preparation of *m*-bromoaniline from benzene has been described in Problem 22.18. All that remains is to write the equation for its conversion to *m*-bromoiodobenzene.

1. NaNO₂, HCl, H₂O
2. KI

m-Bromoaniline   *m*-Bromoiodobenzene

22.20 The final step in the preparation of ethyl *m*-fluorophenyl ketone is shown in the text example immediately preceding this problem; therefore all that is necessary is to describe the preparation of *m*-aminophenyl ethyl ketone.

Ethyl *m*-fluorophenyl
ketone

m-Aminophenyl ethyl
ketone

Ethyl *m*-nitrophenyl
ketone

Recalling that arylamines are normally prepared by reduction of nitroarenes, we see that ethyl *m*-nitrophenyl ketone is a pivotal synthetic intermediate. It is prepared by nitration of ethyl phenyl ketone, which is analogous to nitration of acetophenone, shown in text Section 12.16. The preparation of ethyl phenyl ketone by Friedel–Crafts acylation of benzene is shown in text Section 12.7.

Ethyl *m*-nitrophenyl
ketone

Ethyl phenyl ketone

Reversing the order of introduction of the nitro and acyl groups is incorrect. It is possible to nitrate ethyl phenyl ketone but not possible to carry out a Friedel–Crafts acylation on nitrobenzene, owing to the strong deactivating influence of the nitro group.

22.21 Direct nitration of the prescribed starting material cumene (isopropylbenzene) is not suitable, because isopropyl is an ortho, para-directing substituent and will give the target molecule *m*-nitrocumene as only a minor component of the nitration product. However, the conversion of 4-isopropyl-2-nitroaniline to *m*-isopropylnitrobenzene, which was used to illustrate reductive deamination of arylamines in the text, establishes the last step in the synthesis.

m-Nitrocumene   4-Isopropyl-2-nitroaniline   Cumene

Our task simplifies itself to the preparation of 4-isopropyl-2-nitroaniline from cumene. The following procedure is a straightforward extension of the reactions and principles developed in this chapter.

Cumene   *p*-Nitrocumene   *p*-Isopropylaniline   *p*-Isopropylacetanilide

p-Isopropylacetanilide   4-Isopropyl-2-nitroacetanilide   4-Isopropyl-2-nitroaniline

Reductive deamination of 4-isopropyl-2-nitroaniline by diazotization in the presence of ethanol or hypophosphorous acid yields *m*-nitrocumene and completes the synthesis.

22.22 Chrysoidine can be prepared from the reaction of benzenediazonium ion and *m*-benzenediamine.

Chrysoidine   Benzenediazonium ion   *m*-Benzenediamine

22.23 Amines may be primary, secondary, or tertiary. The $C_4H_{11}N$ primary amines, compounds of the type $C_4H_9NH_2$, and their systematic names are

$$CH_3CH_2CH_2CH_2NH_2 \qquad (CH_3)_2CHCH_2NH_2 \qquad CH_3CHCH_2CH_3 \qquad (CH_3)_3CNH_2$$
$$|$$
$$NH_2$$

Butylamine	Isobutylamine	*sec*-Butylamine	*tert*-Butylamine
(1-butanamine or butan-1-amine)	(2-methyl-1-propanamine or 2-methylpropan-1-amine)	(2-butanamine or butan-2-amine)	(2-methyl-2-propanamine or 2-methylpropan-2-amine)

Secondary amines have the general formula R_2NH. Those of molecular formula $C_4H_{11}N$ are

$$(CH_3CH_2)_2NH \qquad CH_3NCH_2CH_2CH_3 \qquad CH_3NCH(CH_3)_2$$
$$| \qquad\qquad |$$
$$H \qquad\qquad\quad H$$

Diethylamine	N-Methylpropylamine	N-Methylisopropylamine
(N-ethylethanamine)	(N-methyl-1-propanamine or N-methylpropan-1-amine)	(N-methyl-2-propanamine or N-methylpropan-2-amine)

There is only one tertiary amine (R_3N) of molecular formula $C_4H_{11}N$:

$$(CH_3)_2NCH_2CH_3$$

N,N-Dimethylethylamine
(*N,N*-dimethylethanamine)

22.24 (*a*) The name 2-ethyl-1-butanamine designates a four-carbon chain terminating in an amino group and bearing an ethyl group at C-2.

$$CH_3CH_2CHCH_2NH_2$$
$$|$$
$$CH_2CH_3$$

2-Ethyl-1-butanamine

(*b*) The prefix *N*- in *N*-ethyl-1-butanamine identifies the ethyl group as a substituent on nitrogen in a secondary amine.

$$CH_3CH_2CH_2CH_2NCH_2CH_3$$
$$|$$
$$H$$

N-Ethyl-1-butanamine

(*c*) Dibenzylamine is a secondary amine. It bears two benzyl groups on nitrogen.

$$C_6H_5CH_2NCH_2C_6H_5$$
$$|$$
$$H$$

Dibenzylamine

(*d*) Tribenzylamine is a tertiary amine.

$$(C_6H_5CH_2)_3N$$

Tribenzylamine

(*e*) Tetraethylammonium hydroxide contains a quaternary ammonium ion.

$$(CH_3CH_2)_4\overset{+}{N} \ HO^-$$

Tetraethylammonium
hydroxide

(*f*) This compound is a secondary amine; it bears an allyl substituent on the nitrogen of cyclohexylamine.

N-Allylcyclohexylamine

(*g*) Piperidine is a cyclic secondary amine that contains nitrogen in a six-membered ring. *N*-Allylpiperidine is a tertiary amine.

$NCH_2CH=CH_2$

N-Allylpiperidine

(*h*) The compound is the benzyl ester of 2-aminopropanoic acid.

$$\underset{\underset{NH_2}{|}}{CH_3CHCOCH_2C_6H_5}$$

(with O double bonded above the C)

Benzyl 2-aminopropanoate

(*i*) The parent compound is cyclohexanone. The substituent $(CH_3)_2N-$ group is attached to C-4.

4-(*N,N*-Dimethylamino)-
cyclohexanone

(*j*) The suffix -*diamine* reveals the presence of two amino groups, one at either end of a three-carbon chain that bears two methyl groups at C-2.

$$\underset{\underset{CH_3}{|}}{\overset{\overset{CH_3}{|}}{H_2NCH_2CCH_2NH_2}}$$

2,2-Dimethyl-1,3-
propanediamine

22.25 (*a*) A phenyl group and an amino group are trans to each other on a three-membered ring in this compound.

trans-2-Phenylcyclopropylamine
(tranylcypromine)

(*b*) This compound is a tertiary amine. It bears a benzyl group, a methyl group, and a 2-propynyl group on nitrogen.

$$C_6H_5CH_2-N\begin{smallmatrix}CH_3\\CH_2C\equiv CH\end{smallmatrix}$$

N-Benzyl-*N*-methyl-2-propynylamine
(pargyline)

(*c*) The amino group is at C-2 of a three-carbon chain that bears a phenyl substituent at its terminus.

$$C_6H_5CH_2\underset{\underset{NH_2}{|}}{C}HCH_3$$

1-Phenyl-2-propanamine
(amphetamine)

(*d*) Phenylephrine is named systematically as an ethanol derivative.

1-(*m*-Hydroxyphenyl)-
2-(*N*-methylamino)ethanol

22.26 (*a*) There are five isomers of C_7H_9N that contain a benzene ring.

$$C_6H_5CH_2NH_2 \quad C_6H_5NHCH_3$$

Benzylamine   *N*-Methylaniline

o-Methylaniline   *m*-Methylaniline   *p*-Methylaniline

(*b*) Benzylamine is the strongest base because its amine group is bonded to an sp^3-hybridized carbon. Benzylamine is a typical alkylamine; the pK_a of its conjugate acid is 9.3. All the other isomers are arylamines; the pK_a's of their conjugate acids are about 4.

(*c*) The formation of *N*-nitrosoamines on reaction with sodium nitrite and hydrochloric acid is a characteristic reaction of secondary amines. The only C_7H_9N isomer in this problem that is a secondary amine is *N*-methylaniline.

$$C_6H_5NHCH_3 \xrightarrow[\text{HCl, H}_2\text{O}]{\text{NaNO}_2} C_6H_5\underset{\underset{N=O}{|}}{N}CH_3$$

N-Methylaniline   *N*-Methyl-*N*-
nitrosoaniline

(*d*) Ring nitrosation is a characteristic reaction of tertiary arylamines.

Tertiary arylamine   *p*-Nitroso-*N,N*-dialkylaniline

None of the C_7H_9N isomers in this problem is a tertiary amine; hence none will undergo ring nitrosation.

22.27 (*a*) Basicity decreases in proceeding across a row in the periodic table. The increased nuclear charge as one progresses from carbon to nitrogen to oxygen to fluorine causes the electrons to be bound more strongly to the atom and thus less readily shared.

$$H_3\bar{C}: \; > \; H_2\bar{N}: \; > \; H\ddot{O}:^- \; > \; :\ddot{F}:^-$$

Strongest   Weakest
base   base

pK_a
of conjugate   60   36   16   3.4
acid

(*b*) The strongest base in this group is amide ion, H_2N^-, and the weakest base is water, H_2O. Ammonia is a weaker base than hydroxide ion; the equilibrium lies to the left.

$$:NH_3 \; + \; H_2O \; \rightleftharpoons \; \overset{+}{N}H_4 \; + \; OH^-$$

Weaker   Weaker   Stronger   Stronger
base   acid   acid   base

The correct order is

$$H_2\bar{N}: \; > \; H\ddot{O}:^- \; > \; :NH_3 \; > \; H_2\ddot{O}:$$

Strongest   Weakest
base   base

(*c*) These anions can be ranked according to their basicity by considering the respective acidities of their conjugate acids.

Base	Conjugate acid	pK_a of conjugate acid
H_2N^-	NH_3	36
HO^-	H_2O	16
$^-:C{\equiv}N:$	$HC{\equiv}N:$	9.1
$^-O-\overset{+}{N}(=O)O^-$	$HO\overset{+}{N}(=O)O^-$	−1.4

The order of basicities is the opposite of the order of acidities of their conjugate acids.

$$H_2N^- > HO^- > :\bar{C}{\equiv}N: \; > \; NO_3^-$$

Strongest   Weakest
base   base

(*d*) A carbonyl group attached to nitrogen stabilizes its negative charge. The strongest base is the anion that has no carbonyl groups on nitrogen; the weakest base is phthalimide anion, which has two carbonyl groups.

Strongest base   Weakest base

22.28 In evaluating base strengths in each part of this problem, be sure to recall that the strongest base has the weakest conjugate acid (largest pK_a).

(*a*) An alkyl substituent on nitrogen is electron-releasing and base-strengthening; thus methylamine is a stronger base than ammonia. An aryl substituent is electron-withdrawing and base-weakening, and so aniline is a weaker base than ammonia.

$$CH_3NH_2 \quad > \quad NH_3 \quad > \quad C_6H_5NH_2$$

	Methylamine, strongest base:	Ammonia:	Aniline, weakest base:
pK_a of conjugate acid	11.6	9.3	4.6

(*b*) An acetyl group is an electron-withdrawing and base-weakening substituent, especially when bonded directly to nitrogen. Amides are weaker bases than amines, and thus acetanilide is a weaker base than aniline. Alkyl groups are electron-releasing; *N*-methylaniline is a slightly stronger base than aniline.

$$C_6H_5NHCH_3 \quad > \quad C_6H_5NH_2 \quad > \quad C_6H_5NH\overset{\overset{\displaystyle O}{\|}}{C}CH_3$$

	N-methylaniline, strongest base:	Aniline:	Acetanilide, weakest base:
pK_a of conjugate acid	4.9	4.6	−1.0

(*c*) Chlorine substituents are slightly electron-withdrawing, and methyl groups are slightly electron-releasing. 2,4-Dimethylaniline is therefore a stronger base than 2,4-dichloroaniline. Nitro groups are strongly electron-withdrawing, their base-weakening effect being especially pronounced when a nitro group is ortho or para to an amino group because the two groups are then directly conjugated.

	2,4-Dimethylaniline, strongest base:	2,4-Dichloroaniline:	2,4-Dinitroaniline, weakest base:
pK_a of conjugate acid	4.9	2.0	−4.5

(*d*) Nitro groups are more electron-withdrawing than chlorine, and the base-weakening effect of a nitro substituent is greater when it is ortho or para to an amino group than when it is meta to it.

3,4-Dichloroaniline, strongest base:	4-Chloro-3-nitroaniline:	4-Chloro-2-nitroaniline, weakest base:

pK_a of conjugate acid ≈ 3 1.9 -1.0

(*e*) According to the principle applied in part (*a*) (alkyl groups increase basicity, aryl groups decrease it), the order of decreasing basicity is as shown:

$$(CH_3)_2NH \quad > \quad C_6H_5NHCH_3 \quad > \quad (C_6H_5)_2NH$$

Dimethylamine, strongest base:	*N*-Methylaniline:	Diphenylamine, weakest base:

pK_a of conjugate acid   10.7   4.9   0.8

22.29 Nitrogen (a) is the most basic and the most nucleophilic of the three nitrogen atoms of physostigmine and is the one that reacts with methyl iodide.

Physostigmine	Methyl iodide	"Physostigmine methiodide"

The nitrogen that reacts is the one that is a tertiary alkylamine. Of the other two nitrogens, (b) is attached to an aromatic ring and is much less basic and less nucleophilic. The third nitrogen, (c), is an amide nitrogen; amides are less nucleophilic than amines.

22.30 (*a*) Looking at the problem retrosynthetically, it can be seen that a variety of procedures are available for preparing ethylamine from ethanol. The methods by which a primary amine may be prepared include

$CH_3CH_2NH_2$

Gabriel synthesis

$CH_3CH_2N_3$

Reduction of an azide

$$CH_3\overset{\overset{\displaystyle O}{\|}}{C}H$$

Reductive amination

$$CH_3\overset{\overset{\displaystyle O}{\|}}{C}NH_2$$

Reduction of an amide

Two of these methods, the Gabriel synthesis and the preparation and reduction of the corresponding azide, begin with ethyl bromide.

$$CH_3CH_2OH \xrightarrow[\text{or HBr}]{PBr_3} CH_3CH_2Br$$

Ethanol   Ethyl bromide

$$CH_3CH_2Br + \text{[N-Potassiophthalimide]} \longrightarrow \text{[N-Ethylphthalimide]} \xrightarrow{H_2NNH_2} CH_3CH_2NH_2$$

Ethyl bromide   N-Potassiophthalimide   N-Ethylphthalimide   Ethylamine

$$CH_3CH_2Br \xrightarrow{NaN_3} CH_3CH_2N_3 \xrightarrow[\text{2. H}_2\text{O}]{\text{1. LiAlH}_4} CH_3CH_2NH_2$$

Ethyl bromide   Ethyl azide   Ethylamine

To use reductive amination, we must begin with oxidation of ethanol to acetaldehyde.

$$CH_3CH_2OH \xrightarrow[CH_2Cl_2]{\text{PCC or PDC}} CH_3\overset{\overset{\displaystyle O}{\|}}{C}H$$

Ethanol   Acetaldehyde

$$CH_3\overset{\overset{\displaystyle O}{\|}}{C}H \xrightarrow{NH_3,\ H_2,\ Ni} CH_3CH_2NH_2$$

Acetaldehyde   Ethylamine

Another possibility is reduction of acetamide. This requires an initial oxidation of ethanol to acetic acid.

$$CH_3CH_2OH \xrightarrow[\text{H}_2\text{O, heat}]{\text{K}_2\text{Cr}_2\text{O}_7,\ \text{H}_2\text{SO}_4} CH_3CO_2H \xrightarrow[\text{2. NH}_3]{\text{1. SOCl}_2} CH_3\overset{\overset{\text{O}}{\|}}{C}NH_2 \xrightarrow[\text{2. H}_2\text{O}]{\text{1. LiAlH}_4} CH_3CH_2NH_2$$

Ethanol   Acetic acid   Acetamide   Ethylamine

(*b*) Acylation of ethylamine with acetyl chloride, prepared in part (*a*), gives the desired amide.

$$CH_3\overset{\overset{\text{O}}{\|}}{C}Cl\ +\ 2CH_3CH_2NH_2 \longrightarrow CH_3\overset{\overset{\text{O}}{\|}}{C}NHCH_2CH_3\ +\ CH_3CH_2\overset{+}{N}H_3\ Cl^-$$

Acetyl   Ethylamine   *N*-Ethylacetamide   Ethylammonium
chloride   chloride

Excess ethylamine can be allowed to react with the hydrogen chloride formed in the acylation reaction. Alternatively, equimolar amounts of acyl chloride and amine can be used in the presence of aqueous hydroxide as the base.

(*c*) Reduction of the *N*-ethylacetamide prepared in part (*b*) yields diethylamine.

$$CH_3\overset{\overset{\text{O}}{\|}}{C}NHCH_2CH_3 \xrightarrow[\text{2. H}_2\text{O}]{\text{1. LiAlH}_4} CH_3CH_2NHCH_2CH_3$$

N-Ethylacetamide   Diethylamine

Diethylamine can also be prepared by reductive amination of acetaldehyde [from part (*a*)] with ethylamine.

$$CH_3\overset{\overset{\text{O}}{\|}}{C}H\ +\ CH_3CH_2NH_2 \xrightarrow[\text{or NaBH}_3\text{CN}]{\text{H}_2,\ \text{Ni}} CH_3CH_2NHCH_2CH_3$$

Acetaldehyde   Ethylamine   Diethylamine

(*d*) The preparation of *N*,*N*-diethylacetamide is a standard acylation reaction. The reactants, acetyl chloride and diethylamine, have been prepared in previous parts of this problem.

$$CH_3\overset{\overset{\text{O}}{\|}}{C}Cl\ +\ (CH_3CH_2)_2NH \xrightarrow{\text{HO}^-} CH_3\overset{\overset{\text{O}}{\|}}{C}N(CH_2CH_3)_2$$

Acetyl   Diethylamine   *N*,*N*-Diethylacetamide
chloride

(*e*) Triethylamine arises by reduction of *N*,*N*-diethylacetamide or by reductive amination.

$$CH_3\overset{\overset{\text{O}}{\|}}{C}N(CH_2CH_3)_2 \xrightarrow[\text{2. H}_2\text{O}]{\text{1. LiAlH}_4} (CH_3CH_2)_3N$$

N,*N*-Diethylacetamide   Triethylamine

$$CH_3\overset{\overset{\text{O}}{\|}}{C}H\ +\ (CH_3CH_2)_2NH \xrightarrow[\text{NaBH}_3\text{CN}]{\substack{\text{H}_2,\ \text{Ni} \\ \text{or}}} (CH_3CH_2)_3N$$

Acetaldehyde   Diethylamine   Triethylamine

(*f*) Quaternary ammonium halides are formed by reaction of alkyl halides and tertiary amines.

$$CH_3CH_2Br \ + \ (CH_3CH_2)_3N \longrightarrow (CH_3CH_2)_4\overset{+}{N} \ Br^-$$

Ethyl bromide Triethylamine Tetraethylammonium bromide

22.31 (*a*) In this problem a primary alkanamine must be prepared with a carbon chain extended by one carbon. This can be accomplished by way of a nitrile.

$$RCH_2NH_2 \ \Longrightarrow \ RCN \ \Longrightarrow \ RBr \ \Longrightarrow \ ROH$$

$$(R = CH_3CH_2CH_2CH_2{-})$$

The desired reaction sequence is therefore

$$CH_3CH_2CH_2CH_2OH \ \xrightarrow[\substack{\text{or} \\ \text{HBr}}]{PBr_3} \ CH_3CH_2CH_2CH_2Br \ \xrightarrow{NaCN} \ CH_3CH_2CH_2CH_2CN$$

1-Butanol Butyl bromide Pentanenitrile

$$\Big\downarrow \begin{array}{l} \text{1. LiAlH}_4 \\ \text{2. H}_2\text{O} \end{array}$$

$$CH_3CH_2CH_2CH_2CH_2NH_2$$

1-Pentanamine

(*b*) The carbon chain of *tert*-butyl chloride cannot be extended by a nucleophilic substitution reaction; the S$_N$2 reaction that would be required on the tertiary halide would not work. The sequence employed in part (*a*) is therefore not effective in this case. The best route is carboxylation of the Grignard reagent and subsequent conversion of the corresponding amide to the desired primary amine product.

$$(CH_3)_3CCH_2NH_2 \ \Longrightarrow \ (CH_3)_3C\overset{\overset{\displaystyle O}{\|}}{C}NH_2 \ \Longrightarrow \ (CH_3)_3CCO_2H \ \Longrightarrow \ (CH_3)_3CCl$$

The reaction sequence to be used is

$$(CH_3)_3CCl \ \xrightarrow[\substack{\text{2. CO}_2 \\ \text{3. H}_3\text{O}^+}]{\text{1. Mg, diethyl ether}} \ (CH_3)_3CCO_2H$$

tert-Butyl chloride 2,2-Dimethylpropanoic acid

Once the carboxylic acid has been obtained, it is converted to the desired amine by reduction of the corresponding amide.

$$(CH_3)_3CCO_2H \ \xrightarrow[\text{2. NH}_3]{\text{1. SOCl}_2} \ (CH_3)_3C\overset{\overset{\displaystyle O}{\|}}{C}NH_2 \ \xrightarrow[\text{2. H}_2\text{O}]{\text{1. LiAlH}_4} \ (CH_3)_3CCH_2NH_2$$

2,2-Dimethylpropanoic acid 2,2-Dimethylpropanamide 2,2-Dimethyl-1-propanamine

(c) Oxidation of cyclohexanol to cyclohexanone gives a compound suitable for reductive amination.

Cyclohexanol   →   Cyclohexanone   →   N-Methylcyclohexylamine

(d) The desired product is the reduction product of the cyanohydrin of acetone.

Acetone
cyanohydrin

1-Amino-2-methyl-
2-propanol

The cyanohydrin is made from acetone in the usual way. Acetone is available by oxidation of isopropyl alcohol.

Isopropyl
alcohol

Acetone

Acetone
cyanohydrin

(e) The target amino alcohol is the product of nucleophilic ring opening of 1,2-epoxypropane by ammonia. Ammonia attacks the less hindered carbon of the epoxide function.

1,2-Epoxypropane

1-Amino-2-propanol

The necessary epoxide is formed by epoxidation of propene.

Isopropyl
alcohol

Propene

1,2-Epoxypropane

(f) The reaction sequence is the same as in part (e) except that dimethylamine is used as the nucleophile instead of ammonia.

1,2-Epoxypropane
[prepared as in part (e)]

Dimethylamine

1-(N,N-Dimethylamino)-
2-propanol

(g) The key to performing this synthesis is recognition of the starting material as an acetal of acetophenone. Acetals may be hydrolyzed to carbonyl compounds.

2-Methyl-2-phenyl-
1,3-dioxolane
        Acetophenone   1,2-Ethanediol

Once acetophenone has been obtained, it may be converted to the required product by reductive amination.

    Acetophenone   Piperidine   N-(1-Phenylethyl)-
                                  piperidine

22.32 (a) The reaction of alkyl halides with N-potassiophthalimide (the first step in the Gabriel synthesis of amines) is a nucleophilic substitution reaction. Alkyl bromides are more reactive than alkyl fluorides; that is, bromide is a better leaving group than fluoride.

N-Potassiophthalimide   1-Bromo-2-
                         fluoroethane   2-Phthalimidoethyl fluoride

(b) In this example one bromine is attached to a primary and the other to a secondary carbon. Phthalimide anion is a good nucleophile and reacts with alkyl halides by the S_N2 mechanism. It attacks the less hindered primary carbon.

1,4-Dibromopentane

N-(4-Bromopentyl)phthalimide
(only product, 67% yield)

(*c*) Both bromines are bonded to primary carbons, but branching at the adjacent carbon hinders nucleophilic attack at one of them.

N-(4-Bromo-3,3-dimethylbutyl)phthalimide
(only product, 53% yield)

1,4-Dibromo-2,2-dimethylbutane

22.33 (*a*) Amines are basic and are protonated by hydrogen halides.

$$C_6H_5CH_2NH_2 \ + \ HBr \longrightarrow C_6H_5CH_2\overset{+}{N}H_3 \ \ Br^-$$

Benzylamine

Benzylammonium
bromide

(*b*) Equimolar amounts of benzylamine and sulfuric acid yield benzylammonium hydrogen sulfate as the product.

$$C_6H_5CH_2NH_2 \ + \ HOSO_2OH \longrightarrow C_6H_5CH_2\overset{+}{N}H_3 \ \ ^-OSO_2OH$$

Benzylamine   Sulfuric acid

Benzylammonium hydrogen
sulfate

(*c*) Acetic acid transfers a proton to benzylamine.

$$C_6H_5CH_2NH_2 \ + \ CH_3\overset{O}{\overset{\|}{C}}OH \longrightarrow C_6H_5CH_2\overset{+}{N}H_3 \ \ ^-O\overset{O}{\overset{\|}{C}}CH_3$$

Benzylamine   Acetic acid

Benzylammonium acetate

(*d*) Acetyl chloride reacts with benzylamine to form an amide.

$$2C_6H_5CH_2NH_2 \ + \ CH_3\overset{O}{\overset{\|}{C}}Cl \longrightarrow CH_3\overset{O}{\overset{\|}{C}}NHCH_2C_6H_5 \ + \ C_6H_5CH_2\overset{+}{N}H_3 \ \ Cl^-$$

Benzylamine   Acetyl
chloride

N-Benzylacetamide

Benzylammonium
chloride

(*e*) Acetic anhydride also gives an amide with benzylamine.

$$2C_6H_5CH_2NH_2 \ + \ CH_3\overset{O}{\overset{\|}{C}}O\overset{O}{\overset{\|}{C}}CH_3 \longrightarrow CH_3\overset{O}{\overset{\|}{C}}NHCH_2C_6H_5 \ + \ C_6H_5CH_2\overset{+}{N}H_3 \ \ ^-O\overset{O}{\overset{\|}{C}}CH_3$$

Benzylamine   Acetic anhydride

N-Benzylacetamide

Benzylammonium acetate

(*f*) Primary amines react with ketones to give imines.

$$C_6H_5CH_2NH_2 \ + \ CH_3\overset{O}{\overset{\|}{C}}CH_3 \longrightarrow (CH_3)_2C{=}NCH_2C_6H_5$$

Benzylamine   Acetone

N-Isopropylidenebenzylamine

(*g*) These reaction conditions lead to reduction of the imine formed in part (*f*). The overall reaction is reductive amination.

$$C_6H_5CH_2NH_2 \ + \ CH_3\overset{\overset{\displaystyle O}{\|}}{C}CH_3 \xrightarrow{\ H_2, \ Ni\ } (CH_3)_2CHNHCH_2C_6H_5$$

Benzylamine   Acetone   *N*-Isopropylbenzylamine

(*h*) Amines are nucleophilic and bring about the opening of epoxide rings.

$$C_6H_5CH_2NH_2 \ + \ H_2C\!\!-\!\!CH_2 \ \longrightarrow \ C_6H_5CH_2NHCH_2CH_2OH$$
$$\overset{}{\underset{O}{\diagdown\!\diagup}}$$

Benzylamine   Ethylene oxide   2-(*N*-Benzylamino)ethanol

(*i*) In these nucleophilic ring-opening reactions the amine attacks the less sterically hindered carbon of the ring.

$$C_6H_5CH_2NH_2 \ + \ H_2C\!\!-\!\!CHCH_3 \ \longrightarrow \ C_6H_5CH_2NHCH_2\underset{\underset{\textstyle OH}{|}}{C}HCH_3$$
$$\overset{}{\underset{O}{\diagdown\!\diagup}}$$

Benzylamine   1,2-Epoxypropane   1-(*N*-Benzylamino)-2-propanol

(*j*) With excess methyl iodide, amines are converted to quaternary ammonium iodides.

$$C_6H_5CH_2NH_2 \ + \ 3CH_3I \ \longrightarrow \ C_6H_5CH_2\overset{+}{N}(CH_3)_3 \ I^-$$

Benzylamine   Methyl iodide   Benzyltrimethylammonium iodide

(*k*) Nitrous acid forms from sodium nitrite in dilute hydrochloric acid. Nitrosation of benzylamine in water gives benzyl alcohol via a diazonium ion intermediate.

$$C_6H_5CH_2NH_2 \ \xrightarrow[\ H_2O\ ]{NaNO_2, \ HCl} \ C_6H_5CH_2\overset{+}{N}{\equiv}N \ \xrightarrow[\ H_2O\ ]{-N_2} \ C_6H_5CH_2OH$$

Benzylamine   Benzyldiazonium ion   Benzyl alcohol

22.34 (*a*) Aniline is a weak base and yields a salt on reaction with hydrogen bromide.

$$C_6H_5NH_2 \ + \ HBr \ \longrightarrow \ C_6H_5\overset{+}{N}H_3 \ Br^-$$

Aniline   Hydrogen bromide   Anilinium bromide

(*b*) Aniline acts as a nucleophile toward methyl iodide. With excess methyl iodide, a quaternary ammonium salt is formed.

$$C_6H_5NH_2 \ + \ 3CH_3I \ \longrightarrow \ C_6H_5\overset{+}{N}(CH_3)_3 \ I^-$$

Aniline   Methyl iodide   *N,N,N*-Trimethylanilinium iodide

(*c*) Aniline is a primary amine and undergoes nucleophilic addition to aldehydes and ketones to form imines.

$$C_6H_5NH_2 \ + \ CH_3\overset{\overset{\displaystyle O}{\|}}{C}H \ \longrightarrow \ C_6H_5N{=}CHCH_3 \ + \ H_2O$$

Aniline   Acetaldehyde   *N*-Phenylacetaldimine   Water (*d*) When an imine is formed in the presence of hydrogen and a suitable catalyst, reductive amination occurs to give an amine.

$$C_6H_5NH_2 \ + \ CH_3\overset{\overset{O}{\|}}{C}H \ \xrightarrow{\text{H}_2,\,\text{Ni}} \ C_6H_5NHCH_2CH_3$$

| Aniline | Acetaldehyde | *N*-Ethylaniline |

(*e*) Aniline undergoes *N*-acylation on treatment with carboxylic acid anhydrides.

$$2C_6H_5NH_2 \ + \ CH_3\overset{\overset{O}{\|}}{C}O\overset{\overset{O}{\|}}{C}CH_3 \ \longrightarrow \ C_6H_5NH\overset{\overset{O}{\|}}{C}CH_3 \ + \ C_6H_5\overset{+}{N}H_3 \ {}^{-}O\overset{\overset{O}{\|}}{C}CH_3$$

Aniline　　　　Acetic anhydride　　　　Acetanilide　　　　Anilinium acetate

(*f*) Acyl chlorides bring about *N*-acylation of arylamines.

$$2C_6H_5NH_2 \ + \ C_6H_5\overset{\overset{O}{\|}}{C}Cl \ \longrightarrow \ C_6H_5NH\overset{\overset{O}{\|}}{C}C_6H_5 \ + \ C_6H_5\overset{+}{N}H_3 \ Cl^{-}$$

Aniline　　Benzoyl　　　　Benzanilide　　　Anilinium
　　　　　chloride　　　　　　　　　　　　chloride

(*g*) Nitrosation of primary arylamines yields aryl diazonium salts.

$$C_6H_5NH_2 \ \xrightarrow[\text{H}_2\text{O, 0--5°C}]{\text{NaNO}_2,\,\text{H}_2\text{SO}_4} \ C_6H_5\overset{+}{N}\!\equiv\!N\!:\ HSO_4^{-}$$

Aniline　　　　　　　　　　　Benzenediazonium
　　　　　　　　　　　　　　hydrogen sulfate

The replacement reactions that can be achieved by using diazonium salts are illustrated in parts (*h*) through (*n*). In all cases molecular nitrogen is lost from the ring carbon to which it was attached and is replaced by another substituent.

$$C_6H_5\overset{+}{N}\!\equiv\!N\!:\ HSO_4^{-}$$

Benzenediazonium
hydrogen sulfate

(*h*) $\xrightarrow[\text{heat}]{\text{H}^+,\,\text{H}_2\text{O}}$ C$_6$H$_5$OH Phenol

(*i*) $\xrightarrow{\text{CuCl}}$ C$_6$H$_5$Cl Chlorobenzene

(*j*) $\xrightarrow{\text{CuBr}}$ C$_6$H$_5$Br Bromobenzene

(*k*) $\xrightarrow{\text{CuCN}}$ C$_6$H$_5$CN Benzonitrile

(*l*) $\xrightarrow{\text{H}_3\text{PO}_2}$ C$_6$H$_6$ Benzene

(*m*) $\xrightarrow{\text{KI}}$ C$_6$H$_5$I Iodobenzene

(*n*) $\xrightarrow[\text{2. heat}]{\text{1. HBF}_4}$ C$_6$H$_5$F Fluorobenzene

(o) The nitrogens of an aryl diazonium salt are retained on reaction with the electron-rich ring of a phenol. Azo coupling occurs.

$$C_6H_5\overset{+}{N}\equiv N\colon \; HSO_4^- \; + \; C_6H_5OH \longrightarrow C_6H_5N=N-\!\!\!\left\langle\right\rangle\!\!\!-OH$$

<table>
<tr><td>Benzenediazonium
hydrogen sulfate</td><td>Phenol</td><td>p-(Azophenyl)phenol</td></tr>
</table>

(p) Azo coupling occurs when aryl diazonium salts react with N,N-dialkylarylamines.

$$C_6H_5\overset{+}{N}\equiv N\colon \; HSO_4^- \; + \; C_6H_5N(CH_3)_2 \longrightarrow C_6H_5N=N-\!\!\!\left\langle\right\rangle\!\!\!-N(CH_3)_2$$

<table>
<tr><td>Benzenediazonium
hydrogen sulfate</td><td>N,N-Dimethylaniline</td><td>p-(Azophenyl)-N,N-dimethylaniline</td></tr>
</table>

22.35 (a) Amides are reduced to amines by lithium aluminum hydride.

$$\underset{\text{Acetanilide}}{C_6H_5NH\overset{\overset{\textstyle O}{\|}}{C}CH_3} \xrightarrow[\text{2. H}_2\text{O}]{\text{1. LiAlH}_4\text{, diethyl ether}} \underset{N\text{-Ethylaniline}}{C_6H_5NHCH_2CH_3}$$

(b) Acetanilide is strongly activated toward electrophilic aromatic substitution. An acetamido group is ortho, para-directing.

<table>
<tr><td>Acetanilide</td><td>o-Nitroacetanilide</td><td>p-Nitroacetanilide</td></tr>
</table>

(c) Sulfonation of the ring occurs.

<table>
<tr><td>Acetanilide</td><td>p-Acetamidobenzenesulfonic acid</td></tr>
</table>

(*d*) Bromination of the ring takes place.

$$C_6H_5\overset{\overset{\displaystyle O}{\|}}{N}HCCH_3 \quad \xrightarrow[\text{acetic acid}]{Br_2} \quad$$

+ ortho isomer

Acetanilide   *p*-Bromoacetanilide

(*e*) Acetanilide undergoes Friedel–Crafts alkylation readily.

$$C_6H_5\overset{\overset{\displaystyle O}{\|}}{N}HCCH_3 \; + \; (CH_3)_3CCl \quad \xrightarrow{AlCl_3} \quad$$

+ ortho isomer

Acetanilide   *tert*-Butyl   *p-tert*-Butylacetanilide
                  chloride

(*f*) Friedel–Crafts acylation also is easily carried out.

$$C_6H_5\overset{\overset{\displaystyle O}{\|}}{N}HCCH_3 \; + \; CH_3\overset{\overset{\displaystyle O}{\|}}{C}Cl \quad \xrightarrow{AlCl_3} \quad$$

+ ortho isomer

Acetanilide   Acetyl chloride   *p*-Acetamidoacetophenone

(*g*) Acetanilide is an amide and can be hydrolyzed when heated with aqueous acid. Under acidic conditions the aniline that is formed exists in its protonated form as the anilinium cation.

$$C_6H_5\overset{\overset{\displaystyle O}{\|}}{N}HCCH_3 \; + \; H_2O \; + \; HCl \quad \longrightarrow \quad C_6H_5\overset{+}{N}H_3 \; Cl^- \; + \; HO\overset{\overset{\displaystyle O}{\|}}{C}CH_3$$

Acetanilide   Water   Hydrogen   Anilinium   Acetic acid
                      chloride   chloride

(*h*) Amides are hydrolyzed in base.

$$C_6H_5\overset{\overset{\displaystyle O}{\|}}{N}HCCH_3 \; + \; NaOH \quad \xrightarrow{H_2O} \quad C_6H_5NH_2 \; + \; Na^+ \; {}^-O\overset{\overset{\displaystyle O}{\|}}{C}CH_3$$

Acetanilide   Sodium   Aniline   Sodium acetate
                   hydroxide

22.36 (*a*) The reaction illustrates the preparation of a secondary amine by reductive amination.

Cyclohexanone   Cyclohexylamine   Dicyclohexylamine (70%)

(*b*) Amides are reduced to amines by lithium aluminum hydride.

6-Ethyl-6-
azabicyclo[3.2.1]octan-7-one

6-Ethyl-6-
azabicyclo[3.2.1]octane

(*c*) Treatment of alcohols with *p*-toluenesulfonyl chloride converts them to *p*-toluenesulfonates.

3-Phenyl-1-propanol   *p*-Toluenesulfonyl chloride   3-Phenylpropyl *p*-toluenesulfonate

p-Toluenesulfonate is an excellent leaving group in nucleophilic substitution reactions. Dimethylamine is the nucleophile.

$C_6H_5CH_2CH_2CH_2OSO_2$—⟨ ⟩—CH_3 + $(CH_3)_2NH$ ⟶ $C_6H_5CH_2CH_2CH_2N(CH_3)_2$

3-Phenylpropyl *p*-toluenesulfonate   Dimethylamine   *N,N*-Dimethyl-3-phenyl-
1-propanamine (86%)

(*d*) Amines are sufficiently nucleophilic to react with epoxides. Attack occurs at the less substituted carbon of the epoxide.

2-(2,5-Dimethoxyphenyl)oxirane   Isopropylamine   1-(2,5-Dimethoxyphenyl)-2-
(*N*-isopropylamino)ethanol (67%)

(*e*) α-Halo ketones are reactive substrates in nucleophilic substitution reactions. Dibenzylamine is the nucleophile.

Dibenzylamine   1-Chloro-2-propanone   1-(Dibenzylamino)-2-
propanone (87%)

Because the reaction liberates hydrogen chloride, it is carried out in the presence of added base—in this case triethylamine—so as to avoid converting the dibenzylamine to its hydrochloride salt.

(*f*) Quaternary ammonium hydroxides undergo Hofmann elimination when they are heated. A point to be considered here concerns the regioselectivity of Hofmann eliminations: it is the less hindered β proton that is removed by the base giving the less substituted alkene.

trans-1-Isopropenyl-4-methylcyclohexane (98%)   Trimethylamine

Elimination to give   [structure]   does not occur.

(*g*) The combination of sodium nitrite and aqueous acid is a nitrosating agent. Secondary alkylamines react with nitrosating agents to give *N*-nitroso amines as the isolated products.

$$(CH_3)_2CHNHCH(CH_3)_2 \xrightarrow[\text{HCl, H}_2\text{O}]{\text{NaNO}_2} (CH_3)_2CHNCH(CH_3)_2$$

Diisopropylamine   *N*-Nitrosodiisopropylamine (91%)

22.37 (*a*) Catalytic hydrogenation reduces nitro groups to amino groups.

1,2-Diethyl-4-nitrobenzene   3,4-Diethylaniline (93–99%)

(*b*) Nitro groups are readily reduced by tin(II) chloride.

1,3-Dimethyl-2-nitrobenzene   2,6-Dimethylaniline

This reaction is the first step in a synthesis of the drug *lidocaine*.

(*c*) The amino group of arylamines is nucleophilic and undergoes acylation on reaction with chloroacetyl chloride. Chloroacetyl chloride is a difunctional compound—it is both an acyl chloride and an alkyl chloride. Acyl chlorides react with nucleophiles faster than do alkyl chlorides. Therefore, acylation of the amine nitrogen occurs rather than alkylation.

2,6-Dimethylaniline   Chloroacetyl chloride   *N*-(Chloroacetyl)-2,6-dimethylaniline (*d*) The final step in the synthesis of lidocaine is displacement of the chloride by diethylamine from the α-halo amide formed in part (*c*) in a nucleophilic substitution reaction.

N-(Chloroacetyl)-   Diethylamine   Lidocaine
2,6-dimethylaniline

The reaction is carried out with excess diethylamine, which acts as a base to neutralize the hydrogen chloride formed.

(*e*) For use as an anesthetic, lidocaine is made available as its hydrochloride salt. Of the two nitrogens in lidocaine, the amine nitrogen is more basic than the amide.

Lidocaine   Lidocaine hydrochloride

(*f*) Lithium aluminum hydride reduction of amides is one of the best methods for the preparation of amines, including arylamines.

$$C_6H_5NHCCH_2CH_2CH_3 \xrightarrow[\text{2. H}_2\text{O}]{\text{1. LiAlH}_4} C_6H_5NHCH_2CH_2CH_2CH_3$$

N-Phenylbutanamide   N-Butylaniline (92%)

(*g*) Arylamines react with aldehydes and ketones in the presence of hydrogen and nickel to give the product of reductive amination.

$$C_6H_5NH_2 + CH_3(CH_2)_5CH \xrightarrow{\text{H}_2, \text{Ni}} C_6H_5NHCH_2(CH_2)_5CH_3$$

Aniline   Heptanal   N-Heptylaniline (65%)

(*h*) Acetanilide is activated toward electrophilic aromatic substitution. On reaction with chloroacetyl chloride, it undergoes Friedel–Crafts acylation, primarily at its para position.

Acetanilide   Chloroacetyl   p-Acetamidophenacyl chloride
                     chloride   (79–83%)

Acylation, rather than alkylation, occurs. Acyl chlorides are more reactive than alkyl chlorides toward electrophilic aromatic substitution reactions as a result of the more stable intermediate (acylium ion) formed.

(*i*) Reduction with iron in hydrochloric acid is one of the most common methods for converting nitroarenes to arylamines.

4-Bromo-4′-nitrobiphenyl   4-Amino-4′-bromobiphenyl (94%)

(*j*) Primary arylamines are converted to aryl diazonium salts on treatment with sodium nitrite in aqueous acid. When the aqueous acidic solution containing the diazonium salt is heated, a phenol is formed.

4-Amino-4′-bromobiphenyl

H_2O, heat
$(-N_2)$

4-Bromo-4′-hydroxybiphenyl (85%)

(*k*) This problem illustrates the conversion of an arylamine to an aryl chloride by the Sandmeyer reaction.

2,6-Dinitroaniline   2-Chloro-1,3-dinitrobenzene (71–74%)

(*l*) Diazotization of primary arylamines followed by treatment with copper(I) bromide converts them to aryl bromides.

m-Bromoaniline   *m*-Dibromobenzene (80–87%)

(*m*) Nitriles are formed when aryl diazonium salts react with copper(I) cyanide.

o-Nitroaniline   *o*-Nitrobenzonitrile (87%)

(*n*) An aryl diazonium salt is converted to an aryl iodide on reaction with potassium iodide.

2,6-Diiodo-
4-nitroaniline

1,2,3-Triiodo-
5-nitrobenzene
(94–95%)

(*o*) Aryl diazonium fluoroborates are converted to aryl fluorides when heated. Both diazonium salt functions in the starting material undergo this reaction.

4,4′-Bis(diazonio)biphenyl fluoroborate

4,4′-Difluorobiphenyl (82%)

(*p*) Hypophosphorous acid (H_3PO_2) reduces aryl diazonium salts to arenes.

2,4,6-Trinitroaniline

1,3,5-Trinitrobenzene
(60–65%)

(*q*) Ethanol, like hypophosphorous acid, is an effective reagent for the reduction of aryl diazonium salts.

2-Amino-5-
iodobenzoic acid

m-Iodobenzoic acid
(86–93%)

(*r*) Diazotization of aniline followed by addition of a phenol yields a bright-red diazo-substituted phenol. The diazonium ion acts as an electrophile toward the activated aromatic ring of the phenol.

Aniline

Benzenediazonium
hydrogen sulfate

2,3,6-Trimethyl-4-
(phenylazo)phenol (98%)

(*s*) Nitrosation of *N,N*-dialkylarylamines takes place on the ring at the position para to the dialkylamino group.

N,N-Dimethyl-*m*-toluidine

3-Methyl-4-nitroso-*N,N*-dimethylaniline (83%)

22.38 (*a*) 4-Methylpiperidine can participate in intermolecular hydrogen bonding in the liquid phase.

These hydrogen bonds must be broken for individual 4-methylpiperidine molecules to escape into the gas phase. *N*-Methylpiperidine lacks a proton bonded to nitrogen and so cannot engage in intermolecular hydrogen bonding. Less energy is required to transfer a molecule of *N*-methyl-piperidine to the gaseous state, and therefore it has a lower boiling point than 4-methylpiperidine.

N-Methylpiperidine;
no hydrogen bonding possible
to other *N*-methylpiperidine molecules

(*b*) The two products are diastereomeric quaternary ammonium chlorides that differ in the configuration at the nitrogen atom.

4-*tert*-Butyl-*N*-
methylpiperidine

(c) Tetramethylammonium hydroxide cannot undergo Hofmann elimination. The only reaction that can take place is nucleophilic substitution.

| Tetramethylammonium hydroxide | | Trimethylamine | Methanol |

(d) The key intermediate in the reaction of an amine with nitrous acid is the corresponding diazonium ion.

$$CH_3CH_2CH_2NH_2 \xrightarrow[\text{H}_2\text{O}]{\text{NaNO}_2, \text{HCl}} CH_3CH_2CH_2-\overset{+}{N}\equiv N:$$

1-Propanamine   Propyldiazonium ion

Loss of nitrogen from this diazonium ion is accompanied by a hydride shift to form a secondary carbocation.

Propyldiazonium ion   Isopropyl cation   Nitrogen

Capture of isopropyl cation by water yields the major product of the reaction, 2-propanol.

Isopropyl cation   Water   2-Propanol

22.39 Alcohols are converted to *p*-toluenesulfonates by reaction with *p*-toluenesulfonyl chloride. None of the bonds to the chirality center is affected in this reaction.

(*S*)-2-Octanol   *p*-Toluenesulfonyl chloride   (*S*)-1-Methylheptyl *p*-toluenesulfonate (compound A)

Displacement of the *p*-toluenesulfonate leaving group by sodium azide is an S_N2 process and proceeds with inversion of configuration.

(*S*)-1-Methylheptyl *p*-toluenesulfonate (compound A)   (*R*)-1-Methylheptyl azide (compound B)

Reduction of the azide yields a primary amine. A nitrogen–nitrogen bond is cleaved; all the bonds to the chirality center remain intact.

$$\overset{..}{:}\overset{+}{N}=\overset{..}{N}=\overset{..}{N}-\overset{H}{\underset{CH_3}{C}}(CH_2)_5CH_3 \quad \xrightarrow[\text{2. H}_2\text{O, HO}^-]{\text{1. LiAlH}_4} \quad H_2\overset{..}{N}-\overset{H}{\underset{CH_3}{C}}(CH_2)_5CH_3$$

(*R*)-1-Methylheptyl azide   (*R*)-2-Octanamine
(compound B)   (compound C)

22.40 (*a*) The overall transformation can be expressed as RBr → RCH$_2$NH$_2$. In many cases this can be carried out via a nitrile, as RBr → RCN → RCH$_2$NH$_2$. In this case, however, the substrate is 1-bromo-2,2-dimethylpropane, an alkyl halide that reacts very slowly in nucleophilic substitution. Carbon–carbon bond formation with 1-bromo-2,2-dimethylpropane can be achieved more effectively by carboxylation of the corresponding Grignard reagent.

$$(CH_3)_3CCH_2Br \quad \xrightarrow[\substack{\text{2. CO}_2 \\ \text{3. H}_3\text{O}^+}]{\text{1. Mg}} \quad (CH_3)_3CCH_2CO_2H$$

1-Bromo-2,2-   3,3-Dimethylbutanoic
dimethylpropane   acid (63%)

The carboxylic acid can then be converted to the desired amine by reduction of the derived amide.

$$(CH_3)_3CCH_2CO_2H \quad \xrightarrow[\text{2. NH}_3]{\text{1. SOCl}_2} \quad (CH_3)_3CCH_2\overset{O}{\overset{||}{C}}NH_2 \quad \xrightarrow[\text{2. H}_2\text{O}]{\text{1. LiAlH}_4} \quad (CH_3)_3CCH_2CH_2NH_2$$

3,3-Dimethylbutanoic   3,3-Dimethylbutanamide   3,3-Dimethyl-1-butanamine
acid   (51%)   (57%)

The yields listed in parentheses are those reported in the chemical literature for this synthesis.

(*b*) Consider the starting materials in relation to the desired product.

$$H_2C=CH(CH_2)_8CH_2\text{—N}\bigcirc \quad \Longrightarrow \quad H_2C=CH(CH_2)_8\overset{O}{\overset{||}{C}}OH \quad + \quad \underset{\underset{H}{N}}{\bigcirc}$$

N-(10-Undecenyl)pyrrolidine   10-Undecenoic acid   Pyrrolidine

The synthetic tasks are to form the necessary carbon–nitrogen bond and to reduce the carbonyl group to a methylene group. This has been accomplished by way of the amide as a key intermediate.

$$H_2C=CH(CH_2)_8\overset{O}{\overset{||}{C}}OH \quad \xrightarrow[\text{2. pyrrolidine}]{\text{1. SOCl}_2} \quad H_2C=CH(CH_2)_8\overset{O}{\overset{||}{C}}\text{—N}\bigcirc \quad \xrightarrow[\text{2. H}_2\text{O}]{\text{1. LiAlH}_4} \quad H_2C=CH(CH_2)_8CH_2\text{—N}\bigcirc$$

10-Undecenoic acid   *N*-(10-Undecenoyl)pyrrolidine (75%)   *N*-(10-Undecenyl)pyrrolidine (66%)

A second approach converts the starting carboxylic acid to an aldehyde followed by reductive amination.

$$H_2C=CH(CH_2)_8\overset{O}{\overset{||}{C}}OH \quad \xrightarrow[\text{2. H}_2\text{O}]{\text{1. LiAlH}_4} \quad H_2C=CH(CH_2)_8CH_2OH \quad \xrightarrow[\text{CH}_2\text{Cl}_2]{\text{PCC or PDC}} \quad H_2C=CH(CH_2)_8\overset{O}{\overset{||}{C}}H$$

10-Undecenoic acid   10-Undecen-1-ol   10-Undecenal

The reducing agent in the reductive amination process cannot be hydrogen, because that would result in hydrogenation of the double bond. Sodium cyanoborohydride is required.

| 10-Undecenal | Pyrrolidine | | N-(10-Undecenyl)pyrrolidine |

(*c*) It is stereochemistry that determines the choice of which synthetic method to employ in introducing the amine group. The carbon–nitrogen bond must be formed with inversion of configuration at the alcohol carbon. Conversion of the alcohol to its *p*-toluenesulfonate ensures that the leaving group is introduced with exactly the same stereochemistry as the alcohol.

| *cis*-2-Phenoxycyclo-pentanol | *p*-Toluenesulfonyl chloride | | *cis*-2-Phenoxycyclopentyl *p*-toluenesulfonate |

Once the leaving group has been introduced with the proper stereochemistry, it can be displaced by a nitrogen nucleophile suitable for subsequent conversion to an amine.

| *cis*-2-Phenoxycyclopentyl *p*-toluenesulfonate | *trans*-2-Phenoxycyclopentyl azide (90%) | *trans*-2-Phenoxy-cyclopentylamine |

(As actually reported, the azide was reduced by hydrogenation over a palladium catalyst, and the amine was isolated as its hydrochloride salt in 66% yield.)

(*d*) Recognition that the primary amine is derivable from the corresponding nitrile by reduction,

$$\text{C}_6\text{H}_5\text{CH}_2\text{NCH}_2\text{CH}_2\text{CH}_2\text{CH}_2\text{NH}_2 \quad \Longrightarrow \quad \text{C}_6\text{H}_5\text{CH}_2\text{NCH}_2\text{CH}_2\text{CH}_2\text{C}\equiv\text{N}$$
$$\qquad\quad | \qquad\qquad\qquad\qquad\qquad\qquad\qquad\qquad | $$
$$\qquad\quad \text{CH}_3 \qquad\qquad\qquad\qquad\qquad\qquad\qquad \text{CH}_3$$

and that the necessary tertiary amine function can be introduced by a nucleophilic substitution reaction between the two given starting materials suggests the following synthesis.

$$\underset{\overset{|}{\text{CH}_3}}{\text{C}_6\text{H}_5\text{CH}_2\text{NH}} + \text{BrCH}_2\text{CH}_2\text{CH}_2\text{CN} \longrightarrow \underset{\overset{|}{\text{CH}_3}}{\text{C}_6\text{H}_5\text{CH}_2\text{NCH}_2\text{CH}_2\text{CH}_2\text{CN}} \xrightarrow[\text{2. H}_2\text{O}]{\text{1. LiAlH}_4} \underset{\overset{|}{\text{CH}_3}}{\text{C}_6\text{H}_5\text{CH}_2\text{NCH}_2\text{CH}_2\text{CH}_2\text{CH}_2\text{NH}_2}$$

N-Methylbenzylamine   4-Bromobutanenitrile   *N*-Benzyl-*N*-methyl-1,4-butanediamine

Alkylation of *N*-methylbenzylamine with 4-bromobutanenitrile has been achieved in 92% yield in the presence of potassium carbonate as a weak base to neutralize the hydrogen bromide produced. The nitrile may be reduced with lithium aluminum hydride, as shown in the equation, or by catalytic hydrogenation. Catalytic hydrogenation over platinum gave the desired diamine, isolated as its hydrochloride salt, in 90% yield.

(*e*) The overall transformation may be viewed retrosynthetically as follows:

$$ArCH_2N(CH_3)_2 \implies ArCH_2Br \implies ArCH_3 \qquad Ar = NC\!-\!\!\bigcirc\!\!-$$

The sequence that presents itself begins with benzylic bromination with *N*-bromosuccinimide.

$$NC\!-\!\!\bigcirc\!\!-CH_3 \xrightarrow[\substack{\text{benzoyl peroxide, CCl}_4 \\ \text{heat}}]{\text{N-bromosuccinimide}} NC\!-\!\!\bigcirc\!\!-CH_2Br$$

　　　　p-Cyanotoluene　　　　　　　　　　　　　　　　　*p*-Cyanobenzyl bromide

The reaction shown in the equation has been reported in the chemical literature and gave the benzylic bromide in 60% yield. Treatment of this bromide with dimethylamine gives the desired product. (The isolated yield was 83% by this method.)

$$NC\!-\!\!\bigcirc\!\!-CH_2Br \;+\; (CH_3)_2NH \longrightarrow NC\!-\!\!\bigcirc\!\!-CH_2N(CH_3)_2$$

　　p-Cyanobenzyl bromide　　　　Dimethylamine　　　　　　　*p*-Cyano-*N,N*-dimethylbenzylamine

22.41 (*a*) This problem illustrates the application of the Sandmeyer reaction to the preparation of aryl cyanides. Diazotization of *p*-nitroaniline followed by treatment with copper(I) cyanide converts it to *p*-nitrobenzonitrile.

　　　p-Nitroaniline　　　　　　　　　　　　*p*-Nitrobenzonitrile

(*b*) An acceptable pathway becomes apparent when it is realized that the amino group in the product is derived from the nitro group of the starting material. Two chlorines are introduced by electrophilic aromatic substitution, the third by a Sandmeyer reaction.

Two of the required chlorine atoms can be introduced by chlorination of the starting material, *p*-nitroaniline.

　　　p-Nitroaniline　　　　　　　　　　2,6-Dichloro-4-
　　　　　　　　　　　　　　　　　　　　　　　nitroaniline

The third chlorine can be introduced via the Sandmeyer reaction. Reduction of the nitro group completes the synthesis of 3,4,5-trichloroaniline.

2,6-Dichloro-4-
nitroaniline

1. NaNO$_2$, HCl, H$_2$O
2. CuCl

1,2,3-Trichloro-5-
nitrobenzene

reduce

3,4,5-Trichloroaniline

The reduction step has been carried out by hydrogenation with a nickel catalyst in 70% yield.

(c) The amino group that is present in the starting material facilitates the introduction of the bromine substituents, and is then removed by reductive deamination.

p-Nitroaniline

Br$_2$
acetic acid

2,6-Dibromo-4-
nitroaniline
(95%)

1. NaNO$_2$, H$_2$O, H$^+$
2. ethanol

1,3-Dibromo-5-
nitrobenzene
(70%)

Hypophosphorous acid has also been used successfully in the reductive deamination step.

(d) Reduction of the nitro group of the 1,3-dibromo-5-nitrobenzene prepared in the preceding part of this problem gives the desired product. The customary reducing agents used for the reduction of nitroarenes would all be suitable.

1,3-Dibromo-5-nitrobenzene
[prepared from p-nitroaniline as in part (c)]

H$_2$, Ni

3,5-Dibromoaniline
(80%)

(e) The synthetic objective is

p-Acetamidophenol

This compound, known as *acetaminophen* and used as an analgesic to reduce fever and relieve minor pain, may be prepared from p-nitroaniline by way of p-nitrophenol.

p-Nitroaniline   *p*-Nitrophenol   *p*-Acetamidophenol

Any of the customary reducing agents suitable for converting aryl nitro groups to arylamines (Fe, HCl; Sn, HCl; H_2, Ni) may be used. Acetylation of *p*-aminophenol may be carried out with acetyl chloride or acetic anhydride. The amino group of *p*-aminophenol is more nucleophilic than the hydroxyl group and is acetylated preferentially.

22.42 (*a*) Replacement of an amino substituent by a bromine is readily achieved by the Sandmeyer reaction.

o-Anisidine   *o*-Bromoanisole (88–93%)

(*b*) This conversion demonstrates the replacement of an amino substituent by fluorine via the Schiemann reaction.

o-Anisidine   *o*-Methoxybenzenediazonium   *o*-Fluoroanisole
                                    fluoroborate (57%)   (53%)

(*c*) We can use the *o*-fluoroanisole prepared in part (*b*) to prepare 3-fluoro-4-methoxyacetophenone by Friedel–Crafts acylation.

o-Anisidine   *o*-Fluoroanisole   3-Fluoro-4-methoxyacetophenone
                                                                              (70–80%)

Remember from text Section 12.16 that it is the more activating substituent that determines the regioselectivity of electrophilic aromatic substitution when an arene bears two different substituents. Methoxy is a strongly activating substituent; fluorine is slightly deactivating. Friedel–Crafts acylation takes place at the position para to the methoxy group.

(*d*)　The *o*-fluoroanisole prepared in part (*b*) serves nicely as a precursor to 3-fluoro-4-methoxy-benzonitrile via diazonium salt chemistry.

The desired sequence of reactions to carry out the synthesis is

Conversion of *o*-fluoroanisole to 4-amino-2-fluoroanisole proceeds in the conventional way by preparation and reduction of a nitro derivative. Once the necessary arylamine is at hand, it is converted to the nitrile by a Sandmeyer reaction.

(*e*)　Diazotization followed by hydrolysis of the 4-amino-2-fluoroanisole prepared as an intermediate in part (*d*) yields the desired phenol.

22.43 (*a*) The compound is an imine. Retrosynthetically, we disconnect the N=C double bond to see that this particular imine is prepared by the reaction of aniline and *p*-nitrobenzaldehyde.

One of the reactants given in the problem is nitrobenzene, from which we can prepare aniline by reduction.

The other reactant given is *p*-nitrobenzyl alcohol, which can be oxidized to *p*-nitrobenzaldehyde with PCC or PDC in dichloromethane.

Combining aniline and *p*-nitrobenzaldehyde gives the desired imine. An acid catalyst such as a trace of HCl can be used to speed up the reaction.

 (*b*) The most direct way to approach this synthesis is to convert *p*-nitrotoluene to *p*-nitrobenzyl alcohol, then proceed as in part (*a*). Benzylic bromination with *N*-bromosuccinimide (NBS), followed by hydrolysis will work.

The alternative sequence

will not work in this synthesis because the nitro group will also be reduced by lithium aluminum hydride.

22.44 (*a*) The carboxyl group of *p*-aminobenzoic acid can be derived from the methyl group of *p*-methylaniline by oxidation. First, however, the nitrogen must be acylated so as to protect the ring from oxidation.

The sequence of reactions to be used is

p-Methylaniline   p-Methylacetanilide   p-Acetamidobenzoic
acid

p-Aminobenzoic
acid

(b) Attachment of fluoro and propanoyl groups to a benzene ring is required. The fluorine substituent can be introduced by way of the diazonium tetrafluoroborate, the propanoyl group by way of a Friedel–Crafts acylation. Because the fluorine substituent is ortho, para-directing, introducing it first gives the proper orientation of substituents.

Ethyl p-fluorophenyl
ketone

Fluorobenzene

Aniline

Fluorobenzene is prepared from aniline by the Schiemann reaction, shown in Section 22.17. Aniline is prepared from benzene via nitrobenzene. Friedel–Crafts acylation of fluorobenzene has been carried out with the results shown and gives the required ethyl p-fluorophenyl ketone as the major product.

Fluorobenzene   Propanoyl
chloride

Ethyl p-fluorophenyl
ketone (86%)

(c) The synthetic plan is based on the essential step of forming the fluorine derivative from an amine by way of a diazonium salt.

1-Bromo-2-fluoro-
3,5-dimethylbenzene

2,4-Dimethylaniline

The required substituted aniline is derived from *m*-xylene by a standard synthetic sequence.

m-Xylene   1,3-Dimethyl-4-   2,4-Dimethylaniline
                                          nitrobenzene (98%)

1-Bromo-2-fluoro-   2-Bromo-4,6-dimethylaniline
3,5-dimethylbenzene (60%)

(*d*) In this problem two nitrogen-containing groups of the starting material are each to be replaced by a halogen substituent. The task is sufficiently straightforward that it may be confronted directly.

Replace amino group by bromine:

2-Methyl-4-nitro-1-   1-Bromo-2-methyl-4-
naphthylamine   nitronaphthalene (82%)

Reduce nitro group to amine:

1-Bromo-2-methyl-   4-Bromo-3-methyl-
4-nitronaphthalene   1-naphthylamine

Replace amino group by fluorine:

4-Bromo-3-methyl-   1-Bromo-4-fluoro-2-
1-naphthylamine   methylnaphthalene
                                                          (64%)

(*e*) Bromination of the starting material will introduce the bromine substituent at the correct position, that is, ortho to the *tert*-butyl group.

p-tert-Butylnitrobenzene   2-Bromo-1-*tert*-butyl-4-nitrobenzene

The desired product will be obtained if the nitro group can be removed. This is achieved by its conversion to the corresponding amine, followed by reductive deamination.

2-Bromo-1-*tert*-butyl-4-nitrobenzene   3-Bromo-4-*tert*-butylaniline   *o*-Bromo-*tert*-butylbenzene

(*f*) The proper orientation of the chlorine substituent can be achieved only if it is introduced after the nitro group is reduced.

The correct sequence of reactions to carry out this synthesis is shown.

p-tert-Butylnitrobenzene   *p-tert*-Butylaniline   *p-tert*-Butylacetanilide   4-*tert*-Butyl-2-chloroacetanilide

Hydrolysis to remove the acetyl group followed by reductive deamination completes the synthesis.

4-*tert*-Butyl-2-
chloroacetanilide

4-*tert*-Butyl-2-
chloroaniline

m-tert-Butylchlorobenzene

(*g*) The orientation of substituents in the target molecule can be achieved by using an amino group to control the regiochemistry of bromination, then removing it by reductive deamination.

The amino group is introduced in the standard fashion by nitration of an arene followed by reduction. This analysis leads to the synthesis shown.

m-Diethylbenzene

2,4-Diethyl-1-nitrobenzene
(75–80%)

2,4-Diethylaniline
(80–90%)

1-Bromo-3,5-
diethylbenzene (70%)

2-Bromo-4,6-
diethylaniline (40%)

(*h*) In this exercise the two nitrogen substituents are differentiated; one is an amino nitrogen, the other an amide nitrogen. By keeping them differentiated they can be manipulated independently. Remove one amino group completely before deprotecting the other.

4-Amino-2-bromo-6-
(trifluoromethyl)acetanilide

2-Bromo-6-(trifluoromethyl)-
acetanilide (92%)

Once the acetyl group has been removed by hydrolysis, the molecule is ready for introduction of the iodo substituent by way of a diazonium salt.

2-Bromo-6-(trifluoromethyl)-
acetanilide

2-Bromo-6-(trifluoromethyl)-
aniline (69%)

1-Bromo-2-iodo-3-
(trifluoromethyl)benzene
(87%)

(*i*) To convert the designated starting material to the indicated product, both the nitro group and the ester function must be reduced and a carbon–nitrogen bond must be formed. Converting the starting material to an amide gives the necessary carbon–nitrogen bond and has the advantage that amides can be reduced to amines by lithium aluminum hydride. The amide can be formed intramolecularly by reducing the nitro group to an amine, then heating to cause cyclization.

This synthesis is the one described in the chemical literature. Other routes are also possible, but the one shown is short and efficient.

22.45 Weakly basic nucleophiles react with α,β-unsaturated carbonyl compounds by conjugate addition.

Ammonia and its derivatives are very prone to react in this way; thus conjugate addition provides a method for the preparation of β-amino carbonyl compounds.

(a) (CH$_3$)$_2$C=CHCCH$_3$ + NH$_3$ ⟶ (CH$_3$)$_2$CCH$_2$CCH$_3$
(with O above both carbonyls; NH$_2$ on substituted carbon)

4-Methyl-3-penten-2-one   Ammonia   4-Amino-4-methyl-2-pentanone (63–70%)

(b) 2-Cyclohexenone + Piperidine ⟶ 3-Piperidinocyclohexanone (45%)

(c) C$_6$H$_5$CCH=CHC$_6$H$_5$ + HN(morpholine) ⟶ C$_6$H$_5$CCH$_2$CHC$_6$H$_5$ (with morpholino N)

1,3-Diphenyl-2-propen-1-one   Morpholine   3-Morpholino-1,3-diphenyl-1-propanone (91%)

(d) The conjugate addition reaction that takes place in this case is an intramolecular one and occurs in virtually 100% yield.

22.46 The first step in the synthesis is the conjugate addition of methylamine to ethyl acrylate. Two sequential Michael additions take place.

CH$_3$NH$_2$ + H$_2$C=CHCOCH$_2$CH$_3$ ⟶ CH$_3$NHCH$_2$CH$_2$COCH$_2$CH$_3$

Methylamine   Ethyl acrylate

H$_2$C=CHCO$_2$CH$_2$CH$_3$

CH$_3$N(CH$_2$CH$_2$CO$_2$CH$_2$CH$_3$)$_2$

Conversion of this intermediate to the desired *N*-methyl-4-piperidone requires a Dieckmann cyclization followed by decarboxylation of the resulting β-keto ester.

N-Methyl-4-piperidone

Treatment of *N*-methyl-4-piperidone with the Grignard reagent derived from bromobenzene gives a tertiary alcohol that can be dehydrated to an alkene. Hydrogenation of the alkene completes the synthesis.

N-Methyl-4-piperidone Phenylmagnesium bromide

N-Methyl-4-phenylpiperidine (compound A)

22.47 Sodium cyanide reacts with alkyl bromides by the S_N2 mechanism. Reduction of the cyano group with lithium aluminum hydride yields a primary amine. This reveals the structure of mescaline to be 2-(3,4,5-trimethoxyphenyl)ethylamine.

3,4,5-Trimethoxybenzyl bromide

2-(3,4,5-Trimethoxyphenyl)-ethanenitrile

2-(3,4,5-Trimethoxyphenyl)ethylamine (mescaline)

22.48 Reductive amination of a ketone with methylamine yields a secondary amine. Methamphetamine is *N*-methyl-1-phenyl-2-propanamine.

Benzyl methyl ketone   Methylamine

N-Methyl-1-phenyl-2-propanamine (methamphetamine)

22.49 There is no obvious reason why the dimethylamino group in 4-(*N*,*N*-dimethylamino)pyridine should be appreciably more basic than it is in *N*,*N*-dimethylaniline; it is the ring nitrogen of 4-(*N*,*N*-dimethylamino)pyridine that is more basic. Note that protonation of the ring nitrogen permits delocalization of the dimethylamino lone pair and dispersal of the positive charge.

Most stable protonated form of
4-(*N*,*N*-dimethylamino)pyridine

22.50 Only the unshared electron pair on nitrogen that is not part of the π electron cloud of the aromatic system will be available for protonation. Treatment of 5-methyl-γ-carboline with acid will give the salt shown.

5-Methyl-γ-carboline

22.51 The ¹H NMR spectrum of each isomer shows peaks corresponding to five aromatic protons, so compounds A and B each contain a monosubstituted benzene ring. Only four compounds of molecular formula $C_8H_{11}N$ meet this requirement.

$$C_6H_5CH_2NHCH_3 \quad\quad C_6H_5NHCH_2CH_3 \quad\quad C_6H_5\underset{\underset{NH_2}{|}}{C}HCH_3 \quad\quad C_6H_5CH_2CH_2NH_2$$

N-Methylbenzylamine   *N*-Ethylaniline   1-Phenylethylamine   2-Phenylethylamine

Neither ¹H NMR spectrum is consistent with *N*-methylbenzylamine, which would have two singlets due to the methyl and methylene groups. Likewise, the spectra are not consistent with *N*-ethylaniline, which would exhibit the characteristic triplet–quartet pattern of an ethyl group. Although a quartet occurs in the spectrum of compound A, it corresponds to only one proton, not the two that an ethyl group requires. The one-proton quartet in compound A arises from an H—C—CH₃ unit. Compound A is 1-phenyl-ethylamine.

Compound B has an ¹H NMR spectrum that fits 2-phenylethylamine.

22.52 Write the structural formulas for the two possible compounds given in the problem and consider how their ¹³C NMR spectra will differ from each other. Both will exhibit their CH_3 carbon signals at high field, but they differ in the positions of their CH_2 and quaternary carbons. A carbon bonded to nitrogen is more shielded than one bonded to oxygen, because nitrogen is less electronegative than oxygen.

Lower field signal · Higher field signal

$(CH_3)_2C-CH_2NH_2$
|
OH

1-Amino-2-methyl-2-propanol

Higher field signal · Lower field signal

$(CH_3)_2C-CH_2OH$
|
NH_2

2-Amino-2-methyl-1-propanol

In one isomer the lowest field signal is a quaternary carbon; in the other it is a CH_2 group. The spectrum shown in text Figure 22.10 shows the lowest field signal as a CH_2 group. The compound is therefore 2-amino-2-methyl-1-propanol, $(CH_3)_2CCH_2OH$.
|
NH_2

This compound *cannot* be prepared by reaction of ammonia with an epoxide, because in basic solution nucleophiles attack epoxides at the less hindered carbon, and therefore epoxide ring opening will give 1-amino-2-methyl-2-propanol rather than 2-amino-2-methyl-1-propanol.

$(CH_3)_2C-CH_2$ + $\ddot{N}H_3$ $\longrightarrow$ $(CH_3)_2CCH_2NH_2$
$\quad\quad\backslash O \diagup$ $\quad\quad\quad\quad\quad\quad\quad\quad\quad\quad\quad\quad$ |
$\quad\quad\quad\quad\quad\quad\quad\quad\quad\quad\quad\quad\quad\quad\quad\quad\quad\quad\quad$ OH

2,2-Dimethyloxirane　　Ammonia　　　1-Amino-2-methyl-2-propanol

ANSWERS TO INTERPRETIVE PROBLEMS

22.53 A;　**22.54** B;　**22.55** D;　**22.56** A;　**22.57** C;　**22.58** B

SELF-TEST

PART A

A-1. Give an acceptable name for each of the following. Identify each compound as a primary, secondary, or tertiary amine.

$\quad\quad\quad\quad\quad CH_3$
$\quad\quad\quad\quad\quad\;|$
(*a*)　$CH_3CH_2CCH_3$　　(*b*) ⬠—$NHCH_3$　　(*c*) ⬡—$NHCH_2CH_2CH_3$
$\quad\quad\quad\quad\quad\;|$
$\quad\quad\quad\quad\;\;NH_2$　　　　　　　　　　　　　　　　　　Br

A-2. Provide the correct structure of the reagent omitted from each of the following reactions:

(*a*)　$C_6H_5CH_2Br$　$\xrightarrow[\substack{2.\ LiAlH_4 \\ 3.\ H_2O}]{1.\ ?}$　$C_6H_5CH_2NH_2$

(*b*)　$C_6H_5CH_2Br$　$\xrightarrow[\substack{2.\ LiAlH_4 \\ 3.\ H_2O}]{1.\ ?}$　$C_6H_5CH_2CH_2NH_2$

(*c*)　$C_6H_5CH_2Br$　$\xrightarrow[2.\ H_2NNH_2]{1.\ ?}$　$C_6H_5CH_2NH_2$　+ phthalazine-1,4-dione structure

A-3. Provide the missing component (reagent or product) for each of the following:

(a) H_3C—⟨benzene⟩—NH_2 $\xrightarrow[H_2O]{NaNO_2,\ HCl}$?

(b) Product of part (a) $\xrightarrow{CuBr}$?

(c) Product of part (a) $\xrightarrow{?}$ toluene

(d) H_3C—⟨benzene⟩—NH_2 $\xrightarrow{?}$ H_3C—⟨benzene⟩—$\overset{\overset{\displaystyle O}{\|}}{N}HCCH_3$

(e) [ring with $\overset{\overset{\displaystyle O}{\|}}{N}HCCH_3$ and CH_2CH_3 substituents] $\xrightarrow[H_2SO_4]{HNO_3}$?

(f) ⟨benzene⟩—$N(CH_3)_2$ $\xrightarrow{NaNO_2,\ HCl,\ H_2O}$?

(g) ⟨benzene⟩—$NHCH_2CH_3$ $\xrightarrow{NaNO_2,\ HCl,\ H_2O}$?

A-4. Provide structures for compounds A through E in the following reaction sequences:

(a) A $\xrightarrow{CH_3I}$ B $\xrightarrow[H_2O]{Ag_2O}$ C $\xrightarrow{heat}$ $H_2C{=}CHCH_2CH_2\overset{\overset{\displaystyle CH_3}{|}}{N}CH_2CH_3$

(b) [cyclohexanone] $+\ CH_3CH_2NH_2$ $\xrightarrow[CH_3OH]{NaBH_3CN}$ D $\xrightarrow[H_2O]{NaNO_2,\ HCl}$ E

A-5. Give the series of reaction steps involved in the following synthetic conversions:

(a) [benzene ring with $C(CH_3)_3$ at top and I at bottom] from benzene

(b) *m*-Chloroaniline from benzene

(c) $C_6H_5N{=}N$—⟨benzene⟩—$N(CH_3)_2$ from aniline

A-6. *p*-Nitroaniline (A) is less basic than *m*-nitroaniline (B). Using resonance structures, explain the reason for this difference.

A   B

A-7. Identify the strongest and weakest bases among the following:

A   B   C   D

A-8. Write the structures of the compounds A–D formed in the following reaction sequence:

PART B

B-1. Which of the following is a secondary amine?

(*a*) 2-Butanamine

(*b*) *N*-Ethyl-2-pentanamine

(*c*) *N*-Methylpiperidine

(*d*) *N,N*-Dimethylcyclohexylamine

B-2. Which of the following C_8H_9NO isomers is the weakest base?

(*a*) *o*-Aminoacetophenone

(*b*) *m*-Aminoacetophenone

(*c*) *p*-Aminoacetophenone

(*d*) Acetanilide

B-3. Rank the following compounds in order of increasing basicity (weakest → strongest):

1   2   3   4

(*a*) 4 < 2 < 1 < 3

(*b*) 4 < 1 < 3 < 2

(*c*) 4 < 3 < 1 < 2

(*d*) 2 < 1 < 3 < 4

B-4. Which of the following arylamines will *not* form a diazonium salt on reaction with sodium nitrite in hydrochloric acid?

(*a*) *m*-Ethylaniline

(*b*) 4-Chloro-2-nitroaniline

(*c*) *p*-Aminoacetophenone

(*d*) *N*-Ethyl-2-methylaniline

B-5. The amines shown are isomers. Choose the one with the lowest boiling point.

B-6. Which of the following is the strongest acid?

(*a*) $H-\overset{\overset{\displaystyle H}{|}}{\underset{\displaystyle H}{N}}:$

(*b*) [structure]

(*c*) [structure]

(*d*) $H-\overset{\overset{\displaystyle H}{|}}{\underset{\displaystyle H}{N^+}}-H$

(*e*) [structure]

B-7. The reaction

$+$ CH$_3$I (excess) ⟶ ?

gives as the final product

(*a*) A primary amine

(*b*) A secondary amine

(*c*) A tertiary amine

(*d*) A quaternary ammonium salt

B-8. A substance is soluble in dilute aqueous HCl and has a single peak in the region 3200–3500 cm^{-1} in its IR spectrum. Which of the following best fits the data?

(*a*) [structure] —N(CH$_3$)$_2$

(*b*) [structure] —NH$_2$

(*c*) [structure] —NHCH$_3$

(*d*) [structure] —CO$_2$H

B-9. Identify product D in the following reaction sequence:

$$CH_3\underset{\underset{CH_3}{|}}{\overset{\overset{CH_3}{|}}{C}}CH_2CH_2OH \xrightarrow[\text{H}_2\text{O, heat}]{\text{K}_2\text{Cr}_2\text{O}_7,\ \text{H}_2\text{SO}_4} A \xrightarrow{\text{SOCl}_2} B \xrightarrow[\text{(2 mol)}]{\text{(CH}_3)_2\text{NH}} C \xrightarrow[\text{2. H}_2\text{O}]{\text{1. LiAlH}_4,\ \text{diethyl ether}} D$$

(a) $CH_3\underset{\underset{CH_3}{|}}{\overset{\overset{CH_3}{|}}{C}}CH_2C\equiv N$

(b) $CH_3\underset{\underset{CH_3}{|}}{\overset{\overset{CH_3}{|}}{C}}CH_2\overset{\overset{N(CH_3)_2}{|}}{C}HN(CH_3)_2$

(c) $CH_3\underset{\underset{CH_3}{|}}{\overset{\overset{CH_3}{|}}{C}}CH_2\overset{\overset{O}{\|}}{C}N(CH_3)_2$

(d) $CH_3\underset{\underset{CH_3}{|}}{\overset{\overset{CH_3}{|}}{C}}CH_2CH_2N(CH_3)$

(e) $CH_3\underset{\underset{CH_3}{|}}{\overset{\overset{CH_3}{|}}{C}}CH_2\overset{\overset{|}{C}}{}HN(CH_3)_2$ with OH on that carbon

B-10. Which one of the following is the best catalyst for the reaction shown?

$$CH_3(CH_2)_8CH_2Br \xrightarrow[\text{benzene}]{\text{KCN}} CH_3(CH_2)_8CH_2CN$$

(a) phenyl–CH$_2$Cl

(b) phenyl–NH$_2$

(c) phenyl–NHCCH$_3$ (with C=O)

(d) phenyl–$\overset{+}{\text{N}}$H$_3$ Cl$^-$

(e) phenyl–CH$_2\overset{+}{\text{N}}$(CH$_3$)$_3$ Cl$^-$

B-11. What will be the *major* product of each of the two reactions shown?

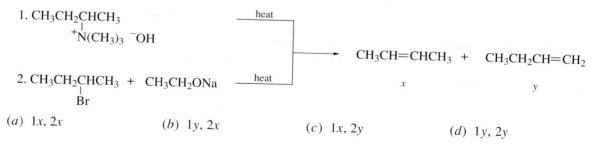

1. $CH_3CH_2\underset{\underset{\overset{+}{N}(CH_3)_3\ ^-OH}{|}}{C}HCH_3$

2. $CH_3CH_2\underset{\underset{Br}{|}}{C}HCH_3 + CH_3CH_2ONa$

$\xrightarrow{\text{heat}}$ $\xrightarrow{\text{heat}}$

$CH_3CH=CHCH_3$ + $CH_3CH_2CH=CH_2$
          x y

(a) 1x, 2x   (b) 1y, 2x   (c) 1x, 2y   (d) 1y, 2y

B-12. Which sequence represents the best synthesis of 4-isopropylbenzonitrile?

$$(CH_3)_2CH-\!\!\!\langle\ \rangle\!\!\!-C\!\equiv\!N$$

4-Isopropylbenzonitrile

(*a*) 1. Benzene + $(CH_3)_2CHCl$, $AlCl_3$; 2. Br_2, $FeBr_3$; 3. KCN

(*b*) 1. Benzene + $(CH_3)_2CHCl$, $AlCl_3$; 2. HNO_3, H_2SO_4; 3. Fe, HCl; 4. NaOH; 5. $NaNO_2$, HCl, H_2O; 6. CuCN

(*c*) 1. Benzene + $(CH_3)_2CHCl$, $AlCl_3$; 2. HNO_3, H_2SO_4; 3. Fe, HCl; 4. NaOH; 5. KCN

(*d*) 1. Benzene + HNO_3, H_2SO_4; 2. $(CH_3)_2CHCl$, $AlCl_3$; 3. Fe, HCl; 4. NaOH; 5. $NaNO_2$, HCl, H_2O; 6. CuCN

(*e*) 1. Benzene + HNO_3, H_2SO_4; 2. Fe, HCl; 3. NaOH; 4. $NaNO_2$, HCl, H_2O; 5. CuCN; 6. $(CH_3)_2CHCl$, $AlCl_3$

B-13. The major products from the following sequence of reactions are

$$(CH_3)_2CHCH_2N(CH_2CH_3)_2 \xrightarrow{CH_3I} \xrightarrow[H_2O]{Ag_2O} \xrightarrow{heat} ?$$

(*a*) $(CH_3)_2CHCH_2NH_2$ + $H_2C\!=\!CH_2$

(*b*) $(CH_3)_2NCH_2CH_3$ + $H_2C\!=\!C(CH_3)_2$

(*c*) $(CH_3)_2CHCH_2\overset{\overset{\displaystyle CH_3}{|}}{N}CH_2CH_3$ + $H_2C\!=\!CH_2$

(*d*) $(CH_3)_3\overset{+}{N}CH_2CH_3$ I^- + $H_2C\!=\!CH_2$

(*e*) None of these combinations of products is correct.

B-14. Which compound yields an *N*-nitroso amine after treatment with nitrous acid ($NaNO_2$, HCl)?

(*a*) ⟨benzene ring⟩$-CH_2NH_2$

(*b*) ⟨benzene ring⟩$-N$⟨piperidine ring⟩

(*c*) ⟨benzene ring⟩$-NHCH_3$

(*d*) H_3C-⟨benzene ring⟩$-NH_2$

(*e*) ⟨benzene ring⟩$-\overset{\overset{\displaystyle O}{\|}}{C}NH_2$

CHAPTER 23

Aryl Halides

SOLUTIONS TO TEXT PROBLEMS

23.1 Four isomers of C_7H_7Cl contain a benzene ring, namely, *o*-, *m*-, and *p*-chlorotoluene and benzyl chloride.

| *o*-Chlorotoluene | *m*-Chlorotoluene | *p*-Chlorotoluene | Benzyl chloride |

Of this group only benzyl chloride is not an aryl halide; its halogen is not attached to the aromatic ring but to an sp^3-hybridized carbon. Benzyl chloride has the weakest carbon–halogen bond, its measured carbon–chlorine bond dissociation enthalpy being only 293 kJ/mol (70 kcal/mol). Homolytic cleavage of this bond produces a resonance-stabilized benzyl radical.

Benzyl chloride → Benzyl radical + Chlorine atom

23.2 (*b*) The negatively charged sulfur in $C_6H_5CH_2\ddot{\underset{\cdot\cdot}{S}}{:}^-$ Na$^+$ is a good nucleophile, which displaces chloride from 1-chloro-2,4-dinitrobenzene.

1-Chloro-2,4-
dinitrobenzene

Benzyl 2,4-
dinitrophenyl sulfide

(*c*) The nitrogen in methylamine has an unshared electron pair and is nucleophilic; it displaces chloride from 1-chloro-2,4-dinitrobenzene.

1-Chloro-2,4-
dinitrobenzene

N-Methyl-2,4-
dinitroaniline

23.3 The most stable resonance structure for the cyclohexadienyl anion formed by reaction of methoxide ion with *o*-fluoronitrobenzene involves the nitro group and has the negative charge on oxygen.

23.4 The positions that are activated toward nucleophilic attack are those that are ortho and para to the nitro group. Among the carbons that bear a bromine leaving group in 1,2,3-tribromo-5-nitrobenzene, only C-2 satisfies this requirement.

1,2,3-Tribromo-
5-nitrobenzene

1,3-Dibromo-2-ethoxy-
5-nitrobenzene

23.5 Nucleophilic addition occurs in the rate-determining step at one of the six equivalent carbons of hexafluorobenzene to give the cyclohexadienyl anion intermediate.

Hexafluorobenzene   Methoxide
                              ion

Cyclohexadienyl anion
intermediate

Elimination of fluoride ion from the cyclohexadienyl anion intermediate restores the aromaticity of the ring and completes the reaction.

Cyclohexadienyl anion intermediate

2,3,4,5,6-Pentafluoroanisole

Fluoride ion

23.6 4-Chloropyridine is more reactive toward nucleophiles than 3-chloropyridine because the anionic intermediate formed by reaction of 4-chloropyridine has its charge on nitrogen. Because nitrogen is more electronegative than carbon, the intermediate is more stable.

4-Chloropyridine

Anionic intermediate (more stable)

3-Chloropyridine

Anionic intermediate (less stable)

23.7 The negative charge of the intermediate is delocalized among the two nitrogen atoms in the ring, the carbon bearing the cyano group, and the nitrogen of the cyano group.

23.8 The aryl halide is incapable of elimination and so cannot form the benzyne intermediate necessary for substitution by the elimination–addition pathway.

2-Bromo-1,3-dimethylbenzene

(No protons ortho to bromine; elimination is impossible.)

23.9 The aryne intermediate from *p*-iodotoluene can undergo addition of hydroxide ion at the position meta to the methyl group or para to it. The two isomeric phenols are *m*- and *p*-methylphenol.

p-Iodotoluene   *m*-Methylphenol   *p*-Methylphenol

23.10 The "triple bond" of benzyne adds to the conjugated diene system of furan.

23.11 (*a*)

m-Chlorotoluene

(*b*)

2,6-Dibromoanisole

(*c*)

p-Fluorostyrene

(*d*)

4,4′-Diiodobiphenyl

(*e*)

2-Bromo-1-chloro-4-
nitrobenzene

(*f*)

1-Chloro-1-phenylethane
(*Note*: This compound
is not an aryl halide.)

(*g*)

p-Bromobenzyl
chloride

(*h*)

2-Chloronaphthalene

(*i*) 1,8-Dichloronaphthalene

(*j*) 9-Fluorophenanthrene

23.12 (*a*) Chlorine is a weakly deactivating, ortho, para-directing substituent.

Chlorobenzene   Acetyl chloride   *o*-Chloroacetophenone   *p*-Chloroacetophenone

(*b*) Bromobenzene reacts with magnesium to give a Grignard reagent.

$$C_6H_5Br \ + \ Mg \ \xrightarrow{\text{diethyl ether}} \ C_6H_5MgBr$$

Bromobenzene   Phenylmagnesium bromide

(*c*) Protonation of the Grignard reagent in part (*b*) converts it to benzene.

$$C_6H_5MgBr \ \xrightarrow[\text{HCl}]{\text{H}_2\text{O}} \ C_6H_6$$

Phenylmagnesium bromide   Benzene

(*d*) Aryl halides react with lithium in much the same way that alkyl halides do, to form organolithium reagents.

$$C_6H_5I \ + \ 2Li \ \xrightarrow{\text{diethyl ether}} \ C_6H_5Li \ + \ LiI$$

Iodobenzene   Lithium   Phenyllithium   Lithium iodide

(*e*) With a base as strong as sodium amide, nucleophilic aromatic substitution by the elimination–addition mechanism takes place.

Bromobenzene $\xrightarrow[\text{NH}_3]{\text{NaNH}_2}$ Benzyne $\xrightarrow[\text{NH}_3]{\text{NaNH}_2}$ Aniline (*f*) The benzyne intermediate from *p*-bromotoluene gives a mixture of *m*- and *p*-methylaniline.

| *p*-Bromotoluene | 4-Methylbenzyne | *m*-Methylaniline | *p*-Methylaniline |

(*g*) Nucleophilic aromatic substitution of bromide by ammonia occurs by the addition–elimination mechanism.

| 1-Bromo-4-nitrobenzene | *p*-Nitroaniline |

(*h*) The bromine attached to the benzylic carbon is far more reactive than the one on the ring and is the one replaced by the nucleophile.

| *p*-Bromobenzyl bromide | *p*-Bromobenzyl cyanide |

(*i*) The aromatic ring of *N,N*-dimethylaniline is very reactive and is attacked by *p*-chlorobenzenediazonium ion.

| *N,N*-Dimethylaniline | *p*-Chlorobenzenediazonium ion | 4-(4′-Chlorophenylazo)-*N,N*-dimethylaniline |

(*j*) Hexafluorobenzene undergoes substitution of one of its fluorines on reaction with nucleophiles such as sodium hydrogen sulfide.

| Hexafluorobenzene | Sodium hydrogen sulfide | 2,3,4,5,6-Pentafluorobenzenethiol |

23.13 (*a*) Because the *tert*-butoxy group replaces fluoride at the position occupied by the leaving group, substitution likely occurs by the addition–elimination mechanism.

o-Fluorotoluene   tert-Butoxide ion   tert-Butyl o-methylphenyl ether

(b) In nucleophilic aromatic substitution reactions that proceed by the addition–elimination mechanism, aryl fluorides react faster than aryl bromides. Because the aryl bromide is more reactive in this case, it must be reacting by a different mechanism, which is most likely elimination–addition.

Bromobenzene   Benzene   tert-Butyl phenyl ether

23.14 (a) Two benzyne intermediates are equally likely to be formed. Reaction with amide ion can occur in two different directions with each benzyne, giving three possible products. They are formed in a 1:2:1 ratio. The asterisk (*) refers to ^{14}C.

Ratio: 1   :   2   :   1

(b) Only one benzyne intermediate is possible, leading to two products in a 1:1 ratio. D refers to ^{2}H (deuterium).

Ratio: 1   :   1

23.15 (*a*) *o*-Chloronitrobenzene is more reactive than chlorobenzene because the cyclohexadienyl anion intermediate is stabilized by the nitro group.

Comparing the rate constants for the two aryl halides in this reaction reveals that *o*-chloronitrobenzene is more than 20 billion times more reactive at 50°C.

(*b*) The cyclohexadienyl anion intermediate is more stable, and is formed faster, when the electron-withdrawing nitro group is ortho to chlorine. *o*-Chloronitrobenzene reacts faster than *m*-chloronitrobenzene. The measured difference is a factor of approximately 40,000 at 50°C.

(*c*) 4-Chloro-3-nitroacetophenone is more reactive, because the ring bears two powerful electron-withdrawing groups in positions where they can stabilize the cyclohexadienyl anion intermediate.

(*d*) Nitro groups activate aryl halides toward nucleophilic aromatic substitution best when they are ortho or para to the leaving group.

is more reactive than

2-Fluoro-1,3-dinitrobenzene

1-Fluoro-3,5-dinitrobenzene

(*e*) The aryl halide with nitro groups ortho and para to the bromide leaving group is more reactive than the aryl halide with only one nitro group.

is more reactive than

1-Bromo-2,4-dinitrobenzene

1,4-Dibromo-2-nitrobenzene

23.16 (*a*) The nucleophile is the lithium salt of pyrrolidine, which reacts with bromobenzene by an elimination–addition mechanism.

Bromobenzene   Lithium   N-Phenylpyrrolidine
                pyrrolidide

(*b*) The nucleophile in this case is piperidine. The substrate, 1-bromo-2,4-dinitrobenzene, is very reactive in nucleophilic aromatic substitution by the addition–elimination mechanism.

1-Bromo-2,4-   Piperidine   N-(2,4-Dinitrophenyl)-
dinitrobenzene   piperidine

(*c*) Of the two bromine atoms, one is ortho and the other meta to the nitro group. Nitro groups activate positions ortho and para to themselves toward nucleophilic aromatic substitution, and so it will be the bromine ortho to the nitro group that is displaced.

1,4-Dibromo-2-   Piperidine   N-(4-Bromo-2-nitrophenyl)-
dinitrobenzene   piperidine

23.17 Because isomeric products are formed by reaction of 1- and 2-bromonaphthalene with piperidine at elevated temperatures, it is reasonable to conclude that these reactions do not involve a common intermediate and hence follow an addition–elimination pathway. Piperidine acts as a nucleophile and substitutes for bromine on the same carbon atom from which bromine is lost.

1-Bromonaphthalene   Piperidine   Compound A

2-Bromonaphthalene   Piperidine   Compound B

When the strong base sodium piperidide is used, reaction occurs by the elimination–addition pathway via a "naphthalyne" intermediate. Only one mode of elimination is possible from 1-bromonaphthalene.

This intermediate can yield both A and B in the addition stage.

Compound A   Compound B

Two modes of elimination are possible from 2-bromonaphthalene:

Compounds A and B   Compound B only

Both naphthalyne intermediates are probably formed from 2-bromonaphthalene because there is no reason to expect elimination to occur only in one direction.

23.18 Reaction of a nitro-substituted aryl halide with a good nucleophile leads to nucleophilic aromatic substitution. Methoxide will displace fluoride from the ring, preferentially at the positions ortho and para to the nitro group.

1,2,3,4,5-Pentafluoro-
6-nitrobenzene
    2,3,4,5-Tetrafluoro-
6-nitroanisole
    2,3,5,6-Tetrafluoro-
4-nitroanisole

23.19 (*a*) This reaction is nucleophilic aromatic substitution by the addition–elimination mechanism.

4-Chloro-3-
nitrotoluene
    4-(Benzylthio)-3-
nitrotoluene

The nucleophile, $C_6H_5CH_2\overset{..}{\underset{..}{S}}:^-$, displaces chloride directly from the aromatic ring. The product in this case was isolated in 57% yield.

(*b*) The nucleophile, hydrazine, will react with 1-chloro-2,4-dinitrobenzene by an addition–elimination mechanism as shown.

1-Chloro-2,4-
dinitrobenzene
    Hydrazine
    2,4-Dinitrophenylhydrazine

The nitrogen atoms of hydrazine each have an unshared electron pair and hydrazine is fairly nucleophilic. The product, 2,4-dinitrophenylhydrazine, is formed in quantitative yield.

(*c*) The problem requires you to track the starting material through two transformations. The first of these is nitration of *m*-dichlorobenzene, an electrophilic aromatic substitution reaction.

m-Dichlorobenzene
    2,4-Dichloro-1-
nitrobenzene

Because the final product of the sequence has four nitrogen atoms ($C_6H_6N_4O_4$), 2,4-dichloro-1-nitrobenzene is an unlikely starting material for the second transformation. Stepwise nucleophilic aromatic substitution of both chlorines is possible but leads to a compound with the wrong molecular formula ($C_6H_7N_3O_2$).

2,4-Dichloro-1-
nitrobenzene

2,4-Diamino-1-
nitrobenzene

To obtain a final product with the correct molecular formula, the original nitration reaction must lead not to a mononitro but to a dinitro derivative. This is reasonable in view of the fact that this reaction is carried out at elevated temperature (120°C).

m-Dichlorobenzene

1,5-Dichloro-2,4-
dinitrobenzene

1,5-Diamino-2,4-dinitrobenzene
($C_6H_6N_4O_4$)

This two-step sequence has been carried out with product yields of 70–71% in the first step and 88–95% in the second step.

(*d*) This problem also involves two transformations, nitration and nucleophilic aromatic substitution. Nitration will take place ortho to chlorine (meta to trifluoromethyl).

1-Chloro-4-
(trifluoromethyl)-
benzene

1-Chloro-2-nitro-4-
(trifluoromethyl)benzene

2-Nitro-4-
(trifluoromethyl)-
anisole

(*e*) The primary alkyl halide is more reactive toward nucleophilic substitution than the aryl halide. A phosphonium salt forms by an S_N2 process.

p-Iodobenzyl bromide

Triphenyl-
phosphine

(*p*-Iodobenzyl)triphenyl-
phosphonium bromide (86%)

(*f*) *N*-Bromosuccinimide (NBS) is a reagent used to substitute benzylic and allylic hydrogens with bromine. The benzylic bromide undergoes S_N2 substitution with the nucleophile, methanethiolate. As in part (*e*), the alkyl halide is more reactive toward substitution than the aryl halide.

2-Bromo-5-
methoxytoluene

2-Bromo-5-methoxybenzyl bromide

2-Bromo-5-methoxybenzyl
methyl sulfide

23.20 The reaction of *p*-bromotoluene with aqueous sodium hydroxide at elevated temperature proceeds by way of a benzyne intermediate.

m-Methylphenol   *p*-Methylphenol

The same benzyne intermediate is formed when *p*-chlorotoluene is the reactant, and so the product ratio must be identical regardless of whether the leaving group is bromide or chloride.

23.21 Dinitration of *p*-chloro(trifluoromethyl)benzene will take place at the ring positions ortho to the chlorine. Compound A is 2-chloro-1,3-dinitro-5-(trifluoromethyl)benzene. Trifluralin is formed by nucleophilic aromatic substitution of chlorine by dipropylamine. Trifluralin is -2,6-dinitro-*N*,*N*-dipropyl -4-(trifluoromethyl) aniline.

p-Chloro-
(trifluoromethyl)-
benzene

2-Chloro-1,3-dinitro-5-
(trifluoromethyl)benzene
(compound A)

2,6-Dinitro-*N*,*N*-Dipropyl-4-
(trifluoromethyl)aniline
(trifluralin)

23.22 *p*-Chlorobenzenethiolate reacts with *p*-nitrobenzyl chloride by an S_N2 process to give compound A. Reduction of the nitro group yields the aniline derivative, compound B. Chlorbenside is then formed by a Sandmeyer reaction in which the diazonium ion is replaced by chlorine.

p-Nitrobenzyl chloride

Sodium
p-chlorobenzenethiolate

p-Chlorophenyl *p*-nitrobenzyl sulfide
(compound A)

Compound A

Compound B

1. NaNO$_2$, HCl
2. CuCl

Chlorbenside

23.23 *p*-Chloro(trifluoromethyl)benzene undergoes nucleophilic substitution by the alkoxide anion to give compound A.

Compound A

Prozac (Fluoxetine hydrochloride) differs from compound A in having an —NHCH$_3$ group in place of —N(CH$_3$)$_2$.

Compound A   Prozac

23.24 Benzyne is formed by loss of nitrogen and carbon dioxide.

Benzenediazonium-2-   Benzyne   Nitrogen   Carbon dioxide
carboxylate

23.25 *o*-Bromofluorobenzene yields benzyne on reaction with magnesium (see text Section 23.9). Triptycene is the Diels–Alder cycloaddition product from the reaction of benzyne with anthracene (compound A). Although anthracene is aromatic, it is able to undergo cycloaddition at the center ring with a dienophile because the adduct retains the stabilization energy of two benzene rings.

o-Bromofluorobenzene   Anthracene   Triptycene
                               (compound A)

23.26 (*a*) Ethoxide ion adds to the aromatic ring to give a cyclohexadienyl anion.

2,4,6-Trinitroanisole   Sodium ethoxide   Meisenheimer complex

(*b*) The same Meisenheimer complex results when ethyl 2,4,6-trinitrophenyl ether reacts with sodium methoxide.

Ethyl 2,4,6-trinitrophenyl ether   Sodium methoxide   Meisenheimer complex

23.27 Methoxide ion may add to 2,4,6-trinitroanisole either at the ring carbon that bears the methoxyl group or at an unsubstituted ring carbon.

2,4,6-Trinitroanisole   A   B

The two Meisenheimer complexes are the sodium salts of the anions shown. It was observed that compound A was the more stable of the two. Compound B was present immediately after adding sodium methoxide to 2,4,6-trinitroanisole but underwent relatively rapid isomerization to compound A.

23.28 (*a*) The first reaction that occurs is an acid–base reaction between diethyl malonate and sodium amide.

$$CH_2(COOCH_2CH_3)_2 \ + \ NaNH_2 \ \longrightarrow \ Na^+ \ :\bar{C}H(COOCH_2CH_3)_2 \ + \ NH_3$$

Diethyl malonate   Sodium amide   Diethyl sodiomalonate   Ammonia

A second equivalent of sodium amide converts bromobenzene to benzyne.

Bromobenzene   Benzyne   Ammonia   Sodium bromide

The anion of diethyl malonate adds to benzyne.

Benzyne   Anion of diethyl
                 malonate

This anion then abstracts a proton from ammonia to give the observed product.

Ammonia   Diethyl 2-phenylmalonate   Amide
                                                   anion

(*b*) The ester is deprotonated by the strong base sodium amide, after which the ester enolate undergoes an elimination reaction to form a benzyne intermediate. Cyclization to the final product occurs by intramolecular attack of the ester enolate on the reactive triple bond of the aryne.

Ethyl 5-(2-chlorophenyl)pentanoate   Ester enolate

Ethyl
1,2,3,4-tetrahydronaphthalene-
1-carboxylate

Aryne intermediate

(*c*) In the presence of very strong bases, aryl halides undergo nucleophilic aromatic substitution by an elimination–addition mechanism. The structure of the product indicates that a nitrogen of the side chain acts as a nucleophile in the addition step.

(*d*) On treatment with base, intramolecular nucleophilic aromatic substitution leads to the observed product.

23.29 Polychlorinated biphenyls (PCBs) are derived from biphenyl as the base structure. It is numbered as shown.

(*a*) There are three monochloro derivatives of biphenyl:

2-Chlorobiphenyl
(*o*-chlorobiphenyl)

3-Chlorobiphenyl
(*m*-chlorobiphenyl)

4-Chlorobiphenyl
(*p*-chlorobiphenyl)

(*b*) The two chlorine substituents may be in the same ring (six isomers):

2,3-Dichlorobiphenyl   2,4-Dichlorobiphenyl   2,5-Dichlorobiphenyl   2,6-Dichlorobiphenyl   3,4-Dichlorobiphenyl   3,5-Dichlorobiphenyl

The two chlorine substituents may be in different rings (six isomers):

2,2'-Dichlorobiphenyl   2,3'-Dichlorobiphenyl   2,4'-Dichlorobiphenyl   3,3'-Dichlorobiphenyl   3,4'-Dichlorobiphenyl   4,4'-Dichlorobiphenyl

There are therefore a total of 12 isomeric dichlorobiphenyls.

(*c*) The number of octachlorobiphenyls will be equal to the number of dichlorobiphenyls (12). In both cases we are dealing with a situation in which eight of the ten substituents of the biphenyl system are the same. We then need to consider how the remaining two may be arranged. In the dichlorobiphenyls described in part (*b*), eight substituents are hydrogen and two are chlorine; in the octachlorobiphenyls, eight substituents are chlorine and two are hydrogen.

(*d*) The number of nonachloro isomers (nine chlorines, one hydrogen) must equal the number of monochloro isomers (one chlorine, nine hydrogens). There are therefore three nonachloro derivatives of biphenyl.

23.30 The principal isotopes of chlorine are ^{35}Cl and ^{37}Cl. A cluster of five peaks indicates that dichlorodiphenyldichloroethane (DDE) contains *four* chlorines.

m/z for $C_{14}H_8Cl_4$

316	^{35}Cl	^{35}Cl	^{35}Cl	^{35}Cl
318	^{35}Cl	^{35}Cl	^{35}Cl	^{37}Cl
320	^{35}Cl	^{35}Cl	^{37}Cl	^{37}Cl
322	^{35}Cl	^{37}Cl	^{37}Cl	^{37}Cl
324	^{37}Cl	^{37}Cl	^{37}Cl	^{37}Cl

The peak at m/z 316 therefore corresponds to a compound $C_{14}H_8Cl_4$ in which all four chlorines are ^{35}Cl. The respective molecular formulas indicate that DDE is the dehydrochlorination product of dichlorodiphenyltrichloroethane (DDT).

$$C_{14}H_9Cl_5 \xrightarrow{\ -HCl\ } C_{14}H_8Cl_4$$

DDT   DDE

The structure of DDT was given in the statement of the problem. This permits the structure of DDE to be assigned.

DDE (only reasonable
dehydrochlorination product of DDT)

ANSWERS TO INTERPRETIVE PROBLEMS

23.31 D;   **23.32** C;   **23.33** D;   **23.34** B;   **23.35** A

SELF TEST

PART A

A-1. Give the product(s) obtained from each of the following reactions:

(a)

$\xrightarrow[\text{NH}_3]{\text{KNH}_2}$? (two products)

(b)

$\xrightarrow[\text{CH}_3\text{OH}]{\text{CH}_3\text{O}^-}$? (monosubstitution)

(c)

$\xrightarrow[\text{NH}_3]{\text{NaNH}_2}$? (two products)

A-2. Draw the structure of the intermediate formed in each reaction of problem A-1.

A-3. Suggest synthetic schemes by which chlorobenzene may be converted into

(a) 2,4-Dinitroanisole (1-methoxy-2,4-dinitrobenzene)

(b) *p*-Isopropylaniline

A-4. Write a mechanism using resonance structures to show how a nitro group directs ortho, para in nucleophilic aromatic substitution.

A-5. What is the cycloaddition product of the following reaction? What is the structure of the short-lived intermediate formed in this reaction?

PART B

B-1. The reaction

most likely occurs by which of the following mechanisms?

(*a*) Addition–elimination   (*c*) Both (*a*) and (*b*)

(*b*) Elimination–addition   (*d*) Neither of these

B-2. Rank the following in order of decreasing rate of reaction with ethoxide ion ($CH_3CH_2O^-$) in a nucleophilic aromatic substitution reaction:

(*a*) $3 > 4 > 1 > 2$ (*c*) $3 > 4 > 2 > 1$

(*b*) $2 > 1 > 4 > 3$ (*d*) $4 > 3 > 2 > 1$

B-3. The reaction

most likely involves which of the following aromatic substitution mechanisms?

(*a*) Addition–elimination   (*c*) Elimination–addition

(*b*) Electrophilic substitution   (*d*) Both (*a*) and (*c*)

B-4. Identify the principal organic product of the following reaction:

(a)

(d)

(b)

(e)

(c)

B-5. Which of the following compounds gives a single benzyne intermediate on reaction with sodium amide?

1   2   3

(a) 1 only   (b) 1 and 3   (c) 3 only   (d) 1 and 2

B-6. Which one of the following compounds can be efficiently prepared by a procedure in which nucleophilic aromatic substitution is the last step?

(a)   (b)   (c)   (d)

B-7. Which one of the following undergoes nucleophilic aromatic substitution at the fastest rate?

(a)   (b)   (c)   (d)   (e)

B-8. What combination of reactants will give the species shown as a reactive intermediate?

(a) 1-Bromo-4-nitrobenzene and NaOH   (d) Bromobenzene and $HONO_2$

(b) 4-Nitrophenol and HBr   (e) Nitrobenzene, Br_2, and water

(c) 4-Nitrophenol, Br_2, and $FeBr_3$

CHAPTER 24

Phenols

SOLUTIONS TO TEXT PROBLEMS

24.1 (*b*) A benzyl group ($C_6H_5CH_2$—) is ortho to the phenolic hydroxyl group in *o*-benzylphenol.

$$OH$$
$$CH_2C_6H_5$$

(*c*) Naphthalene is numbered as shown. 3-Nitro-1-naphthol has a hydroxyl group at C-1 and a nitro group at C-3.

Naphthalene

3-Nitro-1-naphthol

24.2 Intramolecular hydrogen bonding between the hydroxyl group and the ester carbonyl can occur when these groups are ortho to each other.

Methyl salicylate

Intramolecular hydrogen bonds form at the expense of intermolecular ones, and intramolecularly hydrogen-bonded phenols have lower boiling points than isomers in which only intermolecular hydrogen bonding is possible.

24.3 (b) A cyano group withdraws electrons from the ring by resonance. A p-cyano substituent is conjugated directly with the negatively charged oxygen and stabilizes the anion more than does a m-cyano substituent.

p-Cyanophenol is slightly more acidic than m-cyanophenol, the pK_a values being 8.0 and 8.5, respectively.

(c) The electron-withdrawing inductive effect of fluorine will be more pronounced at the ortho position than at the para. o-Fluorophenol ($pK_a = 8.7$) is a stronger acid than p-fluorophenol ($pK_a = 9.9$).

24.4 The text points out that the reaction proceeds by the addition–elimination mechanism of nucleophilic aromatic substitution. Under the strongly basic conditions of the reaction, p-toluenesulfonic acid is first converted to its anion.

p-Toluenesulfonic acid   Hydroxide ion   p-Toluenesulfonate ion   Water

Nucleophilic addition of hydroxide ion gives a cyclohexadienyl anion intermediate.

p-Toluenesulfonate ion   Hydroxide   Cyclohexadienyl anion

Loss of sulfite ion (SO_3^{2-}) gives p-cresol.

Cyclohexadienyl anion   p-Cresol   Sulfite

It is also possible that the elimination stage of the reaction proceeds as follows:

p-Methylphenoxide ion

24.5 The text states that the hydrolysis of chlorobenzene in base follows an elimination–addition mechanism.

Chlorobenzene   Benzyne

Benzyne   Phenol

24.6 (*b*) The reaction is Friedel–Crafts alkylation. Proton transfer from sulfuric acid to 2-methylpropene gives *tert*-butyl cation. Because the position para to the hydroxyl substituent already bears a bromine, the *tert*-butyl cation attacks the ring at the position ortho to the hydroxyl.

4-Bromo-2-
methylphenol

2-Methylpropene

4-Bromo-2-*tert*-butyl-
6-methylphenol
(isolated yield, 70%)

(*c*) Acidification of sodium nitrite produces nitrous acid, which nitrosates the strongly activated aromatic ring of phenols.

2-Isopropyl-5-methylphenol   2-Isopropyl-5-methyl-4-nitrosophenol
                                                        (isolated yield, 87%)

(*d*) Friedel–Crafts acylation occurs ortho to the hydroxyl group.

p-Cresol   Propanoyl   1-(2-Hydroxy-5-methylphenyl)-
                     chloride   1-propanone (isolated yield, 87%)

24.7 (*b*) The hydroxyl group of 2-naphthol is converted to the corresponding acetate ester.

2-Naphthol   Acetic anhydride   2-Naphthyl acetate   Sodium acetate

(*c*) Benzoyl chloride acylates the hydroxyl group of phenol.

Phenol   Benzoyl chloride   Phenyl benzoate   Hydrogen
                                                                                 chloride

24.8 Epoxides are sensitive to nucleophilic ring-opening reactions. Phenoxide ion attacks the less hindered carbon to yield 1-phenoxy-2-propanol.

Phenoxide ion   1,2-Epoxypropane   1-Phenoxy-2-propanol

24.9 The aryl halide must be one that is reactive toward nucleophilic aromatic substitution by the addition–elimination mechanism. *p*-Fluoronitrobenzene is far more reactive than fluorobenzene. The reaction shown yields *p*-nitrophenyl phenyl ether in 92% yield.

Sodium
phenoxide

p-Fluoronitrobenzene

p-Nitrophenyl phenyl ether

24.10 Substituted allyl aryl ethers undergo a Claisen rearrangement similar to the reaction described in text Section 24.13 for allyl phenyl ether. 2-Butenyl phenyl ether rearranges on heating to give *o*-(1-methyl-2-propenyl)phenol.

2-Butenyl phenyl
ether

o-(1-Methyl-2-propenyl)phenol

24.11 (*a*) The parent compound is benzaldehyde. Vanillin bears a methoxy group (CH_3O) at C-3 and a hydroxyl group (HO) at C-4.

Vanillin
(4-hydroxy-3-methoxybenzaldehyde)

(*b*, *c*) Thymol and carvacrol differ with respect to the position of the hydroxyl group.

Thymol
(2-isopropyl-5-methylphenol)

Carvacrol
(5-isopropyl-2-methylphenol)

(*d*) An allyl substituent is $-CH_2CH=CH_2$.

Eugenol
(4-allyl-2-methoxyphenol)

(*e*) Benzoic acid is $C_6H_5CO_2H$. Gallic acid bears three hydroxyl groups, located at C-3, C-4, and C-5.

Gallic acid
(3,4,5-trihydroxybenzoic acid)

(*f*) Benzyl alcohol is $C_6H_5CH_2OH$. Salicyl alcohol bears a hydroxyl group at the ortho position.

Salicyl alcohol
(*o*-hydroxybenzyl alcohol)

24.12 (*a*) The compound is named as a derivative of phenol. The substituents (ethyl and nitro) are cited in alphabetical order with numbers assigned in the direction that gives the lowest number at the first point of difference.

3-Ethyl-4-nitrophenol

(*b*) An isomer of the compound in part (*a*) is 4-ethyl-3-nitrophenol.

4-Ethyl-3-nitrophenol

(*c*) The parent compound is phenol. It bears, in alphabetical order, a benzyl group at C-4 and a chlorine at C-2.

4-Benzyl-2-chlorophenol

(*d*) This compound is named as a derivative of anisole, $C_6H_5OCH_3$. Because multiplicative prefixes (di-, tri-, etc.) are not considered when alphabetizing substituents, isopropyl precedes dimethyl.

4-Isopropyl-2,6-
dimethylanisole

(*e*) The compound is an aryl ester of trichloroacetic acid. The aryl group is 2,5-dichlorophenyl.

2,5-Dichlorophenyl
trichloroacetate

24.13 (*a*) The reaction is an acid–base reaction. Phenol is the acid; sodium hydroxide is the base.

| Phenol (stronger acid) | Sodium hydroxide (stronger base) | Sodium phenoxide (weaker base) | Water (weaker acid) |

(*b*) Sodium phenoxide reacts with ethyl bromide to yield ethyl phenyl ether in a Williamson reaction. Phenoxide ion acts as a nucleophile.

$$C_6H_5ONa \ + \ CH_3CH_2Br \longrightarrow C_6H_5OCH_2CH_3 \ + \ NaBr$$

| Sodium phenoxide | Ethyl bromide | Ethyl phenyl ether | Sodium bromide |

(*c*) *p*-Toluenesulfonates behave much like alkyl halides in nucleophilic substitutions. Phenoxide ion displaces *p*-toluenesulfonate from the primary carbon.

Sodium phenoxide   Butyl *p*-toluenesulfonate   Butyl phenyl ether   Sodium *p*-toluenesulfonate

(*d*) Acid anhydrides react with phenoxide anions to yield aryl esters.

Sodium phenoxide   Acetic anhydride   Phenyl acetate   Sodium acetate (*e*) Acyl chlorides convert phenols to aryl esters.

o-Cresol   Benzoyl chloride   2-Methylphenyl benzoate   Hydrogen
                                                                                              chloride

(*f*) Phenols react as nucleophiles toward epoxides.

m-Cresol   Ethylene oxide   2-(3-Methylphenoxy)ethanol

The reaction as written conforms to the requirements of the problem that a balanced equation be written. Of course, the reaction will be much faster if catalyzed by acid or base, but the catalysts do not enter into the equation representing the overall process.

(*g*) Bromination of the aromatic ring of 2,6-dichlorophenol occurs para to the hydroxy group. The more activating group (—OH) determines the orientation of the product.

2,6-Dichlorophenol   Bromine   4-Bromo-2,6-   Hydrogen
                                                          dichlorophenol   bromide

(*h*) In aqueous solution bromination occurs at all the open positions that are ortho and para to the hydroxyl group.

p-Cresol   Bromine   2,6-Dibromo-4-   Hydrogen
                                                    methylphenol   bromide

(*i*) Hydrogen bromide cleaves ethers to give an alkyl halide and a phenol.

Isopropyl phenyl ether   Hydrogen   Phenol   Isopropyl
                            bromide   bromide

24.14 (*a*) Strongly electron-withdrawing groups, particularly those such as —NO$_2$, increase the acidity of phenols by resonance stabilization of the resulting phenoxide anion. Electron-releasing substituents such as —CH$_3$ exert a very small acid-weakening effect.

2,4,6-Trinitrophenol,
more acidic
(pK_a = 0.4)

2,4,6-Trimethylphenol,
less acidic
(pK_a = 10.9)

2,4,6-Trinitrophenol, also known as picric acid, is a stronger acid by far than 2,4,6-trimethylphenol. All three nitro groups participate in resonance stabilization of the picrate anion.

(*b*) Stabilization of a phenoxide anion is most effective when electron-withdrawing groups are present at the ortho and para positions because these carbons bear most of the negative charge in phenoxide anion.

2,6-Dichlorophenol is therefore expected to be (and is) a stronger acid than 3,5-dichlorophenol.

2,6-Dichlorophenol, more acidic   3,5-Dichlorophenol, less acidic
$(pK_a = 6.8)$ $(pK_a = 8.2)$

(*c*) The same principle is at work here as in part (*b*). A nitro group para to the phenol oxygen is directly conjugated to it and stabilizes the anion better than one at the meta position.

4-Nitrophenol, stronger acid   3-Nitrophenol, weaker acid
$(pK_a = 7.2)$ $(pK_a = 8.4)$

(*d*) A cyano group is strongly electron-withdrawing, and so 4-cyanophenol is a stronger acid than phenol.

4-Cyanophenol, more acidic   Phenol, less acidic
$(pK_a = 8.0)$ $(pK_a = 10)$

There is resonance stabilization of the 4-cyanophenoxide anion.

(*e*) The 5-nitro group in 2,5-dinitrophenol is meta to the hydroxyl group and so does not stabilize the resulting anion as much as does an ortho or a para nitro group.

2,6-Dinitrophenol, more acidic   2,5-Dinitrophenol, less acidic
$(pK_a = 3.7)$ $(pK_a = 5.2)$

24.15 (*a*) The rate-determining step of ester hydrolysis in basic solution is formation of the tetrahedral intermediate.

Because this intermediate is negatively charged, there will be a small effect favoring its formation when the aryl group bears an electron-withdrawing substituent. Furthermore, this intermediate can either return to starting materials or proceed to products.

The proportion of the tetrahedral intermediate that goes on to products increases as the leaving group ArO⁻ becomes less basic. This is strongly affected by substituents; electron-withdrawing groups stabilize ArO⁻. The prediction is that *m*-nitrophenyl acetate undergoes hydrolysis in basic solution faster than phenol. Indeed, this is observed to be the case; *m*-nitrophenyl acetate reacts some ten times faster than does phenol acetate at 25°C.

m-Nitrophenyl acetate
(more reactive)

m-Nitrophenoxide anion
(a better leaving group
than phenoxide because
it is less basic)

(*b*) The same principle applies here as in part (*a*). *p*-Nitrophenyl acetate reacts faster than *m*-nitrophenyl acetate (by about 45%) largely because *p*-nitrophenoxide is less basic and thus a better leaving group than *m*-nitrophenoxide.

Resonance in *p*-nitrophenoxide is particularly effective because the *p*-nitro group is directly conjugated to the oxyanion; direct conjugation of these groups is absent in *m*-nitrophenoxide.

(*c*) The reaction of ethyl bromide with a phenol is an S_N2 reaction in which the oxygen of the phenol is the nucleophile. The reaction is much faster with sodium phenoxide than with phenol because an anion is more nucleophilic than a corresponding neutral molecule.

Faster reaction:

Slower reaction:

(*d*) The answer here also depends on the nucleophilicity of the attacking species, which is a phenoxide anion in both reactions.

The more nucleophilic anion is phenoxide ion, because it is more basic than *p*-nitrophenoxide.

More basic;
better nucleophile

Better delocalization of negative
charge makes this less
basic and less nucleophilic.

Rate measurements reveal that sodium phenoxide reacts 17 times faster with ethylene oxide (in ethanol at 70°C) than does its *p*-nitro derivative.

(*e*) This reaction is electrophilic aromatic substitution. Because a hydroxy substituent is more activating than an acetate group, phenol undergoes bromination faster than does phenyl acetate.

Resonance involving ester group reduces
tendency of oxygen to donate electrons to ring.

24.16 Nucleophilic aromatic substitution by the elimination–addition mechanism is impossible, owing to the absence of any protons that might be abstracted from the substrate. The addition–elimination pathway is available, however.

Hexafluorobenzene

Pentafluorophenol

This pathway is favorable because the cyclohexadienyl anion intermediate formed in the rate-determining step is stabilized by the electron-withdrawing inductive effect of its fluorine substituents.

24.17 (*a*) Allyl bromide is a reactive alkylating agent and converts the free hydroxyl group of the aryl compound (a natural product known as *guaiacol*) to its corresponding allyl ether.

Guaiacol

Allyl bromide

2-Allyloxyanisole (80–90%)

(*b*) Sodium phenoxide acts as a nucleophile in this reaction and is converted to an ether.

Sodium phenoxide

3-Chloro-1,2-propanediol

3-Phenoxy-1,2-propanediol (61–63%)

(*c*) Orientation in nitration is governed by the most activating substituent, in this case the hydroxyl group.

Vanillin

$\xrightarrow[\text{acetic acid, heat}]{\text{HNO}_3}$

4-Hydroxy-3-methoxy-5-nitrobenzaldehyde (83%)

(*d*) Allyl aryl ethers undergo a Claisen rearrangement on heating. Heating *p*-acetamidophenyl allyl ether gave an 83% yield of 4-acetamido-2-allylphenol.

p-Acetamidophenyl allyl ether

$\xrightarrow{\text{heat}}$

4-Acetamido-2-allylphenol

(*e*) The hydroxyl group, as the most activating substituent, controls the orientation of electrophilic aromatic substitution. Bromination takes place ortho to the hydroxyl group.

2-Ethoxy-4-nitrophenol

$+ \text{ Br}_2 \xrightarrow{\text{acetic acid}}$

2-Bromo-6-ethoxy-4-nitrophenol (65%)

(*f*) Oxidation of hydroquinone derivatives (*p*-dihydroxybenzenes) with Cr(VI) reagents is a method for preparing quinones.

2-Chloro-1,4-benzenediol

$\xrightarrow[\text{H}_2\text{SO}_4]{\text{K}_2\text{Cr}_2\text{O}_7}$

2-Chloro-1,4-benzoquinone (88%)

(g) Aryl esters undergo a reaction known as the *Fries rearrangement* on being treated with aluminum chloride, which converts them to acyl phenols. Acylation takes place para to the hydroxyl in this case.

5-Isopropyl-2-methylphenyl acetate	4-Hydroxy-2-isopropyl-5-methylacetophenone (90%)

(h) Nucleophilic aromatic substitution takes place to yield a diaryl ether. The nucleophile is the phenoxide ion derived from 2,6-dimethylphenol.

2,6-Dimethylphenol	*p*-Chloronitrobenzene	2,6-Dimethylphenyl *p*-nitrophenyl ether (82%)

(i) Chlorination with excess chlorine occurs at all available positions that are ortho and para to the hydroxyl group.

2,5-Dichlorophenol	Chlorine	2,3,4,6-Tetrachlorophenol (isolated yield, 100%)	Hydrogen chloride

(j) Amines react with esters to give amides. In the case of a phenyl ester, phenol is the leaving group.

o-Methylaniline	Phenyl salicylate	*N*-(*o*-Methylphenyl)salicylamide (isolated yield, 73–77%)	Phenol

(*k*) Aryl diazonium salts attack electron-rich aromatic rings, such as those of phenols, to give the products of electrophilic aromatic substitution.

2,4,5- Trichlorophenol   Benzenediazonium chloride   2-Benzeneazo-3,4,6-trichlorophenol (80%)

24.18 In the first step *p*-nitrophenol is alkylated on its phenolic oxygen with ethyl bromide.

p-Nitrophenol   Ethyl bromide   Ethyl *p*-nitrophenyl ether

Reduction of the nitro group gives the corresponding arylamine.

Ethyl *p*-nitrophenyl ether   *p*-Ethoxyaniline

Treatment of *p*-ethoxyaniline with acetic anhydride gives phenacetin.

p-Ethoxyaniline   Acetic anhydride   *p*-Ethoxyacetanilide (phenacetin)

24.19 The three parts of this problem make up the series of steps by which *o*-bromophenol is prepared.

(*a*) Because direct bromination of phenol yields both *o*-bromophenol and *p*-bromophenol, it is essential that the para position be blocked prior to the bromination step. In practice, what is done is to disulfonate phenol, which blocks the para and one of the ortho positions.

Phenol   4-Hydroxy-1,3-benzenedisulfonic acid (compound A)

(*b*) Bromination then can be accomplished cleanly at the open position ortho to the hydroxyl group.

Compound A
           5-Bromo-4-hydroxy-1,3-
           benzenedisulfonic acid
           (compound B)

(*c*) After bromination the sulfonic acid groups are removed by acid-catalyzed hydrolysis.

Compound B
           o-Bromophenol
           (compound C)

24.20 Nitration of 3,5-dimethylphenol gives a mixture of the 2-nitro and 4-nitro derivatives.

3,5-Dimethylphenol
      3,5-Dimethyl-2-
      nitrophenol
           3,5-Dimethyl-4-
           nitrophenol

The more volatile compound (compound A), isolated by steam distillation, is the 2-nitro derivative. Intramolecular hydrogen bonding is possible between the nitro group and the hydroxyl group.

Intramolecular hydrogen bonding
in 3,5-dimethyl-2-nitrophenol

The 4-nitro derivative participates in intermolecular hydrogen bonds and has a much higher boiling point; it is compound B.

24.21 The relationship between the target molecule and the starting materials tells us that two processes are required, formation of a diaryl ether linkage and nitration of an aromatic ring. The proper order of carrying out these two separate processes is what needs to be considered.

The critical step is ether formation, a step that is feasible for the reactants shown:

Phenol   p-Chloronitrobenzene   4-Nitrophenyl phenyl ether

The reason this reaction is suitable is that it involves nucleophilic aromatic substitution by the addition–elimination mechanism on a p-nitro-substituted aryl halide. Indeed, this reaction has been carried out and gives an 80–82% yield. A reasonable synthesis would therefore begin with the preparation of p-chloronitrobenzene.

Chlorobenzene   o-Chloronitrobenzene   p-Chloronitrobenzene

Separation of the p-nitro-substituted aryl halide and reaction with phenoxide ion complete the synthesis.
     The following alternative route is less satisfactory:

Phenol   Chlorobenzene   Diphenyl ether

Diphenyl ether   2-Nitrophenyl phenyl ether   4-Nitrophenyl phenyl ether

The difficulty with this route concerns the preparation of diphenyl ether. Direct reaction of phenoxide ion with chlorobenzene is very slow and requires high temperatures because chlorobenzene is a poor substrate for nucleophilic substitution.
     A third route is also unsatisfactory because it, too, requires nucleophilic substitution on chlorobenzene.

Phenol   o-Nitrophenol   p-Nitrophenol

p-Nitrophenol   Chlorobenzene   4-Nitrophenyl phenyl ether

24.22 The overall transformation that needs to be accomplished is

2,3-Dimethoxybenzaldehyde   3-Pentadecylcatechol

A reasonable place to begin is with the attachment of the side chain. The aldehyde function allows for chain extension by a Wittig reaction.

2,3-Dimethoxybenzaldehyde

Hydrogenation of the double bond and hydrogen halide cleavage of the ether functions complete the synthesis.

3-Pentadecylcatechol

Other synthetic routes are of course possible. One of the earliest approaches used a Grignard reaction to attach the side chain.

2,3-Dimethoxybenzaldehyde

The resulting secondary alcohol can then be dehydrated to the same alkene intermediate prepared in the preceding synthetic scheme.

Again, hydrogenation of the double bond and ether cleavage leads to the desired 3-pentadecylcatechol.

24.23 Recall that the Claisen rearrangement converts an aryl allyl ether to an ortho-substituted allyl phenol. The presence of an allyl substituent in the product ortho to an aryl ether thus suggests the following retrosynthesis:

As reported in the literature synthesis, the starting phenol may be converted to the corresponding allyl ether by reaction with allyl bromide in the presence of base. This step was accomplished in 80% yield. Heating the allyl ether yields the *o*-allyl phenol.

The synthesis is completed by methylation of the phenolic oxygen and saponification of the acetate ester. The final three steps of the synthesis proceeded in an 82% overall yield.

24.24 The driving force for this reaction is the stabilization that results from formation of the aromatic ring. A reasonable series of steps begins with protonation of the carbonyl oxygen.

Resonance forms of protonated ketone

Protonated ketone can rearrange by alkyl migration.

Aromatization of this intermediate occurs by loss of a proton.

24.25 Bromination of *p*-hydroxybenzoic acid takes place in the normal fashion at both positions ortho to the hydroxy group.

p-Hydroxybenzoic acid

3,5-Dibromo-4-hydroxybenzoic acid

A third bromination step, this time at the para position, leads to the intermediate shown.

Aromatization of this intermediate occurs by decarboxylation.

2,4,6-Tribromophenol

24.26 Electrophilic attack of bromine on 2,4,6-tribromophenol leads to a cationic intermediate.

2,4,6-Tribromophenol

Loss of the hydroxyl proton from this intermediate generates the observed product.

2,4,4,6-Tetrabromocyclohexadienone

24.27 A good way to approach this problem is to assume that bromine attacks the aromatic ring of the phenol in the usual way, that is, para to the hydroxyl group.

2,4,6-Tri-*tert*-butylphenol

This cation cannot yield the product of electrophilic aromatic substitution by loss of a proton from the ring but can lose a proton from oxygen to give a cyclohexadienone derivative.

4-Bromo-2,4,6-tri-*tert*-
butyl-2,5-cyclohexadienone

This cyclohexadienone is the compound $C_{18}H_{29}BrO$, and the peaks at 1655 and 1630 cm^{-1} in the infrared are consistent with C=O and C=C stretching vibrations. The compound's symmetry is consistent with the observed ^{1}H NMR spectrum; two equivalent *tert*-butyl groups at C-2 and C-6 appear as an 18-proton singlet at δ 1.3, the other *tert*-butyl group is a nine-proton singlet at δ 1.2, and the two equivalent vinyl protons of the ring appear as a singlet at δ 6.9.

24.28 Because the starting material is an acetal and the reaction conditions lead to hydrolysis with the production of 1,2-ethanediol, a reasonable reaction course is

Compound A   1,2-Ethanediol   Compound B

Indeed, dione B satisfies the spectroscopic criteria. Carbonyl bands are seen in the infrared spectrum, and compound B has two sets of protons to be seen in its ^{1}H NMR spectrum. The two vinyl protons are equivalent and appear at low field, δ 6.7; the four methylene protons are equivalent to each other and are seen at δ 2.9.

Compound B is the doubly ketonic tautomeric form of hydroquinone, compound C, to which it isomerizes on standing in water.

Compound B   Compound C
                                   (hydroquinone)

24.29 A reasonable first step is protonation of the hydroxyl oxygen.

Cumene hydroperoxide

The weak oxygen–oxygen bond can now be cleaved, with loss of water as the leaving group.

This intermediate bears a positively charged oxygen with only six electrons in its valence shell. Like a carbocation, such a species is highly electrophilic. The electrophilic oxygen attacks the π system of the neighboring aromatic ring to give an unstable intermediate.

Ring opening of this intermediate is assisted by one of the lone pairs of oxygen and restores the aromaticity of the ring.

The cation formed by ring opening is captured by a water molecule to yield the hemiacetal product.

24.30 (*a*) The molecular formula of the compound ($C_9H_{12}O$) tells us that it has a total of four double bonds and rings (index of hydrogen deficiency = 4). The prominent peak in the IR spectrum is the hydroxyl absorption of an alcohol or a phenol at 3300 cm^{-1}.

Peaks in the δ 110–160 region of the ^{13}C NMR spectrum suggest an aromatic ring, which accounts for six of the nine carbon atoms and all its double bonds and rings. The presence of four peaks in this region, two of which are C and two CH, indicates a para-disubstituted aromatic derivative. That the remaining three carbons are sp^3-hybridized is indicated by the upfield absorptions at δ 15, 26, and 38. None of these carbons has a chemical shift below δ 40, and so none of them can be bonded to the hydroxyl group. Thus the hydroxyl group must be bonded to the aromatic ring. The compound is 4-propylphenol.

4-Propylphenol

(*b*) Once again the molecular formula ($C_9H_{11}BrO$) indicates a total of four double bonds and rings.

The four peaks in the δ 110–160 region of the spectrum, three of which represent CH, suggest a monosubstituted aromatic ring.

The remaining atoms to be accounted for are O and Br. Because all the unsaturations are accounted for by the benzene ring and the IR spectrum lacks any hydroxyl absorption, the oxygen atom must be part of an ether function. The three CH$_2$ groups indicated by the absorptions at δ 32, 35, and 66 in the ^{13}C NMR spectrum allow the compound to be identified as 3-bromopropyl phenyl ether.

3-Bromopropyl phenyl ether

ANSWERS TO INTERPRETIVE PROBLEMS

24.31 A;   **24.32** B;   **24.33** A;   **24.34** C

SELF TEST

PART A

A-1. Which is the stronger acid, *m*-hydroxybenzaldehyde or *p*-hydroxybenzaldehyde? Explain your answer, using resonance structures.

A-2. The cresols are methyl-substituted phenols. Predict the major products to be obtained from the reactions of *o*-, *m*-, and *p*-cresol with dilute nitric acid.

A-3. Give the structure of the product from the reaction of *p*-cresol with propanoyl chloride, $CH_3CH_2\overset{\displaystyle O}{\overset{\|}{C}}Cl$, in the presence of $AlCl_3$. What product is obtained in the absence of $AlCl_3$?

A-4. Provide the structure of the reactant, reagent, or product omitted from each of the following:

(*a*) —OCH(CH₃)₂ $\xrightarrow{\text{HBr}}$? (two products)

(*b*) ? (two compounds) $\longrightarrow$

(*c*) $\xrightarrow[\text{2. } H_3O^+]{\text{1. ?}}$

(*d*) $\xrightarrow[\text{heat}]{\text{HBr}}$? (C_8H_9BrO)

A-5. Provide the structures of compounds A and B in the following sequence of reactions:

$\xrightarrow[\text{K}_2\text{CO}_3]{CH_3CH_2CH=CHCH_2Br}$ A $\xrightarrow{\text{heat}}$ B ($C_{11}H_{14}O$)

A-6. Prepare *p-tert*-butylphenol from *tert*-butylbenzene using any necessary organic or inorganic reagents.

PART B

B-1. Rank the following in order of decreasing acid strength (most acidic → least acidic):

1   2   3   4

(a) $2 > 4 > 1 > 3$

(b) $3 > 1 > 2 > 4$

(c) $1 > 3 > 4 > 2$

(d) $3 > 1 > 4 > 2$

B-2. Rank the following compounds in order of increasing acidity (weakest acid first).

1   2   3

(a) $2 < 3 < 1$

(b) $3 < 2 < 1$

(c) $3 < 1 < 2$

(d) $2 < 1 < 3$

(e) $1 < 2 < 3$

B-3. Which of the following phenols has the largest pK_a value (i.e., is least acidic)?

(a) Cl—⟨⟩—OH

(c) O_2N—⟨⟩—OH

(b) H_3C—⟨⟩—OH

(d) $N \equiv C$—⟨⟩—OH

B-4. Which of the following reactions is a more effective method for preparing phenyl propyl ether?

I: $C_6H_5ONa + CH_3CH_2CH_2Br \longrightarrow$

II: $CH_3CH_2CH_2ONa + C_6H_5Br \longrightarrow$

(a) Reaction I is more effective.

(b) Reaction II is more effective.

(c) Both reactions I and II are effective.

(d) Neither reaction I nor reaction II is effective.

B-5. What reactant gives the product shown on heating with aluminum chloride?

H_3C—⟨⟩—C(=O)—⟨⟩—OH

(a)

(b)

(c)

(d)

B-6. What are the products of the following reaction?

(a)

(b)

(c)

(d)

(e)

B-7. Which of the following sets of reagents, used in the order shown, would enable preparation of p-chlorophenol from p-chloronitrobenzene?

(a) 1. Fe, HCl; 2. NaOH; 3. $NaNO_2$, H_2SO_4; 4. H_3PO_2

(b) 1. Fe, HCl; 2. NaOH; 3. $NaNO_2$, H_2SO_4; 4. H_2O, heat

(c) 1. Fe, HCl; 2. NaOH; 3. $NaNO_2$, H_2SO_4; 4. ethanol

(d) 1. NaOH, heat; 2. HCl

B-8. What is the product obtained by heating the following allylic ether of phenol?

(a)

(b)

(c)

(d)

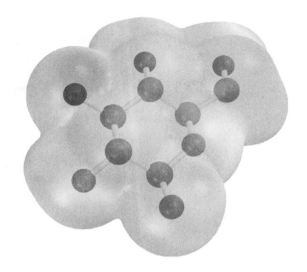

CHAPTER 25

Carbohydrates

SOLUTIONS TO TEXT PROBLEMS

25.1 (*b*) Reorient the three-dimensional representation, putting the aldehyde group at the top and the primary alcohol at the bottom.

$$HOCH_2 \overset{H}{\underset{OH}{-C-}} CHO \quad \xrightarrow{\text{turn } 90°} \quad H\cdots\overset{CHO}{\underset{CH_2OH}{C}}\cdots OH$$

What results is not equivalent to a proper Fischer projection, because the horizontal bonds are directed "back" when they should be "forward." The opposite is true for the vertical bonds. To make the drawing correspond to a proper Fischer projection, we need to rotate it 180° around a vertical axis.

$$H\cdots\overset{CHO}{\underset{CH_2OH}{C}}\cdots OH \quad \longrightarrow \quad HO\overset{CHO}{\underset{CH_2OH}{-C-}}H \quad \text{is equivalent to} \quad HO\overset{CHO}{\underset{CH_2OH}{-\!\!\!-\!\!\!-}}H$$

rotate 180°

Now, having the molecule arranged properly, we see that it is L-glyceraldehyde.

(c) Look at the drawing from a perspective that permits you to see the carbon chain oriented vertically with the aldehyde at the top and the CH_2OH at the bottom. Both groups should point away from you.

When examined from this perspective, the hydrogen is to the left and the hydroxyl to the right with both pointing toward you.

The molecule is D-glyceraldehyde.

25.2 Begin by drawing a perspective view of the molecular model shown in the problem. To view the compound as a Fischer projection, redraw it in an eclipsed conformation.

Staggered conformation   Same molecule in eclipsed conformation

The eclipsed conformation shown, when oriented so that the aldehyde carbon is at the top, vertical bonds back, and horizontal bonds pointing outward from their chirality centers, is readily transformed into the Fischer projection of L-erythrose.

L-Erythrose

25.3 L-Arabinose is the mirror image of D-arabinose, the structure of which is given in text Figure 25.2. The configuration at *each* chirality center of D-arabinose must be inverted to transform it into L-arabinose.

D-(−)-Arabinose   L-(+)-Arabinose

25.4 The configuration at C-5 is opposite to that of D-(+)-glyceraldehyde; it is *S*. This particular carbohydrate therefore belongs to the L series. Comparing it with the Fischer projection formulas of the eight D-aldohexoses reveals it to be in the mirror image of D-(+)-talose; it is L-(−)-talose.

25.5 (*b*) As in part (*a*) the —OH group on the highest numbered chirality center (C-3) is up, and this stereoisomer belongs to the L series. The —OH group at the anomeric carbon is also up, making this the α-furanose form.

α-L-Threofuranose

(*c, d*) The —OH group on the highest numbered chirality center (C-3) is down in both of these structures, and these stereoisomers belong to the D series. For D series carbohydrates the configuration of the anomeric carbon is α if its hydroxyl group is down and β if the hydroxyl group is up.

α-D-Threofuranose   β-D-Threofuranose

25.6 (*b*) The Fischer projection of D-arabinose may be found in text Figure 25.2. The Fischer projection and the eclipsed conformation corresponding to it are

D-Arabinose   Eclipsed conformation   Conformation suitable for
                     of D-arabinose   furanose ring formation

Cyclic hemiacetal formation between the carbonyl group and the C-4 hydroxyl yields the α- and β-furanose forms of D-arabinose.

β-D-Arabinofuranose   α-D-Arabinofuranose

(*c*) The mirror image of D-arabinose [from part (*b*)] is L-arabinose.

D-Arabinose   L-Arabinose   Eclipsed conformation
                                          of L-arabinose

The C-4 atom of the eclipsed conformation of L-arabinose must be rotated 120° in a clockwise sense so as to bring its hydroxyl group into the proper orientation for furanose ring formation.

Original eclipsed conformation
of L-arabinose

Conformation suitable for
furanose ring formation

Cyclization gives the α- and β-furanose forms of L-arabinose.

α-L-Arabinofuranose   β-L-Arabinofuranose

In the L series the anomeric hydroxyl is up in the α isomer and down in the β isomer.

25.7 (*b*) The Fischer projection and Haworth formula for D-mannose are

D-Mannose

β-D-Mannopyranose
(Haworth formula)

The Haworth formula is more realistically drawn as the following chair conformation:

β-D-Mannopyranose

Mannose differs from glucose in configuration at C-2. All hydroxyl groups are equatorial in β-D-glucopyranose; the hydroxyl at C-2 is axial in β-D-mannopyranose.

(*c*) The conformational depiction of β-L-mannopyranose begins in the same way as that of β-D-mannopyranose. L-Mannose is the mirror image of D-mannose.

D-Mannose   L-Mannose   Eclipsed conformation of L-mannose

To rewrite the eclipsed conformation of L-mannose in a way that permits hemiacetal formation between the carbonyl group and the C-5 hydroxyl, C-5 is rotated clockwise 120°.

rotate about
C-4—C-5 bond

β-L-Mannopyranose
(remember, the anomeric
hydroxyl is down in the L series)

Translating the Haworth formula into a proper conformational depiction requires that a choice be made between the two chair conformations shown.

Haworth formula of
β-L-mannopyranose

Less stable chair conformation;
CH_2OH is axial

More stable chair conformation;
CH_2OH is equatorial

(*d*) The Fischer projection formula for L-ribose is the mirror image of that for D-ribose.

D-Ribose   L-Ribose   Eclipsed conformation of L-ribose is
oriented properly for ring closure.

Haworth formula of
β-L-ribopyranose

Of the two chair conformations of β-L-ribose, the one with the greater number of equatorial substituents is more stable.

Less stable chair
conformation of
β-L-ribopyranose

More stable chair
conformation of
β-L-ribopyranose

25.8 The equation describing the equilibrium is

α-D-Mannopyranose
$[\alpha]_D^{20} + 29.3\infty$

Open-chain form of D-mannose

β-D-Mannopyranose
$[\alpha]_D^{20} ñ 17.0\infty$

Let A = percent α isomer; $100 - A$ = percent β isomer. Then

$$A(+29.3°) + (100 - A)(-17.0°) = 100(+14.2°)$$
$$46.3A = 3120$$
$$\text{Percent } \alpha \text{ isomer} = 67\%$$
$$\text{Percent } \beta \text{ isomer} = (100 - A) = 33\%$$

25.9 Based on the anomeric effect the C—Cl bond and the O—CH$_3$ bond are gauche in the most stable conformation of ClCH$_2$OCH$_3$.

25.10 Review carbohydrate terminology by referring to text Table 25.1. A *ketotetrose* is a four-carbon ketose. Writing a Fischer projection for a four-carbon ketose reveals that only one chirality center is present, and thus there are only two ketotetroses. They are enantiomers of each other and are known as D- and L-erythrulose.

D-Erythrulose   L-Erythrulose

25.11 (*b*) Because L-rhamnose is 6-deoxy-L-mannose, first write the Fischer projection formula of D-mannose, and then transform it to its mirror image, L-mannose. Transform the C-6 CH$_2$OH group to CH$_3$ to produce 6-deoxy-L-mannose.

D-Mannose
(from text Figure 25.2)   L-Mannose   6-Deoxy-L-mannose
                                                (L-rhamnose)

25.12 Numbering of the 9-carbon chain begins at the carboxyl group in *N*-acetylneuraminic acid. The compound can be described as a ketose as C-2 is the hemiacetal of a ketone carbonyl. It is also a deoxy sugar as evidenced by the lack of a hydroxyl group on C-3. The highest-numbered chirality center is C-8. It is D; it has the same configuration as D-glyceraldehyde.

N-Acetylneuraminic acid

25.13 Reaction of a carbohydrate with an alcohol in the presence of an acid catalyst gives mixed acetals at the anomeric position.

D-Galactose   Methanol   Methyl   Methyl
                                   α-D-galactopyranoside β-D-galactopyranoside

25.14 Acid-catalyzed addition of methanol to the glycal proceeds by regioselective protonation of the double bond in the direction that leads to the more stable carbocation. The more stable carbocation is the one stabilized by the ring oxygen.

Capture on either face of the carbocation by methanol yields the α and β methyl glycosides.

25.15 The hemiacetal opens to give an intermediate containing a free aldehyde function. Cyclization of this intermediate can produce either the α or the β configuration at this center. The axial and equatorial orientations of the anomeric hydroxyl can best be seen by drawing maltose with the pyranose rings in chair conformations.

β-Configuration of
hemiacetal (equatorial)

Key intermediate formed by
cleavage of hemiacetal

α-Configuration of
hemiacetal (axial)

Only the configuration of the hemiacetal function is affected in this process. The α configuration of the glycosidic linkage remains unchanged.

25.16 The galactitol formed by reduction of D-galactose has a plane of symmetry and is a meso form. It is achiral and thus not optically active.

25.17 Because the groups at both ends of the carbohydrate chain are oxidized to carboxylic acid functions, two combinations of one CH_2OH with one CHO group are possible.

D-Glucose

D-Glucaric acid

L-Gulose

L-Gulose yields the same aldaric acid on oxidation as does D-glucose.

25.18 Hydrolysis of the CN group gives a CO_2H at one end of the carbohydrate chain. The CH_2OH is unchanged at the other. The product is an aldonic acid.

25.19 In analogy with the D-fructose $\rightleftharpoons$ D-glucose interconversion, dihydroxyacetone phosphate and D-glyceraldehyde 3-phosphate can equilibrate by way of an enediol intermediate.

Dihydroxyacetone phosphate   Enediol   D-Glyceraldehyde 3-phosphate

25.20 The structure of methyl α-D-arabinopyranoside indicating the bonds that are cleaved on oxidation with periodic acid was given in the paragraph preceding this problem in the text. Carbon-3 is lost as formic acid. Carbons-2 and 4 become aldehyde functions.

The same changes occur irrespective of the stereochemistry of carbons-2,3, and 4. Therefore, the α-methyl pyranosides of D-ribose, arabinose, xylose, and lyxose all yield the same dialdehyde.

25.21 (*a*) The structure shown in the text is D-(+)-xylose; therefore (−)-xylose must be its mirror image and has the L-configuration at C-4.

D-(+)-Xylose   L-(−)-Xylose

(*b*) Alditols are the reduction products of carbohydrates; xylitol is derived from xylose by conversion of the terminal —CHO to —CH_2OH.

Xylitol

(*c*) Redraw the Fischer projection of D-xylose in its eclipsed conformation.

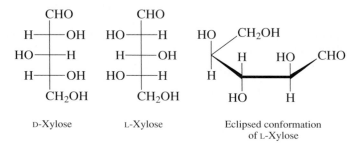

D-Xylose   Eclipsed conformation   Haworth formula of
                               of D-xylose   β-D-xylopyranose

The pyranose form arises by closure to a six-membered cyclic hemiacetal, with the C-5 hydroxyl group undergoing nucleophilic addition to the carbonyl. In the β-pyranose form of D-xylose the anomeric hydroxyl group is up.

The preferred conformation of β-D-xylopyranose is a chair with all the hydroxyl groups equatorial.

is better represented as

Haworth formula of   Chair conformation of
β-D-xylopyranose   β-D-xylopyranose

(*d*) L-Xylose is the mirror image of D-xylose.

D-Xylose   L-Xylose   Eclipsed conformation
                                       of L-Xylose

To construct the furanose form of L-xylose, the hydroxyl at C-4 needs to be brought into the proper orientation to form a five-membered ring.

rotate about
C-3—C-4 bond

The α-anomeric hydroxyl group is up in the L series.

(*e*) Methyl α-L-xylofuranoside is the methyl glycoside corresponding to the structure in part (*d*).

(*f*) Aldonic acids are derived from aldoses by oxidation of the terminal aldehyde to a carboxylic acid.

D-Xylose   D-Xylonic acid

(*g*) Aldonic acids tend to exist as lactones. A δ-lactone has a six-membered ring.

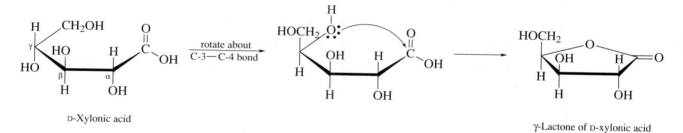

(*h*) A γ-lactone has a five-membered ring.

(*i*) Aldaric acids have carboxylic acid groups at both ends of the chain.

D-Xylose   Xylaric acid

25.22 Begin the problem by converting the Fischer projection of D-xylose to a perspective view. Remember that the horizontal lines of a Fischer projection represent bonds coming toward you, and the vertical lines are going away from you.

Rank the groups attached to each chirality center. Identify each chirality center as either R or S according to the methods described in Chapter 7. Remember that the proper orientation of the lowest ranked group (usually H) is away from you. Two of the chirality centers in D-xylose have the R configuration. The IUPAC name of D-xylose is $(2R,3S,4R)$-2,3,4,5-tetrahydroxypentanal.

25.23 (*a*) Reduction of aldoses with sodium borohydride yields polyhydroxy alcohols called *alditols*. Optically inactive alditols are those that have a plane of symmetry, that is, those that are meso forms. The D-aldohexoses that yield optically inactive alditols are D-allose and D-galactose.

(*b*) All the aldonic acids and their lactones obtained on oxidation of the aldohexoses with bromine are optically active. The presence of a carboxyl group at one end of the carbon chain and a CH_2OH at the other precludes the existence of meso forms.

(*c*) Nitric acid oxidation of aldoses converts them to aldaric acids. The same D-aldoses found to yield optically inactive alditols in part (*a*) yield optically inactive aldaric acids.

(*d*) Aldoses that differ in configuration only at C-2 enolize to the same enediol.

D-Allose   Enediol   D-Altrose

The chirality center at C-2 in the D-aldose becomes sp^2-hybridized in the enediol. The other pairs of D-aldohexoses that form the same enediols are

D-Glucose and D-mannose
D-Gulose and D-idose
D-Galactose and D-talose

25.24 (*a*) To unravel a pyranose form, locate the anomeric carbon and mentally convert the hemiacetal linkage to a carbonyl compound and a hydroxyl function.

Convert the open-chain form to a Fischer projection.

(*b*) Proceed in the same manner as in part (*a*) and unravel the furanose sugar by disconnecting the hemiacetal function.

The Fischer projection is

(*c*) By disconnecting and unraveling as before, the Fischer projection is revealed.

(*d*) Remember in disconnecting cyclic hemiacetals that the anomeric carbon is the one that bears two oxygen substituents.

25.25 (*a*) The L sugars have the hydroxyl group to the left at the highest numbered chirality center in their Fischer projection. The L sugars are the ones in Problem 25.24*a* and *c*.

Highest numbered chirality center is L.

Highest numbered chirality center is L.

(b) Deoxy sugars are those that lack an oxygen substituent on one of the carbons in the main chain. The carbohydrate in Problem 25.24b is a deoxy sugar.

No hydroxyl group at C-5

(c) Branched-chain sugars have a carbon substituent attached to the main chain; the carbohydrate in Problem 25.24c fits this description.

Methyl group attached to main chain ⟶

(d) Only the sugar in Problem 25.24d is a ketose.

Ketone carbonyl in Fischer projection

(e) A furanose ring is a five-membered cyclic hemiacetal. Only the compound in Problem 25.24b is a furanose form.

CH$_2$OH
H————H   O   OH
   H   H
H   H
HO   OH

(*f*) In D sugars, the α-configuration corresponds to the condition in which the hydroxyl group at the anomeric carbon is down. The α-D sugar is that in Problem 25.24*d*.

HOCH$_2$
HO   O
              CH$_2$OH
         HO
HO   OH

Anomeric hydroxyl
is down.

α-Pyranose form of a D-ketose

In the α-L series the anomeric hydroxyl is up. Neither of the L sugars—namely, those of Problem 25.24*a* and *c*—is α; both are β.

25.26 There are seven possible ketopentoses. The ketone carbonyl can be located at either C-2 or C-3. When the carbonyl group is at C-2, there are two chirality centers, giving rise to four stereoisomers (two pairs of enantiomers).

CH$_2$OH	CH$_2$OH	CH$_2$OH	CH$_2$OH
C=O	C=O	C=O	C=O
H——OH	HO——H	HO——H	H——OH
H——OH	HO——H	H——OH	HO——H
CH$_2$OH	CH$_2$OH	CH$_2$OH	CH$_2$OH
D-Ribulose	L-Ribulose	D-Xylulose	L-Xylulose

When the carbonyl group is located at C-3, there are only three stereoisomers, because one of them is a meso form and is superimposable on its mirror image.

CH$_2$OH	CH$_2$OH	CH$_2$OH
HO——H	H——OH	H——OH
C=O	C=O	C=O
H——OH	HO——H	H——OH
CH$_2$OH	CH$_2$OH	CH$_2$OH

25.27 (*a*) Carbon-2 is the only chirality center in D-apiose.

^{1}CHO
H——^{2}——OH
HOCH$_2$——^{3}——OH
       ^{4}CH$_2$OH

D-Apiose

Carbon-3 is not a chirality center; it bears two identical CH$_2$OH substituents.

(*b*) The alditol obtained on reduction of D-apiose retains the chirality center. It is chiral and optically active.

D-Apiose
(optically active)

D-Apiitol
(optically active)

(*c, d*) Cyclic hemiacetal formation in D-apiose involves addition of a CH$_2$OH hydroxyl group to the aldehyde carbonyl.

Three chirality centers occur in the furanose form, namely, the anomeric carbon C-1 and the original chirality center C-2, as well as a new chirality center at C-3.

In addition to the two furanose forms just shown, two more are possible. Instead of the reaction of the CH$_2$OH group that was shown to form the cyclic hemiacetal, the other CH$_2$OH group may add to the aldehyde carbonyl.

two furanose forms just shown

rotate C-3
120° about the
C-2—C-3 bond

25.28 The most reasonable conclusion is that all four are methyl glycosides. Two are the methyl glycosides of the α- and β-pyranose forms of mannose and two are the methyl glycosides of the α- and β-furanose forms.

Methyl
α-D-mannopyranoside

Methyl
β-D-mannopyranoside

Methyl
α-D-mannofuranoside

Methyl
β-D-mannofuranoside

In the case of the methyl glycosides of mannose, comparable amounts of pyranosides and furanosides are formed. The major products are the α isomers.

25.29 (*a*) The α-hydroxy aldehyde unit at the end of the sugar chain is cleaved, as well as all the vicinal diol functions. Four moles of periodic acid are required per mole of D-arabinose. Four moles of formic acid and one mole of formaldehyde are produced

(*b*) The points of cleavage of D-ribose on treatment with periodic acid are as indicated.

Four moles of periodic acid per mole of D-ribose is required. Four moles of formic acid and one mole of formaldehyde are produced.

(*c*) Write the structure of methyl β-D-glucopyranoside so as to identify the adjacent alcohol functions.

Methyl β-D-glucopyranoside

Two moles of periodic acid per mole of glycoside is required. One mole of formic acid is produced.

(*d*) There are two independent vicinal diol functions in this glycoside. Two moles of periodic acid are required per mole of substrate.

25.30 Because gentiobiose undergoes mutarotation, it must have a free hemiacetal group. Formation of two molecules of D-glucose indicates that it is a disaccharide and because that hydrolysis is catalyzed by emulsin, the glycosidic linkage is β. The methylation data, summarized in the following equation, require that the glucose units be present in pyranose forms and be joined by a β(1,6)-glycoside bond.

2,3,4,6-Tetra-*O*-methyl-D-glucose   2,3,4-Tri-*O*-methyl-D-glucose

25.31 The fact that phlorizin is hydrolyzed to D-glucose and compound A by emulsin indicates that it is a β-glucoside in which D-glucose is attached to one of the phenolic hydroxyls of compound B.

$C_{21}H_{24}O_{10} + H_2O$

D-Glucose ($C_6H_{12}O_6$)   Compound A ($C_{15}H_{14}O_5$)

The methylation experiment reveals to which hydroxyl glucose is attached. Excess methyl iodide reacts with all the available phenolic hydroxyl groups, but the glycosidic oxygen is not affected. Thus when the methylated phlorizin undergoes acid-catalyzed hydrolysis of its glycosidic bond, the oxygen in that bond is exposed as a phenolic hydroxyl group.

Compound B

This compound must arise by hydrolysis of

The structure of phlorizin is therefore

25.32 (*a*) The primary alcohol function reacts with triphenylmethyl chloride to give an ether by an S_N1 mechanism.

(*b*) The only two OH groups of methyl α-D-glucopyranoside that can yield an acetal with a structure analogous to *trans*-decalin are those at C-4 and C-6. Periodic acid oxidation of the product involves the OH groups at C-2 and C-3.

25.33 (*a*) Periodic acid cleaves the vicinal diol function of compound A.

Compound B

The CH=O at one end of compound B cannot react with the CH$_2$OH at the other end to give a cyclic hemiacetal because the structure produced would have a highly strained trans fusion between two five-membered rings.

Compound B is an acetal. Acid hydrolysis converts it to acetone and compound C.

Compound C

Oxidation of compound C converts it to the lactone (compound D).

Compound D

25.34 (*a*) Unravel the lactone so as to reveal the carbon chain and its stereochemistry.

The chain extension method adds a carboxyl group to the aldehyde end of an aldose. To determine the aldose, disconnect the carboxyl group and transform the next carbon to an aldehyde.

The aldose subjected to chain extension is D-xylose.

25.35 Like other glycosides, cyanogenic glycosides are cleaved to a carbohydrate and an alcohol on hydrolysis.

(a) In the case of linamarin the alcohol is recognizable as the cyanohydrin of acetone. Once formed, this cyanohydrin dissociates to hydrogen cyanide and acetone.

Linamarin   D-Glucose   Acetone cyanohydrin

$$CH_3CCH_3 \ + \ HCN$$

Acetone   Hydrogen cyanide

(b) Amygdalin undergoes an analogous hydrolytic cleavage to yield the cyanohydrin of benzaldehyde.

Amygdalin   D-Glucuronic acid   Benzaldehyde cyanohydrin

$$C_6H_5CH \ + \ HCN$$

Benzaldehyde   Hydrogen cyanide

25.36 Comparing D-glucose, D-mannose, and D-galactose, it can be said that the configuration of C-2 has a substantial effect on the relative energies of the α- and β-pyranose forms, but that the configuration of C-4 has virtually no effect. With this observation in mind, write the structures of the pyranose forms of the carbohydrates given in each part.

(a) The β-pyranose form of D-gulose is the same as that of D-galactose except for the configuration at C-3.

β-D-Galactopyranose (64% at equilibrium)   β-D-Gulopyranose   α-D-Gulopyranose (1,3 diaxial repulsion between hydroxyl groups)

The axial hydroxyl group at C-3 destabilizes the α-pyranose form more than the β form because of its repulsive interaction with the axially disposed anomeric hydroxyl group. There should be an even higher β/α ratio in D-gulopyranose than in D-galactopyranose. This is so; the observed β/α ratio is 88:12.

(b) The β-pyranose form of D-talose is the same as that of D-mannose except for the configuration at C-4.

β-D-Talopyranose   α-D-Talopyranose   α-D-Mannopyranose
(68% at equilibrium)

Because the configuration at C-4 has little effect on the α- to β-pyranose ratio (compare D-glucose and D-galactose), we would expect that talose would behave very much like mannose and that the α-pyranose form would be preferred at equilibrium. This is indeed the case; the α-pyranose form predominates at equilibrium, the observed α/β ratio being 78:22.

(c) The pyranose form of D-xylose is just like that of D-glucose except that it lacks a CH$_2$OH group.

β-D-Glucopyranose   β-D-Xylopyranose   α-D-Xylopyranose
(64% at equilibrium)

We would expect the equilibrium between pyranose forms in D-xylose to be much like that in D-glucose and predict that the β-pyranose form would predominate. It is observed that the β/α ratio in D-xylose is 64:36, exactly the same as in D-glucose.

(d) The pyranose form of D-lyxose is like that of D-mannose except that it lacks a CH$_2$OH group. As in D-mannopyranose, the α form should predominate over the β.

β-D-Lyxopyranose   α-D-Lyxopyranose   α-D-Mannopyranose
(68% at equilibrium)

The observed α/β distribution ratio in D-lyxopyranose is 73:27.

25.37 (a) The rate-determining step in glycoside hydrolysis is carbocation formation at the anomeric position. The carbocation formed from methyl α-D-fructofuranoside (compound A) is tertiary and therefore more stable than the one from methyl α-D-glucofuranoside (compound B), which is secondary. The more stable a carbocation is, the more rapidly it will be formed.

Faster:

A   Tertiary carbocation

Slower:

B   Secondary carbocation

(*b*) The carbocation formed from methyl β-D-glucopyranoside (compound D) is less stable than the one from its 2-deoxy analog (compound C) and is formed more slowly. It is destabilized by the electron-withdrawing inductive effect of the hydroxyl group at C-2.

Faster:

C   More stable

Slower:

D   Less stable

25.38 D-Altrosan is a glycoside. The anomeric carbon—the one with two oxygen substituents—has an alkoxy group attached to it. Hydrolysis of D-altrosan follows the general mechanism for acetal hydrolysis.

D-Altrose

25.39 Galactose has hydroxyl groups at carbons 2, 3, 4, 5, 6. Ten trimethyl ethers are therefore possible.

2,3,4	2,4,5	3,4,5	4,5,6
2,3,5	2,4,6	3,4,6	
2,3,6	2,5,6	3,5,6	

To find out which one of these is identical with the degradation product of compound A, carry compound A through the required transformations.

Compound A

$$\xrightarrow[\text{Ag}_2\text{O}]{\text{CH}_3\text{I}}$$

Tri-*O*-methyl ether of
compound A

$$\downarrow \begin{array}{l} \text{H}_3\text{O}^+ \\ \text{(acetal hydrolysis)} \end{array}$$

2,3,5-Tri-*O*-methyl-
D-galactose

ANSWERS TO INTERPRETIVE PROBLEMS

25.40 E;   **25.41** F;   **25.42** C;   **25.43** A;   **25.44** D

SELF-TEST

PART A

A-1. The structure of D-erythrose is shown. Draw the following:

$$
\begin{array}{c}
\text{CHO} \\
\text{H}\!-\!\!-\!\text{OH} \\
\text{H}\!-\!\!-\!\text{OH} \\
\text{CH}_2\text{OH}
\end{array}
$$

(*a*) The enantiomer of D-erythrose

(*b*) A diastereomer of D-erythrose

(*c*) The α-furanose form of D-erythrose (use a Haworth formula)

(*d*) The anomer of the structure in part (*c*)

(*e*) Assign the configuration of each chirality center of D-erythrose as either *R* or *S*.

A-2. The structure of D-mannose is

$$
\begin{array}{c}
\text{CHO} \\
\text{HO}\!-\!\!-\!\text{H} \\
\text{HO}\!-\!\!-\!\text{H} \\
\text{H}\!-\!\!-\!\text{OH} \\
\text{H}\!-\!\!-\!\text{OH} \\
\text{CH}_2\text{OH}
\end{array}
$$

D-Mannose

Using Fischer projections, draw the product of the reaction of D-mannose with

(*a*) $NaBH_4$ in H_2O

(*b*) Excess periodic acid

A-3. Referring to the structure of D-arabinose shown, draw the following:

(*a*) The α-pyranose form of D-arabinose

(*b*) The β-furanose form of D-arabinose

(*c*) The β-pyranose form of L-arabinose

$$
\begin{array}{c}
\text{CHO} \\
\text{HO}\!-\!\!-\!\text{H} \\
\text{H}\!-\!\!-\!\text{OH} \\
\text{H}\!-\!\!-\!\text{OH} \\
\text{CH}_2\text{OH}
\end{array}
$$

D-Arabinose

A-4. Using text Figure 25.2, identify the following carbohydrate:

A-5. Write structural formulas for the α- and β-methyl pyranosides formed from the reaction of D-mannose (see Problem A-2 for its structure) with methanol in the presence of hydrogen chloride. How are the two products related—are they enantiomers? Diastereomers?

PART B

B-1. Choose the response that provides the best match between the terms given and the structures shown.

CHO   CHO   CHO CH_2OH

H——OH   HO——H   HO——H   C=O

H——OH   H——OH   HO——H   H——OH

H——OH   H——OH   HO——H   H——OH

CH_2OH CH_2OH CH_2OH CH_2OH

     1   2   3   4

	Diastereomers	Enantiomers
(a)	1, 3, and 4	1 and 3
(b)	1 and 2	1 and 3
(c)	1, 2, and 3	1 and 3
(d)	1 and 4	1 and 2

B-2. A D carbohydrate is

(a) Always dextrorotatory

(b) Always levorotatory

(c) Always the anomer of the corresponding L carbohydrate

(d) None of the above

B-3. Two of the three compounds shown yield the same product on reaction with warm HNO_3. The *exception* is

(a)
CHO

HO——H

H——OH

H——OH

CH_2OH

(b)
CHO

HO——H

HO——H

H——OH

CH_3

(c)
CHO

HO——H

HO——H

H——OH

CH_2OH

(d) None of these (all yield the same product)

B-4. The optical rotation of the α form of a pyranose is +150.7°; that of the β form is +52.8°. In solution an equilibrium mixture of the anomers has an optical rotation of +80.2°. The percentage of the α form at equilibrium is

(a) 28%   (b) 32%   (c) 68%   (d) 72%

B-5. Which of the following represents the anomer of the compound shown?

(a)

(c)

(b)

(d) None of these

B-6. Which of the following aldoses yields an optically inactive substance on reaction with sodium borohydride?

1   2   3   4

(a) 3 only   (b) 1 and 4   (c) 2 and 3   (d) All (1, 2, 3, and 4)

B-7. The structure of D-arabinose is shown in Problem A-3. Which of the following is L-arabinose?

(a)   (b)   (c)   (d)   (e)

B-8. Which one of the statements concerning the equilibrium shown is true?

(*a*) The two structures are enantiomers of each other. They have equal but opposite optical rotations and racemize slowly at room temperature.

(*b*) The two structures are enantiomers of each other. They racemize too rapidly at room temperature for their optical rotations to be measured.

(*c*) The two structures are diastereomers of each other. Their interconversion is called mutarotation.

(*d*) The two structures are diastereomers of each other. Their interconversion does not require breaking and making bonds, only a change in conformation.

(*e*) The two structures are diastereomers of each other. One is a furanose form, the other a pyranose form.

B-9. The configurations of the chirality centers in D-threose (shown) are

(*a*) 2R,3R   (*b*) 2R,3S   (*c*) 2S,3R   (*d*) 2S,3S

CHAPTER 26

Lipids

SOLUTIONS TO TEXT PROBLEMS

26.1 The triacylglycerol shown in text Figure 26.2a, with an oleyl group at C-2 of the glycerol unit and two stearyl groups at C-1 and C-3, yields stearic and oleic acids in a 2 : 1 molar ratio on hydrolysis. A constitutionally isomeric structure in which the oleyl group is attached to C-1 of glycerol would yield the same hydrolysis products.

$$
CH_3(CH_2)_7CH = CH(CH_2)_7\overset{\overset{\displaystyle O}{\|}}{C}O\overset{\overset{\displaystyle CH_2O\overset{\overset{\displaystyle O}{\|}}{C}(CH_2)_{16}CH_3}{|}}{C}H \quad\text{or}\quad CH_3(CH_2)_{16}\overset{\overset{\displaystyle O}{\|}}{C}O\overset{\overset{\displaystyle CH_2O\overset{\overset{\displaystyle O}{\|}}{C}(CH_2)_7CH=CH(CH_2)_7CH_3}{|}}{C}H
$$

$$\Big\downarrow \; 3H_2O$$

$$
CH_3(CH_2)_7CH = CH(CH_2)_7\overset{\overset{\displaystyle O}{\|}}{C}OH \;+\; HO\overset{\overset{\displaystyle CH_2OH}{|}}{\underset{\displaystyle CH_2OH}{C}}H \;+\; 2HO\overset{\overset{\displaystyle O}{\|}}{C}(CH_2)_{16}CH_3
$$

Oleic acid	Glycerol	Stearic acid

26.2 The catalytic hydrogenation intermediate described in the text shows the metal becoming bonded to C-10 of *cis*-9-octadecenoic acid. The metal can also bond to C-9.

$$
\underset{CH_3(CH_2)_7}{\overset{H}{\diagdown}}C = C\underset{(CH_2)_7\overset{\overset{\displaystyle O}{\|}}{C}OR}{\overset{H}{\diagup}} \;+\; H-(M)_n-H \;\;\rightleftharpoons\;\; CH_3(CH_2)_7 - \overset{\overset{\displaystyle H}{|}}{C}H - \overset{\overset{\displaystyle (M)_n-H}{|}}{C}H - (CH_2)_7\overset{\overset{\displaystyle O}{\|}}{C}OR
$$

Ester of cis-9-octadecenoic acid	Hydrogenated metal catalyst surface	Intermediate in catalytic hydrogenation

Dissociation of this intermediate can give rise to esters of 8-decenoic acid as well as 9-decenoic acid.

26.3 The sulfur of acyl carrier protein acts as a nucleophile and attacks the acetyl group of acetyl coenzyme A. The reaction is nucleophilic acyl substitution.

| Acetyl coenzyme A | Acyl carrier protein | Tetrahedral intermediate | S-Acetyl acyl carrier protein | Coenzyme A |

26.4 Conversion of acyl carrier protein–bound tetradecanoate to hexadecanoate proceeds through the series of intermediates shown.

26.5 The structure of L-glycerol 3-phosphate is shown in a Fischer projection. Translate the Fischer projection to a three-dimensional representation.

The order of decreasing sequence rule precedence is

$$HO- > H_2O_3POCH_2- > HOCH_2- > H-$$

When the three-dimensional formula is viewed from a perspective in which the lowest ranked group is away from us, we see

Order of decreasing rank is
clockwise, therefore *R*.

The absolute configuration is *R*.

    The conversion of L-glycerol 3-phosphate to a phosphatidic acid does not affect any of the bonds to the chirality center, nor does it alter the sequence rule ranking of the groups.

The absolute configuration is *R*.

26.6   Cetyl palmitate (hexadecyl hexadecanoate) is an ester in which both the acyl group and the alkyl group contain 16 carbon atoms.

$$CH_3(CH_2)_{14}\overset{\overset{\text{O}}{\|}}{C}O(CH_2)_{15}CH_3$$

Hexadecyl hexadecanoate

26.7   The biosynthesis of PGE$_2$ is described in the text. The fatty acid precursor of PGE$_2$ is *cis,cis,cis,cis*-5,8,11,14-icosatetraenoic acid, better known by its common name arachidonic acid.

cis,cis,cis,cis-5,8,11,14-Icosatetraenoic
acid (arachidonic acid)

PGE$_1$, shown in text Figure 26.4, has one less double bond than PGE$_2$. By analogy, the fatty acid precursor to PGE$_1$ also has one less double bond.

cis,cis,cis-8,11,14-Icosatrienoic acid

26.8   We are told that LTA$_4$ is an epoxide. The thiol group of glutathione (abbreviated GSH in the following reaction) acts as a nucleophile and opens the epoxide ring. In abbreviated form the reaction can be viewed as

Reasoning backwards from the structure of leukotriene C$_4$ (LTC$_4$) in the text, we can deduce the structure of the epoxide leukotriene A$_4$ (LTA$_4$).

Leukotriene A$_4$ (LTA$_4$)

26.9 Isoprene units are fragments in the carbon skeleton. Functional groups and multiple bonds are ignored when structures are examined for the presence of isoprene units.

α-Phellandrene (two equally correct answers):

or

Menthol (same carbon skeleton as α-phellandrene but different functionality):

or

Citral:

α-Selinene is shown in text Section 26.7.

Farnesol:

Abscisic acid:

Cembrene (two equally correct answers):

or

Vitamin A:

26.10 β-Carotene is a tetraterpene because it has 40 carbon atoms. The tail-to-tail linkage is at the midpoint of the molecule and connects two 20-carbon fragments.

Tail-to-tail link between isoprene units

26.11 Isopentenyl diphosphate acts as an alkylating agent toward farnesyl diphosphate. Alkylation is followed by loss of a proton from the carbocation intermediate, giving geranylgeranyl diphosphate. Hydrolysis of the diphosphate yields geranylgeraniol.

Farnesyl diphosphate   Isopentenyl diphosphate

−H⁺

Geranylgeranyl diphosphate

H₂O

Geranylgeraniol

26.12 Borneol, the structure of which is given in text Figure 26.6, is a secondary alcohol. Oxidation of borneol converts it to the ketone camphor.

Borneol   Camphor

Reduction of camphor with sodium borohydride gives a mixture of stereoisomeric alcohols, of which one is borneol and the other isoborneol.

Camphor   Borneol   Isoborneol

26.13 Figure 26.7 in the text describes the distribution of ^{14}C (denoted by *) in citronellal biosynthesized from acetate enriched with ^{14}C in its methyl group.

If, instead, acetate enriched with ^{14}C at its carbonyl carbon were used, exactly the opposite distribution of the ^{14}C label would be observed.

When $^{14}CH_3CO_2H$ is used, C-2, C-4, C-6, C-8, and both methyl groups of citronellal are labeled. When $CH_3{}^{14}CO_2H$ is used, C-1, C-3, C-5, and C-7 are labeled.

26.14 Start with squalene 2,3-epoxide, number the carbons of the main chain, and track these carbons through the biosynthetic pathway outlined in Mechanism 26.2.

(*b*) The hydrogens that migrate in step 4 are those attached to C(14) and C(18) of squalene 2,3-epoxide.

(*c*) The carbons at the C,D ring junction of cholesterol are C(14) and C(15) of squalene 2,3-epoxide. The methyl group that is eventually attached to C(14) was originally attached to C(15) and migrated to C(14) in step 4.

(*d*) In the conversion of lanosterol to cholesterol in step 5, both methyl groups at C(2) are lost, as well as the methyl group at C(15). The methyl group at C(15) was originally attached to C(10) of squalene 2,3-epoxide.

26.15 By analogy with Problem 26.13, we can determine the distribution of ^{14}C (denoted by *) in squalene 2,3-epoxide. Track the ^{14}C labels through the biosynthetic scheme shown in Problem 26.14 to locate their position in cholesterol.

Squalene 2,3-epoxide   Steps 1-5   Cholesterol

26.16 By analogy to the reaction in which 7-dehydrocholesterol is converted to vitamin D_3, the structure of vitamin D_2 can be deduced from that of ergosterol.

7-Dehydrocholesterol   Ergosterol

light   light

Vitamin D_3 Vitamin D_2

26.17 Crocetin and crocin both have the same 20-carbon conjugated polyene unit. The isoprene units are shown by the wavy lines in the structure below. Note the tail-to-tail link between isoprene units in the center.

Tail-to-tail link

R = H: Crocetin
R = gentiobiose: Crocin

The six-membered ring bearing the aldehyde group in both safranal and picrocrocin is derived from two isoprene units.

Safranal

Picrocrocin

26.18 (*a*) Fatty acid biosynthesis proceeds by the joining of acetate units.

Acetyl coenzyme A   Acetoacetyl coenzyme A   Palmitoyl coenzyme A

Thus, the even-numbered carbons will be labeled with ^{14}C when palmitic acid is biosynthesized from $^{14}CH_3CO_2H$ (^{14}C is represented as ∗).

$$\overset{*}{C}H_3CH_2\overset{*}{C}H_2CH_2\overset{*}{C}H_2CH_2\overset{*}{C}H_2CH_2\overset{*}{C}H_2CH_2\overset{*}{C}H_2CH_2\overset{*}{C}H_2CH_2\overset{*}{C}H_2CO_2H$$

(*b*) As noted in text Section 26.6, arachidonic acid is the biosynthetic precursor of PGE_2. The distribution of the ^{14}C label in PGE_2 biosynthesized from $^{14}CH_3CO_2H$ reflects the fatty acid origin of the prostaglandins.

PGE_2

(*c*) The biosynthetic pathway leading to prostacyclin I_2 (PGI_2), like the pathway leading to PGE_2, begins with arachidonic acid.

PGI_2

(*d*) Limonene is a monoterpene, biosynthesized from acetate by way of mevalonate and isopentenyl diphosphate.

Acetic acid   Isopentenyl   Limonene
                                     diphosphate

(*e*) The distribution of the ^{14}C label in β-carotene becomes evident once its isoprene units are identified.

β-Carotene

26.19 The isoprene units in the designated compounds are shown by disconnections in the structural formulas.

(*a*) Ascaridole:

or

(*b*) Dendrolasin:

(*c*) γ-Bisabolene

or

(*d*) α-Santonin

(*e*) Tetrahymanol

Tail-to-tail linkage
of isoprene units

...OH

CH₃

CH₃
CH₃

H₃C   CH₃   CH₃

H₃C   CH₃

26.20 Of the four isoprene units of cubitene, three of them are joined in the usual head-to-tail fashion, but the fourth one is joined in an irregular way.

Irregular linkage of this
isoprene unit to remainder
of molecule

26.21 (*a*) Cinerin I is an ester, the acyl portion of which is composed of two isoprene units, as follows:

Cinerin I

(*b*) Hydrolysis of cinerin I involves cleavage of the ester unit.

Cinerin I   (+)-Chrysanthemic
acid

Chrysanthemic acid has the constitution shown in the equation. Its stereochemistry is revealed by subsequent experiments.

(+)-Chrysanthemic
acid

(−)-Caronic acid   Acetone

Because caronic acid is optically active, its carboxyl groups must be trans to each other. (The cis stereoisomer is an optically inactive meso form.) The structure of (+)-chrysanthemic acid must therefore be either the following or its mirror image.

The carboxyl group and the 2-methyl-1-propenyl side chain must be trans to each other.

26.22 (*a*) Hydrolysis of phrenosine cleaves the glycosidic bond. The carbohydrate liberated by this hydrolysis is D-galactose.

Phrenosine is a β-glycoside of D-galactose.

(*b*) The species that remains on cleavage of the galactose unit has the structure

The two substances, sphingosine and cerebronic acid, that are formed along with D-galactose arise by hydrolysis of the amide bond.

Sphingosine   Cerebronic acid

26.23 (*a*) Catalytic hydrogenation over Lindlar palladium converts alkynes to cis alkenes.

9-Octadecynoic acid   (Z)-9-Octadecenoic acid (74%)
(stearolic acid)   (oleic acid)

(*b*) Carbon–carbon triple bonds are converted to trans alkenes by reduction with lithium and ammonia.

9-Octadecynoic acid   (E)-9-Octadecenoic acid (97%)
(stearolic acid)   (elaidic acid)

(*c*) The carbon–carbon double bond is hydrogenated readily over a platinum catalyst. Reduction of the ester function does not occur.

$$(Z)\text{-}CH_3(CH_2)_7CH{=}CH(CH_2)_7\overset{\overset{\displaystyle O}{\|}}{C}OCH_2CH_3 \xrightarrow{\text{H}_2, \text{Pt}} CH_3(CH_2)_{16}\overset{\overset{\displaystyle O}{\|}}{C}OCH_2CH_3$$

Ethyl (*Z*)-9-octadecenoate
(ethyl oleate)

Ethyl octadecanoate (91%)
(ethyl stearate)

(*d*) Lithium aluminum hydride reduces the ester function but leaves the carbon–carbon double bond intact.

$$(Z)\text{-}CH_3(CH_2)_5\underset{\underset{\displaystyle OH}{|}}{C}HCH_2CH{=}CH(CH_2)_7\overset{\overset{\displaystyle O}{\|}}{C}OCH_3 \xrightarrow[\text{2. H}_2\text{O}]{\text{1. LiAlH}_4} (Z)\text{-}CH_3(CH_2)_5\underset{\underset{\displaystyle OH}{|}}{C}HCH_2CH{=}CH(CH_2)_7CH_2OH \ + \ CH_3OH$$

Methyl (*Z*)-12-hydroxy-9-octadecenoate
(methyl ricinoleate)

(*Z*)-9-Octadecen-1,12-diol (52%)

Methanol

(*e*) Epoxidation of the double bond occurs when an alkene is treated with a peroxy acid. The reaction is stereospecific; substituents that are cis to each other in the alkene remain cis in the epoxide.

$$(Z)\text{-}CH_3(CH_2)_7CH{=}CH(CH_2)_7COOH \ + \ C_6H_5\overset{\overset{\displaystyle O}{\|}}{C}OOH \longrightarrow$$

Oleic acid

Peroxybenzoic acid

cis-9,10-Epoxyoctadecanoic acid
(62–67%)

Benzoic acid

(*f*) Acid-catalyzed hydrolysis of the epoxide yields a diol; its stereochemistry corresponds to net anti dihydroxylation of the double bond of the original alkene.

$$\xrightarrow{\text{H}_3\text{O}^+}$$

cis-9,10-Epoxyoctadecanoic acid

9,10-Dihydroxyoctadecanoic acid

The product is chiral but is formed as a racemic mixture containing equal amounts of the 9*R*,10*R* and 9*S*,10*S* stereoisomers when the starting epoxide is racemic.

(*g*) Dihydroxylation of carbon–carbon double bonds with osmium tetraoxide proceeds with syn addition of hydroxyl groups.

$$(Z)\text{-}CH_3(CH_2)_7CH{=}CH(CH_2)_7COOH \xrightarrow[\text{2. H}^+]{\text{1. OsO}_4, (CH_3)_3COOH, HO^-}$$

9,10-Dihydroxyoctadecanoic acid
(70%)

The product is chiral but is formed as a racemic mixture containing equal amounts of the 9*R*,10*S* and 9*S*,10*R* stereoisomers.

(*h*) Hydroboration–oxidation gives syn hydration of carbon–carbon double bonds with a regioselectivity opposite to Markovnikov's rule. The reagent attacks the less hindered face of the double bond of α-pinene.

Isopinocampheol (79%)

(*i*) The starting alkene in this case is β-pinene. As in the preceding exercise with α-pinene, diborane adds to the bottom face of the double bond.

cis-Myrtanol (81%)

(*j*) The starting material is an acetal. It undergoes hydrolysis in dilute aqueous acid to give a ketone.

(95% yield)

26.24 (*a*) There are no direct methods for the reduction of a carboxylic acid to an alkane. A number of indirect methods that may be used, however, involve first converting the carboxylic acid to an alkyl bromide via the corresponding alcohol.

$$CH_3(CH_2)_{16}CO_2H \xrightarrow[\text{2. }H_2O]{\text{1. LiAlH}_4} CH_3(CH_2)_{16}CH_2OH \xrightarrow[\text{or PBr}_3]{\text{HBr, heat}} CH_3(CH_2)_{16}CH_2Br$$

Octadecanoic acid   1-Octadecanol   1-Bromooctadecane

Once the alkyl bromide is in hand, it may be converted to an alkane by conversion to a Grignard reagent followed by addition of water.

$$CH_3(CH_2)_{16}CH_2Br \xrightarrow[\text{diethyl ether}]{\text{Mg}} CH_3(CH_2)_{16}CH_2MgBr \xrightarrow{H_2O} CH_3(CH_2)_{16}CH_3$$

1-Bromooctadecane   Octadecane

Other routes are also possible. For example, E2 elimination from 1-bromooctadecane followed by hydrogenation of the resulting alkene will also yield octadecane.

(b) Retrosynthetic analysis reveals that the 18-carbon chain of the starting material must be attached to a benzene ring.

$$C_6H_5-(CH_2)_{17}CH_3 \implies C_6H_5- \quad + \quad -(CH_2)_{17}CH_3$$

1-Phenyloctadecane

The desired sequence may be carried out by a Friedel–Crafts acylation, followed by Clemmensen or Wolff–Kishner reduction of the ketone.

$$CH_3(CH_2)_{16}CO_2H \xrightarrow{SOCl_2} CH_3(CH_2)_{16}\overset{O}{\overset{\|}{C}}Cl \xrightarrow[AlCl_3]{benzene} C_6H_5-\overset{O}{\overset{\|}{C}}(CH_2)_{16}CH_3$$

Octadecanoic acid · Octadecanoyl chloride · 1-Phenyl-1-octadecanone

$$\downarrow \; Zn(Hg), HCl$$

$$C_6H_5-CH_2(CH_2)_{16}CH_3$$

1-Phenyloctadecane

(c) First examine the structure of the target molecule 3-ethylicosane.

$$CH_3(CH_2)_{16}\underset{\underset{CH_2CH_3}{|}}{CH}CH_2CH_3$$

Retrosynthetic analysis reveals that two ethyl groups have been attached to a C_{18} unit.

$$CH_3(CH_2)_{16}\underset{\underset{CH_2CH_3}{|}}{CH}-CH_2CH_3 \implies CH_3(CH_2)_{16}\underset{|}{CH}- \quad + \quad 2CH_3CH_2-$$

The necessary carbon–carbon bonds can be assembled by the reaction of an ester with two moles of a Grignard reagent.

$$CH_3(CH_2)_{16}\overset{O}{\overset{\|}{C}}OCH_2CH_3 + 2CH_3CH_2MgBr \xrightarrow[2.\ H_3O^+]{1.\ diethyl\ ether} CH_3(CH_2)_{16}\underset{\underset{CH_2CH_3}{|}}{\overset{\overset{OH}{|}}{C}}CH_2CH_3$$

Ethyl octadecanoate (from octadecanoic acid and ethanol) · Ethylmagnesium bromide · 3-Ethyl-3-icosanol

With the correct carbon skeleton in place, all that is needed is to convert the alcohol to the alkene. This can be accomplished by dehydration and reduction.

$$CH_3(CH_2)_{16}\overset{OH}{\underset{CH_2CH_3}{C}}CH_2CH_3 \xrightarrow[\text{heat}]{H_2SO_4} CH_3(CH_2)_{16}\underset{CH_2CH_3}{C}=CHCH_3 + CH_3(CH_2)_{15}CH=\underset{CH_2CH_3}{C}CH_2CH_3$$

3-Ethyl-3-icosanol   3-Ethyl-2-icosene   3-Ethyl-3-icosene

$$\xrightarrow{H_2,\ Pt}$$

$$CH_3(CH_2)_{16}\underset{CH_2CH_3}{CH}CH_2CH_3$$

3-Ethylicosane

(d) Icosanoic acid contains two more carbon atoms than octadecanoic acid.

$$CH_3(CH_2)_{18}CO_2H \implies CH_3(CH_2)_{16}CH_2Br + \overset{..}{C}H_2CO_2H$$

Icosanoic acid

A reasonable approach utilizes a malonic ester synthesis.

$$CH_3(CH_2)_{16}CH_2Br + CH_2(CO_2CH_2CH_3)_2 \xrightarrow{NaOCH_2CH_3} CH_3(CH_2)_{16}CH_2CH(CO_2CH_2CH_3)_2$$

1-Bromooctadecane   Diethyl malonate   Diethyl 2-octadecylmalonate
[prepared as in part (a)]

1. HO⁻, H₂O
2. H₃O⁺
3. heat

$$CH_3(CH_2)_{16}CH_2CH_2CO_2H$$

Icosanoic acid

(e) Lithium aluminum hydride reduction of octadecanamide gives the corresponding amine. The amide can be prepared from octadecanoic acid.

$$CH_3(CH_2)_{16}CO_2H \xrightarrow[\text{2. NH}_3]{\text{1. SOCl}_2} CH_3(CH_2)_{16}\overset{O}{\overset{||}{C}}NH_2 \xrightarrow[\text{2. H}_2O]{\text{1. LiAlH}_4} CH_3(CH_2)_{16}CH_2NH_2$$

Octadecanoic acid   Octadecanamide   1-Octadecanamine

(f) Chain extension can be achieved via cyanide displacement of bromine from 1-bromooctadecane. Reduction of the cyano group completes the synthesis.

$$CH_3(CH_2)_{16}CH_2Br \xrightarrow{KCN} CH_3(CH_2)_{16}CH_2C\equiv N \xrightarrow[\text{2. H}_2O]{\text{1. LiAlH}_4} CH_3(CH_2)_{16}CH_2CH_2NH_2$$

1-Bromooctadecane   Nonadecanenitrile   1-Nonadecanamine
[from part (a)]

26.25 First acylate the free hydroxyl group with an acyl chloride.

Treatment with aqueous acid brings about hydrolysis of the acetal function.

The two hydroxyl groups of the resulting diol are then esterified with 2 moles of the second acyl chloride.

26.26 The overall transformation

requires converting the alcohol function to some suitable leaving group, followed by substitution by an appropriate nucleophile.

3-Methyl-3-buten-1-ol   4-Bromo-2-methyl-1-butene   3-Methyl-3-butenyl methyl sulfide

As reported in the literature, the alcohol was converted to its corresponding *p*-toluenesulfonate and this substance was then used as the substrate in the nucleophilic substitution step to produce the desired sulfide in 76% yield.

26.27 The first transformation is an intramolecular aldol condensation. This reaction was carried out under conditions of base catalysis.

6-Methyl-2,5-heptanedione   (Not isolated)   3-Isopropyl-2-cyclopentenone (71%)

The next step is reduction of a ketone to a secondary alcohol. Lithium aluminum hydride is suitable; it reduces carbonyl groups but leaves the double bond intact.

3-Isopropyl-2-
cyclopentenone

3-Isopropyl-2-
cyclopenten-1-ol (97%)

Conversion of an alkene to a cyclopropane can be accomplished by using the Simmons–Smith reagent (iodomethylzinc iodide).

3-Isopropyl-2-
cyclopenten-1-ol

5-Isopropylbicyclo[3.1.0]hexan-2-ol
(66%)

Oxidation of the secondary alcohol to the ketone can be accomplished with any of a number of oxidizing agents. The chemists who reported this synthesis used chromic acid.

5-Isopropylbicyclo-
[3.1.0]hexan-2-ol

5-Isopropylbicyclo[3.1.0]hexan-2-one
(89%)

A Wittig reaction converts the ketone to sabinene.

5-Isopropylbicyclo-
[3.1.0]hexan-2-one

Sabinene
(70%)

26.28 The first step is a 1,4 addition of hydrogen bromide to the conjugated diene system of isoprene.

Hydrogen
bromide

2-Methyl-
1,3-butadiene

1-Bromo-3-
methyl-2-butene

This is followed by Markovnikov addition of hydrogen bromide to the remaining double bond.

1-Bromo-3-methyl-2-butene   Hydrogen bromide

1,3-Dibromo-3-methylbutane

26.29 A reasonable mechanism is protonation of the isolated carbon–carbon double bond, followed by cyclization.

α-Ionone

−H⁺

β-Ionone

26.30 The double bond has a tendency to become conjugated with the carbonyl group. Two mechanisms are more likely than any others under conditions of acid catalysis. One of these involves protonation of the double bond followed by loss of a proton from C-4.

The other mechanism proceeds by enolization followed by proton-induced double-bond migration.

26.31 See the June 1995 issue of the *Journal of Chemical Education,* pages 541–542, for the solution to this problem.

ANSWERS TO INTERPRETIVE PROBLEMS

26.32 A;   **26.33** D;   **26.34** B;   **26.35** B;   **26.36** C;   **26.37** A

SELF-TEST

PART A

A-1. Write a balanced chemical equation for the basic hydrolysis of tristearin.

A-2. Both waxes and fats are lipids that contain the ester functional group. In what way do the structures of these lipids differ?

A-3. Classify each of the following isoprenoid compounds as a monoterpene, a diterpene, and so on. Indicate with dashed lines the isoprene units that make up each structure.

(*a*) α-Pinene:

(*c*) Abietic acid:

(*b*) Caryophyllene:

A-4. Propose a series of synthetic steps to carry out the preparation of oleic acid [(Z)-9-octadecenoic acid] from compound A. You may use any necessary organic or inorganic reagents.

Compound A

A-5. Write a mechanism for the biosynthetic pathway by which limonene is formed from geranyl diphosphate.

Geranyl diphosphate   Limonene

PART B

B-1. A major component of a lipid bilayer is

(*a*) A triacylglycerol such as tristearin

(*b*) Phosphatidylcholine, also known as lecithin

(*c*) A sterol such as cholesterol

(*d*) A prostaglandin such as PGE$_1$

B-2. Compare the following two triacylglycerols:

$$
\begin{array}{ll}
CH_2O_2CC_{17}H_{35} & CH_2O_2CC_{17}H_{35} \\
CHO_2CC_{17}H_{35} & CHO_2CC_{17}H_{31} \\
CH_2O_2CC_{17}H_{35} & CH_2O_2CC_{17}H_{31}
\end{array}
$$

A   B

(*a*) The melting point of A will be higher.

(*b*) The melting point of B will be higher.

(*c*) The melting points of A and B will be the same.

(*d*) No comparison of melting points can be made.

B-3. Lanosterol, a biosynthetic precursor of cholesterol, exists naturally as a single enantiomer. How many *possible* stereoisomers having the lanosterol skeleton are there?

Lanosterol

(*a*) 7   (*b*) 64   (*c*) 128   (*d*) 256

B-4. The compound whose carbon skeleton is shown, known as selinene, is found in celery.

Selinene is an example of a

(*a*) Monoterpene   (*b*) Diterpene   (*c*) Sesquiterpene   (*d*) Triterpene

B-5. Which of the following correctly represents the isoprenoid units of selinene?

(*a*)

(*c*) Both of these are acceptable

(*b*)

(*d*) Neither of these is acceptable

B-6. What is the distribution of radioactive carbon (^{14}C) in isopentenyl diphosphate biosynthesized from acetic acid labeled with ^{14}C at its carboxyl carbon ($CH_3\overset{*}{C}O_2H$)? ^{14}C is indicated by an asterisk (*) in the structures.

(*a*) OPP

(*d*) OPP

(*b*) OPP

(*e*) OPP

(*c*) OPP

CHAPTER 27

Amino Acids, Peptides, and Proteins

SOLUTIONS TO TEXT PROBLEMS

27.1 (*b*) L-Cysteine is the only amino acid in text Table 27.1 that has the *R* configuration at its chirality center.

$$H_3\overset{+}{N}\!-\!\!\!\!\overset{\displaystyle CO_2^-}{\underset{\displaystyle CH_2SH}{|}}\!\!\!\!-\!H \equiv \overset{H \ \ \overset{+}{N}H_3}{\underset{HSCH_2}{C\!-\!CO_2^-}} \equiv \overset{HSCH_2 \ \ H}{\underset{H_3\overset{+}{N}}{C\!-\!CO_2^-}}$$

L-Cysteine

The order of decreasing sequence rule precedence is

$$H_3\overset{+}{N}\!-\!\! > \ HSCH_2\!-\!\! > \ -\!CO_2^- > \ H\!-\!$$

When the molecule is oriented so that the lowest ranked atom (H) is held away from us, the order of decreasing precedence traces a clockwise path.

$$\overset{\displaystyle CH_2SH}{\underset{H_3\overset{+}{N} \qquad CO_2^-}{}}$$

Clockwise; therefore *R*

The reason why L-cysteine has the *R* configuration and all the other L-amino acids have the *S* configuration lies in the fact that the —CH_2SH substituent is the only side chain that outranks —CO_2^- according to the sequence rule. Remember, rank order is determined by atomic number at the first point of difference, and —C—S outranks —C—O. In all the other amino acids —CO_2^- outranks the substituent at the chirality center. The reversal in the Cahn–Ingold–Prelog descriptor comes not from any change in the spatial arrangement of substituents at the chirality center but rather from a reversal in the relative ranks of the carboxylate group and the side chain.

(*c*) The order of decreasing sequence rule precedence in L-methionine is

$$H_3\overset{+}{N}— \quad > \quad —CO_2^- \quad > \quad —CH_2CH_2SCH_3 \quad > \quad H—$$

Sulfur is one atom further removed from the chirality center, and so C—O outranks C—C—S.

$$H_3\overset{+}{N}—\overset{\displaystyle CO_2^-}{\underset{\displaystyle CH_2CH_2SCH_3}{|—H}} \quad \equiv \quad CH_3SCH_3CH_2 \quad \overset{H \; \overset{+}{NH_3}}{\underset{}{\diagdown \; C—CO_2^-}}$$

The absolute configuration is *S*.

27.2 The amino acids in text Table 27.1 that have more than one chirality center are isoleucine and threonine. The chirality centers are marked with an asterisk in the structural formulas shown.

$$CH_3CH_2\overset{*}{C}H—\overset{*}{C}HCO_2^- \qquad CH_3\overset{*}{C}H—\overset{*}{C}HCO_2^-$$
$$\underset{\displaystyle CH_3 \; {}^+NH_3}{||} \qquad\qquad \underset{\displaystyle OH \; {}^+NH_3}{||}$$

Isoleucine   Threonine

27.3 The isoelectric point of an amino acid is midway between the pK_a value of the zwitterion and its conjugate acid. For cysteine the equilibria are

$$HSCH_2\underset{\displaystyle {}^+NH_3}{\overset{}{C}HCO_2H} \;\overset{\substack{\text{p}K_{a1}\\=1.96}}{\rightleftharpoons}\; HSCH_2\underset{\displaystyle {}^+NH_3}{\overset{}{C}HCO_2^-} \;\overset{\substack{\text{p}K_a\,(\text{side chain})\\=8.18}}{\rightleftharpoons}\; {}^-SCH_2\underset{\displaystyle {}^+NH_3}{\overset{}{C}HCO_2^-} \;\overset{\substack{\text{p}K_{a2}\\=10.28}}{\rightleftharpoons}\; {}^-SCH_2\underset{\displaystyle NH_2}{\overset{}{C}HCO_2^-}$$

Conjugate acid   Zwitterion
of zwitterion

The isoelectric point of cysteine is the average of the pK_{a1} and pK_a (side chain), or 5.07.

27.4 The zwitterion form of tyrosine is shown

$$HO—\!\!\left\langle \text{—}\bigcirc\text{—} \right\rangle\!\!—CH_2\underset{\displaystyle {}^+NH_3}{\overset{}{C}HCO_2^-}$$

Zwitterion form
of tyrosine

As the pH is raised, protons are lost from both the ammonium group and the phenolic hydroxyl group to form a dianion.

$$^-O—\!\!\left\langle \text{—}\bigcirc\text{—} \right\rangle\!\!—CH_2\underset{\displaystyle NH_2}{\overset{}{C}HCO_2^-}$$

Dianion form
of tyrosine

27.5 To convert 3-methylbutanoic acid to valine, a leaving group must be introduced at the α carbon prior to displacement by ammonia. This is best accomplished by bromination under the conditions of the Hell–Volhard–Zelinsky reaction.

$$(CH_3)_2CHCH_2CO_2H \xrightarrow[\text{or } Br_2, PCl_3]{Br_2, P} (CH_3)_2CHCHCO_2H \xrightarrow{NH_3} (CH_3)_2CHCHCO_2^-$$

with Br under the central structure, and ⁺NH₃ under the valine structure.

3-Methylbutanoic acid	2-Bromo-3-methylbutanoic acid	Valine

Valine has been prepared by this method. The Hell–Volhard–Zelinsky reaction was carried out in 88% yield, but reaction of the α-bromo acid with ammonia was not very efficient, valine being isolated in only 48% yield in this step.

27.6 In the Strecker synthesis an aldehyde is treated with ammonia and a source of cyanide ion. The resulting amino nitrile is hydrolyzed to an amino acid.

$$(CH_3)_2CHCH \xrightarrow{NH_3 \atop HCN} (CH_3)_2CHCHC\equiv N \xrightarrow[\text{2. HO}^-]{1.\ H_3O^+,\ heat} (CH_3)_2CHCHCO_2^-$$

with O (double bond) above the first CH, NH₂ under the central structure, and ⁺NH₃ under the valine structure.

2-Methylpropanal	2-Amino-3-methylbutanenitrile	Valine

As actually carried out, the aldehyde was converted to the amino nitrile by treatment with an aqueous solution containing ammonium chloride and potassium cyanide. Hydrolysis was achieved in aqueous hydrochloric acid and gave valine as its hydrochloride salt in 65% overall yield.

27.7 The alkyl halide with which the anion of diethyl acetamidomalonate is treated is 2-bromopropane.

$$CH_3CNHCH(CO_2CH_2CH_3)_2 \ +\ (CH_3)_2CHBr \xrightarrow[CH_3CH_2OH]{NaOCH_2CH_3} CH_3CNHC(CO_2CH_2CH_3)_2$$

with O (double bond) above each carbonyl C, and CH(CH₃)₂ under the product structure.

Diethyl acetamidomalonate	2-Bromopropane	Diethyl acetamidoisopropylmalonate

This is the difficult step in the synthesis; it requires a nucleophilic substitution of the S_N2 type involving a secondary alkyl halide. Competition of elimination with substitution results in only a 37% observed yield of alkylated diethyl acetamidomalonate.

Hydrolysis and decarboxylation of the alkylated derivative are straightforward and proceed in 85% yield to give valine.

$$CH_3CNHC(CO_2CH_2CH_3)_2 \xrightarrow[\text{heat}]{HBr,\ H_2O} H_3\overset{+}{N}C(CO_2H)_2 \xrightarrow[-CO_2]{heat} H_3\overset{+}{N}CHCO_2^-$$

with O (double bond) above the first carbonyl, CH(CH₃)₂ under each structure.

Diethyl acetamidoisopropylmalonate	2-Aminoisopropylmalonic acid	Valine

The overall yield of valine (31%) is the product of 37% × 85%.

27.8   Ninhydrin is the hydrate of a triketone and is in equilibrium with it.

Hydrated form of ninhydrin   Triketo form of ninhydrin

An amino acid reacts with this triketone to form an imine.

Triketo form of ninhydrin   α-Amino acid   Imine

This imine then undergoes decarboxylation.

The anion that results from the decarboxylation step is then protonated. The product is shown as its diketo form but probably exists as an enol.

Hydrolysis of the imine function gives an aldehyde and a compound having a free amino group.

This amine then reacts with a second molecule of the triketo form of ninhydrin to give an imine.

Proton abstraction from the neutral imine gives its conjugate base, which is a violet dye.

Violet dye

27.9 The carbon that bears the amino group of 4-aminobutanoic acid corresponds to the α carbon of an α-amino acid, glutamic acid.

$$\underset{\overset{|}{^{+}NH_3}}{CH_2CH_2CH_2CO_2^{-}} \quad \text{arises by decarboxylation of} \quad \underset{\overset{|}{^{+}NH_3}}{^{-}O_2CCHCH_2CH_2CO_2^{-}}$$

4-Aminobutanoic acid   Glutamic acid

27.10 In transamination the ketone carbonyl of an α-keto acid and the α-amine group of an amino acid effectively trade places by the steps outlined in text Mechanism 27.2. Thus the transamination reaction of α-ketoglutaric acid with L-aspartic acid yields L-glutamic acid and oxaloacetic acid.

$$^{-}O_2CCH_2CH_2\overset{\overset{O}{\|}}{C}CO_2^{-} + \underset{\overset{|}{^{+}NH_3}}{^{-}O_2CCH_2CHCO_2^{-}} \longrightarrow \underset{\overset{|}{^{+}NH_3}}{^{-}O_2CCH_2CH_2CHCO_2^{-}} + ^{-}O_2CCH_2\overset{\overset{O}{\|}}{C}CO_2^{-}$$

α-Ketoglutaric   L-Aspartic acid   L-Glutamic acid   Oxaloacetic
acid   acid

27.11 (*b*) Alanine is the N-terminal amino acid in Ala-Phe. Its carboxyl group is joined to the nitrogen of phenylalanine by a peptide bond.

$$\underset{\overset{|}{CH_3}}{\overset{+}{H_3N}CH\overset{\overset{O}{\|}}{C}}-\underset{\overset{|}{CH_2C_6H_5}}{NHCHCO_2^{-}} \quad AF$$

Alanine   Phenylalanine

(*c*) The positions of the amino acids are reversed in Phe-Ala. Phenylalanine is the N terminus and alanine is the C terminus.

$$\underset{\overset{|}{C_6H_5CH_2}}{\overset{+}{H_3N}CH\overset{\overset{O}{\|}}{C}}-\underset{\overset{|}{CH_3}}{NHCHCO_2^{-}} \quad FA$$

Phenylalanine   Alanine

(*d*) The carboxyl group of glycine is joined by a peptide bond to the amino group of glutamic acid.

$$\overset{+}{H_3N}CH_2\overset{\overset{O}{\|}}{C}-NHCHCO_2^- \qquad GE$$
$$\qquad\qquad\qquad | $$
$$\qquad\qquad\qquad CH_2CH_2CO_2^-$$

Glycine　　　Glutamic acid

The dipeptide is written in its anionic form because the carboxyl group of the side chain is ionized at pH 7. Alternatively, it could have been written as a neutral zwitterion with a $CH_2CH_2CO_2H$ side chain.

(*e*) The peptide bond in Lys-Gly is between the carboxyl group of lysine and the amino group of glycine.

$$\overset{+}{H_3N}CH\overset{\overset{O}{\|}}{C}-NHCH_2CO_2^- \quad KG$$
$$\qquad\quad |$$
$$\overset{+}{H_3N}CH_2CH_2CH_2CH_2$$

Lysine　　　Glycine

The amino group of the lysine side chain is protonated at pH 7, and so the dipeptide is written here in its cationic form. It could have also been written as a neutral zwitterion with the side chain $H_2NCH_2CH_2CH_2CH_2$.

(*f*) Both amino acids are alanine in D-Ala-D-Ala. The fact that they have the D configuration has no effect on the constitution of the dipeptide.

$$\overset{+}{H_3N}CH\overset{\overset{O}{\|}}{C}-NHCHCO_2^- \quad D\text{-}A\text{-}D\text{-}A$$
$$\qquad |\qquad\qquad |$$
$$\qquad CH_3\qquad CH_3$$

Alanine　　Alanine

27.12 (*b*) When amino acid residues in a dipeptide are indicated without a prefix, it is assumed that the configuration at the α carbon atom is L. For all amino acids except cysteine, the L configuration corresponds to *S*. The stereochemistry of Ala-Phe may therefore be indicated for the zigzag conformation as shown.

The L configuration corresponds to *S* for each of the chirality centers in Ala-Phe.

(*c*) Similarly, Phe-Ala has its substituent at the N-terminal amino acid directed away from us, whereas the C-terminal side chain is pointing toward us, and the L configuration corresponds to *S* for each chirality center.

(*d*) There is only one chirality center in Gly-Glu. It has the L (or *S*) configuration.

$$H_3\overset{+}{N} \cdots \overset{\displaystyle O}{\underset{\displaystyle N}{\parallel}} \cdots \overset{\displaystyle CH_2CH_2CO_2^-}{\underset{\displaystyle H}{\overset{\displaystyle H}{\cdots}}} CO_2^-$$

(*e*) In order for the N-terminal amino acid in Lys-Gly to have the L (or *S*) configuration, its side chain must be directed away from us in the conformation indicated.

$$H_3\overset{+}{N} \cdots \qquad \overset{\displaystyle O}{\parallel} \qquad$$
$$H_3\overset{+}{N}CH_2CH_2CH_2CH_2\ \ H \quad \overset{\displaystyle N}{\underset{\displaystyle H}{}} CO_2^-$$

(*f*) The configuration at both α-carbon atoms in D-Ala-D-Ala is exactly the reverse of the configuration of the chirality centers in parts (*a*) through (*e*). Both chirality centers have the D (or *R*) configuration.

$$H_3\overset{+}{N} \cdots \overset{\displaystyle O}{\underset{\displaystyle H}{\underset{\displaystyle CH_3}{}}} \overset{H\ \ CH_3}{\underset{\displaystyle N}{\underset{\displaystyle H}{}}} CO_2^-$$

27.13 Figure 27.5 in the text gives the structure of leucine enkephalin. Methionine enkephalin differs from it only with respect to the C-terminal amino acid. The amino acid sequences of the two pentapeptides are

  Tyr-Gly-Gly-Phe-Leu   Tyr-Gly-Gly-Phe-Met

  Leucine enkephalin   Methionine enkephalin

The peptide sequence of a polypeptide can also be expressed using the one-letter abbreviations listed in text Table 27.1. Methionine enkephalin becomes YGGFM.

27.14 The structure of oxytocin is given in Figure 27.6. The amino acids present are Cys, Tyr, Ile, Gln, Asn, Pro, Leu, and Gly-NH$_2$ (the amide of glycine). The number of negative charges is zero because the C-terminal amino acid (Gly-NH$_2$) is an amide, not a carboxylic acid, and none of the other amino acids has an acidic side chain. The N-terminal amino acid (cysteine) has a positively charged nitrogen. Therefore, the net charge of oxytocin is +1.

27.15 Twenty-four tetrapeptide combinations are possible for the four amino acids alanine (A), glycine (G), phenylalanine (F), and valine (V). Remember that the order is important; AG is not the same peptide as GA. Using the one-letter abbreviations for each amino acid the possibilities are

AGFV	AGVF	AFGV	AFVG	AVGF	AVFG
GAFV	GAVF	GFAV	GFVA	GVFA	GVAF
FAGV	FAVG	FVAG	FVGA	FGAV	FGVA
VAGF	VAFG	VGAF	VGFA	VFAG	VFGA

27.16 Chymotrypsin cleaves a peptide selectively at the carboxyl group of amino acids that have aromatic side chains. The side chain of phenylalanine is a benzyl group, C$_6$H$_5$CH$_2$—. If the dipeptide isolated after treatment with chymotrypsin contains valine (V) and phenylalanine (F), its sequence must be VF.

Valine   Phenylalanine   Valinylphenylalanine (VF)

The possible sequences for the unknown tetrapeptide are VFAG and VFGA.

27.17 The Edman degradation removes the N-terminal amino acid, which is identified as a phenylthiohydantoin (PTH) derivative. The first Edman degradation of Val-Phe-Gly-Ala gives the PTH derivative of valine; the second gives the PTH derivative of phenylalanine.

Val-Phe-Gly-Ala $\xrightarrow[\text{Edman degradation}]{\text{first}}$ Phe-Gly-Ala + [PTH derivative with C_6H_5, S, N, O, HN, $CH(CH_3)_2$] $\xrightarrow[\text{Edman degradation}]{\text{second}}$ Gly-Ala + [PTH derivative with C_6H_5, S, N, O, HN, $CH_2C_6H_5$]

27.18 Leucine and isoleucine are isomers. Therefore, they have the same molecular weight and the same m/z value.

27.19 Lysine has two amino groups. Both amino functions are converted to amides on reaction with benzyloxycarbonyl chloride.

$H_2NCHCO_2^-$ (with $H_2NCH_2CH_2CH_2CH_2$) $+ 2C_6H_5CH_2OCCl$ (O double bond) $\longrightarrow$ $C_6H_5CH_2OCNHCH_2CH_2CH_2CH_2$ (with $C_6H_5CH_2OCNHCHCO_2H$ above, O double bonds)

27.20 The anion formed from the reaction of 9-fluorenylmethoxycarbonylalanine and ammonia dissociates to give dibenzofulvene, carbon dioxide, and alanine.

CH_2—O—C—NHCHCH$_3$ (with O double bond and CO_2^-) $\longrightarrow$ [dibenzofulvene, CH_2] $+ CO_2 + H_2NCHCH_3$ (with CO_2^-)

Anion of 9-fluorenyl-methoxycarbonylalanine   Dibenzofulvene   Carbon dioxide   Alanine

27.21 The peptide bond of Ala-Leu connects the carboxyl group of alanine and the amino group of leucine. We therefore need to protect the amino group of alanine and the carboxyl group of leucine.

Protect the amino group of alanine as its benzyloxycarbonyl derivative.

$H_3\overset{+}{N}CHCO_2^-$ (with CH_3) $+ C_6H_5CH_2OCCl$ (O double bond) $\longrightarrow$ $C_6H_5CH_2OCNHCHCO_2H$ (O double bond, with CH_3)

Alanine   Benzyloxycarbonyl chloride   Z-Protected alanine

Protect the carboxyl group of leucine as its benzyl ester.

$H_3\overset{+}{N}CHCO_2^-$ (with $(CH_3)_2CHCH_2$) $+ C_6H_5CH_2OH$ $\xrightarrow[\text{2. HO}^-]{\text{1. H}^+, \text{heat}}$ $H_2NCHCO_2CH_2C_6H_5$ (with $(CH_3)_2CHCH_2$)

Leucine   Benzyl alcohol   Leucine benzyl ester

Coupling of the two amino acids is achieved by N,N'-dicyclohexylcarbodiimide (DCCI)-promoted amide bond formation between the free amino group of leucine benzyl ester and the free carboxyl group of Z-protected alanine.

$$
\underset{\substack{\text{Z-Protected alanine}}}{\underset{\underset{\text{CH}_3}{|}}{\text{C}_6\text{H}_5\text{CH}_2\text{OCNHCHCO}_2\text{H}}}
\;+\;
\underset{\substack{\text{Leucine benzyl ester}}}{\underset{\underset{(\text{CH}_3)_2\text{CHCH}_2}{|}}{\text{H}_2\text{NCHCOCH}_2\text{C}_6\text{H}_5}}
\xrightarrow{\text{DCCI}}
\underset{\substack{\text{Protected dipeptide}}}{\underset{\underset{\text{CH}_3 \quad\;\; \text{CH}_2\text{CH(CH}_3)_2}{|\qquad\quad |}}{\text{C}_6\text{H}_5\text{CH}_2\text{OCNHCHCNHCHCOCH}_2\text{C}_6\text{H}_5}}
$$

Both the benzyloxycarbonyl protecting group and the benzyl ester protecting group may be removed by hydrogenolysis over palladium. This step completes the synthesis of Ala-Leu.

$$
\underset{\substack{\text{Protected dipeptide}}}{\underset{\underset{\text{CH}_3 \quad\;\; \text{CH}_2\text{CH(CH}_3)_2}{|\qquad\quad |}}{\text{C}_6\text{H}_5\text{CH}_2\text{OCNHCHCNHCHCOCH}_2\text{C}_6\text{H}_5}}
\xrightarrow{\text{H}_2,\,\text{Pd}}
\underset{\substack{\text{Ala-Leu}}}{\underset{\underset{\text{CH}_3 \quad\;\; \text{CH}_2\text{CH(CH}_3)_2}{|\qquad\quad |}}{\overset{+}{\text{H}_3}\text{NCHCNHCHCO}_2^{-}}}
$$

27.22 As in the DCCI-promoted coupling of amino acids, the first step is the addition of the Z-protected amino acid to DCCI to give an O-acylisourea.

$$
\underset{\substack{\text{Z-Protected}\\\text{amino acid}}}{\underset{\underset{\text{R}}{|}}{\text{ZNHCHCOH}}}
\;+\;
\underset{\text{DCCI}}{\text{C}_6\text{H}_{11}\text{N}=\text{C}=\text{NC}_6\text{H}_{11}}
\longrightarrow
\underset{\substack{O\text{-Acylisourea}}}{\underset{\underset{\text{R}}{|}}{\text{ZNHCHCO}-\text{C}\underset{\searrow \text{NHC}_6\text{H}_{11}}{\overset{\nearrow \text{NC}_6\text{H}_{11}}{}}}}
$$

This O-acylisourea is attacked by p-nitrophenol to give the p-nitrophenyl ester of the Z-protected amino acid.

$$
\text{O}_2\text{N}-\!\!\bigcirc\!\!-\ddot{\text{O}}\text{H}
\;+\;
\underset{\underset{\text{R}}{|}}{\text{ZNHCHC}-\text{O}-\text{C}\underset{\searrow \text{NHC}_6\text{H}_{11}}{\overset{\nearrow \text{NC}_6\text{H}_{11}}{}}}
\longrightarrow
\text{O}_2\text{N}-\!\!\bigcirc\!\!-\underset{\underset{\text{R}}{|}}{\text{OCCHNHZ}}
\;+\;
\text{C}_6\text{H}_{11}\text{NHCNHC}_6\text{H}_{11}
$$

27.23 To add a leucine residue to the N terminus of the ethyl ester of Z-Phe-Gly, the benzyloxycarbonyl protecting group must first be removed. This can be accomplished by hydrogenolysis.

$$
\underset{\substack{\text{Z-Protected ethyl ester of Phe-Gly}}}{\underset{\underset{\text{C}_6\text{H}_5\text{CH}_2}{|}}{\text{C}_6\text{H}_5\text{CH}_2\text{OCNHCHCNHCH}_2\text{COCH}_2\text{CH}_3}}
\xrightarrow{\text{H}_2,\,\text{Pd}}
\underset{\substack{\text{Phe-Gly ethyl ester}}}{\underset{\underset{\text{C}_6\text{H}_5\text{CH}_2}{|}}{\text{H}_2\text{NCHCNHCH}_2\text{COCH}_2\text{CH}_3}}
$$

The reaction shown has been carried out in 100% yield. Alternatively, the benzyloxycarbonyl protecting group may be removed by treatment with hydrogen bromide in acetic acid. This latter route has also been reported in the chemical literature and gives the hydrobromide salt of Phe-Gly ethyl ester in 82% yield.

Once the protecting group has been removed, the ethyl ester of Phe-Gly is allowed to react with the *p*-nitrophenyl ester of Z-protected leucine to form the protected tripeptide. Hydrogenolysis of the Z-protected tripeptide gives Leu-Phe-Gly as its ethyl ester.

$$C_6H_5CH_2O\overset{O}{\overset{||}{C}}NH\underset{(CH_3)_2CHCH_2}{\overset{|}{C}}H\overset{O}{\overset{||}{C}}O—\langle\text{benzene}\rangle—NO_2 \quad + \quad H_2N\underset{C_6H_5CH_2}{\overset{|}{C}}H\overset{O}{\overset{||}{C}}NHCH_2\overset{O}{\overset{||}{C}}OCH_2CH_3$$

p-Nitrophenyl ester of Z-protected leucine	Phe-Gly ethyl ester

$$C_6H_5CH_2O\overset{O}{\overset{||}{C}}NH\underset{(CH_3)_2CHCH_2}{\overset{|}{C}}H\overset{O}{\overset{||}{C}}NH\underset{CH_2C_6H_5}{\overset{|}{C}}H\overset{O}{\overset{||}{C}}NHCH_2\overset{O}{\overset{||}{C}}OCH_2CH_3$$

Z-protected Leu-Phe-Gly ethyl ester

$$\downarrow\ H_2,\ Pd$$

$$H_2N\underset{(CH_3)_2CHCH_2}{\overset{|}{C}}H\overset{O}{\overset{||}{C}}NH\underset{CH_2C_6H_5}{\overset{|}{C}}H\overset{O}{\overset{||}{C}}NHCH_2\overset{O}{\overset{||}{C}}OCH_2CH_3$$

Leu-Phe-Gly ethyl ester

27.24 Amino acid residues are added by beginning at the C terminus in the Merrifield solid-phase approach to peptide synthesis. Thus the synthesis of Phe-Gly requires glycine to be anchored to the solid support. Begin by protecting glycine as its *tert*-butoxycarbonyl (Boc) derivative.

$$(CH_3)_3CO\overset{O}{\overset{||}{C}}Cl \quad + \quad H_3\overset{+}{N}CH_2CO_2^- \quad \longrightarrow \quad (CH_3)_3CO\overset{O}{\overset{||}{C}}NHCH_2CO_2H$$

tert-Butoxycarbonyl chloride	Glycine	Boc-protected glycine

The protected glycine is attached via its carboxylate anion to the solid support.

$$(CH_3)_3CO\overset{O}{\overset{||}{C}}NHCH_2CO_2H \quad \xrightarrow[\text{2. ClCH}_2\text{–resin}]{\text{1. HO}^-} \quad (CH_3)_3CO\overset{O}{\overset{||}{C}}NHCH_2\overset{O}{\overset{||}{C}}OCH_2—\text{resin}$$

Boc-protected glycine

The amino group of glycine is then exposed by removal of the protecting group. Typical conditions for this step involve treatment with hydrogen chloride in acetic acid.

$$(CH_3)_3CO\overset{O}{\overset{||}{C}}NHCH_2\overset{O}{\overset{||}{C}}OCH_2—\text{resin} \quad \xrightarrow[\text{acetic acid}]{\text{HCl}} \quad H_2NCH_2\overset{O}{\overset{||}{C}}OCH_2—\text{resin}$$

Boc-protected, resin-bound glycine	Resin-bound glycine

To attach phenylalanine to resin-bound glycine, we must first protect the amino group of phenylalanine. A Boc protecting group is appropriate.

$$(CH_3)_3COCOCl \quad + \quad H_3\overset{+}{N}CHCO_2^{\bar{n}} \quad \longrightarrow \quad (CH_3)_3COCNHCHCO_2H$$
$$\underset{CH_2C_6H_5}{}$$

tert-Butoxycarbonyl chloride	Phenylalanine	Boc-protected phenylalanine

Peptide bond formation occurs when the resin-bound glycine and Boc-protected phenylalanine are combined in the presence of DCCI.

$$(CH_3)_3COCNHCHCO_2H \quad + \quad H_2NCH_2COCH_2\text{—resin} \quad \xrightarrow{DCCI} \quad (CH_3)_3COCNHCHCNHCH_2COCH_2\text{—resin}$$
$$CH_2C_6H_5 \qquad\qquad\qquad\qquad\qquad\qquad\qquad\qquad\qquad CH_2C_6H_5$$

Boc-protected phenylalanine	Resin-bound glycine	Boc-protected, resin-bound Phe-Gly

Remove the Boc group with HCl and then treat with HBr in trifluoroacetic acid to cleave Phe-Gly from the solid support.

$$(CH_3)_3COCNHCHCNHCH_2COCH_2\text{—resin} \xrightarrow[\text{2. HBr, trifluoroacetic acid}]{\text{1. HCl, acetic acid}} H_3\overset{+}{N}CHCNHCH_2CO_2^-$$
$$CH_2C_6H_5 \qquad\qquad\qquad\qquad\qquad\qquad\qquad\qquad CH_2C_6H_5$$

Boc-protected, resin-bound Phe-Gly		Phe-Gly

27.25 Examining the structure of a single strand of the pleated sheet made up of Ala-Ala-Ala-Ala reveals that the methyl groups alternate up, down, up, down.

Ala-Ala-Ala-Ala

27.26 A similar salt bridge to the one shown in text Table 27.4 can form between lysine and glutamic acid.

Lysine	Glutamic acid

27.27 The protonated form of imidazole represented by structure A is stabilized by delocalization of the lone pair of one of the nitrogens. The positive charge is shared by both nitrogens.

A

The positive charge in structure B is localized on a single nitrogen. Resonance stabilization of the type shown in structure A is not possible, thus structure A is the more stable protonated form.

B

27.28 Disconnections reveal the biosynthetic precursors of penicillin to be cysteine and valine.

Penicillin   Cysteine   Valine

27.29 (*a*) The amino acid in the problem is valine [$R = CH(CH_3)_2$]. A is the zwitterion of valine; B is its conjugate base. The pK_a for deprotonation of the ammonium ion, from text Table 27.2, is 9.62. We can use the Henderson-Hasselbalch equation (text Section 19.4) to determine the ratio of the zwitterion (A) to the conjugate base (B).

$$pH = pK_a + \log \frac{[\text{conjugate base}]}{[\text{acid}]}$$

$$pH = pK_a + \log \frac{[B]}{[A]}$$

$$\log \frac{[B]}{[A]} = pH - pK_a$$

$$\log \frac{[B]}{[A]} = 7.00 - 9.62 = -2.62$$

$$\frac{[B]}{[A]} = 10^{-2.62} = 2.40 \times 10^{-3}$$

$$\frac{[A]}{[B]} = \frac{1}{2.40 \times 10^{-3}} = 417$$

(*b*) The maximum concentration of the zwitterion (A) is found at the isoelectric point, pI. The pH where the concentration of A is at a maximum is 5.96.

27.30 Sulfanilic acid contains both an acidic sulfonic acid functional group and a basic amino group. The structure of the zwitterion is:

$$H_3\overset{+}{N}-\underset{}{\text{⟨benzene ring⟩}}-SO_3^-$$

Zwitterion of
sulfanilic acid

27.31 From the molecular formulas and net charges given in the text the structures of citrulline, ornithine, and putrescine can be determined.

Citrulline

Ornithine

Putrescine

27.32 The following outlines a synthesis of β-alanine in which conjugate addition to acrylonitrile plays a key role.

$$H_2C=CHC\equiv N \xrightarrow{NH_3} H_2NCH_2CH_2C\equiv N \xrightarrow[\text{heat}]{H_2O,\ HO^-} H_3\overset{+}{N}CH_2CH_2CO_2^-$$

Acrylonitrile   3-Aminopropanenitrile   β-Alanine

Addition of ammonia to acrylonitrile has been carried out in modest yield (31–33%). Hydrolysis of the nitrile group can be accomplished in the presence of either acids or bases. Hydrolysis in the presence of $Ba(OH)_2$ has been reported in the literature to give β-alanine in 85–90% yield.

27.33 (*a*) The first step involves alkylation of diethyl malonate by 2-bromobutane.

$$CH_3CH_2\overset{}{\underset{Br}{CHCH_3}} + {:}\overset{-}{C}H(COOCH_2CH_3)_2 \longrightarrow CH_3CH_2\overset{}{\underset{CH_3}{CHCH}}(COOCH_2CH_3)_2$$

2-Bromobutane   Anion of diethyl malonate   Compound A

In the second step of the synthesis, compound A is subjected to ester saponification. Following acidification, the corresponding diacid (compound B) is isolated.

$$CH_3CH_2\overset{}{\underset{CH_3}{CHCH}}(COOCH_2CH_3)_2 \xrightarrow[\text{2. HCl}]{\text{1. KOH}} CH_3CH_2\overset{}{\underset{CH_3}{CHCH}}(COOH)_2$$

Compound A   Compound B $(C_7H_{12}O_4)$

Compound B is readily brominated at its α-carbon atom by way of the corresponding enol form.

Compound B   Enol form   Compound C
($C_7H_{11}BrO_4$)

When compound C is heated, it undergoes decarboxylation to give an α-bromo carboxylic acid.

$$CH_3CH_2CH-C(COOH)_2 \xrightarrow{\text{heat}} CH_3CH_2CH-CHCOOH + CO_2$$
$$\quad\quad | \quad\quad |$$
$$\quad\quad CH_3 \ Br \quad\quad\quad\quad\quad CH_3 \ Br$$

Compound C   Compound D   Carbon dioxide

Treatment of compound D with ammonia converts it to isoleucine by nucleophilic substitution.

$$CH_3CH_2CH-CHCO_2H + NH_3 \longrightarrow CH_3CH_2CH-CHCO_2^-$$
$$\quad\quad | \quad\quad | \quad\quad\quad\quad\quad\quad\quad\quad | \quad\quad |$$
$$\quad\quad CH_3 \ Br \quad\quad\quad\quad\quad\quad\quad\quad CH_3 \ {}^+NH_3$$

Compound D   Isoleucine (racemic)

(*b*) The procedure just described can be adapted to the synthesis of other amino acids. The group attached to the α-carbon atom is derived from the alkyl halide used to alkylate diethyl malonate. Benzyl bromide (or chloride or iodide) would be appropriate for the preparation of phenylalanine.

$$C_6H_5CH_2Br \longrightarrow C_6H_5CH_2CHCO_2^-$$
$$\quad\quad\quad\quad\quad\quad\quad\quad\quad\quad |$$
$$\quad\quad\quad\quad\quad\quad\quad\quad\quad\quad {}^+NH_3$$

Benzyl bromide   Phenylalanine
(racemic)

27.34 Acid hydrolysis of the triester converts all its ester functions to free carboxyl groups and cleaves both amide bonds.

The hydrolysis product is a substituted derivative of malonic acid and undergoes decarboxylation on being heated. The product of this decarboxylation is aspartic acid (in its protonated form under conditions of acid hydrolysis).

$$\quad\quad CH_2COOH \quad\quad\quad\quad\quad\quad CH_2COOH$$
$$\quad\quad | \quad\quad\quad\quad\quad\quad\quad\quad\quad\quad\quad |$$
$$H_3\overset{+}{N}-C(COOH)_2 \xrightarrow{\text{heat}} H_3\overset{+}{N}-CHCOOH + CO_2$$

Aspartic acid   Carbon dioxide

Aspartic acid is chiral, but is formed as a racemic mixture, so the product of this reaction is not optically active. The starting triester is achiral and cannot give an optically active product when it reacts with optically inactive reagents.

27.35 (*a*) Nitration occurs ortho to the ortho, para directing, strongly activating OH group.

Tyrosine → 3-Nitrotyrosine

(*b*) The phenolic OH is alkylated with allyl bromide to give an *O*-allyl ether. In the second step, a methyl ester is converted to an amide by reaction with ammonia. In the following structural formulas the Boc protecting group is shown in detail.

$C_{18}H_{25}NO_5$

NH_3, CH_3OH

$C_{17}H_{24}N_2O_4$

(*c*) With both chlorine atoms in ortho-para relationships with three nitro groups, the aryl halide is especially reactive to nucleophilic aromatic substitution. The molecular formulas of the products indicate both halogens are replaced by glycines.

$C_{10}H_7N_5Na_2O_{10}$

HCl | H_2O

$C_{10}H_9N_5O_{10}$

27.36 The amino acids leucine, phenylalanine, and serine each have one chirality center.

$$\underset{\overset{|}{\underset{+}{N}H_3}}{RCHCO_2^-}$$

Leucine: $R = (CH_3)_2CHCH_2$
Phenylalanine: $R = C_6H_5CH_2$
Serine: $R = HOCH_2$

When prepared by the Strecker synthesis, each of these amino acids is obtained as a racemic mixture containing 50% of the D enantiomer and 50% of the L enantiomer.

$$\underset{}{\overset{O}{\overset{||}{RCH}}} + NH_3 + HCN \longrightarrow \underset{\overset{|}{NH_2}}{RCHC\equiv N} \xrightarrow[\text{heat}]{H_2O, H^+} \underset{\overset{|}{\underset{+}{N}H_3}}{RCHCO_2^-}$$

Chiral, but racemic

Thus, preparation of the tripeptide Leu-Phe-Ser will yield a mixture of 2^3 (eight) stereoisomers.

D-Leu-D-Phe-D-Ser	L-Leu-L-Phe-L-Ser
D-Leu-D-Phe-L-Ser	L-Leu-L-Phe-D-Ser
D-Leu-L-Phe-D-Ser	L-Leu-D-Phe-L-Ser
D-Leu-L-Phe-L-Ser	L-Leu-D-Phe-D-Ser

27.37 Bradykinin is a nonapeptide but contains only five different amino acids. Three of the amino acid residues are proline, two are arginine, and two are phenylalanine. Five peaks will appear on the strip chart after amino acid analysis of bradykinin.

Arg-Pro-Pro-Gly-Phe-Ser-Pro-Phe-Arg $\longrightarrow$ 2Arg + 3Pro + Gly + 2Phe + Ser

27.38 Asparagine and glutamine each contain an amide function in their side chain. Under the conditions of peptide bond hydrolysis that characterize amino acid analysis, the side-chain amide is also hydrolyzed, giving ammonia.

$$\underset{\overset{|}{\underset{+}{N}H_3}}{\overset{O}{\overset{||}{H_2NCCH_2CHCO_2^-}}} + H_2O \longrightarrow NH_3 + \underset{\overset{|}{\underset{+}{N}H_3}}{\overset{O}{\overset{||}{HOCCH_2CHCO_2^-}}}$$

Asparagine   Water   Ammonia   Aspartic acid

$$\underset{\overset{|}{\underset{+}{N}H_3}}{\overset{O}{\overset{||}{H_2NCCH_2CH_2CHCO_2^-}}} + H_2O \longrightarrow NH_3 + \underset{\overset{|}{\underset{+}{N}H_3}}{\overset{O}{\overset{||}{HOCCH_2CH_2CHCO_2^-}}}$$

Glutamine   Water   Ammonia   Glutamic acid

27.39 (*a*) 1-Fluoro-2,4-dinitrobenzene reacts with the amino group of the N-terminal amino acid in a nucleophilic aromatic substitution reaction of the addition–elimination type.

$$O_2N-\overset{NO_2}{\underset{}{\bigcirc}}-F + \underset{(CH_3)_2CHCH_2}{\overset{+}{\underset{}{H_3NCHC}}}\overset{O}{\overset{||}{}}NHCH_2\overset{O}{\overset{||}{C}}NH\underset{CH_2OH}{CHCO_2^-} \longrightarrow \underset{(CH_3)_2CHCH_2}{O_2N-\overset{NO_2}{\underset{}{\bigcirc}}-NHCHC}\overset{O}{\overset{||}{}}NHCH_2\overset{O}{\overset{||}{C}}NH\underset{CH_2OH}{CHCO_2H}$$

1-Fluoro-2,4-dinitrobenzene   Leu-Gly-Ser   DNP-Leu-Gly-Ser

(b) Hydrolysis of the product in part (a) cleaves the peptide bonds. Leucine is isolated as its 2,4-dinitrophenyl (DNP) derivative, but glycine and serine are isolated as the free amino acids.

DNP-Leu-Gly-Ser   → hydrolysis →   DNP-Leu   + $H_3\overset{+}{N}CH_2CO_2^-$ + $H_3\overset{+}{N}CHCO_2^-$

Gly   Ser

(c) Phenyl isothiocyanate is a reagent used to identify the N-terminal amino acid of a peptide by the Edman degradation. The N-terminal amino acid is cleaved as a phenylthiohydantoin (PTH) derivative, the remainder of the peptide remaining intact.

Ile-Glu-Phe   1. $C_6H_5N{=}C{=}S$ 2. HBr, nitromethane →   PTH derivative of isoleucine   +   Glu-Phe

(d) Benzyloxycarbonyl chloride reacts with amino groups to convert them to amides. The only free amino group in Asn-Ser-Ala is the N terminus. The amide function of asparagine does not react with benzyloxycarbonyl chloride.

Asn-Ser-Ala   + $C_6H_5CH_2OCCl$ Benzyloxycarbonyl chloride → Z-Asn-Ser-Ala

(e) The Z-protected tripeptide formed in part (d) is converted to its C-terminal p-nitrophenyl ester on reaction with p-nitrophenol and N,N'-dicyclohexylcarbodiimide (DCCI).

$$\underset{\text{Z-Asn-Ser-Ala}}{C_6H_5CH_2O\overset{O}{\overset{\|}{C}}NHCHC\overset{O}{\overset{\|}{}}NHCHC\overset{O}{\overset{\|}{}}NHCHCO_2H}\quad+\quad\underset{p\text{-Nitrophenol}}{O_2N-\!\!\!\!\bigcirc\!\!\!\!-OH}$$

with side chains: CH_2 ($C{=}O$, NH_2); CH_2OH; CH_3

$$\downarrow \text{DCCI}$$

$$C_6H_5CH_2O\overset{O}{\overset{\|}{C}}NHCHC\overset{O}{\overset{\|}{}}NHCHC\overset{O}{\overset{\|}{}}NHCHC\overset{O}{\overset{\|}{}}O-\!\!\!\!\bigcirc\!\!\!\!-NO_2$$

with side chains: CH_2 ($C{=}O$, NH_2); CH_2OH; CH_3

Z-Asn-Ser-Ala *p*-nitrophenyl ester

(*f*) The *p*-nitrophenyl ester prepared in part (*e*) is an "active" ester. The *p*-nitrophenyl group is a good leaving group and can be displaced by the amino nitrogen of valine ethyl ester to form a new peptide bond.

$$C_6H_5CH_2O\overset{O}{\overset{\|}{C}}NHCHC\overset{O}{\overset{\|}{}}NHCHC\overset{O}{\overset{\|}{}}NHCHC\overset{O}{\overset{\|}{}}O-\!\!\!\!\bigcirc\!\!\!\!-NO_2\quad+\quad H_2NCHC\overset{O}{\overset{\|}{}}OCH_2CH_3$$

with side chains: CH_2 ($C{=}O$, NH_2); CH_2OH; CH_3 | $CH(CH_3)_2$

Z-Asn-Ser-Ala *p*-nitrophenyl ester   Valine ethyl ester

$$\downarrow$$

$$C_6H_5CH_2O\overset{O}{\overset{\|}{C}}NHCHC\overset{O}{\overset{\|}{}}NHCHC\overset{O}{\overset{\|}{}}NHCHC\overset{O}{\overset{\|}{}}NHCHC\overset{O}{\overset{\|}{}}OCH_2CH_3$$

with side chains: CH_2 ($C{=}O$, NH_2); CH_2OH; CH_3; $CH(CH_3)_2$

Z-Asn-Ser-Ala-Val ethyl ester

(*g*) Hydrogenolysis of the Z-protected tetrapeptide ester formed in part (*f*) removes the Z protecting group.

$$C_6H_5CH_2O\overset{O}{\overset{\|}{C}}NHCHC\overset{O}{\overset{\|}{}}NHCHC\overset{O}{\overset{\|}{}}NHCHC\overset{O}{\overset{\|}{}}NHCHC\overset{O}{\overset{\|}{}}OCH_2CH_3 \quad\xrightarrow{\text{H}_2,\text{ Pd}}\quad H_2NCHC\overset{O}{\overset{\|}{}}NHCHC\overset{O}{\overset{\|}{}}NHCHC\overset{O}{\overset{\|}{}}NHCHC\overset{O}{\overset{\|}{}}OCH_2CH_3$$

with side chains: CH_2 ($C{=}O$, NH_2); CH_2OH; CH_3; $CH(CH_3)_2$

Z-Asn-Ser-Ala-Val ethyl ester   Asn-Ser-Ala-Val ethyl ester

27.40 (*a*) Trypsin catalyzes the hydrolysis of peptide bonds involving the carboxyl group of a lysine (K) or arginine (R) residue. The primary sequence of a peptide is written from the N terminus on the left to the C terminus on the right, so the cleavages are to the right of K and R.

$$\text{AQDDYR} \xi \text{YIHFLTQHYDAK} \xi \text{PKGR} \xi \text{NDEYCFNMMK}$$

Trypsin-catalyzed cleavage occurs at the points indicated

(*b*) Chymotrypsin-catalyzed cleavage is selective for peptide bonds involving the carboxyl groups (that is, to the right) of amino acids with aromatic side chains: phenylalanine (F), tyrosine (Y), and tryptophan (W).

$$\text{AQDDY} \xi \text{RY} \xi \text{IHF} \xi \text{LTQHY} \xi \text{DAKPKGRNDEY} \xi \text{CF} \xi \text{NMMK}$$

Chymotrypsin-catalyzed cleavage occurs at the points indicated

27.41 Somatostatin is a tetradecapeptide and so is composed of 14 amino acids. The fact that Edman degradation gave the PTH derivative of alanine identifies this as the N-terminal amino acid. A major piece of information is the amino acid sequence of a hexapeptide obtained by partial hydrolysis:

$$\text{Ala-Gly-Cys-Lys-Asn-Phe}$$

Using this as a starting point and searching for overlaps with the other hydrolysis products gives the entire sequence.

```
Ala-Gly-Cys-Lys-Asn-Phe
        Asn-Phe-Phe-Trp-Lys
                Phe-Trp
                    Lys-Thr-Phe
                        Thr-Phe-Thr-Ser-Cys
                            Thr-Ser-Cys
Ala-Gly-Cys-Lys-Asn-Phe-Phe-Trp-Lys-Thr-Phe-Thr-Ser-Cys
 1   2   3   4   5   6   7   8   9  10  11  12  13  14
```

The disulfide bridge in somatostatin is between cysteine 3 and cysteine 14. Thus, the primary structure is

$$\begin{array}{c} \text{Lys-Asn-Phe-Phe-Trp-Lys} \\ \nearrow \qquad\qquad\qquad\qquad | \\ \text{Ala-Gly-Cys} \\ \searrow \qquad\qquad\qquad\qquad \\ \text{S-S-Cys-Ser-Thr-Phe-Thr} \end{array}$$

27.42 It is the C-terminal amino acid that is anchored to the solid support in the preparation of peptides by the Merrifield method. Refer to the structure of oxytocin in Figure 27.6 of the text and note that oxytocin, in fact, has no free carboxyl groups; all the acyl groups of oxytocin appear as amide functions. Thus, the carboxyl terminus of oxytocin has been modified by conversion to an amide.

There are three amide functions of the type $\overset{\overset{\text{O}}{\|}}{\text{C}}\text{NH}_2$, two of which belong to side chains of asparagine and glutamine, respectively. The third amide belongs to the C-terminal amino acid, glycine,

—NHCH$_2\overset{\overset{\text{O}}{\|}}{\text{C}}$OH, which in oxytocin has been modified so that it appears as —NHCH$_2\overset{\overset{\text{O}}{\|}}{\text{C}}NH_2$. Therefore, attach glycine to the solid support in the first step of the Merrifield synthesis. The carboxyl group can be modified to the required amide after all the amino acid residues have been added and the completed peptide is removed from the solid support.

ANSWERS TO INTERPRETIVE PROBLEMS

27.43 A;   **27.44** C;   **27.45** B;   **27.46** D;   **27.47** C;   **27.48** A

SELF-TEST

PART A

A-1. Give the structure of the reactant, reagent, or product omitted from each of the following:

(*a*)   ? $\xrightarrow[\substack{2.\ H_3O^+,\ heat \\ 3.\ neutralize}]{1.\ NH_4Cl,\ NaCN}$ $\overset{\overset{+}{N}H_3}{\underset{|}{C_6H_5CH_2CHCO_2^-}}$

(*b*) $C_6H_5CH_2O\overset{O}{\overset{||}{C}}Cl$ + valine $\xrightarrow[2.\ H^+]{1.\ HO^-,\ H_2O}$?

(*c*) Boc-Phe + $H_2NCH_2CO_2CH_2CH_3$ $\xrightarrow{?}$ $\underset{\underset{CH_2C_6H_5}{|}}{Boc-NHCH\overset{O}{\overset{||}{C}}NHCH_2CO_2CH_2CH_3}$

A-2. Give the structure of the derivative that would be obtained by treatment of Phe-Ala with Sanger's reagent (1-fluoro-2,4-dinitrobenzene) followed by hydrolysis.

A-3. Outline a sequence of steps that would allow the following synthetic conversions to be carried out:

(*a*) $\overset{\overset{+}{N}H_3}{\underset{|}{(CH_3)_2CHCH_2CHCO_2^-}}$ (leucine) from $CH_3\overset{O}{\overset{||}{C}}NHCH(CO_2CH_2CH_3)_2$

(*b*) Leu-Val from leucine and $\overset{\overset{+}{N}H_3}{\underset{|}{(CH_3)_2CHCHCO_2^-}}$ (valine)

A-4. The carboxypeptidase-catalyzed hydrolysis of a pentapeptide yielded phenylalanine (Phe). One cycle of an Edman degradation gave a derivative of leucine (Leu). Partial hydrolysis yielded the fragments Leu-Val-Gly and Gly-Ala among others. Deduce the structure of the peptide.

A-5. Consider the following compound:

(*a*) What kind of peptide does this structure represent? (For example, dipeptide)

(*b*) How many peptide bonds are present?

(c) Give the name for the N-terminal amino acid.

(d) Give the name for the C-terminal amino acid.

(e) Using three-letter abbreviations, write the sequence.

A-6. Consider the tetrapeptide Ala-Gly-Phe-Leu. What are the products obtained from each of the following? Be sure to account for all the amino acids of the peptide.

(a) Treatment with 1-fluoro-2,4-dinitrobenzene followed by hydrolysis in concentrated HCl at 100°C

(b) Treatment with chymotrypsin

(c) Treatment with carboxypeptidase

(d) Reaction with benzyloxycarbonyl chloride

PART B

B-1. Which phrase correctly completes the statement?
Except for glycine, which is achiral, all the amino acids present in proteins...

(a) are chiral, but racemic

(b) are meso forms

(c) have the L configuration at their α carbon

(d) have the R configuration at their α carbon

(e) have the S configuration at their α carbon

B-2. Assume that a particular amino acid has an isoelectric point of 6.0. In a solution of pH 1.0, which of the following species will predominate?

(a) $\overset{R}{\underset{|}{H_3\overset{+}{N}CHCO_2H}}$ (b) $\overset{R}{\underset{|}{H_2NCHCO_2H}}$ (c) $\overset{R}{\underset{|}{H_3\overset{+}{N}CHCO_2^-}}$ (d) $\overset{R}{\underset{|}{H_2NCHCO_2^-}}$

B-3. Which of the species in problem B-2 will predominate at pH 6.0?

B-4. Which of the species in problem B-2 will predominate at pH 11?

B-5. Which of the following reagents would be combined in the synthesis of Phe-Ala?

[In phenylalanine (Phe), R in the generalized amino acid formula $\overset{R}{\underset{|}{H_2NCHCO_2H}}$ is $CH_2C_6H_5$, and in alanine (Ala) it is CH_3.]

1. $\overset{CH_3}{\underset{|}{ZNHCHCO_2H}}$ 2. $\overset{CH_3}{\underset{|}{H_2NCHCO_2CH_2C_6H_5}}$

3. $\overset{CH_2C_6H_5}{\underset{|}{ZNHCHCO_2H}}$ 4. $\overset{CH_2C_6H_5}{\underset{|}{H_2NCHCO_2CH_2C_6H_5}}$

(a) 1 and 2   (b) 1 and 4   (c) 2 and 3   (d) 3 and 4

B-6. What are the products obtained following treatment of Ser-Tyr-Val-Ala with chymotrypsin?

(*a*) Serine + Tyr-Val-Ala   (*d*) Ser-Tyr-Val + Alanine

(*b*) Ser-Tyr + Valine + Alanine   (*e*) Serine + Tyrosine + Val-Ala

(*c*) Ser-Tyr + Val-Ala

B-7. The first cycle of the Edman degradation of the tetrapeptide Gly-Ala-Ile-Leu would give a PTH derivative of

(*a*) Glycine   (*b*) Alanine   (*c*) Isoleucine   (*d*) Leucine

CHAPTER 28

Nucleosides, Nucleotides, and Nucleic Acids

SOLUTIONS TO TEXT PROBLEMS

28.1 The enol form of cytosine has a double bond between nitrogen at position 1 (see text Table 28.1 for the numbering scheme) in the ring and the carbon at position 2:

28.2 The resonance form of guanine that has a ring with an electronic structure analogous to benzene arises from amide resonance:

Resonance form with six-membered ring analogous to benzene

28.3 Caffeine and theobromine are both purines. Caffeine lacks $H-N-C=O$ units so it cannot enolize. Two constitutionally isomeric enols are possible for theobromine.

28.4 The structure of adenosine is shown in text Table 28.2. 3′-Deoxyadenosine (cordycepin) has an H replacing the OH group on C-3 of the ribose portion of the molecule.

28.5 The structure of cytidine is shown in text Table 28.2. In 2′-deoxycytidine-3′-monophosphate an H has replaced the OH at C-2 of the ribose portion of the molecule. In addition, C-3 of the ribose portion is bonded to a phosphate group.

28.6 Removal of a proton by the base from the 3′-hydroxyl group generates an anion that can act as a nucleophile on phosphorus and form the ring in cyclic-AMP.

28.7 When equations are added, species that appear on both the left and the right of the arrow are canceled. Thus adding the hydrolysis of ATP to equation 1 does, indeed, produce equation 2.

Hydrolysis of ATP:

$$ATP + H_2O \longrightarrow ADP + HPO_4^{2-}$$

Equation 1:

Equation 2:

28.8 The reaction of ammonia with γ-glutamyl phosphate to give glutamine is analogous to the reaction of ammonia with an acid anhydride to yield an amide. As noted in the text, γ-glutamyl phosphate is a mixed anhydride of glutamic acid and phosphoric acid. Addition of ammonia to the carbonyl carbon is followed by elimination to give glutamine and phosphate ion.

γ-Glutamyl phosphate　　　Ammonia

Phosphate ion　　　Glutamine

28.9 Figure 28.1 in the text shows a trinucleotide of 2′-deoxy-D-ribose. If the pentose were D-ribose instead, an OH group would be found on C-2 of the ribose portion of each nucleotide segment. The arrows in the following structures point to the OH groups not found in nucleotides formed from 2′-deoxy-D-ribose.

AUG　　　　　　　GUA

28.10 The ratio of the purine guanine to the pyrimidine cytosine in DNA is 1:1. Thus the guanine content in turtle DNA is the same as the cytosine content, 21.3%.

28.11 Each nucleosome is 146 base pairs long and is separated from the next nucleosome by a linker of about 50 base pairs. Thus a gene with 10,000 base pairs would contain approximately 50 nucleosomes.

28.12 By changing the coding sequence of text Figure 28.12 from AUGGCU to AUGUCU, the second amino acid becomes serine instead of alanine. The serine tRNA sequence that is complementary to the UCU sequence of mRNA is AGA.

28.13 Uracil is a pyrimidine base found in RNA. Its complement in both DNA and RNA is the purine adenine.

28.14 The numbering of the ring in uracil and its derivatives parallels that in pyrimidine.

Pyrimidine   Uracil   5-Fluorouracil

28.15 (*a, b*) Oxygen is more electronegative than nitrogen, and we would expect the OH proton to be more acidic than the NH proton. As the following scheme illustrates, the conjugate base of the keto form and of the enol form are resonance contributors to the same anion.

28.16 Purine has the molecular formula $C_5H_4N_4$. Because we are told that uric acid is a purine having the formula $C_5H_4N_4O_3$ and has no C—H bonds, it is reasonable to presume that three C=O groups have replaced the C—H groups of purine.

Purine   Uric acid

28.17 Purine and its numbering system are as shown:

In nebularine, D-ribose in its furanose form is attached to position 9 of purine. The stereochemistry at the anomeric position is β.

9-β-D-Ribofuranosylpurine
(nebularine)

28.18 The problem states that vidarabine is the arabinose analog of adenosine. Arabinose and ribose differ only in their configuration at C-2.

Adenosine   Vidarabine

28.19 Evaluating each of the three protonated forms of adenine will reveal which is the most stable. The most basic nitrogen will be the one that forms the most stable conjugate acid. *Hint:* You might find it helpful to review amine basicity (particularly the discussion of imidazole) in text Section 22.4.

Unstable; disrupts
aromaticity of five-
membered ring

Not the most stable; disrupts
delocalization of arylamine
unshared electron pair

Stable; positive charge
shared by two nitrogen
atoms as shown by
resonance structures

28.20 Nucleophilic aromatic substitution occurs when 6-chloropurine reacts with hydroxide ion by an addition–elimination pathway.

The enol tautomerizes to give hypoxanthine.

Hypoxanthine

28.21 Nitrous acid reacts with aromatic primary amines to yield diazonium ions.

Adenosine

Treatment of the diazonium ion with water yields a phenol. Tautomerization gives inosine.

Inosine

28.22 The carbon atoms of the ribose portion of a nucleoside are numbered as follows:

A 5′-nucleotide has a phosphate group attached to the C-5′ hydroxyl.

Inosinic acid

28.23 (*a*) The value of $\Delta G^{\circ\prime}$ for the phosphorylation of α-D-glucopyranose is –23 kJ/mol. Reactions with a negative change in free energy are exergonic.

(*b*) Enzymes are catalysts in biological systems. Catalysts speed up a reaction, but do not affect $\Delta G^{\circ\prime}$.

(*c*) To find the value of $\Delta G^{\circ\prime}$ for the reaction of α-D-glucopyranose with inorganic phosphate, add the phosphorylation reaction given in the problem to the *reverse* of the hydrolysis of ATP to ADP. The free energy change for the conversion of ADP to ATP (from text Section 28.4) is –(–31 kJ/mol) = +31 kJ/mol.

Phosphorylation:

ATP + [glucopyranose] → [glucopyranose phosphate] + ADP $\Delta G^{\circ\prime}$ = –23 kJ

Reverse of ATP Hydrolysis:

$$\text{ATP} + \text{HPO}_4^{2-} \longrightarrow \text{ATP} + \text{H}_2\text{O} \quad \Delta G^{\circ\prime} = +31 \text{ kJ}$$

Sum:

The reaction of α-D-glucopyranose with inorganic phosphate has a free energy change, $\Delta G^{\circ\prime} = +8$ kJ. The reaction is endergonic.

28.24 All the bases in the synthetic messenger RNA prepared by Nirenberg were U; therefore, the codon is UUU. By referring to the codons in text Table 28.3, we see that the UUU codes for phenylalanine. A polypeptide in which all the amino acid residues were phenylalanine was isolated in Nirenberg's experiment.

ANSWERS TO INTERPRETIVE PROBLEMS

28.25 C;   **28.26** D;   **28.27** B;   **28.28** A;   **28.29** B;   **28.30** A

SELF TEST

PART A

A-1. How does the nucleoside adenosine found in RNA differ from that found in DNA?

A-2. The standard free energy change ($\Delta G^{\circ\prime}$) for the reaction of glycerol with inorganic phosphate ion to give glycerol 1-phosphate is +9.2 kJ.

Using the free energy change for the conversion of ATP to ADP (text Section 28.4), calculate the free energy change for the reaction:

Problems A-3 through A-7 refer to the following nucleotide segments of a DNA strand.

> (*a*) A-A-A-G-G-T-C-C-C-G-T-A
>
> (*b*) T-A-C-T-C-G-C-G-G-A-T-G

A-3. Write the nucleotide sequence for the complementary DNA strand of each segment.

A-4. Write the mRNA nucleotide sequence that would be produced by transcription of each DNA segment.

A-5. List the codons present in each mRNA segment.

A-6. Determine the amino acid sequence that would be formed from each mRNA nucleotide segment.

A-7. What are the anticodons corresponding to each nucleotide segment in Problem A-5?

PART B

B-1. Which of the following are purines?

1. Adenine	2. Cytosine	3. Guanine	4. Thymine

(*a*) 1, 2   (*b*) 1, 3   (*c*) 1, 4   (*d*) 2, 3   (*e*) 2, 4

B-2. Which of the following are pyrimidines?

1. Adenine	2. Cytosine	3. Guanine	4. Thymine

(*a*) 1, 2   (*b*) 1, 3   (*c*) 1, 4   (*d*) 2, 3   (*e*) 2, 4

B-3. Which of the following is present in RNA but not DNA?

(*a*) Adenine   (*b*) Uracil   (*c*) Guanine   (*d*) Thymine

B-4. Which of the following is present in DNA but not RNA?

(*a*) Adenine   (*b*) Uracil   (*c*) Guanine   (*d*) Thymine

B-5. Which of the following structures is correct for adenine?

B-6. Choose the response that contains all of the correct statements, and no incorrect ones. Complete hydrolysis of a nucleic acid yields:

1. A mixture of heterocyclic bases
2. A mixture of amino acids
3. An aldohexose, either glucose or deoxyglucose
4. Phosphate ion
5. An aldopentose, either ribose or deoxyribose

(a) 1, 3, 4   (b) 2, 3, 4   (c) 1, 2, 5   (d) 2, 4, 5   (e) 1, 4, 5

B-7. The base pairs present in DNA are:

(a) A···G and C···T   (d) A···A and C···C
(b) A···C and G···T   (e) A···A and T···T
(c) A···T and G···C

B-8. How many bases constitute a codon in messenger RNA?

(a) 1   (b) 2   (c) 3   (d) 4   (e) more than 4

B-9. In which of the following is a sequence of bases termed an *anticodon* located?

(a) DNA   (c) Ribosomal RNA
(b) Messenger RNA   (d) Transfer RNA

CHAPTER 29

Synthetic Polymers

SOLUTIONS TO TEXT PROBLEMS

29.1 (*b*) Methyl methacrylate is two words, thus its polymer name is *poly(methyl methacrylate)*. The repeating unit follows the pattern shown in the text.

Poly(methyl methacrylate)

29.2 Formaldehyde can form a cyclic trimer, known as trioxane.

$$3H_2C{=}O \longrightarrow$$

Formaldehyde   Trioxane

29.3 Nucleophilic acyl substitution follows an addition–elimination pathway (text Section 20.3). Nucleophilic addition by 1,4-benzenediamine to the carbonyl group of terephthaloyl chloride gives the tetrahedral intermediate shown.

1,4-Benzenediamine   Terephthaloyl chloride   Tetrahedral intermediate

29.4 The product of the reaction contains all of the atoms of the starting materials. This is an example of an *addition* reaction.

29.5 (*b*) Azobisisobutyronitrile (AIBN) undergoes homolytic bond cleavage as shown in the text to give nitrogen and two 1-cyano-1-methylethyl radicals. These radicals initiate polymerization of styrene by addition to the vinyl group.

29.6 The terminal free radical of the growing polymer chain can react with the free-radical initiator, in this case an alkoxy radical.

29.7 As shown in text Mechanism 29.1, four-carbon branches result from a strain-free six-membered chair cyclohexane-like transition state. Transition states leading to chains shorter than four carbons would have angle strain which would increase ΔH of activation. Transition states leading to chains longer than four carbons require restricting the motion of more atoms and are more ordered. This would make ΔS of activation more negative.

29.8 As noted in the text, the number of chains is equal to the number of molecules of butyllithium used. Thus if less butyllithium is present, there will be fewer chains and each one will be longer. In other words, the average chain length is a function of the ratio of moles of styrene to moles of butyllithium.

29.9 Hydroxide ion adds to methyl 2-cyanoacrylate to give an anion stabilized by delocalization of the negative charge to oxygen and nitrogen.

29.10 A lactam is a cyclic amide (text Section 20.14). ε-Caprolactam has a 7-membered ring.

ε-Caprolactam

29.11 1,3-Benzenediamine reacts with 1,3-benzenedicarboxylic acid to give the repeating unit of Nomex.

1,3-Benzenediamine

1,3-Benzenedicarboxylic acid

Repeating unit of Nomex

29.12 The polymer made from ε-caprolactone is a polyester.

ε-Caprolactone

Repeating unit of poly(ε-caprolactone)

29.13 The reaction of bisphenol A (as its disodium salt) with phosgene ($Cl_2C=O$) is an example of nucleophilic acyl substitution, and follows an addition–elimination pathway.

29.14 If the "polymeric diol" used in the polymer shown in the text were derived from 1,2-epoxypropane its structure would be as shown:

1,2-Epoxypropane

"Polymeric diol"

29.15 Nylon 11 is a condensation polymer as water is released in the polymerization process, as shown below.

$$H_3\overset{+}{N}(CH_2)_{10}\overset{\overset{\displaystyle O}{\|}}{C}O^- \longrightarrow \left[NH(CH_2)_{10}\overset{\overset{\displaystyle O}{\|}}{C}\right]_n + H_2O$$

11-Aminoundecanoic acid   Nylon 11   Water

As the polymer grows, each chain has two growth points, a free amino group at one end and a free carboxyl at the other. The starting material reacts rapidly to form oligomers, which can then react together to form the polymer. Nylon 11 is a step-growth polymer.

29.16 The protein biosynthesis process described in text Figure 28.12 extends the protein by one amino acid at a time on the end of the chain. It is a chain-growth process. The process converts an ester to an amide, so an alcohol is formed as each additional amino acid is added to the growing chain. The protein is a condensation polymer.

29.17 3-Hydroxynonanoic acid will polymerize to form a polyester with a hexyl group side chain in the repeating unit.

$$CH_3(CH_2)_5\underset{\underset{\displaystyle OH}{|}}{C}HCH_2\overset{\overset{\displaystyle O}{\|}}{C}OH \longrightarrow \left[O\underset{\underset{\displaystyle (CH_2)_5CH_3}{|}}{C}HCH_2\overset{\overset{\displaystyle O}{\|}}{C}\right]_n$$

3-Hydroxynonanoic acid   Polyester repeating unit

29.18 Polyvinylene is a polymer of acetylene. Its source-based name is polyacetylene.

$$HC\equiv CH \longrightarrow$$

Acetylene   Polyvinylene

29.19 The lactone used to prepare the polymer shown is 3-propanolide, better known by its common name β-propiolactone.

$$\left[OCH_2CH_2\overset{\overset{\displaystyle O}{\|}}{C}\right]_n \Longrightarrow$$

3-Propanolide (β-Propiolactone)

29.20 The monomers used to prepare the polymer in Kodel fibers are 1,4-benzenedicarboxylic acid (terephthalic acid) and cyclohexane-1,4-dimethanol.

1,4-Benzenedicarboxylic acid
(terephthalic acid)

Cyclohexane-1,4-dimethanol

29.21 The alkene that would be most suited for cationic polymerization is the one that forms the most stable carbocation when protonated. Of those in the problem, 1,3-butadiene is the best candidate.

$$H_2C=CHCH=CH_2$$

1,3-Butadiene

29.22 An alkene suitable for anionic polymerization is one which forms a relatively stable anion. Thus alkenes with electron-withdrawing groups attached to the double bond are the best candidates. Of the alkenes in the problem, acrylonitrile is most suited for anionic polymerization.

$$H_2C=CHC\equiv N$$

Acrylonitrile

29.23 (*a*) The reagents given in the problem [$TiCl_4$, $(CH_3CH_2)_3Al$] are early Ziegler-Natta catalysts, and initiate polymerization by formation of a coordination complex.

(*b*) Organic peroxides readily undergo homolytic bond cleavage of the O–O bond, and initiate polymerization by a free-radical mechanism.

(*c*) Boron trifluoride (BF_3) is a powerful Lewis acid and initiates polymerization by a cationic mechanism.

29.24 *p*-Methoxystyrene contains an electron-donating methoxy group, and thus the anion formed from this compound would be less stable, and formed more slowly, than that formed from styrene. As a result, the anionic polymerization of styrene is more rapid.

p-Methoxystyrene

Less stable anion; formed more slowly

Styrene

More stable anion; formed more rapidly

29.25 The monomer that forms the more stable anion will react faster and should be used first to form the living polymer. Thus acrylonitrile should be polymerized first, and styrene added second to form the copolymer.

Acrylonitrile   Living polymer

29.26 The repeating unit of poly(vinyl butyral) is an acetal, formed from the reaction of the diol starting polymer and an aldehyde. Compound A is butanal

Butanal
(compound A)

29.27 Polymerization of ethylene under Ziegler–Natta conditions yields a linear polymer of methylene (CH_2) groups. Adding a small quantity of 1-hexene will result in introduction of butyl side chains into the polymer.

Ethylene   1-Hexene   Ethylene–1-hexene
copolymer

29.28 (*a*) The formation of bisphenol A begins by electrophilic aromatic substitution at the para position of phenol to form a tertiary alcohol. The electrophile is the protonated form of acetone [$(CH_3)_2C=O$].

A second mole of phenol then reacts with this alcohol to form bisphenol A.

Bisphenol A

(*b*) Bisphenol B differs from bisphenol A in that one of the methyl groups has been replaced by an ethyl group.

Bisphenol B

29.29 In the presence of an acid catalyst the oxygen of ethylene oxide is protonated in the polymerization process.

Ethylene oxide

Poly(ethylene oxide)

Under basic conditions ethylene oxide is attacked directly by hydroxide ion.

Ethylene oxide

Poly(ethylene oxide)

29.30 (*a*) The reaction of phenol and formaldehyde is an example of electrophilic aromatic substitution. Hydroxyl is a strongly activating ortho-para director. The electrophile is the protonated form of formaldehyde.

Phenol   Protonated form   *o*-(Hydroxymethyl)-
                of formaldehyde   phenol

(*b*) The two aromatic rings become linked following a second electrophilic aromatic substitution. The electrophile is the protonated form of *p*-(hydroxymethyl)phenol.

29.31 In the second polymerization step a molecule of formaldehyde adds to the formaldehyde-boron trifluoride complex.

ANSWERS TO INTERPRETIVE PROBLEMS

29.32 C;   **29.33** B;   **29.34** A;   **29.35** D;   **29.36** B;   **29.37** B;   **29.38** C

SELF-TEST

PART A

A-1. Briefly describe the difference between a chain-growth and a step-growth polymer.

A-2. (*a*) What monomer of molecular formula C_2H_4O gives the polymer with the repeating unit shown?

$$\left[O-CH_2CH_2 \right]_n$$

(*b*) What is the source-based name of this polymer?

A-3. Bisphenol A is used to form the polycarbonate polymer known as *Lexan,* as shown in the reaction below. Is this polymer formed in an addition process, or a condensation?

Bisphenol A   Phosgene   Polycarbonate repeating unit

A-4. What is the monomer used to prepare poly(vinyl acetate)?

A.5. Poly(vinyl acetate) is used to prepare poly(vinyl alcohol) by the reaction:

Poly(vinyl acetate)   Methanol   Poly(vinyl alcohol)   Methyl acetate

Why is it impractical to make poly(vinyl alcohol) from vinyl alcohol?

PART B

B-1. Polystyrene (shown) is prepared from styrene ($C_6H_5CH=CH_2$) by a(an) __I__ process. Polystyrene is a __II__ polymer.

Repeating unit of
polystyrene

(*a*) I: addition;   II: step-growth   (*c*) I: condensation;   II: step-growth

(*b*) I: addition;   II: chain-growth   (*d*) I: condensation;   II: chain-growth

B-2. What would be the repeating unit of a polymer made from 4-butanolide?

4-Butanolide

B-3. Which of the following polymer fragments would be *least* durable in an aqueous environment (i.e., least able to withstand exposure to water without degradation)?

B-4. One type of epoxy resin is obtained by the base-catalyzed reaction of bisphenol A with epichlorohydrin. The mechanism of the first step of the process is shown.

Conjugate base of
bisphenol A

From this information deduce the structure of epichlorohydrin.

(a) $HOCH_2CHCH_2Cl$
              |
            OH

(c) $H_2C{-}CHCH_2Cl$ (with epoxide O bridging)

(b) H_2CCCH_2Cl (with C=O and Cl substituent)

(d) $H_2C{=}CHCH_2Cl$

APPENDIX

Answers to the Self-Tests

CHAPTER 1

A-1. (*a*) P; $1s^22s^22p^63s^23p^3$ (*b*) S^{2-}; $1s^22s^22p^63s^23p^6$

A-2. (*a*)

$$:\ddot{N}=C=\ddot{S}:$$

Formal charge: −1   0   0   Net charge: −1

(*b*)

$$:O\equiv N-\ddot{\ddot{O}}:$$

Formal charge: +1   +1   −1   Net charge: +1

(*c*)

$$-1\ \ :\ddot{\ddot{O}}:$$
$$|$$
$$HC=NH_2$$

Formal charge: 0   +1   Net charge: 0

A-3. (*a*)

$$:N\equiv C-\ddot{\ddot{S}}:$$

Formal charge: 0   0   −1   Net charge: −1

(*b*)

$$:\ddot{O}=N=\ddot{O}:$$

Formal charge: 0   +1   0   Net charge: +1

(*c*)

$$0\ \ :\ddot{O}$$
$$||$$
$$HC-\ddot{N}H_2$$

Formal charge: 0   0   Net charge: 0

The more stable Lewis structures are

(*a*) $^-:\ddot{N}=C=\ddot{S}:$ (*b*) $:\ddot{O}=\overset{+}{N}=\ddot{O}:$ (*c*)
$$:\ddot{O}$$
$$||$$
$$HC-\ddot{N}H_2$$

A-4. (*a*)

$$H$$
$$|$$
$$H-C-\ddot{N}-H$$
$$|\ \ \ |$$
$$H\ \ H$$

(*b*)

$$H\ \ H$$
$$|\ \ \ |$$
$$H-C-C=\ddot{\ddot{O}}:$$
$$|$$
$$H$$

A-5. (*a*) $C_{12}H_{20}O$

(*c*) $C_{14}H_{24}O$

(*b*) $C_{10}H_{22}$

(*d*) C_9H_6BrN

A-6. $CH_3CH_2\ddot{O}:^-$ + $H-\ddot{N}H_2$ ⇌ $CH_3CH_2\ddot{O}H$ + $^-\ddot{N}H_2$ $K_{eq} = 10^{-20}$

      Base   Acid   Stronger   Stronger
                                      acid   base

A-7.

A-8. $:\ddot{O}-C\equiv N:$

Formal
charge: −1   0   0   Net charge: −1

A-9. (*a*)

Tetrahedral   Trigonal   Tetrahedral
              planar
        Bent

(*b*) Pyramidal;   Yes, it is polar.

A-10. (*a*) Linear   (*b*) Linear   (*c*) Bent

A-11. (*a*) D   (*c*) None   (*e*) None   (*g*) A
         (*b*) A, B   (*d*) B   (*f*) A, D   (*h*) C

A-12.

A-13. (*a*)

More
stable

(*b*)

More
stable

B-1. (*b*)   **B-2.** (*b*)   **B-3.** (*c*)   **B-4.** (*d*)
B-5. (*a*)   **B-6.** (*b*)   **B-7.** (*d*)   **B-8.** (*b*)
B-9. (*d*)   **B-10.** (*e*)   **B-11.** (*d*)   **B-12.** (*b*)
B-13. (*d*)   **B-14.** (*c*)   **B-15.** (*c*)

CHAPTER 2

A-1. $CH_3CH_2CH_2CH_2$— $CH_3CH_2CHCH_3$
                                                              |

Common: *n*-Butyl   *sec*-Butyl
Systematic: Butyl   1-Methylpropyl

$$CH_3CHCH_2— \qquad CH_3C—$$

Common: Isobutyl   *tert*-Butyl
Systematic: 2-Methylpropyl   1,1-Dimethylethyl

A-2. (*a*) 28 (8 C—C; 20 C—H)   (*b*) 27 (9 C—C; 18 C—H)

A-3. (*a*) Oxidized   (*b*) Neither   (*c*) Neither   (*d*) Reduced

A-4. (*a*) $CH_3CHCHCHCH_3$ with CH_3CHCH_3 and CH_3 CH_3 substituents   (*b*) Six methyl groups, three isopropyl groups

A-5. (*a*) 3,4-Dimethylheptane   (*b*) (1,2-Dimethylpropyl)cyclohexane

A-6.

	Primary	Secondary	Tertiary
(*a*)	4	3	2
(*b*)	3	5	3

A-7. (*a*) 1,3-Dimethylbutyl; secondary
(*b*) 1,1-Diethylpropyl; tertiary
(*c*) 2,2-Diethylbutyl; primary

A-8. $CH_3CHCHCH_2CH_3 \equiv C_7H_{16}$ $C_7H_{16} + 11O_2 \longrightarrow 7CO_2 + 8H_2O$

A-9.

Cyclopentane   Methylcyclobutane   Ethylcyclopropane   1,1-Dimethylcyclopropane   1,2-Dimethylcyclopropane

A-10. (*a*)

(*b*)

(2-Methylbutyl)cyclohexane

(*c*)

3-Ethyl-2,3-
dimethylhexane

4-Ethyl-3-methylheptane

A-11. (*a*) $CH_3CH_2CH_2CH_2CH_2CH_2CH_2CH_3$

(*c*) $(CH_3)_2CHCHCH(CH_3)_2$
                             |
                            CH_3

(*b*) $(CH_3)_3CC(CH_3)_3$

(*d*) $(CH_3)_3CC(CH_3)_3$

A-12.

2,2-Dimethylpentane

2,4-Dimethylpentane

3,3-Dimethylpentane

2,3-Dimethylpentane

3-Ethylpentane

A-13. 10,049 kJ/mol

A-14. (*a*) $11\sigma; 1\pi$ (*b*) $9\sigma; 2\pi$ (*c*) $12\sigma; 4\pi$ (*d*) $13\sigma; 4\pi$

A-15. (*a*) $H_3C-CH=CH-CH_3$

            sp^3 sp^2 sp^3

(*c*) All carbons are sp^2.

(*b*) $H-C\equiv C-CH_2-CH_3$

          sp sp^3

(*d*)

B-1. (*a*)	**B-2.** (*d*)	**B-3.** (*d*)	**B-4.** (*c*)
B-5. (*b*)	**B-6.** (*a*)	**B-7.** (*c*)	**B-8.** (*c*)
B-9. (*a*)	**B-10.** (*a*)	**B-11.** (*b*)	**B-12.** (*e*)
B-13. (*d*)	**B-14.** (*d*)	**B-15.** (*d*)	**B-16.** (*c*)
B-17. (*a*)			

CHAPTER 3

A-1.

Gauche   Anti

A-2. (*a*)

(Eclipsed)

(*b*)

A-3. $(CH_3)_3CCH_2C(CH_3)_3$ = 2,2,4,4-tetramethylpentane

A-4.

A-5.

A-6.

A-7. (*a*) C   (*b*) A and B   (*c*) D   (*d*) A

A-8.

More stable

A-9. *cis*-1-Ethyl-3-methylcyclohexane has the lower heat of combustion.

A-10. Tricyclic; $C_{10}H_{16}$

A-11. The form of the curve more closely resembles ethane than butane.

A-12.

B-1. (*d*)	**B-2.** (*b*)	**B-3.** (*c*)	**B-4.** (*a*)
B-5. (*c*)	**B-6.** (*a*)	**B-7.** (*d*)	**B-8.** (*e*)
B-9. (*c*)	**B-10.** (*e*)	**B-11.** (*b*)	**B-12.** (*a*)
B-13. (*d*)	**B-14.** (*c*)	**B-15.** (*b*)	

CHAPTER 4

A-1. Alcohol, alkene, ester, ketone

A-2. (*a*) *trans*-1-Bromo-3-methylcyclopentane

(*b*) 2-Ethyl-4-methyl-1-hexanol

A-3. (*a*) $ICH_2\overset{\underset{\displaystyle CH_3}{|}}{\underset{\underset{\displaystyle Cl}{|}}{C}}CH_2CH_2CH_2CH_2CH_3$ (*b*)

A-4. (*a*) **Functional class:**   1-ethyl-3-methylbutyl alcohol
       Substitutive: 5-methyl-3-hexanol
   (*b*) **Functional class:**   1,1,2-trimethylbutyl chloride
       Substitutive: 2-chloro-2,3-dimethylpentane

A-5. (*a*) $CH_3CH_2CH_2Cl$ (*b*) $CH_3CH_2\overset{\underset{\displaystyle OH}{|}}{C}(CH_3)_2$

A-6. (a) Three

(b, c)

$$\underset{\substack{\text{(Most stable)}}}{\overset{\begin{array}{c}CH_3 \ \ CH_3\\ | \ \ \ \ \ | \end{array}}{CH_3\overset{\bullet}{C}CH_2CHCH_3}} \qquad \underset{\substack{}}{\overset{\begin{array}{c}CH_3 \ \ CH_3\\ | \ \ \ \ \ | \end{array}}{CH_3CH\overset{\bullet}{C}HCHCH_3}} \qquad \underset{\substack{\text{(Least stable)}}}{\overset{\begin{array}{c}CH_3 \ \ CH_3\\ | \ \ \ \ \ | \end{array}}{\overset{\bullet}{C}H_2CHCH_2CHCH_3}}$$

A-7.

$$\underset{\begin{array}{c}|\\CH_3\end{array}}{\overset{\begin{array}{c}CH_3\\|\end{array}}{CH_3CCH_3}} + Cl_2 \longrightarrow \underset{\begin{array}{c}|\\CH_3\end{array}}{\overset{\begin{array}{c}CH_3\\|\end{array}}{CH_3CCH_2Cl}} + HCl$$

A-8.   Br· +   ⟶   HBr +

+ Br₂ ⟶   + Br·

A-9. $\Delta H° = -57 \text{ kJ} (-13.5 \text{ kcal})$

A-10. (a) $(CH_3)_3C-\overset{..}{\underset{..}{O}}H + H-Br \rightleftharpoons (CH_3)_3C-\overset{+}{O}H_2 + Br^-$

$(CH_3)_3C-\overset{+}{O}H_2 \rightleftharpoons H_2O + (CH_3)_3C^+$

$(CH_3)_3C^+ + :\overset{..}{\underset{..}{Br}}:^- \longrightarrow (CH_3)_3C-Br$

(b)

(c) Water is displaced directly from the oxonium ion of 1-butanol by bromide ion. A primary carbocation is not involved.

$$:\overset{..}{\underset{..}{Br}}:^- \qquad CH_3CH_2CH_2CH_2-\overset{+}{O}H_2 \longrightarrow H_2O + CH_3CH_2CH_2CH_2Br$$

A-11. (a) 3-Methyl-3-pentanol   (b) Fluorine (F₂)

(c) Ethyl radical, $CH_3\overset{\bullet}{C}H_2$ (d) Cl_2

B-1. (e)	**B-2.** (c)	**B-3.** (b)	**B-4.** (c)
B-5. (e)	**B-6.** (c)	**B-7.** (d)	**B-8.** (a)
B-9. (c)	**B-10.** (d)	**B-11.** (c)	**B-12.** (e)
B-13. (a)	**B-14.** (c)	**B-15.** (c)	**B-16.** (c)

CHAPTER 5

A-1.   (*a*)   2,4,4-Trimethyl-2-pentene or 2,4,4-trimethylpent-2-ene

       (*b*)   (*E*)-3,5-Dimethyl-4-octene or (*E*)-3,5-dimethyloct-4-ene

       (*c*)   (*E*)-2,7-Dibromo-3-(2-methylpropyl)-2-heptene or (*E*)-2,7-dibromo-3-(2-methylpropyl)hept-2-ene

       (*d*)   5-Methyl-4-hexen-3-ol or 5-methylhex-4-en-3-ol

A-2.   (*a*)

2,3-Dimethyl-2-pentene

       (*b*)

5-Chloro-2-methyl-1-hexene

       (*c*)

1,6-Dimethylcyclohexene

       (*d*)

4-Methyl-4-penten-2-ol

A-3.   (*a*)

        1   2   3   4   5   6

       (*b*) Isomer 5   (*c*) Isomers 1 and 4   (*d*) Isomers 2 and 3

A-4.   Two sp^2 C atoms; four sp^3 C atoms; three sp^2–sp^3 σ bonds

A-5.   (*a*)   (*b*)   (*c*)

A-6.

    (*Z*)-3-Methyl-3-hexene   (*E*)-3-Methyl-3-hexene

A-7.   (*a*) $H_2C{=}CCH_2CH_2CH_3$ + $(CH_3)_2C{=}CHCH_2CH_3$ (major)

                 |
               CH_3

       (*b*)   +   (major)

       (*c*) $(CH_3)_2CHCHCH(CH_3)_2$ (*d*) $H_2C{=}CC(CH_3)_3$

                       (X = Cl, Br, I)

A-8. $CH_3CH_2CCH_2Br$ with CH_3 substituents

A-9. Step 1: Protonation

Step 2: Dissociation

Step 3: Deprotonation

A-10.

(major)

A-11.

A-12. Cis isomer:

Trans isomer:

The trans isomer will react faster because its most stable conformation (with the isopropyl group equatorial) has an axial Cl able to undergo E2 elimination.

A-13. Rearrangement (hydride migration) occurs to form a more stable carbocation.

A-14.

3-Ethyl-4,4-dimethyl-2-pentene

A   B

B-1. (*c*)   **B-2.** (*c*)   **B-3.** (*d*)   **B-4.** (*c*)
B-5. (*a*)   **B-6.** (*b*)   **B-7.** (*a*)   **B-8.** (*a*)
B-9. (*a*)   **B-10.** (*d*)   **B-11.** (*b*)   **B-12.** (*c*)
B-13. (*a*)   **B-14.** (*c*)   **B-15.** (*a*)

CHAPTER 6

A-1. Five;

3,4-Dimethyl-
1-pentene

2,3-Dimethyl-
2-pentene

2,3-Dimethyl-
1-pentene

(*E*)-3,4-Dimethyl-
2-pentene

(*Z*)-3,4-Dimethyl-
2-pentene

A-2. (*a*) (CH₃)₂CCH₂CH₃

(*c*)

(*b*) HBr, peroxides

(*d*)

A-3. (*a*)

$\xrightarrow[\text{heat}]{\text{H}_2\text{SO}_4 \text{ (conc)}}$

$\xrightarrow[\text{2. H}_2\text{O}_2, \text{ HO}^-]{\text{1. B}_2\text{H}_6}$

(*b*)

$$CH_3CH_2\underset{\overset{|}{Cl}}{C}HCH(CH_3)_2 \xrightarrow[\text{CH}_3\text{OH}]{\text{NaOCH}_3} CH_3CH_2CH{=}C(CH_3)_2 \xrightarrow{\overset{\overset{O}{\|}}{CH_3COOH}} CH_3CH_2\overset{\overset{O}{\diagdown\diagup}}{CH{-}C}(CH_3)_2$$

(*c*) $(CH_3)_3C\underset{\overset{|}{Br}}{C}HCH_3 \xrightarrow[\text{CH}_3\text{OH}]{\text{CH}_3\text{O}^-\text{Na}^+} (CH_3)_3CCH{=}CH_2 \xrightarrow[\text{Peroxides}]{\text{HBr}} (CH_3)_3CCH_2CH_2Br$

A-4. **Initiation:** $ROOR \xrightarrow[\text{or heat}]{\text{light}} 2RO\cdot$

$RO\cdot + HBr \longrightarrow ROH + Br\cdot$

Propagation: $Br\cdot + CH_3CH_2CH{=}CH_2 \longrightarrow CH_3CH_2\overset{\cdot}{C}HCH_2Br$

$CH_3CH_2\overset{\cdot}{C}HCH_2Br + HBr \longrightarrow CH_3CH_2CH_2CH_2Br + Br\cdot$

A-5.

(*E*)-2-Butene

A-6.

A-7. **Step 1:** Protonation to form a carbocation

Step 2: Nucleophilic addition of chloride ion

A-8. $H_2C{=}\underset{\overset{|}{CH_3}}{C}CH_2CH_3$ or $(CH_3)_2C{=}CHCH_3 \xrightarrow{\text{HCl}} CH_3\underset{\overset{\underset{|}{CH_3}}{}}{\overset{\overset{|}{Cl}}{C}}CH_2CH_3$

2-Methyl-1-butene　　　　　2-Methyl-2-butene　　　　　2-Chloro-2-methylbutane

A-9.

A-10.

A-11. (CH₃)₂CHC=CH₂ →(HBr) (CH₃)₂CHCCH₃ →(NaOCH₂CH₃ / CH₃CH₂OH) (CH₃)₂C=C(CH₃)₂

(CH₃)₂CHC=CH₂

$$(CH_3)_2CHC{=}CH_2 \xrightarrow{HBr} (CH_3)_2CH\overset{CH_3}{\underset{Br}{C}}CH_3 \xrightarrow[CH_3CH_2OH]{NaOCH_2CH_3} (CH_3)_2C{=}C(CH_3)_2$$

A   B   C

C $\xrightarrow[2.\ H_2O,\ Zn]{1.\ O_3}$ (CH₃)₂C=O   (2 moles)

B-1. (c)   **B-2.** (a)   **B-3.** (c)   **B-4.** (d)
B-5. (d)   **B-6.** (e)   **B-7.** (b)   **B-8.** (b)
B-9. (b)   **B-10.** (b)   **B-11.** (a)   **B-12.** (e)
B-13. (e)

CHAPTER 7

A-1. (a) 1 and 2, both achiral; identical
  (b) 3 and 4, both chiral; enantiomers
  (c) 5 chiral, 6 achiral (meso); diastereomers
  (d) 7 and 8, both chiral; diastereomers
  (e) 9 and 10, both chiral; diastereomers

A-2. 3: (R)-2-Chlorobutane   4: (S)-2-Chlorobutane

5:           6:

7: (2S,3R)-2,3-Dibromopentane   8: (2R,3R)-2,3-Dibromopentane
9: (2E,5R)-5-Chloro-2-hexene   10: (2Z,5S)-5-Chloro-2-hexene

A-3. (a) Three; meso form is possible.   (c) Four; no meso form possible.
  (b) Eight; no meso form possible.

A-4. (a)

(b)

(c)

A-5. Chiral stereoisomers:

and

(2S,3S)-2,3-
Dichlorobutane

(2R,3R)-2,3-
Dichlorobutane

Meso stereoisomer (achiral);
plane of symmetry indicated
with dashed line

meso-2,3-Dichlorobutane

A-6. (a) $[\alpha] = -31.2°$ (b) 30% S

A-7. (a)

(b)

Meso form
(only stereoisomer)

(c)

A-8. (a) (2S,3S)-1,3-Dibromo-2-chlorobutane

(b) (R)-1-Ethylcyclohex-2-enol

A-9. Two: (2R,3S)-2-bromo-3-chlorobutane and (2S,3S)-2-bromo-3-chlorobutane; they are diastereomers.

A-10.

Racemic mixture

B-1. (*c*) **B-2.** (*c*) **B-3.** (*b*) **B-4.** (*d*)

B-5. (*b*) **B-6.** (*c*) **B-7.** (*d*) **B-8.** (*d*)

B-9. (*b*) **B-10.** (*c*) **B-11.** (*d*) **B-12.** (*d*)

B-13. (*e*) **B-14.** (*b*)

CHAPTER 8

A-1. (*a*) $CH_3CH_2CH_2CH_2OCH_2CH_3$

(*e*)

(*b*)

(X = OTs, Br, I)

(*f*)

(*c*)

(*g*)

(*d*)

A-2. $(CH_3)_2CHO^- Na^+ + CH_3CH_2CH_2Br$

A-3. (*a*)

(*b*)

A-4. **Step 1:** Ionization to form a secondary carbocation

Step 2: Rearrangement by methyl migration to form a more stable tertiary carbocation

$$\underset{\underset{CH_3}{|}}{\overset{\overset{CH_3}{|}}{CH_3\overset{+}{C}-CHCH_3}} \longrightarrow \underset{\underset{CH_3}{|}}{\overset{\overset{CH_3}{|}}{CH_3\overset{+}{C}-CHCH_3}}$$

Step 3: Capture of the carbocation by water, followed by deprotonation

$$\underset{\underset{CH_3}{|}}{\overset{\overset{CH_3}{|}}{CH_3\overset{+}{C}-CHCH_3}} \xrightarrow{\text{H}_2\text{O}} \underset{\underset{CH_3}{|}}{\overset{\overset{\overset{+}{H_2O}\ \ CH_3}{|\ \ \ \ |}}{CH_3C-CHCH_3}} \xrightarrow{-\text{H}^+} \underset{}{\overset{\overset{OH}{|}}{(CH_3)_2C-CH(CH_3)_2}}$$

A-5. (a) $(CH_3)_3CBr \xrightarrow{CH_3OH} (CH_3)_3COCH_3$
S_N1, unimolecular substitution; rate $= k[(CH_3)_3CBr]$

(b)   ⬡—Cl $\xrightarrow{NaN_3}$ ⬡—N$_3$

S_N2, bimolecular substitution; rate $= k[C_6H_{11}Cl][NaN_3]$

A-6. (a) Sodium iodide is soluble in acetone, whereas the byproduct of the reaction, sodium bromide, is not. According to Le Châtelier's principle, the reaction will shift in the direction that will replace the component removed from solution, in this case toward product.

(b) Protic solvents such as water form hydrogen bonds to anionic nucleophiles, thus stabilizing them and decreasing their nucleophilic strength. Aprotic solvents such as DMSO do not solvate anions very strongly, leaving them more able to express their nucleophilic character.

A-7.

OH	O⁻ Na⁺		
⬡	⬡	CH$_3$CH$_2$OH	CH$_3$CH$_2$Br
A	B	C	D

A-8.

A-9. Dissociation to give a secondary carbocation

$$CH_3CH_2CH_2CHCH(CH_3)_2 \longrightarrow CH_3CH_2CH_2\overset{+}{C}HCH(CH_3)_2$$
$$\overset{|}{Br}$$

Rearrangement by hydride migration to give a tertiary carbocation

$$CH_3CH_2CH_2\overset{+}{C}HC(CH_3)_2 \longrightarrow CH_3CH_2CH_2CH_2\overset{+}{C}(CH_3)_2$$
$$\overset{|}{H}$$

Capture of the carbocation by water to give product

$$CH_3CH_2CH_2CH_2\overset{+}{C}(CH_3)_2 \ + \ :\overset{..}{O}H_2 \ \xrightarrow{(-H^+)} \ \overset{OH}{\underset{|}{CH_3CH_2CH_2CH_2C(CH_3)_2}}$$

B-1. *(b)*	**B-2.** *(c)*	**B-3.** *(d)*	**B-4.** *(c)*
B-5. *(d)*	**B-6.** *(a)*	**B-7.** *(c)*	**B-8.** *(d)*
B-9. *(c)*	**B-10.** *(a)*	**B-11.** *(a)*	**B-12.** *(c)*
B-13. *(c)*	**B-14.** *(c)*		

CHAPTER 9

A-1. *(a)* 4,5-Dimethyl-2-hexyne

(b) 4-Ethyl-3-propyl-1-heptyne

(c) 6,6-Dimethylcyclodecyne

A-2. *(a)* $\overset{Cl}{\underset{|}{CH_3CH_2CH_2C}}=CH_2$

(b) $\overset{Cl}{\underset{|}{CH_3CH_2CH_2\underset{|}{C}CH_3}}$
$\overset{}{\underset{Cl}{}}$

(c) $H_2O, H_2SO_4, HgSO_4$

(d) $\underset{H}{\overset{H_3C}{}}C=C\underset{H}{\overset{CH_3}{}}$

(e) $(CH_3)_2CHC\equiv CH$

(f) Na, NH$_3(\ell)$

(g) $\underset{Cl}{\overset{H_3C}{}}C=C\underset{CH_2CH_3}{\overset{Cl}{}}$

(h) $\overset{CH_3}{\underset{|}{CH_3CH_2CH_2CHCO_2H}} \ + \ CH_3CH_2CO_2H$

A-3. Reaction (2) is effective; the desired product is formed by an S$_N$2 reaction.

$$\overset{CH_3}{\underset{|}{CH_3CH_2CH_2CHC}}\equiv CNa \ + \ CH_3I \ \longrightarrow \ \overset{CH_3}{\underset{|}{CH_3CH_2CHC}}\equiv CCH_3 \ + \ NaI$$

Reaction (1) is not effective, owing to E2 elimination from the secondary bromide.

$$\overset{Br}{\underset{|}{CH_3CH_2CHCH_3}} \ + \ CH_3C\equiv CNa \ \longrightarrow \ CH_3CH=CHCH_3 \ + \ CH_3C\equiv CH \ + \ NaBr$$

A-4. (*a*) $CH_3CH_2Br \xrightarrow[\text{DMSO}]{\text{KOC(CH}_3)_3} H_2C=CH_2 \xrightarrow{Br_2} BrCH_2CH_2Br$

$BrCH_2CH_2Br \xrightarrow[\text{2. H}_2\text{O}]{\text{1. NaNH}_2, \text{NH}_3} HC\equiv CH \xrightarrow[\text{2. CH}_3\text{CH}_2\text{Br}]{\text{1. NaNH}_2} HC\equiv CCH_2CH_3$

(*b*) $HC\equiv CCH_2CH_3 \xrightarrow{\text{NaNH}_2} NaC\equiv CCH_2CH_3 \xrightarrow{\text{CH}_3\text{CH}_2\text{Br}} CH_3CH_2C\equiv CCH_2CH_3$

(*c*) $H_2C=CHCH_2CH_3 \xrightarrow{Cl_2} \overset{\overset{\displaystyle Cl}{|}}{ClCH_2CHCH_2CH_3} \xrightarrow[\text{2. H}_2\text{O}]{\text{1. 3NaNH}_2, \text{NH}_3} HC\equiv CCH_2CH_3$

$HC\equiv CCH_2CH_3 \xrightarrow{\text{As in part (}b\text{)}} CH_3CH_2C\equiv CCH_2CH_3$

(*d*) $HC\equiv CH \xrightarrow{\text{NaNH}_2} HC\equiv CNa \xrightarrow{(CH_3)_2CHCH_2Br} HC\equiv CCH_2CH(CH_3)_2$

$HC\equiv CHCH_2CH(CH_3)_2 \xrightarrow[\text{HgSO}_4]{\text{H}_2\text{O, H}_2\text{SO}_4} \overset{\overset{\displaystyle O}{||}}{CH_3CCH_2CH(CH_3)_2}$

A-5.

(*E*)-2-Heptene

A-6. $(CH_3)_3CC\equiv CH$ $(CH_3)_3CC\equiv C\!:^{-} Na^{+}$

      A   B

$CH_3CH_2CH_2OH$ $CH_3CH_2CH_2Br$

      C   D

A-7. **E:** $HC\equiv CCH_2CH_3$ **F:** $CH_3CH_2C\equiv CCH_2CH_3$

A-8.

B-1. (*a*)	**B-2.** (*c*)	**B-3.** (*a*)	**B-4.** (*d*)
B-5. (*b*)	**B-6.** (*b*)	**B-7.** (*e*)	**B-8.** (*c*)
B-9. (*b*)	**B-10.** (*a*)	**B-11.** (*d*)	**B-12.** (*b*)

CHAPTER 10

A-1. $H_2C=CHCH_2CH=CH_2$ $H_2C=CHCH=CHCH_3$

                                             (Conjugated)

$\overset{\displaystyle H_2C=CHC=CH_2}{\underset{\displaystyle \underset{|}{CH_3}}{}}$ $\left.\begin{array}{l} H_2C=C=CHCH_2CH_3 \\ H_2C=C=C(CH_3)_2 \\ CH_3CH=C=CHCH_3 \end{array}\right\}$ Allenes

(Conjugated)

A-2.

$H_2C=CH$ $C=C$ CH_3 / H / H

(3Z)-1,3-Pentadiene

$H_2C=CH$ $C=C$ H / H / CH_3

(3E)-1,3-Pentadiene

$H_2C=CHC=CH_2$
CH_3

2-Methyl-1,3-butadiene

A-3.

A-4. (*a*)

$$CH_3CH=CHCHCHCH_3 \quad + \quad CH_3CHCH=CHCHCH_3$$

Br / Br (1,2-Addition)

Br / Br (1,4-Addition)

(*b*) $H_2C=CHCHCH_3$ (Cl) $+$ $ClCH_2CH=CHCH_3$

(*c*)

(*d*) (NBS), heat

(*e*)

A-5. (cannot adopt the required *s*-cis conformation)

A-6.

A-7. A: B:

A-8.

B-1. (*b*) **B-2.** (*c*) **B-3.** (*a*) **B-4.** (*c*)

B-5. (*a*) **B-6.** (*d*) **B-7.** (*a*) **B-8.** (*a*)

B-9. (*a*) **B-10.** (*d*)

CHAPTER 11

A-1. (*a*) *m*-Bromotoluene (*c*) *o*-Chloroacetophenone

  (*b*) 2-Chloro-3-phenylbutane (*d*) 2,4-Dinitrophenol

A-2. (*a*) (*b*) (*c*) (*d*)

A-3. (*a*) (*b*)

(10 π electrons) (14 π electrons)

A-4. (*a*) Eight π electrons. No, the substance is not aromatic.

  (*b*) 6 π electrons. Yes, it is aromatic.

  (*c*) 14 π electrons. Yes, it is aromatic.

A-5.

A-6. (*a*)

  (*d*) $Na_2Cr_2O_7$, H_2SO_4, H_2O, heat

  (*b*) $C_6H_5CH_2X$ (X = Cl, Br, I, OTs)

  (*e*)

  (*c*)

  (*f*) $C_6H_5\overset{OH}{\underset{Br}{CHCHCH_3}}$

A-7. (I) $C_6H_5CH{=}CHCH_3$ $\xrightarrow{\text{HBr}}$ $C_6H_5\overset{\overset{\displaystyle Br}{|}}{C}HCH_2CH_3$

(II) $C_6H_5CH_2CH_2CH_3$ $\xrightarrow[\substack{\text{light}\\ \text{(or NBS, heat)}}]{\text{Br}_2}$ $C_6H_5\overset{\overset{\displaystyle Br}{|}}{C}HCH_2CH_3$

A-8.

A-9.

A-10. $(CH_3)_3C{-}\langle\text{benzene ring}\rangle{-}CH_2CH_3$

B-1. (*c*)	**B-2.** (*c*)	**B-3.** (*a*)	**B-4.** (*b*)
B-5. (*a*)	**B-6.** (*d*)	**B-7.** (*b*)	**B-8.** (*d*)
B-9. (*b*)	**B-10.** (*d*)	**B-11.** (*a*)	**B-12.** (*b*)
B-13. (*c*)	**B-14.** (*d*)	**B-15.** (*c*)	

CHAPTER 12

A-1.

A-2. (*a*) Slower

(*c*) Slower

(*b*) Faster

A-3. (*a*) NO_2^+ (*b*) $Br{-}\overset{+}{B}r{-}\overset{-}{F}eBr_3$ (*c*) SO_3

A-4. (*a*)

(*b*)

(*c*) $C_6H_5\overset{O}{\overset{\|}{C}}Cl$, $AlCl_3$ or $C_6H_5\overset{O}{\overset{\|}{C}}O\overset{O}{\overset{\|}{C}}C_6H_5$, $AlCl_3$

(*d*)

(*e*)

A-5. (*a*)

(*b*)

(*c*)

(*d*) CH_3CH_2—

A-6. (*a*)

(+ ortho isomer)

(*b*)

(c) [benzaldehyde] —N₂H₄, KOH, heat / or Zn(Hg), HCl→ [toluene] —(CH₃C)₂O / AlCl₃→ [4-methylacetophenone]

(d) [benzene] —H_2SO_4, heat→ [benzenesulfonic acid, SO_3H] —Br_2 / $FeBr_3$→ [3-bromobenzenesulfonic acid]

(e) [benzene] —$(CH_3)_2CHCl$ / $AlCl_3$→ $(CH_3)_2CH$—[phenyl] —HNO_3 / H_2SO_4→ $(CH_3)_2CH$—[phenyl]—NO_2

A-7. (a) [toluene, CH_3] —$Na_2Cr_2O_7$ / H_2O, H_2SO_4, heat→ [benzoic acid, CO_2H] —HNO_3 (excess) / H_2SO_4→ [3,5-dinitrobenzoic acid, CO_2H, O_2N, NO_2]

(b) $(CH_3)_2CH$—[phenyl]—NO_2 —Br_2 / light→ $(CH_3)_2C$(Br)—[phenyl]—NO_2 —$NaOCH_2CH_3$→ $H_2C=C(CH_3)$—[phenyl]—NO_2

[Prepared from benzene as in problem A-6(e)]

(c) [toluene, CH_3] —$(CH_3)_2CHCCl$, O / $AlCl_3$→ [4-isobutyryltoluene, CH_3, $CCH(CH_3)_2$, O] —Cl_2 / $FeCl_3$→ [CH_3, Cl, $CCH(CH_3)_2$, O] —Zn(Hg), HCl→ [CH_3, Cl, $CH_2CH(CH_3)_2$]

A-8. CH_3O—[phenyl] —HNO_3 / H_2SO_4→ CH_3O—[phenyl]—NO_2 (+ ortho) —Br_2 / $FeBr_3$→ CH_3O—[phenyl](Br)—NO_2

B-1. (c)	**B-2.** (b)	**B-3.** (c)	**B-4.** (b)
B-5. (a)	**B-6.** (b)	**B-7.** (c)	**B-8.** (b)
B-9. (c)	**B-10.** (a)	**B-11.** (e)	**B-12.** (c)
B-13. (c)	**B-14.** (c)	**B-15.** (c)	

CHAPTER 13

A-1. 1: 6.10 ppm   3: 200 MHz
2: 1305 Hz   4: 0.00 ppm

A-2. (*a*) Two signals BrCH$_2$CH$_2$CH$_2$Br
  a b a

  a: triplet b: pentet

(*b*) Two signals   CH$_3$CH$_2$CCH$_2$CH$_3$ (with Cl above and Cl below the central C)
  a b b a

  a: triplet b: quartet

(*c*) Three signals, all singlets

A-3. **A:** CH$_3$COC(CH$_3$)$_3$ (ketone C=O)   **B:** CH$_3$OCC(CH$_3$)$_3$ (ester C=O)

A-4. (*a*) C$_6$H$_5$—CH$_2$CCH$_2$CH$_3$ (with C=O)

(*c*) HC(COCH$_2$CH$_3$)$_3$ (with C=O)

(*b*) (CH$_3$)$_2$C—C(CH$_3$)$_2$ (with HO and OH)

(*d*) (CH$_3$)$_2$C—C≡N (with OH)

A-5. Seven signals:

g—C$_6$H$_4$—CCH$_2$CH$_3$ (with C=O; positions d, c b a, f, e)

a: δ 10–30
b: δ 20–40
c: δ 190–220
d–g: δ 110–175

A-6. Pentane: three signals; 2-methylbutane: four signals; 2,2-dimethylpropane: two signals

A-7. 2,3-Dimethylbutane: (CH$_3$)$_2$CHCH(CH$_3$)$_2$

B-1. (*d*)	**B-2.** (*a*)	**B-3.** (*b*)	**B-4.** (*b*)
B-5. (*b*)	**B-6.** (*a*)	**B-7.** (*b*)	**B-8.** (*a*)
B-9. (*c*)	**B-10.** (*a*)	**B-11.** (*c*)	**B-12.** (*c*)
B-13. (*a*)	**B-14.** (*a*)	**B-15.** (*d*)	

CHAPTER 14

A-1. (*a*)

cyclohexyl–X + 2Li ⟶ cyclohexyl–Li + LiX

(X = Cl, Br, I)

(*b*) (CH$_3$)$_3$CBr + Mg ⟶ (CH$_3$)$_3$CMgBr

(*c*) 2C$_6$H$_5$CH$_2$Li + CuX ⟶ (C$_6$H$_5$CH$_2$)$_2$CuLi + LiX

(X = Cl, Br, I)

A-2. (*a*) $(C_6H_5)_2\overset{\underset{\displaystyle |}{OH}}{C}CH_3$ + CH_3CH_2OH

(*d*)

(*b*) $(CH_3)_2CHCH_2D$

(*e*) $H_3C\cdots$ $\cdots H$ (cyclopropane with Br, Br top and H, CH₃ bottom)

(*c*) $H_3C\!\!-\!\!\bigcirc\!\!-\!\!CH_2OH$

A-3. (*a*) $C_6H_5\overset{\underset{\displaystyle }{O}}{\overset{||}{C}}H$ + $(CH_3)_3CMgX$ and C_6H_5MgX + $(CH_3)_3C\overset{\underset{\displaystyle }{O}}{\overset{||}{C}}H$

(X = Cl, Br, I)   (X = Cl, Br, I)

(*b*) $CH_3CH_2CH_2\overset{\underset{\displaystyle }{O}}{\overset{||}{C}}H$ + $CH_3CH_2CH_2CH_2MgX$ and $CH_3CH_2CH_2CH_2\overset{\underset{\displaystyle }{O}}{\overset{||}{C}}H$ + $CH_3CH_2CH_2MgX$

(X = Cl, Br, I)   (X = Cl, Br, I)

A-4. (*a*) $(CH_3CH_2CH_2)_2CuLi$ (*c*) CH_2I_2, Zn(Cu)

(*b*) $(CH_3)_2CHMgX$
(X = Cl, Br, I)

A-5. Solvents A, B, and E are suitable; they are all ethers. Solvents C and F have acidic hydrogens and will react with a Grignard reagent. Solvent D is an ester which will react with a Grignard reagent.

A-6. $CH_3(CH_2)_3OH$ $\xrightarrow[\text{or HBr}]{\text{PBr}_3}$ $CH_3(CH_2)_3Br$ $\xrightarrow{\text{2Li}}$ $CH_3(CH_2)_3Li$ + LiBr

$2CH_3(CH_2)_3Li$ + CuBr $\longrightarrow$ $(C_4H_9)_2CuLi$ $\xrightarrow{CH_3(CH_2)_3Br}$ $CH_3(CH_2)_6CH_3$

A-7. (I) $(CH_3)_2CH\overset{\underset{\displaystyle }{O}}{\overset{||}{C}}CH_3$ + CH_3MgBr

(II) $CH_3\overset{\underset{\displaystyle }{O}}{\overset{||}{C}}CH_3$ + $(CH_3)_2CHMgBr$ $\left.\right\}$ $\xrightarrow[\text{2. H}_3O^+]{\text{1. diethyl ether}}$ $(CH_3)_2CH\overset{\underset{\displaystyle |}{OH}}{C}(CH_3)_2$

(III) $(CH_3)_2CHCO_2CH_3$ + $2CH_3MgBr$

A-8. $C_6H_5CH_2CH_3$ $\xrightarrow[\text{peroxides, heat}]{\text{NBS}}$ $C_6H_5\overset{\underset{\displaystyle }{Br}}{\overset{|}{C}}HCH_3$ $\xrightarrow[\text{diethyl ether}]{\text{Mg}}$ $C_6H_5\overset{\underset{\displaystyle }{CH_3}}{\overset{|}{C}}HMgBr$

$C_6H_5\overset{\underset{\displaystyle }{CH_3}}{\overset{|}{C}}HMgBr$ $\xrightarrow[\text{2. H}_3O^+]{\text{1. }CH_3\overset{O}{\overset{||}{C}}H}$ $C_6H_5\overset{\underset{\displaystyle |}{CH_3}}{\underset{\underset{\displaystyle OH}{|}}{C}}HCHCH_3$

A-9. (*a*)

OH
|
C₆H₅—C—CH₂CH₃
|
CH₃

(The structure shows a phenyl group attached to a carbon bearing OH, CH₂CH₃, and CH₃)

(*c*)

Cyclopentane ring with OH and C≡CCH₃ substituents

(*b*)

Structure: (CH₃CH₂)CH(CH₃)—CH(OH)—CH₃ drawn as a branched chain with OH

B-1. (*c*)   **B-2.** (*a*)   **B-3.** (*d*)   **B-4.** (*a*)

B-5. (*e*)   **B-6.** (*c*)   **B-7.** (*b*)   **B-8.** (*a*)

B-9. (*e*)   **B-10.** (*b*)   **B-11.** (*b*)   **B-12.** (*b*)

CHAPTER 15

A-1. (*a*)

Cyclohexanone (six-membered ring with C=O)

(*d*) OsO_4, $(CH_3)_3COOH$, $(CH_3)_3COH$, HO^-

(*e*) NaSH

(*b*) $C_6H_5\overset{O}{\overset{\|}{C}}OCH_2CH_3$

(*c*) 1. B_2H_6;   2. H_2O_2, HO^-

A-2. (*a*) $C_6H_5CH_2\overset{O}{\overset{\|}{C}}H$

(*c*) $(C_6H_5CH_2CH_2)_2O$

(*d*) $K_2Cr_2O_7$, H^+, H_2O, heat

(*b*) $CH_3\overset{O}{\overset{\|}{C}}Cl$, pyridine; or

$(CH_3\overset{O}{\overset{\|}{C}})_2O$, pyridine; or CH_3CO_2H, H^+

A-3. (*a*) $(CH_3)_2CHO^- Na^+$

(*e*) $(CH_3)_2CHOSO_2$—⟨benzene ring⟩—CH_3

(*b*) $(CH_3)_2C{=}O$

(*f*) CH_3CH_2—⟨benzene ring⟩—$\overset{O}{\overset{\|}{C}}OCH(CH_3)_2$

(*c*) $(CH_3)_2C{=}O$

(*g*) $CH_3\overset{O}{\overset{\|}{C}}OCH(CH_3)_2$

(*d*) $CH_3\overset{O}{\overset{\|}{C}}OCH(CH_3)_2$

A-4. (I) $(CH_3)_2CHBr$ + Mg $\longrightarrow$ $(CH_3)_2CHMgBr$ $\xrightarrow[\text{2. H}_3\text{O}^+]{\text{1. } H_2C\overset{O}{-}CH_2}$ $(CH_3)_2CHCH_2CH_2OH$

(II) $(CH_3)_2CHCH_2Br$ + Mg $\longrightarrow$ $(CH_3)_2CHCH_2MgBr$

$(CH_3)_2CHCH_2MgBr$ $\xrightarrow[\text{2. H}_3\text{O}^+]{\text{1. } H_2C=O}$ $(CH_3)_2CHCH_2CH_2OH$

A-5. (a)

(b)

(c)

A-6. (a) PCC or PDC in CH_2Cl_2

(b) $Na_2Cr_2O_7$, H^+, H_2O, heat

(c) 1. $LiAlH_4$; 2. H_2O

(d) OsO_4, $(CH_3)_3COOH$, $(CH_3)_3COH$, HO^-

A-7.

A   B   C

A-8. (a) $(CH_3)_2C=CHCH_3$ $\xrightarrow[\text{2. H}_2\text{O}_2, \text{HO}^-]{\text{1. B}_2\text{H}_6}$ $(CH_3)_2CH\overset{OH}{\underset{}{C}}HCH_3$ $\xrightarrow[\text{CH}_2\text{Cl}_2]{\text{PDC}}$ $(CH_3)_2CH\overset{O}{\underset{}{C}}CH_3$

(b)

(c)

$C_6H_5CH_3$ $\xrightarrow[\text{peroxides, heat}]{\text{NBS}}$ $C_6H_5CH_2Br$ $\xrightarrow{\text{Mg}}$ $C_6H_5CH_2MgBr$ $\xrightarrow[\text{2. H}_3\text{O}^+]{\text{1. } H_2C\overset{O}{-}CH_2}$ $C_6H_5CH_2CH_2CH_2OH$

$\downarrow$ $\begin{matrix} K_2Cr_2O_7 \\ H^+, H_2O \\ heat \end{matrix}$

$C_6H_5CH_2CH_2CO_2CH_2CH_3$ $\xleftarrow[\text{H}^+]{\text{CH}_3\text{CH}_2\text{OH}}$ $C_6H_5CH_2CH_2CO_2H$

B-1. (e)   **B-2.** (d)   **B-3.** (c)   **B-4.** (c)

B-5. (b)   **B-6.** (b)   **B-7.** (a)   **B-8.** (a)

B-9. (d)   **B-10.** (a)   **B-11.** (b)   **B-12.** (d)

B-13. (e)   **B-14.** (a)   **B-15.** (c)

CHAPTER 16

A-1. $CH_3OCH_2CH_2CH_3$ $CH_3OCH(CH_3)_2$ $CH_3CH_2OCH_2CH_3$

        Methyl propyl ether   Isopropyl methyl ether   Diethyl ether

A-2. (a)

(b)

(c) $C_6H_5CHCH_2OH$ (with I substituent)

(d)

(e) $C_6H_5SCH_2CH_3$

(f) $C_6H_5\overset{O^-}{\underset{+}{S}}CH_2CH_3$

A-3. (a)

A-4. $CH_3CH_2OH \xrightarrow[\text{heat}]{H_2SO_4} H_2C{=}CH_2 \xrightarrow{CH_3COOH} H_2C\text{-}CH_2$ (epoxide)

$CH_3CH_2OH \xrightarrow{Na} CH_3CH_2O^- Na^+ \xrightarrow[CH_3CH_2OH]{H_2C-CH_2 \text{ (epoxide)}} CH_3CH_2OCH_2CH_2OH$

A-5. (a)

(b)

A-6. A: B:

A-7.

A-8.

B-1. (*a*)   **B-2.** (*a*)   **B-3.** (*c*)   **B-4.** (*d*)

B-5. (*d*)   **B-6.** (*e*)   **B-7.** (*a*)   **B-8.** (*c*)

B-9. (*d*)   **B-10.** (*d*)

CHAPTER 17

A-1. (*a*) 3,4-Dimethylhexanal

(*b*) 2,2,5-Trimethylhexan-3-one

(*c*) *trans*-4-Bromo-2-methylcyclohexanone

(*d*) 4-Methyl-3-penten-2-one

A-2.

(*a*)   (*b*)   (*c*)

A-3. (*a*)

(*b*) NH$_2$OH

(*c*) (CH$_3$)$_2$CHCH + HOCH$_2$CH$_2$CH$_2$OH

(*d*) (C$_6$H$_5$)$_3$P$-$CHCH$_2$CH$_3$

(*e*)

(*f*) CH$_3$CH$_2$CH$_2$CH(OCH$_2$CH$_3$)$_2$

(*g*) C$_6$H$_5$CCH$_2$CH$_3$ + (CH$_3$)$_2$NH

(*h*) CH$_3$CH$_2$CH$_2$COH

A-4.　(*a*)　$(C_6H_5)_3\overset{+}{P}-CH_2CH(CH_3)_2$　Br^-　　$(C_6H_5)_3\overset{+}{P}-\overset{\bullet\bullet}{\overset{-}{C}}HCH(CH_3)_2$

　　　　　　　　　　　　A　　　　　　　　　　　　　　　　　B

　　　　　$C_6H_5CH{=}CHCH(CH_3)_2$

　　　　　　　　　　　C

　　(*b*)　$C_6H_5CH_2\overset{\overset{\displaystyle O}{||}}{C}CH_3$　　$C_6H_5CH_2O\overset{\overset{\displaystyle O}{||}}{C}CH_3$

　　　　　　　　D　　　　　　　　　E

A-5.　(*a*)　(1) CH_3MgI;　(2) H_3O^+;　(3) H_2SO_4, heat

　　(*b*)　$(C_6H_5)_3\overset{+}{P}-\overset{\bullet\bullet}{\overset{-}{C}}H_2$　[from $(C_6H_5)_3P$ + CH_3I $\longrightarrow$ $\xrightarrow{C_4H_9Li}$]

　　(*c*)　$HOCH_2CH_2OH$, H^+(cat), heat

　　(*d*)　$CH_3\overset{\overset{\displaystyle O}{||}}{C}OOH$

A-6.　(*a*)

　　(*b*)

　　　　　　　　　　$\equiv$　　$CH_3\overset{\overset{\displaystyle O}{||}}{C}CH_2CH_2CH_2\overset{\overset{\displaystyle OH}{|}}{\underset{\underset{\displaystyle CH_3}{|}}{C}}CH_2OH$

A-7.　(*a*)

CH_3CH_2I + $(C_6H_5)_3P$ $\longrightarrow$ $(C_6H_5)_3\overset{+}{P}-CH_2CH_3$ I^- $\xrightarrow{C_4H_9Li}$ $(C_6H_5)_3\overset{+}{P}-\overset{\bullet\bullet}{\overset{-}{C}}HCH_3$

$(CH_3)_2C{=}O$ + $(C_6H_5)_3\overset{+}{P}-\overset{\bullet\bullet}{\overset{-}{C}}HCH_3$ $\longrightarrow$ $(CH_3)_2C{=}CHCH_3$ $\xrightarrow{CH_3\overset{\overset{\displaystyle O}{||}}{C}OOH}$ $(CH_3)_2C\overset{\displaystyle O}{\overset{\diagdown\diagup}{-}}CHCH_3$

　　(*b*)

(*c*)

A-8.

A-9.

B-1. (*c*)	**B-2.** (*d*)	**B-3.** (*a*)	**B-4.** (*c*)
B-5. (*b*)	**B-6.** (*b*)	**B-7.** (*a*)	**B-8.** (*b*)
B-9. (*e*)	**B-10.** (*c*)	**B-11.** (*c*)	**B-12.** (*c*)
B-13. (*d*)	**B-14.** (*e*)	**B-15.** (*a*)	**B-16.** (*c*)

CHAPTER 18

A-1. (*a*)   (*c*)

(*b*)

A-2.

$$C_6H_5CH_2CH{=}C\overset{\overset{\displaystyle O}{\|}}{C}H \qquad (CH_3CH_2)_2CHCH_2\overset{\overset{\displaystyle O}{\|}}{C}CH_2CH_3$$
$$\underset{\displaystyle C_6H_5}{|}$$

    A   B

A-3.

$$CH_3CH_2\overset{\overset{\displaystyle OH}{|}}{C}H\overset{}{C}H\overset{\overset{\displaystyle O}{\|}}{C}H \qquad CH_3\overset{\overset{\displaystyle OH}{|}}{C}H\overset{}{C}HC(CH_3)_2$$
$$\underset{\displaystyle CH_3}{|} \qquad\qquad\quad \underset{\displaystyle CH_3}{|}\;\; \underset{\displaystyle HC{=}O}{}$$

$$CH_3CH_2\overset{\overset{\displaystyle OH}{|}}{C}HC(CH_3)_2 \qquad CH_3\overset{\overset{\displaystyle OH}{|}}{C}H\overset{}{C}H\overset{\overset{\displaystyle O}{\|}}{C}H$$
$$\underset{\displaystyle HC{=}O}{} \qquad\qquad\qquad \underset{\displaystyle CH_3}{|}\;\; \underset{\displaystyle CH_3}{|}$$

A-4.

$$CH_3CH_2OH \xrightarrow[\text{CH}_2\text{Cl}_2]{\text{PCC}} CH_3\overset{\overset{\displaystyle O}{\|}}{C}H$$

$$2CH_3\overset{\overset{\displaystyle O}{\|}}{C}H \xrightarrow[\text{H}_2\text{O}]{\text{NaOH}} CH_3\overset{\overset{\displaystyle OH}{|}}{C}HCH_2\overset{\overset{\displaystyle O}{\|}}{C}H \xrightarrow[\substack{\text{or 1. LiAlH}_4 \\ \text{2. H}_2\text{O}}]{\text{NaBH}_4,\ \text{CH}_3\text{OH}} CH_3\overset{\overset{\displaystyle OH}{|}}{C}HCH_2CH_2OH$$

A-5.

Note: the order of the two steps may be reversed.

A-6. (*a*) $CH_3CH_2CH_2\overset{}{C}H\overset{\overset{\displaystyle O}{\|}}{C}H$
$$\qquad\qquad\quad \underset{\displaystyle Br}{|}$$

(*b*) $CH_3CH_2CH_2CH_2\overset{\overset{\displaystyle OH}{|}}{C}H\overset{}{C}H\overset{\overset{\displaystyle O}{\|}}{C}H$
$$\qquad\qquad\qquad\qquad\quad \underset{\displaystyle CH_2CH_2CH_3}{|}$$

(*c*)

(*d*)

A-7.

$$CH_3\overset{}{C}H\overset{\overset{\displaystyle O}{\|}}{C}H + {:}\overset{..}{O}H \rightleftharpoons CH_3\overset{..}{C}H\overset{\overset{\displaystyle O}{\|}}{C}H \underset{H_2C=O}{\rightleftharpoons} CH_3\overset{}{C}H\overset{\overset{\displaystyle O}{\|}}{C}H \underset{HO}{\rightleftharpoons} CH_3\overset{}{C}H\overset{\overset{\displaystyle O}{\|}}{C}H$$
$$\underset{\displaystyle H}{|} \qquad\qquad\qquad\qquad\qquad\qquad\qquad\qquad\qquad \underset{\displaystyle H_2\overset{..}{C}{-}\overset{..}{\underset{..}{O}}{:}}{} \qquad\quad H \qquad \underset{\displaystyle CH_2OH}{}$$

A-8. ⌬—CHO + $CH_3CH_2CCH_2CH_3$ (with C=O)

B-1. (*a*)	**B-2.** (*c*)	**B-3.** (*b*)	**B-4.** (*d*)
B-5. (*a*)	**B-6.** (*c*)	**B-7.** (*c*)	**B-8.** (*e*)
B-9. (*c*)	**B-10.** (*b*)	**B-11.** (*a*)	**B-12.** (*a*)

CHAPTER 19

A-1. (*a*) 4-Methyl-5-phenylhexanoic acid

    (*b*) Cyclohexanecarboxylic acid

    (*c*) 3-Bromo-2-ethylbutanoic acid

A-2. 4-Phenylbutanoic acid is $C_6H_5CH_2CH_2CH_2CO_2H$.

$$C_6H_5CH_2CH_2CH(CO_2H)_2 \xrightarrow{\text{heat}}$$

$$C_6H_5CH_2CH_2CH_2Br \xrightarrow[\text{2. H}^+,\text{ H}_2\text{O, heat}]{\text{1. CN}^-}$$

$$C_6H_5CH_2CH_2CH_2Br \xrightarrow[\substack{\text{2. CO}_2 \\ \text{3. H}_3\text{O}^+}]{\text{1. Mg}}$$

A-3. $C_6H_5CH_2CO_2H + CH_3CH_2OH \xrightarrow{\text{H}^+\text{ (cat)}} C_6H_5CH_2COCH_2CH_3 + H_2O$

A-4. $(CH_3)_2CHCH_2\overset{\text{Br}}{\underset{|}{C}}HCO_2H$ $(CH_3)_2CHCH_2\overset{\text{I}}{\underset{|}{C}}HCO_2H$ $C_6H_5CC(CH_3)_2$ (with C=O and CO_2H)

             A   B   C

A-5. (*a*) $\xrightarrow{\text{Mg}}$ ⌬—MgBr

    (*b*) $\xrightarrow[\text{2. H}_2\text{O}]{\text{1. LiAlH}_4} CH_3CH_2CH_2CH_2OH$

    (*c*) $\xrightarrow{\text{Br}_2,\text{ P}} CH_3CH_2\overset{\text{Br O}}{\underset{}{C}}HCOH$

    (*d*) $CH_3CH_2\overset{\text{CN}}{\underset{\text{OH}}{C}}H \xrightarrow[\text{heat}]{\text{H}^+,\text{ H}_2\text{O}}$

A-6. $CH_3CH_2\overset{}{\underset{\text{Br}}{C}}HCO_2H$

A-7.
$$\underset{\underset{OCH_2CH_2CH_2CH_3}{|}}{\overset{\overset{OH}{|}}{H C O H}}$$

A-8.

$$CH_3\overset{O}{\overset{\|}{C}}OH \;\underset{H^+}{\rightleftharpoons}\; CH_3\overset{\overset{+}{O}H}{\overset{\|}{C}}OH + H\ddot{O}CH_3 \;\rightleftharpoons\; CH_3\underset{\underset{+}{H\overset{\cdot\cdot}{O}CH_3}}{\overset{\overset{OH}{|}}{C}}OH \;\underset{-H^+}{\rightleftharpoons}\; CH_3\underset{OCH_3}{\overset{\overset{OH}{|}}{C}}OH$$

$$CH_3\underset{OCH_3}{\overset{\overset{OH}{|}}{C}}OH \;\underset{H^+}{\rightleftharpoons}\; CH_3\underset{OCH_3}{\overset{\overset{\ddot{O}H}{|}}{C}}\overset{+}{O}H_2 \;\rightleftharpoons\; CH_3\overset{\overset{+}{O}H}{\overset{\|}{C}}OCH_3 + H_2O \;\underset{-H^+}{\rightleftharpoons}\; CH_3\overset{O}{\overset{\|}{C}}OCH_3$$

B-1. (b)	**B-2.** (a)	**B-3.** (c)	**B-4.** (d)
B-5. (c)	**B-6.** (d)	**B-7.** (c)	**B-8.** (d)
B-9. (e)	**B-10.** (c)	**B-11.** (e)	

CHAPTER 20

A-1. (a) Propyl butanoate　　　　　　　　　(c) 4-Methylpentanoyl chloride

　　　(b) N-Methylbenzamide

A-2. (a) $C_6H_5\overset{O}{\overset{\|}{C}}O\overset{O}{\overset{\|}{C}}C_6H_5$　　　　(b) $CH_3\overset{O}{\overset{\|}{C}}NHCH\underset{CH_3}{\overset{|}{}}CH_2CH_3$　　　(c) $C_6H_5\overset{O}{\overset{\|}{C}}OC_6H_5$

A-3. (a) $SOCl_2$　　　　　　　　　　(b) $C_6H_5\overset{O}{\overset{\|}{C}}OCH_3$

A-4. (a) CH_3CO_2H + —OH

　　　(b) $C_6H_5\overset{O}{\overset{\|}{C}}O$—

　　　(c)

(d) $CH_3CH_2\overset{O}{\overset{\|}{C}}N(CH_3)_2$ + CH_3CH_2OH

(e) H_3C——CO_2H + $CH_3\overset{+}{N}H_3\;HSO_4^-$

A-5.

A-6.

A-7. (*a*) $CH_3\underset{\overset{|}{OH}}{\overset{\overset{|}{OH}}{C}}OCH_3$ (*b*) $CH_3\underset{\overset{|}{NH_2}}{\overset{\overset{||}{O}\ \ \overset{|}{OH}}{C}}CH_3$

A-8.

A-9.

Toluene   Benzoic acid

Benzoic acid   Benzyl alcohol

Benzoic acid   Benzyl alcohol   Benzyl benzoate

A-10. The compound is 2-chloropropanamide.

$$CH_3CHC\overset{\displaystyle O}{\underset{\displaystyle Cl}{\diagup}}\diagdown NH_2$$

2-Chloropropanamide

The compound may be prepared from propanoic acid as shown.

$$CH_3CH_2CO_2H \xrightarrow[P]{Cl_2} CH_3\underset{Cl}{CH}CO_2H \xrightarrow{SOCl_2} CH_3\underset{Cl}{CH}C\overset{O}{\diagdown}Cl \xrightarrow{NH_3} CH_3\underset{Cl}{CH}\overset{O}{C}NH_2$$

| Propanoic acid | 2-Chloropropanoic acid | 2-Chloropropanoyl chloride | 2-Chloropropanamide |

B-1. (*a*) **B-2.** (*b*) **B-3.** (*b*) **B-4.** (*c*)

B-5. (*d*) **B-6.** (*c*) **B-7.** (*d*) **B-8.** (*d*)

B-9. (*b*) **B-10.** (*a*) **B-11.** (*d*) **B-12.** (*b*)

B-13. (*b*)

CHAPTER 21

A-1. (*a*) $CH_3CH_2CH_2\overset{O}{\overset{||}{C}}\underset{\overset{|}{CH_2CH_3}}{CH}\overset{O}{\overset{||}{C}}OCH_2CH_3$

(*d*) $(CH_3CH_2O_2C)_2CHCH_2CH_2\overset{O}{\overset{||}{C}}OCH_2CH_3$

(*b*) $C_6H_5CH_2\overset{O}{\overset{||}{C}}OCH_2CH_3$

(*e*) $CH_3\overset{O}{\overset{||}{C}}\underset{\overset{|}{CH_2C_6H_5}}{CH}\overset{O}{\overset{||}{C}}OCH_2CH_3$

(*c*)   +

(*f*) 1. HO⁻, H₂O
2. H₃O⁺
3. heat

(*g*) $CH_3\overset{O}{\overset{||}{C}}CH_2CH_2CO_2H$

A-2. (*a*) $CH_3CH_2O\overset{O}{\overset{||}{C}}(CH_2)_4\overset{O}{\overset{||}{C}}OCH_2CH_3$

A

B

C

(*b*) $CH_3CH_2CH_2\overset{O}{\overset{||}{C}}\underset{\overset{|}{CH_2CH_3}}{CH}\overset{O}{\overset{||}{C}}OCH_2CH_3$

$CH_3CH_2CH_2\overset{O}{\overset{||}{C}}CH_2CH_2CH_3$

E

D

A-3. (*a*)

$$CH_3CCH_2COCH_2CH_3 \xrightarrow[\text{2. BrCH}_2\text{COCH}_2\text{CH}_3]{\text{1. NaOCH}_2\text{CH}_3} CH_3CCHCOCH_2CH_3 \xrightarrow[\substack{\text{2. H}_3\text{O}^+ \\ \text{3. heat}}]{\text{1. HO}^-,\ \text{H}_2\text{O}} CH_3CCH_2CH_2COH$$

with the $CH_2COCH_2CH_3$ substituent (with carbonyl O) on the middle carbon of the product.

(*b*)

$$C_6H_5CCH_3 + CH_3CH_2OCOCH_2CH_3 \xrightarrow{\text{NaOCH}_2\text{CH}_3} C_6H_5CCH_2COCH_2CH_3$$

$$\xrightarrow[\text{2. H}_2\text{C}=\text{CHCCH}_3]{\text{1. NaOCH}_2\text{CH}_3} C_6H_5CCHCOCH_2CH_3$$

with $CH_2CH_2CCH_3$ substituent.

$$\downarrow \substack{\text{1. HO}^-,\ \text{H}_2\text{O} \\ \text{2. H}_3\text{O}^+ \\ \text{3. heat}}$$

$$C_6H_5CCH_2CH_2CH_2CCH_3$$

A-4.

$$2CH_3CH_2COCH_2CH_3 \rightleftharpoons CH_3CH_2COCH_2CH_3 + CH_3CH_2COCH_2CH_3$$

$$\longrightarrow CH_3CHCOCH_2CH_3$$
$$CH_3CH_2C-OCH_2CH_3$$

$$\downarrow {}^-CH_3CH_2O^-$$

$$CH_3CHCOCH_2CH_3 \quad CH_3CH_2C=O$$

$$\xleftarrow{-H^+}$$

$$CH_3CCOCH_2CH_3 \quad CH_3CH_2C=O$$

A-5. α-Deprotonation of the Claisen condensation product is necessary for completion of the reaction. The condensation product of ethyl 3-methylbutanoate can be deprotonated under the reaction conditions; the product from condensation of ethyl 2-methylbutanoate cannot.

$$2CH_3CHCH_2COCH_2CH_3 \rightleftharpoons CH_3CHCHCOCH_2CH_3$$

Ethyl 3-methylbutanoate (with CH$_3$ and with CH$_3$CHCH$_2$C=O substituents)

$$\xrightarrow{-H^+} CH_3CHCCOCH_2CH_3$$

$$2CH_3CH_2CHCOCH_2CH_3 \rightleftharpoons CH_3CH_2CCOCH_2CH_3$$

Ethyl 2-methylbutanoate — Claisen product cannot enolize

B-1. (*b*)   **B-2.** (*d*)   **B-3.** (*c*)   **B-4.** (*b*)

B-5. (*c*)   **B-6.** (*c*)   **B-7.** (*b*)   **B-8.** (*d*)

CHAPTER 22

A-1. (*a*) 1,1-Dimethylpropylamine, 2-methyl-2-butanamine or 2-methylbutan-2-amine; primary

   (*b*) *N*-Methylcyclopentylamine or *N*-methylcyclopentanamine; secondary

   (*c*) *m*-Bromo-*N*-propylaniline; secondary

A-2. (*a*) NaN_3 (*b*) KCN   (*c*)

A-3. (*a*)

   (*b*)

   (*c*) H_3PO_2 or CH_3CH_2OH

   (*d*) $(CH_3C)_2O$ or CH_3CCl

   (*e*)

   (*f*)

   (*g*)

A-4. (*a*)

   A   B   C

   (*b*)

   D   E

A-5. (*a*) C_6H_6 $\xrightarrow[\text{AlCl}_3]{(\text{CH}_3)_3\text{CCl}}$ [structure with $C(CH_3)_3$] $\xrightarrow[\text{H}_2\text{SO}_4]{\text{HNO}_3}$ [structure with $C(CH_3)_3$ and NO_2] $\xrightarrow[\text{2. NaOH}]{\text{1. Sn, HCl}}$ [structure with $C(CH_3)_3$ and NH_2]

$\downarrow$ NaNO$_2$, HCl / H$_2$O

[structure with $C(CH_3)_3$ and I] $\xleftarrow{\text{KI}}$ [structure with $C(CH_3)_3$ and N_2^+ Cl^-]

(*b*) C_6H_6 $\xrightarrow[\text{H}_2\text{SO}_4]{\text{HNO}_3}$ [structure with NO_2] $\xrightarrow{\text{Cl}_2, \text{FeCl}_3}$ [structure with NO_2 and Cl] $\xrightarrow[\text{2. NaOH}]{\text{1. Sn, HCl}}$ [structure with NH_2 and Cl]

(*c*) $C_6H_5NH_2$ $\xrightarrow[\text{H}_2\text{O}]{\text{NaNO}_2, \text{HCl}}$ $C_6H_5N_2^+$ Cl^- $\xrightarrow{C_6H_5N(CH_3)_2}$ $C_6H_5N=N-$[ring]$-N(CH_3)_2$

A-6. In the para isomer, resonance delocalization of the electron pair of the amine nitrogen involves the nitro group.

[resonance structures]

A-7. Strongest base: C, an alkylamine; Weakest base: D, a lactam (cyclic amide)

A-8.

A: [structure with NHCCH$_3$ (O), $C(CH_3)_3$]
B: [structure with NH$_2$, $C(CH_3)_3$]
C: [structure with NH$_2$, Cl, Cl, $C(CH_3)_3$]
D: [structure with Br, Cl, Cl, $C(CH_3)_3$]

| | A | B | C | D |

B-1. (*b*)   **B-2.** (*d*)   **B-3.** (*c*)   **B-4.** (*d*)

B-5. (*c*)   **B-6.** (*e*)   **B-7.** (*d*)   **B-8.** (*c*)

B-9. (*d*)   **B-10.** (*e*)   **B-11.** (*b*)   **B-12.** (*b*)

B-13. (*c*)   **B-14.** (*c*)

CHAPTER 23

A-1. (a)

(c)

(b)

A-2. (a)

(b)

(c)

A-3. (a)

(b)

(+ ortho isomer)   (+ meta isomer)

A-4.

The mechanism for para substitution is similar.

A-5. **Product:**   **Intermediate:**

B-1. (*a*)   **B-2.** (*a*)   **B-3.** (*c*)   **B-4.** (*d*)

B-5. (*b*)   **B-6.** (*a*)   **B-7.** (*a*)   **B-8.** (*a*)

CHAPTER 24

A-1. *p*-Hydroxybenzaldehyde is the stronger acid. The phenoxide anion is stabilized by conjugation with the aldehyde carbonyl.

A-2.

o-Cresol

m-Cresol

p-Cresol

A-3.

(Friedel–Crafts acylation)

(Esterification)

A-4. (*a*) C₆H₅—OH + BrCH(CH₃)₂

(*c*) CO_2, 125°C, 100 atm

(*b*)

+ BrCH₂CH(CH₃)₂

(*d*)

A-5.

A   B

A-6.

B-1. (*d*)	**B-2.** (*d*)	**B-3.** (*b*)	**B-4.** (*a*)
B-5. (*c*)	**B-6.** (*b*)	**B-7.** (*b*)	**B-8.** (*c*)

CHAPTER 25

A-1. (*a*)

```
        CHO
HO —————— H
HO —————— H
       CH₂OH
```
L-Erythrose

(*d*)

β-D-Erythrofuranose

(*b*)

```
        CHO
H —————— OH
HO —————— H
       CH₂OH
```
L-Threose

or

```
        CHO
HO —————— H
H —————— OH
       CH₂OH
```
D-Threose

(*e*)

```
   R    CHO
H —————— OH
H —————— OH
   R   CH₂OH
```

(*c*)

α-D-Erythrofuranose

A-2. (*a*)

```
       CH₂OH
HO —————— H
HO —————— H
H —————— OH
H —————— OH
       CH₂OH
```

(*b*) $5HCO_2H + H_2C{=}O$

A-3. (*a*)

(*c*)

(*b*)

A-4. β-D-Idopyranose (β-pyranose form of D-idose)

A-5. The products are diastereomers.

| D-Mannose | Methanol | | Methyl α-D-mannopyranoside | Methyl β-D-mannopyranoside |

B-1. (*b*)   **B-2.** (*d*)   **B-3.** (*b*)   **B-4.** (*a*)

B-5. (*c*)   **B-6.** (*c*)   **B-7.** (*c*)   **B-8.** (*c*)

B-9. (*c*)

CHAPTER 26

A-1.

Tristearin

A-2. Fats are triesters of glycerol. A typical example is tristearin, shown in the preceding problem. A wax is usually a mixture of esters in which the alkyl and acyl group each contain 12 or more carbons. An example is hexadecyl hexadecanoate (cetyl palmitate).

$$\overset{\overset{\displaystyle O}{\overset{\displaystyle \|}{}}}{C_{15}H_{31}C-O-C_{16}H_{33}}$$

A-3. (*a*) Monoterpene;

(*b*) Sesquiterpene;

(*c*) Diterpene;

A-4.

$$CH_3(CH_2)_7 \qquad (CH_2)_7CO_2H$$
$$C=C$$
$$H \qquad\qquad H$$

Oleic acid

$$HC\equiv C(CH_2)_7CH \xrightarrow{NaNH_2} Na^+ \ ^-\!:C\equiv C(CH_2)_7CH \xrightarrow{CH_3(CH_2)_6CH_2Br} CH_3(CH_2)_7C\equiv C(CH_2)_7CH$$

$$\downarrow H_3O^+$$

$$\begin{array}{c} CH_3(CH_2)_7 \qquad (CH_2)_7CO_2H \\ C=C \\ H \qquad\qquad H \end{array} \xleftarrow[\text{Lindlar Pd}]{H_2} CH_3(CH_2)_7C\equiv C(CH_2)_7CO_2H \xleftarrow[H_2SO_4,\ H_2O]{Na_2Cr_2O_7} CH_3(CH_2)_7C\equiv C(CH_2)_7CH$$

A-5.

Geranyl diphosphate

Limonene

B-1. (*b*) **B-2.** (*a*) **B-3.** (*c*) **B-4.** (*c*)

B-5. (*a*) **B-6.** (*a*)

CHAPTER 27

A-1. (*a*) $C_6H_5CH_2CH$ (with =O)   (*b*) $C_6H_5CH_2OCNHCHCO_2H$ (with =O on C, $CH(CH_3)_2$ below)   (*c*) DCCI

A-2. O_2N—(benzene ring with NO_2)—$NHCHCO_2H$, $CH_2C_6H_5$

A-3. (*a*)

$$CH_3\overset{O}{\overset{\|}{C}}NHCH(CO_2CH_2CH_3)_2 \xrightarrow[\text{ethanol}]{NaOCH_2CH_3} CH_3\overset{O}{\overset{\|}{C}}NH\ddot{\overset{\ \ }{C}}(CO_2CH_2CH_3)_2 \xrightarrow{(CH_3)_2CHCH_2Br} CH_3\overset{O}{\overset{\|}{C}}NHC(CO_2CH_2CH_3)_2$$

$$CH_2CH(CH_3)_2$$

$$\downarrow \begin{array}{l}\text{1. } H_3O^+ \\ \text{2. heat}\end{array}$$

$$\overset{+}{H_3}NCHCO_2^-$$
$$CH_2CH(CH_3)_2$$

(*b*) Leu-Val $\equiv$ $\overset{+}{H_3}N\underset{CH_2CH(CH_3)_2}{\overset{O}{\overset{\|}{CHC}}}-NH\underset{}{\overset{CH(CH_3)_2}{CH}}CO_2^-$

N-Protect leucine: $C_6H_5CH_2O\overset{O}{\overset{\|}{C}}Cl$ + $\overset{+}{H_3}NCHCO_2^-$ $\xrightarrow[\text{2. } H^+]{\text{1. NaOH, } H_2O}$ $C_6H_5CH_2O\overset{O}{\overset{\|}{C}}NHCHCO_2H$

$$CH_2CH(CH_3)_2 \qquad\qquad CH_2CH(CH_3)_2$$

(Z-Leu)

C-Protect valine: $C_6H_5CH_2OH$ + $\overset{+}{H_3}NCHCO_2^-$ $\xrightarrow{H^+}$ $\overset{+}{H_3}NCH\overset{O}{\overset{\|}{C}}OCH_2C_6H_5$

$$CH(CH_3)_2 \qquad\qquad CH(CH_3)_2$$

Couple: Z-Leu + $H_2NCH\overset{O}{\overset{\|}{C}}OCH_2C_6H_5$ $\xrightarrow{DCCI}$ $C_6H_5CH_2O\overset{O}{\overset{\|}{C}}NHCHC\overset{O}{\overset{\|}{}}NHCH\overset{CH(CH_3)_2}{C}OCH_2C_6H_5$

$$CH(CH_3)_2 \qquad\qquad (CH_3)_2CHCH_2 \qquad O$$

Deprotect: $C_6H_5CH_2O\overset{O}{\overset{\|}{C}}NHCHC\overset{O}{\overset{\|}{}}NHCH\overset{CH(CH_3)_2}{C}OCH_2C_6H_5$ $\xrightarrow{H_2, Pd}$ $\overset{+}{H_3}NCHC\overset{O}{\overset{\|}{}}NHCH\overset{CH(CH_3)_2}{}CO_2^-$

$$(CH_3)_2CHCH_2 \qquad O \qquad\qquad CH_2CH(CH_3)_2$$

A-4. Leu-Val-Gly-Ala-Phe

A-5. (*a*) Pentapeptide   (*c*) Serine   (*e*) Ser-Ala-Leu-Phe-Gly

   (*b*) Four   (*d*) Glycine

A-6. (*a*)

O_2N—(ring with NO_2)—$NHCH\overset{O}{\overset{\|}{C}}OH$ + $\overset{+}{H_3}NCH_2CO_2^-$ + $\overset{+}{H_3}NCHCO_2^-$ + $\overset{+}{H_3}NCHCO_2^-$

$$CH_3 \qquad\qquad\qquad\qquad CH_2C_6H_5 \qquad CH_2CH(CH_3)_2$$

DNP-Ala   Gly   Phe   Leu

(*b*) $\overset{+}{H_3}NCH\overset{O}{\overset{\|}{C}}NHCH_2\overset{O}{\overset{\|}{C}}NHCHCO_2^-$ + $\overset{+}{H_3}NCHCO_2^-$

$$CH_3 \qquad\quad CH_2C_6H_5 \qquad\qquad CH_2CH(CH_3)_2$$

Ala-Gly-Phe   Leu

(c) Same as part b; Ala-Gly-Phe + Leu

$$(d) \quad C_6H_5CH_2O\overset{O}{\overset{||}{C}}NHCHC\overset{O}{\overset{||}{C}}NHCH_2\overset{O}{\overset{||}{C}}NHCHC\overset{O}{\overset{||}{C}}NHCHCO_2H$$

with substituents CH_3 and $CH_2CH(CH_3)_2$, $CH_2C_6H_5$

Z-Ala-Gly-Phe-Leu

B-1. (c) **B-2.** (a) **B-3.** (c) **B-4.** (d)
B-5. (c) **B-6.** (c) **B-7.** (a)

CHAPTER 28

A-1. In RNA the sugar is ribose. In DNA the sugar is 2-deoxyribose; the OH on C-2 of ribose is replaced by an H.

A-2. −21.8 kJ

A-3. Write the symbol for the complementary bases opposite each nucleotide segment in the original strand. Recall that the complementary base pairs in DNA are A···T and G···C.

(a) Original strand: —A—A—A—G—G—T—C—C—C—G—T—A—
Complementary strand: —T—T—T—C—C—A—G—G—G—C—A—T—

(b) Original strand: —T—A—C—T—C—G—C—G—G—A—T—G—
Complementary strand: —A—T—G—A—G—C—G—C—C—T—A—C—

A-4. The pairing of RNA bases with the DNA strand is the same as DNA base-pairing, except that uracil (U) replaces thymine (T) in RNA.

(a) Original strand: —A—A—A—G—G—T—C—C—C—G—T—A—
mRNA strand: —U—U—U—C—C—A—G—G—G—C—A—U—

(b) Original strand: —T—A—C—T—C—G—C—G—G—A—T—G—
mRNA strand: —A—U—G—A—G—C—G—C—C—U—A—C—

A-5. The codons are triplets of nucleotides in the mRNA strand.

(a) UUU   CCA   GGG   CAU

(b) AUG   AGC   GCC   UAC

A-6. Each mRNA codon codes for a specific tRNA and each tRNA delivers a particular amino acid to the growing protein. The codons and the amino acids for which they code are listed in text Table 28.3.

(a) Codon	Amino Acid	(b) Codon	Amino Acid
UUU	Phenylalanine	AUG	Methionine
CCA	Proline	AGC	Serine
GGG	Glycine	GCC	Alanine
CAU	Histidine	UAC	Tyrosine

The amino acid sequences are:
(a) Phe-Pro-Gly-His

(b) Met-Ser-Ala-Tyr

A-7. An anticodon is a group of three bases in a particular section of tRNA complementary to the bases of each mRNA codon. There is a different tRNA for each amino acid.

(*a*) Codons:	UUU	CCA	GGG	CAU
Anticodons:	AAA	GGU	CCC	GUA
(*b*) Codons:	AUG	AGC	GCC	UAC
Anticodons:	UAC	UCG	CGG	AUG

B-1. (*b*)	**B-2.** (*e*)	**B-3.** (*b*)	**B-4.** (*d*)
B-5. (*d*)	**B-6.** (*e*)	**B-7.** (*c*)	**B-8.** (*c*)
B-9. (*d*)			

CHAPTER 29

A-1. In a chain-growth process monomers add one by one to the end of the growing polymer chain. Each chain has only one growth point. In a step-growth process each chain has at least two growth points. The monomer is consumed quickly to form oligomers, and these react with each other to form the polymer.

A-2. (*a*) The repeating unit is formed by connecting the end atoms of the monomer. Thus the repeating unit shown arose from polymerization of ethylene oxide.

Ethylene oxide

(*b*) The source-based name of the polymer is poly(ethylene oxide).

A-3. Reaction of one mole of bisphenol A with one mole of phosgene produces two moles of hydrogen chloride (HCl). The reaction is a condensation process.

A-4. The monomer used to prepare poly(vinyl acetate) is the alkene vinyl acetate.

Poly(vinyl acetate)   Vinyl acetate

A-5. Poly(vinyl alcohol) cannot be prepared directly from vinyl alcohol as vinyl alcohol is an enol. The enol form is less stable than the keto form, and the major species present is acetaldehyde.

Vinyl alcohol   Acetaldehyde

B-1. (*b*)	**B-2.** (*a*)	**B-3.** (*b*)	**B-4.** (*c*)